International Workshop on Discovery Physics at the LHC

(Kruger2012)

AA002740

Journal of Physics: Conference Series Volume 455

Kruger, South Africa
3-7 December 2012

ISBN: 978-1-62993-093-0
ISSN: 1742-6588

Printed from e-media with permission by:

Curran Associates, Inc.
57 Morehouse Lane
Red Hook, NY 12571

CURRAN ASSOCIATES INC. proceedings.com

Some format issues inherent in the e-media version may also appear in this print version.

Printed by Curran Associates, Inc. (2012)

For permission requests, please contact the Institute of Physics
at the address below.

Institute of Physics
Dirac House, Temple Back
Bristol BS1 6BE UK

Phone: 44 1 17 929 7481
Fax: 44 1 17 920 0979

techtracking@iop.org

Additional copies of this publication are available from:

Curran Associates, Inc.
57 Morehouse Lane
Red Hook, NY 12571 USA
Phone: 845-758-0400
Fax: 845-758-2634
Email: curran@proceedings.com
Web: www.proceedings.com

TABLE OF CONTENTS

Journal of Physics: Conference Series **455** (2013) 011001 doi:10.1088/1742-6596/455/1/011001

Preface

The second conference on ‘Discovery Physics at the LHC’ was held on 3–7 December 2012 at the Kruger Gate Hotel in South Africa. In total there were 110 participants from Armenia, Belgium, Brazil, Canada, Czech Republic, France, Germany, Greece, Israel, Italy, Norway, Poland, USA, Russia, Slovakia, Spain, Sweden, United Kingdom, Switzerland and South Africa.

The latest results from the Large Hadron Collider, Brookhaven National Laboratory, Jefferson Laboratory and BABAR experiments, as well as the latest theoretical insights were presented.

Set against the backdrop of the majestic Kruger National Park a very stimulating conference with many exchanges took place. The proceedings reflect the high standard of the conference.

The financial contributions from the National Institute for Theoretical Physics (NITHeP), the SA-CERN programme, the UCT-CERN Research Centre, the University of Johannesburg, the University of the Witwatersrand and iThemba Labs—Laboratory for Accelerator Based Science are gratefully acknowledged.

Jean Cleymans
Chair of the Local Organizing Committee

Local Organizing Committee
Oana Boeriu
Jean Cleymans
Simon H Connell
Alan S Cornell
William A Horowitz
Andre Peshier
Trevor Vickey
Zeblon Z Vilakazi

Journal of Physics: Conference Series **455** (2013) 011001 doi:10.1088/1742-6596/455/1/011001

Group Picture

Peer review statement

All papers published in this volume of *Journal of Physics: Conference Series* have been peer reviewed through processes administered by the proceedings Editors. Reviews were conducted by expert referees to the professional and scientific standards expected of a proceedings journal published by IOP Publishing.

Journal of Physics: Conference Series **455** (2013) 012001 doi:10.1088/1742-6596/455/1/012001

Status of the LHC

Mike Lamont
CERN, Geneva, Switzerland
E-mail: Mike.Lamont@cern.ch

Abstract.
A brief overview of LHC operations over the last 3 years is provided. Luminosity performance has been satisfactory and the factors that have been exploited are outlined. Availability and operational efficiency are discussed. Finally a brief overview of the planned long shutdown is given and estimates of the potential post shutdown performance briefly enumerated.

1. Introduction

The LHC re-started initial commissioning with beam at the end of 2009. Since then the LHC has had three years of operations as summarized in table 1. Essentially, 2010 was devoted to commissioning and establishing confidence in the critical machine protection system. At this stage the basics were sorted out well, laying the foundation for what followed. 2011 saw the start of exploitation and the exploration of performance limits. Pushing these limits, outlined below, allowed the instantaneous luminosity performance to be increased throughout 2011. 2012 was a production year at an increased centre of mass (COM) energy of 8 TeV. Again limits were pushed revealing some interesting issues at high bunch and total beam intensity.

The integrated luminosity performance over the 3 years can be regarded as good, with the LHC delivering enough integrated luminosity to enable ATLAS and CMS to annonce the discovery of a/the Higgs boson on July 4th 2012.

Table 1. LHC operations 2010 - 2012

Year	Overview	COM energy	Integrated luminosity [fb^{-1}]
2010	Commissioning	7 TeV	0.04
2011	Exploring limits	7 TeV	6.1
2012	Production	8 TeV	23.1

One of the main features of operations in 2011 and 2012 was the use of high bunch intensity with 50 ns bunch spacing. As discussed below, this gave good instantaneous luminosity performance but at the cost of high pile-up to the high luminosity experiments. In 2011 the mean number of collision per crossing (mu) was around 12 with Poisson tails up to approximately 20.

Journal of Physics: Conference Series **455** (2013) 012001 doi:10.1088/1742-6596/455/1/012001

In 2012 mu had increased to around 30 events/crossing at the beginning of a fill with tails up to approximately 40.

Besides the delivery of high instantaneous and integrated luminosity to ATLAS and CMS, the LHC team was also able deliver physics to a number of other users.

- 2010 and 2011 saw lead-lead ion runs which delivered 9.7 and 166 μb^{-1} respectively at an energy of 3.5Z TeV. Here the clients were ALICE, ATLAS and CMS.
- Luminosity levelling at around 4×10^{32} $cm^{-2}s^{-1}$ via transverse separation,with a tilted crossing angle to make life difficult, enabled LHCb to collect 1.2 and 2.2 fb^{-1} in 2011 and 2012 respectively.
- ALICE enjoyed some sustained proton-proton running in 2012 at around 5×10^{30} $cm^{-2}s^{-1}$ with collisions between enhanced satellite bunches and the main bunches.
- There was a successful $\beta^* = 1$ km run for TOTEM and ALFA. With t_{min} of approximately 0.0004 GeV^2 this was the first LHC measurement in Coulomb-Nuclear Interference region.
- The three years operational period culminated in successful proton-lead run at the start of 2013. Here the clients were ALICE, ATLAS, CMS and LHCb.

2. Performance

Basic variations on the equation for the luminosity of a collider are shown in (1) and (2) .

$$\mathcal{L} = \frac{N^2 k_b f}{4\pi\sigma_x^*\sigma_y^*} F \tag{1}$$

Assuming round beams and equal values of the beta function for both beams in both planes, this may be expressed:

$$\mathcal{L} = \frac{N^2 k_b f \gamma}{4\pi\epsilon_n\beta^*} F \tag{2}$$

Here:

- N is the number of particles per bunch
- k_b is the number of bunches
- f is the revolution frequency
- γ is the usual relativistic factor
- σ_x^* and σ_y^* are the horizontal and vertical beam sizes at the interaction point
- ϵ_n is the normalized emittance
- β^* is the value of the beta function at the interaction point
- F is the geometrical reduction factor arising from the crossing angle

The second variation (2) nicely illustrates the parameters that the LHC has exploited in the last 3 years in the search of instantaneous luminosity performance. The corresponding values for these parameters at the peak performance of the LHC (so far) are shown in table 2. The design report values are shown for comparison. Remembering that the beam size is naturally larger at lower energy, it can be seen that the LHC has achieved 77% of design luminosity at 4 sevenths of the design energy with a β^* of 0.6 m (cf. design value of 0.55 m) with half nominal number of bunches.

Table 2. Performance related parameter overview

Parameter	Value in 2012	Design value
Beam energy [TeV]	4	7
β^* in IP 1,2,5,8 [m]	0.6,3.0,0.6,3.0	0.55
Bunch spacing [ns]	50	25
Number of bunches	1374	2808
Average bunch intensity [protons per bunch]	1.6 - 1.7×10^{11}	1.15×10^{11}
Normalized emittance at start of fill [mm.mrad]	2.5	3.75
Peak luminosity [$cm^{-2}s^{-1}$]	7.7×10^{33}	1×10^{34}
Max. mean number of events per bunch crossing	$\approx$40	19
Stored beam energy [MJ]	$\approx$140	362

One of the main reasons for the impressive luminosity has been the excellent beam quality delivered by the injectors. As shown in table 3 the injector complex has succeed in delivering beam with significantly more protons per bunch than nominal with lower emittances than nominal. This is particularly significant for the 50 ns beam.

Happily the LHC has proven capable of absorbing these brighter beams, notably from a beam-beam perspective. This fact has lead to the LHC choosing to operate with 50 ns in both 2011 to 2012 and pushing hard at this bunch spacing. The clear cost has been increased pile-up for the high luminosity experiments which they have successfully learnt to deal with.

Table 3. 2012 values of beam parameters at exit of SPS

Bunch spacing [ns]	Protons per bunch	Emittance [μm]
50	1.7×10^{11}	1.8
25	1.2×10^{11}	2.7

In brief the LHC has achieved good luminosity performance via the following.

- Exploiting the important advantage that high bunch intensities bring (with luminosity proportional to N_b^2). Here the bunch intensity has been up to 150% of nominal with the 50 ns bunch spacing.
- The normalized emittance going into collisions has been around 2.5 mm.mrad i.e. 67% of nominal. Again this is thanks to very good injector performance and ability to conserve the emittance through the Booster, PS, and SPS. Some systematic blow-up at injection and in the ramp is seen in the LHC.
- It has proved possible to squeeze to a β^* of 60 cm thanks to measurement of a very good aperture in the interaction regions (credit to alignment, respect of mechanical tolerances, optics measurement and correction, and orbit correction).
- The total intensity has reached 2.2×10^{14} i.e. 70% of nominal. Here a fully trustworthy machine protection has been instrumental in providing the confidence to routinely deal with 140 MJ beams.

Operational efficiency has also been good and occasionally excellent as illustrated in table 4.

Table 4. Integrated performance highlights

Max. luminosity delivered in one fill	237 pb^{-1}
Max. luminosity delivered in 7 days	1.35 fb^{-1}
Longest time in stable beams (2012)	22.8 hours
Longest time in stable beams over 7 days	91.8 hours (55%)

3. Overview of machine characteristics

The performance described above is on the back of some excellent system performance and some fundamental characteristics of the LHC.

- There is excellent single beam lifetime and on the whole the LHC enjoys excellent vacuum conditions.
- There is excellent field quality, coupled with good correction of non-non-linearities. Certainly dynamic aperture appears not to be an issue.
- There is low tune modulation, low power converter ripple, and low RF noise.
- Head-on beam-beam is not a limitation although long range has to taken reasonably seriously with enough separation at the long range encounters guaranteed by sufficiently large crossing angles.
- Collective effects have been seen with the high bunch intensities. Single and coupled bunch instabilities have been surpressed using a range of tools (high chromaticity, Landau damping octupoles and transverse feedback).

Very good understanding of the beam physics and a good level of operational control has been established.

- The linear optics is well measured and remarkably close to the machine model. The bare β beating is acceptable and has been corrected to excellent [1].
- There is a very good magnetic model which includes dynamic effects. This a major contributory factor to excellent agreement with the optics model. It has also allowed the once feared persistent current decay and snapback to be tamed.
- There is better than expected aperture due excellent alignment and respect of mechanical tolerances.
- The β^* reach has been established and exploited. Reduction has been pursued aggressively, exploiting: the better than specified available aperture; tight collimator settings; and very good stability and reproducibility.

The complex operational cycle is now well established and is robust.

- The pre-cycle, injection process, 450 GeV machine, ramp, squeeze, collide are largely sequencer driven. The devils is in the details but on the whole the process is well mastered. There is generally good beam lifetime throughout the whole process.
- A strict pre-cycling regime means the magnetic machine is remarkably reproducible. This is reflected in the optics, orbit, collimator set-up, tune and chromaticity. Importantly orbit stability (or the ability to consistently correct back to a reference) means that collimator set-up remains good for a year's run [2].

- Operations is unpinned by superb performance of machine protection and associated systems. This includes the beam interlock system, the beam dump system, the beam loss monitors, and the collimation system. There is rigorous machine protection follow-up, qualification, and monitoring; all non-conformities are examined rigorously. The importance of this to the success of the LHC so far can not be over stressed - there has been a move from commissioning to real confidence in under two years.

Availability has, in general, been pretty good considering the size, complexity and operating principles of the LHC. Issues, outlined below, have seen vigorous follow-up and consolidation has been performed. A outline of 2012's availability is shown in table 5. The percent of scheduled physics time spent in "Stable beams" in 2012 was around 36% of a total scheduled time for proton-proton physics of around 200 days. This is encouraging for a machine only 3 years into it's operational lifetime. The machine is performing well and a huge amount of experience and understanding has been gained. There is good system performance, excellent tools, and reasonable availability following targeted consolidation. This is the legacy for post long shutdown 1 operation.

Table 5. LHC availability 2012

Mode	% of scheduled time
Access	14%
Setup	28%
Beam in	15%
Ramp and squeeze	8%
Stable beams	36%

4. Issues

There have inevitably been a number of challenges arising during the exploitation of the LHC. A brief outline is provided below together with potential mitigation measures.

Initially single event effects (SEEs) caused by beam induced radiation to tunnel electronics was a serious cause of inefficiency. However this problem had been foreseen and its impact was considerably reduced following sustained program of mitigation measures coordinated by the R2E (Radiation to Electronics) team. The success of their efforts are impressive. There were several shielding campaigns prior to the 2011 run including relocations "on the fly" and equipment upgrades. The 2011/12 Christmas stop saw some "early" relocation and additional Shielding and further equipment upgrades. This has resulted in the reduction of premature dumps from $\approx$12 per fb^{-1} to $\approx$3 per fb^{-1} in 2012, going a long way to helping the efficiency of integrated luminosity delivery.

UFOs (Unidentified Falling Objects) have now been exquisitely well studied and simulated [3]. There were occasional dumps in 2012 following adjustment of BLM thresholds at the appropriate time-scales (the beam loss spike caused by a UFO is typically of order 1 ms). With the increase in energy to 6.5 TeV and the proposed move to 25 ns there is potentially serious problem with the UFOs become harder (energy) and potentially more frequent (25 ns). Investigations have continued and potentially encouraging results from the 2013 quench test program are noted [3].

Beam induced heating has been an issue and essentially all cases have been local and in some way due to non-conformities either in design or installation. The guilty parties have been clearly enumerated [4]. Design problems have effected the injection dumps (TDI) and the mirror assemblies of the synchrotron radiation telescopes. Installation problem have occurred in a low number of vacuum assemblies.

Instabilities were an interesting problem that dogged operations through 2012. Although never debilitating there were times when they cut into operational efficiency. It should be noted that these problems paralleled a gentle push in bunch intensity with the peak going into stable beams reaching around 1.7×10^{11} protons per bunch i.e. ultimate bunch intensity. Cofactors included increased impedance from tight collimator settings; smaller than nominal emittance; and operation with low chromaticity during the first half of the run.

The final issue to be discussed here is that of electron cloud. Although this has not been a serious issue with the 50 ns beam, there are potential problems with the 25 ns foreseen for post LS1 operation.

There were 3.5 days of scrubbing with 25 ns beams at 450 GeV between 6 and 9 December 2012. The tests saw regular filling of the ring with up to 2748 bunches with a total intensity per beam of up to 2.7×10^{14}. Scrubbing effects in the arcs saw quite rapid conditioning observed in the first stages. The secondary electron yield (SEY) evolution significantly slows down during the last scrubbing fills (more than expected by estimates from laboratory measurements and simulations) and preliminary conclusions [5] are that an electron cloud free environment with 25 ns beam after scrubbing at 450 GeV seem not be reachable in acceptable time. Operation with high heat load and electron cloud density (with blow-up) seems to be unavoidable with a corresponding slow intensity ramp-up. (In 2015 following the warm-up and opening of the entire ring to atmosphere, the SEY and vacuum conditions will be reset and initial re-conditioning will be required.)

5. Long shutdown 1 (LS1)

The primary aim of LS1 is the consolidation of the superconducting splices in the magnet interconnects following the incident of 2008. This will allow the current in the main dipole and quadrupole circuits to be increased to the 6.5 and then 7 TeV level. Besides this a huge amount of maintenance and other consolidation work is to be performed. Key LS1 work packages are outlined below.

- Measure all splices and repair the defective ones. Repeat for the 6 splices per interconnect (2 splices for the main dipoles and 4 for the 2 main quadrupole circuits). There are approximately 1700 interconnects in the machine.
- Consolidate interconnects with a new design (clamp, shunt)
- Finish installation of pressure release mechanisms on cryostats not yet so equipped
- Magnet consolidation: the exchange of weak cryo-magnets
- Consolidation of the current lead feed-boxes (DFBAs)
- Measures to further reduce SEE (R2E) - the R2E project is aiming for 0.3 premature dumps per fb^{-1} via a combination of: equipment relocations at 4 LHC points; additional shielding; and critical system upgrades (QPS, FGC)
- Install collimators with integrated button BPMs (tertiary collimators and a few secondary collimators)
- Extensive experiments consolidation and upgrades
- Plus a lot of other maintenance work covering cryogenics, quench protection, electrical infrastructure, cooling and ventilation, Radio Frequency, beam dump absorber and magnet, change of dump switches (radiation), electron cloud mitigations

6. Post LS1

6.1. Energy

The magnets coming from the damaged sector 34 do not show degradation of performance, and the best estimates to train the LHC (with large errors) are: 30 quenches to reach 6.25 TeV; and 100 quenches to reach 6.5 TeV [6]. With two quenches/day this would mean between 2 to 5 days of training per sector. The proposed plan is to try to reach 6.5 TeV in four sectors in mid 2014. Based on that experience, the decision can then be made to go at 6.5 TeV or step back to 6.25 TeV.

6.2. 50 versus 25 ns

50 ns offers:

- lower total beam current;
- higher bunch intensity (at the cost of having to wrestle with beam instablities);
- lower emittance.

However the perhaps crippling cost at 6.5 TeV is the very high pile-up which will certainly required levelling to be operationally useful. On the other hand 25 ns has a number of negative points which include:

- more long range collisions: requiring a larger crossing angle and thus higher β^*;
- higher emittance;
- seriously more electron cloud with the need for scrubbing;
- higher UFO rate;
- higher injected bunch train intensity;
- higher total beam current.

However, the push will be to go for 25 ns to avoid the inefficiencies and cost of high pile-up.

6.3. Beam from the injectors LS1 to LS2

The bunch spacings and associated performance on offer from the injectors post LS1 are shown in table 6. 50 ns proved a good choice in 2011 and 2012 opening the way to an increased number of bunches and the excellent performance in terms of emittance and bunch intensity. The best that was taken into collisions in 2012 was around 1.7 $\times 10^{11}$ protons per bunch with an emittance of around 2.5 μm going into collision. Further imaginative developments on the PS side have lead to the creation of the so-called BCMS (Batch Compression and (bunch) Merging and (bunch) Splittings) scheme [7] which offer remarkably low emittance coupled with healthy bunch intensity as shown in table 6.

This scheme should open the way to well above nominal performance in the post LS1 era. Major upgrades of the injectors, including the increase of the Booster to PS transfer energy, and the connection of LINAC4 to the Booster will only take place during LS2. These upgrades should open the way in their turn to ultimate LHC performance (i.e. peak luminosity of order 2.3×10^{34} cm^{-2}s^{-1}).

6.4. Potential performance

The potential performance for four scenarios enumerated in table 6 are shown in table 7. The estimates assume:

- a beam energy of 6.5 TeV;
- a 1.1 ns bunch length (nominal);

Journal of Physics: Conference Series **455** (2013) 012001 doi:10.1088/1742-6596/455/1/012001

Table 6. Post LS1 beam parameters at the exit of SPS

Scheme	Protons per bunch [10^{11}]	Emittance exit SPS [μm]	Emittance into collisions [μm]
25 ns	1.15	2.8	3.75
25 ns BCMS	1.15	1.4	1.9
50 ns	1.65	1.7	2.3
50 ns BCMS	1.6	1.2	1.6

- a scheduled 150 days of proton physics with a reasonable optimistic availability (Hübner factor ≈ 0.2);
- 85 mb visible cross-section.

It should be noted that the 50 ns options necessitates the use of a levelling scheme of some sort, as yet unproven operationally.

Table 7. Post LS1 performance estimates - usual caveats apply

Scheme	Number of bunches	Protons per bunch [10^{11}]	β_x^*[cm]/β_y^*[cm]/ half crossing angle [μrad]	Emittance [μm]	Peak luminosity	Iile-up	Int. lumi fb^{-1}
25 ns	2760	1.15	55/43/189	3.75	9.3e33	25	24
25 ns BCMS	2520	1.15	45/43/149	1.9	1.7e34	52	45
50 ns	1380	1.6	42/43/136	2.3	1.6e34 level to 0.8e34	87 level to 44	40
50 ns BCMS	1260	1.6	38/43/115	1.6	2.3e34 level to 0.8e34	138 level to 44	40

From table 7 one notes the following.

- The nominal 25 ns parameters gives more-or-less nominal luminosity at 6.5 TeV as might be expected.
- The BCMS 25 ns scheme gives a healthy 1.7×10^{34}cm^{-2}s^{-1} with peak <mu> of around 50 with 83% of the nominal intensity.
- The now operational 50 ns scheme gives a virtual luminosity of 1.6×10^{34} cm^{-2}s^{-1} but with a pile-up of over 70 making levelling mandatory.
- The BCMS 50 ns scheme gives a virtual luminosity of 2.3×10^{34} cm^{-2}s^{-1} with a pile-up of over 100 making levelling even more mandatory.

7. Long term plans

An outline of the tentative 10 year plan is presented below. Some adjustments will surely be made over the coming years. The results from operation at high energy will naturally be important and could well lead, to, say, a delay of LS2.

- **2015 - 2017** Physics operation, initially at 6.5 TeV. One might hope for a peak luminosity in the region of 1.7×10^{34} cm^{-2}s^{-1}. 2015 will be a re-commissioning year and likely to deliver less integrated luminosity than a nominal year at 6.5 TeV.
- **2018** Long shutdown 2 - here the main focus will the injector complex upgrade including connection of LINAC4 to the booster.
- **2019 - 2021** Given the increased performance of the injectors, it might possible to approach the ultimate performance of the LHC i.e. a luminosity of 2.3×10^{34} cm^{-2}s^{-1}.
- **2022** Long shutdown 3: this is essentially for the HL-LHC upgrade which includes the installation of new inner triplet magnet assemblies.

8. Conclusions

The LHC has shown the results of excellent design, construction, and installation. Operations is now well bedded in and both instantaneous and integrated luminosity performance is healthy. The injector complex has performed very well and delivered 50 ns beam with high bunch intensities and low emittance. Machine availability has been reasonable for a machine the size and complexity of the LHC.

The LHC carries forward a wealth of experience from operation at 3.5 and 4 TeV, and is anticipating operation at 6.5 TeV in 2015 following a two year shutdown. There are potential issues. Measures to address and mitigate these are under examination.

Acknowledgments

The success of the LHC represents a huge international effort at all phases: design, construction, installation, commissioning and operations.

References

[1] R. Tomas Garcia, Optics and dynamic aperture at 4 and 6.5 TeV, Proceedings of the LHC Beam Operation workshop - Evian 2012 (to be published).
[2] B. M. Salvachua Ferrando, Cleaning and collimator operation - outlook, Proceedings of the LHC Beam Operation workshop - Evian 2012 (to be published).
[3] T. Baer, UFOs, Proceedings of the LHC Beam Operation workshop - Evian 2012 (to be published).
[4] B. Salvant, Beam induced RF heating, Proceedings of the LHC Beam Operation workshop - Evian 2012 (to be published).
[5] G. Iadarola, Electron Cloud and Scrubbing, Proceedings of the LHC Beam Operation workshop - Evian 2012 (to be published).
[6] E. Todesco, What is the maximum reasonable energy?, Proceedings of Chamonix 2012 workshop on LHC performance, CERN-ATS-2012-069.
[7] R. Steerenberg, Post LS1 25 ns and 50 ns options from the injectors, Proceedings of the LHC Beam Operation workshop - Evian 2012 (to be published).

Journal of Physics: Conference Series **455** (2013) 012002 doi:10.1088/1742-6596/455/1/012002

The charm of hot matter - charmonium and open charm measurements in Pb–Pb collisions with ALICE at the LHC

Anton Andronic, for the ALICE Collaboration
EMMI & GSI Research Division, Planckstr. 1, 64291 Darmstadt, Germany
E-mail: A.Andronic@gsi.de

Abstract. We present selected results obtained by the ALICE Collaboration concerning charmed hadron and quarkonium production in Pb–Pb collisions at the LHC.

1. Introduction

The goal of high-energy nucleus-nucleus collisions is to produce and characterize a state of nuclear (QCD) matter at (energy) densities well above the nuclear ground state ($\varepsilon_0 \simeq 0.15$ GeV/fm^3). At high densities and/or at high temperatures one expects [1, 2] that quarks are no longer confined in protons and neutrons but move freely over distances larger than the size of the nucleon ($\simeq 1$ fm$=10^{-15}$ m). Employing Quantum Chromo-Dynamics calculations on a space-time lattice, a deconfinement phase transition for an energy density of about 1 GeV/fm^3 was predicted (see [3] for an early review). The characterization of quark-gluon matter in terms of its equation of state (EoS, relating pressure to energy) and of its transport properties (such as viscosity) and delineating its phase diagram is a major ongoing research effort in collisions of nuclei at high energies.

Heavy quarks (charm and bottom) are a prominent part of the so-called "hard probes", observables characterized by an energy (mass) scale much larger than the temperature of the deconfined medium (of a few hundred MeV). These observables are produced in primary hard scattering processes and their yields are calculable in elementary hadronic collisions with perturbative QCD techniques [4, 5, 6, 7]. The production of hard probes in nucleus-nucleus (AA) collisions is quantified by the nuclear modification factor: $R_{\rm AA} = \frac{{\rm d}^2N_{\rm AA}/{\rm d}p_{\rm T}{\rm d}y}{\langle N_{\rm coll}\rangle \cdot {\rm d}^2N_{\rm pp}/{\rm d}p_{\rm T}{\rm d}y}$, where ${\rm d}^2N/{\rm d}p_{\rm T}{\rm d}y$ denotes the transverse momentum ($p_{\rm T}$) and rapidity (y) differential yield of a given observable measured in AA or pp collisions and $\langle N_{\rm coll}\rangle$ is the average number of nucleon-nucleon collisions over the given centrality interval of AA collisions; $\langle N_{\rm coll}\rangle$ is calculated using the Glauber model [8]. Also calculated within this geometric model is the number of participating nucleons, $\langle N_{\rm part}\rangle$, used in the following for study of centrality dependence of $R_{\rm AA}$.

The $p_{\rm T}$ spectra of hadrons carrying light quarks and their azimuthal distributions with respect to the reaction plane (defined by the beam axis and the impact parameter of the colliding nuclei) exhibit features of collective flow and are described well by hydrodynamical models (see a recent review in [9]). In non-central collisions, the initial approximately elliptic shape (in the transverse plane) of the overlap zone of the colliding nuclei leads, through the initial gradients of the energy density (or pressure), to anisotropic spatial (azimuthal angular) emission of hadrons. This is quantified by the second order (quadrupole) Fourier coefficient $v_2 = \langle\cos(2\phi)\rangle$, where ϕ is the azimuthal angle with respect to the reaction plane.

The study of (deconfined) QCD matter has entered a new era in year 2010 with the advent of Pb–Pb collisions at the Large Hadron Collider (LHC), delivering the largest ever collision energy, $\sqrt{s_{\mathrm{NN}}}$=2.76 TeV, more than a factor of 10 larger than previously available. We present selected results obtained by the ALICE Collaboration concerning charmed hadron and quarkonium production. The ALICE detector is described in [10]. The results presented below correspond to minimum-bias Pb-Pb run of the year 2010 and to the high-statistics Pb-Pb run of the year 2011. An integrated luminosity $L_{int} \simeq$30 μb^{-1} was collected in 2011 with centrality triggers for the ALICE Central Barrel at midrapidity ($|y| < 0.9$), while $L_{int} \simeq$70 μb^{-1} was collected with (di)muon triggers in the forward Muon Spectrometer ($2.5 < y < 4$).

2. Open charm hadrons

Produced very early in the collision, at times $t \simeq 1/2m_q$, heavy quarks experience the full evolution of the hot fireball of partons. At variance with light quarks, which can be thermally produced, heavy quarks maintain their identity through the hot stage of the collision. They help to address the question if parton energy loss (by gluon radiation) exhibits the expected quark mass pattern. The theoretical expectation is that heavy quarks (charm and bottom) lose less energy (by gluon radiation) compared to lighter (up, down, strange) quarks [11]. The related question, often asked, is whether heavy quarks thermalize alongside the light quarks and gluons. The measurements at the LHC were eagerly awaited in order to establish in a new energy regime the observation at RHIC, at a center of mass energy per nucleon pair $\sqrt{s_{\mathrm{NN}}}$ = 200 GeV, of heavy quark energy loss and collective flow [12].

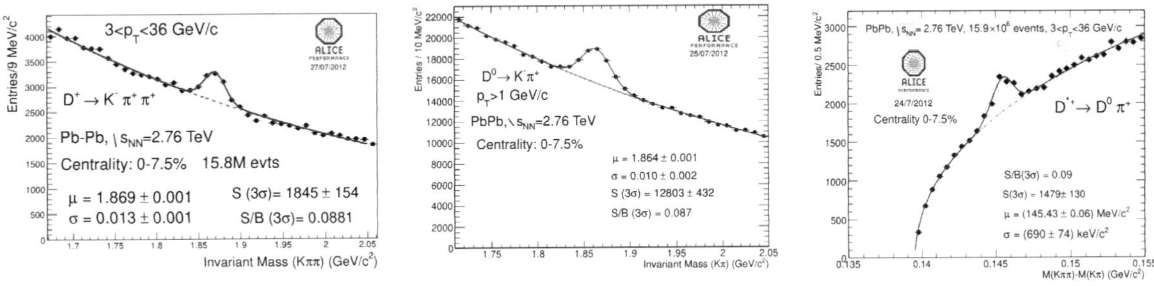

Figure 1. Invariant mass distributions for charmed meson candidates for central (0-7.5% centrality) Pb-Pb collisions. Particles and antiparticles are included.

The production of hadrons carrying open charm is studied in ALICE via the direct reconstruction of hadronic decays [13] at midrapidity, $|y| < 0.5$, and via the measurement of decay electrons [14] (at midrapidity) or muons [15] (at forward rapidity, $2.5 < y < 4$). Invariant mass distributions for charmed meson candidates are shown in Fig. 1 for central (0-7.5% centrality) Pb-Pb collisions. Owing to the very good resolution of track impact parameter in the transverse plane (65 μm for $p_{\mathrm{T}} = 1$ GeV/c and reaching asymptotically 20 μm) and to the good hadron identification performance, a very good signal to background ratio S/B of about 10% is reached for various D-meson species. The very good momentum resolution of ALICE at low p_{T} leads to invariant mass resolutions around 10 MeV/c^2 for the D^+ and D^0 mesons.

The transverse momentum dependence of the nuclear modification factor R_{AA} of D mesons in central Pb-Pb collisions is shown in Fig. 2 [14]. The left panel shows the measurements for three D-meson species, D^+, D^0, and D^{*+}, while the right panel shows the average values of these

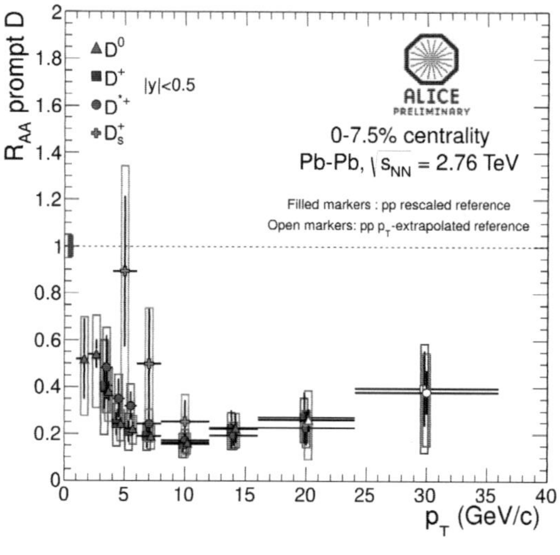

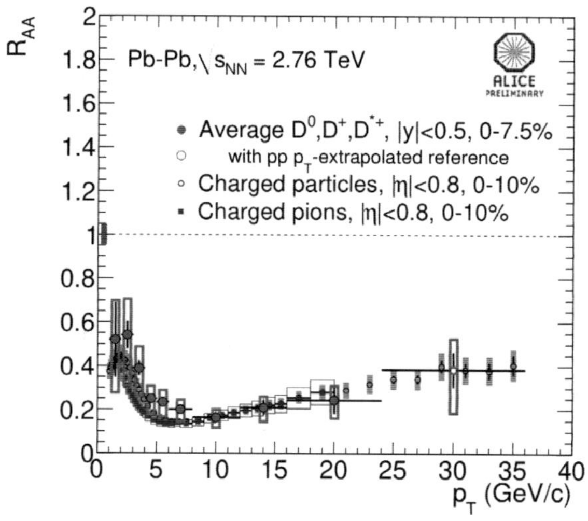

Figure 2. The transverse momentum dependence of the nuclear modification factor of prompt D mesons (particles and antiparticles) in central (0-7.5% centrality) Pb–Pb collisions (left panel). The right panel shows the average values of R_{AA} for D^+, D^0, and D^{*+} meson species in comparison to data for all charged particles and for charged pions (for 0-10% centrality).

species in comparison to data for all charged particles and for charged pions (which are for 0-10% centrality). The similar R_{AA} values of D mesons and of pions (or charged particles), well below unity, indicate that the energy loss of charm quarks is of similar magnitude compared to that of lighter quarks or gluons. The first measurement of R_{AA} for D_s mesons in heavy-ion collisions [16] is also shown in Fig. 2 (left panel). In the highest measured p_T bin (8-12 GeV/c), the R_{AA} for D_s mesons is compatible with that of non-strange charmed mesons. At lower p_T, the R_{AA} of D_s mesons seems to increase, but with the current statistical and systematic uncertainties no conclusion can be drawn on the expected enhancement of D_s mesons with respect to non-strange D mesons [17].

The transverse momentum dependence of the elliptic flow coefficient v_2 of charmed mesons [14] is shown in Fig. 3 together with that of inclusive charged hadrons for the centrality interval 30-50%. Despite the large uncertainties, the challenging measurement of v_2 for D mesons shows a significant anisotropy, suggesting that charm quarks may take part in the collective motion built up at the quark level in the deconfined stage.

The R_{AA} and v_2 results are confronted with model predictions in Fig. 4. The theoretical models [18, 19, 20, 21, 22, 23, 24] implement parton energy loss via elastic (collisional) and radiative mechanisms. The models are able to describe the features of the data, but a quantitative description of both energy loss and elliptic flow remains challenging for all models. As noted in [24], an increase of R_{AA} as a function of p_T is generic for a constant fractional parton energy loss and for a parton spectrum of power-law form, with the power exponent increasing as a function of p_T.

A further challenge for theory is the consistent description of energy loss for light [25] and heavy quarks [13], illustrated in Fig. 5 (left panel). Partonic energy loss models achieve a good description at high p_T while the low p_T region is not described well. The data suggest $R_{AA}^D > R_{AA}^\pi$, as expected. Further hints on the quark flavor dependence of the nuclear modification factor are given in Fig. 5 (right panel) for intermediate p_T values (where the R_{AA} values exhibit a shallow minimum, see Fig. 4). The ALICE data on light flavors [25] and charmed

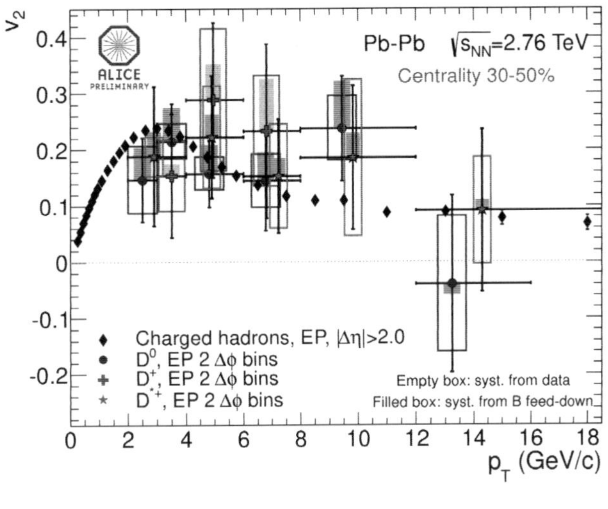

Figure 3. The transverse momentum dependence of the elliptic flow coefficient v_2 of charmed mesons for the centrality interval 30-50%. The data for inclusive charged hadrons is included for comparison.

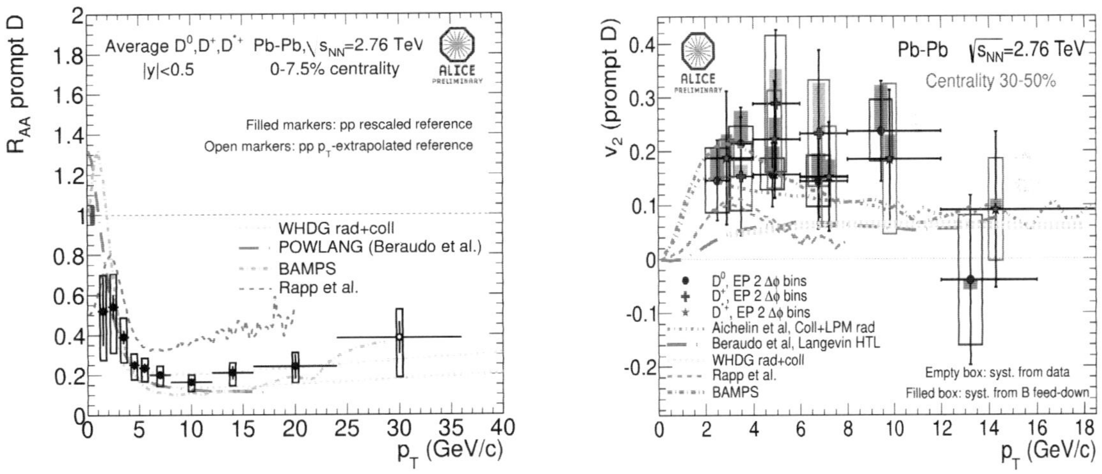

Figure 4. Transverse momentum dependence of R_{AA} (left panel, central collisions) and v_2 (right panel, 30-50% centrality) of D mesons. The mesurements are compared to model predictions (see text).

hadrons [14] are compared to the measurements of primordial J/ψ and J/ψ from decays of B hadrons [26]. Although the details of such a comparison need to be carefully considered [24], the data suggest the expected hierarchy of flavor dependence of in-medium parton energy loss. A further challenge to theory is the extraction of the heavy quark (momentum) diffusion coefficient, a quantity which was recently calculated within lattice QCD [27].

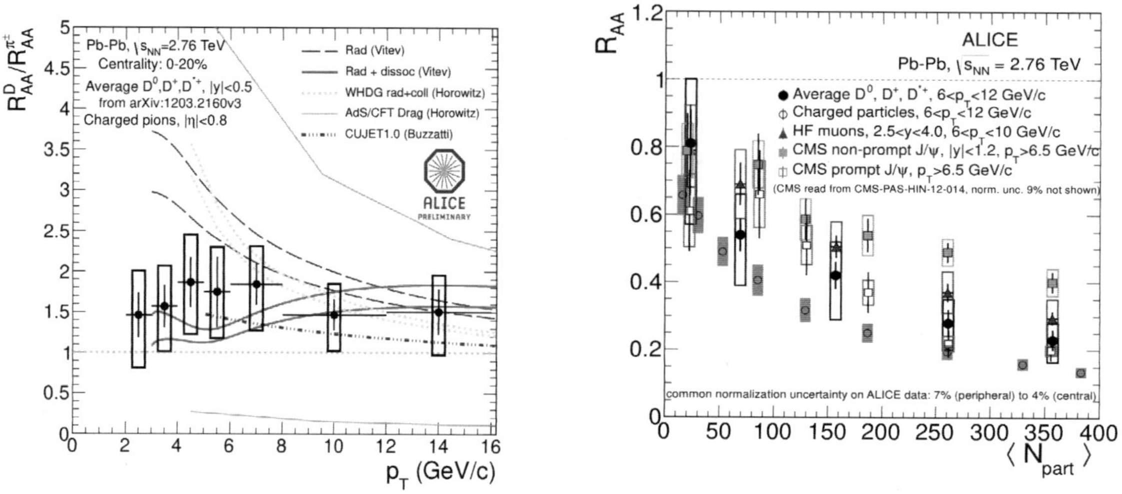

Figure 5. Left panel: transverse momentum dependence of the ratio of R_{AA} of D mesons and pions, data and model predictions. Right panel: centrality dependence of R_{AA} for hadrons carrying light and heavy quarks (see text).

3. Quarkonium

Among the various suggested probes of deconfinement, charmonium plays a distinctive role. It is the first hadron for which a clear mechanism of suppression in the QGP was proposed early on, based on the color analogue of Debye screening [28] (see [29] for a recent review). A suppression compared to pp collisions was observed in pA collisions [30], which was understood as a destruction of the pre-resonant $c\bar{c}$ state by the nucleons of the colliding nuclei. The measurements in Pb-Pb at the SPS ($\sqrt{s_{\rm NN}}$ = 17.3 GeV) [31] and in Au-Au at RHIC [32] demonstrated an "anomalous" suppression, attributed to the dense and hot QCD matter. The theoretical interpretation was a sequential suppression [33, 34] of various charmonium states as a function of energy density (temperature) of the QGP. In this picture, the J/ψ meson survives in the QGP and only the contribution from the $\psi(2S)$ and χ_c charmonium states (leading to about 10% and 30% of the total J/ψ yield, respectively) vanishes as a consequence of the melting of $\psi(2S)$ and $\chi_{c1,2}$ states. As appealing as this interpretation is, it was questioned by the non-zero $\psi(2S)$ yield measured at the SPS [35] and also later by the re-evaluation of cold nuclear matter suppression derived from p-A measurements [30].

The $\psi(2S)$ yield relative to the yield of J/ψ was noted earlier [36, 37] to correspond, for central collisions, to the chemical freeze-out temperature derived from fits to other hadron abundances. This led to the idea of statistical hadronization of charm quarks in nucleus-nucleus collisions [37]. In this model, the charm quarks produced in initial hard collisions thermalize in the QGP and are "distributed" into hadrons at chemical freeze-out. All charmonium states are assumed to be fully melted (or, more precisely, not formed at all) in the QGP and produced, together with all other hadrons, exclusively at chemical freeze-out. This model [38, 39] (see also a recent review [40]) has gained support from experimental data at RHIC [32, 41]. The model predicted a notably different pattern of J/ψ production at the LHC. Depending on the charm production cross section, even an enhanced production relative to pp collisions could be expected at the LHC in central Pb–Pb collisions [39] (see [42] for further predictions). The statistical model predicted [43] an energy-independent relative production ratio $\psi(2S)$/J/ψ about 4 times smaller in (central) AA collisions compared to pp.

Proposed at the same time as the statistical hadronization model, the idea of kinetic recombination of charm and anti-charm quarks in the QGP [44] is an alternative quarkonium production mechanism. In this model (see [45, 46] for recent results), a continuous dissociation and regeneration of charmonium takes place in the QGP over its entire lifetime. Besides the charm production cross section, this model has as input parameters the time dependence of the temperature of the fireball as well as relevant cross sections and assumptions on the melting scenarios of charmonium states. Important observables, like the transverse momentum dependence of production yields and of elliptic flow can be calculated within the transport (kinetic) models. It predicted a rather small J/ψ regeneration component at RHIC and a sizable one at the LHC.

The measurement of J/ψ production in Pb-Pb collisions at the LHC was expected to provide a definitive answer on the question of (re)generation. The first data, measured at high-p_{T} by ATLAS [47] and CMS [48], showed a pronounced suppression with respect to pp collisions. Subsequently, it was seen that this J/ψ suppression is of the same magnitude as that of open-charm hadrons [13], as shown above (Fig. 4, right panel). This may indicate that the high-p_{T} charm quarks that form either D or J/ψ mesons had the same dynamics including a thermalization stage and a late hadronization.

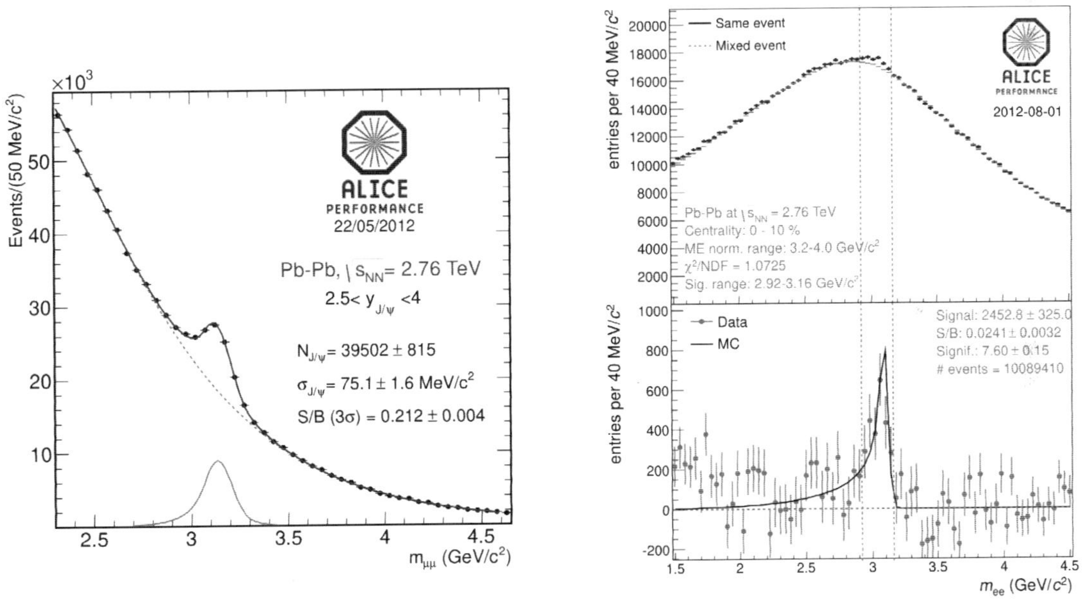

Figure 6. Invariant mass distributions measured for the $\mu^+\mu^-$ channel (Muon Spectrometer, 2.5< y <4, right panel; the dashed line is a fit to the background) and e^+e^- channel (Central Barrel, $|y|$ <0.9, right panel; the lower panel is the signal obtained after the mixed-event background subtraction, with the line the signal form obtained from Monte Carlo simulations).

Quarkonium is measured in ALICE via its dielectron channel at midrapidity and via the dimuon channel at forward rapidity, respectively, as shown in Fig. 6 (see [49, 50] for the details of the analysis). A clear difference to the RHIC results [32, 41] was seen in the first LHC measurement of the overall (inclusive in p_{T}) production, performed in ALICE [49]. In this case, where low p_{T} J/ψ production is dominant, less nuclear suppression (larger R_{AA} value) is seen at the LHC both at forward rapidity [51] and at mid-rapidity [52], as shown in Fig. 7. Corroborated with the CMS data [48], the result indicates an enhanced production of J/ψ at low-p_{T} compared to high p_{T}.

Journal of Physics: Conference Series **455** (2013) 012002 doi:10.1088/1742-6596/455/1/012002

At the LHC, the estimated energy density is at least a factor of 3 larger than at RHIC [53], leading to an initial temperature most likely above the one required for J/ψ dissociation. Therefore, one can conclude that the production mechanism of J/ψ and charmonium in general at the LHC is determined (to a large extent) by regeneration in QGP or by generation at chemical freeze-out.

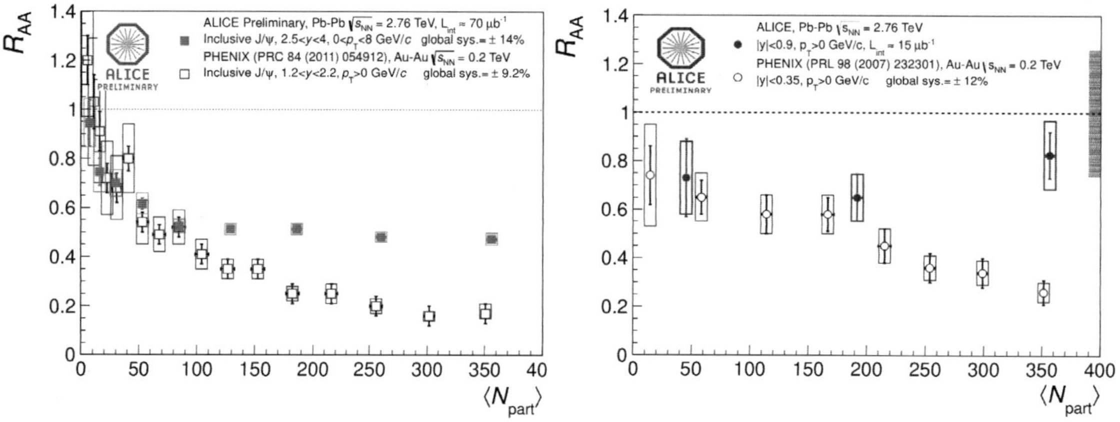

Figure 7. Centrality dependence of the nuclear modification factor for inclusive J/ψ production. The ALICE measurement (preliminary) at the LHC is compared to the PHENIX data at RHIC. The two panels show the data at forward rapidity (left) and at mid-rapidity (right).

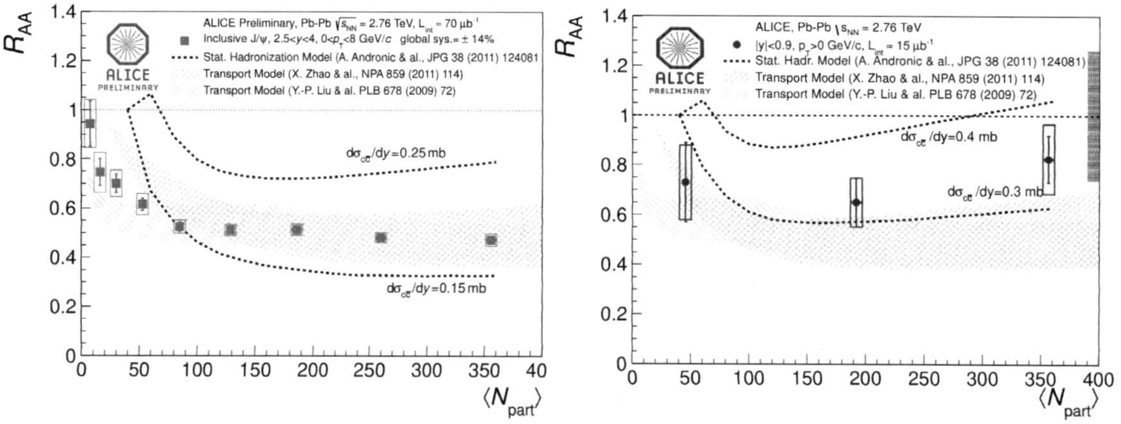

Figure 8. Centrality dependence of the nuclear modification factor for J/ψ at all momenta in comparison to theoretical models, for forward rapidity (left panel) and mid-rapidity (right panel).

Indeed, both the statistical hadronization [42] and transport [45, 46] models reproduce the data [49], as seen in Fig. 8. Based on these observations, the J/ψ production can be considered a probe of QGP as initially proposed [28], but may not be a "thermometer" of the medium [34]. Within the statistical model, the charmonium states become probes of the phase boundary between QGP and hadron phase. This extends with a heavy quark the family of quarks employed for the determination of the hadronization temperature (via the conjectured connection to the chemical freeze-out temperature extracted from fits of statistical model calculations to hadron abundances).

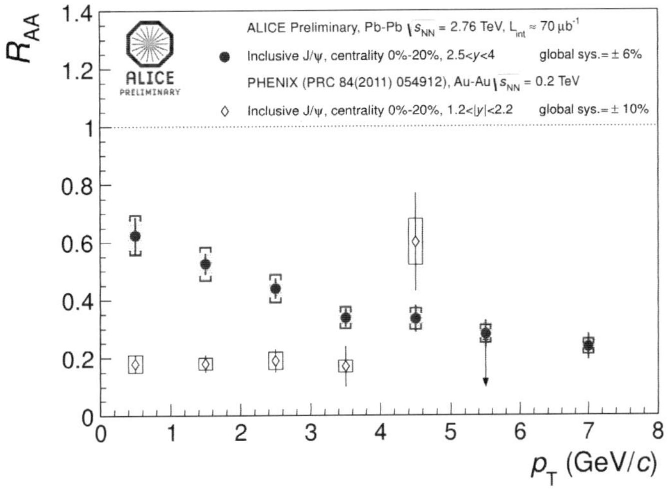

Figure 9. . Transverse momentum dependence of the nuclear modification factor for J/ψ at forward rapidity for the centrality range 0-20%. The preliminary ALICE data at the LHC are compared to measurements at RHIC by PHENIX.

The transverse momentum dependence of the nuclear modification factor, shown in Fig. 9, is, at the LHC, dramatically different than the one measured at RHIC. At low-p_T the nuclear suppression is significantly reduced (i.e. larger R_{AA}) at the LHC [51] compared to RHIC [32]. Transport model predictions reproduce the data quantitatively, as can be seen in Fig. 10 (with model of ref. [46]; the open bands represent the yield due to regeneration). In current models [45, 46], about half of the low-p_T J/ψ yield in Pb-Pb collisions at $\sqrt{s_{NN}} = 2.76$ TeV is produced by the recombination of charm quarks in QGP, while the rest is due to primordial production.

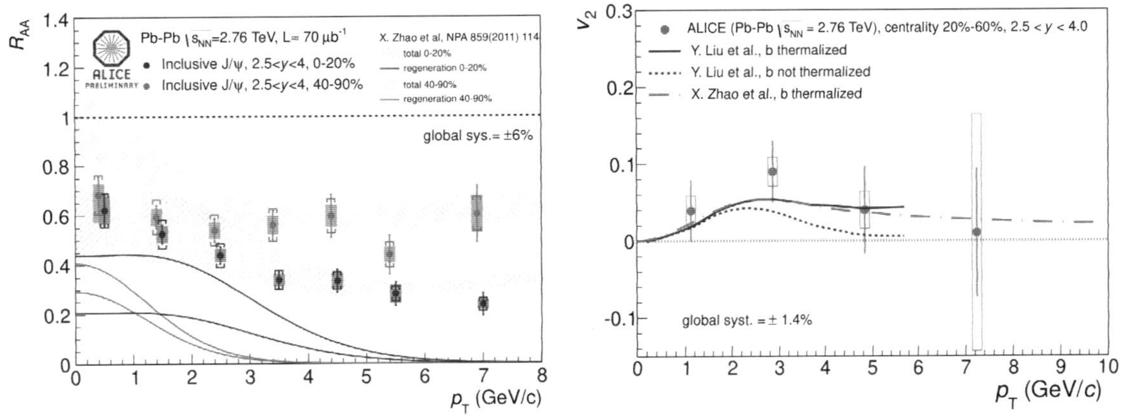

Figure 10. Nuclear modification factor (left panel, for centrality ranges 0-20% and 40-90%) and elliptic flow (right panel, for centrality range 20-60%) of J/ψ mesons as a function of transverse momentum. The ALICE data is compared with parton transport model predictions.

Both the kinetic and the statistical hadronization models require thermalization of the charm quarks in QGP. As a consequence they will follow the collective behavior of the bulk QGP and their flow will be reflected in that of charmed hadrons and quarkonia. Indeed, elliptic flow of J/ψ at LHC energies was predicted within a transport model [45]. The first measurement at the LHC [54], shown in Fig. 10 (right), provides a tantalizing hint of a non-zero elliptic flow of J/ψ (the significance of non-zero flow for the centrality range 20-60% is 2.3σ in the p_T range 2-4 GeV/c). The preliminary ALICE data is consistent with the expectation from transport models, but precision data is needed in order to be able to extract information on the QGP properties and on the amount of J/ψ produced via regeneration.

The picture outlined above for the charmonium production, extracted by comparison of data and model predictions, remains to be tested by precision measurements at the LHC, which will place further constraints to models. A generic feature expected within (re)generation models is an increase of R_{AA} at the higher energy ($\sqrt{s_{NN}}$ = 5.1 TeV) of collisions expected in the LHC after the current shutdown period. In particular, both the statistical and the transport models employ as input parameter the $c\bar{c}$ production cross section $\sigma_{c\bar{c}}$. The sensitivity of the calculations to $\sigma_{c\bar{c}}$ is rather large (see Fig. 8). A precision measurement of $\sigma_{c\bar{c}}$ in Pb-Pb collisions, within reach with the proposed ALICE upgrade [55], will place an important constraint to models. A precision measurement of excited states of charmonium, in particular of $\psi(2S)$, for which current data [56] are rather imprecise, could allow to discriminate between the regeneration [46] and statistical production [43] mechanisms.

4. Summary and outlook

A wealth of data has been measured by ALICE on charmed hadrons and charmonium in Pb-Pb collisions at the LHC. Parton energy loss is strong also for charm quarks and its quantitative description together with the observed elliptic flow in theoretical models remains a challenge. Charmonium exhibits features of production via regeneration in the hot deconfined medium or via statistical hadronization at the phase boundary. Disentangling these two pictures is the next challenge for the experiments and theory.

Acknowledgments

I would like to thank the organizers for the invitation to talk at this special workshop at the close proximity to the magnificent Kruger Park and in particular many thanks to Simon Connell for a special strategy to obtain an unlikely visa at the Johannesburg airport.

References

[1] Collins J C and Perry M 1975 *Phys. Rev. Lett.* **34** 1353
[2] Cabibbo N and Parisi G 1975 *Phys. Lett. B* **59** 67–69
[3] Karsch F 2002 *Lect. Notes Phys.* **583** 209–249 (*Preprint* hep-lat/0106019)
[4] Abelev B *et al.* (ALICE Collaboration) 2012 *JHEP* **1201** 128 (*Preprint* 1111.1553)
[5] Kniehl B, Kramer G, Schienbein I and Spiesberger H 2012 *Eur. Phys. J. C* **72** 2082 (*Preprint* 1202.0439)
[6] Cacciari M, Frixione S, Houdeau N, Mangano M L, Nason P *et al.* 2012 *JHEP* **1210** 137 (*Preprint* 1205.6344)
[7] Maciula R and Szczurek A 2013 (*Preprint* 1301.3033)
[8] Miller M L, Reygers K, Sanders S J and Steinberg P 2007 *Ann. Rev. Nucl. Part. Sci.* **57** 205–243 (*Preprint* nucl-ex/0701025)
[9] Heinz U W and Snellings R 2013 (*Preprint* 1301.2826)
[10] Aamodt K *et al.* (ALICE Collaboration) 2008 *JINST* **3** S08002
[11] Dokshitzer Y L and Kharzeev D 2001 *Phys. Lett. B* **519** 199–206 (*Preprint* hep-ph/0106202)
[12] Adare A *et al.* (PHENIX Collaboration) 2011 *Phys. Rev. C* **84** 044905 (*Preprint* 1005.1627)
[13] Abelev B *et al.* (ALICE Collaboration) 2012 *JHEP* **1209** 112 (*Preprint* 1203.2160)
[14] del Valle Z C (ALICE Collaboration) 2012 (*Preprint* 1212.0385)
[15] Abelev B *et al.* (ALICE Collaboration) 2012 *Phys. Rev. Lett.* **109** 112301 (*Preprint* 1205.6443)
[16] Innocenti G M (ALICE Collaboration) 2012 (*Preprint* 1210.6388)
[17] He M, Fries R J and Rapp R 2013 *Phys. Rev. Lett.* **110** 112301 (*Preprint* 1204.4442)
[18] Gossiaux P, Bierkandt R and Aichelin J 2009 *Phys. Rev. C* **79** 044906 (*Preprint* 0901.0946)
[19] Greco V, van Hees H and Rapp R 2012 *AIP Conf.Proc.* **1422** 117–126 (*Preprint* 1110.4138)
[20] Sharma R, Vitev I and Zhang B W 2009 *Phys. Rev. C* **80** 054902 (*Preprint* 0904.0032)

[21] Alberico W, Beraudo A, De Pace A, Molinari A, Monteno M *et al.* 2011 *Eur. Phys. J. C* **71** 1666 (*Preprint* 1101.6008)
[22] Buzzatti A and Gyulassy M 2012 *Phys. Rev. Lett.* **108** 022301 (*Preprint* 1106.3061)
[23] Uphoff J, Fochler O, Xu Z and Greiner C 2012 (*Preprint* 1208.1970)
[24] Horowitz W 2012 (*Preprint* 1210.8330)
[25] Abelev B *et al.* (ALICE Collaboration) 2013 *Phys. Lett. B* **720** 52 (*Preprint* 1208.2711)
[26] Chatrchyan S *et al.* (CMS Collaboration) 2012 *Physics Analysis Note* **CMS-PAS-HIN-12-014**
[27] Banerjee D, Datta S, Gavai R and Majumdar P 2012 *Phys. Rev. D* **85** 014510 (*Preprint* 1109.5738)
[28] Matsui T and Satz H 1986 *Phys. Lett. B* **178** 416
[29] Kluberg L and Satz H 2009 (*Preprint* 0901.3831)
[30] Arnaldi R *et al.* (NA60 Collaboration) 2012 *Phys. Lett. B* **706** 263–267 (*Preprint* 1004.5523)
[31] Alessandro B *et al.* (NA50 Collaboration) 2005 *Eur. Phys. J. C* **39** 335–345 (*Preprint* hep-ex/0412036)
[32] Adare A *et al.* (PHENIX Collaboration) 2007 *Phys. Rev. Lett.* **98** 232301 (*Preprint* nucl-ex/0611020)
[33] Digal S, Petreczky P and Satz H 2001 *Phys. Rev. D* **64** 094015 (*Preprint* hep-ph/0106017)
[34] Karsch F, Kharzeev D and Satz H 2006 *Phys. Lett. B* **637** 75–80 (*Preprint* hep-ph/0512239)
[35] Alessandro B *et al.* (NA50 Collaboration) 2007 *Eur. Phys. J. C* **49** 559–567 (*Preprint* nucl-ex/0612013)
[36] Sorge H, Shuryak E V and Zahed I 1997 *Phys. Rev. Lett.* **79** 2775–2778 (*Preprint* hep-ph/9705329)
[37] Braun-Munzinger P and Stachel J 2000 *Phys. Lett. B* **490** 196–202 (*Preprint* nucl-th/0007059)
[38] Andronic A, Braun-Munzinger P, Redlich K and Stachel J 2007 *Nucl. Phys. A* **789** 334–356 (*Preprint* nucl-th/0611023)
[39] Andronic A, Braun-Munzinger P, Redlich K and Stachel J 2007 *Phys. Lett. B* **652** 259–261 (*Preprint* nucl-th/0701079)
[40] Braun-Munzinger P and Stachel J 2009 (*Preprint* 0901.2500)
[41] Adare A *et al.* (PHENIX Collaboration) 2011 *Phys. Rev. C* **84** 054912 (*Preprint* 1103.6269)
[42] Andronic A, Braun-Munzinger P, Redlich K and Stachel J 2011 *J. Phys. G* **38** 124081 (*Preprint* 1106.6321)
[43] Andronic A, Beutler F, Braun-Munzinger P, Redlich K and Stachel J 2009 *Phys. Lett. B* **678** 350–354 (*Preprint* 0904.1368)
[44] Thews R L, Schroedter M and Rafelski J 2001 *Phys. Rev. C* **63** 054905 (*Preprint* hep-ph/0007323)
[45] Liu Y P, Qu Z, Xu N and Zhuang P F 2009 *Phys. Lett. B* **678** 72–76 (*Preprint* 0901.2757)
[46] Zhao X and Rapp R 2011 *Nucl. Phys. A* **859** 114–125 (*Preprint* 1102.2194)
[47] Aad G *et al.* (ATLAS Collaboration) 2011 *Phys. Lett. B* **697** 294–312 (*Preprint* 1012.5419)
[48] Chatrchyan S *et al.* (CMS Collaboration) 2012 *JHEP* **1205** 063 (*Preprint* 1201.5069)
[49] Abelev B *et al.* (ALICE Collaboration) 2012 *Phys. Rev. Lett.* **109** 072301 (*Preprint* 1202.1383)
[50] Aamodt K *et al.* (ALICE Collaboration) 2011 *Phys. Lett. B* **704** 442–455 (*Preprint* 1105.0380)
[51] Suire C (ALICE collaboration) 2012 *Hard Probes* (*Preprint* 1208.5601)
[52] Arsene I C (ALICE Collaboration) 2012 *Nucl.Phys.A* (*Preprint* 1210.5818)
[53] Chatrchyan S *et al.* (CMS Collaboration) 2012 *Phys. Rev. Lett.* **109** 152303 (*Preprint* 1205.2488)
[54] Abbas E *et al.* (ALICE Collaboration) 2013 (*Preprint* 1303.5880)
[55] Musa L and Safarik K 2012 Letter of intent for the upgrade of the alice experiment Tech. Rep. CERN-LHCC-2012-012. LHCC-I-022 CERN Geneva
[56] Arnaldi R (ALICE Collaboration) 2012 (*Preprint* 1211.2578)

Hadronic resonance production in pp and Pb–Pb collisions at LHC with the ALICE experiment

A. Badalá[1] and H. Oeschler[2] for the ALICE collaboration

[1] INFN-Sezione di Catania, V. S. Sofia 64, Catania, I95125, Italy

[2] Institut für Kernphysik, Technische Universität Darmstadt, Darmstadt, Germany

E-mail: Angela.Badala@ct.infn.it

Abstract. Resonances, with their lifetimes comparable to that of the fireball, can be used to estimate the time span and hadronic interaction cross section in the phase between chemical and kinetic freeze-out. In pp collisions, measurements of resonances provide an important baseline for heavy-ion data and allow for the tuning of QCD-inspired particle production models. The ALICE collaboration measured $K^*(892)^0$ and $\phi(1020)$ production, both in pp collisions at $\sqrt{s}$=7 TeV and in Pb–Pb collisions at $\sqrt{s_{NN}}$=2.76 TeV. The inelastic yield of $\Sigma(1385)^{\pm}$ in pp collision at $\sqrt{s}$=7 TeV was also measured. Transverse momentum spectra, ratios to stable particles and a comparison between central and peripheral production are shown.

1. Introduction

The study of resonance production plays an important role both in elementary and in heavy ion collisions. In pp and e^+e^- collisions, it contributes to the understanding of hadron production [1, 2] as the decay products of resonances represent a large fraction of the final state particles. In addition, it provides a reference for tuning event generators inspired by Quantum Chromodynamics (QCD) such as PHOJET [3] and PYTHIA [4]. Hadronic resonances are a sensitive probe of the dynamical evolution of the fireball. Due to their short lifetime (a few fm/c) a significant fraction decays during the evolution from chemical to kinetic freeze-out and their hadronic daughters interact with the medium during the fireball expansion [5, 6, 7, 8]. In particular, products of their hadronic decay may rescatter reducing the measured resonance signal. Resonances may also be regenerated through collisions of hadrons. The competition between resonance-generating processes and rescattering, and therefore the ratio of resonance yields to non-resonance yield is governed by the lifetime and the temperature of the hadronic medium. Thermal models [7, 9, 10, 11] predict particle ratios as function of the chemical freeze-out temperature (T_{ch}) and the time between chemical and thermal freeze-out. Particularly interesting is the comparison of ϕ and K^* production, considering the different lifetimes (about a factor 10) of the two resonances. Due to the large lifetime (44 fm/c) the ϕ is expected to decay outside the hot and dense interacting medium.

2. Experimental setup and data analysis

The results reported in this paper refer to analyses carried out using a sample of minimum-bias pp data at $\sqrt{s}$=7 TeV (60 to 250 millions events, for the different resonances analyzed) and of minimum-bias Pb–Pb data at $\sqrt{s_{NN}}$=2.76 TeV (about 10 millions events), collected during 2010. The events are selected with a primary vertex within 10 cm of the detector center.

2.1. Experimental setup

The ALICE detector [12, 13] provides extensive particle tracking and identification in the central pseudorapidity range ($|\eta| \leq 0.9$) as well as muon tracking and identification at forward angles ($-2.5 > \eta > -4$). For the analysis described in this paper, only the central barrel has been used. The central tracking and particle identification detectors include, from the innermost outwards, the Inner Tracking System (ITS), the Time Projection Chamber (TPC) and the Time of Flight array (TOF). The central detectors are embedded in a 0.5 T solenoidal field. The moderate field, together with a low material budget permits the reconstruction of low p_T tracks. Furthermore, two forward scintillator hodoscopes (VZERO) placed along the beam direction at -0.9 m and 3.3 m on either side of the interaction point, which cover the pseudorapidity regions $-3.7 < \eta < -1.7$ and $2.8 < \eta < 5.1$, were used for minimum-bias triggering and for rejecting beam-gas interactions.

The Inner Tracking System (ITS) is a silicon detector that surrounds the interaction point, with six layers between radii 3.9 cm to 43 cm from the beam axis. The two innermost layers, based on silicon pixels (SPD), are also used as an on-line trigger and to reconstruct the collision vertex with a resolution better than 100 μm.

The TPC [14] provides track reconstruction with up to 159 three-dimensional space points per track in a cylindrical active volume of about 90 m^3. The standard tracking used in this analysis combines the information from the ITS and TPC. The momentum resolution of the TPC is in the range 1-7% for pions with 1<p_T<10 GeV/c. Furthermore it provides very good resolution in the distance of closest approach to the vertex (impact parameter resolution in the transverse direction is <100 μm for p_T>1 GeV/c) and hence an excellent separation of primary and secondary particles.

The TPC identifies particles via the specific energy loss dE/dx determined with a truncated-mean procedure. It achieves a resolution ranging from 5-6.5% (tracks with 159 clusters - mean over all the reconstructed tracks) in pp and Pb–Pb collisions. Identification is achieved by calculating the difference between the measured energy loss and the one expected for different mass hypotheses. A selection on this difference, normalized to the resolution σ_{TPC}, is optimized for each analysis and depends in general on the signal to background ratio and on the transverse momentum. The TPC dE/dx measurements allow pions to be separated from kaons for momenta up to p ~0.7 GeV/c, while the proton/antiproton band starts to overlap with the pion/kaon band at p ~1 GeV/c. The electron/positron dE/dx crosses the other bands at various momenta.

In order to ensure high efficiency and good dE/dx resolution and to minimize the contamination from secondaries and fakes, tracks were required to have at least 70 reconstructed clusters in the TPC. To improve the resolution (<1% at p_T~1 GeV/c), tracks were accepted only in the range $|\eta| < 0.8$ (i.e. well within the TPC acceptance) and with $p_T \geq$ 0.15 GeV/c.

The Time-of-Flight Detector (TOF) is an array of multi-gap resistive-plate chambers placed at a radius of 370 to 399 cm. Particles are identified by the difference between the measured time-of-flight and the one expected from a given particle (π, K, p). The selection is expressed in units of the estimated resolution σ_{TOF} for each track, which has a mean value of 160 ps and 85 ps in pp and Pb–Pb collisions, respectively. The TOF allows pions and kaons to be unambiguously identified up to $p \sim$ 1.5-2.0 GeV/c. The two mesons can be distinguished from (anti)protons up to $p \sim$ 2.5 GeV/c. For the analyses described in this paper the start time of the collision (event time zero) is measured by the T0 detector, an array of Cherenkov counters located at +350 cm and -70 cm along the beam line. For events in which the T0 signal is not present, it is estimated using the particle arrival times at the TOF or the averaged collision time observed in the fill.

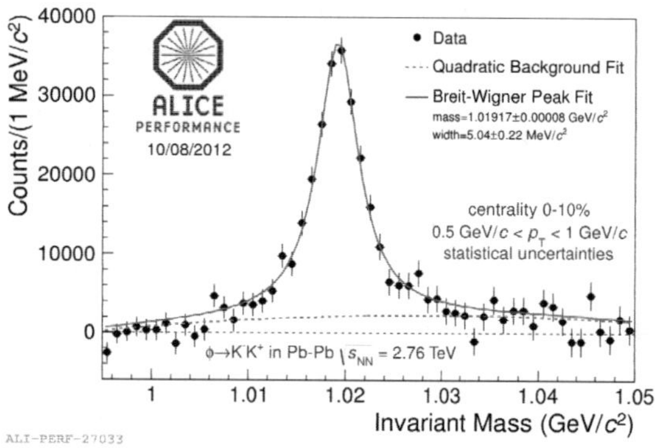

Figure 1. The K^+K^- invariant mass distribution in Pb–Pb collisions at $\sqrt{s_{\rm NN}}$=2.76 TeV. The fitting function is the sum of a Breit-Wigner function and a polynomial.

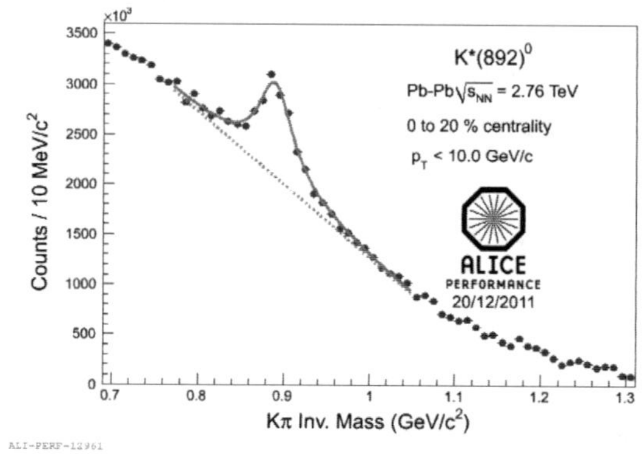

Figure 2. The $K\pi$ invariant mass distribution in Pb–Pb collisions at $\sqrt{s_{\rm NN}}$=2.76 TeV. The fitting function is the sum of a Breit-Wigner function and a polynomial.

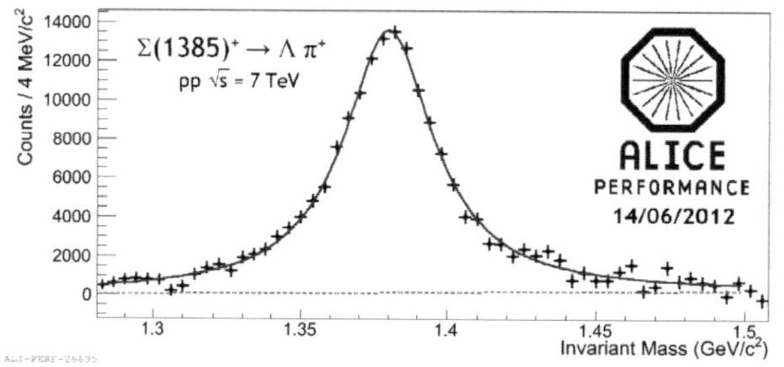

Figure 3. The $\Lambda\pi^+$ invariant mass distribution in pp collisions at $\sqrt{s}$=7 TeV after background subtraction. The solid line is the result of the combined fit: a Breit-Wigner function plus a polynomial. The dashed line describes the residual background.

2.2. Raw yield extraction and $p_{\rm T}$ spectrum

Resonances are identified by their main hadronic decay ($K^* \longrightarrow \pi^\pm$+$K^\mp$, $\phi \longrightarrow K^+$+K^-, $\Sigma^{*\pm} \longrightarrow \Lambda\pi^\pm$). Due to their very short lifetimes, decay products cannot be distinguished from particles coming from the primary vertex. Their yield is obtained by computing the invariant mass spectrum of all primary candidates (tracks or hyperons) and then subtracting a combinatorial background. This was performed by the event-mixing or the like-sign technique. The signal, after subtracting the combinatorial background, was then fitted with a Breit-Wigner plus a polynomial for the residual background. A Voigtian function (convolution of Breit-Wigner function and Gaussian) was used in pp collisions for the extraction of the ϕ raw yield. For the Σ^*, the residual background originating from correlated $\Lambda\pi$ pairs coming from $\Lambda(1520)$ decay, has been estimated by Monte-Carlo simulations and subtracted before fitting the invariant mass spectrum. Some examples of invariant mass spectra are presented in Figs.1,2 and 3. The mass and width of analyzed resonances are close to the PDG values. In particular, in Pb-Pb collisions no mass shift nor broadening has been observed for the $\phi(1020)$, nor for $K^*(892)^0$, similar to what is observed in pp collisions.

In order to extract the total production yield, the raw counts were corrected for the decay branching ratio and for the losses due to geometrical acceptance and detector efficiency, which was determined by Monte-Carlo simulation using PYTHIA or HIJING generators, for pp and

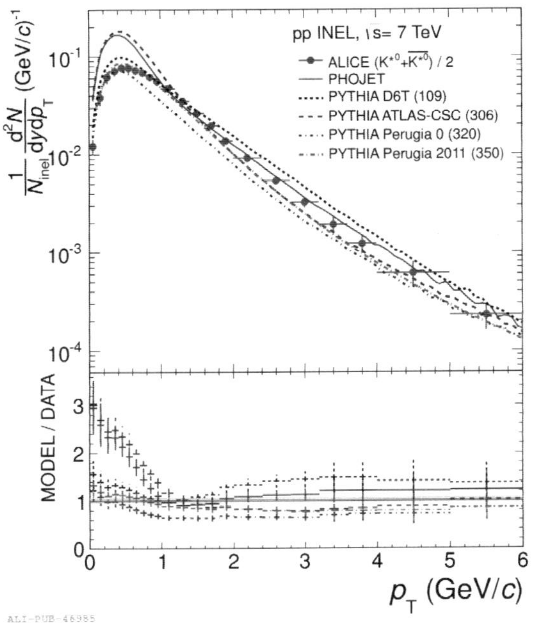

Figure 4. Comparison of the K* $p_{\rm T}$ spectrum in inelastic pp collisions at $\sqrt{s}$=7 TeV with PHOJET and PYTHIA tunes [15].

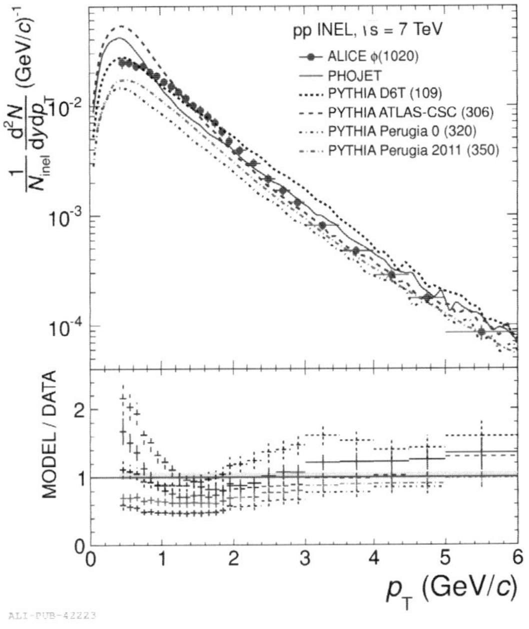

Figure 5. Comparison of the ϕ $p_{\rm T}$ spectrum in inelastic pp collisions at $\sqrt{s}$=7 TeV with PHOJET and PYTHIA tunes [15].

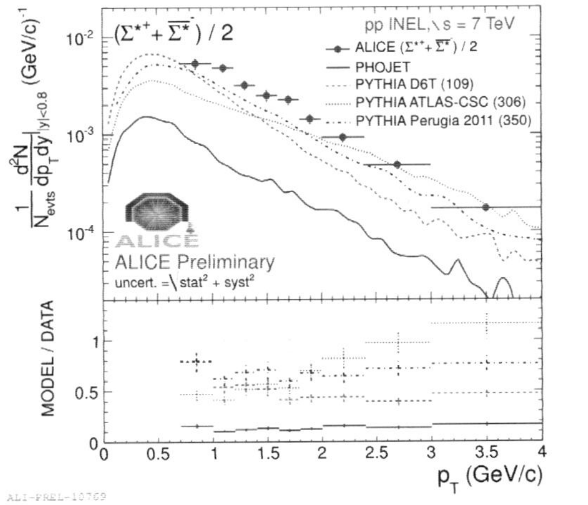

Figure 6. Comparison of the $(\Sigma^{*+} + \overline{\Sigma^{*-}})/2$ $p_{\rm T}$ spectrum in inelastic pp collisions at $\sqrt{s}$=7 TeV with PHOJET and PYTHIA tunes.

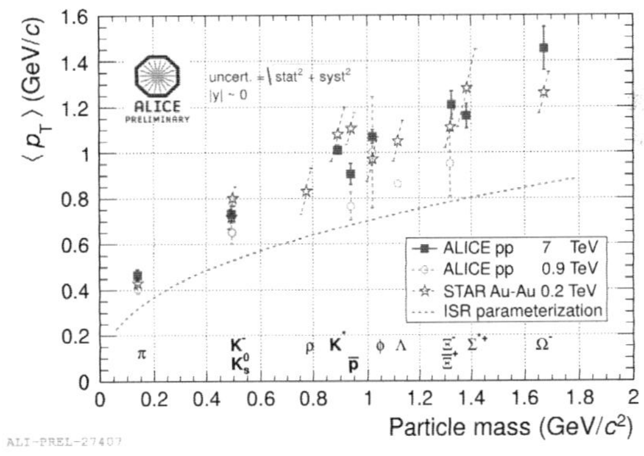

Figure 7. Mean transverse momentum $\langle p_{\rm T}\rangle$ as a function of the particle mass. Au-Au data at $\sqrt{s_{\rm NN}}$=0.2 TeV [22]; ALICE data for pp collisions at $\sqrt{s}$=0.9 TeV [23, 24]; ALICE data at $\sqrt{s}$=7 TeV [25].

Pb–Pb data, respectively. Final yields were obtained normalizing to the number of inelastic collisions (pp collisions) or to the number of analyzed events in a given centrality range (Pb–Pb collisions). The trigger efficiency was also taken into account. The analyses of $K^*(892)^0$ and $\phi(1020)$ mesons in pp collisions at $\sqrt{s}$=7 TeV are described in detail in [15].

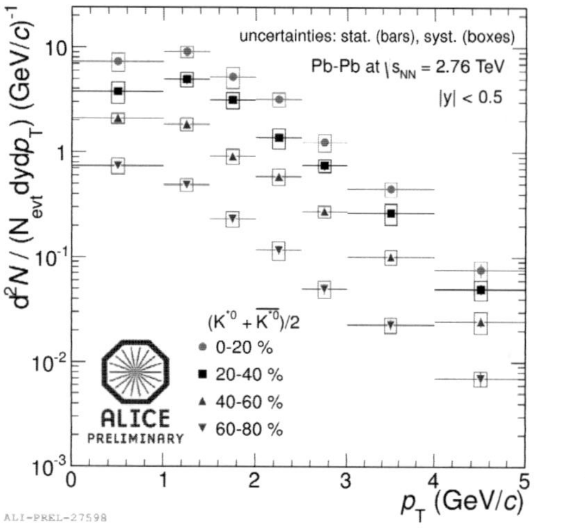

Figure 8. K* p_T spectra in Pb–Pb collisions at $\sqrt{s_{NN}}$=2.76 TeV at different centralities.

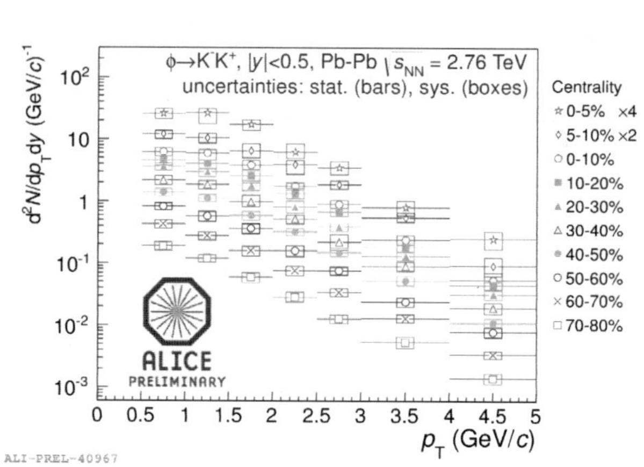

Figure 9. $\phi(1020)$ p_T spectra in Pb–Pb collisions at $\sqrt{s_{NN}}$=2.76 TeV at different centralities.

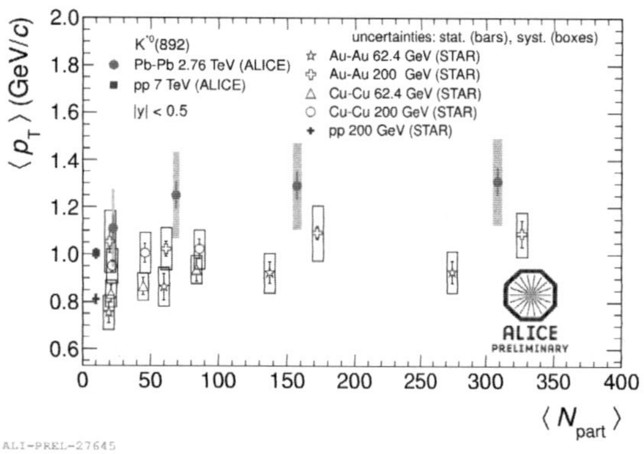

Figure 10. Mean transverse momentum of K* meson as a function of the mean number of participants for STAR [31] and ALICE data.

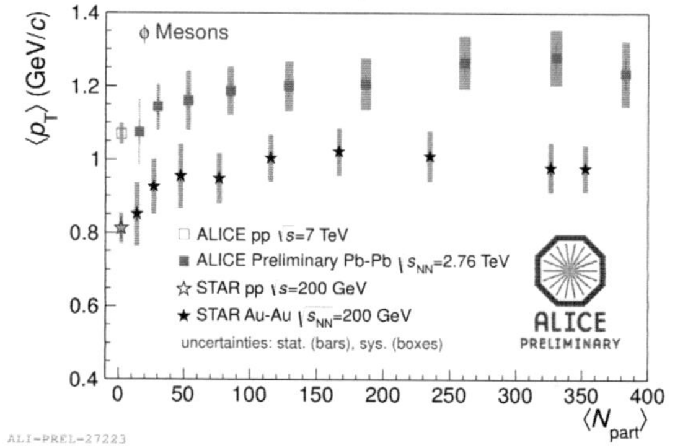

Figure 11. Mean transverse momentum of $\phi(1020)$ meson as a function of the mean number of participants for STAR [20, 28] and ALICE data [15].

3. Results

3.1. pp collisions at $\sqrt{s}$=7 TeV

Figures 4, 5 and 6 show the K* [1], $\phi(1020)$ and $(\Sigma^{*+} + \overline{\Sigma^{*+}})/2$ spectra with a comparison to a number of PYTHIA [16, 17, 18] tunes and PHOJET [3]. The best agreement is found for the recent PYTHIA Perugia 2011 tune. In fact there is a good agreement of the model to the data for the K*. It reproduces only the high p_T part (p_T>3 GeV/c) of the ϕ spectrum. Very poor agreement is found for the baryonic resonance.

The spectra have been fitted by a Tsallis function [19] and the extracted n values (6.2±0.07±0.8 and 6.7±0.20±0.4 for K* and ϕ, respectively) are similar to those quoted by the STAR experiment at RHIC for the ϕ measured in pp collisions at 200 GeV ($n = 8.3 \pm 1.2$) [20].

[1] We denote by K* the average of K^{*0} and $\overline{K^{*}}^{0}$.

In contrast, the slope parameters (254±2±18 and 272±4±11 MeV for K* and ϕ, respectively) are significantly higher than the values obtained at RHIC, $T = 202 \pm 14 \pm 11$ MeV for ϕ [21]. For the ϕ the inelastic yield increases proportionally to the charged particle multiplicity from 0.9 to 7 TeV. The mean transverse momenta $\langle p_T \rangle$ for K*, $\phi(1020)$ and $\Sigma(1385)^{\pm}$ are presented in Fig. 7 as a function of the particle mass. They follow the trend observed for stable particles indicating a common production mechanism. In this figure the $\langle p_T \rangle$ for different system and various collision energies (ALICE pp data at 0.9 and 7 TeV, STAR Au-Au data at $\sqrt{s_{NN}}$=0.2 TeV) are also shown.

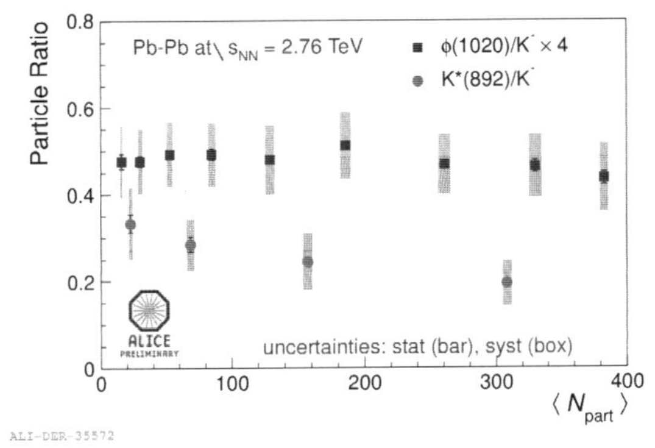

Figure 12. K*/K, ϕ/K ratios as a function of mean number of participants in Pb–Pb collisions at $\sqrt{s_{NN}}$=2.76 TeV.

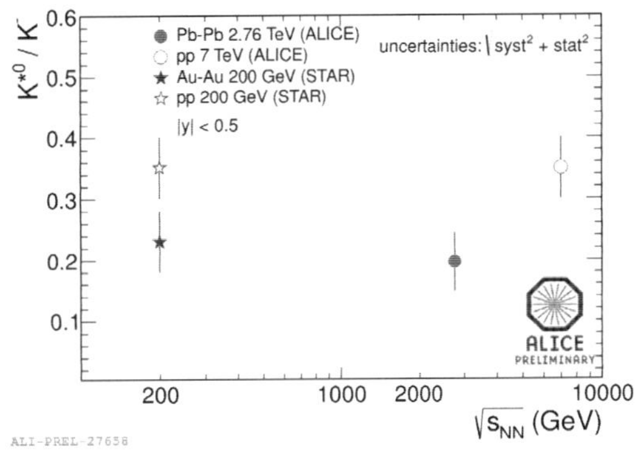

Figure 13. K*/K ratio as a function of the collision energy. Open symbols are for pp data, closed symbols for ion-ion data [31, 15].

3.2. *Pb–Pb collisions at $\sqrt{s_{NN}}$=2.76 TeV*

The transverse momentum spectra for K* and ϕ in several event centrality bins are shown in Figs. 8 and 9. The mean transverse momentum as a function of the mean number of participant $\langle N_{part} \rangle$ at LHC and at RHIC energies for K* and $\phi(1020)$ is shown in Figs. 10 and 11, respectively. We note that the $\langle p_T \rangle$ measured in pp collisions at $\sqrt{s}$=7 TeV is equal to the value measured in Pb–Pb peripheral collisions at $\sqrt{s_{NN}}$=2.76 TeV. For the ϕ, the $\langle p_T \rangle$ at LHC energies is larger than the one at RHIC energies. This is consistent with a stronger radial flow at LHC than RHIC. In fact, a global blast-wave fit of π, K, p shows a 10% increase in $\langle \beta_T \rangle$ with respect to RHIC [29], for central collisions.

In heavy ion collisions, the yields for stable and long-lived hadrons reflect the thermodynamic conditions (temperature, chemical potentials) at freeze-out, whereas the yield for short-lived resonances can be modified by final-state interactions inside the hot and dense reaction zone [5, 30]. Particularly interesting is the comparison of $\phi(1020)$ and K* production, considering the different lifetimes (about a factor of 10) of the two resonances.

While the ϕ/K ratio is independent of the collision centrality, the K*/K ratio decreases with increasing centrality (Fig.12). Figure 13 shows that the K*/K ratio in pp collisions is the same at RHIC and at the LHC. In contrast, this ratio decreases in heavy ion collisions and this effect seems to be larger at the LHC energies. The difference in K* and ϕ production could be related to the interaction of K* with hadronic medium, which does not affect the ϕ yield due to its long lifetime. Figure 14 shows the decrease of the K*/K ratio with respect to the system size represented by $(dN_{ch}/d\eta)^{1/3}$ for the different data sets of different collision systems and energies. The fact that they all fall on a common line could indicate that the observed decrease is related to the radial extent of the fireball. The decrease could be caused by the pion rescattering mechanism ($\sigma(\pi,\pi)$), which destroyes the pion-kaon correlation of the K* decay products. However, there is also the possibility that the ratio K*/K changes from chemical to kinetic freeze-out just by the variation of the temperature.

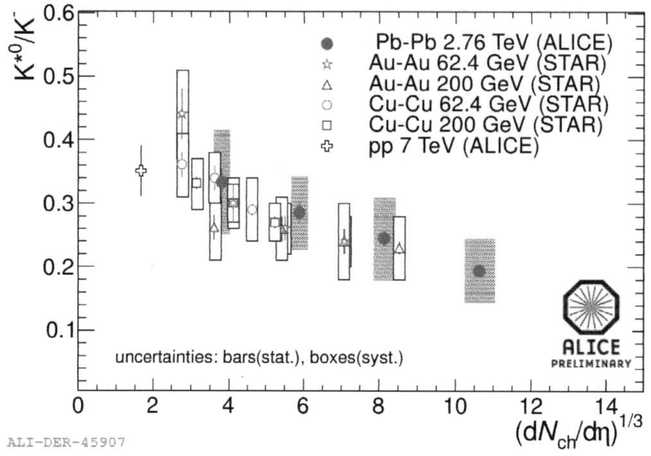

Figure 14. K*/K ratio as a function of $(dN_{\rm ch}/d\eta)^{1/3}$. STAR data are from [31].

4. Conclusions

The hadronic resonance $K^*(892)^0$, $\phi(1020)$ and $\Sigma(1385)^{\pm}$ have been measured in pp collisions at $\sqrt{s}$=7 TeV and in Pb–Pb collisions at $\sqrt{s_{\rm NN}}$=2.76 TeV for different event centrality bins by the ALICE experiment at the LHC. Transverse momentum spectra of $K^*(892)^0$, $\phi(1020)$ and $\Sigma(1385)^{\pm}$ measured in pp collisions have been compared to PHOJET and different PYTHIA tunes. Only the $K^*(892)^0$ is reproduced by Perugia 2011 PYTHIA tune. None of tunes gives a satisfactory description of the $\phi(1020)$ and $\Sigma(1385)^{\pm}$ data.

In proton-proton collisions the mean transverse momenta increase with the collision energy and they follow the trend of the stable particles.

While the ϕ/K is rather flat versus centrality, the K*/K ratio decreases with centrality. It is shown that this decrease is related to the radial extension of the fireball.

References

[1] Aguilar-Benitez M et al. (LEBC-EHS Collaboration) 1991 Inclusive particle production in 400 GeV/*c* pp-interactions *Z. Phys.* C **50** 405
[2] Albrecht H et al. 1994 Inclusive production of K*(892), $\rho^0(770)$, and $\omega(780)$ mesons in the Υ energy region *Z. Phys.* C **61** 1
[3] Engel R 1995 Photoproduction within the two component dual parton model. Amplitudes and cross-sections *Z. Phys.* C **66** 203; Engel R and Ranft J 1996 Hadronic photon-photon interactions at high-energies *Phys. Rev.* D **54**, 4244.
[4] Sjöstrand T, Mrenna S and Skands P 2006 PYTHIA 6.4 physics and manual *J. High Energy Phys.* **05** 026
[5] Bleicher M and Aichelin J 2002 Strange resonance production: Probing chemical and thermal freezeout in relativistic heavy ion collisions *Phys. Lett.* B **530** 81
[6] Bleicher M and Stöcker H 2004 Dynamics and freeze-out of hadron resonances at RHIC *J. Phys. Nucl. Part. Phys.* G. **30** S111
[7] Markert C et al. 2002 Strange hadron resonances: freeze-out probes in heavy-ion collisions *Proceedings* of PASI 2002 and *hep-ph*/0206260
[8] Vogel S and Bleicher M 2005 Resonance absorption and regeneration in relativistic heavy ion collisions *Proceedings* of Nucl. Phys. Winter Meeting 2005 in Bormio and *nucl-th*/0505027v1

[9] Andronic A et al. 2009 Thermal hadron production in relativistic nuclear collisions: the hadron mass spectrum, the horn, and the QCD phase transition *Phys. Lett.* B **673** 142

[10] Andronic A et al. 2012 The statistical model in Pb-Pb collisions at the LHC *arXiv* 1210.7724v1

[11] Torrieri G and Rafelski J 2001 Strange hadron resonances as a signature of freeze-out dynamics *Phys. Lett.* B **509** 239

[12] Aamodt K et al. (ALICE Collaboration) 2008 The ALICE experiment at the CERN LHC *J. Instrum.* **3** S08002

[13] Aamodt K et al. (ALICE Collaboration) 2010 Alignment of the ALICE Inner Tracking System with cosmic-ray tracks *J. Instrum.* **5** P03003

[14] Alme J et al. 2010 The ALICE TPC, a large 3-dimensional tracking device with fast readout for ultra-high multiplicity events *Nucl. Instrum. Meth. in Phys. Res.* A **622** 316

[15] Abelev B et al. (ALICE Collaboration) 2012 Production of $K^*(892)^0$ and $\phi(1020)$ in pp collisions at $\sqrt{s}$=7 TeV *Eur. Phys. J.* C **72** 2183

[16] Field R 2008 Physics at the Tevatron *Acta Phys. Pol.* B **39** 2611; Field R 2008 Studying the underlying events at CDF and the LHC *Proceedings* of the First International Workshop on Multiple Partonic Interactions at the LHC, (MPI08), Perugia(Italy) 27-31 October(2008),pag. 12 *arXiv*:1003.4220

[17] Buttar C et al. 2004 Simulations of minimum bias events and the underlying event, MC tuning and predictions for the LHC *Acta Phys. Pol.* B **35** 433

[18] Skands P Z 2010 Tuning Monte Carlo Generators: The Perugia Tunes *Phys. Rev.* D **82** 074018

[19] Tsallis C 1988 Possible generalization of Boltzmann-Gibbs statistics *J. Stat. Phys.* **52** 479

[20] Abelev B I et al. (STAR Collaboration) 2009 Measurements of ϕ meson production in relativistic heavy-ion collisions in the BNL Relativistic Heavy Ion Collider (RHIC) *Phys. Rev.* C **79** 064903

[21] Adams J et al. (STAR Collaboration) 2005 $K(892)^*$ resonance production in Au+Au and p+p collisions at $\sqrt{s_{NN}}$=200 GeV *Phys. Rev.* C **71** 064902

[22] Abelev B I et al. (STAR Collaboration) 2007 Strange particle production in p+p collisions at $\sqrt{s}$= 200-GeV *Phys. Rev.* C **75** 064901

[23] Aamodt K et al. (ALICE Collaboration) 2011 Strange particle production in proton-proton collisions at $\sqrt{s}$=0.9 TeV with ALICE at the LHC *Eur. Phys. J.* C **71** 1594

[24] Aamodt K et al. (ALICE Collaboration) 2011 Production of pions, kaons and protons in pp collisions at $\sqrt{s}$=900 GeV with ALICE at the LHC *Eur. Phys. J.* C **71** 1655

[25] Abelev B et al. (STAR Collaboration) 2012 Multi-strange baryon production in pp collisions at $\sqrt{s}$=7 TeV with ALICE *Phys. Lett.* B **712** 309

[26] Abelev B et al. (STAR Collaboration) 2006 Strange baryon resonance production in $\sqrt{s_{\rm NN}}$=200 GeV p+p and Au+Au collisions *Phys. Rev. Lett.* **97** 132301

[27] Abelev B et al. (STAR Collaboration) 2009 Systematic measurements of identified particle spectra in pp, d+Au, and Au+Au collisions at the STAR detector *Phys. Rev.* C **79** 034909

[28] Adams J et al. (STAR Collaboration) 2005 ϕ meson production in Au+Au and p+p collisions at $\sqrt{s_{\rm NN}}$=200 GeV *Phys. Lett.* B **612** 181

[29] Abelev B et al. (ALICE Collaboration) 2012 Pion, kaon, and proton production in central Pb-Pb collisions at $\sqrt{s_{\rm NN}}$=2.76 TeV *Phys. Rev. Lett.* **109** 252301

[30] Markert C 2005 Resonance production in heavy ion collisions *J. Phys. Nucl. Part. Phys.* G **31** S897

[31] Aggarwal M M et al. (STAR Collaboration) 2011 K^{*0} production in Cu+Cu and Au+Au collisions at $\sqrt{s_{\rm NN}}$=62.4 GeV and 200 GeV *Phys. Rev.* C **84** 034909

Journal of Physics: Conference Series **455** (2013) 012004 doi:10.1088/1742-6596/455/1/012004

Overview of results from ALICE at the CERN LHC

Panagiota Foka for the ALICE Collaboration

GSI Helmholtzzentrum für Schwerionenforschung GmbH, Planckstr. 1, 64291 Darmstadt, Germany

E-mail: yiota.foka@cern.ch

Abstract. ALICE, a general-purpose heavy-ion detector at the CERN Large Hadron Collider, has collected data from pp, p–Pb and Pb–Pb collisions. The results show intriguing properties of the produced matter in Pb–Pb collisions at $\sqrt{s_{\rm NN}} = 2.76$ TeV demonstrating that the created system is larger, hotter and denser compared to the one created in heavy-ion collisions at lower energies and it still behaves like an almost perfect, strongly interacting liquid. The high energy and luminosity at LHC open up the possibility to explore high-$p_{\rm T}$ and high-mass observables as probes of the medium at the new energy regime.

1. Introduction

QCD calculations on the Lattice predict that, at certain critical conditions of energy density and temperature, normal nuclear matter undergoes a phase transition to a deconfined state of quarks and gluons known as Quark-Gluon Plasma (QGP), where chiral symmetry is also restored [1, 2]. Experimentally, ultra-relativistic heavy-ion collisions make it possible to reach and exceed the critical energy density (with latest estimates being close to 0.5 GeV/fm^3 [1]) thus allowing to study, in the laboratory, strongly interacting matter under extreme conditions and the details of the QCD phase transition. After intense theoretical and experimental research at SPS and RHIC the first experimental results from the LHC heavy-ion program [3, 4] confirmed the RHIC observations [5] and provided additional evidence of the existence of this new state of matter at the new energy regime [6]. Hence, the focus now turns towards the detailed characterization of the properties of the QGP [7].

ALICE, designed as a multi-purpose detector [8], measures a multitude of observables characterizing the different stages of the collision with the aim to deduce the properties of the created system in terms of its equation of state and transport properties [9]. A short overview of ALICE experimental results, mainly from the analysis of Pb–Pb collisions at $\sqrt{s_{\rm NN}} = 2.76$ TeV, is presented.

2. Bulk properties

The properties of the bulk matter created in heavy-ion collisions and its dynamic evolution is studied by measuring the characteristics of the majority of particles with momenta below a few GeV/c. They include multiplicity distributions, which can be related to the initial energy density, yields and momentum spectra of identified particles, which are determined by the conditions at and shortly after hadronisation, and particle correlations, which measure the space-time evolution of the source and its collective transport properties.

Journal of Physics: Conference Series **455** (2013) 012004 doi:10.1088/1742-6596/455/1/012004

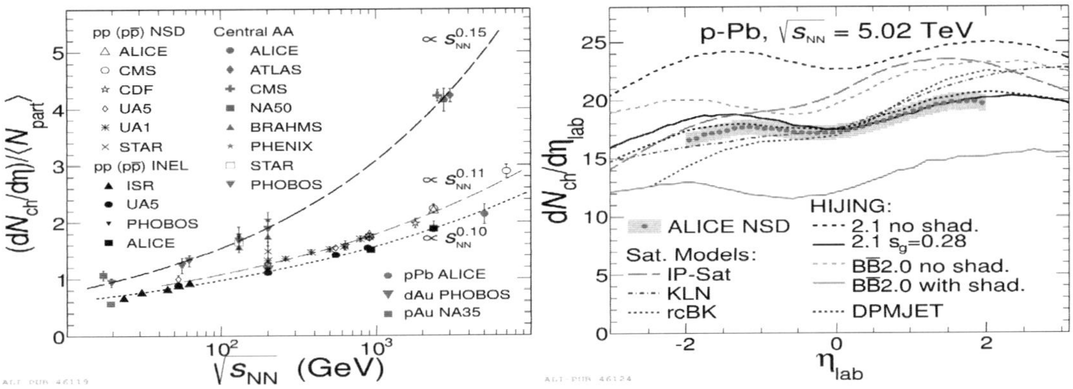

Figure 1. (Left) Charged particle pseudorapidity density, $dN_{ch}/d\eta$, per participant as a function of $\sqrt{s_{NN}}$ for nucleus–nucleus, proton–nucleus and proton–proton collisions. (Right) Charged particle pseudorapidity density, $dN_{ch}/d\eta_{lab}$, in NSD p–Pb collisions at $\sqrt{s_{NN}}$ = 5.02 TeV compared to theoretical predictions.

Multiplicity Particle production at low-p_T is a non-perturbative process and cannot be calculated based on first principles (QCD). The first measurements of particle multiplicities at LHC were eagerly awaited to constrain model predictions which varied widely even after the RHIC data became available. Within days after the first ion collisions a charged particle pseudorapidity density of $dN_{ch}/d\eta \approx 1600$ was published by ALICE for the 5% most central Pb–Pb collisions in the central pseudorapidity $|\eta| < 0.5$ [10]. Compared to the RHIC measurement of $dN_{ch}/d\eta \approx 690$ for the 5% most central Au–Au collisions at $\sqrt{s_{NN}} = 200$ GeV/c [11] shows an increase of ≈ 2.3.

Figure 1–left presents the charged particle pseudorapidity density per participant measured in pp, p–Pb and Pb–Pb collisions by the LHC experiments [12, 13, 14] compared to nucleon-nucleon and nucleus-nucleus data at lower energies. The energy dependence of $dN_{ch}/d\eta$ shows that particle production in Pb–Pb at LHC is no longer compatible with a logarithmic dependence with $\sqrt{s_{NN}}$, as it was true for the data up to the top RHIC energy [15], but follows a power law with an exponent of 0.15. The proton–nucleus and proton–proton data show a less steep dependence on energy than the Pb–Pb data and they can both be described by a power law with slightly different exponents of 0.10 and 0.11, respectively.

To shed light into the different production mechanisms p–Pb studies are of particular interest because of their potential to disentangle final-state effects characteristic of the formation of hot QCD matter, from initial-state effects already present in cold nuclear matter. Figure 1–right shows the $dN_{ch}/d\eta_{lab}$, measured by ALICE, in Non-Single Diffractive (NSD) p–Pb collisions at $\sqrt{s_{NN}}$ = 5.02 TeV [16]. The forward–backward asymmetry between the proton and lead hemisphere is clearly visible. The measurement is compared to particle production models [17, 18, 19, 20, 21] that describe similar measurements in other collision systems; see for details [16]. Most models which include shadowing [21] or saturation [17, 20], approximately get the right value of $dN_{ch}/d\eta_{lab}$. HIJING 2.1 [21], where the gluon shadowing parameter $s_g = 0.28$ was tuned to describe the RHIC data [22, 15], describes the pseudorapidity distribution quite well.

Energy density The measured $dN_{ch}/d\eta$ can be related to the initial energy density of the created system using the formula proposed by Bjorken [23]. The energy density reached in central Pb–Pb collisions of about $\epsilon = 15$ GeV/fm^3 [24] is almost three times higher than the one reported at RHIC [5] and well above the critical energy density required for the predicted phase transition.

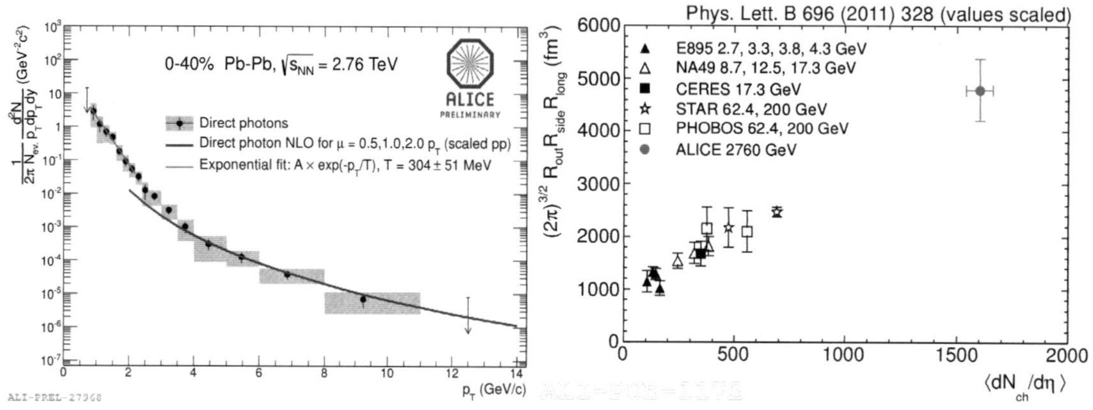

Figure 2. (Left) Direct-photon p_T-spectrum with NLO prediction and exponential fit at the low-p_T region. (Right) Freeze-out volume of the source as measured by identical pion interferometry compared to measurements at lower energies.

Initial temperature The relative increase of energy density from RHIC to LHC implies an initial temperature at LHC of $\approx$ 300 MeV for central Pb–Pb collisions. This is confirmed by the measurement of direct thermal photons which are emitted at the very initial stage of the collision. Figure 2–left shows the p_T spectrum of direct photons measured by ALICE using the measurement of γ conversions for the 40% most central Pb–Pb collisions [25]. At $p_T > 4$ GeV/c the spectrum is well reproduced by the NLO pQCD prediction for pp collisions, scaled by the number of binary collisions. However, below 2 GeV/c there is an excess attributed to thermal photons. An exponential fit in the p_T range 0.8–2.2 GeV/c yields an inverse slope parameter $T = (304 \pm 51)$ MeV, where the quoted error includes both the statistical and systematic uncertainties. The LHC value is about 40% higher than that measured in a similar analysis by PHENIX [26].

Source size and lifetime The space-time evolution of the expanding system has been studied using identical pion interferometry techniques (known as Hanbury-Brown-Twiss correlations [27]) in 5% central Pb–Pb collisions showing that the freeze-out volume (the size of matter when strong interactions cease) is 5000 fm^3, two times larger than the one measured at RHIC, and its total lifetime (the time between collisions and freeze-out) is 10 fm/c, 30% larger than at RHIC. The extracted volume, plotted as a function of charged particle multiplicity, together with the world data, shows a linear monotonic increase, seen in Fig. 2–right, where the freeze-out volume is given as the product of a geometrical factor and the HBT-radii (called $R_{\text{long}}, R_{\text{side}}, R_{\text{out}}$) [28]. Extrapolation to zero $\text{d}N_{\text{ch}}/\text{d}\eta$ shows that the source size coincides with the initial volume of a Pb nucleus ($\approx$ 800 fm^3) and the source lifetime, not shown here, vanishes [6].

In general, and as expected, the system created in central Pb–Pb collisions at $\sqrt{s_{\text{NN}}} =$ 2.76 TeV at LHC is characterized by higher energy density and temperature, bigger volume at freeze-out and longer life-time, than the one created at the highest RHIC energies in Au–Au collisions at $\sqrt{s_{\text{NN}}} = 200$ GeV. Further details of its properties are reflected at the spectra and correlations of different particle species.

2.1. Identified particle spectra and yields

The energies available at the LHC open up the possibility for detailed measurements at an extended p_T range. In ALICE p_T spectra of identified particles, shown in Fig. 3, were measured

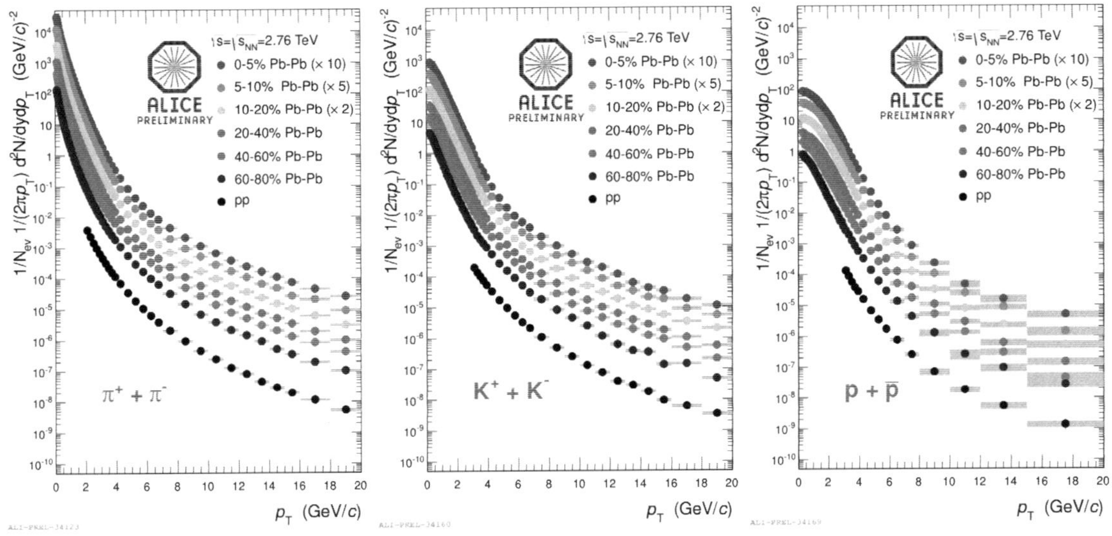

Figure 3. Transverse momentum (p_T) spectra of identified particles in Pb–Pb and pp collisions up to 20 GeV/c using different particle identification techniques.

by various techniques, and for the first time up to $p_T = 20$ GeV/c for π, K and p including the TPC dE/dx measurements at the relativistic rise above 5 GeV/c [29]. Measurements at RHIC have shown that at low $p_T < 2$ GeV/c, the dynamics of bulk matter can be described by Relativistic Hydrodynamic Models (RHM), at high $p_T > 8$ GeV/c, the spectra reflect the interaction of partons from hard scatterings with the medium while at intermediate p_T, the interplay of soft and hard processes is reflected.

Even at LHC energies about 95% of particles are produced with p_T below ≈ 1.5 GeV/c reflecting the dynamics of the expanding system which shows collective behaviour that can be described in terms of hydrodynamics [30].

Transverse momentum spectra at low-p_T In particular, the p_T distributions at low-p_T reflect the conditions at kinetic freeze–out, where particle momenta are fixed. Details on the ALICE measurements of low-p_T spectra of identified charged hadrons (π, K and p) are presented in this conference [31]. The spectra, measured in the p_T range $0.1 < p_T < 4.5$ GeV/c at mid–rapidity ($|y| < 0.5$), for the 5% most central Pb–Pb collisions [32], are harder than the ones measured in central Au–Au collisions at $\sqrt{s_{NN}} = 200$ GeV at RHIC [33, 34] reflecting the stronger radial flow at LHC. A blast-wave fit of the spectra [35] yields the values of the kinetic freeze–out temperature $T_{kin} = 96 \pm 10$ MeV, similar to the one at RHIC, and of the collective radial flow velocity $\langle \beta_T \rangle = 0.65 \pm 0.02$, 10% higher than the one at RHIC. The ALICE data compared to hydrodynamical calculations [36, 37, 38] show better agreement when interactions during the hadronic phase are taken into account.

Particle yields and thermal models The conditions at chemical freeze–out, where particle abundances are fixed, are characterized by the chemical freeze–out temperature (T_{ch}) and baryochemical potential (μ_B) and are determined from measured particle yields via thermal model calculations. The ALICE measurements for the 20% most central Pb–Pb collisions, presented in this conference [31], are compared to a thermal model prediction [39] with chemical freeze-out temperature $T_{ch} = 164$ MeV and baryochemical potential $\mu_B = 1$ MeV, very similar to the prediction of $T_{ch} = 170$ MeV given in [40]. The measured yields of p and Λ's are significantly lower than those predicted by the model calculations and the ones measured at RHIC. When adjusting T_{ch} to describe the data, a much lower temperature $T_{\mathrm{ch}} = 152$ MeV is obtained and the fit does not reproduce the data well (χ^2=39.6 per 9 d.o.f.) [41]. A possible explanation for these deviations from the thermal model predictions may be interactions in the hadronic phase. In particular, large cross sections for baryon–antibaryon annihilation can be the origin of the lower yields of some baryons [42, 43]. This is also supported by ALICE femtoscopic measurements of $\mathrm{p\bar{p}}$ and $\Lambda\bar{\Lambda}$ correlations [44, 45].

2.2. Azimuthal anisotropies

Measurements of azimuthal particle anisotropies probe collective phenomena characteristic of a bulk system such as the one expected to be created in heavy–ion collisions [46]. In non–central collisions, pressure gradients, developed in the overlap region of the two colliding nuclei, transform, through interactions between the produced particles, the initial spatial anisotropy into an observed momentum anisotropy, leading to an anisotropic particle distribution $\mathrm{dN}/\mathrm{d}\varphi$. This anisotropy is usually quantified via a Fourier expansion of the azimuthal distribution of particles [47], φ, with the Fourier (or flow) coefficients v_n dependent on p_{T} and given by:

$$\mathsf{v}_n = \langle \cos\left[n(\varphi - \Psi_n)\right]\rangle, \tag{1}$$

where n is the order of the flow harmonic, φ the azimuthal angle of the particle and Ψ_n the azimuthal angle of the initial state spatial plane of symmetry of harmonic n. The second Fourier coefficient, v_2, is called elliptic flow with $\Psi_2 \approx \Psi_{RP}$ where Ψ_{RP} is the reaction-plane angle, defined by the beam direction and the impact parameter plane. It has been extensively studied as it was first suggested in [48] as a measure of collective phenomena of bulk matter. It was contrasted to thermal motion, a superposition of independent pp collisions, where particle momenta are not correlated relative to the reaction plane.

Higher order odd harmonics had usually been neglected because they were expected to be zero due to symmetry reasons. However, the statistical nature of individual nucleon–nucleon collisions can lead to highly irregular shapes of the reaction zone and the corresponding initial energy and pressure distributions [49, 50] resulting in event-by-event fluctuations of the elliptic flow direction and magnitude and other v_n coefficients. Different experimental methods are used to measure the symmetry plane angles and the different v_n coefficients, via two- and multi-particle correlations [51, 52, 53] each one of them being sensitive to different effects, thus allowing a comprehensive study of non-flow contributions and fluctuations.

Properties of the medium and hydrodynamic models The first ALICE measurements [3] confirmed hydrodynamic predictions and indicated that the system created in Pb–Pb collisions at LHC still behaves as a strongly interacting almost perfect liquid with minimal shear viscosity to entropy ratio, η/s, like the one at RHIC [54, 55, 56, 57, 58]. Additional constrains on η/s were obtained studying v_2 as a function of centrality and p_{T} for different particle species as well as from measurements of v_3. Comparison with models yields η/s in the range 1–2.5 $\times$ $1/4\pi$ (with $\hbar = k_B = 1$) [46], very close to the lower bound conjectured by AdS/CFT for a good relativistic quantum fluid [59], where the uncertainty in the determination of η/s comes mainly from the definition of the initial conditions. Resent results [46] show that IP-Glasma initial conditions [60] and average values for η/s of 0.2 for Pb–Pb collisions at the LHC and 0.12 for Au–Au collisions at top RHIC energies, provide a good description of all currently available data.

v_n measurements in extended pseudorapidity ALICE extended the azimuthal anisotropy measurements both in pseudorapidity and p_{T} [61, 62]. At mid-rapidity five harmonics were measured while v_2 and v_3 were measured up to $\eta = 5$. Figure 4–left shows the pseudorapidity dependence of v_2 and v_3 up to $\eta = 5$ for very central (0–5%) and peripheral (50–60%) collisions. The anisotropic flow is almost independent of pseudorapidity for $|\eta| < 2$ and decreases at

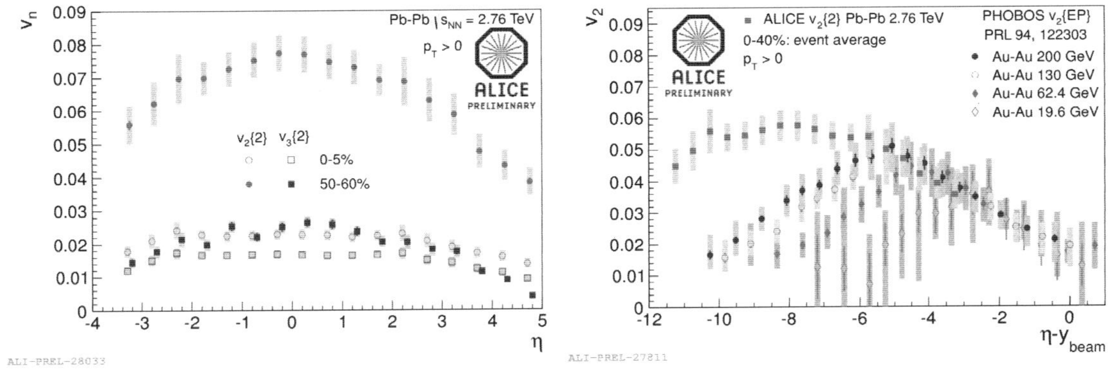

Figure 4. (Left) v_2 and v_3 pseudorapidity dependence for very central (0–5)%) and peripheral (60–80)% Pb–Pb collisions. (Right) v_2 longitudinal scaling over two orders of magnitude in collision energy.

larger rapidities. The integrated v_2 increases by 20–30% at mid-rapidity and about 30% at forward rapidity relative to RHIC, at the higher limit of predictions, but still in agreement with hydrodynamic model calculations. v_3 shows a weak centrality dependence with similar magnitude for the very central and peripheral collisions. For the 5% most central collisions v_2 exhibits a similar trend and magnitude as v_3. These measurements indicate that v_2 is driven by geometry while v_3 is driven by initial state fluctuations, which are also generating a non-zero v_2 for the very central collisions. The elliptic flow measurements in the extended pseudorapidity range were performed with two- as well as four-particle cumulant methods allowing to study flow fluctuations up to very forward pseudorapidities [61].

Extending the measurement to $\eta = 5$ allows testing the limiting fragmentation picture first observed at RHIC; namely, v_2 was found to be independent of collision energy when studied in the rest frame of one of the colliding nuclei. The ALICE v_2 measurement, shown in Fig. 4–right, together with the RHIC data, confirms that this longitudinal scaling holds from the lowest RHIC energy $\sqrt{s_{\rm NN}} = 19.6$ GeV up to the TeV regime at LHC $\sqrt{s_{\rm NN}} = 2.76$ TeV [63].

v_n measurements in extended p_T ALICE also extended the measurement of the charged particle anisotropies (v_2, v_3 and v_4) up to $p_T \approx 20$ GeV/c as shown in Fig. 5 [62]. The fourth order harmonic anisotropy was measured with respect to the second and fourth order event planes, $v_4(\Psi_2)$ and $v_4(\Psi_4)$. The difference between the two is totally due to fluctuations in the fourth order harmonic flow and as such provides important constraints on the physics and origin of the flow fluctuations. Significant non-zero v_2 and v_3 were found up to the highest measured transverse momenta. At $p_T > 10$ GeV/c the elliptic flow results are well described by a model [64] which takes into account collision and radiative energy loss in an expanding medium.

In general, the shape of the p_T-differential anisotropic flow suggests the existence of several regions in the transverse-momentum space with distinctly different underlying physics processes. It is now widely accepted that at $p_T < 1$–2 GeV/c the flow pattern is mostly determined by hydrodynamical flow exhibiting a typical "mass splitting" [65]. At higher momenta, $p_T > 10$ GeV/c, the anisotropy seems to be defined by the energy loss and one would expect little particle type dependence in this region. The intermediate p_T region is less understood.

v_2 of identified particles Hydrodynamic model calculations predict a characteristic dependence of v_2 on the particle mass at low values of transverse momentum ($p_T < 2$ GeV/c) induced by the collective radial expansion of the system, which being cumulative throughout the whole collision,

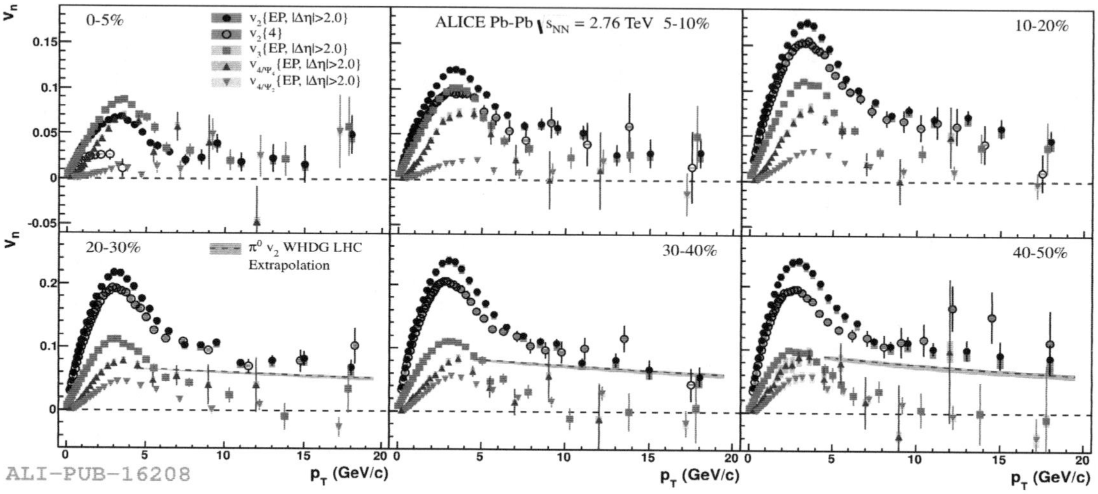

Figure 5. p_T-differential event anisotropies, v_2, v_3, v_4 measured for different centrality classes of Pb–Pb collisions up to $p_T \approx 20$ GeV/c.

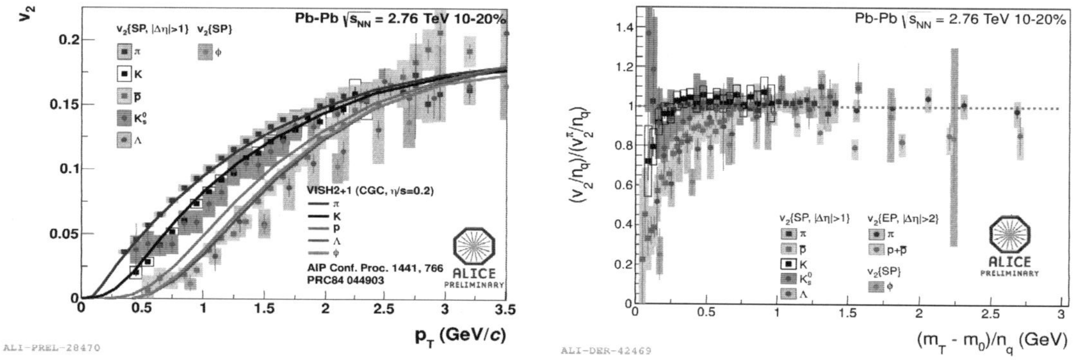

Figure 6. (Left) The p_T-differential v_2 for different particle species. (Right) The dependence of v_2/n_q on the scaled kinetic energy. In both plots, the 10–20% centrality bin is shown.

has a significant contribution from the partonic phase. However, the hadronic rescattering that follows might mask the information of the early stage. Hence, particles with small hadronic cross section could be more accurate probes of the early stage of the collision.

Figure 6–left shows the p_T-differential v_2 for different particle species (π, K, K^0_s, $\bar{p}$, Λ, ϕ) measured in the 10–20% centrality range. At $p_T \leq 2$ GeV/c, a clear mass ordering is seen attributed to collective radial flow that shifts heavier particles to higher p_T. Remarkable is the flow of the ϕ-meson which has a mass similar to the one of the proton. While at low-p_T the v_2 of ϕ is similar to the one of the protons, at high-p_T follows the trend of mesons. The measurements are compared to hydrodynamical calculations [36] using $\eta/s = 0.2$ and the Color Glass Condensate (CGC) initial conditions, shown by the solid curves. The model reproduces the p_T dependence of v_2 for π and K up to $p_T \approx 2$ GeV/c but overestimates v_2 for heavier particles. However, including a hadronic phase improves the agreement with the data.

Quark number scaling At RHIC it was observed that all baryons exhibit similar flow which differs from the flow of mesons, approximately by a factor 3:2 [66]. These findings led to the interpretation that hadron formation is dominated by quark coalescence at the end of the partonic evolution [67]. ALICE studied quark scaling at different centrality ranges and different representations. Figure 6–right shows the v_2 of identified particles in the same centrality range, 10–20%, scaled by the number of constituent quarks, n_q ($n_q = 2$ and 3 for mesons and baryons, respectively), as a function of the transverse kinetic energy, $KE_T = (m_t - m_0)$, scaled by n_q, and divided by the polynomial fit to the pion elliptic flow. In this representation quark scaling is broken below 1 GeV while at higher p_T holds at the level of 20%.

3. Hard processes and medium induced effects

The high energy and luminosity at LHC lead to large cross sections and production rates for hard processes, i.e. those involving high-momenta or high-mass scales. Such probes, originating from hard parton scatterings with large momentum transfer at the initial stage of the collision, are calculable with pQCD in elementary collisions and can be used to probe the created medium as they are expected to interact with it prior to hadronisation.

In particular, the energy of the partons is reduced through collisional energy loss and medium–induced gluon radiation. Based on QCD, it is predicted that gluons lose a larger amount of energy than quarks because of the dependence of the energy loss on the colour coupling factor [68]. In addition, the in–medium effects are expected to be further reduced for heavy quarks due to a reduced small angle gluon radiation known as the "dead–cone effect" [69].

On the other hand, the "J/Ψ suppression" was one of the first "QGP signatures" [70] and a pattern of "sequential melting" of quarkonia bound states was predicted to signal deconfinement due to colour screening [71, 72]. However, at LHC, due to the abundant production of c and b quarks a recombination effect may be expected during the collision history [73, 74, 75] or at hadronization [76].

Hence, the details of the production and propagation of high-p_T and high-mass probes are studied to explore the mechanisms of parton-energy loss and deconfinement in the medium. Experimentally, several techniques are used to investigate suppression effects. One can compare the relative production of single particles or fully reconstructed jets in nuclear collisions to expectations from superposition of independent nucleon–nucleon collisions or study angular particle correlations. The suppression is usually quantified by the nuclear modification factor, R_{AA}, which is the ratio of yields in A–A collisions to the pp cross section scaled by the number of binary collisions,

$$R_{AA} = \frac{(1/N^{AA}_{evt.})d^2N^{AA}_{ch}//dp_T d\eta}{\langle N_{coll}\rangle(1/N^{pp}_{evt.})d^2N^{pp}_{ch}/dp_T d\eta}, \quad (2)$$

where the number of binary nucleon–nucleon collisions $\langle N_{coll}\rangle$, is given by the product of the nuclear overlap function T_{AA} calculated from the Glauber model [77], and the inelastic NN cross–section σ^{NN}_{inel}. In the absence of nuclear effects R_{AA} is unity by construction.

3.1. High–p_T particles and jets

Charged particles R_{AA} in Pb–Pb collisions A large suppression of high-p_T hadrons was first observed at RHIC [5], in the 5% most central Au–Au collisions at $\sqrt{s_{NN}} = 200$ GeV relative to pp, whereas particles that do not interact strongly, e.g. photons, are not modified. The reported suppression is a factor of 5 at p_T = 5–6 GeV/c and independent of p_T up to 20 GeV/c. At LHC, the suppression is confirmed, the accessible p_T range is significantly extended, and the measurements include particles, such as Z, W and isolated photons, which show no modification [78, 79, 80].

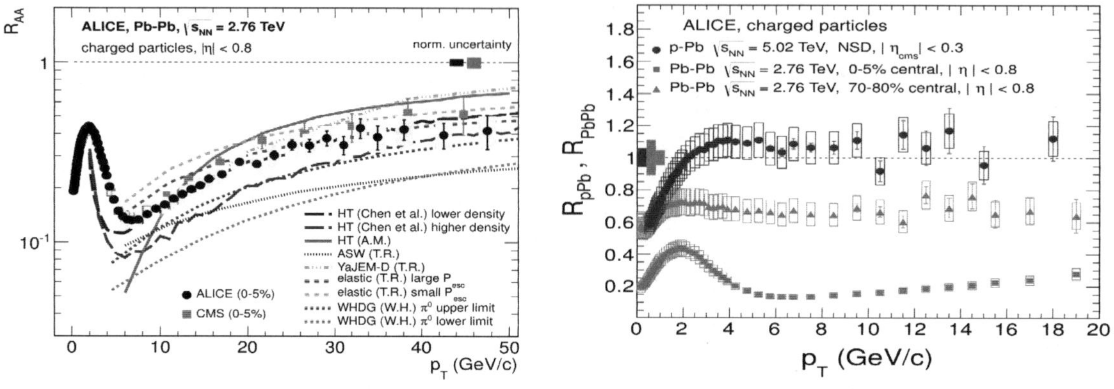

Figure 7. (Left) The nuclear modification factor $R_{\rm AA}$ for the 5% most central Pb–Pb collisions at $\sqrt{s_{\rm NN}}$ = 2.76 TeV, compared to CMS data and model calculations. (Right) The nuclear modification factor $R_{\rm pPb}$ for p–Pb collisions at $\sqrt{s_{\rm NN}}$ = 5.02 TeV compared to $R_{\rm PbPb}$ for the 5% most central and 70–80% peripheral Pb–Pb collisions.

Figure 7–left shows the charged particles $R_{\rm AA}$ for the 5% most central Pb–Pb collisions [81]. The ALICE results (full circles), measured at mid-rapidity, are in good agreement with the CMS measurement [82] (full squares) within the uncertainties. At $p_{\rm T} \approx 2$ GeV/c a pronounced maximum is observed which is interpreted as another manifestation of collective radial flow [6]. At higher $p_{\rm T}$ $R_{\rm AA}$ exhibits a minimum, reaching ≈ 0.13 at $p_{\rm T}$ = 5–7 GeV/c, a slightly lower value than the one at RHIC, where $R_{\rm AA} \approx 0.2$. Above 7 GeV/c, $R_{\rm AA}$ increases to ≈ 0.4 at $p_{\rm T}>$ 30 GeV/c and remains flat up to more than 100 GeV/c (as measured by CMS [82]). The measurements are compared to model calculations that used RHIC data to calibrate the medium density and implement different energy-loss mechanisms [83]. Several of them can reproduce qualitatively the increase of $R_{\rm AA}$ with increasing $p_{\rm T}$. This rise can be understood as a decrease of the parton fractional energy loss with increasing $p_{\rm T}$, reflecting the weak energy dependence of pQCD radiative energy loss on parton energy.

The ALICE $R_{\rm AA}$ measurement for π, K and p, presented in this conference [31], show that while at low $p_{\rm T}$ a hierarchy of suppression is observed, reflecting strong radial flow, at $p_{\rm T}$ above $\approx$ 8–10 GeV/c, the suppression seems to be the same for the different particle species indicating that the effect of the medium is the same for all light-flavoured quarks.

Charged particles $R_{\rm AA}$ in p–Pb collisions As a control experiment to clarify intial state effects the data from the p–Pb pilot run at $\sqrt{s_{\rm NN}}$ = 5.02 TeV are used [84] and the nuclear modification factor is compared to Pb–Pb results as shown in Fig. 7–right. The charged particles $p_{\rm T}$ distribution is measured around mid-rapidity in the $p_{\rm T}$ range $0.5 < p_{\rm T} < 20$ GeV/c. The $R_{\rm pPb}$, shows a suppression of charged particle production in p–Pb collisions for $p_{\rm T} < 2$ GeV/c followed by a rise to $R_{\rm pPb} \approx 1.1$ for $p_{\rm T} \approx 4$ GeV/c and is consistent with unity for $p_{\rm T} > 6$ GeV/c. The observed trends qualitatively resemble the $R_{\rm dAu}$ at RHIC although, the enhancement shown by ALICE data is less than that observed at RHIC. At low-$p_{\rm T}$ the suppression might be related to parton shadowing or saturation while the rise at $p_{\rm T} \approx 4$ GeV/c is interpreted as originating from multiple scattering in the initial phase of the collision and is known as Cronin effect.

The $R_{\rm pPb}$ is compared to the Pb–Pb central (0–5%) and peripheral (70–80%) collisions. Exhibiting a rather good $N_{\rm part}$ scaling for $p_{\rm T} > 5$ GeV/c the p–Pb data demonstrate that the strong suppression of hadron production at high $p_{\rm T}$ observed in central Pb–Pb collisions at LHC is a fingerprint of the hot quark-gluon matter created in Pb–Pb collisions.

Journal of Physics: Conference Series **455** (2013) 012004 doi:10.1088/1742-6596/455/1/012004

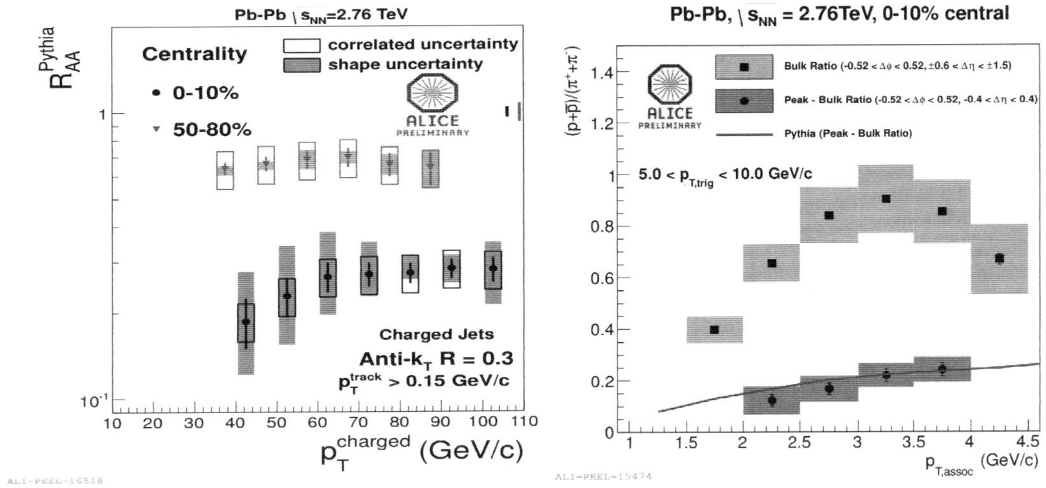

Figure 8. (Left) The nuclear modification factor for jets, $R_{\rm AA}^{\rm Pythia}$, with the reference jet spectrum extracted from Pythia Perugia0, for the 10% most central and 50–80% peripheral Pb–Pb collisions [85]. (Right) Ratio of proton to pion spectra as function of $p_{\rm T}$ for the "bulk" and "peak-bulk" angular regions in $\Delta\eta$–$\Delta\phi$ space compared to Pythia.

Reconstructed jets R_{AA} Similarly, the properties of the medium are reflected in the modification of fully reconstructed jets. The ALICE measurement of the jet nuclear modification factor $R_{\rm AA}^{\rm Pythia}$, shown in Fig. 8–left, are based on charged particles reconstructed with the TPC and the ITS [85] while the reference jet spectrum was extracted from Pythia and verified with a small sample of pp data at the same energy. The measurement covers the low-$p_{\rm T}$ region, down to $p_{\rm T} \approx 30$ GeV/c and extends up to $p_{\rm T} \approx 110$ GeV/c. The suppression pattern is similar to the one observed for single particles. The jets $R_{\rm AA}^{\rm Pythia}$ exhibits a strong suppression for central events and is gradually increasing for more peripheral collisions.

Particle composition in jet-like structures To study the jet hadronchemistry and the effects of the interplay of soft and hard processes the particle composition in jet-like structures is investigated studying the p/π ratio [86, 44]. Jet-like structures are selected via particle correlations in $\Delta\eta$ (difference in pseudorapidity) and $\Delta\phi$ (difference in azimuthal angle) relative to a trigger particle (with $p_{\rm T}$ in the 5–10 GeV/c range). The p/π ratio is defined in the following areas. (a) "bulk": a region far from the trigger particle (b) "peak": particles close in $\Delta\eta$–$\Delta\phi$ to the trigger particle (c) "jet" ("peak"–"bulk"): the region obtained subtracting "peak-bulk" after correcting the "peak" p/π ratio for the contribution of the underlying event.

Figure 8–right shows the ratio of the p/π spectra in the two angular regions, "bulk" and "jet" ("peak–bulk") as a function of $p_{\rm T}$. It is found that in the "near-side" of the "jet" region, the p/π ratio is consistent with the expectation from pp estimated by Pythia while in the "bulk" region it is compatible with the one obtained for non-triggered events of same centrality (in the $p_{\rm T}$ range 1.5–4.5 GeV/c where the analysis was performed) i.e. the ratio increases by factor 3–4 compared to pp. This suggests that there is no significant medium-induced modification of jet particle ratios and that the enhancement of the p/π ratio observed in Pb–Pb minimum bias events is a result of bulk processes and not jet fragmentation. The detailed understanding of the influence of the medium on jet fragmentation requires further studies.

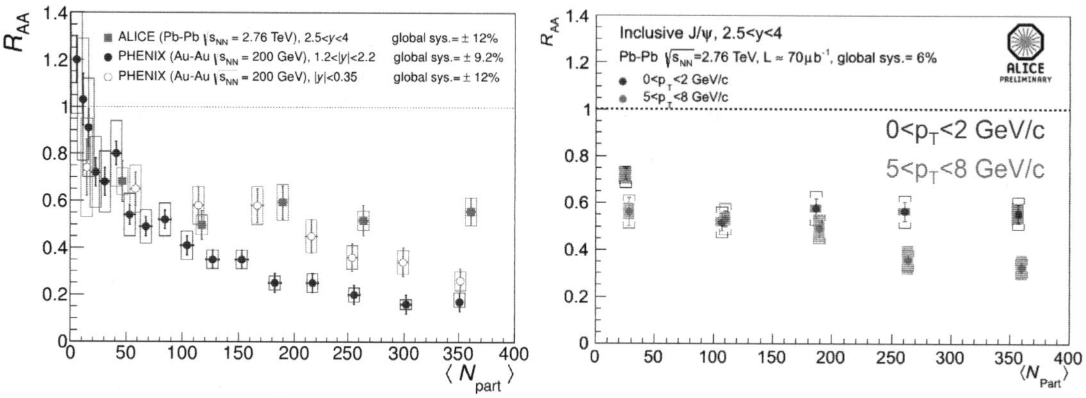

Figure 9. The centrality dependence of the nuclear modification factor for inclusive J/Ψ, measured at forward rapidity ($2.5 < y < 4$) via the $\mu^+\mu^-$ channel down to $p_{\rm T} = 0$ compared to RHIC data (left) and measured for low- and high-$p_{\rm T}$ regions (right).

3.2. Heavy flavour

The abundant production of c and b quarks at LHC (about 100 pairs in a central collision) makes possible highly differential measurements which can probe the details of the underlying physics processes such as deconfinement, energy loss, thermalisation and hadronisation mechanisms. After the first observations of heavy-quark energy loss and collective flow at RHIC [87] ALICE performs a comprehensive study of quarkonia production in nuclear and elementary collisions with emphasis on their dependence on different kinematic variables, measuring them in various decay channels and a wide kinematic range, as detailed in this conference [88] and [89, 90, 91].

Deconfinement and quarkonium suppression Figure 9–left shows the $R_{\rm AA}$ for inclusive J/Ψ as a function of the mean number of participants [92]. The ALICE measurement (full squares) was performed via the $\mu^+\mu^-$ channel at forward rapidity ($2.5 < y < 4$), and covers low-$p_{\rm T}$ down to $p_{\rm T} = 0$. It shows little centrality dependence, and gives a centrality integrated $R_{\rm AA}^{0\%-80\%} = 0.545$ ± 0.032 (stat.) ± 0.083 (syst.). The ALICE $R_{\rm AA}$ is compared with the RHIC measurements at forward (filled points) and central (open points) rapidity [93]. For central collisions, the ALICE data show less suppression than at RHIC at forward rapidity. The difference becomes smaller but still remains significant comparing the ALICE data to the RHIC mid-rapidity data [92]. The ALICE measurements at mid-rapidity show similar trends. For details of the analysis at mid-rapidity see [92, 94] and at forward rapidity [95, 96]. Further differential measurements, as presented in Fig. 9–right, show that a smaller suppression is observed at low-$p_{\rm T}$ ($p_{\rm T} < 2$ GeV/c) than at high-$p_{\rm T}$ ($5 < p_{\rm T} < 8$ GeV/c) especially in more central collisions.

Such results are qualitatively in agreement with recombination scenarios, where regeneration effects are expected to be important at low $p_{\rm T}$ and mid-rapidity for central collisions; further details on the measurements and comparisons with models are detailed in this conference [88].

D-mesons R_{AA} and suppression hierarchy Based on the QCD predictions on energy loss, see section 3.1, pions, mostly originating from gluons should be more suppressed than charmed particles and those than particles containing beauty, leading to a hierarchy of suppressions with $R^B_{AA} > R^D_{AA} > R^{charged}_{AA}$. Figure 10–left presents the $R_{\rm AA}$ for prompt D-mesons calculated as the average of the relevant factors for $\rm D^0$, $\rm D^+$ and $\rm D^{*+}$ mesons [97] for the 20% most central Pb–Pb collisions measured at mid-rapidity [98, 99]. A suppression factor of 3–4 is observed corresponding to a minimum of $R_{\rm AA} \approx 0.25$ at $p_{\rm T}$= 5–6 GeV/c.

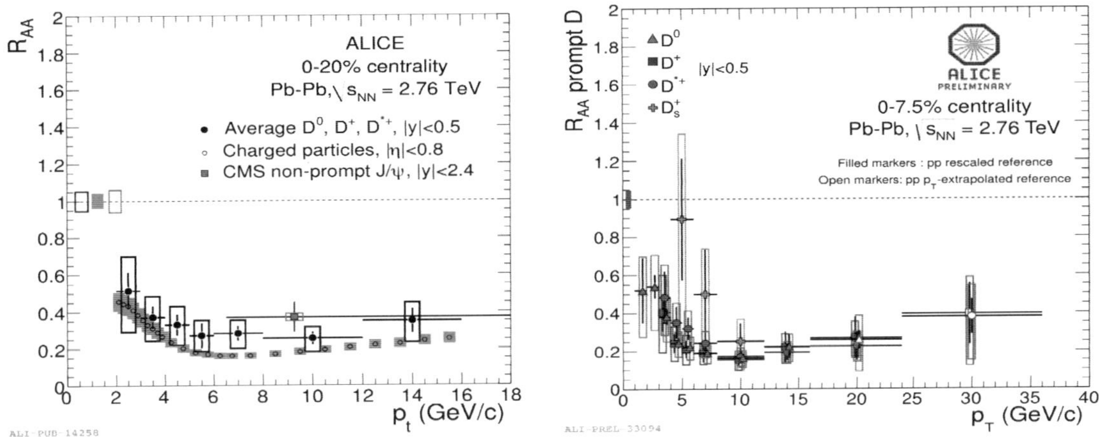

Figure 10. (Left) The nuclear modification factor, R_{AA}, for the average D–meson for the 20% central Pb–Pb collisions compared to charged particles and non–prompt J/Ψ from CMS [101]. (Right) The R_{AA} of D_s compared to the one of D^0, D^+ and D^{*+} for the 0–7.5% centrality range.

To test the predicted hierarchy of suppression, the results are compared to the R_{AA} of charged particles measured also at mid-rapidity, and found to be very similar. At $p_{\mathrm{T}} < 8$ GeV/c the average R_{AA} for prompt D-mesons is slightly higher than the charged-particle R_{AA} (however, still within the systematic uncertainties) showing a weak indication that $R_{\mathrm{AA}}^{D} > R_{\mathrm{AA}}^{charged}$. At higher p_{T} the D-mesons R_{AA} is similar to the one of charged particles (mostly light hadrons) [100], also supported by higher statistics measurements up to $p_{\mathrm{T}} \approx 36$ GeV/c [98, 99]. The ALICE results are compared to the CMS measurements [101] for non–prompt J/Ψ mesons with $p_{\mathrm{T}} > 6.5$ GeV/c, mostly originating from B decays, where a weaker suppression is seen suggesting that $R_{\mathrm{AA}}^{B} > R_{\mathrm{AA}}^{D}$ [102]. While, within the current statistical errors, the measurements indicate that the R_{AA} of non-prompt J/Ψ is larger than that of prompt D-mesons, and the later larger than the one of light hadrons, the current precision of the measurements prevents to conclude with respect to the expected colour charge and mass hierarchy of parton energy loss [90].

Comparisons of the heavy-flavour suppression factor R_{AA} in the most central class to NLO MNR calculations with EPS09 shadowing parametrizations [90] show that the heavy flavour suppression cannot be explained with only nuclear shadowing at $p_{\mathrm{T}} > 4$ GeV/c. Detailed studies based on the p–Pb data at 5.02 TeV will clarify the issues of the initial state effects.

D_s R_{AA} and hadronisation mechanism The ALICE measurements of D-mesons are complemented with the first measurement of D_s in heavy-ion collisions [103], of particular interest since it contains c and s quarks and can probe details of the hadronisation mechanism. Figure 10–right shows that in the 8–12 GeV/c p_{T} range the D_s R_{AA} is compatible with that of non-strange charmed mesons exhibiting a suppression factor 4–5 above 8 GeV/c. At lower p_{T} the D_s R_{AA} seems to increase relative to the one of the D^0. However, the current uncertainties preclude any conclusion on the predicted enhancement of D_s mesons relative to non-strange D-mesons [104].

v_2 of D-mesons and J/Ψ Further insight on the properties of the medium can be obtained investigating azimuthal anisotropies of heavy-flavoured hadrons. If heavy quarks interact strongly with the medium, heavy-flavoured hadrons should inherit the medium azimuthal anisotropy. Measurements of heavy-flavoured particles v_2 at low-p_{T} can provide information on the degree of thermalisation while at high-p_{T} on the path-length dependence of energy loss.

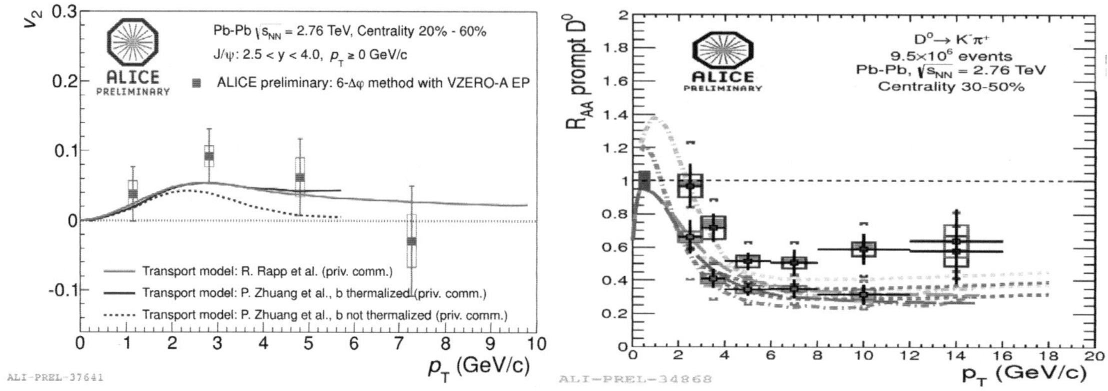

Figure 11. (Right) The p_T–differential v_2 measurements of inclusive J/Ψ measured at forward rapidity. (Left) R_{AA} of D^0 for 30–50% central Pb–Pb collisions measured in the in- and out- of plane azimuthal angle regions compared to model predictions.

Figure 11–left presents the first measurement of J/Ψ v_2 at LHC [105]. The p_T–differential v_2 of inclusive J/Ψ measured at forward rapidities for the 20–60% centrality range shows a non–zero v_2 value suggesting that charm quarks may follow the collective behaviour of the bulk system. A similar measurement for prompt D–mesons at mid-rapidity, for the 30–50% centrality range, in the p_T range $2 < p_T < 6$ GeV/c, show a finite v_2 value with a significance of 3σ compatible within errors with that of the light hardons [90]. The measurement of the D^0 R_{AA} in- and out-of plane azimuthal regions for the 30–50% centrality class, shown in Fig. 11–right [106], indicates larger suppression in the out-of-plane azimuthal region as expected due to the longer path length. Comparisons with models show that at low-p_T, both v_2 and energy loss could explain the results while at high-p_T energy loss seems to be the most probable scenario. A challenge for theory is the simultaneous description of the R_{AA} and v_2 and a consistent description of energy loss for light and heavy quarks [88]. In summary, the measurements of v_2 and R_{AA} are sensitive to the medium transport properties. However, more precise measurements covering low-p_T are required to resolve the contributions of different underlying mechanisms.

4. Summary

ALICE, with unique capabilities on tracking and particle identification, delivered a comprehensive set of results characterising the different stages of Pb–Pb collisions. The created system shows the expected features extrapolating from RHIC thus confirming hydrodynamic and thermal models in the new energy regime. Precision measurements are advancing providing detailed information on its initial conditions and transport properties. The analysis of the p–Pb pilot run already demonstrates that the observed features in the Pb–Pb collisions are not a cold matter effect but the fingerprint of the created quark-gluon plasma.

References

[1] S. Borsanyi, JHEP **1011**, (2010) 077.
[2] H. Satz, Rep. Prog. Phys. **63**, (2000) 1511;
S.A. Bass, M. Gyulassy, H. Stöcker, W. Greiner, J. Phys. **G25**, (1999) R1;
E.V. Shuryak, Phys. Rep. **115**, (1984) 151;
J. Cleymans, R.V. Gavai, E. Suhonen, Phys. Rep. **130**, (1986) 217.
[3] K. Aamodt *et al.* [ALICE Collaboration], Phys. Rev. Lett. **105**, (2010) 252302.
K. Aamodt *et al.* [ALICE Collaboration], Phys. Rev. Lett. **107**, (2011) 032301.
[4] G. Aad *et al.* [ATLAS Collaboration], Phys. Rev. Lett. **105**, (2010) 252303.
K. Aamodt *et al.* [ALICE Collaboration], Phys. Lett. **B696**, (2011) 30.

[5] I. Arsene *et al.* [BRAHMS Collaboration], Nucl. Phys. **A757**, (2005) 1.
K. Adcox *et al.* [PHENIX Collaboration], Nucl. Phys. **A757**, (2005) 184.
B. B. Back *et al.* [PHOBOS Collaboration], Nucl. Phys. **A757**, (2005) 28.
J. Adams *et al.* [STAR Collaboration], Nucl. Phys. **A757**, (2005) 102.
[6] B. Muller, J. Schukraft and B. Wyslouch, arXiv:1202.3233 [hep-ex].
[7] QM 2012 Proceedings to be published in Nucl. Phys. A
[8] K. Aamodt *et al.* [ALICE Collaboration], JINST **3**, (2008) S08002.
[9] K. Aamodt *et al.* [ALICE Collaboration], J. Phys. **G30**, (2004) 1517; and J. Phys. **G32**, (2006) 1295.
[10] K. Aamodt *et al.* [ALICE Collaboration], Phys. Rev. Lett. **105**, (2010) 252301.
[11] S. S. Adler *et al.* [PHENIX Collaboration], Phys. Rev. **C71**, (2005) 034908.
[12] K. Aamodt *et al.* [ALICE Collaboration], Phys.Rev.Lett. 106:032301 (2011), 1012.1657.
[13] G. Aad *et al.* [ATLAS Collaboration], (2011), 1108.6027.
[14] S. Chatrchyan *et al.*, [CMS Collaboration], JHEP 1108:141 (2011), 1107.4800.
[15] B. Alver *et al.*, [PHOBOS Collaboration], Phys.Rev. **C83**, 024913 (2011), arXiv:1011.1940 [nucl-ex].
[16] B. Abelev et al [ALICE Collaboration] 2013 Phys Rev Lett **110** 032301.
[17] A. Dumitru, D. E. Kharzeev, E. M. Levin, and Y. Nara, Phys.Rev. **C85**, 044920 (2012), arXiv:1111.3031.
[18] G. Barnafoldi, J. Barrette, M. Gyulassy, P. Levai, and V. Topor Pop, Phys.Rev. **C85**, 024903 (2012), arXiv:1111.3646 [nucl-th].
[19] P. Tribedy and R. Venugopalan, Phys.Lett. **B710**, 125 (2012), arXiv:1112.2445 [hep-ph].
[20] J.L. Albacete, A. Dumitru, H. Fujii, and Y. Nara, (2012), arXiv:1209.2001 [hep-ph].
[21] R. Xu, W.-T. Deng, and X.-N. Wang, (2012), arXiv:1204.1998 [nucl-th].
[22] B.Back *et al.*, [PHOBOS Collaboration], Phys.Rev.Lett. **93**, 082301 (2004), ar xiv:nucl-ex/0311009 [nucl-ex].
[23] J. Bjorken, Phys. Rev. **D27**, (1983) 140.
[24] K. Krajczar (for the CMS Collaboration), J. Phys. **G38**, (2011) 124041.
[25] M. Wilde [ALICE Collaboration], QM 2012 Proceedings to be published in Nucl. Phys. A.
[26] A.M. Adare *et al.* [PHENIX Collaboration], Phys. Rev. Lett. 104, 132301 (2010).
[27] M.A. Lisa, S. Pratt, R. Soltz, U. Wiedemann, Ann.Rev.Nucl.Part.Sci. 55:357 (2005), nucl-ex/0505014.
[28] K. Aamodt *et al.* [ALICE Collaboration], Phys.Lett. B696:328 (2011), 1012.4035.
[29] M. Ivanov [ALICE Collaboration], QM 2012 Proceedings to be published in Nucl. Phys. A.
[30] P. Kolb and U. Heinz, arXiv:nucl-th/0305084
[31] R. Preghenella [ALICE Collaboration], these proceedings.
[32] B. Abelev *et al.* [ALICE Collaboration], arXiv:1208.1974 [hep-ex].
[33] B. Abelev *et al.* [STAR Collaboration], Phys. Rev. **C79**, (2009) 034909.
[34] S. S. Adler *et al.* [PHENIX Collaboration, Phys. Rev. **C69**, (2004) 034909.
[35] E. Schnedermann, J. Sollfrank, U. W. Heinz, Phys. Rev. **C48**, (1993) 2462.
[36] C. Shen *et al.*, Phys.Rev. **C84**, (2011) 044903.
[37] Y. Karpenko and Y. Sinyukov, J.Phys. **G38**, (2011)124059.
Y. Karpenko, Y. Sinyukov, and K. Werner, nucl-th/1204.5351.
[38] P. Bozek, Phys. Rev. **C85**, (2012) 034901.
[39] A. Andronic, P. Braun-Munzinger, and J. Stachel, Phys. Lett. B673, 142 (2009).
[40] J. Cleymans, I. Kraus, H. Oeschler, K. Redlich, and S. Wheaton, Phys. Rev. C 74, 034903 (2006).
[41] L. Milano [ALICE Collaboration], QM 2012 Proceedings to be published in Nucl. Phys. A.
[42] F. Becattini *et al.*, arXiv:1203.5302[nucl-th] (2012).
[43] J. Steinheimer, J. Aichelin, and M. Bleicher, arXiv:1203.5302[nucl-th] (2012).
[44] A.M. Adare [ALICE Collaboration], QM 2012 Proceedings to be published in Nucl. Phys. A.
[45] M. Szymanski [ALICE Collaboration], QM 2012 Proceedings to be published in Nucl. Phys. A.
[46] U. W. Heinz and R. Snellings R 2013 (Preprint 1301.2826).
[47] S. Voloshin, Y. Zhang, Z. Phys. **C70**, (1996) 665.
A. M. Poskanzer and S. Voloshin, Phys. Rev. **C58**, (1998) 1671.
[48] J.Y. Ollitrault, Phys.Rev. D46:229 (1992).
[49] Sorensen P, J.Phys.G 37:094011 (2010), 1002.4878.
[50] Alver B, Roland G, Phys.Rev. C81:054905 (2010), 1003.0194.
[51] A. M. Poskanzer, S.A. Voloshin, Phys.Rev. C58:1671 (1998), nucl-ex/9805001.
[52] S. A. Voloshin, A. M. Poskanzer, R. Snellings, Landolt-Boernstein: Relativistic Heavy Ion Physicsvolume 1/23 (Springer-Verlag, 2010), pp. 5–54, 0809.2949.
[53] R. Snellings, New J.Phys. 13:055008 (2011), 1102.3010.
[54] B.B. Back *et al.* [PHOBOS Collaboration], Nucl. Phys. A757:28 (2005), nucl-ex/0410022.
[55] I. Arsene *et al.* [BRAHMS Collaboration], Nucl. Phys. A757:1 (2005), nucl-ex/0410020.
[56] K. Adcox, *et al.*, [PHENIX Collaboration], Nucl. Phys. A757:184 (2005), nucl-ex/0410003.

[57] J. Adams *et al.* [STAR Collaboration], Nucl. Phys. A757:102 (2005), nucl-ex/0501009.
[58] Müller B, Nagle JL, Ann.Rev.Nucl.Part.Sci. 56:93 (2006), nucl-th/0602029.
[59] P. Kovtun, D. Son, A. Starinets, Phys.Rev.Lett. 94:111601 (2005), hep-th/0405231.
[60] B. Schenke, P. Tribedy, R. Venugopalan Phys.Rev.Lett. 108:252301 (2012); Phys.Rec. C86:034908 (2012).
[61] S.A. Voloshin [ALICE Collaboration], QM 2012 Proceedings to be published in Nucl. Phys. A.
[62] B. Abelev *et al.* [ALICE Collaboration], ar xiv:1205.5761 [nucl-ex].
[63] B.B. Back *et al.* [PHOBOS Collaboration], Phys. Rev. Lett. **94** 122303 (2005) [nucl-ex/040602)
[64] W. A. Horowitz and M. Gyulassy, Nucl. Phys. A **872**, 265 (2011).
[65] P. Huovinen, P. F. Kolb, U. W. Heinz, P. V. Ruuskanen, S. A. Voloshin, Phys. Lett. B **503**, 58 (2001).
[66] B. Abelev *et al.* [STAR Collaboration], Phys. Rev. Lett. **99**, (2007) 112301.
[67] S. A. Voloshin, Nucl. Phys. A **715**, 379 (2003);
D. Molnar and S. A. Voloshin, Phys. Rev. Lett. **91**, 092301 (2003)
[68] A. Majumder, M. van Leeuwen, Prog. Part. Nucl. Phys. ́textbfA66, (2011) 41.
[69] Y. L. Dokshitzer, D. Kharzeev, Phys. Lett. **B519**, (2001) 199.
[70] T. Matsui and H. Satz 1986 Phys. Lett. B **178** 416.
[71] S. Digal, P. Petreczky and H. Satz (2001) Phys. Rev. **D64** 094015 (Preprint hep-ph/010617)
[72] F. Karsch, D. Kharzeev and H. Satz 2006 Phys. Lett. B **637** 75-80 (Preprint hep-ph/0512239)
[73] R. L. Thews, M. Schroedter and J. Rafelski 2001 Phys. Rev. C **63** 054905 (Preprint hep-ph/0007323
[74] Y.P. Liu, Z. Qu, N. Xu and P. F. Zhuang 2009 Phys. Lett. B **678** 72-76 (Preprint 0901.2757)
[75] X. Zhao and R. Rapp 2011 Nucl. Phys. A **859** 114-125 (Preprint 1102.2194)
[76] P. Braun-Munzinger and J. Stachel 2000 Phys. Lett. B **490** 196-202 (preprint nucl-th/0007059)
[77] M. Muller *et al.*, Annu. Rev. Nucl. Part. Sci. **57**, (2007) 205.
[78] S. Chatrchyan *et al.* [CMS Collaboration] PRL 106 (2011) 212301.
[79] S. Chatrchyan *et al.* [CMS Collaboration] PLB 715 (2012) 66.
[80] S. Chatrchyan *et al.* [CMS Collaboration] PLB 710 (2012) 256.
[81] B. Abelev *et al.* [ALICE Collaboration], arXiv:1208.2711 [hep-ex].
[82] S. Chatrchyan *et al.* [CMS Collaboration], Eur. Phys. J. **C72**, (2012) 1945.
[83] W. A. Horowitz and M. Gyulassy, Nucl. Phys. **A872**, (2011) 265.
C. A. Salgado and U. A. Wiedemann, Phys. Rev. **D68**, (2003) 014008.
X. F. Chen *et al.*, Phys. Rev. **C84**, (2011) 034902.
A. Majumder and B. Muller, Phys. Rev. Lett. **105**, (2010) 252002.
T. Renk *et al.*, Phys. Rev. **C84**, (2011) 014906.
T. Renk, Phys. Rev. **C83**, (2011) 024908.
[84] B. Abelev *et al* [ALICE Collaboration], arXiv:1210.4520 (nucl-ex)
[85] M. Verweij (for the ALICE Collaboration), Nucl. Phys. **A**, (2012) 1.
[86] K. Safarik [ALICE Collaboration], QM 2012 Proceedings to be published in Nucl. Phys. A.
[87] A. Adare *et al* [PHENIX Collaboration] 2011 Phys. Rev. C **84** 044905 (preprint 1005.1627)
[88] Andronic, these proceedings.
[89] B. Abelev *et al.* [ALICE Collaboration], arXiv:1203.2160 [nucl-ex].
[90] Z.C. del Valle [ALICE Collaboration] 2012 (Preprint 1212.0385)
[91] B. Abelev *et al.* [ALICE Collaboration] 2012 Phys. Rev. Lett. **109** 112301 (Preprint 1205.6443)
[92] B. Abelev *et al.* [ALICE Collaboration], Phys.Rev.Lett. **109**, (2012) 072301.
[93] A. Adare *et al.* [PHENIX Collaboration], Phys. Rev. Lett., **98**, (2007) 232301.
A. Adare *et al.* [PHENIX Collaboration], Phys. Rev., **C84**, (2011) 054912.
[94] I.C. Arsene [ALICE Collaboration] 2012 Nucle. Phys. A (Preprint 1210.5818)
[95] K. Aamodt *et al* [ALICE Collaboration] 2011 Phys. Lett. B **704** 442-455 (Preprint 1105.0380)
[96] C. Suire [ALICE Collaboration] 2012 Hard Probes (Preprint 1208.5601)
[97] B. Abelev *et al.* [ALICE Collaboration], arXiv:1203.2160 [nucl-ex].
[98] Z. Conesa del Valle [ALICE Collaboration], QM 2012 Proceedings to be published in Nucl. Phys. A.
[99] A. Grelli [ALICE Collaboration], QM 2012 Proceedings to be published in Nucl. Phys. A.
[100] B. Abelev *et al* [ALICE Collaboration] 2013 Phys. Lett. B **720** 52 (Preprint 1208.2711)
[101] S. Chatrchyan (for the CMS Collaboration), JHEP **1205** (2012) 063.
[102] S. Chatrchyan *et al* [CMS Collaboration] 2012 Physics Analysis Note **CMS-PAS-HIN-12-014**
[103] G.M. Innocenti [ALICE Collaboration] 2012 (Preprint 1210.6388)
[104] M. He, R. J. Fries and R. Rapp 2012 (Preprint 1204.4442)
I. Kuznetsova, J. Rafelski, Eur.Phys.J.C51:113-133,2007.
[105] E. Abbas *et al.* [ALICE Collaboration] 2013 (Preprint 1303.5880)
[106] D. Caffarri [ALICE Collaboration], QM 2012 Proceedings to be published in Nucl. Phys. A. arXiv:1212.0786

Journal of Physics: Conference Series **455** (2013) 012005 doi:10.1088/1742-6596/455/1/012005

Strangeness production in ALICE

Domenico Elia, for the ALICE Collaboration

INFN Sezione di Bari, via Orabona 4, 70126 Bari, Italy

E-mail: Domenico.Elia@ba.infn.it

Abstract. The ALICE experiment at the LHC has measured the production of strange and multi-strange hadrons both in Pb–Pb and pp collisions at unprecedented beam energies. Transverse momentum spectra at mid-rapidity for K^0_S mesons, Λ, Ξ^- and Ω^- baryons and their anti-particles in Pb–Pb collisions at $\sqrt{s_{NN}} = 2.76$ TeV are presented in different centrality intervals. Particles are reconstructed via their weak decay topologies into charged particles only over a wide momentum range, exploiting the tracking and particle identification capabilities of the ALICE detector. The corresponding production yields are compared with those observed in pp collisions and lower energy heavy-ion reactions. In particular, results on baryon-to-meson ratio, strangeness enhancements, elliptic flow and nuclear modification factors are discussed. The comparison of strange particle spectra in pp collisions at $\sqrt{s} = 7$ TeV with PYTHIA Perugia-2011 predictions is also presented.

1. Introduction

Measuring strange and multi-strange particle production in relativistic heavy-ion interactions is a unique tool to investigate the properties of the hot and dense matter created in the collision, as there is no net strangeness content in the initially colliding nuclei. In particular, an enhanced production of strange particles in A–A compared to pp interactions was one of the earliest proposed signature of a deconfined Quark-Gluon Plasma (QGP) [1, 2].

Measurements in pp collisions serve not only as baseline for those in heavy-ion reactions but also as a valuable handle on the different production mechanisms at play. The low material budget in the central rapidity region and the complementary particle identification techniques allow ALICE to measure strange particle spectra over a wide transverse momentum (p_T) range: from low to medium p_T (< 6 GeV/c) where soft interactions are expected to be the main source of particle production, to the high p_T region where pQCD-calculable processes dominate.

Following a brief description of the ALICE detector and the analysis techniques used to reconstruct the strange particle decays (section 2), the strange hadron p_T spectra measured in pp collisions at $\sqrt{s} = 7$ TeV are presented in section 3 and compared with predictions from Monte Carlo generators. The fourth section deals with the main results in Pb–Pb collisions at $\sqrt{s_{NN}} = 2.76$ TeV: baryon to meson ratios as a function of p_T for different centrality intervals and compared to pp results, strangeness enhancements as a function of $\langle N_{part} \rangle$ and their excitation function from SPS to LHC, strange particle elliptic flow and nuclear modification factors.

2. Strange decay reconstruction with the ALICE detector

The ALICE detector was designed to study heavy-ion physics at the LHC. It consists of a central barrel with a large solenoid providing a 0.5 T field for tracking and particle identification, plus a dimuon spectrometer equipped with its own dipole magnet giving a 0.7 T field, and other forward detectors for triggering and centrality selection. Tracking and vertexing are performed using the Inner Tracking System (ITS), consisting of six layers of silicon detectors, and the Time Projection Chamber (TPC). The two innermost layers of the ITS and the VZERO detector (scintillation hodoscopes placed on either side of the interaction region) are used for triggering. The VZERO also provides the centrality determination in Pb–Pb collisions. A complete description of the ALICE sub-detectors can be found in [3].

Results presented in this contribution are based on a sample of about 130×10^6 pp collisions at $\sqrt{s} = 7$ TeV and about 30×10^6 Pb–Pb collisions at $\sqrt{s_{\rm NN}} = 2.76$ TeV, both collected in 2010. Strange hadrons are measured through the reconstruction of their weak decays into charged particles only. K^0_S, Λ and $\overline{\Lambda}$ candidates are found by combining two charged tracks into V-shaped decays (V0 topology), while for multi-strange baryons, a selected V0 (Λ candidate) is further combined with a third track (bachelor candidate). Candidates are required to satisfy topological and kinematic restrictions. In addition, each of the daughter tracks is checked for compatibility with the pion, kaon or proton hypotheses using their energy loss in the TPC. Further details on the candidate selection and signal extraction procedure can be found in [4, 5]. A correction factor, which includes both detector acceptance and reconstruction efficiency, is determined for each particle species as a function of $p_{\rm T}$, using PYTHIA [6] and HIJING [7] for pp and Pb–Pb respectively. To reduce the required event statistics, enriched events with additional injected particles have been used for Pb–Pb and for multi-strange baryons in pp.

3. Results in pp collisions

The transverse momentum spectra for K^0_S, Λ and $\overline{\Lambda}$ in pp collisions at $\sqrt{s} = 7$ TeV are shown in Fig. 1. The $\overline{\Lambda}/\Lambda$ ratio is compatible with unity over the measured $p_{\rm T}$ range. The integrated production yields are obtained by fitting the spectra with the Tsallis function: the fraction in the unmeasured region is about 15% for Λ and negligible for K^0_S.

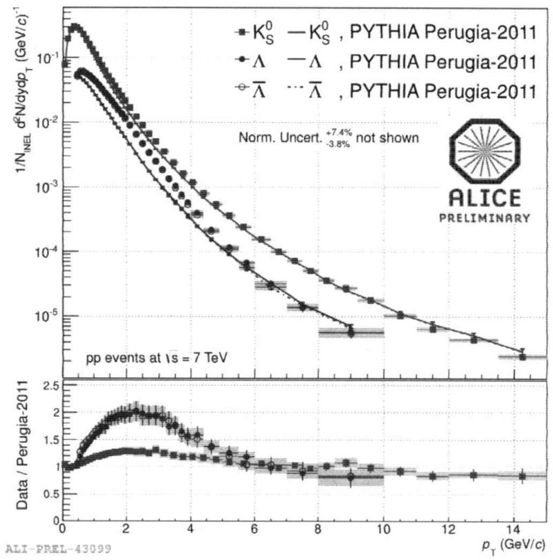

Figure 1. Transverse momentum spectra for Λ, $\overline{\Lambda}$ and K^0_S (symbols) in pp collisions at $\sqrt{s} = 7$ TeV, compared to PYTHIA Perugia-2011 predictions (lines).

The spectra have been compared with predictions from PYTHIA Perugia-2011 [8, 9], which provides the best description for the multi-strange baryons [5] and has been tuned to the measured charged particle multiplicity in pp collisions at $\sqrt{s} = 7$ TeV. The agreement is rather good in the high $p_{\rm T}$ region (> 6 GeV/c), while in the soft part PYTHIA underestimates the spectra by approximately 20% for $\rm K^0_S$ and more than a factor of two for Λ.

A larger disagreement between PYTHIA Perugia-2011 and data can be seen in the left panel of Fig. 2 for the multi-strange baryon spectra. While being the best available tune, PYTHIA Perugia-2011 underestimates up to a factor of 2 and 5 the Ξ and Ω spectra respectively, in the intermediate $p_{\rm T}$ region. At high $p_{\rm T}$ the discrepancy decreases for Ξ^- and $\overline{\Xi}^+$ when entering the fragmentation regime, while no conclusion can be draw for Ω^- and $\overline{\Omega}^+$ with the available statistics. In the right panel of Fig. 2, the multi-strange baryon yields and $\langle p_{\rm T}\rangle$ are shown for ALICE data at $\sqrt{s} = 0.9$ TeV and 7 TeV [4, 5] and STAR at $\sqrt{s} = 0.2$ TeV [10]: it can be seen that PYTHIA Perugia-2011 roughly reproduces the energy dependence.

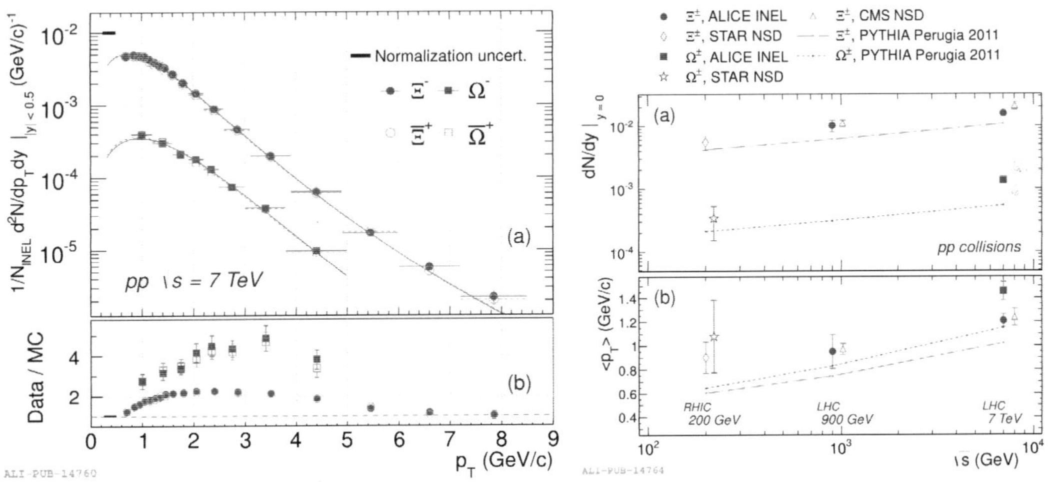

Figure 2. Left: Transverse momentum spectra for Ξ^- and Ω^- baryons (full symbols) and their anti-particles (open symbols) in pp collisions at $\sqrt{s} = 7$ TeV, compared to PYTHIA Perugia-2011 predictions (lines). Right: Multi-strange baryon yields (a) and $\langle p_{\rm T}\rangle$ (b) as a function of centre-of-mass energy in pp collisions.

4. Results in Pb–Pb collisions

A broad range of measurements has been obtained in Pb–Pb collisions at $\sqrt{s_{\rm NN}} = 2.76$ TeV. Strange and multi-strange hadron spectra have been measured in different centrality intervals according to the fraction of the total inelastic cross-section. The definition of the event centrality is based on the sum of the amplitudes measured in the VZERO detectors, as described in [11].

4.1. Strange baryon to meson ratio

An enhanced baryon to meson ratio in the intermediate $p_{\rm T}$ region (around 2.5 GeV/c) was first observed in $\sqrt{s_{\rm NN}} = 130$ and 200 GeV Au–Au collisions at RHIC [12]. The result of these measurements cannot be explained within the fragmentation model, where baryons are energetically more difficult to be produced than mesons. Alternative hadronization mechanisms have been invoked, such as quark coalescence: this could explain qualitatively the results in A–A but unlikely be at work in pp due to the low phase space density in the final state.

Results from the ALICE measurements are illustrated in Fig. 3. The left panel shows the $\Lambda/\rm K^0_S$ ratio as a function of $p_{\rm T}$ for several centrality classes in Pb–Pb collisions at

$\sqrt{s_{\mathrm{NN}}} = 2.76$ TeV and for minimum bias pp collisions at $\sqrt{s} = 0.9$ and 7 TeV. While in pp the ratio stays about the same at different energies (also compared to RHIC data), a strong centrality dependence is observed in Pb–Pb: the value increases with centrality up to its maximum of about 1.5 at $p_{\mathrm{T}} \sim 3$ GeV/c for the most central events. The behaviour appears to be specific to A–A collisions and is reproduced by the EPOS model [13]. The maximum value of the Λ/K^0_S ratio as a function of $\langle N_{\mathrm{part}} \rangle$ is shown in the right panel of Fig. 3 to increase with centrality and with the centre-of-mass energy of the colliding system.

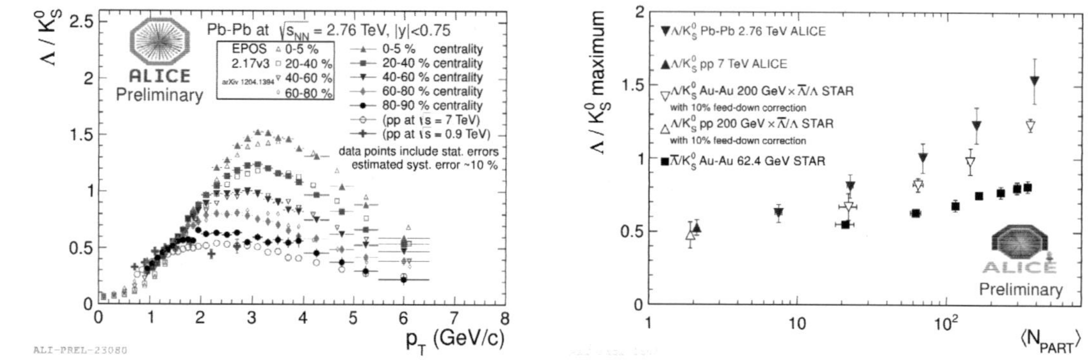

Figure 3. Left: Centrality dependence of Λ/K^0_S as a function of p_{T} in Pb–Pb collisions at $\sqrt{s_{\mathrm{NN}}} = 2.76$ TeV, compared with minimum bias pp collisions at $\sqrt{s} = 0.9$ and 7 TeV. Right: Maximum value of the Λ/K^0_S ratio as a function of the average number of participating nucleons $\langle N_{\mathrm{part}} \rangle$, for different collision systems and energies.

4.2. Strangeness enhancements

The enhancement of strange and multi-strange particle production in heavy-ion collisions compared to pp was one of the earliest proposed signatures for the QGP [1, 2]. In the original studies, it was argued that in a deconfined state, the system of free quarks and gluons would quickly find chemical equilibrium describable in terms of a grand canonical ensemble, implying a considerable increase in strange quark phase space occupancy compared to the observed strangeness production levels in pp interactions. Indeed, both at SPS [15, 16] and RHIC [17] the large production yields observed in A–A, especially for multi-strange baryons, could hardly be understood as coming from a hadronic phase and would require instead a faster equilibration.

The strangeness enhancements are defined as ratios of the strange particle yields measured in Pb–Pb collisions, normalized to the mean number of participant nucleons $\langle N_{\mathrm{part}} \rangle$, to the corresponding quantities in pp interactions at the same energy. The ALICE results for Λ ($|S|$=1), Ξ^- ($|S|$=2) and $\Omega^- + \overline{\Omega}^+$($|S|$=3) are shown in the left panel of Fig. 4 as a function of $\langle N_{\mathrm{part}} \rangle$. The pp reference values were obtained by interpolating ALICE data at two energies ($\sqrt{s} = 0.9$ and 7 TeV [5, 14]) for Λ and Ξ^-, and STAR data at 200 GeV [10] and ALICE data at 7 TeV for $\Omega^- + \overline{\Omega}^+$. The enhancements are larger than unity for all the particles and increase with $\langle N_{\mathrm{part}} \rangle$. They also increase with the strangeness content of the particle, showing the hierarchy already observed at lower energies and consistent with the picture of enhanced $s\bar{s}$ pair production in a hot and dense partonic medium. In the right panel of Fig. 4 the ALICE results are compared with lower energy measurements performed at SPS and RHIC: the enhancements are seen to decrease with increasing centre-of-mass energy, continuing the same trend first established at the SPS and then confirmed going from SPS to RHIC energies. However, since the hyperon to pion ratios have been shown to be constant with centrality and the charged particle multiplicity does not scale with $\langle N_{\text{part}} \rangle$, further studies are currently ongoing to quantify the specific enhancements of strange particles out of the general relative increase of multiplicity at mid-rapidity.

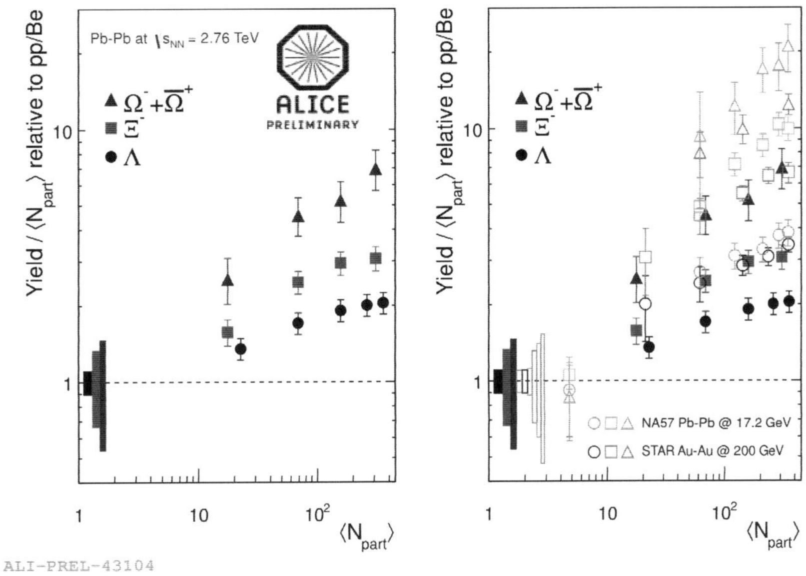

Figure 4. Left: Strangeness enhancements for Λ, Ξ^- and $\Omega^- + \overline{\Omega}^+$ as a function of the mean number of participants $\langle N_{\text{part}} \rangle$. Boxes on the dashed line at unity indicate statistical and systematic uncertainties on the pp reference. Right: Comparison with lower energy measurements at SPS and RHIC (hollow symbols).

4.3. Strange particle elliptic flow

The elliptic flow (v_2) coefficient is connected to one of the most informative observables for understanding the dynamics of the heavy-ion collisions and the properties of the created hot and dense matter [18]. Since collectivity is cumulative throughout the evolution of the collision, both partonic and hadronic stages should contribute to the observed transverse flow. To estimate the relative contributions to the elliptic flow from early deconfined partonic interactions and later hadronic rescatterings in the system evolution, measurements of the v_2 for strange and particularly multi-strange particles are crucial due to their small hadronic cross-section.

Fig. 5 shows the K^0_S, Λ and ϕ v_2 as a function of p_T for 10-20% (left panel) and 40-50% (right panel) centrality Pb–Pb collisions at $\sqrt{s_{NN}} = 2.76$ TeV, compared with charged pions, kaons and protons. At low p_T (< 2.5 GeV/c), a clear mass ordering of the v_2 is observed, i.e. lower v_2 for heavier hadrons than that for lighter ones. The relative mass-splitting of v_2 also increases with centrality: this can be explained by a stronger radial flow effect for more central collisions. In Fig. 6 the measurements for multi-strange baryons confirm the mass ordering at low p_T, which is also reproduced by calculations from the VISH2+1 viscous hydrodynamic model [19].

4.4. Strange particle nuclear modification factors

One of the most striking differences between the A–A and pp collisions first observed at RHIC was the suppression of the hadron production at high p_T in central Au–Au collisions at $\sqrt{s_{NN}} = 200$ GeV when compared to expectations from a geometrical superposition of nucleon-nucleon interactions [20, 21]. The effect, confirmed by the charged-particle nuclear modification

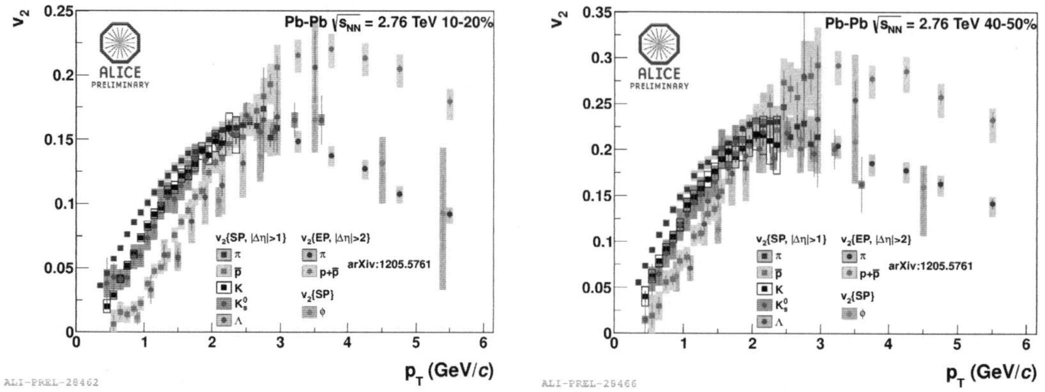

Figure 5. p_T differential v_2 for K^0_S, Λ and ϕ compared with charged pions, kaons and protons, measured for 10-20% (left) and 40-50% (right) centrality Pb–Pb collisions at $\sqrt{s_{NN}} = 2.76$ TeV.

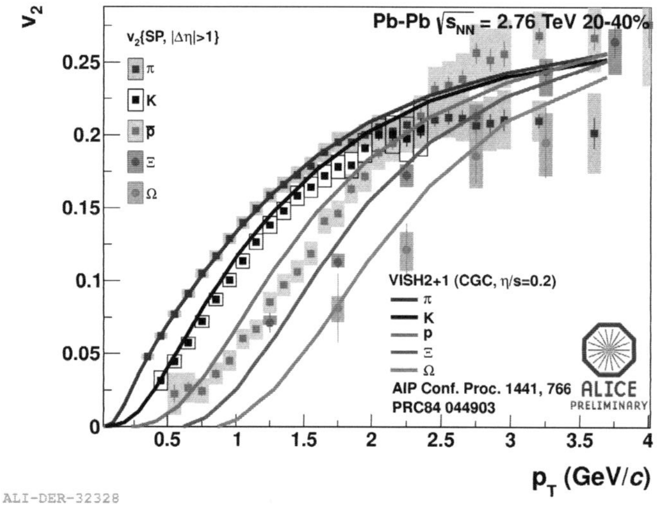

Figure 6. p_T differential v_2 for identified particles in 20-40% centrality Pb–Pb collisions at $\sqrt{s_{NN}} = 2.76$ TeV, compared to VISH2+1 viscous hydrodynamic model calculations.

factor (R_{AA}) as a function of p_T measured in Pb–Pb collisions at the LHC [22], is interpreted as the result of the energy loss suffered by partons when traversing the hot and dense matter created in the ultra-relativistic heavy-ion collision. The larger suppression observed at LHC compared to that at RHIC can be explained by the higher density of the medium created in collisions at higher centre-of-mass energy.

An important point to be further studied is the meson/baryon and flavour dependence of the parton energy loss. This is partially addressed by the measurements reported in Fig. 7. The left panel shows that the suppression of K^0_S and Λ are similar to that of unidentified charged hadrons for $p_T > 6$ GeV/c in the 5% most central Pb–Pb collisions at $\sqrt{s_{NN}} = 2.76$ TeV. This is also the case for K^0_S and charged particles at low p_T, where Λ instead show a significantly higher R_{AA}, probably related to the baryon enhancement at intermediate transverse momenta and then to alternative hadronization mechanisms. Higher R_{AA} values and even reduced differences are seen for more peripheral events, as shown in the right panel of Fig. 7.

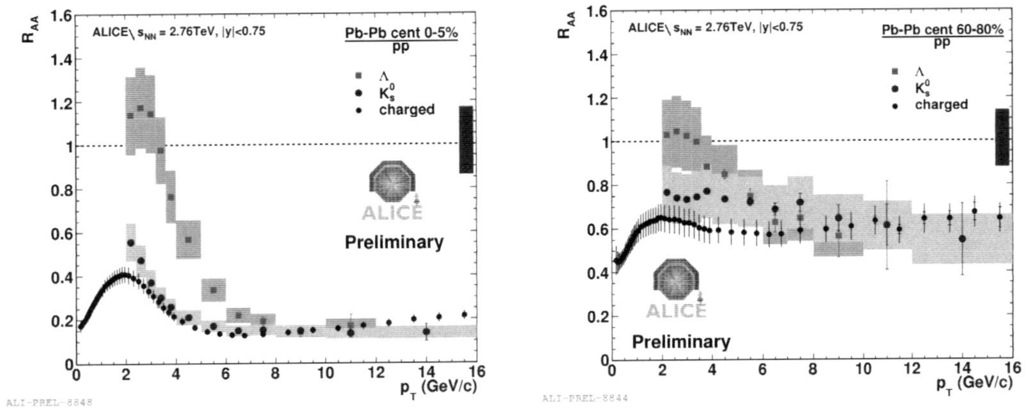

Figure 7. Nuclear modification factors for K^0_S, Λ and charged particles as a function of p_T in 0-5% (left) and 60-80% (right) centrality Pb–Pb collisions at $\sqrt{s_{NN}} = 2.76$ TeV.

5. Conclusions

ALICE measurements confirm that particles containing s-quarks are a valuable tool for the investigation of the strongly interacting matter created in Pb–Pb collisions at the LHC energy. Results on baryon anomaly, strangeness enhancements, elliptic flow and high p_T suppression have been presented and discussed. In addition, strange and multi-strange hadron spectra measured in pp collisions have been shown to play a relevant role not only as a baseline for the Pb–Pb measurements but also as constrain to the Monte Carlo generator models.

References

[1] J. Rafelski and B. Müller, Phys. Rev. Lett. **48**, 1066 (1982).
[2] P. Koch, B. Müller and J. Rafelski, Phys. Rep. **142**, 167 (1986).
[3] K. Aamodt *et al.*, JINST **3**, S08002 (2008).
[4] K. Aamodt *et al.*, Eur. Phys. J. C **71**, 1594 (2011).
[5] B. Abelev *et al.*, Phys. Lett. B **712**, 309 (2012).
[6] T. Sjostrand, S. Mrenna and P.Z. Skands, JHEP **05**, 026 (2006).
[7] X.-N. Wang and M. Gyulassy, Phys. Rev. D **44**, 3501 (1991).
[8] P.Z. Skands, Phys. Rev. D **82**, 074018 (2010).
[9] R. Skands, arXiv:1010.3558v1, (2010).
[10] B. I. Abelev *et al.*, Phys. Rev. C **75**, 064901 (2007).
[11] K. Aamodt *et al.*, Phys. Rev. Lett. **106**, 032301 (2011).
[12] M.A.C. Lamont *et al.*, Eur. Phys. J. C **49**, 35 (2007).
[13] K. Werner *et al.*, arXiv:1204.1394v1, (2012).
[14] K. Aamodt *et al.*, Eur. Phys. J. C **71**, 1594 (2011).
[15] F. Antinori *et al.*, J. Phys. G **32**, 427 (2006).
[16] F. Antinori *et al.*, J. Phys. G **37**, 045105 (2010).
[17] B. I. Abelev *et al.*, Phys. Rev. C **77**, 044908 (2008).
[18] R. Snellings, New J. Phys. **13**, 055008 (2011).
[19] U.W. Heinz, C. Shen and H. Song, AIP Conf. Proc. **1441**, 766 (2012).
[20] K. Adcox *et al.*, Phys. Rev. Lett. **88**, 022301 (2002).
[21] C. Adler *et al.*, Phys. Rev. Lett. **89**, 202301 (2002).
[22] K. Aamodt *et al.*, Phys. Lett. B **696**, 30 (2011).

International Workshop on Discovery Physics at the LHC (Kruger2012) IOP Publishing
Journal of Physics: Conference Series **455** (2013) 012006 doi:10.1088/1742-6596/455/1/012006

Coherent photoproduction of J/ψ at forward rapidity in Pb–Pb collisions with the ALICE detector

J. G. Contreras (for the ALICE Collaboration)

Faculty of Nuclear Sciences and Physical Engineering, Czech Technical University in Prague, Prague, Czech Republic

Abstract. We present the measurement of coherent photoproduction of J/ψ at forward rapidity in Pb–Pb collisions with the ALICE detector at the LHC.

1. The LHC as a photon–ion collider

The interaction of energetic beams of Pb ions at the LHC generates not only hadronic Pb-Pb collisions, but also γ–Pb and γ–γ collisions. This is because an accelerated charge can be viewed as a source of quasi–real photons, as first proposed by Fermi [2, 3] with the ultra–relativistic formulation provided by Weizsäcker [4] and Williams [5]. These quasi–real photons may interact either with an ion or with another quasi–real photon.

Since hadronic interactions have short range, hadronic Pb–Pb collisions with an impact parameter larger than the sum of the nuclear radii are strongly suppressed. In contrast, collisions involving quasi real photons occur at very large impact parameters. Such processes are called ultra peripheral collisions (UPC).

The intensity of the photon flux depends on the square of the ion electric charge. The photons are quasi–real and when they are emitted coherently from the whole nucleus, their wavelength is restricted to be larger than the nuclear radius. This implies that in the direction perpendicular to that of the ion beam the transverse momentum is of the order of 30 MeV, while in the longitudinal direction, taking into account the boost, the photon energy, in the laboratory reference system, may reach up to 40 GeV for Pb–Pb collisions at $\sqrt{s_{\mathrm{NN}}} = 2.76$ TeV. The LHC energy during the 2011 ion run.

These quasi–real photons may interact with a full ion, in which case the interaction is called coherent. In the other extreme the interaction may be with a single nucleon in the nuclei, in which case the interaction is called incoherent. The physics of UPC is discussed in a number of reviews [6, 7]. In this contribution the measurement of the coherent production of a J/ψ in UPC at large rapidities with the ALICE detector is presented. This measurement was carried out using the ALICE forward muon arm, and was recently published in [1]. A study of the coherent production of ρ^0 vector mesons in UPC can also be found in these proceedings [8].

2. Coherent production of J/ψ in UPC

One of the most interesting processes in UPC is the coherent photoproduction of vector mesons composed of heavy quarks. These processes are expected to probe the nuclear gluon distribution, which is not well constrained by the pre-LHC measurements, especially at low values of Bjorken

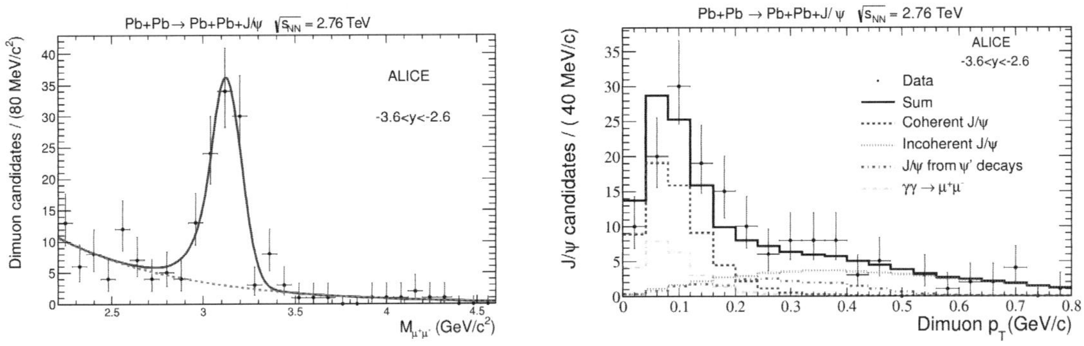

Figure 1. Dimuon invariant mass (left) and transverse momentum (right) distributions for selected events. The transverse momentum distribution is presented only for dimuons in the mass range between 2.8 GeV/c^2 and 3.4 GeV/c^2. For details see the text.

x [9]. Due to the coherent coupling of the photon to the lead ion the transverse momentum of the vector meson is restricted to be ≈ 60 MeV/c for Pb–Pb collisions at $\sqrt{s_{\rm NN}} = 2.76$ TeV, while its rapidity depends on the photon energy, k, as $y = \ln(2k/M)$, where M is the mass of the vector meson.

In Pb–Pb collisions there are two possible sources of photons at a given rapidity, because both nuclei could radiate the photon. Since the photon energy spectrum is steep, the contribution from the the low energy photon dominates the cross section. At rapidity $y = -3$ the two contributions correspond to center of mass energy, in the γ–Pb system, of $W_{\gamma\rm Pb} = 20$ GeV and $W_{\gamma\rm Pb} = 400$ GeV respectively. Taking into account the difference in the flux and in the corresponding cross sections, the contribution from the higher energy is approximately 4% of that at the lower energy.

In pQCD models the coherent photoproduction of the J/ψ occurs as follows: the quasi-real photon fluctuates into a quark–antiquark pair which interacts with the nuclei to produce the J/ψ through the exchange of two gluons at the same Bjorken x. The value of the rapidity of the vector meson and the Bjorken x variable are related by $x = Me^{\pm y}/\sqrt{s_{\rm NN}}$, so that at $y = -3$ values of $x \approx 10^{-2}$ and $x \approx 10^{-5}$ are probed, where the dominant contribution to the process comes from the higher x value.

3. The ALICE detector

The ALICE detector [10] consists of a central barrel with a uniform magnetic field of 0.5 T, and a muon spectrometer in the forward direction. The muon spectrometer measures muons with full azimuthal acceptance and within $-4.0 < \eta < -2.5$, where η denotes the pseudorapidity. The muon spectrometer is made of an absorber of ten interaction lengths, followed by five tracking stations made of two planes of cathode pad MWPC. A third MWPC station is situated inside a dipole magnet which has 3 Tm integrated magnetic field. The spectrometer also has a triggering system to select muon candidates with a transverse momentum larger than a programmable threshold, set to 1 GeV/c for the measurement described in this contribution. The trigger system consist of four planes of resistive plate chambers (RPC) placed after the MWPC and behind an iron wall of 7.2 interaction lengths, which absorbs secondary punch–through hadrons from the front absorber and low momentum muons from π and K weak decays.

Other detectors used in the measurement presented here are the VZERO and the ZDC. The VZERO consists of two arrays of 32 scintillator tiles each, covering the ranges $2.8 < \eta < 5.1$ (VZERO-A) and $-3.7 < \eta < -1.7$ (VZERO-C). The time resolution of each array is better than

Table 1. Summary of the contributions to the systematic uncertainty for the J/ψ photoproduction cross section measurement.

Source	Value
Theoretical uncertainty in $\sigma_{\gamma\gamma}$	20%
Coherent signal extraction	$^{+9}_{-14}$%
Reconstruction efficiency	6%
RPC trigger efficiency	5%
J/ψ acceptance calculation	3%
Two-photon e^+e^- background	2%
Branching ratio	1%
Total	$^{+24}_{-26}$%

1 ns. They contribute to the level–0 trigger of ALICE. In addition two zero degree calorimeters (ZDC) located at $\pm$ 116 m from the interaction point are used to veto hadronic interactions by measuring neutrons in the very forward direction.

4. Cross section for the coherent production of J/ψ in UPC at high rapidities

The J/ψ vector meson has been measured through its decay into a pair of muons. The signature is then two muons with opposite charge from a J/ψ vector meson decay at high rapidities and with low transverse momentum, in an otherwise empty detector. The only irreducible background comes from the continuum production of muon pairs in γ–γ collisions.

This measurement used an integrated luminosity of 55 μb^{-1} of Pb–Pb data recorded by ALICE in 2011. A dedicated UPC trigger selected events with at least one muon track in the muon spectrometer, with transverse momentum larger than 1 GeV/c, at least one hit in the VZERO–C, and no activity on the VZERO-A.

The offline event selection is described in detail in [1]. The selection ensured that muon tracks are not from beam–gas interactions and that they are within the acceptance of both the muon spectrometer and the VZERO–C. Accepted events contained only two muon tracks with opposite charges. These two muon tracks were added to form a dimuon, required to have transverse momentum less than 300 MeV/c and rapidity within $-3.6 < y < -2.6$. Rejection of the remaining beam–gas background and low multiplicity hadronic interactions was performed using VZERO timing and the ZDC respectively. The data set passing these selection criteria contained 117 events within the mass range (2.8,3.4) GeV/c^2.

The mass distribution for the selected dimuons and the transverse momentum distribution for dimuons within the mass range (2.8,3.4) GeV/c^2 are shown in Figure 1. The left panel shows the mass distribution for the selected events. A peak around the J/ψ mass is clearly seen. Also shown is a fit of an exponential plus a Crystal Ball parameterization. The parameters of the minimum likelihood fit are consistent with expectations from the production of J/ψ vector mesons and the di–muon continuum. The right panel shows the distribution of transverse momentum for events in the mass range (2.8,3.4) GeV/c^2. The histograms show the contributions expected from the Starlight MC [11] for coherent and incoherent J/ψ production, feed–down from $\psi(2S)$ and the contribution from the continuum. At low transverse momentum the coherent production dominates. The yield of J/ψ with transverse momentum less than 300 MeV/c is estimated to be 96 ± 12 (stat) ± 6 (sys) with a 11 ± 6 % contribution from feed down and a 12^{+14}_{-4} % contribution from incoherent processes as estimated from a comparison of data with MC expectations.

To normalize the J/ψ cross–section, the cross section for the continuum production of dimuons ($\sigma_{\gamma\gamma}$) was used.

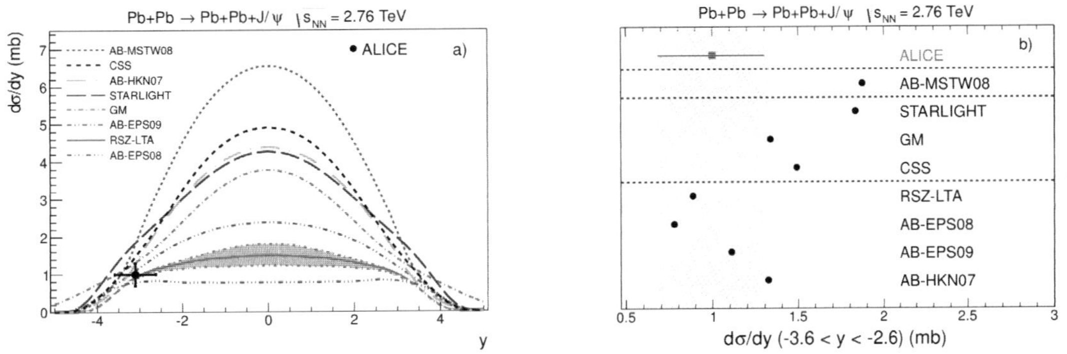

Figure 2. Comparison between the measurement and available theoretical calculations.

$$\frac{d\sigma_{\rm coh}}{dy} = \frac{1}{\rm BR} \cdot \frac{N_{\rm coh}}{N_{\gamma\gamma}} \cdot \frac{({\rm Acc \ x}\ \epsilon)_{\gamma\gamma}}{({\rm Acc \ x}\ \epsilon)_{\rm coh}} \frac{\sigma_{\gamma\gamma}}{\Delta y},$$

where BR is the branching ratio of $J/\psi \rightarrow \mu^+\mu^-$, (Acc x ϵ) are the factors to correct for the detector efficiency and acceptance of the coherent and continuum samples, Δy is the width of the rapidity range.

The coherent cross section at $\sqrt{s_{NN}} = 2.76$ TeV for $-3.6 < y < -2.6$ was measured to be

$$\frac{d\sigma_{\rm coh}}{dy} = 1.00 \pm 0.18({\rm stat})^{+0.24}_{-0.26}({\rm sys})\ {\rm mb}$$

where the contributions to the systematic error are shown in Table 1. The dominant contribution is the theoretical uncertainty in $\sigma_{\gamma\gamma}$ followed by the extraction of the coherent contribution to the mass and transverse momentum distributions.

5. Comparison to theory predictions

The measured cross section has been compared to five recent theoretical calculations [12–16]. The comparison is shown in Figure 2. STARLIGHT [12], uses a GVDM coupled to a Glauber approach to link the γA to the γp cross section, where the later is obtained from a parameterization of HERA data. GM [13] is based on the color dipole model, where the scattering amplitude depends on the nuclear profile and the dipole nucleon cross section, which is taken from the IIM model which incorporates saturation. The CSS [16] model uses a Glauber approach and the color dipole nucleon amplitude based on the unintegrated gluon distribution of the proton. The AB models [15] use the LO pQCD amplitude scaled to fit $\gamma + p \rightarrow J/\psi + p$ data. For the gluon distribution, AB–MSTW08 assumes no nuclear effects. The other AB models incorporate nuclear effects according to the EPS08, EPS09 or HKN07 prescriptions. Finally, RSZ-LTA [14] is based on the LO pQCD amplitude for two gluon exchange where the nuclear gluon density incorporates shadowing computed in the leading twist approximation.

From Figure 2 it can be concluded that models which incorporate shadowing – with the scale fixed near the mass of the J/ψ – are closer to data. This data will help to constraint and improve the theoretical predictions such that the uncertainty on the nuclear gluon distribution can be reduced.

Acknowledgments
I would like to thank the organizers of KRUGER12 for the opportunity to present this work, and for a wonderful meeting. This work has been partially supported by the project LK11209 from the MSMT of the Czech Republic

References
[1] B. Abelev et al. [ALICE Collaboration], 2013 *Phys. Lett.* B **718** 1273-1283; arXiv:1209.3715 [nucl-ex].
[2] E. Fermi 1924 *Z. Phys.* **29** 315–327.
[3] E. Fermi 1925 *Nuovo Cim* **2** 143–158. Translation available in `http://arxiv.org/abs/hep-th/0205086`.
[4] C. F. Weizsäcker 1934 *Z. Phys.* **88** 612–625.
[5] E. J. Williams 1934 *Phys. Rev.* **45** 729–730.
[6] A. J. Baltz et al. 2008 *Phys. Rept* **458** 1–171.
[7] C. A. Bertulani, S. R. Klein and J. Nystrand, 2005 *Ann. Rev. Nucl. Part. Sci.* **55** 271–310.
[8] K. Skjerdal, this proceedings.
[9] K. J. Eskola, H. Paukkunen and C. A. Salgado 2009 *JHEP* **0904** 065.
[10] K. Aamodt et al. [ALICE Collaboration] 2008 *JINST* **3** S08002.
[11] STARLIGHT website, http://starlight.hepforge.org/.
[12] S.R. Klein and J. Nystrand 1999 *Phys. Rev.* C **60** 014903.
[13] V.P. Goncalves and M.V.T. Machado 2011 *Phys. Rev.* C **84** 011902.
[14] V. Rebyakova, M. Strikman and M. Zhalov 2012 *Phys. Lett.* B **710** 647.
[15] A. Adeluyi and C.A. Bertulani 2012 *Phys. Rev.* C **85** 044904.
[16] A. Cisek, W. Schafer and A. Szczurek 2012 *Phys. Rev.* C **86** 014905.

Journal of Physics: Conference Series **455** (2013) 012007 doi:10.1088/1742-6596/455/1/012007

(Anti-)matter and hyper-matter production at the LHC with ALICE

Nicole Martin[1,2] **(for the ALICE Collaboration)**

[1] ExtreMe Matter Institute EMMI and Research Division, GSI Helmholtzzentrum für Schwerionenforschung, Darmstadt, Germany

[2] Institut für Kernphysik, Technische Universität Darmstadt, Darmstadt, Germany

E-mail: n.martin@gsi.de

Abstract. ALICE is the experiment at the CERN LHC dedicated to the investigation of high energy nucleus–nucleus collisions. Its excellent particle identification capabilities allow for the measurement of rarely produced particles, like hypernuclei and light nuclei (and their anti-particles). We present here results from Pb–Pb collisions at a center of mass energy of $\sqrt{s_{\rm NN}}$ = 2.76 TeV per nucleon–nucleon pair. In particular the measurement of $^3\overline{\rm He}$ and $^4\overline{\rm He}$ is discussed. In addition the reconstruction of (anti-)hypertritons via their mesonic decay ($^3_\Lambda{\rm H} \rightarrow ^3{\rm He} + \pi$) is presented. Searches for even lighter hypermatter states, i.e. $\Lambda\Lambda$ (also known as H-Dibaryon) and Λn bound states, are dicussed. The results are compared with thermal model predictions.

1. Introduction

The measurement of light (anti-)nuclei and (anti-)hypernuclei, created in heavy-ion collisions is challenging as the production probability decreases with increasing mass. The production yields of these particles can for example be predicted by thermal models based on an analysis of lighter particles [1, 2]. The chemical freeze-out temperature T, the volume V and the baryo-chemical potential μ_B are the only three parameters in this approach. If particle ratios instead of yields are calculated the volume cancels out and the yield ratios, as for example $\frac{^n\overline{\rm He}}{^n{\rm He}}$, scale with $\exp(-2n\mu_B/T)$. Therefore, the fit to the data of lighter particles provides constraints for the baryo-chemical potential μ_B and the chemical freeze-out temperature T. The production yields depend on the center-of-mass energies $\sqrt{s_{\rm NN}}$ of the colliding nuclei. Figure 1 shows thermal model predictions for particle ratios as a function of $\sqrt{s_{\rm NN}}$ together with experimental data from AGS, SPS and RHIC [1]. A fit to the experimental data at the highest energy available before the startup of the LHC, namely $\sqrt{s_{\rm NN}}$ = 200 GeV per nucleon–nucleon pair, gives: $\mu_B = 24 \pm 2$ MeV and T = 162 ± 2 MeV. The analysis can be extended by investigating also light nuclei and hypernuclei, which are discussed in section 4. The lightest anti-hypernucleus, the anti-hypertriton $^3_{\overline{\Lambda}}\overline{\rm H}$ was first measured by the STAR Collaboration [3]. The experimental ratios measured by STAR can be compared with the theoretical predictions from a thermal model: the experimental ratios of $^3{\rm He}/^3\overline{\rm He}$ and $^3_\Lambda{\rm H}/^3_{\overline{\Lambda}}\overline{\rm H}$ are in good agreement with the model predictions, but the measured ratios of $^3{\rm He}/^3_\Lambda{\rm H}$ and $^3\overline{\rm He}/^3_{\overline{\Lambda}}\overline{\rm H}$ are about two standard deviations larger then the predicted ratios [1]. To further investigate this discrepancy, additional experimental data is needed. The energy reached in heavy-ion collisions at the LHC provide the opportunity to measure such particles in unprecedented abundances. Up to now in two periods in the years 2010 and 2011, each roughly four weeks long, data from Pb–Pb collisions at $\sqrt{s_{\rm NN}} = 2.76\,\text{TeV}$ per nucleon–nucleon pair were recorded. The ALICE detector system [4] is well suited to measure the rarely produced states created in Pb–Pb collisions due to its unique particle identification capabilities. Exotic states like a possible Λn bound state or the H-Dibaryon are also investigated. The H-Dibaryon is a bound state of *uuddss* ($\Lambda\Lambda$) and was first predicted by Jaffe in a bag model calculation [5]. Recent lattice QCD calculations [6, 7] also suggest a bound state, with binding energies in the range 13-50 MeV/c. A chiral extrapolation of these lattice calculations to a physical pion mass resulted in a H-Dibaryon unbound by either 13 ± 14 MeV/c [8] or it is lies close to the Ξp threshold [9]. This renewed the interest in the experimental searches for the H-Dibaryon. The results related to these searches of such exotic bound states are presented in section 5.

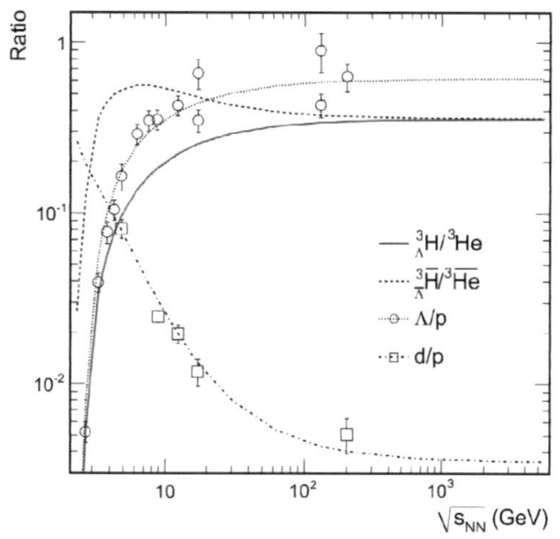

Figure 1. Thermal model predictions for particle ratios as a function of the center of mass energy per nucleon–nucleon pair ($\sqrt{s_{NN}}$) for central Pb–Pb collisions [1].

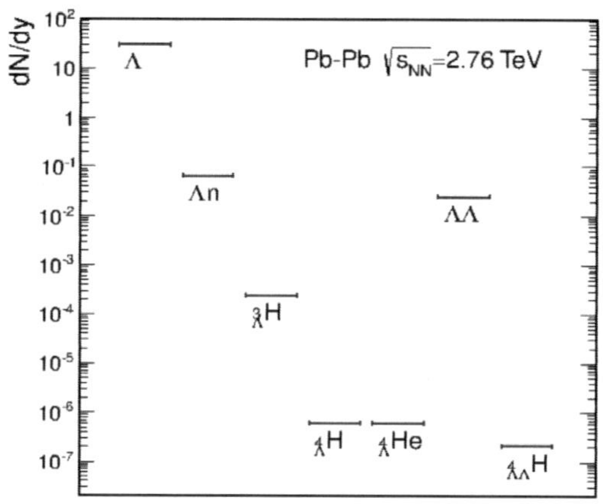

Figure 2. Shown here are predictions for the production yield dN/dy from thermal model calculations for various particles for central Pb–Pb collisions. [10].

2. The ALICE detector system

The ALICE detector is one of the four large experiments at the Large Hadron Collider (LHC) at the European Organization for Nuclear Research (CERN) in Geneva, Switzerland. Besides Pb–Pb collisions, proton-proton and p-Pb collisions are investigated. The ALICE detector is in total 26 meters long, 16 meters high and 16 meters wide. It weighs 10,000 tons and is located at the Interaction Point 2 of the LHC. It has a central barrel which holds most of the detector systems, a single arm backward muon spectrometer and two Zero Degree Calorimeters (ZDC) to distinguish between central and peripheral collisions. The four detectors in the central part, which all cover a pseudo-rapidity interval of $|\eta| \leq 0.9$ over the full azimuth, are the Inner Tracking System (ITS) and the Time Projection Chamber (TPC) for tracking, a Time Of Flight (TOF) detector for clear π/K and K/p separation at intermediate momenta and the Transition Radiation Detector (TRD) for electron identification. A schematic view of the ALICE detector is shown in figure 3.

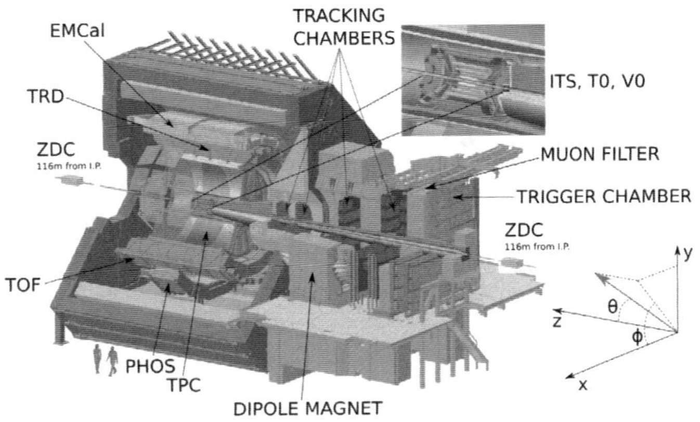

Figure 3. The ALICE detector system together with the ALICE coordinate system: right-handed orthogonal Cartesian system with the point of origin at the interaction point. The positive x,y,z-axes of the Cartesian coordinate system are shown together with the spherical coordinates r, ϕ and θ.

Particle identification
The precise particle identification (PID) and continous tracking from very low (100 MeV/c) to very high p_{T} (20 GeV/c) is a unique speciality of the ALICE detector. The PID used in the analysis described in this proceedings takes advantage of three different techniques: the specific energy loss ($\mathrm{d}E/\mathrm{d}x$), which is described by the Bethe-Bloch formula [11], measured with the TPC is used to identify particles in particular at low and high momenta (relativistic rise). This allows the identification of all particle species, from the lightest (electron) to the heaviest (anti-nuclei). The resolution of the TPC ranges from $\sigma \approx 5.2$ % in pp collisions to 6.5% (2010), respectively 7.2% (2011), in central Pb–Pb collisions. For intermediate p_{T} the TOF detector is used. The TOF measurement has a resolution of $\sigma \approx 85$ ps in Pb–Pb collisions and 120 ps in pp collisions. The TOF performance allows a 2σ p/K-separation up to a momentum of 5 GeV/c. Figure 4 shows the velocity β measured with the TOF detector as a function of the momentum p of the particle for Pb–Pb collisions. In addition unstable particles are identified via their decay topologies.

3. Anti-nuclei

Light anti-nuclei such as anti-protons $\overline{p}$, anti-deuterons $\overline{d}$, anti-tritons $\overline{t}$ and anti-helium $^{3}\overline{\mathrm{He}}$ are produced even in pp collisions in a significant amount. More nuclei than anti-nuclei are detected, but in principle the anti-baryon to baryon ratio at the LHC energies is close to unity thus also light nuclei and their corresponding anti-nuclei are expected to have similar yields. On the matter side a substantial background is created via knockout from the material, whereas anti-particles suffer only from annihilation when detector material is crossed. This is demonstrated in figures 5 and 6. They show the distance of closest approach (DCA) to the primary vertex in the xy-plane (DCA_{XY}) for different DCA in the z direction (DCA_Z). Whereas the number of deuterons increases with increasing DCA_Z, because of the deuterons which are knocked out from the detector material, the distributions for anti-deuteron stay the same. The knockout is a significant problem at low p_{T}. Therefore anti-matter studies benefit from a cleaner sample.

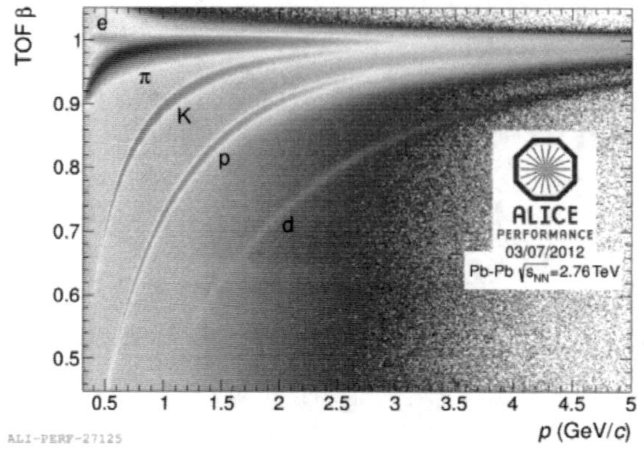

Figure 4. The velocity β measured with the TOF detector as function of the momentum p of the particle.

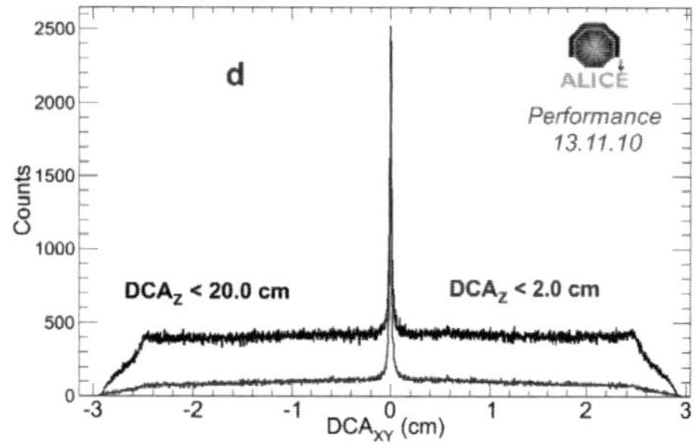

Figure 5. DCA$_{XY}$ distribution for different DCA$_Z$ values for deuterons.

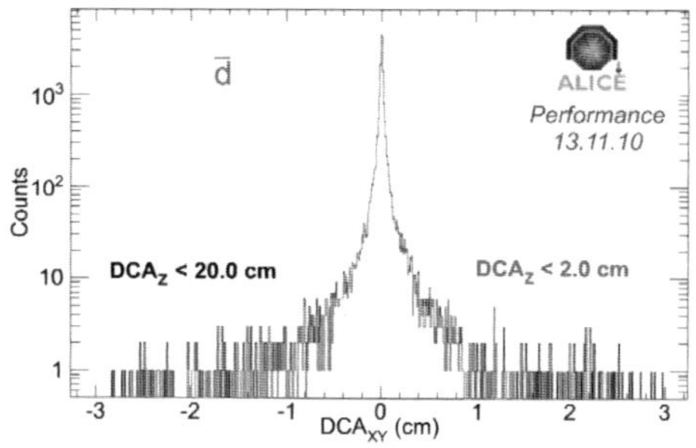

Figure 6. DCA$_{XY}$ distribution for different DCA$_Z$ values for anti-deuterons.

Anti-helium-4 $^4\overline{\mathrm{He}}$

For the identification of the $^4\overline{\mathrm{He}}$ we first selected all events, which had at least one particle with a dE/dx corresponding to a $^3\overline{\mathrm{He}}$ or a higher mass. This offline "trigger" is indicated in figure 7 with a dashed line. Then we identified $^4\overline{\mathrm{He}}$ candidates in a 3σ range around the theoretical Bethe-Bloch curve. But this is not sufficient enough, especially at higher rigidity $\frac{p}{z}$ (the momentum p of the particle divided by its charge number z) where the two Bethe-Bloch curves of $^4\overline{\mathrm{He}}$ and $^3\overline{\mathrm{He}}$ are very similar. Therefore we use in addition the information from the TOF detector to calculate the $\frac{m^2}{z^2}$. The result is shown in the inlay in figure 7. The particles with the ratio of m^2 to z^2 which corresponds to $^4\overline{\mathrm{He}}$ are indicated with larger (red) markers in the dE/dx vs. rigidity figure. We identified 10 $^4\overline{\mathrm{He}}$ using the TPC and TOF, corresponding to $23 \cdot 10^6$ events of a trigger mix (central, semi-central and minimum bias). The first $^4\overline{\mathrm{He}}$ nuclei were recently observed by the STAR Collaboration at RHIC [12].

4. (Anti-)Hypertriton

The investigation of hypernuclei has a long tradition in nuclear physics. They were discovered in the 1950s by M. Danysz and J. Pniewski in a photographic emulsion exposed to cosmic rays [13]. Hypernuclei are nuclei, in which at least one hyperon is bound in addition to the normal nucleons. All hyperons are unstable, even if they are bound in a nucleus. Nuclei of mass 5 only exist in the form of hypernuclei $\left(^5_\Lambda\mathrm{He}\right)$. Therefore hypernuclei give access

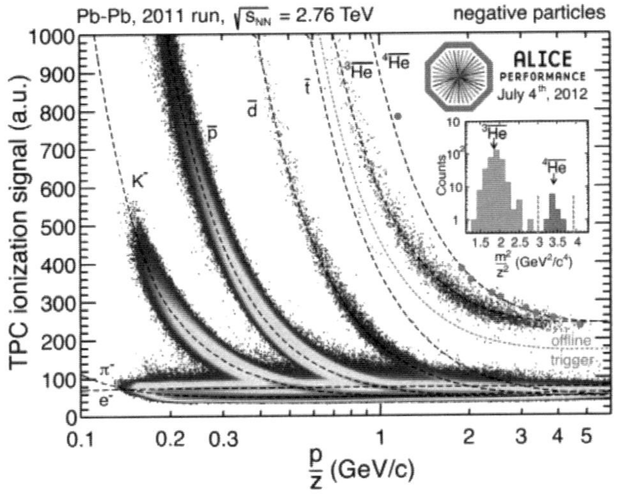

Figure 7. TPC $\mathrm{d}E/\mathrm{d}x$ spectrum for negative particles after a selection of events that contain at least one $^3\overline{\mathrm{He}}$ or $^4\overline{\mathrm{He}}$ candidate for the 2011 data. The inlet shows the $\frac{m^2}{z^2}$ distribution for this pre-selected data. The 10 $^4\overline{\mathrm{He}}$ nuclei clearly identified by TPC and TOF are indicated as red dots.

to the next higher mass (anti-)nucleus, which will be measurable in heavy-ion collisions. The (anti-)hypertriton has a mass of $2.991 \pm 0.001 \pm 0.002\,\mathrm{GeV}/c^2$ and a decay length of $c\tau = 5.5^{+2.7}_{-1.4} \pm 0.8$ cm [3]. This is a similar decay length as the one of the free Λ particle ($c\tau \approx 7.8\,\mathrm{cm}$). We identify the (anti-)hypertriton candidates via their decay daughters $(\pi, ^3\mathrm{He})$. We first look for light nuclei $(^3\mathrm{He}, ^3\overline{\mathrm{He}})$ which are daughter tracks originating from decay vertices. They are identified via the TPC $\mathrm{d}E/\mathrm{d}x$ information. If the second daughter is identified as a pion, we reconstruct the invariant mass of the pair. The resulting invariant mass distributions for the Pb–Pb data corresponding to $23 \cdot 10^6$ events of a trigger mix (central, semi-central and minimum bias) are shown in figure 8 and figure 9. The signal of the (anti-)hypertriton has a mean mass of $2.990 \pm 0.001\,\mathrm{GeV}/c^2$ $(2.993 \pm 0.001\,\mathrm{GeV}/c^2)$ and a width of $3.35 \pm 0.7\,\mathrm{MeV}/c^2$ $(2.0 \pm 1.2\,\mathrm{MeV}/c^2)$.

5. Exotic bound states

We investigate two possible bound states: the Λn bound state and the H-Dibaryon ($\Lambda\Lambda$).

Λn bound state

We search for a possible Λn bound state, which decays into a deuteron and a pion, in the data set of Pb–Pb collisions from 2010 ($13.8 \cdot 10^6$ min. bias events). The HypHI experiment at GSI observes a signal, compared to mixed events, in the $d + \pi^-$ which they connect to the possible existence of a new bound state [14]. We investigate here only the $\overline{\Lambda\mathrm{n}}$ bound state, because of the lower background compared to the corresponding particle. To estimate the efficiency for the detection of the $\overline{\Lambda\mathrm{n}}$ Monte Carlo sample has been produced, where the $\overline{\Lambda\mathrm{n}}$ was generated flat in rapidity y and in transverse momentum p_T. We assume the lifetime to be that of the Λ particle. As we do not know the true shape of the p_T spectrum we estimate it from the extrapolation of blast-wave fits for π, K, p done at the same energy. Blast-wave functions describe the shape of p_T spectra in heavy-ion collisions quite well, since the blast-wave describes the thermal part and the radial flow visible in these spectra. The resulting blast-wave is normalised to unity

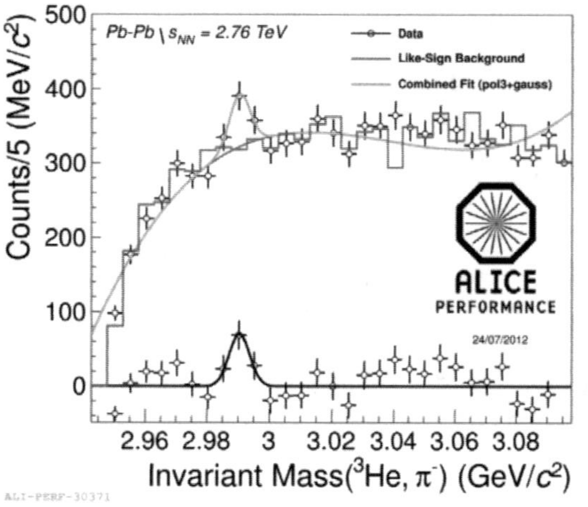

Figure 8. Invariant mass distribution of $^{3}\mathrm{He} + \pi^{-}$.

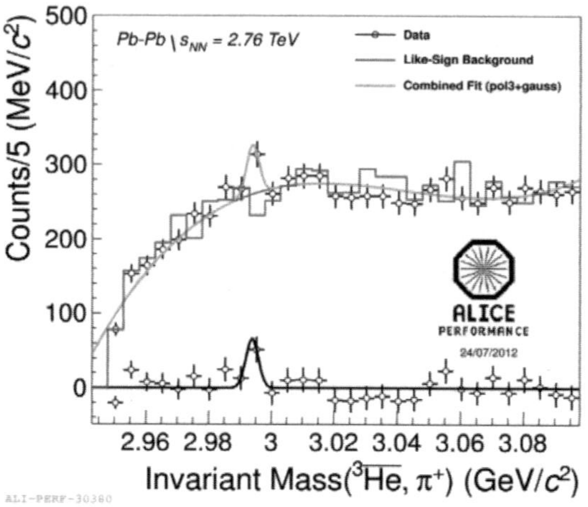

Figure 9. Invariant mass distribution of $^{3}\overline{\mathrm{He}} + \pi^{+}$.

and convoluted with the correction factor (efficiency × acceptance) to get a weighted efficiency. The unknown p_{T} shape is the main source of uncertainty. therefore, we use different functions for the systematics (limiting cases: blast-wave of deuteron and ^{3}He). To estimate the number of $\overline{\Lambda\mathrm{n}}$ particles, which we expect to measure, we need in addition also the branching ratio and the yield $\mathrm{d}N/\mathrm{d}y$. The branching ratio depends on the binding energy shown in figure 10. The production yield $\mathrm{d}N/\mathrm{d}y$ is taken from thermal model predictions, see figure 2. The resulting number of $\overline{\Lambda\mathrm{n}}$ particles which we expect is $N_{\overline{\Lambda\mathrm{n}},rec} = N_{events} \cdot Acc \times Eff \cdot BR \cdot \frac{\mathrm{d}N}{\mathrm{d}y} \cdot \mathrm{d}y \approx 4000$. For the data analysis we follow a similar strategy as for the hypertriton. We search for displaced decay vertices. If we can identifiy one of the daughter tracks as a deuteron and the second daughter as a pion, we calculate the invariant mass of these candidates. The result is shown in figure 12 (black points) together with the systematic error (grey band) and the peak which we expect from the Monte Carlo estimation and its corresponding error. So far we have no visible signal. From this non observation we can set an upper limit: $\mathrm{d}N/\mathrm{d}y \leq 1.5 \cdot 10^{-3}\,(99\,\%CL)$. This is a factor of 10 different to the thermal model input $\mathrm{d}N/\mathrm{d}y = 1.65 \cdot 10^{-2}$. However, the model describes the yields of the lighter hadrons and the hypertriton measured by STAR at $\sqrt{s_{NN}} = 200$ GeV correctly within uncertainties.

H-Dibaryon ($\Lambda\Lambda$)

For the H-Dibaryon we investigate two possible cases. In the first case the mass of the H-Dibaryon m_{H} is below the $\Lambda\Lambda$ threshold, such that the H-Dibaryon would be weakly bound and a measurable decay channel would be $\mathrm{H} \rightarrow \Lambda + p + \pi$. The mass of the H-Dibaryon would then lie between $2.2\,\mathrm{GeV}/c^2 < m_{\mathrm{H}} < 2.231\,\mathrm{GeV}/c^2$. In the second case the mass of the H-Dibaryon m_{H} is above the $\Lambda\Lambda$ threshold, such that the H-Dibaryon would be a resonant state and a measurable decay channel would be $\mathrm{H} \rightarrow \Lambda + \Lambda$. The mass of the H-Dibaryon would then lie above $m_{\mathrm{H}} > 2.231\,\mathrm{GeV}/c^2$. The analysis strategy for the H-Dibaryon is very similar as for the $\overline{\Lambda\mathrm{n}}$ bound state described above, except that here a second V0 type decay particle is involved. The result is shown in figure 13. Again we want to estimate the number of H-Dibaryons, which we expect from the current statistics from the Pb–Pb run of 2010 ($13.8 \cdot 10^6$ min. bias events). Therefore, we needed again estimates of the branching ratio and the yield $\mathrm{d}N/\mathrm{d}y$. The branching ratio as function of the binding energy is shown in figure 11. The production yield $\mathrm{d}N/\mathrm{d}y$ is

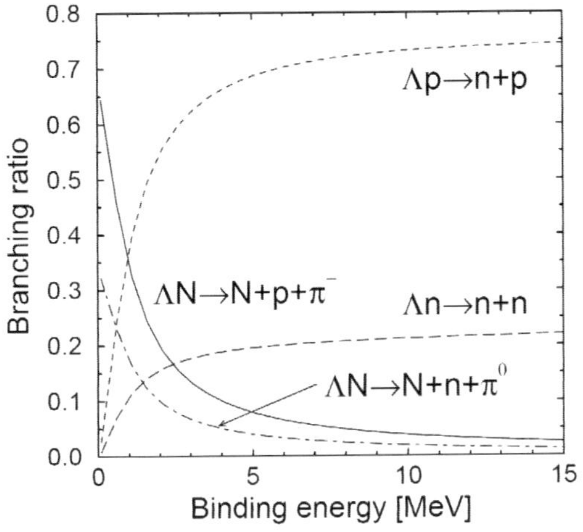

Figure 10. Theory predictions for the branching ratios for different decay channels of the Λn and the Λp [15].

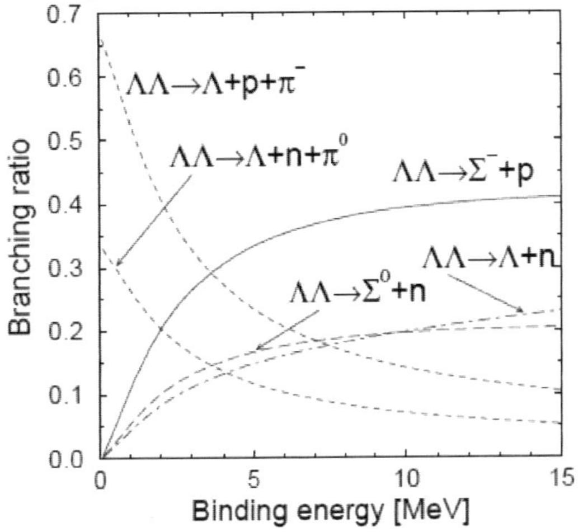

Figure 11. Theory predictions for the branching ratios for different decay channels of the H-Dibaryon [16].

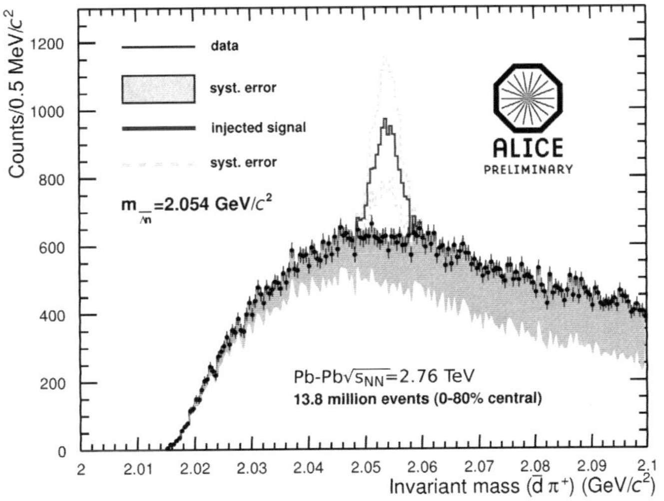

Figure 12. Invariant mass distribution for $d + \pi^+$. The figure also show the expected signal for the $\overline{\Lambda n}$ bound state.

again taken from the thermal model predictions, see figure 2. This results in 211 strongly bound H-Dibaryons which we should detect and 1350 lightly bound particles. The invariant mass distribution for the data is shown in figure 13 together with the two peaks, which we obtained from the Monte Carlo simulations. Again we see no hint of a signal from the data. From the non observation we can again obtain upper limits. For a strongly bound H we can set it to $\mathrm{d}N/\mathrm{d}y \leq 8.4 \cdot 10^{-4}$ $(99\,\%CL)$ and for a lightly bound H $\mathrm{d}N/\mathrm{d}y \leq 2 \cdot 10^{-4}$ $(99\,\%CL)$. Compared to the thermal model prediction $\mathrm{d}N/\mathrm{d}y = 3.1 \cdot 10^{-3}$ a discrepancy of a factor 10 is observed.

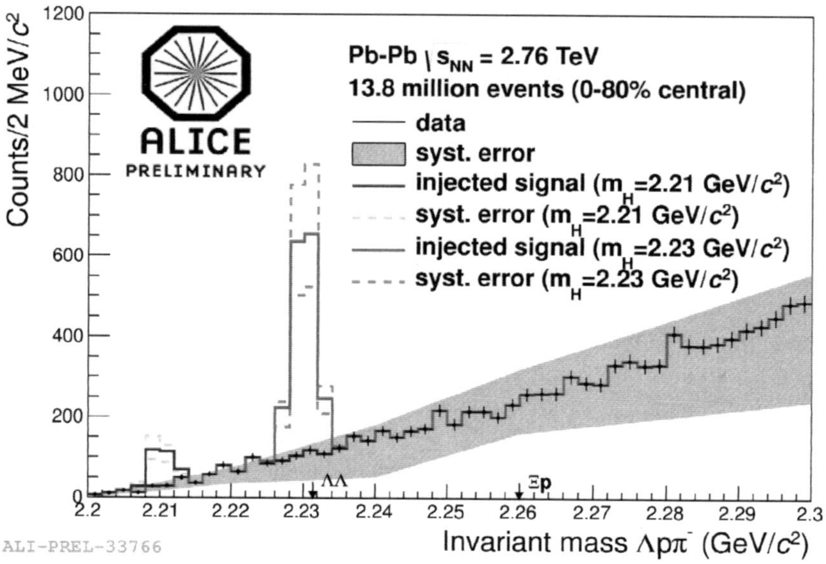

Figure 13. Invariant mass distribution for $\Lambda + p + \pi^-$. The figure also shows the expected signal for the two possible H-Dibaryons, which we investigated.

6. Summary and Outlook

We have shown that by combining the different particle identification techniques, e.g. TPC dE/dx and TOF, the ALICE detector is well suited for the detection of stable, weakly and strongly decaying particle species up to high masses. We have observed 10 $^4\overline{\text{He}}$ nuclei. The (anti-)hypertriton signal has also been observed and the work on the p_T spectra and the measurement of the lifetime is ongoing. Upper limits for the $\overline{\Lambda\text{n}}$ bound state and for two bound cases for the H-Dibaryon were obtained. Final conclusion on the existence of these exotic particles will be possible with larger statistics, as planned within the ALICE upgrade studies.

References

[1] Andronic A, Braun-Munzinger P, Stachel J and Stöcker H 2011 *Phys. Lett.* B **697** 203
[2] Cleymans J, Kabana S, Kraus I, Oeschler H, Redlich K and Sharma N 2011*Phys. Rev.* C **84** 054916
[3] Abelev B I *et al* (STAR Collaboration) 2010 *Science* **328** 58
[4] Aamodt K *et al* (ALICE Collaboration) 2008 *JINST* **3** S08002
[5] Jaffe R 1977 *Phys. Rev. Lett.* **38** 195 and erratum ibid 1977 **38** 617
[6] Inoue T, Ishii N, Aoki S, Doi T, Hatsuda T, Ikeda Y, Murano K, Nemura H and Sasaki K (HAL QCD Collaboration) 2011 *Phys. Rev. Lett.* **106** 162001
[7] Beane S R *et al* (NPLQCD Collaboration) 2011 *Phys. Rev. Lett.* **106** 162002
[8] Shanahan P E, Thomas A W and Young R D 2011 *Phys. Rev. Lett.* **107** 092004
[9] Haidenbauer J and Meißner U-G 2011 *Phys. Lett.* B **706** 100
[10] Andronic A private communication
[11] Bethe H 1930 *Annalen der Physik* **397** 325
[12] Agakishiev H *et al* (STAR Collaboration) 2011 *Nature* **473** 353
[13] Danysz M and Pniewski J 1953 *Bull. Acad. Pol. Sci.* III **1** 42;
Danysz M and Pniewski J 1953 *Phil. Mag.* **44** 348
[14] http://www.bnl.gov/hhi/files/talks/TakehikoSaito.pdf, as shown 1.3.2012 at the Riken-BNL workshop on "Hyperon-Hyperon Interactions and Searches for Exotic Di-Hyperons in Nuclear Collisions"
[15] Schaffner-Bielich J private communication
[16] Schaffner-Bielich J, Mattiello R and Sorge H 2000 *Phys. Rev. Lett.* **84** 4305

Journal of Physics: Conference Series **455** (2013) 012008 doi:10.1088/1742-6596/455/1/012008

Charged particle production in Pb-Pb and p-Pb collisions at ALICE

Chiara Oppedisano for the ALICE Collaboration
INFN Sezione di Torino, V. P. Giuria 1, 10125 Torino (Italy)
E-mail: Chiara.Oppedisano@cern.ch

Abstract. Global event observables are of fundamental importance in characterizing the properties of the strongly interacting medium created in heavy-ion collisions. The ALICE experiment at LHC measured charged particle multiplicity distributions in Pb-Pb collisions at $\sqrt{s_{NN}} = 2.76$ TeV in a wide pseudo-rapidity range ($-5 < \eta < 5.5$) exploiting different techniques. The production of high transverse momentum particles in Pb-Pb relative to pp is studied to characterize the in-medium energy loss. The pseudorapidity distribution of charged particles and the nuclear modification factor for Pb-Pb collisions at $\sqrt{s_{NN}} = 2.76$ TeV and for p-Pb collisions at $\sqrt{s_{NN}} = 5.02$ TeV are discussed. The ALICE results are compared to experimental results at lower energies and to theoretical predictions.

1. Introduction

Heavy ion collisions are the best experimental tool to study the phase transition to the Quark Gluon Plasma state predicted by lattice QCD. The RHIC experimental programme has established in the last decade that the hot medium formed in Au-Au collisions at $\sqrt{s_{NN}} = 200$ GeV behaves as a perfect liquid [1] [2]. The ALICE (A Large Ion Collider Experiment) experiment at LHC has studied Pb-Pb collisions at $\sqrt{s_{NN}} = 2.76$ TeV extending the experimental reach towards extreme energy densities. The ALICE experimental programme includes the study of proton-proton interactions as a benchmark for the observables studied in nucleus-nucleus collisions. Moreover proton-nucleus collisions are of crucial importance to disentangle Cold Nuclear Matter (CNM) effects and as a tool to assess QCD in the low x and high gluon density domain reached at LHC energies [3].

Global event properties allow the characterization of the collisions by describing the initial state and the following dynamical evolution. In nucleus-nucleus collisions the centrality is usually defined through the impact parameter b, the distance between the centers of the colliding nuclei in the plane transverse to their direction. The centrality is directly related to the geometry of the collisions and hence to the number of participating nucleons (N_{part}) and to the number of binary collisions (N_{coll}). The observables are usually studied as a function of centrality: the more central the collision, the higher the energy density. The particle multiplicity produced in A-A or p-A collisions is linked to the initial energy density and it can provide constraints on the particle production mechanisms, allowing the discrimination of the soft ($\sim N_{part}$) from the hard ($\sim N_{coll}$) regime [4].

The medium formed can also be characterized by studying the probes that propagate through it. High transverse momentum partons, produced during the very early stages of the collisions

Journal of Physics: Conference Series **455** (2013) 012008 doi:10.1088/1742-6596/455/1/012008

in initial hard scattering, are an ideal tool to study the properties of the medium. Hard partons lose energy by interacting with partons from the medium or by medium-induced gluon radiation [5] [6]. The nuclear modification factor (R_{AA}), defined as the ratio of the particle yield in A-A over p-p yield scaled by the number of collisions, characterizes the in-medium effects:

$$R_{AA} = \frac{d^2 N_{AA}/dp_T d\eta}{< N_{coll} > d^2 N_{pp}/dp_T d\eta}. \quad (1)$$

In the hard regime, where particle production scales with the number of binary collisions, R_{AA} is expected to approach unity if the nucleus-nucleus interactions are a simple superposition of N_{coll} hadron collisions. Modifications due to the presence of a hot and dense medium result in R_{AA} different from unity.

In the following, results on the charged particle multiplicity and on the charged hadron nuclear modification factor, measured in Pb-Pb collisions at $\sqrt{s_{NN}} = 2.76$ TeV and in p-Pb interactions at $\sqrt{s} = 5.02$ TeV, are presented and compared to results at lower energies and to predictions.

2. Detectors

ALICE is the dedicated heavy ion experiment at LHC. The experimental apparatus is constituted of a central ($|\eta| < 0.9$) barrel placed on a solenoidal magnetic field where different detectors track and identify particles down to $p_T = 200$ MeV/c, a forward ($2.5 < \eta < 4$) muon spectrometer to detect muons from heavy-quarkonia decays and very forward detectors which characterize the collision and provide the trigger. A comprehensive description of the experimental apparatus can be found in [7]. For the results that will be discussed in the following, four detectors are considered:

- the Silicon Pixel Detector (SPD): the 2 innermost layers of the Inner Tracking System (ITS), consisting of two cylindrical layers of hybrid silicon pixel assemblies covering $\eta < 2.0$ and $\eta < 1.4$ for the inner and outer layers, respectively. Besides the trigger, the SPD provides the interaction vertex and the charged particle multiplicity at midrapidity.
- the VZERO counters: two arrays of 32 scintillator tiles placed at $z = 3.3\ m$ and $z = -0.9\ m$ from the nominal interaction point (IP), covering the full azimuth within $2.8 < \eta < 5.1$ (VZERO-A) and $-3.7 < \eta < -1.7$. (VZERO-C).
- the Forward Multiplicity Detector (FMD): a silicon strip detector composed of three sub-detectors, each made by an inner and an outer ring, covering 2π in azimuth and $-3.4 < \eta < -1.7$ (FMD3), $1.7 < \eta < 5.0$ (FMD1-2) in pseudorapidity.
- the Zero Degree Calorimeters (ZDC): quartz fiber calorimeters placed at about 113 m from the IP, each set consisting of a neutron (ZN) and proton (ZP) ZDC. They detect the energy carried in the forward direction by nucleons that do not interact (spectators). The ZDC are used in Pb-Pb collisions to determine the centrality, to reject the background due to purely electromagnetic interactions and to identify satellite collisions.

3. Pb-Pb results

3.1. Charged particle multiplicity

The charged particle multiplicity, measured by ALICE in Pb-Pb collisions at midrapidity $|\eta| < 0.9$, has been compared to the existing measurements with different colliding systems [8] [9]. The charged particle multiplicity per participant pair at LHC is ~ 2.2 times the value measured at RHIC and a factor of 1.9 larger than in Non Single Diffractive (NSD) pp interactions at $\sqrt{s_{NN}} = 2.36$ TeV (see fig. 1). This measurement provides some basic constraints on theoretical models predicting particle production. The logarithmic increase of multiplicity as a function of

$\sqrt{s}$, valid for data up to the highest RHIC energy [10], had to be replaced by a power law growth $\sim s^{0.15}$. The ALICE measurement, together with the ATLAS [11] and CMS [12] results, draw

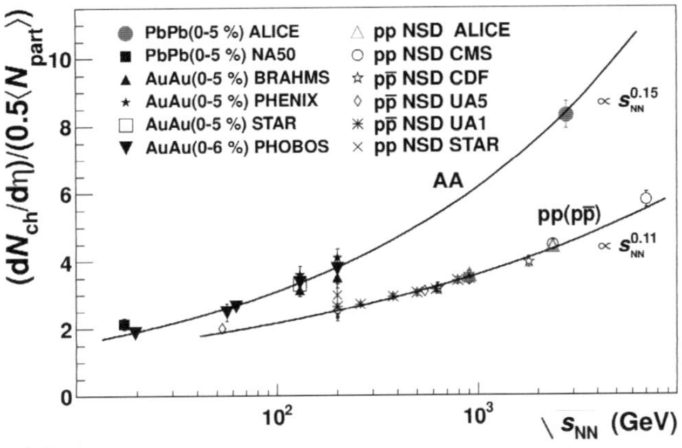

Figure 1. Charged particle multiplicity per participant pair as a function of $\sqrt{s}$ for different colliding systems [8].

a coherent picture that shows a very weak dependence on centrality going from 0.2 to 2.76 TeV (see fig. 2). Such a behavior is reproduced only by models that include saturation [13].

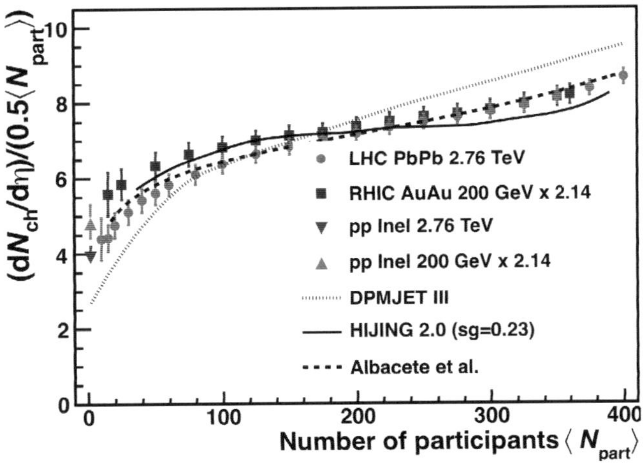

Figure 2. Charged particle multiplicity per participant pair as a function of centrality (expressed here in number of participating nucleons). LHC and RHIC data show the same weak centrality dependence [13].

ALICE extended the multiplicity measurement to forward rapidities, covering nearly 10 units of pseudorapidity, using the VZERO and the FMD. Since these detectors do not allow the rejection of secondary particles, a special technique has been developed to avoid large Monte Carlo dependent corrections and to extend the rapidity coverage. Interactions of the main Pb bunch with the satellite bunches from the other beam are selected, to identify the interaction vertices for which the produced particles only cross the beam pipe before reaching the VZERO and FMD detectors. In this way, the secondary production and the corresponding correction are minimized. The satellite vertices are selected using the timing information provided by the neutron ZDCs that measure the arrival time of the emitted neutrons with a resolution of about 0.5 ns. The sum (interaction time) and the difference (Z_{vertex}) of the arrival time on the two ZN are simultaneously used to select the interactions vertices of interest (see fig. 3).

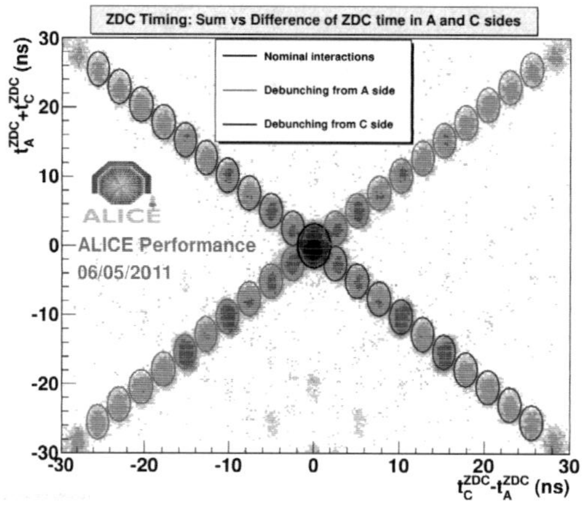

Figure 3. Sum vs. difference of the timing measured by the two ZN detectors positioned on both sides relative to ALICE IP. Interactions between the main bunches are at (0, 0). Interactions of main-satellite bunches are displaced by 2.5 ns with the corresponding vertices spatially displaced by 37.5 cm.

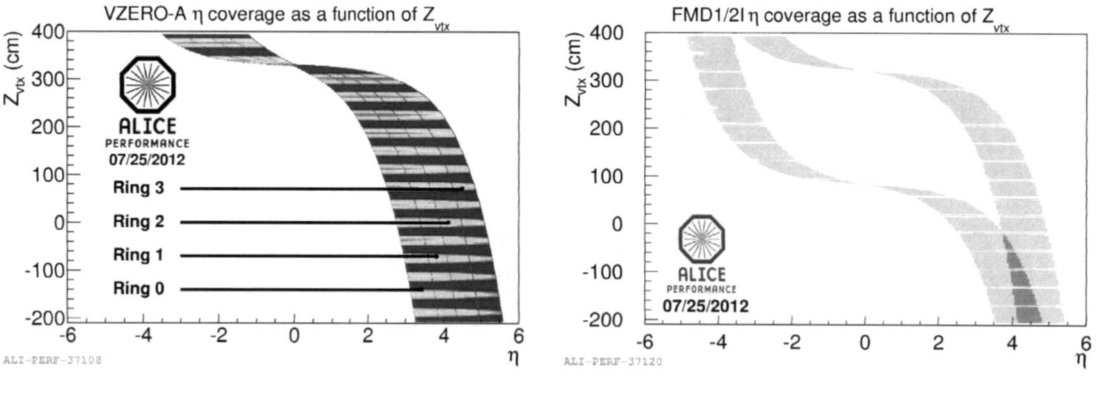

Figure 4. VZERO (left) and FMD (right) pseudorapidity coverage for the satellite vertices.

In fig. 4 the pseudorapidity coverage for the two detectors is shown for the different satellite vertices used in the analysis. Monte Carlo simulations are used to relate the VZERO signal amplitude to the true charged particle multiplicity. The VZERO absolute response is calibrated in the η range where VZERO and SPD detectors overlap. Finally, the charged particle multiplicity is extracted using the ratio from different vertex positions in order to cancel out contributions due to secondaries. The FMD detects the energy deposited by crossing particles, a statistical approach is used to determine the charged particle multiplicity. The absolute measurement is then corrected for the secondary contribution estimated through a Monte Carlo simulation. The FMD and VZERO measurements agree within 10% and they are both in agreement with the SPD results in the overlapping region. The combined results are shown in fig. 5. The centrality classes have been defined using the ZDC since their distance from the IP makes their response weakly dependent on the interaction vertex position (the same is not true for other estimators such as the VZERO or the SPD multiplicity). The distributions are then symmetrized between positive and negative pseudorapidities and fitted with a double Gaussian. The results are compared to predictions from different models: predictions based on Color Glass Condensate [14] [15] seem to reproduce the data around midrapidity but are only available over a limited pseudorapidity range. AMPT [16] and UrQMD [17] models provide a reasonable description of the forward region but they are unable to describe the overall magnitude and the shape on the whole η range.

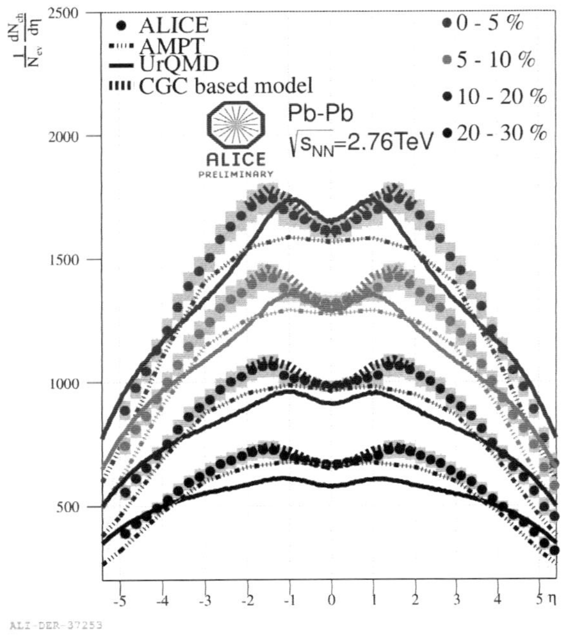

Figure 5. Combined $dN_{ch}/d\eta$ distributions in 4 centrality bins compared to model predictions.

In fig. 6 the charged particle distribution is plotted in the rest frame of one of the colliding nuclei for LHC and RHIC energies. This measurement is consistent with the validity of longitudinal scaling [19] within the errors, arising mainly from the extrapolation of the charged particle pseudorapidity density from the measured region to the rapidity region of the projectile.

3.2. Nuclear modification factor R_{AA}

The nuclear modification factor for charged particles has been measured in Pb-Pb collisions at $\sqrt{s} = 2.76$ TeV [20]. While a weak parton energy loss is observed in peripheral events, the high p_T particle production is strongly suppressed for central events. The pattern of R_{AA} as a function of p_T is qualitatively similar to that observed at RHIC (see fig. 7). At LHC the production is more suppressed, providing evidence for strong parton energy loss in a very dense medium.

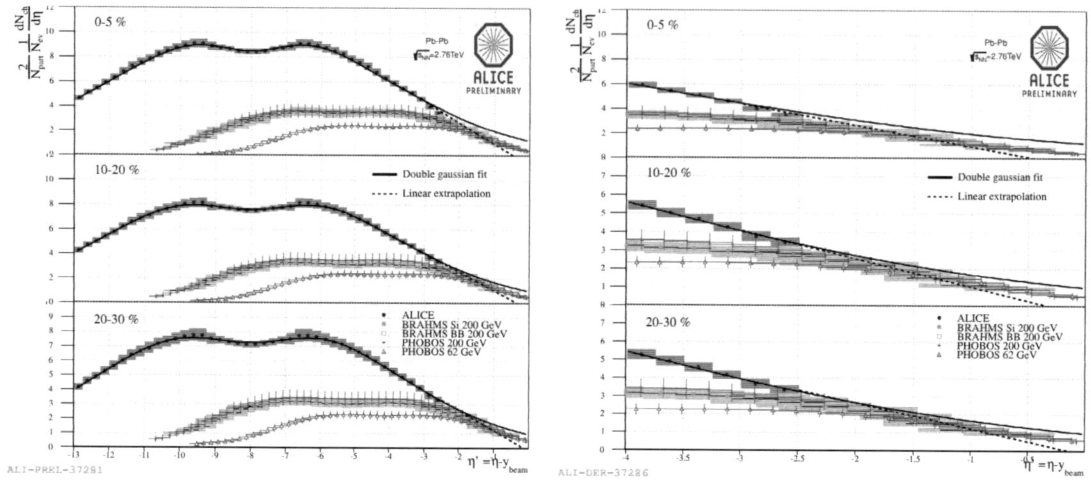

Figure 6. The charged-particle pseudorapidity density distribution per participating nucleon pair for three centrality bins shown in the rest frame of one of the projectiles (the projectile η region is zoomed in the right plots).

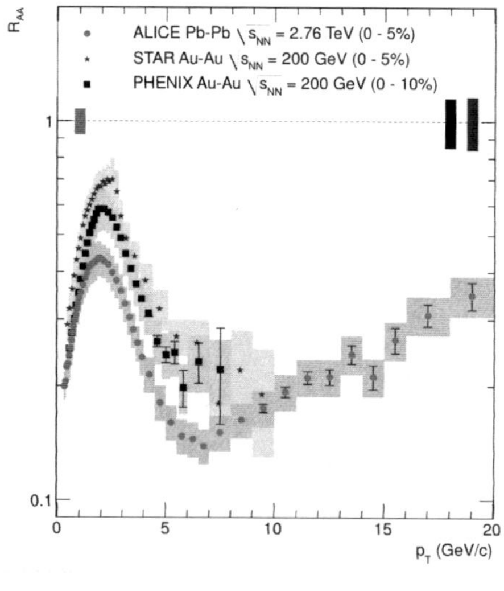

Figure 7. R_{AA} as a function of transverse momentum in most central bin compared to RHIC results [20].

4. p-Pb results

4.1. Charged particle multiplicity distributions

Particle production in p-Pb collisions is sensitive to the nuclear effects in the initial state. Therefore p-A interactions are an essential tool to discriminate between initial (already present in p-A interactions) and final (presumably accessible only in A-A collisions) state effects. ALICE measured the charged particle multiplicity in NSD p-A interactions with $|\eta_{LAB}| < 2$ [21]. The charged particle multiplicity is compared with several models (see fig. 8). Most of the models including either shadowing or saturation are able to predict the result within 20%. DPMJET [22] and HIJING [23] with gluon shadowing are closer to the data. The charged particle multiplicity per participant (compared to existing data in fig. 8) is consistent with inelastic pp collisions interpolated to the same energy. This measurement has provided the first constraint on the description of particle production in proton-nucleus collisions at LHC energies.

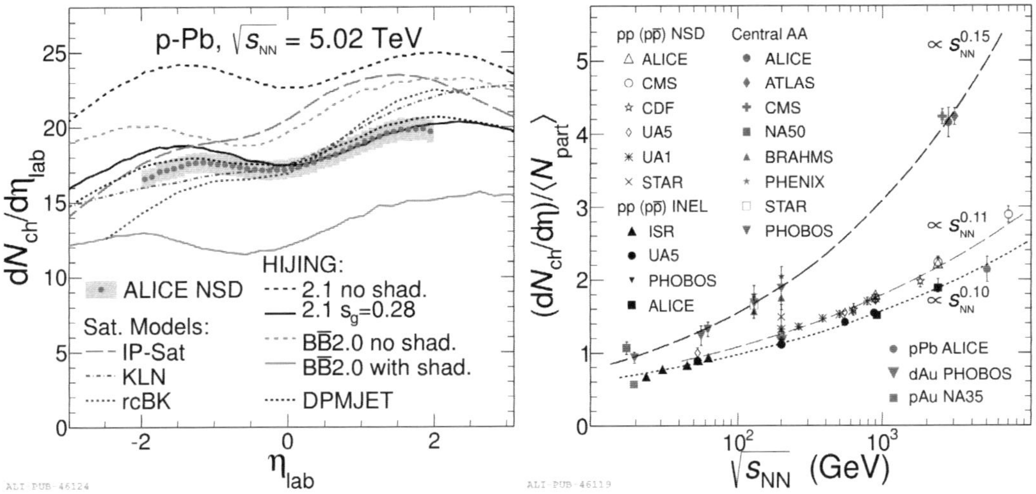

Figure 8. Left: charged particle pseudorapidity distribution measured in NSD p-A collisions. Right: Multiplicity per participant as a function of beam energy for different systems [21].

4.2. Nuclear modification factors R_{pA}

The nuclear modification factor in p-A collisions is a crucial tool to assess whether the initial state of the colliding system plays a role in the suppression of high transverse momentum particles observed in central Pb-Pb collision. R_{pPb} as a function of p_T is shown in fig. 9, compared with R_{AA} for top 5% central and for peripheral Pb-Pb collisions. R_{pPb} is compatible with unity for hard processes ($p_T > 2$ GeV/c), providing evidence that the strong suppression in central A-A collisions is due to the hot and dense medium formed in the collision. The Cronin enhancement observed at RHIC has a lower magnitude at LHC. The centrality dependence of R_{pPb} will help in understanding the cold nuclear matter effects and will provide constraints for the theoretical models describing particle production.

5. Conclusions

ALICE has measured the charged particle pseudorapidity distribution in Pb-Pb collisions in more than 10 units of pseudorapidity. The $\sqrt{s}$ dependence follows a power law. A weak centrality dependence of produced multiplicity is observed, indicating a strong shadowing of nuclear Parton Distribution Functions. Longitudinal scaling seems to persist at LHC energies. Currently no available model is able to describe ALICE data over the whole η range. A strong suppression of high transverse momentum particle production has been observed in Pb-Pb collisions, indicating that a dense and opaque medium is formed in A-A collisions at LHC.

First p-Pb data have provided an insight into cold nuclear matter effects. R_{pPb} is found to be compatible with unity, providing evidence that the suppression observed in Pb-Pb collisions is the fingerprint of an in-medium effect.

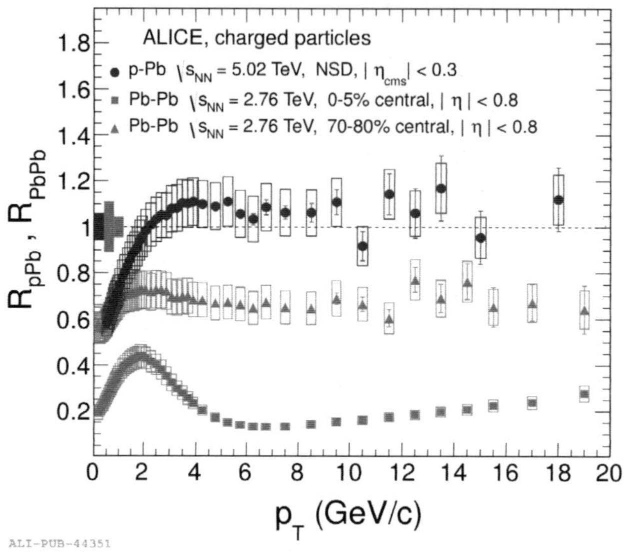

Figure 9. R_{pPb} (blue circles) compared to R_{AA} for top 5% central (red squares) and a more peripheral (green triangles) bin [24].

References

[1] PHENIX Collaboration, 2005 *Nucl.Phys.A* **757** 184
[2] STAR COllaboration, 2005 *Nucl.Phys.A* **757** 102
[3] C. Salgado et al., 2012 *J. Phys. G* **39** 015010
[4] D. Kharzeev et al., 2005 *Nucl. Phys. A* **747** 609
[5] M. Gyulassy et al. 1990 *Phys. Lett. B* **243** 432
[6] R. Baier et al. 1997 *Nucl. Phys. B* **484** 265
[7] ALICE Collaboration 2008 *JINST* **3** S08002
[8] ALICE Collaboration 2010 *Phys. Rev. Lett.* **105** 252301
[9] ALICE Collaboration 2011 *Phys. Rev. Lett.* **106** 032301
[10] PHOBOS Collaboration, 2011 *Phys.Rev. C* **83** 024913
[11] ATLAS Collaboration 2012 *Physics Letters B* **710** 363
[12] CMS Collaboration 2011 *Journal of High Eenergy Physics* **08** 141
[13] B. Muller et al. 2012 *Ann.Rev.Nucl.Part.Sci.* **62** 361
[14] J.L. Albacete and A. Dumitru 2010 arXiv:1011.5161
[15] J.L. Albacete at el. 2011 *J. Phys. Conf. Ser.* **316** 012011
[16] Zi-Wei Lin et al. *Physical Review C* **72** 064901
[17] M. Mitrovski et al. 2009 *Phys.Rev. C* **79** 044901
[18] PHOBOS Collaboration 2011 *Phys. Rev. C* **83** 024913
[19] J. Benecke et al. 1969 *Phys. Rev.* **188** 2159
[20] ALICE Collaboration 2011 *Physics Letters B* **696** 30
[21] ALICE Collaboration 2013 *Physical Review Letters* **110** 032301
[22] R. Engel et al. 1995 *Physical Review D* **52** 1459
[23] X.-N. Wang and M. Gyulassy 1991 *Physical Review D* **44** 3501
[24] ALICE Collaboration 2013 *Phys. Rev. Lett.* **110** 082302

Journal of Physics: Conference Series **455** (2013) 012009 doi:10.1088/1742-6596/455/1/012009

Light-flavour identified charged-hadron production in pp and Pb–Pb collisions at the LHC

Roberto Preghenella for the ALICE Collaboration
Centro Studi e Ricerche e Museo Storico della Fisica "Enrico Fermi", Rome, Italy
Sezione INFN, Bologna, Italy
E-mail: preghenella@bo.infn.it

Abstract.
Thanks to the unique detector design adopted to fulfill tracking and particle-identification (PID) requirements (e.g. low momentum cut-off and low material budget), the ALICE experiment provides significant information about hadron production both in pp and Pb–Pb collisions. In particular, the p_{T}-differential and integrated production yields of identified particles play a key role in the study of the collective and thermal properties of the matter formed in high-energy heavy-ion collisions. Furthermore, the production of high-p_{T} particles provides insights into the property of the hot medium created in such collisions and the in-medium energy-loss mechanisms.

Transverse momentum spectra of $\pi^{\pm}$, $\mathrm{K}^{\pm}$, p and $\bar{\mathrm{p}}$ are measured at mid-rapidity ($|y| < 0.5$) over a wide momentum range, from $\sim$ 100 MeV/c up to $\sim$ 20 GeV/c. The measurements are performed exploiting the $\mathrm{d}E/\mathrm{d}x$ in silicon and gas, the time-of-flight and the ring-imaging Cherenkov particle-identification techniques, which will be briefly reviewed in this report. The current results on light-flavour charged-hadron production will be presented for pp collisions at $\sqrt{s} = 0.9$, 2.76 and 7 TeV and for Pb–Pb collisions at $\sqrt{s_{\mathrm{NN}}} = 2.76$ TeV. Integrated production yields, transverse momentum spectra and particle ratios in pp are discussed as a function of the collision energy and compared to previous experiments and commonly-used Monte Carlo models. Pb–Pb collisions at the LHC feature the highest radial flow ever observed and an unexpectedly low p/π production ratio. The results are presented as a function of collision centrality and compared to RHIC data in Au–Au collisions at $\sqrt{s_{\mathrm{NN}}} = 200$ GeV and predictions from thermal and hydrodynamic models. The nuclear modification factor (R_{AA}) of identified hadrons will also be discussed and compared to unidentified charged particles and theoretical predictions. This is observed to be identical for all particle species at high-p_{T}.

1. Introduction

ALICE (A Large Ion Collider Experiment) is a general-purpose heavy-ion detector at the CERN LHC (Large Hadron Collider). It has been designed in order to fulfill the requirements to track and identify particles from very low ($\sim$100 MeV/c) up to quite high ($\sim$100 GeV/c) transverse momenta in an environment with large charged-particle multiplicities as in the case of central lead-lead (Pb–Pb) collisions at the LHC.

The ALICE experiment, shown in Figure 1, consists of a central-barrel detector and several forward detector systems. The central system covers the mid-rapidity region ($|\eta| \leq 0.9$) over the full azimuthal angle. It is installed inside a large solenoidal magnet providing a moderate magnetic field of 0.5 T. It includes a six-layer high-resolution inner-tracking system (ITS), a large-volume time-projection chamber (TPC) and electron and charged-hadron

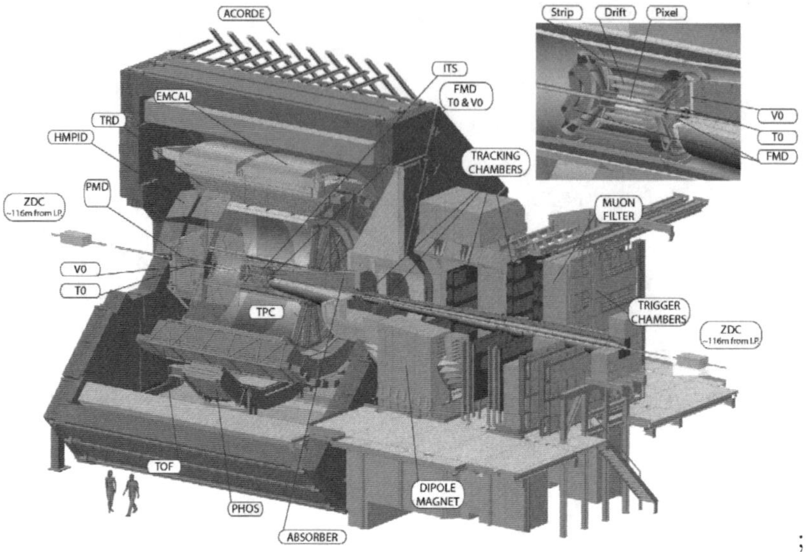

Figure 1. Schematic layout of the ALICE detector with its main subsystems.

identification detectors which exploit transition-radiation (TRD) and time-of-flight (TOF) techniques, respectively. Small-area systems for high-p_{T} particle-identification (HMPID), photon and neutral-meson measurements (PHOS) and jet reconstruction (EMCal) complement the central barrel. Thanks to these unique features the experiment is able to identify hadrons in a wide momentum range by employing different detection systems and techniques, as discussed in Section 2. The detectors which cover larger rapidity regions include a single-arm muon spectrometer covering the pseudorapidity range $-4.0 \leq \eta \leq -2.4$ and several smaller detectors (VZERO, TZERO, FMD, ZDC, and PMD) for triggering, multiplicity measurements and centrality determination. A detailed description of the ALICE detector layout and of its subsystems can be found in [1].

Since November 2009 when the first collisions at the LHC occurred, the ALICE experiment has collected proton-proton data at several centre-of-mass energies ($\sqrt{s}$ = 0.9, 2.76, 7 and 8 TeV). During the first two LHC heavy-ion runs, in 2010 and 2011, the detector recorded Pb–Pb collisions at a centre-of-mass energy per nucleon pair of $\sqrt{s_{\mathrm{NN}}}$ = 2.76 TeV and could profit from of an integrated luminosity of about 10 $\mu\mathrm{b}^{-1}$ and 100 $\mu\mathrm{b}^{-1}$, respectively. It is worth stressing the outstanding performance of the LHC complex: the instant luminosity exceeded $10^{26}\mathrm{cm}^{-2}\mathrm{s}^{-1}$ in the second run, higher than the design value. Proton-lead (p–Pb) collision data were also collected with the ALICE detector. This occurred during a short run performed in September 2012 in preparation for the main p–Pb run at the beginning of 2013. Eight pairs of bunches collided in the ALICE interaction region, providing a luminosity of about $8 \times 10^{25}\mathrm{cm}^{-2}\mathrm{s}^{-1}$. The configuration (4 TeV protons colliding with fully stripped ^{208}Pb ions at 82 × 4 TeV) resulted in interactions at $\sqrt{s_{\mathrm{NN}}}$ = 5.02 TeV in the nucleon-nucleon centre-of-mass system, which moves with a rapidity of $\Delta y_{\mathrm{NN}} = 0.465$.

2. Particle identification

In this section the main particle-identification (PID) detectors relevant to the analyses presented in this paper are briefly discussed. A detailed review of the ALICE experiment and of its PID capabilities can be found in [2]. The ITS is the innermost detector system, a six-layer silicon detector located at radii between 4 and 43 cm. Four of the six layers provide specific energy loss $\mathrm{d}E/\mathrm{d}x$ measurements and are used for particle identification in the non-relativistic ($1/\beta^2$)

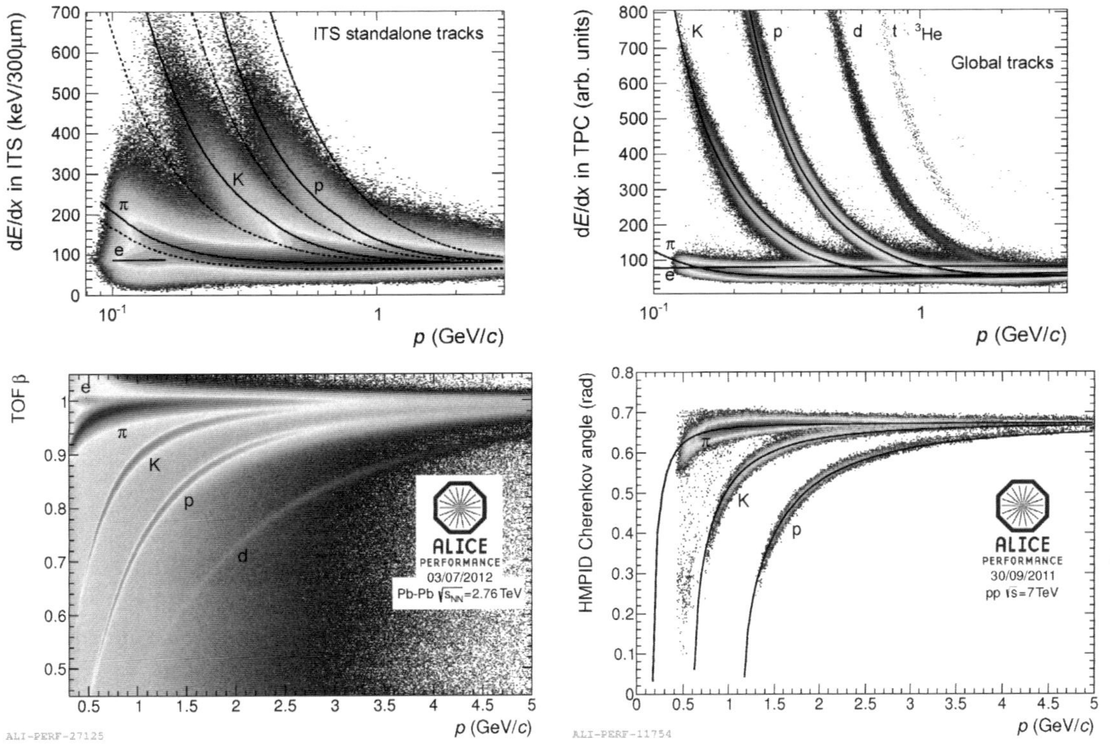

Figure 2. Energy loss $\mathrm{d}E/\mathrm{d}x$ in the ITS (top-left) and in the TPC (top-right). The continuous curves represent the Bethe-Bloch parametrization. (bottom-left) particle velocity β measured with TOF as a function of momentum. (bottom-right) Cherenkov angle measured in the HMPID as a function of the track momentum.

region. By using the ITS as a standalone tracker it is possible to reconstruct and identify low-momentum particles (below 200 MeV/c) not reaching the main tracking systems (Figure 2 top-left). The TPC is the main central-barrel tracking detector of ALICE. It provides three-dimensional hit information and specific energy-loss measurements with up to 159 samples. With the measured particle momentum and $\langle \mathrm{d}E/\mathrm{d}x \rangle$ the particle type can be determined by comparing the measurements with the Bethe-Bloch expectation (Figure 2 top-right). The TOF detector is a large-area array of Multigap Resistive Plate Chambers (MRPC) and covers the central pseudorapidity region ($|\eta| < 0.9$, full azimuth). Particle identification is performed by matching momentum and trajectory-length measurements performed by the tracking system with the time-of-flight information provided by the TOF system. The overall time-of-flight resolution is measured to be about 85 ps in Pb–Pb collisions (about 120 ps in pp collisions) and it is determined by the time resolution of the detector itself and by the start-time resolution (Figure 2 bottom-left). The HMPID detector consists of seven identical proximity focusing RICH (Ring Imaging Cherenkov) counters. Photon detection is performed using the proportional multiwire chambers coupled to pad-segmented CsI photocathode. Particle identification is obtained by means of the measurement of the Cherenkov angle allowing the separation of pions and kaons between 1 GeV/c and 3 GeV/c and protons from 1.5 GeV/c up to 5 GeV/c (Figure 2).

The transverse momentum spectra of primary $\pi^{\pm}$, $\mathrm{K}^{\pm}$, p and $\bar{\mathrm{p}}$ are measured at mid-rapidity ($|y| < 0.5$) combining the techniques and detectors described above. Primary particles are defined as prompt particles produced in the collision and all decay daughters, except products

Journal of Physics: Conference Series **455** (2013) 012009 doi:10.1088/1742-6596/455/1/012009

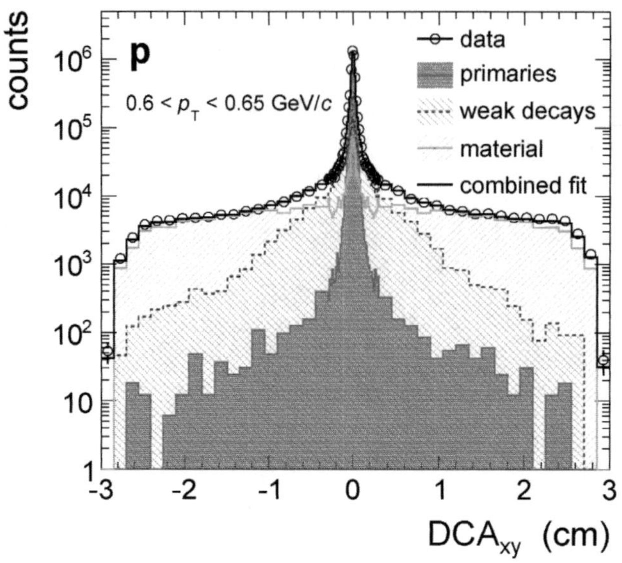

Figure 3. Transverse DCA (DCA_{xy}) of protons in the range between 0.6 GeV/c and 0.65 GeV/c in 0-5% most central Pb–Pb collisions together with the Monte Carlo templates which are fitted to the data.

from weak decays of strange particles. The contribution from the feed-down of weakly-decaying particles to $\pi^{\pm}$, p and $\bar{\text{p}}$ and from protons emitted from secondary interactions with material are subtracted by fitting the data using Monte Carlo templates of the DCA[1] distributions (Figure 3). Particles can also be identified in ALICE via their characteristics decay topology or invariant mass. This, combined with the direct identification of the decay daughters, allows one to reconstruct weakly-decaying particles and hadronic resonances with a good signal-to-background ratio.

3. Light-flavour hadron production

ALICE has measured the production yields of primary charged pions, kaon and (anti)protons in a wide momentum range and in several colliding systems. The measurements have been performed in proton-proton collisions at several centre-of-mass energies ($\sqrt{s}$ = 0.9, 2.76 and 7 TeV) and in Pb–Pb collisions at $\sqrt{s_{\mathrm{NN}}}$ = 2.76 TeV as a function of collision centrality.

In Pb–Pb collisions the transverse momentum p_{T} distributions and yields are compared to previous results at RHIC and expectations from hydrodynamic and thermal model. The results obtained for central Pb–Pb collisions are shown in Figure 4 and 5. The spectral shapes are harder than those observed at RHIC, indicating an increase of the radial flow velocity with the centre-of-mass energy. The radial flow at the LHC is found to be about 10% higher than at RHIC energy. The hydrodynamic models shown in Figure 4 give for central collisions a fair description of the data, describing the experimental spectra within 20%. This supports a hydrodynamic interpretation of the transverse momentum distribution in central collisions at the LHC. Further details of this anaysis and the models used to compare with the data can be found in [5].

While the K/π integrated production ratio is observed to be in line with lower energy measurements and predictions from the thermal model, both the p/π and the Λ/π ratios are

[1] Distance of Closest Approach to the reconstructed primary vertex.

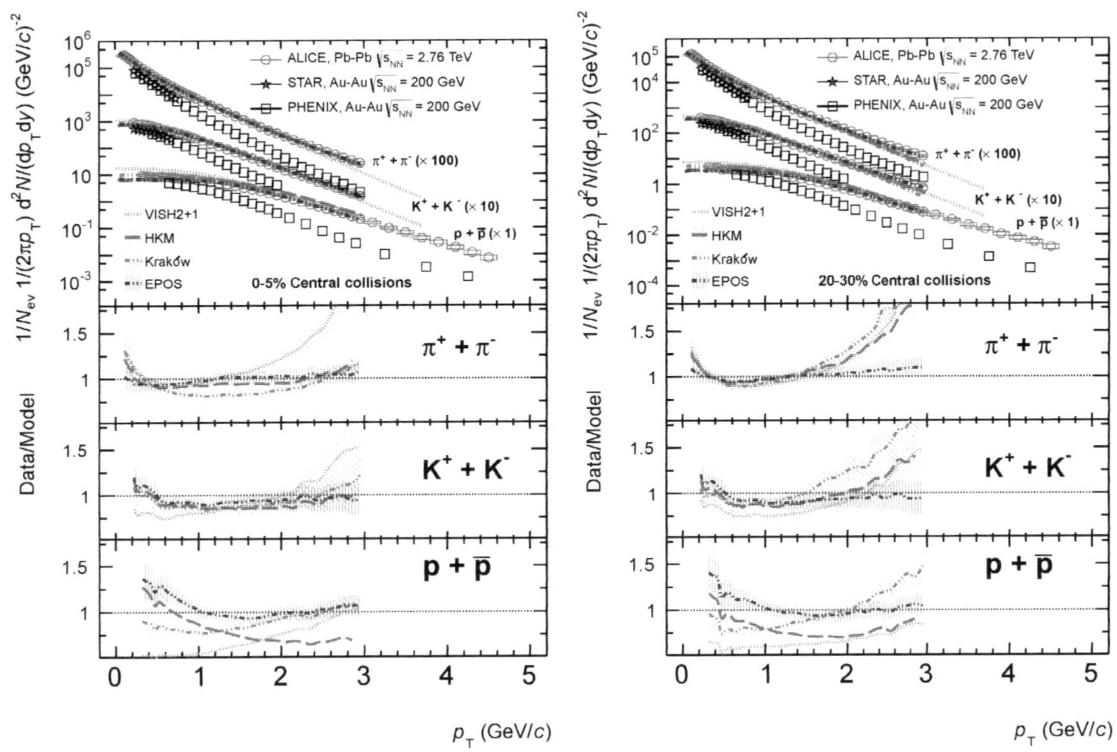

Figure 4. Transverse momentum distributions of the sum of positive and negative pions, kaons and protons for central Pb–Pb collisions. The results are compared to RHIC data and hydrodynamic models.

lower than those at RHIC and significantly lower (a factor $\sim$ 1.5 and 1.35, respectively) than predictions. A possible explanation of these deviations from the thermal-model predictions may be re-interactions in the hadronic phase due to large cross sections for antibaryon-baryon annihilation [6, 7, 8].

The p_T-dependent yield of charged kaons and protons normalized to charged pions are shown in Figure 6 for pp collisions at $\sqrt{s}$ = 2.76 and 7 TeV where they are compared with theoretical model predictions. Within the experimental uncertainties no energy dependence is observed in the data. The observed production ratios are not reproduced by NLO calculations [9]. PYTHIA Monte Carlo generator [10] underpredicts the proton-to-pion ratio at intermediate p_T.

Pion, kaon and (anti)proton production in Pb–Pb collisions have been compared to that in pp collisions and all show a suppression pattern which is similar to that of inclusive charged hadrons at high momenta (p_T above $\simeq$ 10 GeV/c), as shown in Figure 7 and 8 [11]. This seems to suggest that the dense medium formed in Pb–Pb collisions does not affect the fragmentation. A similar conclusion can be drawn from the proton-to-pion ratio measured in Pb–Pb collisions (Figure 9 and 10): for intermediate momenta (3–7 GeV/c) it exibits a relatively strong enhancement, by a factor of 3 compared to that in pp collisions at $p_T \approx$ 3 GeV/c and returns to the value in pp collisions at higher momenta (p_T above $\simeq$ 10 GeV/c) [11]. A similar observation was also reported for the Λ/K^0_s ratio and possible explanations have been proposed which include particle production via quark recombination [12].

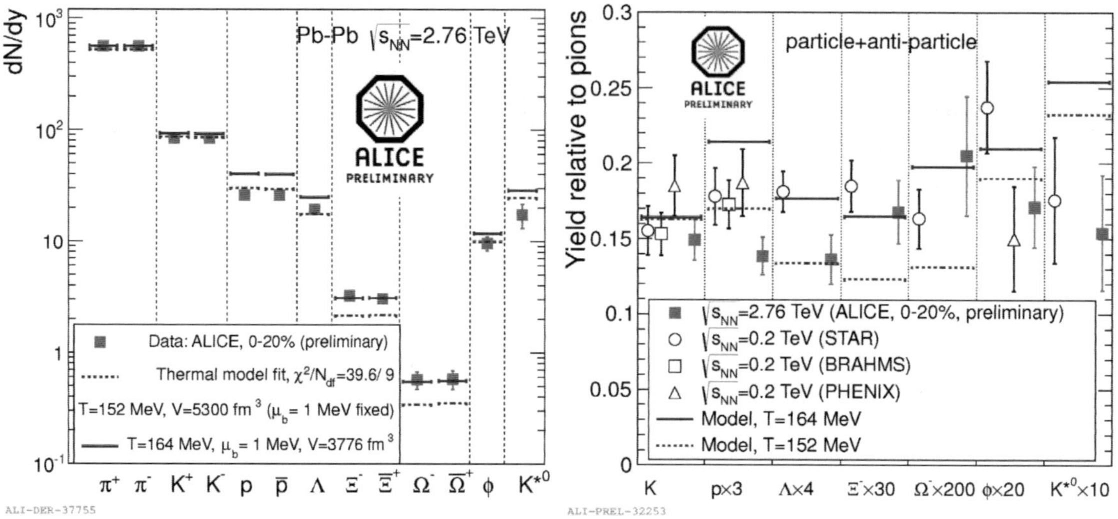

Figure 5. (left) The thermal fit of ALICE data showing data and model for the best fit. (right) Mid-rapidity particle ratios compared to RHIC results and predictions from thermal models for central Pb–Pb collisions at the LHC.

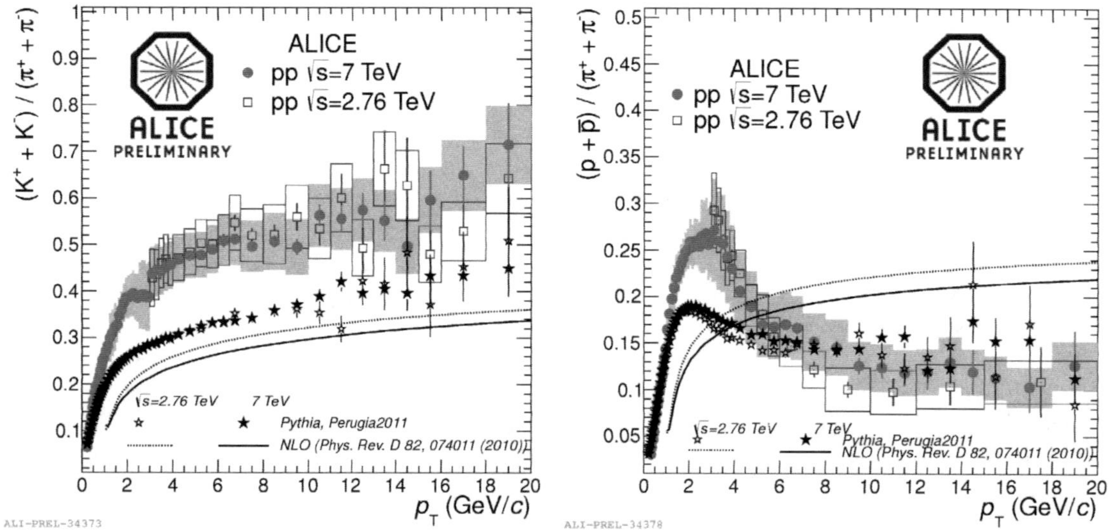

Figure 6. K/π (top) and p/π production ratio in pp collisions compared with PYTHIA Monte Carlo predictions and NLO calculations.

4. Transverse momentum distribution in proton-lead collisions

Particle production in proton-lead (p–Pb) collisions allows one to study and understand QCD at low parton fractional momentum x and high gluon density. It is moreover expected to be sensitive to nuclear effects in the initial state. For this reason p–Pb measurements provide an essential reference tool to discriminate between initial and final state effects and they are crucial for the studies and the understanding of deconfined matter created in nucleus-nucleus collisions.

The measurement of the transverse momentum p_T distributions of charged particle in p–Pb collisions were reported already in [13]. It was previously shown that the production of

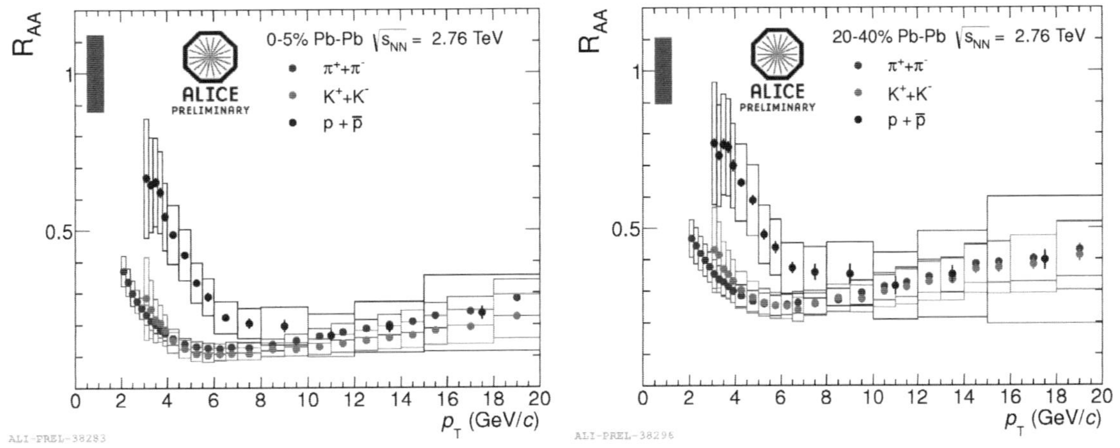

Figure 7. Charged pion, kaon and (anti)proton nuclear modification factor R_{AA} as a function of p_T for Pb–Pb collisions at $\sqrt{s_{NN}} = 2.76$ TeV for centrality 0-5% (left) and 20-40% (right). Statistical (vertical error bars) and systematic (gray and colored boxes) are shown for the charged pion RAA. The gray boxes contains the common systematic error related to pp normalization to INEL and N_{coll}.

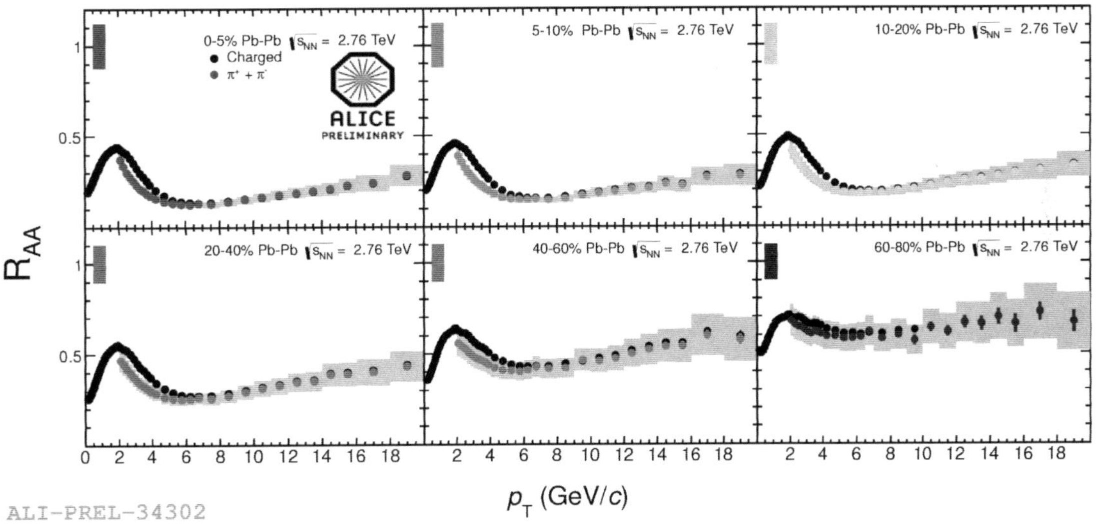

Figure 8. Nuclear modification factor R_{AA} of charged pions compared to the R_{AA} of unidentified charged particles as a function of p_T for different centrality classes. Statistical (vertical error bars) and systematic (gray and colored boxes) errors are shown for the charged pions. The colored boxes contain the common systematic uncertainty related to the number of binary collisions and the pp normalization to INEL. Only statistical errors are shown for the unidentified charged R_{AA}.

charged hadrons in central Pb–Pb collisions at the LHC is strongly suppressed [14, 15]. The suppression remains substantial up to 100 GeV/c and is also seen in reconstructed jets [16]. Proton-lead collisions provide a control experiment to clearly establish whether the initial state of the colliding nuclei plays a role in the observed high-p_T hadron production in Pb–Pb collisions. In

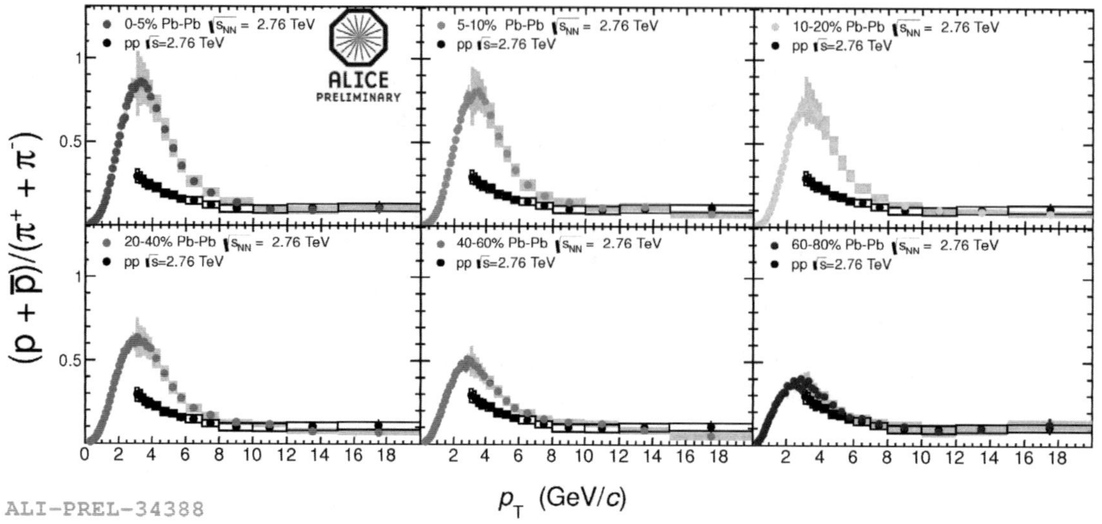

Figure 9. Proton-to-pion ratio in Pb–Pb collisions compared with pp results at 2.76 TeV in several centrality bins. In Pb–Pb the statistical and systematic uncertainties are displayed as error bars and gray bands, respectively. In pp the systematic uncertainty is displayed in black squares.

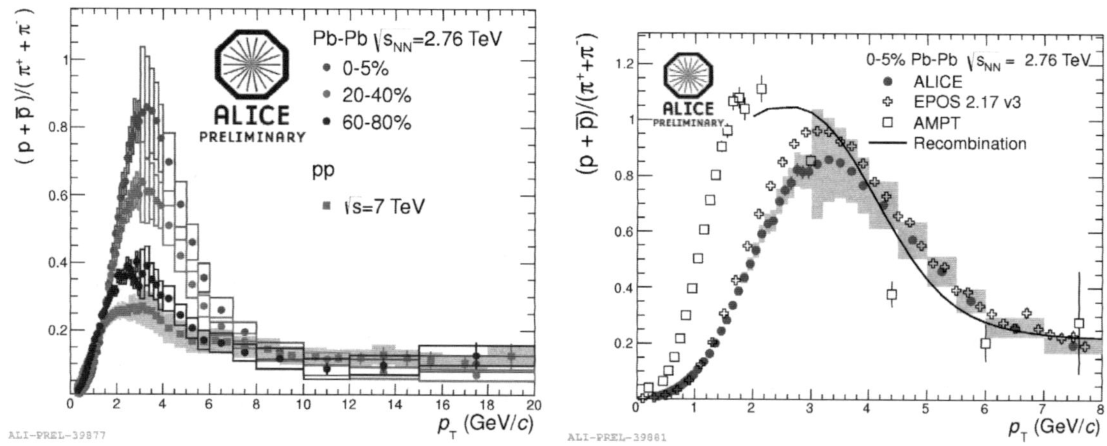

Figure 10. (left) Proton-to-pion ratio in Pb–Pb collisions in several centrality bins compared with pp results at 2.76 TeV. (right) Proton-to-pion ratio in central Pb–Pb collisions compared with models.

order to quantify nuclear effects, the p_T-differential yield relative to the proton-proton reference, the so-called nuclear modification factor, is calculated. The nuclear modification factor is expected to be unity for hard processes which exhibit binary collision scaling. This has been recently confirmed in Pb–Pb collisions at the LHC by the measurements of direct photon [17], Z^0 [18] and $W^{\pm}$ [19], observables which are not affected by hot QCD matter. In Figure 11 the measurement of the nuclear modification factor in p–Pb collisions R_{pPb} is compared to that in central (0-5% centrality) and peripheral (70–80%) Pb–Pb collisions R_{PbPb}. R_{pPb} is observed to be consistent with unity for transverse momenta higher than about 2 GeV/c. This important

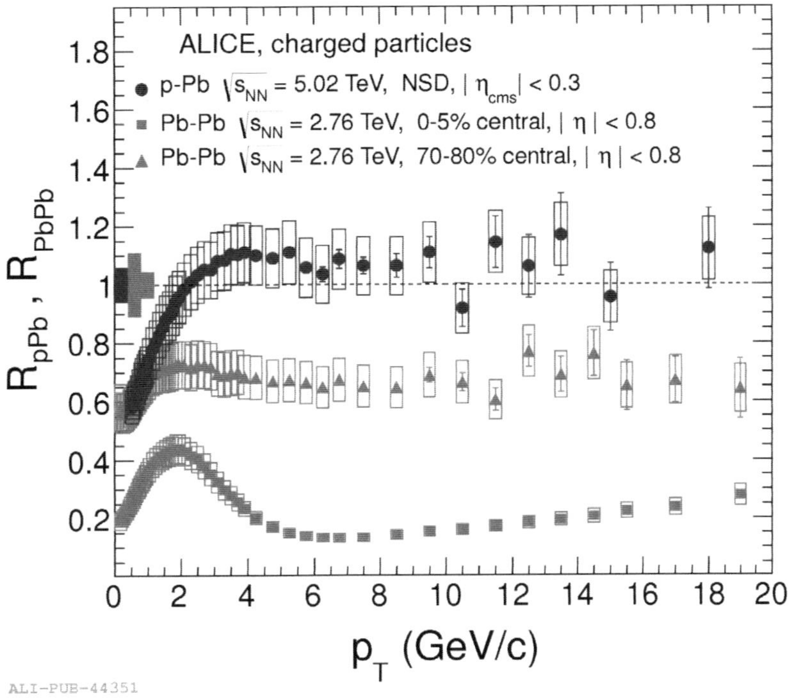

Figure 11. The nuclear modification factor of charged particles as a function of transverse momentum in NSD p–Pb collisions at $\sqrt{s_{NN}} = 5.02$ TeV compared to measurements in central (0-5%) and peripheral (70-80%) Pb–Pb collisions at $\sqrt{s_{NN}} = 2.76$ TeV.

measurement demonstrates that the strong suppression observed in central Pb–Pb collisions at the LHC is not due to an initial-state effect, but it is a final state effect related to the hot matter created in high-energy heavy-ion reactions.

5. Summary and conclusions

ALICE has obtained so far a wealth of physics results both from the analysis of proton-proton collision data and from the first two LHC heavy-ion runs.

The transverse momentum spectra of $\pi^{\pm}$, $K^{\pm}$, p and $\bar{p}$ have been measured with ALICE in several colliding systems and energies at the LHC demonstrating the excellent PID capabilities of the experiment. Data in pp collisions show no evident $\sqrt{s}$ dependence in hadron production ratios. In Pb–Pb collisions $\bar{p}/\pi^-$ integrated ratio is significantly lower than statistical model predictions with a chemical freeze-out temperature $T_{ch} \simeq 160 - 170$ MeV. The average transverse momenta and the transverse momentum spectra indicate a ~10% stronger radial flow than at RHIC energies.

In the intermediate transverse momentum $p_{\rm T}$ region, an enhancement of the baryon-to-meson ratio is observed. The maximum of the ratio is shifted to higher $p_{\rm T}$ with respect to RHIC measurements. The results of the measurements of charged pion, kaon, protons and antiproton production at high $p_{\rm T}$ were also presented. The nuclear modification factors for these species are similar in magnitude, suggesting that the medium does not significantly affect fragmentation.

First results from a short pilot run with proton-lead beams have been also reported. The measurement of the charged-particle transverse momentum $p_{\rm T}$ spectra and nuclear modification factor in p–Pb collisions at $\sqrt{s_{NN}} = 5.02$ TeV, covering $0.5 < p_{\rm T} < 20$ GeV/c, show a nuclear modification factor consistent with unity for $p_T > 2$ GeV/c. This measurement indicates that the strong suppression of hadron production at high p_T observed at the LHC in Pb–Pb collisions is not due to an initial-state effect, but is the fingerprint of jet quenching in hot QCD matter.

References

[1] K. Aamodt *et al.* [ALICE Collaboration], "The ALICE experiment at the CERN LHC", JINST **3** (2008) S08002.
[2] G. Alessandro, (Ed.) *et al.* [ALICE Collaboration], "ALICE: Physics performance report, volume II", J. Phys. G **32** (2006) 1295.
[3] B. Abelev *et al.* [ALICE Collaboration], "Pseudorapidity density of charged particles *p*-Pb collisions at $\sqrt{s_{NN}} = 5.02$ TeV", arXiv:1210.3615 [nucl-ex].
[4] B. Alver, M. Baker, C. Loizides and P. Steinberg, "The PHOBOS Glauber Monte Carlo", arXiv:0805.4411 [nucl-ex].
[5] B. Abelev *et al.* [ALICE Collaboration], "Pion, Kaon, and Proton Production in Central Pb–Pb Collisions at $\sqrt{s_{NN}} = 2.76$ TeV", Phys. Rev. Lett. **109** (2012) 252301 [arXiv:1208.1974 [hep-ex]].
[6] J. Steinheimer, J. Aichelin and M. Bleicher, "Non-thermal p/π ratio at LHC as a consequence of hadronic final state interactions", arXiv:1203.5302 [nucl-th].
[7] F. Becattini, M. Bleicher, T. Kollegger, M. Mitrovski, T. Schuster and R. Stock, "Hadronization and Hadronic Freeze-Out in Relativistic Nuclear Collisions", Phys. Rev. C **85** (2012) 044921 [arXiv:1201.6349 [nucl-th]].
[8] Y. Pan and S. Pratt, "Baryon Annihilation in Heavy Ion Collisions", arXiv:1210.1577 [nucl-th].
[9] R. Sassot, P. Zurita and M. Stratmann, "Inclusive Hadron Production in the CERN-LHC Era", Phys. Rev. D **82** (2010) 074011 [arXiv:1008.0540 [hep-ph]].
[10] P. Z. Skands, Phys. Rev. D **82** (2010) 074018 [arXiv:1005.3457 [hep-ph]].
[11] A. Ortiz Velasquez [ALICE Collaboration], "Production of pions, kaons and protons at high p_T in $\sqrt{s_{NN}} =$ 2.76 TeV Pb-Pb collisions", arXiv:1210.6995 [hep-ex].
[12] R. J. Fries, B. Muller, C. Nonaka and S. A. Bass, "Hadronization in heavy ion collisions: Recombination and fragmentation of partons", Phys. Rev. Lett. **90** (2003) 202303 [nucl-th/0301087].
[13] B. Abelev *et al.* [ALICE Collaboration], "Transverse Momentum Distribution and Nuclear Modification Factor of Charged Particles in *p*-Pb Collisions at $\sqrt{s_{NN}} = 5.02$ TeV", arXiv:1210.4520 [nucl-ex].
[14] K. Aamodt *et al.* [ALICE Collaboration], "Suppression of Charged Particle Production at Large Transverse Momentum in Central Pb–Pb Collisions at $\sqrt{s_{NN}} = 2.76$ TeV", Phys. Lett. B **696** (2011) 30 [arXiv:1012.1004 [nucl-ex]].
[15] B. Abelev *et al.* [ALICE Collaboration], "Centrality Dependence of Charged Particle Production at Large Transverse Momentum in Pb–Pb Collisions at $\sqrt{s_{\rm NN}} = 2.76$ TeV", [arXiv:1208.2711 [hep-ex]].
[16] G. Aad *et al.* [ATLAS Collaboration], "Measurement of the jet radius and transverse momentum dependence of inclusive jet suppression in lead-lead collisions at $\sqrt{s_{NN}} = 2.76$ TeV with the ATLAS detector", arXiv:1208.1967 [hep-ex].
[17] S. Chatrchyan *et al.* [CMS Collaboration], "Measurement of isolated photon production in *pp* and PbPb collisions at $\sqrt{s_{NN}} = 2.76$ TeV", Phys. Lett. B **710** (2012) 256 [arXiv:1201.3093 [nucl-ex]].
[18] S. Chatrchyan *et al.* [CMS Collaboration], "Study of Z boson production in PbPb collisions at nucleon-nucleon centre of mass energy = 2.76 TeV", Phys. Rev. Lett. **106** (2011) 212301 [arXiv:1102.5435 [nucl-ex]].
[19] S. Chatrchyan *et al.* [CMS Collaboration], "Study of *W* boson production in PbPb and *pp* collisions at $\sqrt{s_{NN}} = 2.76$ TeV", Phys. Lett. B **715** (2012) 66 [arXiv:1205.6334 [nucl-ex]].

International Workshop on Discovery Physics at the LHC (Kruger2012) IOP Publishing
Journal of Physics: Conference Series **455** (2013) 012010 doi:10.1088/1742-6596/455/1/012010

Photoproduction of ρ^0 in ultra–peripheral nuclear collisions at ALICE

Kyrre Skjerdal, for the ALICE collaboration

Department of Physics and Technology, University of Bergen, P.O. Box 7803, 5020 Bergen, Norway

E-mail: kyrre.skjerdal@cern.ch

Abstract. Photoproduction of ρ^0 mesons in ultra-peripheral Pb+Pb collisions has been studied by the ALICE Collaboration at the CERN LHC. The strong photon flux associated with relativistic charged nuclei leads to a very large cross section for exclusive photoproduction of ρ^0 meson in interactions of the type $Pb + Pb \rightarrow Pb + Pb + \rho^0$. For a ρ^0 produced at mid-rapidity at the LHC, the photon-nucleus center of mass energy is higher than in any previous experiment.

The ALICE detector is a general purpose detector dedicated to study heavy–ion collisions. ALICE has excellent performance in the low p_T region, and can reconstruct charged particle tracks with 0.1 GeV/c $\leq p_T \leq$ 100 GeV/c. In this analysis all tracks were required to be within ALICE's central barrel. Analysis of data from the first heavy ion run at the LHC in 2010 will be discussed in this paper.

1. Introduction to ultra–peripheral collisions

Ultra–peripheral collisions are collisions between hadrons, they can be protons or nuclei, where they geometrically miss each other. This implies that the impact parameter is larger than two times the radii of the colliding hadrons, see Figure 1. For light vector meson production in Pb–Pb collisions at LHC energy, the median impact parameter is usually in the range 200 – 300 fm. Because such impact parameters are much larger than the short range of the strong interaction, the interactions will be mediated by the electromagnetic field. The electromagnetic field of a moving charged particle can be treated as a flux of virtual photons. This model is referred to as the Weizäcker–Williams method. The intensity of the electromagnetic field, and therefore the number of virtual photons in the cloud is proportional to Z^2, where Z is the charge of the particle [1].

One can divide ultra-peripheral collisions into two categories: two-photon and photonuclear interactions. The photonuclear interactions can be further subdivided into coherent and incoherent interactions. In coherent photonuclear interactions, the photon interacts with the whole target nucleus coherently. In most cases both nuclei will remain intact after the interaction. For photonuclear interactions a photon from one nucleus is interacting with the other nucleus and, because the photon has spin $J^P = 1^-$, a vector meson can be produced. The transverse momentum of the resulting particles will be determined by the nuclear form factor, the average transverse momentum will be $\langle p_T \rangle \sim 60$ MeV/c.

In an incoherent interaction a photon from one nucleus interacts with a single nucleon in the target nucleus. This will cause the target nucleus to break up. The transverse momentum is

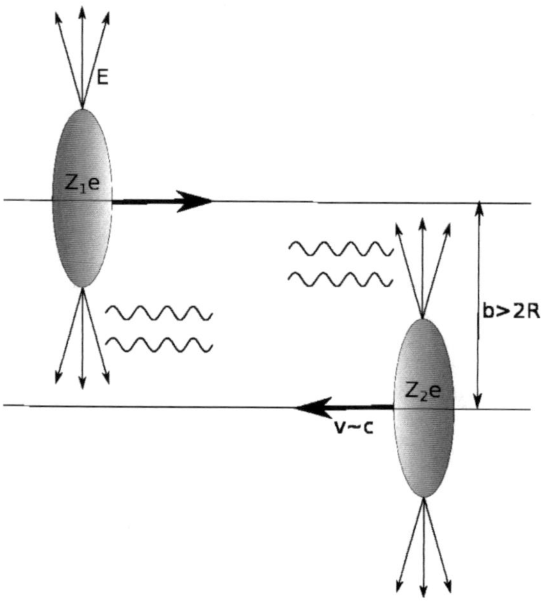

Figure 1. The impact parameter, b, is larger than two times the radius, R, of the colliding hadrons. The photon flux from each nucleus is proportional to Z^2.

higher in incoherent interactions than in coherent interactions, $\langle p_T \rangle \sim 400$ MeV/c for the ρ^0 [6].

In two–photon interactions one photon from each of the two colliding nuclei collides and produces a fermion pair. An example is given by $Pb + Pb \rightarrow Pb + Pb + e^+e^-$, where the interaction is $\gamma + \gamma \rightarrow e^+e^-$.

For vector meson production in photonuclear coherent interactions the event will contain only two tracks, and the detector will otherwise be empty. There will be one positive and one negative track, and the transverse momentum of the track pair will be $\sum p_T \lesssim 100$ MeV/c [6, 9].

2. Model predictions

The center–of–mass energy per nucleon in Pb–Pb collisions at the LHC in the 2010 and 2011 Pb–Pb runs was $\sqrt{s_{NN}} = 2.76$ TeV. This corresponds to a Lorentz factor of $\gamma_L = 1470$ for each beam in the center of mass system. The ρ^0 is a broad resonance with mass $M_\rho = 775$ MeV/c^2 and width $\Gamma_\rho = 149$ MeV/c^2. At mid–rapidity, $y = 0$, the mass 775 MeV/c^2 corresponds to a mean photon energy of $E_\gamma = 775$ MeV$/2 = 338$ MeV. The equivalent γ–proton center–of–mass energy is then

$$W_{\gamma p} = \sqrt{4E_\gamma m_p \gamma_L} = 43.2 \text{ GeV}. \tag{1}$$

This is three times higher than at RHIC, and higher than in any fixed–target experiment.

Three different models predict the cross section for ρ^0 production at LHC energies. The model by Frankfurt, Strikman, and Zhalov (GGM (*Gribov–Glauber Model*)) [3, 4] uses a generalized vector dominance model in the Gribov–Glauber approach. It includes non–diagonal transitions, where the photon fluctuates to a ρ', but appear as a ρ^0 after scattering off the target nucleus. The cross section $\sigma(\rho + \text{nucleon})$ from the Donnachie–Landshoff model, which is in agreement with HERA and lower energy data, is used as input for the photonuclear calculation.

The model developed by Gonçalves and Machado (GM) [5] is based on the color dipole model in combination with saturation from a Color Glass Condensate approach. Starlight [6, 7] is a Monte Carlo event generator developed by Klein and Nystrand. Experimental data for $\gamma + p \rightarrow \rho^0 + p$ is used in combination with a Glauber model, neglecting the elastic scattering. The predictions from the different models can be seen in Figure 2.

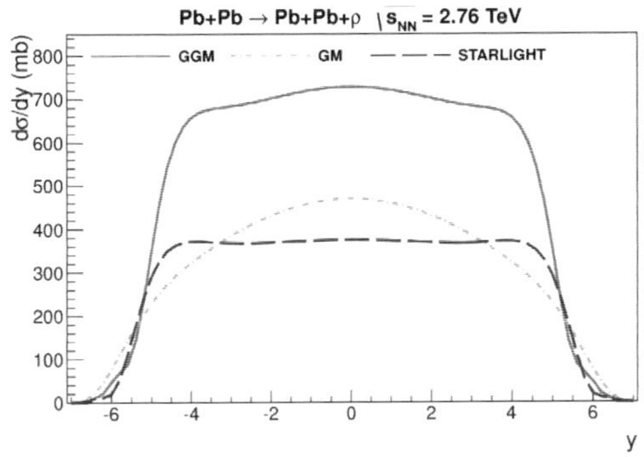

Figure 2. Model predictions for ρ^0 photoproduction at central rapidities ($y = 0$), for the models GGM [3, 4], shown with a full (——) red line, GM [5], shown with a yellow chained (— · —) line and Starlight [6, 7], shown with a blue dashed (- - - -) line. (Color online.)

3. The ALICE detector

The ALICE detector (Figure 3) is a general purpose detector at the CERN LHC [8]. Its main goal is to study ultra–relativistic heavy–ion collisions. It consists of a central barrel, a forward muon arm, and some other smaller forward detectors. The central barrel has a pseudorapidity acceptance of $|\eta| < 0.9$ and a transverse momentum acceptance of $p_T > 100$ MeV/c.

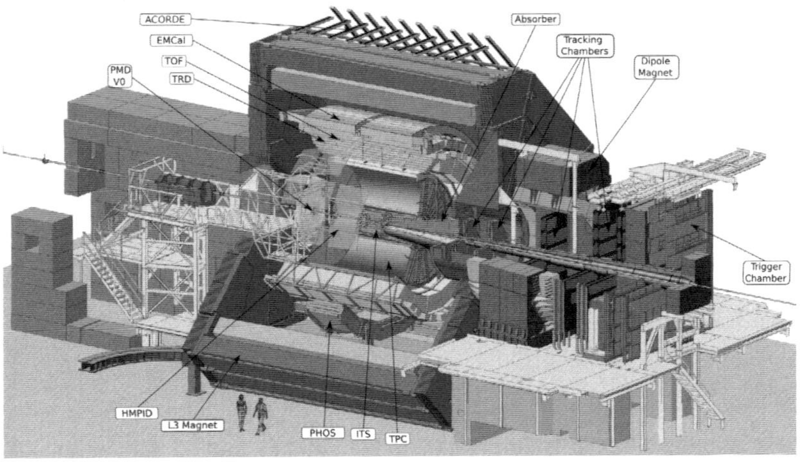

Figure 3. The ALICE detector

For this analysis mainly the central barrel is used. The Inner Tracking System (ITS) and the Time Projection Chamber (TPC) are used for tracking and particle identification. The ITS consists of six layers of silicon detectors; the two innermost layers are silicon pixel detectors (SPD), the two next are silicon drift detectors (SDD), and the two outermost layers are silicon strip detectors (SSD). The ALICE TPC, a cyllindrical gaseous detector with a diameter of 500 cm and a length of 510 cm, is the main tracking device in ALICE. The 557,568 readout pads can provide up to 159 ionization samples for track reconstruction, which can be used to calculate the energy loss ($\mathrm{d}E/\mathrm{d}x$) of the track. The dE/dx of a track is calculated as the truncated mean of the $\mathrm{d}E/\mathrm{d}x$ of the clusters associated with the track. The truncated mean is used in order to reduce the fluctuations in cluster energies resulting from the Landau tail [13]. For triggering the SPD, the Time–of–Flight detector (TOF) and the VZERO counters are used. The TOF detector, which surrounds the TPC, is composed of multigap resistive plate chambers which provide an intrinsic resolution of approximately 80 ps. The VZERO counters, on each side of the interaction point, consists of 32 tiles of scintilators. They are in this analysis used to define rapidity gaps. The VZERO–A has an acceptance of $2.8 \leq \eta \leq 5.1$, and the VZERO–C has an acceptance of $-3.7 \leq \eta \leq -1.7$. To count neutrons from nuclear break up, the Zero Degree Calorimeters (*ZDC*), which are hadronic calorimeters, located ± 116 meters on each side of the interaction point, are used.

4. Analysis of 2010 Pb–Pb data

4.1. Data sets and cuts

For this analysis data recorded during the 2010 Pb–Pb run, at center–of mass energy $\sqrt{s_{NN}} =$ 2.76 TeV, are used. In the first part of the run, when the luminosity was low, a trigger which required at least two hits in the TOF detector was used. Later in the run, when the luminosity was higher, a tighter trigger definition was needed. A trigger which required at least two hits in the TOF detector, at least two hits in the Si–pixel detectors and no activity in the VZERO counters was implemented for the last part of the run.

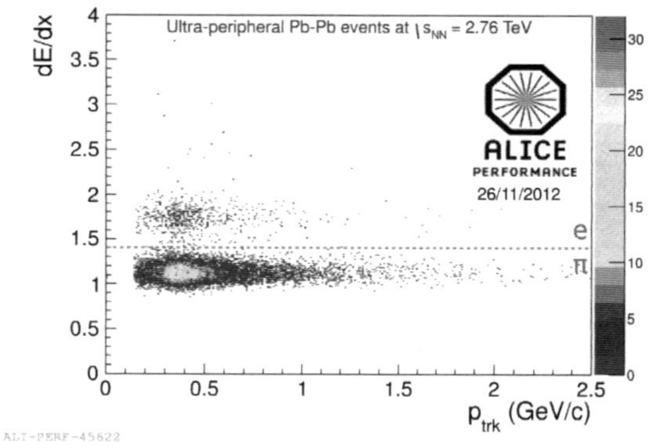

Figure 4. Energy loss in the TPC plotted versus the momentum of the track, for ρ^0 candidates

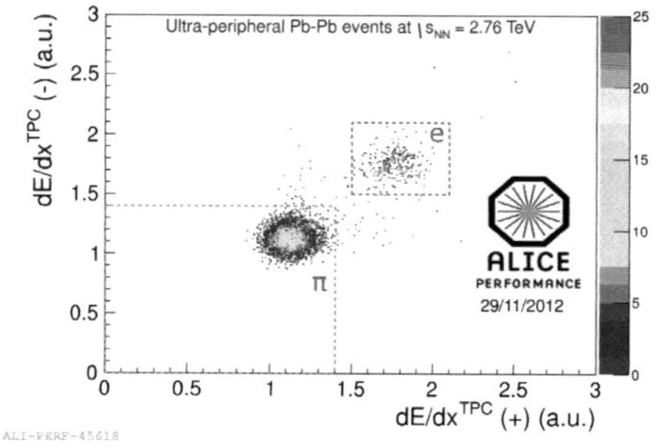

Figure 5. Energy loss of the positively charged particles vs. the energy loss of the negatively charged particles in the TPC, for ρ^0 candidates.

Based on the characheristics of UPC events, a set of cuts are applied to the data at analysis level. The events are required to satisfy one of the two UPC triggers at hardware level. The event must have a primary vertex, where the z–position (along the beam axis) is within 10 cm from the center. There must be exactly two accepted ITS+TPC tracks, and the VZERO detectors should be empty. The two tracks are required to produce ionization in the TPC consistent with pions (Figure 4 and 5). The rapidity of the mother particle should be $|y_{pair}| < 0.5$ and the transverse momentum required to be below $p_T^{pair} < 150$ MeV/c, to get the coherent events. In the end it was required that the tracks have opposite charge, and the like sign background (less than 2%) was subtracted. The cuts, and the number of events surviving each of them, are listed in Table 1 and Table 2.

4.2. Acc $\times$ Eff correction

To correct for acceptance and efficiency a flat invariant mass $\pi^+\pi^-$ simulation is used. The flat simulation is chosen because the shape of the ρ^0 peak would give low statistics for the correction at the tails of the distribution. It is assumed that the $\pi^+\pi^-$ pairs are emitted from a transversely polarized parent, as expected for coherently produced ρ^0's. The generated $\pi^+\pi^-$ events are processed by the ALICE simulation and reconstruction framework using the Geant

Table 1. Number of events surviving the cuts for the TOF–trigger.

Cut	Events left
Triggered events	1,332,041
Primary vertex	850,409
Two accepted tracks	47,978
$\lvert V_z \rvert < 10$ cm	43,413
VZERO veto	8,848
PID cut	7,588
$\lvert y \rvert < 0.5$	5,887
$p_T^{pair} < 150$ MeV/c	2,749
Unlike sign pairs	2,699
Like sign pairs	50

Table 2. Number of events surviving the cuts for the SPD+TOF+VZERO vetoed–trigger.

Cut	Events left
Triggered events	121,487
Primary vertex	103,480
Two accepted tracks	26,217
$\lvert V_z \rvert < 10$ cm	24,020
VZERO veto	17,567
PID cut	15,377
$\lvert y \rvert < 0.5$	11,928
$p_T^{pair} < 150$ MeV/c	6,195
Unlike sign pairs	6,101
Like sign pairs	94

transport to simulate the detector response. These simulated events are then passed through the same analysis as is used on the data. The $(Acc \times Eff)$ is defined as the ratio of the number of reconstructed to generated selected events in the rapidity interval $|y| < 0.5$ and $p_T < 150$ MeV/c as a function of invariant mass in the range $2m_\pi < M_{inv} < 1.5$ GeV/c^2.

4.3. Fitting the invariant mass distribution

The invariant mass distribution is corrected for $(Acc \times Eff)$ and fitted with a Breit–Wigner function with continuum correction (Equation 2).

$$\frac{d\sigma}{dM_{\pi\pi}} = \left| A \frac{\sqrt{M_{\pi\pi} M_{\rho^0} \Gamma(M_{\pi\pi})}}{M_{\pi\pi}^2 - M_{\rho^0}^2 + i M_{\rho^0} \Gamma(M_{\pi\pi})} + B \right|^2 \tag{2}$$

where

$$\Gamma(M_{\pi\pi}) = \Gamma_0 \cdot (M_{\rho^0}/M_{\pi\pi}) \times [(M_{\pi\pi}^2 - 4m_\pi^2)/(M_{\rho^0}^2 - 4m_\pi^2)]^{3/2} \tag{3}$$

is the momentum dependent width of the ρ meson, A is the amplitude of the Breit–Wigner function and B is the amplitude of the direct non–resonant $\pi^+\pi^-$ production.This function has previously been used by the STAR [9] and H1 [10] collaborations.

The fitted invariant mass distribution is shown in linear scale in the left panel of Figure 6, and in logaritmic scale in the right panel. The mass of $M_{\rho^0} = 767.8 \pm 3.5$ MeV/c^2 and the width of $\Gamma_{\rho^0} = 154.1 \pm 8.7$ MeV/c^2, are compatible with the PDG values of $M = 775.49 \pm 0.34$ MeV/c^2 for the mass, and $\Gamma = 149.1 \pm 0.8$ MeV/c^2 for the width.

4.4. Subtraction of incoherent contribution

The transverse momentum cut $p_T < 150$ MeV/c will leave mostly the coherent events, but also some incoherent events will remain. To account for this, one has to find the fraction of incoherent events with $p_T < 150$ MeV/c. To do this the Starlight [6, 7] particle generator is used. The detector response is also simulated, and the simulated particles are reconstructed with the ALICE analysis framework [11].

The data points and the simulated distributions are shown in Figure 7. A coherent and an incoherent simulated sample was generated. The two samples were scaled to fit the data. The contribution from incoherent events under the coherent peak ($p_T < 150$ MeV/c) is found to be $\sim 7\%$. The coherent peak appears to be slightly narrower in data than in the simulation. This

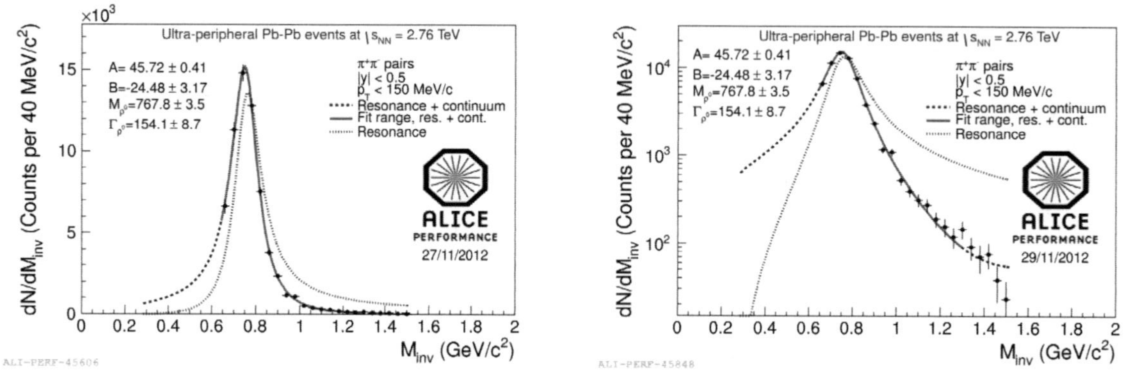

Figure 6. The invariant mass distribution fitted with a Breit–Wigner function with continuum correction. Linear (left) and logartimic (right) scale. The data points are marked with a full circle (●), the blue dashed line (- - - -) is the resonance plus the continuum, the full green line (——) is the resonance plus the continuum in the range used to make the fit and the red dotted line (· · · · · ·) is only the resonance. Statistical errors are shown. (Color online.)

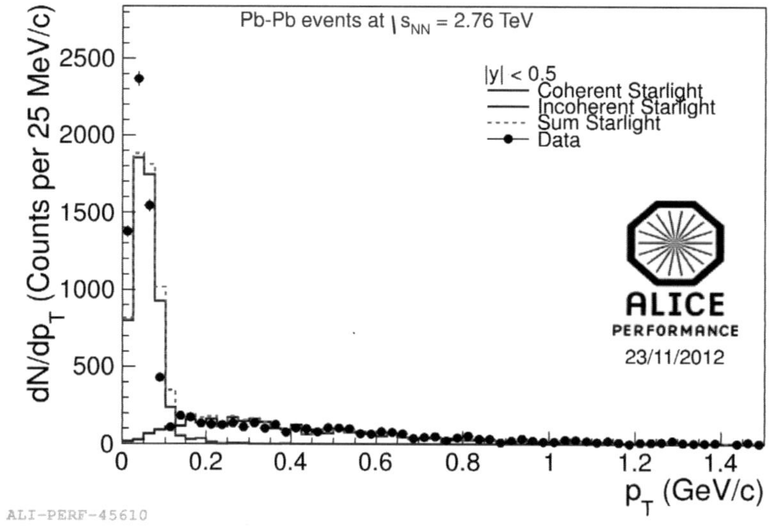

Figure 7. The transverse momentum distribution compared to Starlight predictions for coherent and incoherent photoproduced ρ^0's. The data points are marked with a full circle (●), the simulated coherent production is marked with a red full line (——), the simulated incoherent production is marked with a full blue line (——), and the sum of the incoherent and coherent simulation is marked with a dashed green line (- - - -). (Color online.)

could be a detector effect or related to the implementation of the nuclear form factor in the Monte Carlo.

4.5. Nuclear break up

Exchange of additional photons may lead to coherent vector meson production in coincidence with nuclear break up [12]. To count the number of neutrons emitted from the nuclei, the ZDC neutron counters, located at 116 meters on each side of the interaction point, are used.

Figure 8 shows the energy deposited in each of the ZDCs, for triggered events with two accepted tracks, and the z–position of the primary vertex within 10 cm from the center. The peaks corresponding to zero and one neutron in the ZDCs can be clearly identified. The peak centered around $E = 1380$ GeV, which is the beam energy per nucleon, represents one neutron detected.

In the final selection one distinguishes between the two cases of zero neutrons and one or more neutrons in the ZDC. The separation between no neutrons and one or more neutrons is defined from the minimum between the first two peaks in Figure 8. As expected, the energy at this minimum corresponds to half the beam energy per nucleon. This distinction will make it possible to meassure the cross section for photoproduction of ρ^0 with and without nuclear break up seperately, and this can be compared with model predictions [12].

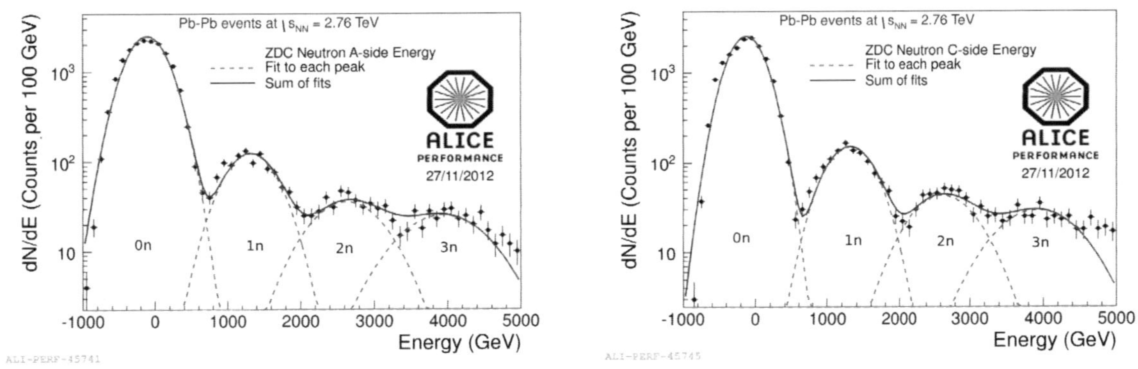

Figure 8. Energy deposited in the ZDC on the A–side (left) and the C–side (right). Each peak is fitted with a Gaussian. The first peak represent zero neutrons, the second peak one neutron and so on. Because a pedestal value is subtracted from the signal, the energy goes below zero. The data points are marked with a full circle (●), the fit to each peak each marked with a green dashed line (- - - -), and the sum of the fits is marked with a red full line (——). (Color online.)

5. Conclusions and outlook

The exclusive photonuclear production of ρ^0 has been studied by the ALICE collaboration. The analysis cuts developed for ultra–peripheral collisions allow a clear separation of the signal, with a background contribution estimated from the like-sign contribution of less than 2% (cf. Tables 1 and 2). The invariant mass and transverse momentum distributions generally agree with the expectations, although the coherent peak in the p_T distribution appears slightly narrower in data. The measurement of the cross section at mid-rapidity will allow a better understanding of ρ^0 photoproduction.

References

[1] Baltz A et al 2008 *Physics Reports* **458(1-3)** 1-171
[2] Bertulani C A, Klein S R and Nystrand J 2005 *Annu. Rev. Nucl. Part. Sci.* **55** 271-310
[3] Frankfurt L, Strikman M and Zhalov M 2002 *Phys. Lett.* B **537** 51-61
[4] Frankfurt L, Strikman M and Zhalov M 2003 *Phys. Rev.* C **67** 034901
[5] Gonçalves V P and Machado M V T 2011 *Phys. Rev.* C **84** 011902
[6] Klein S R and Nystrand J 1999 *Phys. Rev.* C **60** 014903
[7] `http://starlight.hepforge.org`
[8] The ALICE Collaboration 2008 *Journal of Intrumentation* **3(08)** S08002
[9] Abelev B I et al 2008 *Phys. Rev.* C **77** 034910
[10] Aid S et al 1996 *Nucl. Phys.* B **463** 3

[11] http://aliceinfo.cern.ch/Offline
[12] Baltz A, Klein S R and Nystrand J 2002 *Phys. Rev. Lett* **89** 012301
[13] Kalweit A 2011 *J. Phys. G: Nucl. Part. Phys.* **38** 124073

Heavy Ion Physics with ATLAS

Zvi Citron on behalf of the ATLAS Collaboration

Weizmann Institute of Science, Rehovot Israel

E-mail: zvi.citron@weizmann.ac.il

Abstract. An overview of the ATLAS results from Pb+Pb collisions at $\sqrt{s_{NN}} = 2.76$ TeV is presented. ATLAS has measured properties of the bulk particle production including extensive and systematic measurements of the azimuthal particle distributions and correlations. The results for hard probes include both single jet and di-jet measurements, Z bosons, photons, and high p_{T} inclusive charged tracks as well as particles identified as coming from heavy flavor decays. Taken together these results provide a compelling picture of the interaction of hard particles in the dense QCD medium. Results shown are from the $\sim 10\ \mu\mathrm{b}^{-1}$ of minimum bias data recorded in the 2010 LHC heavy ion run, as well as from $\sim 0.15\ \mathrm{nb}^{-1}$ sampled in the 2011 LHC heavy ion run.

1. Introduction

Lead-lead collisions at $\sqrt{s_{NN}} = 2.76$ TeV in the Large Hadron Collider (LHC) provide the opportunity to study strongly interacting matter at the highest temperatures achieved in the laboratory. The ATLAS experiment has a robust heavy-ion program to take advantage of this opportunity. In the 2010 and 2011 LHC runs, yielding approximately $10\ \mu\mathrm{b}^{-1}$ and $0.15\ \mathrm{nb}^{-1}$, respectively, the ATLAS experiment has made a set of measurements that form an emerging picture of the hot dense matter created in a heavy ion collision. These measurements include bulk properties of the system - charged particle multiplicity [1] and extensive measurements of the azimuthal particle distributions and correlations [2, 3] - as well as hard probes such as photons [4, 5], W [6] and Z bosons[7, 8], high transverse momentum, p_{T}, charged tracks [9], single and di-jet measurements [10, 11, 12, 13], and muons from heavy flavor decays [14].

The ATLAS detector [15] at the LHC covers nearly the entire solid angle around the collision point. It consists of an inner tracking detector surrounded by a thin superconducting solenoid, electromagnetic and hadronic calorimeters, and a muon spectrometer incorporating three superconducting toroid magnet systems. The inner-detector system (ID) is immersed in a 2 T axial magnetic field and provides charged particle tracking in the pseudorapidity range $|\eta| < 2.5$. The high-granularity silicon pixel detector covers the vertex region and is surrounded by the silicon microstrip tracker and transition radiation tracker. The calorimeter system covers the range $|\eta| < 4.9$. Within the region $|\eta| < 3.2$, electromagnetic calorimetry is provided by barrel and endcap high-granularity lead liquid-argon (LAr) calorimeters, with an additional thin LAr presampler covering $|\eta| < 1.8$. Forward calorimeters (FCal) are located in the range $3.1 < |\eta| < 4.9$. The muon spectrometer comprises separate trigger and high-precision tracking chambers that measure the deflection of muons in a magnetic field generated by superconducting air-core toroids. The precision tracking chamber system covers the region $|\eta| < 2.7$ with trigger coverage in the range $|\eta| < 2.4$.

2. Particle Flow

An important observable used to understand the hot dense medium is the azimuthal anisotropy of hadron emission. At low p_{T} (less than approximately 4 GeV), this anisotropy is believed to result from a pressure-driven anisotropic expansion of the created matter, with more hadrons emitted in the direction of the largest pressure gradients. The observed azimuthal anisotropy is customarily expressed as a Fourier series in azimuthal angle ϕ:

$$dN/d\phi = N\left(1 + 2\sum_n v_n cos\left(n(\phi - \Phi_n)\right)\right) \tag{1}$$

where Φ_n is the n-th order reaction plane and v_n are the amplitudes of the Fourier expansion. In typical non-central heavy ion collisions where the nuclear overlap region has an "elliptic" shape (or quadrupole asymmetry) on average, the azimuthal anisotropy is expected to be dominated by the v_2 component. However, in addition to this the positions of the nucleons in the overlap region can fluctuate to create matter distributions with additional shape components, such as dipole (n = 1), triangular (n = 3), and higher asymmetries. Figure 1 shows the azimuthal distribution of particles in three events. The sizable fluctuations seen in the figure are consistent with a hydrodynamic description of the evolution of the medium.

For a more thorough understanding of the bulk properties of the system, we may measure directly the event-by-event v_2, v_3, and v_4 components of the system as shown in Figure 2. Beyond the v_n distributions themselves in several p_{T} bins, the lower panels of the figure shows the distributions scaled such that each p_{T} bin's mean matches the mean of the $p_{\mathrm{T}}>$ 0.5 GeV bin. Following this scaling, there is strikingly good agreement in the distributions at different p_{T}. This suggests that the medium's hydrodynamic response to the initial geometry is largely independent of particle p_{T}.

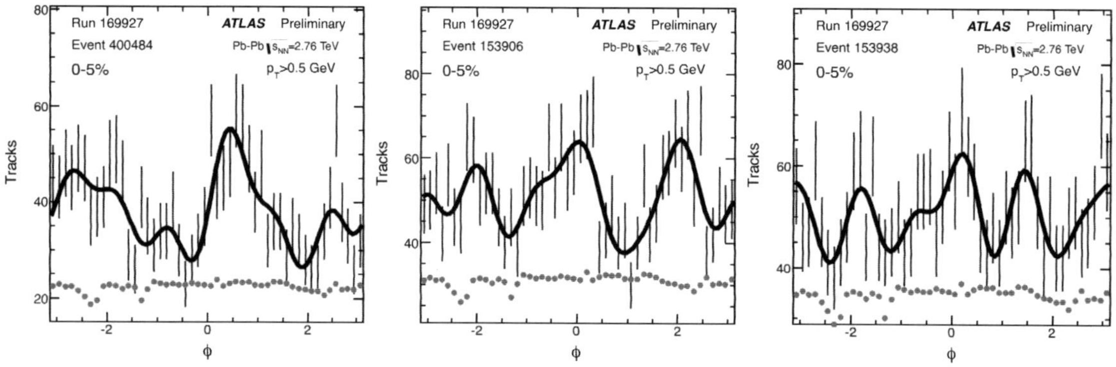

Figure 1. Single track ϕ distributions for three events (from left to right) from the 0-5% centrality class. The bars indicate the foreground distributions, the solid curves indicate a Fourier parameterization including the first six harmonics: $\mathrm{dN}/\mathrm{d}\phi = A(1 + 2\sum_{i=1}^{6} c_n \cos n(\phi - \psi_n))$ and the red solid points indicate the acceptance functions attributable to detector effects. Tracks with $p_{\mathrm{T}}>$ 0.5 GeV are used. Figure is from reference [3].

As shown in Figure 3 the flow components may also be measured using two particle correlations. The analysis employs a large $\Delta\eta$ requirement of $2 < |\Delta\eta| < 5$, to prevent contamination from non-flow physics. Considering all the Fourier components up to n=6 allows a precise reconstruction of the two particle $\Delta\phi$ correlations. This implies that the shape observed in two particle correlations at low p_{T} can be largely attributed to geometric and collective phenomena without invoking jet quenching mechanisms.

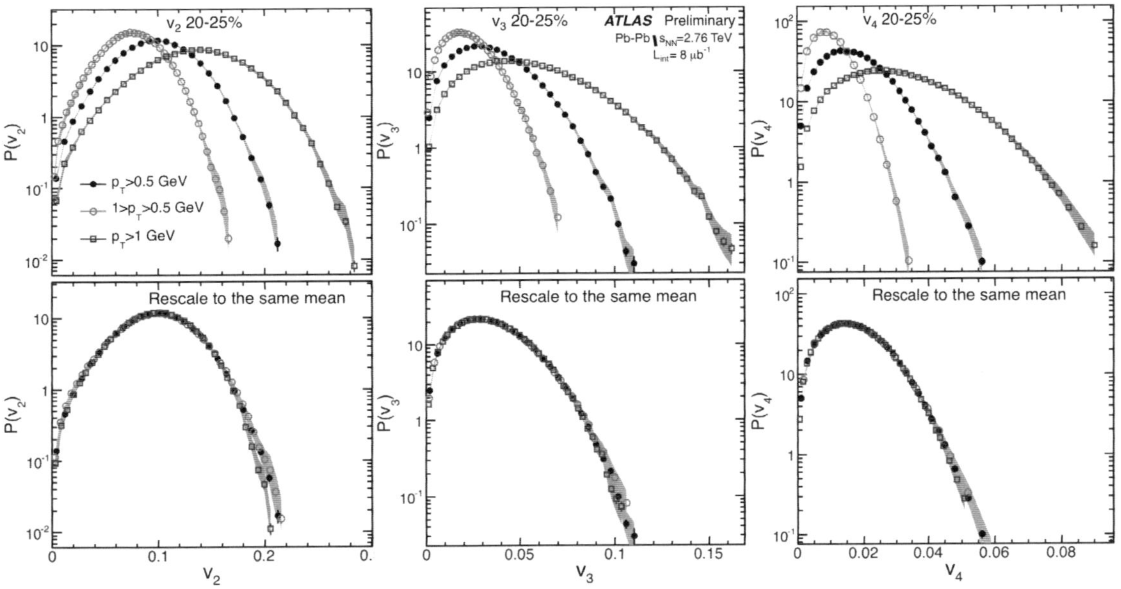

Figure 2. Top panels: The unfolded distributions for v_n in the 20-25% centrality interval for charged particles in p_T>0.5 GeV, 1.0>p_T>0.5 GeV and p_T>1 GeV. Bottom panels: same distributions but rescaled horizontally so the v_n values match that for p_T>0.5 GeV. The shaded bands represent the systematic uncertainties on the v_n-shape. Figure is from reference [3].

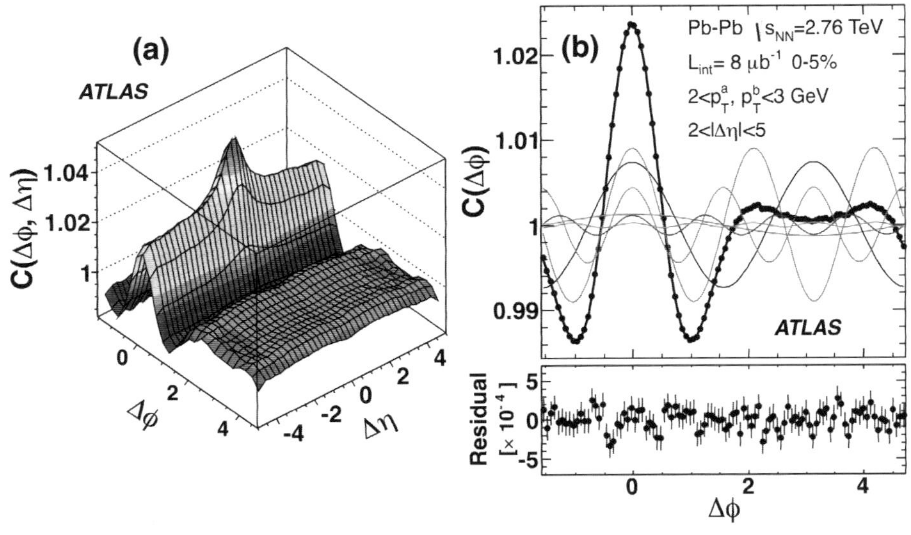

Figure 3. The steps involved in the extraction of v_n values for 2-3 GeV fixed-p_T correlations in the 0-5% centrality interval: (a) two-dimensional correlation function, (b) the one-dimensional $\Delta\phi$ correlation function for $2 < |\Delta\eta| < 5$. Figure is from reference [2].

Taken together these and other measurements allow us to begin to form a coherent picture of the medium geometry and evolution.

3. Penetrating Probes

High momentum particles allow us to probe the properties of the dense medium created in heavy ion collisions, and how color sensitive objects interact with it. Inclusive single particle measurements as well as two particle correlations have suggested that jets are being "quenched" in the medium, and much effort has been made to understand the mechanisms of this quenching. The suppression of particle production may be quantified using the nuclear modification factor, R_{CP}:

$$R_{\mathrm{CP}} = \frac{\langle N_{\mathrm{coll}}\rangle(P)}{\langle N_{\mathrm{coll}}\rangle(C)} \frac{(1/N_{evt,C})d^2N_C/d\eta dp_{\mathrm{T}}}{(1/N_{evt,P})d^2N_P/d\eta dp_{\mathrm{T}}} \tag{2}$$

where C and P refer to central and peripheral event classes, respectively, and $\langle N_{\mathrm{coll}}\rangle$ is the mean number of binary collisions in a given centrality class. The left panel of Figure 4 shows the R_{CP} of inclusive charged particles measured by ATLAS [9], and as expected from previous results it displays a clear suppression of particles in central Pb+Pb events. The right panel shows the nuclear modification factor of muons from heavy flavor decay. The figure shows that although the heavy flavor yield is suppressed in central events compared to peripheral events just as for charged particles, the shape of the modification in p_{T} and the magnitude of the suppression may hint at some differences.

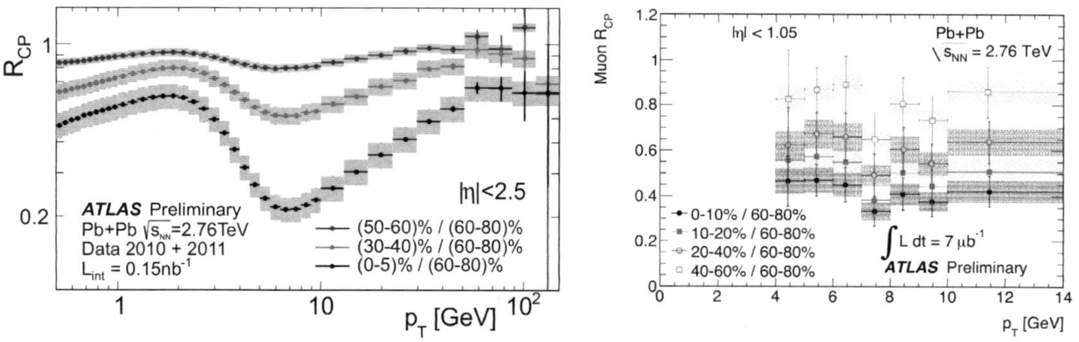

Figure 4. Left: R_{CP} measured in $|\eta| < 2.5$ in three centrality combinations: with 0-5%, 30-40% and 50-60% as numerators and a common peripheral sample 60-80% as denominator. The statistical uncertainties are shown with vertical lines and the overall systematic uncertainty at each point is shown with gray boxes. Figure is from reference [9]. Right: Muons from heavy flavor decays R_{CP} as a function of p_{T} for different centrality bins. The bars represent both statistical and systematic uncertainties. The contribution of the systematic uncertainties for $\langle N_{\mathrm{coll}}\rangle$ and efficiency, which are fully correlated between p_{T} bins, are indicated by the shaded boxes. Figure is from reference [14].

To go beyond the single particle measurement and look more closely at the jets themselves, ATLAS has made the first direct observation of jet quenching by measuring the imbalance of di-jet energies [13]. The imbalance is expressed in terms of the asymmetry, A_J, of two jets:

$$A_J = \frac{E_{T1} - E_{T2}}{E_{T1} + E_{T2}}, \quad \Delta\phi > \frac{\pi}{2} \tag{3}$$

where the jets are reconstructed by an anti-k_t algorithm [16] with distance parameters R=0.4. Figure 5 shows the asymmetry and $\Delta\phi$ distributions for Pb+Pb, p+p, and simulated events. In the more peripheral Pb+Pb events as in p+p and simulation, A_J is peaked at zero implying no relative modification of the jet energy, *i.e.* no quenching. However in more central Pb+Pb events this is no longer the case, showing that there is a relative quenching of one of the two jets due to the presence of the dense medium. Despite this quenching of the jet energy, $\Delta\phi$ distributions remain consistent even in the most central collisions. The energy imbalance is limited in that it

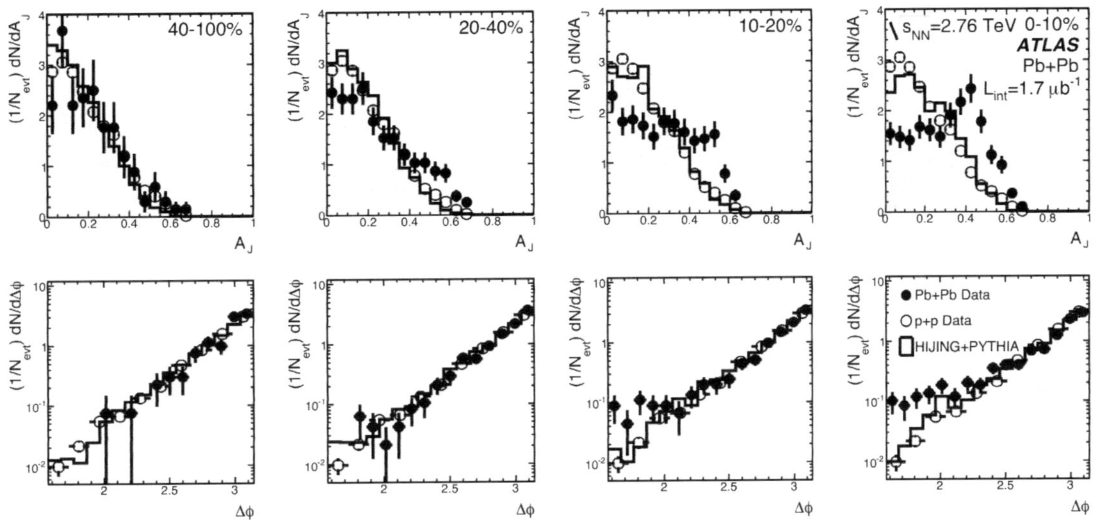

Figure 5. Top: Di-jet asymmetry distributions for data (points) and unquenched HIJING with superimposed PYTHIA dijets (solid yellow histograms), as a function of collision centrality (left to right from peripheral to central events). Proton-proton data from $\sqrt{s_{NN}} = 7$ TeV, analyzed with the same jet selection, is shown as open circles. Bottom: Distribution of $\Delta\phi$, the azimuthal angle between the two jets, for data and HIJING+PYTHIA, also as a function of centrality. Figure is from reference [13].

reflects only the relative modification a two jet event, but does not reflect any modification of the leading jet's energy nor the overall jet production rate. To learn more about jet quenching, in addition to the jet energy imbalance one may measure the nuclear modification factor, R_{CP}, of the fully reconstructed jets rather than inclusive single particles [10]. Figure 6 shows the R_{CP} of jets as a function of N_{part}, clearly demonstrating the suppression of the overall jet yield in addition to the imbalance seen in Figure 5. The suppression observed is independent of the jet p_{T} within the experimental uncertainties.

Further understanding of how jets interact with the medium is provided by considering the azimuthal distribution of jets. The di-jet energy asymmetry and jet R_{CP} show that there is significant quenching, however this is averaged over all medium lengths the jets may be probing. In order to directly examine the path length dependence of jet modification we may measure the distribution of jets relative to the reaction plane of the collision, which as discussed in the previous section reflects the anisotropy of the medium. Similarly defined to the v_2 used to measure the flow, v_2^{jet} measures anisotropy of jets relative to the amount of medium they traverse. Figure 7 shows v_2^{jet} as a function of p_{T}.

Figures 5 - 7 clearly indicate jet quenching due to the medium; to study the mechanisms of the quenching we may measure the structure of the reconstructed jets themselves. A possible mechanism of jet suppression may involve radiated gluons from the leading parton being shunted out of the jet cone due to multiple scattering with the medium, which would lead to a dependence on the jet cone size of the jet R_{CP}. Figure 8 plots the relative R_{CP} of jets with different distance parameters, R, relative to R=0.2 jets. Although the uncertainties are sizable there is a significant

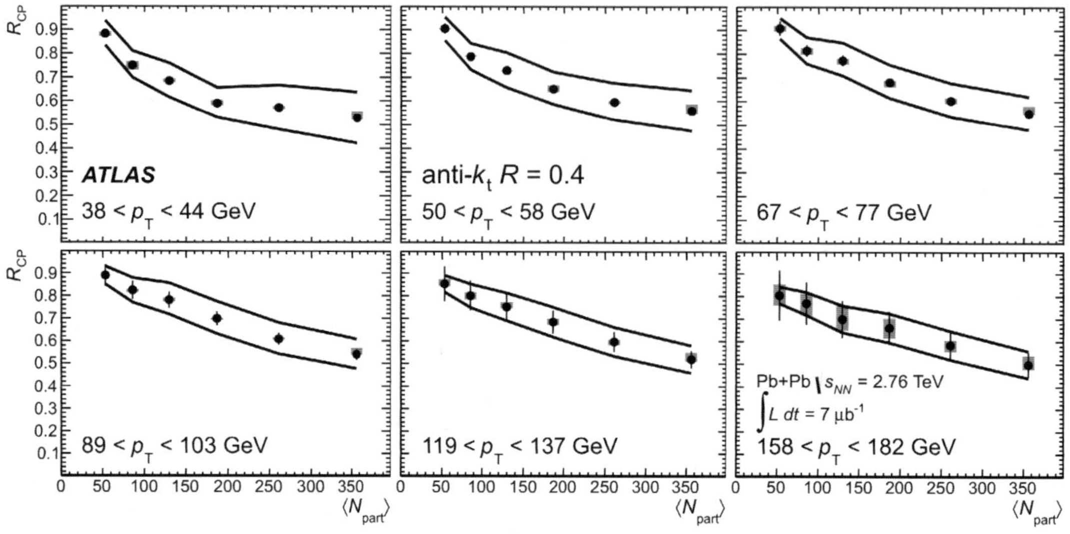

Figure 6. Unfolded R_{CP} values as a function of N_{part} for R = 0.4 anti-k$_t$ jets in six p_{T} bins. The error bars indicate statistical errors from the unfolding; the shaded boxes indicate point-to-point systematic errors that are only partially correlated. The solid lines indicate systematic errors that are fully correlated between all points. The horizontal errors indicate systematic uncertainties on N_{part}. Figure is from reference [10].

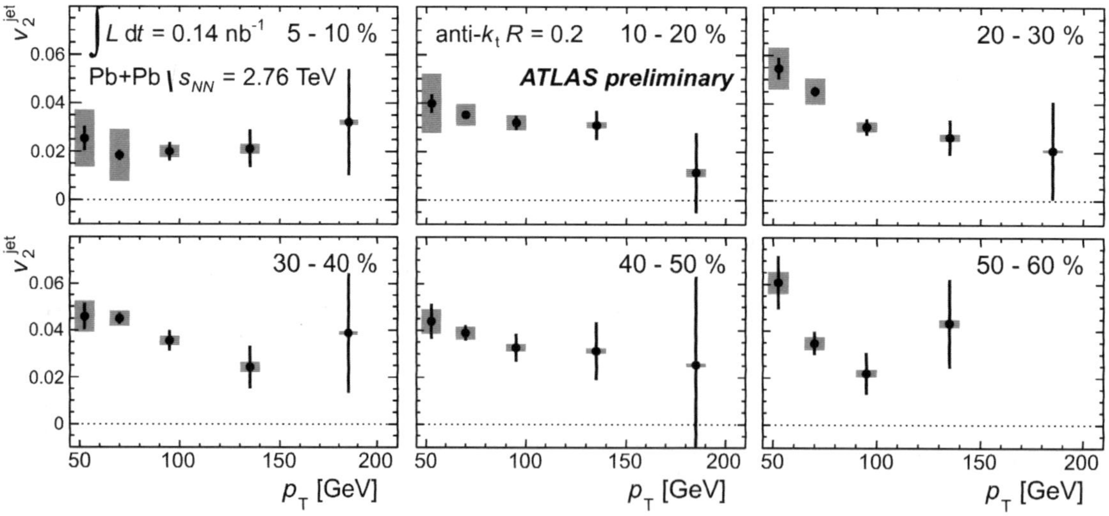

Figure 7. v_2^{jet} as a function of p_{T} for several centrality bins. The error bars on the points indicate statistical uncertainties while the shaded bands represent systematic uncertainties. Figure is from reference [12].

increase with greater R, consistent with a scenario in which some suppression in a smaller jet radius due to multiple scattering of radiated gluons is mitigated by looking in a larger jet radius.

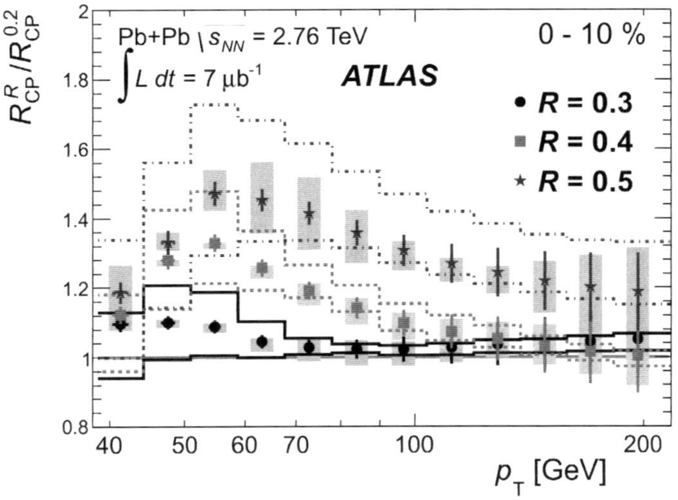

Figure 8. Ratios of R_{CP} values between R = 0.3, 0.4 and 0.5 jets and R =0.2 jets as a function of p_T in the 0-10% centrality bin. The Error bars show statistical uncertainties (see text). The shaded boxes indicate partially correlated systematic errors. The lines indicate systematic errors that are fully correlated between different p_T bins. Figure is from reference [10].

To further understand the structure of the jets and possible mechanisms of quenching, we may consider the fragmentation function of the jets (following background and unfolding):

$$D(z) = \frac{p_T{}^{particle}}{p_T{}^{jet}} cos(\sqrt{\Delta\eta^2 + \Delta\phi^2}) \tag{4}$$

and the ratio:

$$R_{D(z)} = \frac{D(z)|_{central}}{D(z)|_{peripheral}} \tag{5}$$

where particle refers to reconstructed tracks inside the reconstructed jet cone. The ratio $R_{D(z)}$ reflects the modification of the jet substructure with centrality where there is suppression. Figure 9 shows $R_{D(z)}$ for several centralities, and indicates that within the experimental uncertainties the leading particle at high D(z) is unmodified but rather that only lower momentum particles at mid D(z) are.

In contrast to the modification of color sensitive objects discussed above, color neutral bosons are expected to be unmodified in the medium. To confirm this as well as the $\langle N_{\rm coll}\rangle$ scaling assumptions as in Equation 2, we measure photons and Z bosons as shown in Figures 10 and 11. The boson yields are compared to pQCD models of production that include no modification due to the presence of the medium or to the nuclear parton distribution functions, and show good agreement.

The color neutral bosons provide an unbiased baseline from which to measure jet modification in the medium. In a boson+jet event the boson allows direct access to the initial hard scattering which produced both the boson and the jet, and is thus a useful baseline from which to evaluate the jet modification. We may consider the azimuthal distribution of boson+ jet pairs and the momentum fraction $p_T^{\rm jet}/p_T{}^{boson}$ (analogous to the di-jet quantities shown in Figure 5). As in the di-jet case the azimuthal distribution is consistent between central Pb+Pb events and an unmodified pQCD model, however selecting azimuthally correlated events there is a reduction

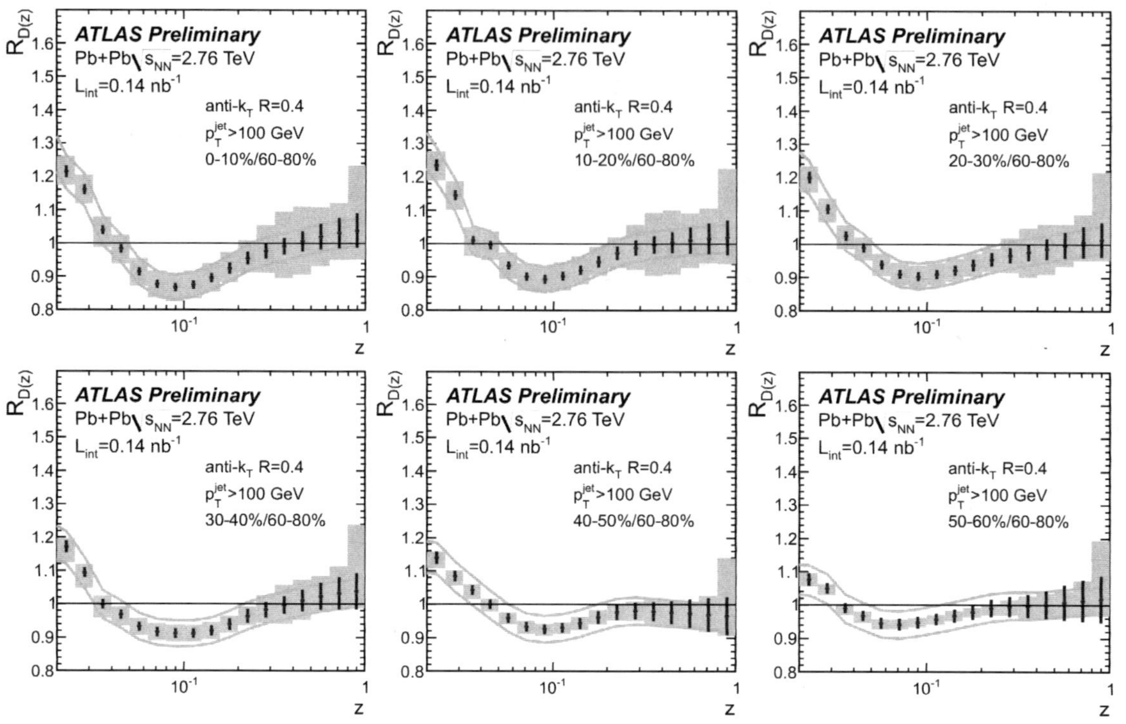

Figure 9. Ratios of $D(z)$ for six bins in collision centrality to those in peripheral (60-80%) collisions, $R_{D(z)}$, for $R = 0.4$ jets. The error bars on the data points indicate statistical uncertainties while the shaded bands indicate systematic uncertainties that are uncorrelated or partially correlated between points. The solid lines indicate systematic uncertainties that are 100% correlated between points. Figure is from reference [11].

in the mean p_T of the jet compared to the boson p_T. The means of the $p_T^{jet}/p_T{}^{boson}$ and the boson + jet yield normalized per the number of bosons are shown in Figures 12 and 13 for both the data and simulation. Although the uncertainties are large (especially in the Z boson case) these measurements show a qualitative result consistent with jet quenching and hold promise for future precision studies.

References

[1] ATLAS Collaboration, Phys. Lett. B**710**, (2012) 363-382
[2] ATLAS Collaboration, Phys. Rev. C**86**, 014907 (2012)
[3] ATLAS Collaboration, ATLAS-CONF-2012-114 (2012) https://cdsweb.cern.ch/record/1472935
[4] ATLAS Collaboration, ATLAS-CONF-2012-051 (2012) https://cdsweb.cern.ch/record/1451913
[5] ATLAS Collaboration, ATLAS-CONF-2012-121 (2012) https://cdsweb.cern.ch/record/1473135
[6] ATLAS Collaboration, Phys. Lett. B**697**, (2011) 294-312
[7] ATLAS Collaboration, Phys. Rev. Lett. **110**, (2013) 02231
[8] ATLAS Collaboration, ATLAS-CONF-2012-119 (2012) https://cds.cern.ch/record/1472941
[9] ATLAS Collaboration, ATLAS-CONF-2011-079 (2011) https://cdsweb.cern.ch/record/1355702
[10] ATLAS Collaboration, Phys. Lett. B**719**, (2013) 220-241
[11] ATLAS Collaboration, ATLAS-CONF-2012-115 (2012) http://cds.cern.ch/record/1472936
[12] ATLAS Collaboration, ATLAS-CONF-2012-116 (2012) http://cds.cern.ch/record/1472938
[13] ATLAS Collaboration, Phys. Rev. Lett. **105**, (2010) 252303
[14] ATLAS Collaboration, ATLAS-CONF-2012-050 (2012) https://cdsweb.cern.ch/record/1451883

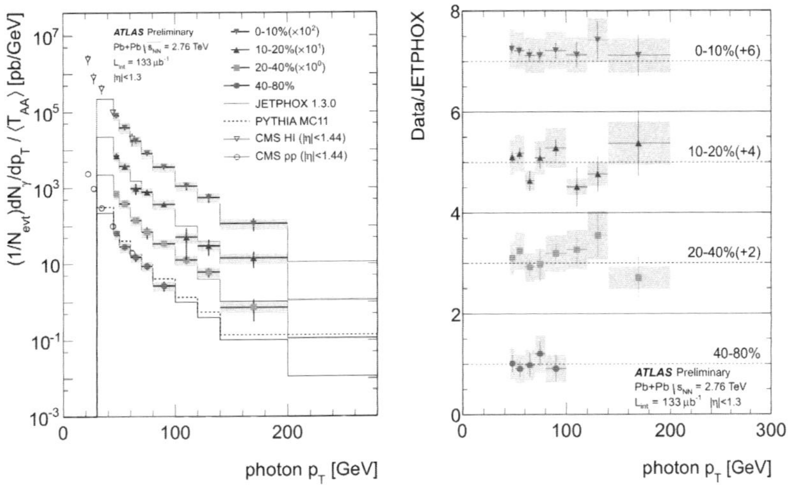

Figure 10. Efficiency corrected yields of isolated photons in $|\eta| < 1.3$ (left), divided by JETPHOX 1.3 predictions with the same isolation requirement (right). Statistical uncertainties are shown by the bars. Systematic uncertainties on the photon yields are combined and shown by the yellow bands. Figure is from reference [5].

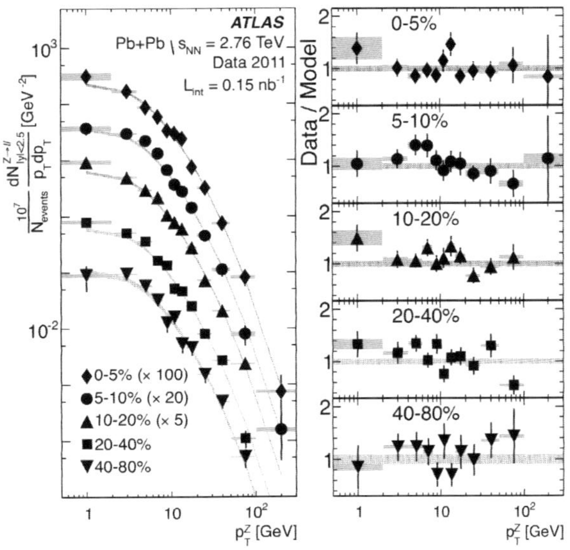

Figure 11. Left: corrected per-event p_T^Z spectra of measured Z bosons in five centrality classes. The data are compared to a PYTHIA simulation normalized to the NNLO p+p cross section and scaled by $\langle T_{AA} \rangle$, shown as bands. Right: ratios of the data to the model in each centrality class. Bars represent statistical uncertainties, boxes represent systematic uncertainties, and bands represent the normalization uncertainty. Figure is from reference [7].

[15] ATLAS Collaboration, Journal of Instrumentation **3**, (2008) no. 08, S08003
[16] M. Cacciari, G. P. Salam, and G. Soyez JHEP **0804**, (2008) 063.

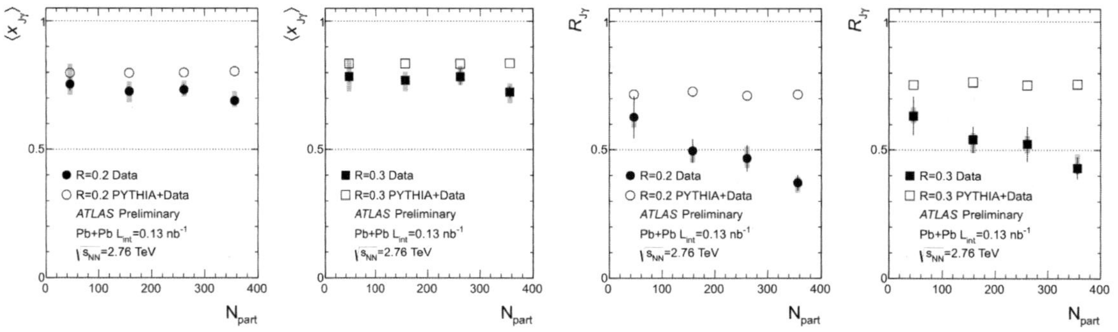

Figure 12. Left: The mean energy fraction $\langle x_{J_\gamma}\rangle = \langle p_T^{\rm jet}/p_T{}^{photon}\rangle$ from fully corrected and unfolded distributions calculated as a function of jet radius ($R = 0.2$ and $R = 0.3$), and for each radius as a function of $N_{\rm part}$. Right: The integrated yield of jets per photon R_{J_γ}. The kinematic cuts are photon $60 < p_T^\gamma < 90$ GeV, $p_T^{\rm jet} > 25$ GeV and $\Delta\phi_{J_\gamma} > 7\pi/8$. The error bars represent statistical errors, while the grey bands indicate the systematic uncertainties. The data are compared to PYTHIA+Data events, calculated for true jets and photons, for simulated events (with data overlay) with a reconstructed photon passing analysis selections. The figures are from reference [5].

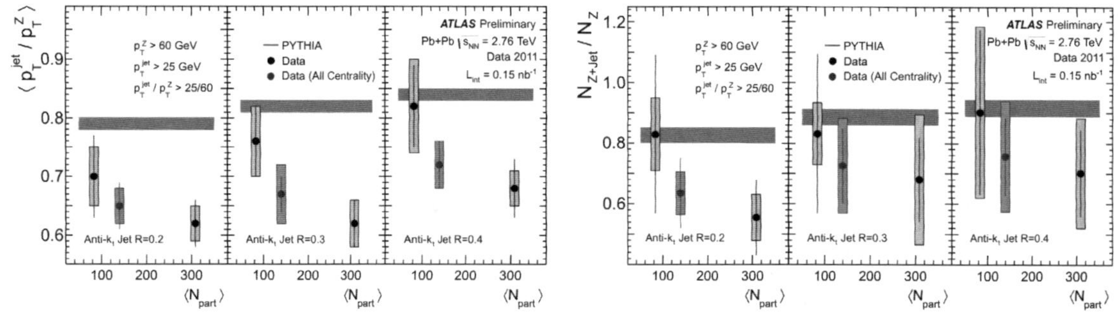

Figure 13. Left: The extracted means from the $p_T^{\rm jet}/p_T^Z$ distributions of the fully corrected data plotted as a function of $\langle N_{\rm part}\rangle$, for each of the three considered jet cone sizes. Right: The fraction of events with a Z boson ($p_T^Z > 60$ GeV) that also have a jet reconstructed ($p_T^{\rm jet} > 25$ GeV, $p_T^{\rm jet}/p_T^Z > 25/60$) as a function of $\langle N_{\rm part}\rangle$. The data are compared to a baseline of PYTHIA, which does not contain any jet energy loss mechanisms. Bars represent statistical uncertainties, and shaded boxes systematic uncertainties. The systematic uncertainties are largely correlated across the $N_{\rm part}$ bins. The width of the PYTHIA band represents its uncertainty. The points at $N_{\rm part} = 140$ refer to 0-80% centrality, and therefore are not independent relative to the points at $N_{\rm part} = 83$, 309. The figures are from reference [8].

Searches for Supersymmetry and Exotics phenomena with the ATLAS detector

Andrée Robichaud-Véronneau
on behalf of the ATLAS collaboration
E-mail: `andree.robichaud-veronneau@cern.ch`

Abstract. Searches for new physics beyond the Standard Model are performed using the p-p LHC data collected with the ATLAS detector in 2011 at 7 TeV and in 2012 at 8 TeV centre-of-mass energy. Various production modes for supersymmetric (SUSY) particles are considered: strong production of squarks and gluinos, weak production of sleptons and gauginos, R-parity violating decays and long-lived particles. No significant deviation from the Standard Model processes is observed and limits are set on different flavours of SUSY, such as cMSSM/mSUGRA, pMSSM, GMSB and AMSB. Searches for exotic models for BSM physics are performed, probing for a large variety of well-motivated models, such as heavy gauge resonances, excited fermions, extra-dimensions, exotic Higgs models and new quarks. As in the case of SUSY searches, no excess is observed above the SM background and limits are obtained for the models mentioned above.

1. Introduction

The Standard Model of particle physics is still to this day the best description of matter and forces in our universe. With the recent discovery of a new Higgs-like boson at the LHC [1], the Standard Model is closer to being completely experimentally verified. It is however an incomplete theory and many new models are currently being developed to overcome the Standard Model's missing ingredients, such as a dark matter candidate particle. Precision measurements of various parameters of the Standard Model, which are more sensitive to new physics phenomena, continue to orient the searches for physics beyond the Standard Model (BSM).

Among the theories currently envisaged as BSM candidates, one theory is of particular interest, Supersymmetry (SUSY). Developed in the 1960s, it introduces new particles, one superpartner (or sparticle) for each existing SM particle. As no SUSY particles have been observed yet, SUSY is believed to be spontaneously broken and consequently, the masses of the superpartners are believed to be higher than their SM counterparts. SUSY is believed to solve several of the current problems with the Standard Model: the hierarchy problem, by cancelling loop contributions to the Higgs mass, as well as providing a dark matter candidate.

In the following, highlights from the searches for SUSY are presented, followed by results from a wide variety of BSM searches in ATLAS [2].

2. Supersymmetry searches

In order to include the SUSY breaking into the observable world, various models are developed, among which the simplest one consistent with the Standard Model is known as the Minimal SuperSymmetric Model (MSSM). The MSSM contains 105 parameters to be determined experimentally, which allows for a large tuning. To make it more manageable, models can be derived from the MSSM with extra constraints to reduce the number of free parameters. This can be achieved by specifying the mechanisms for SUSY breaking, leading to models such as the Gravity-Mediated Supersymmetry Breaking (mSUGRA), also known as the cosmological MSSM (cMSSM), the Gauge Mediated Supersymmetry Breaking (GMSB) and the Anomaly Mediated Supersymmetry Breaking (AMSB). More recently, a new approach was used and phenomenological constraints were added to the MSSM, based on current experimental observations, leading to the phenomenological MSSM (pMSSM). In order to satisfy baryon and lepton number conservation, a new parity is introduced for SUSY models, called R-parity. it assigns a quantum number of 1 for SM particle and -1 for SUSY particles, hence creating a Lightest Supersymmetric Particle (LSP) which cannot decay into Standard Model particles and escapes detection. In certain models known as R-parity violating (RPV), new terms are added to the Lagrangian that also conserve baryon and lepton numbers.

The searches for SUSY in ATLAS are divided into main categories: strong production, searching for squarks and gluinos, weak production, looking for sleptons and gauginos; long-lived particles, which are stable massive particles that cross the detector at a low velocity, and RPV production, where the LSP is allowed to decay to SM particles. Several of these searches were recently updated with 8 TeV results, as pointed out in the following.

A search for coloured sparticles is performed using 8 TeV data [3] looking for squarks and gluinos decaying to SM particles and the LSP, in this case the neutralino, as described in Eq. 1.

$$\tilde{q} \to q\tilde{\chi}_1^0, \tilde{g} \to q\bar{q}\tilde{\chi}_1^0 \tag{1}$$

Events are required to have greater or equal to 2-6 jets and missing transverse energy ($E_{\mathrm{T}}^{\mathrm{miss}}$), while excluding leptons. The Standard Model processes which could fake this signature come from the production of vector bosons with jets ($W \to \ell\nu$, $Z \to \nu\bar{\nu}$), top pair production ($t\bar{t}$) associated with jets and QCD multijets. The analysis is divided into 12 signal regions covering 5 jet multiplicities. No signal is observed outside the bounds of the SM, and limits on the mSUGRA model are set. The squarks and gluinos mass is set to be greater than 1500 GeV. In the case of simplified models [4] where the LSP is massless, limits of $m_{\tilde{g}} > 1100$ GeV and $m_{\tilde{q}} > 730$ GeV are determined.

Events with coloured sparticles are also looked for in the high jet multiplicity region, for models where $m_{\tilde{q}} > m_{\tilde{g}}$ [5], using 8 TeV data. In this case, the events are selected using $E_{\mathrm{T}}^{\mathrm{miss}}/\sqrt{H_{\mathrm{T}}} > 4$ GeV in 6 signal regions based on jet multiplicity (from ≥ 6 to ≥ 9 jets) and jet transverse momentum (p_{T}). The main background events are produced by QCD multijets, $t\bar{t}$ decaying leptonically and W/Z associated with jets. Exclusion limits are determined as no signal is distinguishable from the background, with $m_{\tilde{g}} > 1.0$ TeV for a simplified model where $m_{\tilde{\chi}_1^0} > 300$ GeV.

In models with small fine-tuning, the stop mass is expected to be close to the top mass. The discovery of a low mass Higgs boson has direct consequences on the scale of the MSSM Higgs mass parameter, better known as the "μ term". It is therefore expected that μ should be close to the electroweak scale, with the direct consequence that Higgsinos, Gauginos, and third generations squarks should be light. Searches for these sparticles are then a direct measure of SUSY naturalness.

A search for direct production of sbottom quarks is performed [6], where the sbottom quarks are expected to decay to a b-quark and the LSP. Events with 2 b-tagged jets and $E_{\mathrm{T}}^{\mathrm{miss}}$ are searched in 7 TeV data. The contransverse mass (m_{CT}), as defined in Eq. 2, is used to control the $t\bar{t}$ background. Other backgrounds are the vector boson production with associated heavy flavour jets. No deviation from the SM background is observed, and exclusion limits are set on simplified models, such that $m_{\tilde{b}} > 490$ GeV for a massless LSP and $m_{\tilde{\chi}_1^0} < 180$ GeV for $m_{\tilde{b}} = 400$ GeV.

$$m_{CT} = \sqrt{(E_{\mathrm{T}}(b_1) + E_{\mathrm{T}}(b_2))^2 - (p_{\mathrm{T}}(b_1) - p_{\mathrm{T}}(b_2))^2} \quad (2)$$

Similarly, searches for direct stop production are conducted and combined to form a common exclusion in the $m_{\tilde{\chi}_1^0}$-$m_{\tilde{t}}$ plane. Five searches, completed using the 7 TeV dataset, look for stop decay, either through $\tilde{t} \rightarrow t\tilde{\chi}_1^0$ or $\tilde{t} \rightarrow b\tilde{\chi}_1^{\pm}$ and their several decay channels [7–11]. The combined exclusion limits are shown in Fig. 1. A mixing angle $\theta_i = 0.985$[1] is considered for the signal model generation.

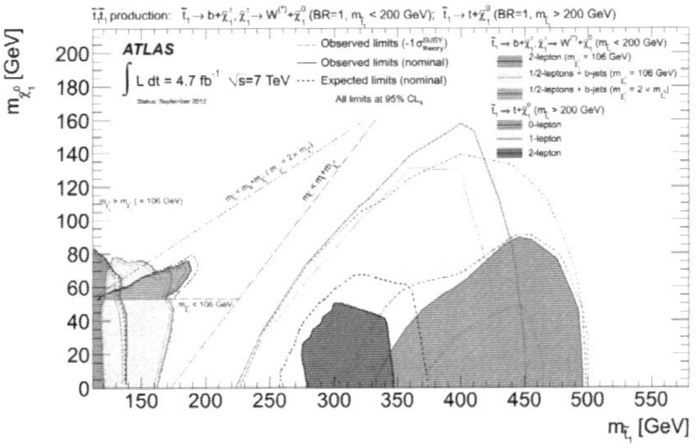

Figure 1. Combined exclusion limits for the direct stop production using 7 TeV data [12].

Third generation squarks can also be produced through decays of directly pair-produced gluinos. The gluino-mediated stop production is tested by four different analyses [5, 13–15] using the 8 TeV dataset and is summarized in a common exclusion limit in the $m_{\tilde{\chi}_1^0}$-$m_{\tilde{g}}$ plane, as shown in Fig. 2.

The gluino-mediated sbottom production can be probed likewise in 8 TeV data using events with three b-tagged jets [14]. The reach in the $m_{\tilde{\chi}_1^0}$-$m_{\tilde{g}}$ plane is improved significantly from the 7 TeV analysis, represented by the green solid line in Fig. 3. New exclusions of $m_{\tilde{g}} < 1240$ GeV for $m_{\tilde{\chi}_1^0} < 200$ GeV and $m_{\tilde{\chi}_1^0} < 570$ GeV for $m_{\tilde{g}} = 1100$ GeV are obtained.

Searches for directly produced weak production are also performed using the 7 TeV dataset. Decays of directly produced left-handed sleptons ($\tilde{\ell}^{\pm} \rightarrow \ell^{\pm}\tilde{\chi}_1^0$) are probed for in events with two leptons and $E_{\mathrm{T}}^{\mathrm{miss}}$ in the final state [16]. Significant improvement is obtained over LEP limits obtained for right-handed smuons, as shown in Fig. 4. Using the same signature, direct production of gauginos are searched for, focusing on the LSP and $\tilde{\chi}_1^{\pm}$. These gauginos decay through intermediate particles, either vector bosons or sleptons. Depending on the decay channel that the intermediate particles take, these processes can lead to higher multiplicity of leptons for which specific searches are designed [17]. For the ≥ 3 leptons searches, the production of $\tilde{\chi}_2^{\pm}$ is also considered. The main backgrounds for direct gauginos searches are $t\bar{t}$, W/Z associated with jets, diboson production and non-isolated leptons. The results for two lepton multiplicities are shown in Fig. 5.

Models with RPV terms added to the Lagrangian [18] allow for LSP decays, thus creating event signatures with large lepton multiplicities as well as fully hadronic three-jet resonances

[1] Corresponding to a mixing matrix of $\begin{pmatrix} \tilde{t_1} \\ \tilde{t_2} \end{pmatrix} = \begin{pmatrix} 0.55 & -0.83 \\ 0.83 & 0.55 \end{pmatrix} \begin{pmatrix} \tilde{t_R} \\ \tilde{t_L} \end{pmatrix}$, resulting in a mostly right handed mixture.

Journal of Physics: Conference Series **455** (2013) 012012 doi:10.1088/1742-6596/455/1/012012

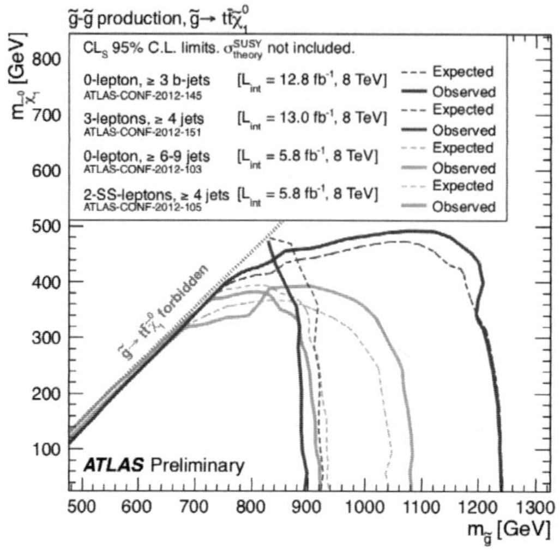

Figure 2. Combined exclusion limits for the gluino-mediated stop production using 8 TeV data [12].

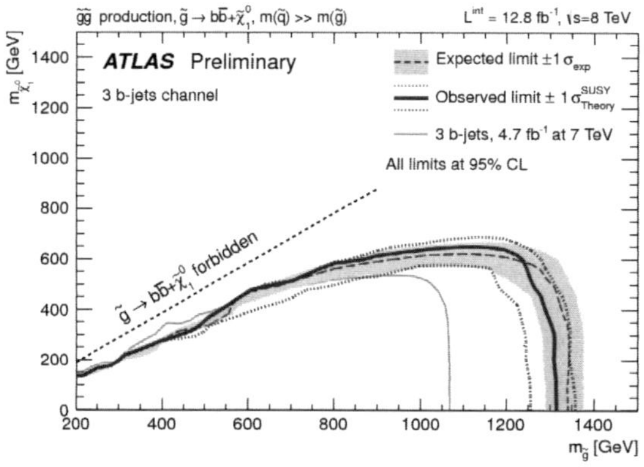

Figure 3. Exclusion limit for the gluino-mediated sbottom production using 8 TeV data [14].

from gluino decay. Also, the existence of Stable Massive particles [19] can be detected as long-lived particle passing through the ATLAS detector, much like the muons. As such, they require completely different trigger and timing paradigm than the standard SUSY analyses.

3. Exotic phenomena searches

Although SUSY is well motivated, other models are developed to account for missing pieces of the Standard Model and describe the physics beyond the SM. There is a large variety of models tested using LHC data. In the following, a selection of interesting results is presented, including searches done using the 8 TeV dataset, testing production of heavy gauge bosons (Z',W'), excited fermions ($f^* \to f\gamma$), extra-dimensions (ADD, UED, RS), exotic Higgs models (H^{++}) and new quarks (fourth generation, vector-like quarks), among others.

A search for heavy resonances, here Z', is performed in the dilepton channel using 8 TeV data [20]. The search is performed for both electron and muon channels. No significant excess is observed over the dilepton invariant mass background. Limits are set on various E6 Z' models.

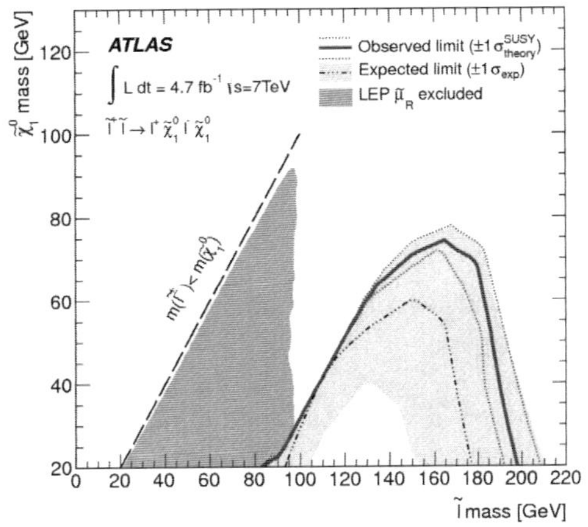

Figure 4. Exclusion limits for direct left-handed sleptons production using 7 TeV data [16].

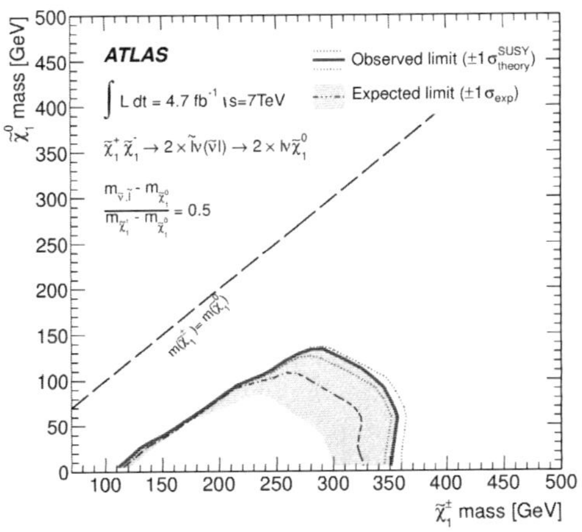

(a)

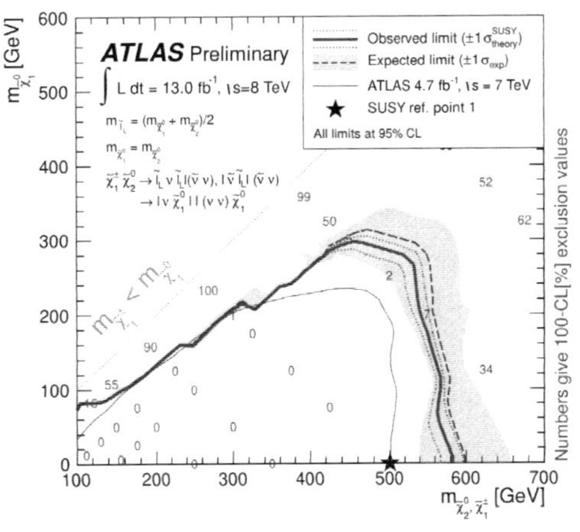

(b)

Figure 5. Exclusion limits for direct gaugino production in (a) the two lepton channel [16] and (b) the three inclusive lepton channel [17] using 7 TeV data.

For all models, a Z' with a mass below 2 TeV is excluded. In the special case of the Sequential Standard Model (SSM), the Z' mass is restricted to be greater than 2.49 TeV, as shown in Fig. 6. Data events considered in this analysis include the highest energy dimuon event, with an invariant mass of 1.258 TeV, the two back-to-back muons entering the detector with a p_T of 289 and 274 GeV respectively.

New heavy resonances can also arise in highly energetic dijet events. In this case, the analysis is used to test quark compositeness, by searching for decays of excited quarks through a gluon ($q^* \rightarrow q\gamma$) in 7 [21] and 8 TeV [22] data. The data is observed to be consistent with the expected background for the dijet invariant mass distribution. Exclusion limits are set on the excited quark masses, of 3.84 TeV, as shown in Fig. 7. The highest energy event recorded by the ATLAS detector in 8 TeV data was found in this analysis, with an invariant mass of 4.69 TeV for

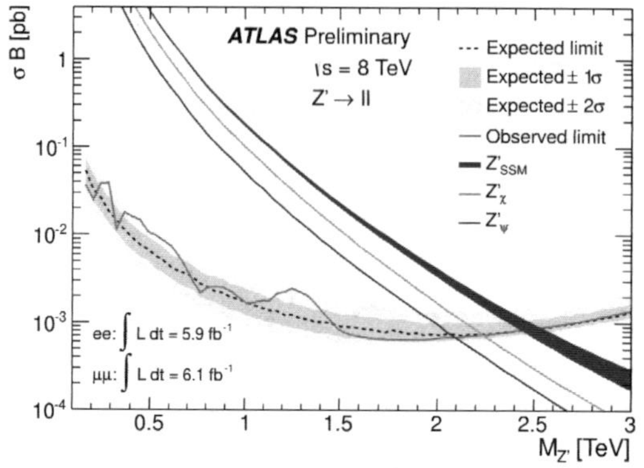

Figure 6. Combined dielectron and dimuon exclusion limits for the cross-section times branching fraction $\sigma \times \mathcal{B}$ using 8 TeV data [20].

two back-to-back jets of $p_T = 2.19$ and 2.29 TeV respectively. This event had $E_T^{miss} = 47$ GeV.

The LHC is a $t\bar{t}$ factory, which is also a good channel to look for heavy resonances. A search for Z' and Randall-Sundstrum (RS) Kaluza-Klein (KK) gluons using the leptons+jets decay channel for top decay, $t\bar{t} \to \mathrm{W}(\ell\nu)\mathrm{bW(qq)b}$ is done using 7 TeV data [23] (a search for fully hadronic top decays [24] is also used but not presented here). The analysis is performed using two complementary reconstruction schemes. The resolved reconstruction looks for the individual hadronic jets from the top quark decay, each with small radius, while the boosted reconstruction searches for one large radius jets, as the hadronic top decay products are expected to be collimated for highly-boosted top quarks. The combination of these two schemes allows for an extended reach in $t\bar{t}$ invariant mass. No significant fluctuation above the predicted background is observed. Exclusion limits for the production of leptophobic topcolor Z' with mass < 1.7 TeV

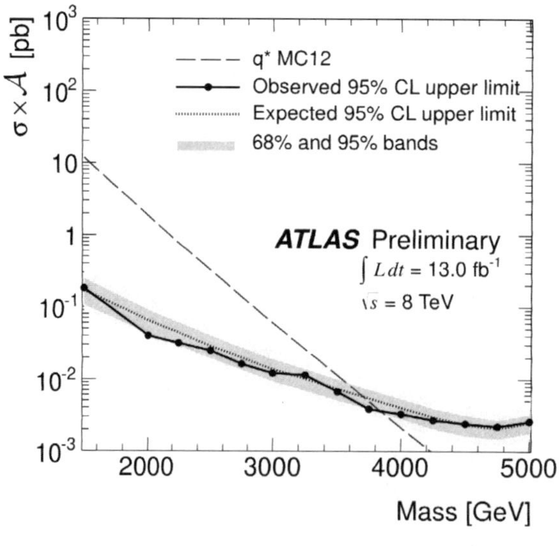

Figure 7. Exclusion limits on the cross-section times acceptance $\sigma \times \mathcal{A}$ for excited quarks production using 8 TeV data [22].

and of RS KK gluon excitations with mass < 1.9 TeV are set, as shown in Fig. 8.

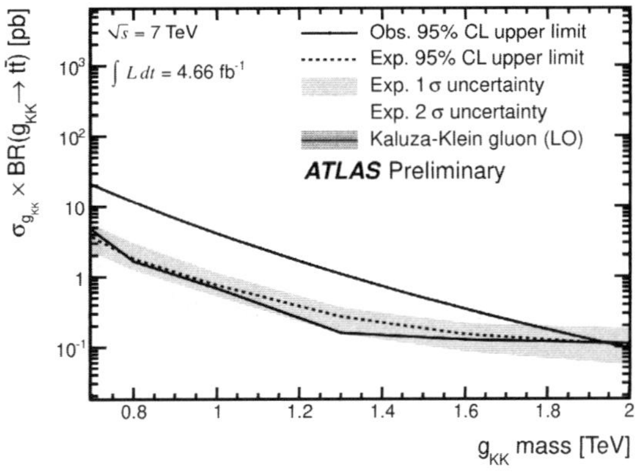

Figure 8. Exclusion limits on $\sigma \times \mathcal{B}$ for KK gluon production using 7 TeV data [23].

Leptons, just like quarks, are theorised to exhibit compositeness. A search in 8 TeV ATLAS data is performed to probe for lepton compositeness [25] with scale Λ in decays through photons, $\ell^* \to \ell\gamma$. The analysis is perfomed for both electron and muon channels, looking for excess in the $m_{\ell\ell\gamma}$ distribution, which is sensitive to all ℓ^* possible widths. No significant departure from the expected background is recorded. For a value of $\Lambda = m_{\ell^*}$, an exclusion limit of $m_{\ell^*} < 2.2$ TeV is determined.

Diboson resonances can be used to probe for bulk RS models by looking for the decay of a spin-2 RS graviton (G^*). This analysis [26] searches through 8 TeV data for ZZ decays to $\ell^+\ell^-$qq, which provides better rejection of the QCD multijets background than the fully hadronic channel. In the high mass regime, events are selected with one large jet resulting from the merging of the two boosted jets from the Z decay. No resonant feature is observed on top of the smoothly decreasing background. Exclusion limits of $m^*_G < 850$ GeV for a dimensionless coupling parameter $\kappa/\bar{m}_{Pl} = 1.0$ is obtained.

Large extra-dimensions, in their ADD flavour, can leave a striking signature in the ATLAS detector by means of an energetic monojet and $E_{\mathrm{T}}^{\mathrm{miss}}$ emerging from the decay of KK gravitons. This event topology can also be used to search for dark matter candidates, or WIMPs arising from a GMSB scenario. A search for such event signatures [27] is performed using the 8 TeV data and good agreement is observed between the experimental measurements and the Standard Model background expectation. An exclusion limit is set on the fundamental Planck scale $M_D < 2.5$ TeV for various number of extra-dimensions n. For a GMSB scenario where $m_{\tilde{q}} = m_{\tilde{g}}$, the best gravitino limits to date are obtained for various values of the squark mass, as shown in Fig. 9.

Searches for ADD extra-dimension models are also performed in electromagnetic channels, such as $\ell^+\ell^-$ and $\gamma\gamma$ [28, 29]. In the dileptonic channel, a non-resonant search (using the total observed number of events) is performed using a similar selection as for the heavy resonances search while in the diphoton case, two isolated γ are required. Since no deviation from the Standard Model background is observed, the two channels are combined to obtain more stringents limits for the string scale $M_S > 2.6$-4.2 TeV and for the mass of the lightest graviton $M_G < 1.03(2.23)$ TeV for $\kappa/\bar{m}_{Pl} = 0.01$ (0.1).

A search for top+jet resonance is performed using 7 TeV ATLAS data in the $t\bar{t}$+jets channel [30]. The lepton+jets mode is used for the $t\bar{t}$ decay. While the Tevatron observed a 3.4σ deviation in the $t\bar{t}$ forward-backward asymmetry A_{FB} [31, 32], ATLAS measures no

Journal of Physics: Conference Series **455** (2013) 012012 doi:10.1088/1742-6596/455/1/012012

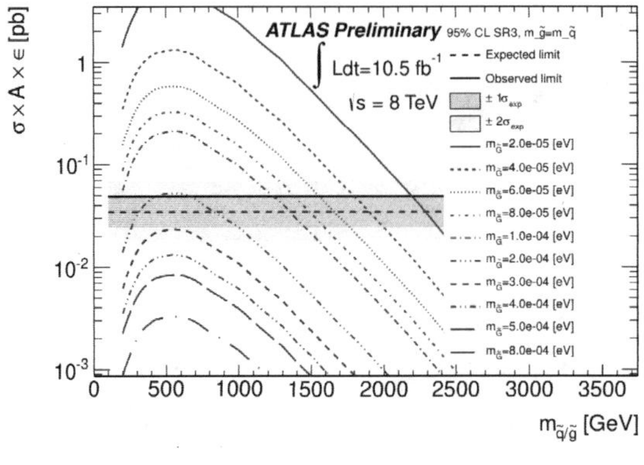

Figure 9. Exclusion limits on $\sigma \times \mathcal{A}$ for gravitinos as a function of the squark/gluino mass using 8 TeV data [27].

deviation from the Standard Model expectations. An exclusion limit on the mass of the new resonance (for both ϕ and W' and coupling parameter g_R=1) of 430 GeV is determined.

Exotic models which include a Higgs triplet can lead to the production of doubly charged Higgs ($H^{\pm\pm}$). They are searched for using 7 TeV ATLAS data in their decay to two pairs of same-sign-same-flavour leptons $\ell^{\pm\pm}\ell'^{\pm\pm}$ (ee, $e\mu$, $\mu\mu$) [33]. Background processes include both prompt ($t\bar{t}$+W/Z, W/Z) and non-prompt (Heavy Flavour production, conversion) sources of leptons. No excess is measured on top of the background and decay-channel dependent limits are set. Masses for doubly charged Higgs of $m_{H^{\pm\pm}} < 409$ (398,375) GeV for ee (μ, μ,$e\mu$) are excluded.

Finally, a search for heavy quarks is performed using 7 TeV ATLAS data [34], probing for pair production of b' and $T_{5/3}$ heavy quarks. A complex signature is considered ($\ell^+\ell^+b\bar{b}q\bar{q}qq$) and is searched for in events with two same-sign leptons, one b-tagged jet and two light jets. After selection 5.6 events are expected and four are measured, therefore exclusion limits are computed. Masses for b' and $T_{5/3} < 0.67$ TeV are excluded. A search for the 4th generation top partner, t', is performed with the same dataset [35]. The t' is looked for in its $t'\bar{t}' \to WbWb$ decay channel. The lepton+jets final state is used to perform this measurement, where no deviation is observed from the Standard Model background. An exclusion limit of 656 GeV is obtained for the mass of the t', when considering BR($t' \to Wb$) = 1.

4. Conclusion

The search for new phenomena beyond the Standard Model is a very active field at the ATLAS experiment, of which only a subset of results is presented here. An impressive variety of Supersymmetry and Exotics models have been considered for searches, amounting to 23 (10) public conference notes with 7 (8) TeV data, and 49 papers with the 2010-2011 dataset for SUSY searches and 48 (6) public conference notes with 7 (8) TeV data, and 52 papers with the 2010-2011 dataset for Exotics searches. These results allowed us to push the masses of the 1st and 2nd generation squarks and gluinos even further above the TeV mark while naturalness arguments are driving the searches for 3rd generation squarks and gauginos. At the same time, the limits on exotic particle masses, coupling parameters and cross-sections are being pushed higher than ever. So far, the data is very much still agreeing with the Standard Model, but this is just the beginning. While the analyses presented above and many more are being updated to 8 TeV, new results with the upgraded LHC energy of 13 TeV will shed more light on the high energy regime of particle physics.

5. References

[1] ATLAS collaboration. Observation of a new particle in the search for the Standard Model Higgs boson with the ATLAS detector at the LHC. *Physics Letters B*, 716(1):1 – 29, 2012.

[2] ATLAS Collaboration. The ATLAS Experiment at the CERN Large Hadron Collider. *Journal of Instrumentation*, 3(08):S08003, 2008.

[3] Search for squarks and gluinos with the ATLAS detector using final states with jets and missing transverse momentum and 5.8 fb^{-1} of $\sqrt{s}$=8 TeV proton-proton collision data. Technical Report ATLAS-CONF-2012-109, CERN, Geneva, Aug 2012.

[4] D. Alves et al. Simplified models for lhc new physics searches. *Journal of Physics G: Nuclear and Particle Physics*, 39(10):105005, 2012.

[5] Search for new phenomena using large jet multiplicities and missing transverse momentum with ATLAS in 5.8 fb^{-1} of $\sqrt{s}$ = 8 TeV proton-proton collisions. Technical Report ATLAS-CONF-2012-103, CERN, Geneva, Aug 2012.

[6] Search for scalar bottom pair production in final states with missing transverse momentum and two b-jets in pp collisions at $\sqrt{s}$=7 TeV with the ATLAS Detector. Technical Report ATLAS-CONF-2012-106, CERN, Geneva, Aug 2012.

[7] ATLAS collaboration. Search for Direct Top Squark Pair Production in Final States with One Isolated Lepton, Jets, and Missing Transverse Momentum in $\sqrt{s}$=7 TeV *pp* Collisions Using 4.7fb^{-1} of ATLAS Data. *Phys. Rev. Lett.*, 109:211803, Nov 2012.

[8] ATLAS collaboration. Search for a Supersymmetric Partner to the Top Quark in Final States with Jets and Missing Transverse Momentum at $\sqrt{s}$=7 TeV with the ATLAS Detector. *Phys. Rev. Lett.*, 109:211802, Nov 2012.

[9] ATLAS collaboration. Search for light top squark pair production in final states with leptons and b-jets with the ATLAS detector in proton-proton collisions. *Physics Letters B*, 720(1-3):13 – 31, 2013.

[10] ATLAS collaboration. Search for a heavy top-quark partner in final states with two leptons with the ATLAS detector at the LHC. *Journal of High Energy Physics*, 2012, 2012.

[11] ATLAS collaboration. Search for light scalar top-quark pair production in final states with two leptons with the ATLAS detector in $\sqrt{s}$=7 TeV proton-proton collisions. *The European Physical Journal C*, 72, 2012.

[12] https://twiki.cern.ch/twiki/bin/view/AtlasPublic/SupersymmetryPublicResults.

[13] Search for Supersymmetry in final states with two same-sign leptons, jets and missing transverse momentum with the ATLAS detector in pp collisions at $\sqrt{s}$=8 TeV. Technical Report ATLAS-CONF-2012-105, CERN, Geneva, Aug 2012.

[14] Search for gluino pair production in final states with missing transverse momentum and at least three b-jets using 12.8 fb^{-1} of pp collisions at $\sqrt{s}$ = 8 TeV with the ATLAS Detector. Technical Report ATLAS-CONF-2012-145, CERN, Geneva, Nov 2012.

[15] Search for supersymmetry using events with three leptons, multiple jets, and missing transverse momentum in 13.0 fb^{-1} of pp collisions with the ATLAS detector at $\sqrt{s}$ = 8 TeV. Technical Report ATLAS-CONF-2012-151, CERN, Geneva, Nov 2012.

[16] ATLAS collaboration. Search for direct slepton and gaugino production in final states with two leptons and missing transverse momentum with the ATLAS detector in pp collisions at $\sqrt{s}$ = 7 TeV collected by the ATLAS detector. *Physics Letters B*, 718(3):879 – 901, 2013.

[17] Search for direct production of charginos and neutralinos in events with three leptons and missing transverse momentum in 13.0 fb^{-1} of pp collisions at $\sqrt{s}$=8 TeV with the ATLAS detector. Technical Report ATLAS-CONF-2012-154, CERN, Geneva, Nov 2012.

[18] Search for Supersymmetry in events with four or more leptons in 13 fb^{-1} pp collisions at $\sqrt{s}$ = 8 TeV with the ATLAS detector. Technical Report ATLAS-CONF-2012-153, CERN, Geneva, Nov 2012.

[19] ATLAS collaboration. Search for long-lived, heavy particles in final states with a muon and multi-track displaced vertex in proton-proton collisions at $\sqrt{s}$ = 7 TeV with the ATLAS detector. *Physics Letters B*, 719(45):280 – 298, 2013.

[20] Search for high-mass dilepton resonances in 6.1/fb of pp collisions at $\sqrt{s}$= 8 TeV with the ATLAS experiment. Technical Report ATLAS-CONF-2012-129, CERN, Geneva, Sep 2012.

[21] ATLAS collaboration. Search for new physics in the dijet mass distribution using 1 fb^{-1} of pp collision data at $\sqrt{s}$ =7 TeV collected by the ATLAS detector. *Physics Letters B*, 708(12):37 – 54, 2012.

[22] Search for New Phenomena in the Dijet Mass Distribution updated using 13.0 fb^{-1} of *pp* Collisions at $\sqrt{s}$ = 8 TeV collected by the ATLAS Detector. Technical Report ATLAS-CONF-2012-148, CERN, Geneva, Nov 2012.

[23] A search for ttbar resonances in the lepton plus jets final state using 4.66 fb^-1 of pp collisions at $\sqrt{s}$ = 7 TeV. Technical Report ATLAS-CONF-2012-136, CERN, Geneva, Sep 2012.

[24] ATLAS collaboration. Search for resonances decaying into top-quark pairs using fully hadronic decays in pp

collisions with ATLAS at $\sqrt{s}$=7 TeV. *Journal of High Energy Physics*, 2013(1):1–50, 2013.

[25] Search for excited electrons and muons with 13 fb^{-1} of proton–proton collisions at $\sqrt{s}$ = 8 TeV with the ATLAS detector. Technical Report ATLAS-CONF-2012-146, CERN, Geneva, Nov 2012.

[26] Search for resonant ZZ production in the ZZ $\rightarrow$ llqq channel with the ATLAS detector using 7.2 fb^{-1} of $\sqrt{s}$ = 8 TeV pp collision data. Technical Report ATLAS-CONF-2012-150, CERN, Geneva, Nov 2012.

[27] Search for New Phenomena in Monojet plus Missing Transverse Momentum Final States using 10fb^{-1} of pp Collisions at $\sqrt{s}$=8 TeV with the ATLAS detector at the LHC. Technical Report ATLAS-CONF-2012-147, CERN, Geneva, Nov 2012.

[28] ATLAS collaboration. Search for contact interactions and large extra dimensions in dilepton events from *pp* collisions at $\sqrt{s}$=7 TeV with the ATLAS detector. *Phys. Rev. D*, 87:015010, Jan 2013.

[29] ATLAS Collaboration. Search for extra dimensions in diphoton events from proton-proton collisions at $\sqrt{s}$ = 7 TeV in the ATLAS detector at the LHC. *New Journal of Physics*, 15(4):043007, 2013.

[30] ATLAS collaboration. Search for resonant top quark plus jet production in $t\bar{t}$+jets events with the ATLAS detector in *pp* collisions at $\sqrt{s}$=7 TeV. *Phys. Rev. D*, 86:091103, Nov 2012.

[31] T. Aaltonen et al. Evidence for a mass dependent forward-backward asymmetry in top quark pair production. *Phys. Rev. D*, 83:112003, Jun 2011.

[32] V. M. Abazov et al. Forward-backward asymmetry in top quark-antiquark production. *Phys. Rev. D*, 84:112005, Dec 2011.

[33] ATLAS Collaboration. Search for doubly charged Higgs bosons in like-sign dilepton final states at $\sqrt{s}$ = 7 TeV with the ATLAS detector. *The European Physical Journal C*, 72(12), 2012.

[34] Search for exotic same-sign dilepton signatures (b' quark, $T_{5/3}$ and four top quarks production) in 4.7/fb of pp collisions at $\sqrt{s}$=7 TeV with the ATLAS detector. Technical Report ATLAS-CONF-2012-130, CERN, Geneva, Sep 2012.

[35] ATLAS collaboration. Search for pair production of heavy top-like quarks decaying to a high- W boson and a b quark in the lepton plus jets final state at $\sqrt{s}$ = 7 TeV with the ATLAS detector. *Physics Letters B*, 718(4–5):1284 – 1302, 2013.

Search for the SM Higgs boson in the diboson decay modes with the ATLAS detector

German Carrillo-Montoya, on behalf of the ATLAS Collaboration
University of the Witwatersrand. Conference proceedings, International Workshop on Discovery Physics at the LHC 2012, Kruger South Africa
E-mail: montoya@cern.ch

Abstract. A Higgs boson search in the $H \to \gamma\gamma$, $H \to ZZ^{(*)} \to 4\ell$ and $H \to WW^{(*)} \to \ell\nu\ell\nu$ decay modes has been performed using proton-proton collisions collected in 2011 and 2012 with the ATLAS detector. Updated results on the observed resonance using up to 4.7 fb^{-1} of $\sqrt{s}$ = 7 TeV and 13.0 fb^{-1} of $\sqrt{s}$ = 8 TeV data including first measurement of its properties are presented and discussed.

1. Introduction

The Standard Model (SM) of particle physics [1–4] has been extensively probed by many experiments in the last decades. In the SM, the Brout-Englert-Higgs mechanism is the source of electroweak symmetry breaking and results in the appearance of the Higgs boson [5–7]. This note describes the updated results of the search of the Higgs boson in the diboson decay modes with the ATLAS detector. Section 2 provides a short description of the production mechanism that create the Higgs boson in proton-proton colliders. Sec. 3 give a short description of the ATLAS detector. The results of the $H \to \gamma\gamma$ channel [8, 9] using 4.8 fb^{-1} of $\sqrt{s}$ =7 TeV and 5.9 fb^{-1} of $\sqrt{s}$ =8 TeV data are described in Sec. 4, while the results obtained in the $H \to ZZ^{(*)} \to 4\ell$ search [8, 10] using an integrated luminosity of 4.8 fb^{-1} of $\sqrt{s}$ =7 TeV and 5.8 fb^{-1} of $\sqrt{s}$ =8 TeV are shown in Sec. 5. Finally the updated measurements of the $H \to WW^{(*)} \to \ell\nu\ell\nu$ analysis [11, 12] where 13.0 fb^{-1} of $\sqrt{s}$ =8 TeV data was used are given in Sec. 6.

2. Higgs boson production at the LHC

In $p-p$ collisions the Higgs boson can be created through different production mechanisms. The fusion of gluons through a quark loop is the dominant process given the high gluon luminosities, then the vector bosons fusion (VBF) emitted from incoming quarks with a distinct signature of two high transverse momentum ($p_{\rm T}$) jets, the so called associated production or Higgs-strahlung where a W or a Z boson is present in the final state. Finally, with a much smaller cross section, the Higgs boson can also be produced accompanied by a top quark pair. In Fig. 1 Feynman diagrams are shown for the three most important processes. The production cross sections and decay branching fractions vary as function of the Higgs boson mass hypothesis ($m_{\rm H}$). In Tab. 1 the production cross section for the gluon-gluon-fusion (ggF) and VBF are shown for $p-p$ collisions at $\sqrt{s}$ =7 and $\sqrt{s}$ =8 TeV. The state for the art high order electroweak, QCD, gluon re-summation, interference and other corrections are taken from [13]. The large instantaneous

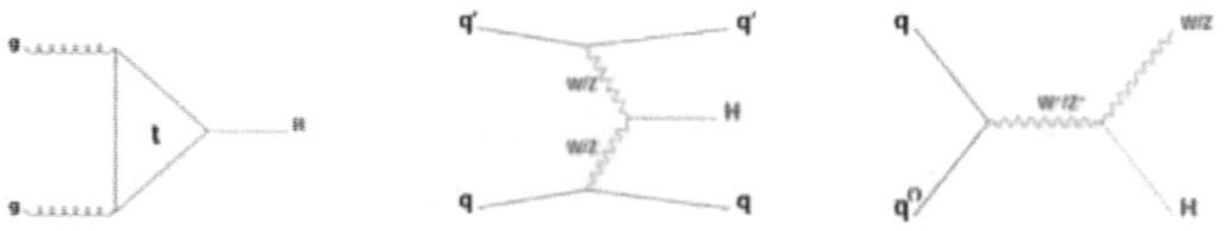

Figure 1. Feynman diagrams for the leading order Higgs boson production mechanisms at the LHC. Left: Gluon-gluon fusion trough and top quark loop. Centre: Vector boson fusion. Right: Associated production with a vector boson (W/Z).

luminosities at the LHC imply a large number of soft collisions per bunch crossing in addition to the hard scattering (event pile-up), this effects are included in the simulations used in the different analyses, a more comprehensive set of details of the simulation chain can be found in [8].

Table 1. Higgs boson production cross sections at the LHC for the ggF and VBF production mechanism at $\sqrt{s}$ =7 and $\sqrt{s}$ =8 TeV centre of mass energies [13].

$\sqrt{s}$ [TeV]	σ_{ggF} [pb]	σ_{VBF} [pb]	$\sigma_{pp\to H}$ [pb]
7	15.3	1.2	17.5
8	19.5	1.6	22.3

3. The ATLAS detector

The ATLAS detector [14] is a multi-purpose particle physics detector with forward-backward symmetric cylindrical geometry[1]. The inner tracking detector (ID) covers $|\eta| < 2.5$ and consists of a silicon pixel detector, a silicon micro-strip detector, and a transition radiation tracker (TRT). The ID is surrounded by a thin superconducting solenoid providing a 2 T axial magnetic field. A high-granularity lead/liquid-argon (LAr) sampling calorimeter measures the energy and the position of electromagnetic showers within $|\eta| < 3.2$. LAr sampling calorimeters are also used to measure hadronic showers in the end-cap ($1.5 < |\eta| < 3.2$) and forward ($3.1 < |\eta| < 4.9$) regions, while an iron/scintillator tile calorimeter measures hadronic showers in the central region ($|\eta| < 1.7$). The muon spectrometer (MS) surrounds the calorimeters and consists of three large superconducting air-core toroid magnets, each with eight coils, a system of precision tracking chambers ($|\eta| < 2.7$), and fast trigger chambers. A three-level trigger system selects events to be recorded for offline analysis.

4. $H \to \gamma\gamma$

The search of the Higgs boson in the diphoton final state is perform in the mass range between 110 and 150 GeV. The smoothly falling background in the invariant $m_{\gamma\gamma}$ is mainly made of SM $\gamma\gamma$, production, with subleading contributions from $\gamma + jets$, $jet - jet$ and Drell-Yan processes, where jets (or e) are misidentified as photons.

4.1. Selection

Photons are reconstructed from electromagnetic clusters using a neural-net algorithm in the 2011 data, and a cut based optimised to minimised the impact of pile-up in 2012. The photon candidates are required to have transverse energy E_{T} >40 and 30 GeV and to be in the fiducial

[1] The ATLAS experiment uses a right-handed coordinate system with its origin at the nominal interaction point. The z-axis is along the beam pipe, the x-axis points to the centre of the LHC ring and the y-axis is defined as pointing upwards. Polar coordinates (r,ϕ) are used in the transverse plane, ϕ being the azimuthal angle around the beam pipe. The pseudo-rapidity η is defined as $\eta = -\ln[\tan(\theta/2)]$ where θ is the polar angle.

volume. Details of the photon reconstruction and the object selection are given in [9]. To enhanced the sensitivity, ten mutually exclusive categories are created based on mass resolution, or the presence of two jets compatible with the VBF production topology. Table 2 list various categories as well as the number of expected signal events in the $\sqrt{s}$ =7 and $\sqrt{s}$ =8 TeV data sets.

Table 2. Number of events in the data (N_D) and expected number of signal events (N_S) for $m_{\rm H}$=126.5 GeV from the $H \rightarrow \gamma\gamma$ analysis, for each category in the mass range 100-160 GeV. The mass resolution FWHM is also given for the $\sqrt{s}$ =8 TeV data. The Higgs boson production cross section multiplied by the branching ratio into two photons ($\sigma \times BR(H \rightarrow \gamma\gamma)$) is listed for $m_{\rm H}$=126.5 GeV. The statistical uncertainties on N_S and FWHM are less than 1% [8].

$\sqrt{s}$	7 TeV		8 TeV		
$\sigma \times B(H \rightarrow \gamma\gamma)$ [fb]		39		50	FWHM
Category	$N_{\rm D}$	$N_{\rm S}$	$N_{\rm D}$	$N_{\rm S}$	[GeV]
Unconv. central, low $p_{\rm Tt}$	2054	10.5	2945	14.2	3.4
Unconv. central, high $p_{\rm Tt}$	97	1.5	173	2.5	3.2
Unconv. rest, low $p_{\rm Tt}$	7129	21.6	12136	30.9	3.7
Unconv. rest, high $p_{\rm Tt}$	444	2.8	785	5.2	3.6
Conv. central, low $p_{\rm Tt}$	1493	6.7	2015	8.9	3.9
Conv. central, high $p_{\rm Tt}$	77	1.0	113	1.6	3.5
Conv. rest, low $p_{\rm Tt}$	8313	21.1	11099	26.9	4.5
Conv. rest, high $p_{\rm Tt}$	501	2.7	706	4.5	3.9
Conv. transition	3591	9.5	5140	12.8	6.1
2-jet	89	2.2	139	3.0	3.7
All categories (inclusive)	23788	79.6	35251	110.5	3.9

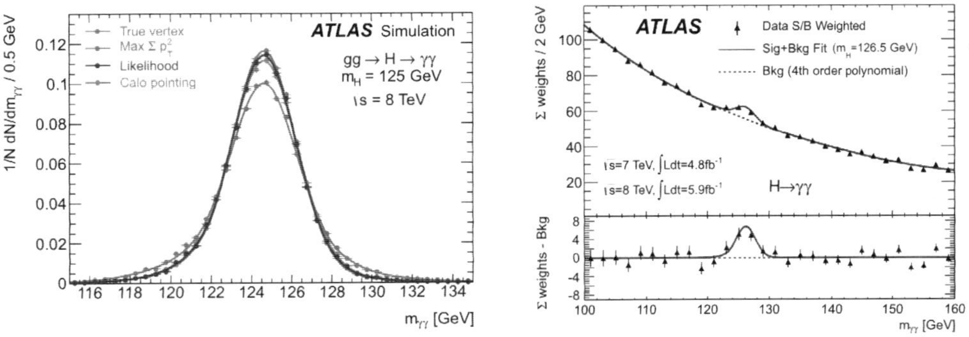

Figure 2. Left: Distribution of the expected diphoton mass for $H \rightarrow \gamma\gamma$ signal events as a function of the algorithm used to determine the longitudinal vertex position of the hard-scattering event. The use of the calorimeter information, labelled as "Calo pointing" is fully adequate to reach the optimal achievable mass resolution labelled as "True vertex". The likelihood described in [8], combining this information with the primary vertex information from the tracking, provides similar mass resolution. Right: The weighted distribution of invariant mass of diphoton candidates for the combined $\sqrt{s}$ = 7 TeV and $\sqrt{s}$ = 8 TeV data samples. The weight w_i for category i from [1, 10] is defined to be $\ln(1+S_i/B_i)$, where S_i is 90% of the expected signal for $m_{\rm H}$ = 126.5 GeV, and B_i is the integral, in a window containing S_i, of a background-only fit to the data. The values S_i/B_i have only a mild dependence on $m_{\rm H}$. The result of a fit to the data of the sum of a signal component fixed to $m_{\rm H}$=126.5 GeV and a background component described by a fourth-order Bernstein polynomial is superimposed. The bottom inset displays the residuals of the data with respect to the fitted background component [8].

4.2. Results

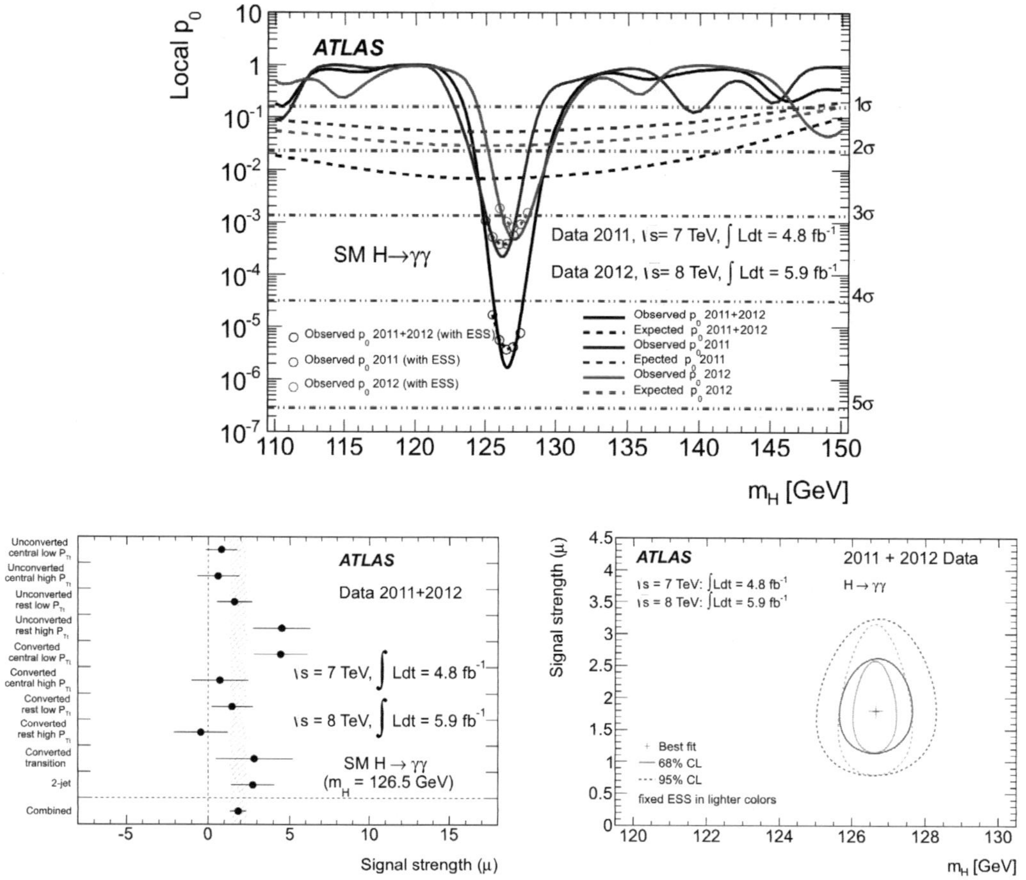

Figure 3. Top: Expected and observed local p_0 values for a SM Higgs boson as a function of the hypothesised Higgs boson mass for the combined analysis and for the $\sqrt{s}=7$ TeV and $\sqrt{s}=8$ TeV data samples separately. The observed p_0 including the effect of the photon energy scale uncertainty on the mass position is included via pseudo-experiments and shown as open circles. Bottom left: Best fit value for the signal strength in the different categories at $m_{\mathrm{H}}=$ 126.5GeV for the combined $\sqrt{s}=7$ TeV and $\sqrt{s}=8$ TeV data samples. The blue band corresponds to the error of the combined result. Bottom right: Confidence intervals contours in the (μ, m_{H}) plane. The 68% and 95% CL contours are for the combined $\sqrt{s}=7$ TeV and $\sqrt{s}=8$ TeV analysis. The light lines indicate the effect of holding constant at their best-fit values the nuisance parameters which describe the energy scale systematic (ESS) uncertainties in the likelihood function [8].

The $\sqrt{s}=7$ and $\sqrt{s}$ =8 TeV combined distribution of the invariant mass, $m_{\gamma\gamma}$, is shown in Fig. 2 (Left). The presence of the excess is statistically quantified by performing an unbinned likelihood fit as described in [9]. Figure 3 (Top) shows the local p_0 value including the possible effect of energy scale systematics. The $\sqrt{s}$ =7 TeV p_0 corresponds to 3.5σ, while the $\sqrt{s}$ =8 TeV p_0 corresponds to a 3.4σ and their combination to 4.5σ. The best fitted mass is 126.5 GeV and its signal strength, $\hat{\mu}$ =1.8±0.5. Figure 3 (Bottom left) gives the strength fitted values for the various categories. The effect of the energy scale systematic on the measured mass can be seen in Fig. 3 (Bottom right). Compatible with the SM Higgs boson prediction within 2σ.

Journal of Physics: Conference Series **455** (2013) 012013 doi:10.1088/1742-6596/455/1/012013

5. $H \to ZZ^{(*)} \to 4\ell$

The $H \to ZZ^{(*)} \to 4\ell$ analysis explores a mass range between 110 and 600 GeV. The good lepton momentum resolution and the clean signature of four charged leptons ($\ell = e, \mu$) coming from the Z bosons decay allows a powerful signal to background separation. The dominant background source is the SM production of $ZZ^{(*)}$, which is an irreducible background. The production of a Z together with a pair of light or heavy jets is the second most important background. A small contribution from top quark pair production is also expected.

5.1. Selection

Leptons are required to be within the fiducial volume and to have a transverse momentum, p_{T}, of at least 20,15,10 GeV for the three leading leptons, and at least 7(6) GeV for the final muon(electron). The lepton quadruplet must satisfy certain criteria that ensures they come from a $ZZ^{(*)}$ pair. Details of the required selections are given in [10]. Lepton isolation and impact parameter requirements are also imposed to reject both misidentified and leptons coming from semileptonic heavy flavour decays. Reducible backgrounds are measured using enriched control samples by relaxing/reversing quality requirements on the subleading lepton pairs, and the irreducible process is constraint during the fit by using the complete invariant $m_{4\ell}$ spectra including the so called single resonant $Z \to 4\ell$ process.

In Fig. 4 (Left) it is shown the invariant mass of the four lepton system for the combined data set. Table 3 shows the expected and observed number of events for the signal and different background components in a window of 5 GeV around $m_{4\ell}$ =125 GeV. The distribution of the candidates in the m_{12} (formed by the lepton pair closest to the on-shell Z boson mass), and m_{34} (invariant mass of the second lepton pair) plane is shown in Fig. 4 (Right).

5.2. Results

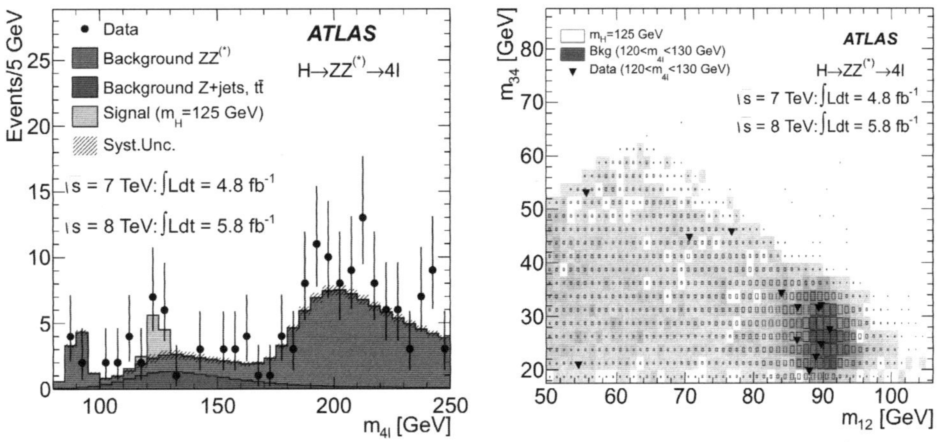

Figure 4. Left: The distribution of the four-lepton invariant mass, $m_{4\ell}$, for the selected candidates, compared to the background expectation in the 80 to 250 GeV mass range, for the combination of the $\sqrt{s} = 7$ TeV and $\sqrt{s} = 8$ TeV data. The signal expectation for a SM Higgs with m_{H}=125 GeV is also shown. Right: Distribution of the m_{34} versus the m_{12} invariant mass, before the application of the Z-mass constrained kinematic fit, for the selected candidates in the $m_{4\ell}$ range 120 to 130 GeV. The expected distributions for a SM Higgs with m_{H}=125 GeV (the sizes of the boxes indicate the relative density) and for the total background (the intensity of the shading indicates the relative density) are also shown [8].

Sensitivity is increased by splitting the analysis into four different lepton flavour combinations $(m_{12}; m_{34})$: $4\mu, 2\mu 2e, 2e2\mu$, and $4e$ where s/b and background compositions are different. The local p_0-values for the independent datasets and their combination as function of $m_{\rm H}$ are shown in Fig. 5 (Left). The likelihood fit of the combined $\sqrt{s}$ = 7 TeV and $\sqrt{s}$ = 8 TeV data indicates an excess of 3.6σ. In Fig. 5 (Right) the likelihood ratio contour plot in the $\hat{\mu}, m_{\rm H}$ is shown, the best fitted signal strength corresponds to 1.2±0.6, fully compatible with the SM within 1σ.

Table 3. The numbers of expected signal ($m_{\rm H}$=125 GeV) and background events, together with the numbers of observed events in the data, in a window of size ±5 GeV around 125 GeV, for the combined $\sqrt{s}$ = 7 TeV and $\sqrt{s}$ = 8 TeV data. [8].

	Signal	$ZZ^{(*)}$	Z + jets, $t\bar{t}$	Observed
4μ	2.09±0.30	1.12±0.05	0.13±0.04	6
$2e2\mu/2\mu 2e$	2.29± 0.33	0.80±0.05	1.27±0.19	5
$4e$	0.90±0.14	0.44±0.04	1.09±0.20	2

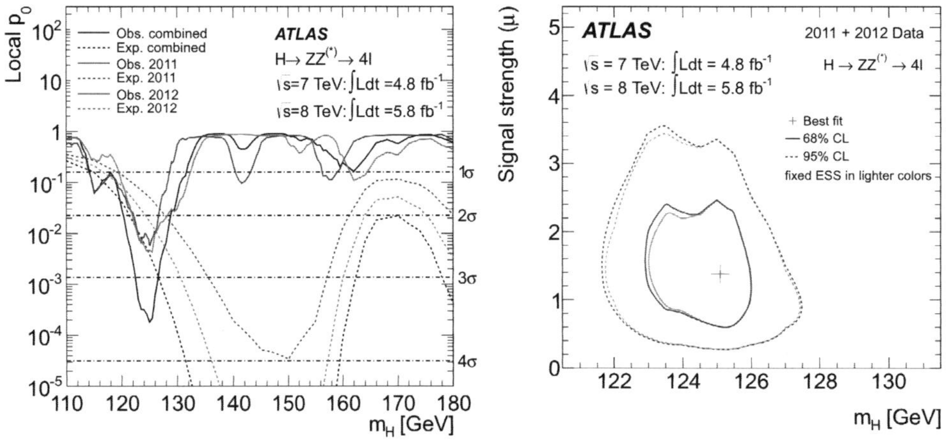

Figure 5. Left: The observed local p_0 for the combination of the 2011 and 2012 data sets (solid black line); the $\sqrt{s}$ = 7 TeV and $\sqrt{s}$ = 8 TeV data results are shown in solid lines (blue and red, respectively). The dashed curves show the expected median local p_0 for the signal hypothesis when tested at the corresponding $m_{\rm H}$. The horizontal dashed lines indicate the p_0 values corresponding to local significances of 1σ, 2σ, 3σ and 4σ. Right: Best fit values for μ and $m_{\rm H}$, and likelihood ratio contours that, in the asymptotic limit, correspond to 68% and 95% level contours in the $(\mu, m_{\rm H})$ plane. The light lines indicate the effect of holding constant at their best-fit values the nuisance parameters which describe the energy scale systematic uncertainties in the likelihood function [8].

6. $H \to WW^{(*)} \to \ell\nu\ell\nu$

The $H \to WW^{(*)} \to \ell\nu\ell\nu$ is the most sensitive analysis in the intermediate mass region (130-180 GeV). To properly normalise the complex background composition in the two lepton plus missing transverse energy ($E_{\rm T}^{miss}$) final state the analysis is divided according to the possible different flavour combinations and exclusive jet multiplicities. The results shown here correspond to an update of [11], where the opposite flavour final states in the 0- and 1-jet categories are revisited using 13 fb^{-1} of $\sqrt{s}$ =8 TeV data.

6.1. Selection

In the zero jet category, the dominant background process is the SM production of $WW^{(*)}$, it is followed by the top quark backgrounds and Drell-Yan process. Subleading contributions of $W + jets$, WZ and $W\gamma$ are also considered. The one jet category has similar processes, but the top quark backgrounds become more important, and some compositions also change like the larger relative presence of $Z \to \tau\tau$. The jet multiplicity and E_{T}^{miss} relative to the objects in the event are shown in Fig. 6.

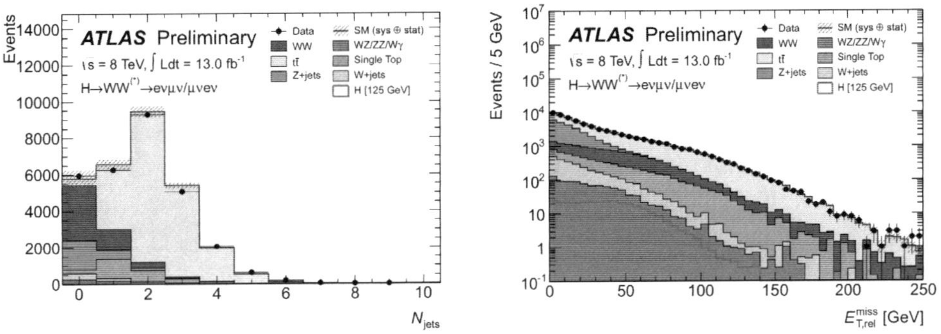

Figure 6. $E_{\mathrm{T}}^{\mathrm{miss,rel}}$ (Left) and multiplicity of jets (Right) for events satisfying the pre-selection criteria described in [12]. No $E_{\mathrm{T}}^{\mathrm{miss,rel}}$ requirement is applied in the $E_{\mathrm{T}}^{\mathrm{miss,rel}}$ distribution. Jets are required to fulfil $p_{\mathrm{T}} > 25$ GeV for $|\eta| < 2.5$ and $p_{\mathrm{T}} > 30$ GeV for $2.5 < |\eta| < 4.5$. The lepton channels are combined. The hashed area indicates the total uncertainty on the background prediction. The WW and top backgrounds are scaled to use the normalisation derived from the corresponding control regions described in the text. The expected signal for a SM Higgs boson with m_{H}=125GeV is superimposed [12].

A series of tight electron a muon identification criteria keeping leptons down to 25 GeV (15 GeV) in p_T for the leading (subleading) objects are applied. The event selection selects candidates compatible with two oppositely charged leptons each coming form a leptonic W decay, where the W and produced from a scalar object. Details about the object and event selection can be found in [12].

All the main backgrounds are normalised using control regions. The control regions are defined by selections similar to those used for the signal region but with some criteria reversed or modified to obtain signal-depleted samples enriched in a particular background. Some control regions have significant contributions from backgrounds other than the targeted one, which introduces dependencies among the background estimates. These correlations are taken into account in the WW control region where the top and $W + jets$ backgrounds are subtracted using their respective measurements. See [11] and [12] for full details on the background control samples and estimates.

6.2. Results

The presence of two neutrinos forbids the reconstruction of the Higgs boson invariant mass, therefore a transverse mass, m_{T}, is used as final discriminant variable. Figure 7 shows the m_{T} for the four different categories. The expected and observed number of events in the signal region for the zero and one jet categories are displayed in Tab. 4 As is was done for the channels previously presented, a statistical procedure is performed to quantify the size of the observed excess. Due to the poor resolution, driven by its final state, the $H \to WW^{(*)} \to \ell\nu\ell\nu$ channel has very little sensitivity to the Higgs mass hypothesis, other than the actual total signal yield.

Table 4. The observed numbers of events compared to the expectation from signal ($m_{\rm H}$125 GeV) and background after the full event selection, including a cut on the transverse mass of $0.75m_{\rm H} < m_{\rm T} < m_{\rm H}$. Data-derived normalisations of the backgrounds are included wherever applicable, as described in [12]. The uncertainties shown include both the statistical and systematic contributions, although the uncertainties on the control-region-derived backgrounds (WW, top in the H+1-jet analysis) do not include the uncertainty from the subtraction of other processes to the control region. The total uncertainty on the predicted background is calculated accounting for the correlations among the predictions, so the total does not correspond exactly to the sum in quadrature of the individual contributions. All numbers are summed over lepton flavours. [12].

	Signal	WW	$WZ/ZZ/W\gamma$	$t\bar{t}$	Single top	Drell-Yan	W+jets	Tot Bkg	Obs.
$H+0$ jet	45±9	242±32	26±4	16±2	11±2	4±3	34±17	334±28	423
$H+1$ jet	18±6	40±22	10±2	37±13	13±7	2±1	11±6	114±18	141

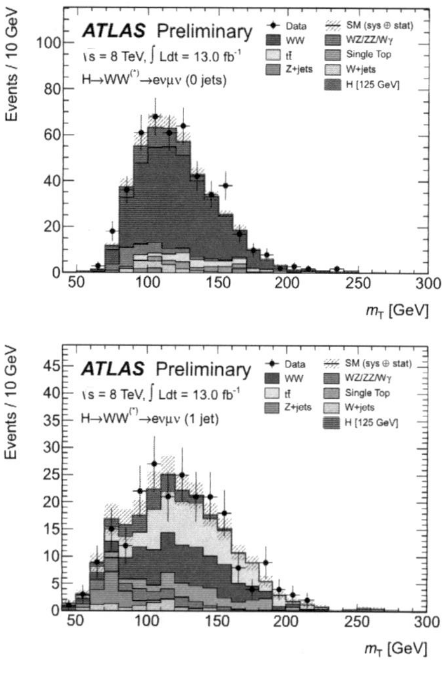

Figure 7. Transverse mass distribution in the zero jet (Top) and one jet (Bottom) channels, for events satisfying all criteria. The plots on the left show the events with a leading electron and the plots on the right show the events with a leading muon. The expected signal for a SM Higgs boson with $m_{\rm H}$ =125 GeV is added on top of the estimated total background. The W+jets background is estimated from data and WW, top and $Z \to \tau\tau$ backgrounds are scaled to use the normalisation derived from the corresponding control regions as described in the text. The hashed area indicates the total uncertainty on the total prediction [12].

In Fig. 8 (Left), the local p_0 is shown. At 125 GeV, an excess of 2.6σ is observed, with a signal strength $\hat{\mu}$ =1.5±0.6. The profiled likelihood contours of $m_{\rm H}$ and $\hat{\mu}$ for the three channels are shown in Fig. 8 (Right). All measured quantities are compatible among each other, as well as with a SM Higgs boson within 2σ.

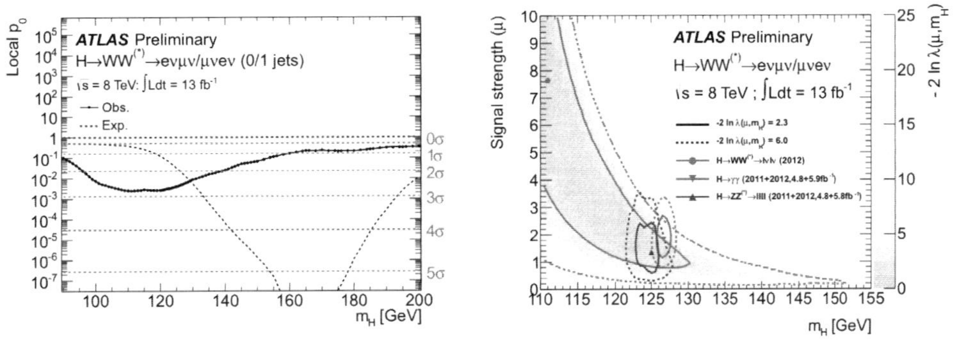

Figure 8. Left: observed (solid line) probability for the background-only scenario as a function of m_{H}. The dashed line shows the corresponding expectation for the signal+background hypothesis at the given value of m_{H}. Right: Approximate 68% and 95% two-dimensional likelihood ($\lambda(\mu, m_{\mathrm{H}})$) contours in the ($\mu, m_{\mathrm{H}}$) plane for the $H \to WW^{(*)} \to e\nu\mu\nu/\mu\nu e\nu$ zero jet and one jet data results with an integrated luminosity of 13 fb^{-1} at $\sqrt{s}$ =8 TeV. The results obtained for the $H \to ZZ^{(*)} \to 4\ell$ and $H \to \gamma\gamma$ analyses are overlaid, The yellow shading shows the $-\ln\lambda(\mu, m_{\mathrm{H}})$ values for $H \to WW^{(*)} \to e\nu\mu\nu/\mu\nu e\nu$ [12].

6.3. Summary

The updated results in the search for the SM Higgs boson in the $H \to \gamma\gamma$, $H \to ZZ^{(*)} \to 4\ell$ and $H \to WW^{(*)} \to \ell\nu\ell\nu$ decay modes with the ATLAS detector are presented. An excess compatible with the production of this particle is observed in the three cases, with local significances corresponding to 4.5σ in the diphoton final state, 3.6σ in the four lepton case, and 2.6σ in the $WW^{(*)}$ mode.

References

[1] S. L. Glashow, *Partial symmetries of weak interactions*, Nucl. Phys. **B 22** (1961) 579.
[2] S. Weinberg, *A model of leptons*, Phys. Rev. Lett. **19** (1967) 1264.
[3] A. Salam, *Elementary particle theory*. Almqvist and Wiksells, Stockholm, 1968.
[4] G. 't Hooft and M. Veltman, *Regularization and Renormalization of Gauge Fields*, Nucl. Phys. **B44** (1972) 189.
[5] F. Englert and R. Brout, *Broken symmetry and the mass of gauge vector mesons*, Phys. Rev. Lett. **13** (1964) 321–323.
[6] P. W. Higgs, *Broken symmetries and the masses of gauge bosons*, Phys. Rev. Lett. **13** (1964) 508–509.
[7] G. S. Guralnik, C. R. Hagen, and T. W. B. Kibble, *Global conservation laws and massless particles*, Phys. Rev. Lett. **13** (1964) 585–587.
[8] ATLAS Collaboration, *Observation of a new particle in the search for the standard model Higgs boson with the ATLAS detector at the LHC*, Phys. Lett. **B716** (2012) 1–29, `arXiv:1207.7214 [hep-ex]`.
[9] ATLAS Collaboration, *Search for the Standard Model Higgs boson in the diphoton decay channel with 4.9 fb^{-1} of pp collision data at $\sqrt{s} = 7$ TeV with ATLAS*, Phys. Rev. Lett. **108** (2012) 111803.
[10] ATLAS Collaboration, *Search for the Standard Model Higgs boson in the decay channel $H \to ZZ^{(*)} \to 4\ell$ with 4.8 fb^{-1} of pp collision data at $\sqrt{s} = 7$ TeV with ATLAS*, Phys. Lett. **B 710** (2012) 383.
[11] ATLAS Collaboration, *Search for the Standard Model Higgs boson in the $H \to WW^{(*)} \to \ell\nu\ell\nu$ decay mode with 4.7 fb^{-1} of ATLAS data at $\sqrt{s} = 7$ TeV*, Physics Letters B **716** (2012) no. 1, 62 – 81, `arXiv:1206.0756 [hep-ex]`.
[12] ATLAS Collaboration, *Update of the $H \to WW^{(*)} \to e\nu\mu\nu$ analysis with 13.0 fb^{-1} of $\sqrt{s} = 8$ TeV data collected with the ATLAS detector*, 2012. `http://cdsweb.cern.ch/record/1493601`.
[13] LHC Higgs cross section working group, S. Dittmaier, C. Mariotti, G. Passarino, and R. Tanaka (Eds.), *Handbook of LHC Higgs Cross Sections: 2. Differential distributions*, 2012. `arXiv:1201.3084 [hep-ph]`. CERN-2011-002.
[14] ATLAS Collaboration, *The ATLAS experiment at the CERN large hadron collider*, JINST **3** (2008) S08003.

Journal of Physics: Conference Series **455** (2013) 012014 doi:10.1088/1742-6596/455/1/012014

Top-quark production in ATLAS

Danilo Enoque Ferreira de Lima

Kelvin Building, University of Glasgow, Glasgow G12 8QQ, UK

E-mail: dferreir@mail.cern.ch

Abstract. Measurements of the top-quark production cross-sections in proton-proton collisions with the ATLAS detector at the Large Hadron Collider are presented. The measurement require no, one or two electrons or muons in the final state (hadronic, single lepton or dilepton channel). In addition, the decay modes with tau leptons are presented (channels with tau leptons). Differential measurements of $t\bar{t}$ final states are presented, in particular, measurements that are able to constrain the modelling of additional parton radiation. Measurements of single top-quark production in the t- and Wt-channels are presented and determination of the CKM matrix element $|V_{tb}|$ is discussed. In addition, the s-channel production is explored and limits on exotic production in single top-quark processes are discussed. This also includes the search for flavour changing neutral currents and the search for additional W' bosons in the s-channel.

1. Introduction

The top quark is the heaviest particle in the Standard Model [1]. It decays into a W-boson and a b-quark almost exclusively. This characteristic has been explored in the measurement of its production cross-section by analysing the final states produced by the W-boson and the b-quark.

The Large Hadron Collider (LHC) [2] is a 27 km circumference synchroton, which has a set of particle detectors which measure the final state particles in proton-proton collisions. The ATLAS [3] experiment is a general purpose detector in the LHC, which has many subdetectors, such as the Inner Detector, the Calorimeters and the Muon Spectrometer. A 2 T magnetic field bends the outgoing charged particles' trajectory. The Inner Detector reconstructs the tracks of charged particles, which are used to measure their momentum and charge with an $|\eta| < 2.5$ coverage. The Calorimeter measures the energy of the particles with a coverage of $|\eta| < 4.9$ and, in the region of overlap with the Inner Detector, the electromagnetic calorimeter is finely granular, aiming at precision measurements for electrons and photons. The Muon Spectrometer is specifically designed to measure the momenta of muon tracks, with a coverage of $|\eta| < 2.7$. The ATLAS detector has many physics goals, such as testing the Standard Model and searching for new physics. It is designed to have 4π solid angle coverage.

In ATLAS, the top quark may be produced in top-antitop pairs or without its antiparticle in the "single-top channel", through dominant strong and electroweak production. In the top-pair channel the final states can be classified into dilepton, if both top-quarks generate leptons in the final state; single-lepton, if only one top-quark decay product includes a lepton; or hadronic, if both top-quarks final state particles are quarks.

Results for the single-top production are also shown, with a measurement of its production in the t-channel, evidence for Wt-channel and a search for the s-channel.

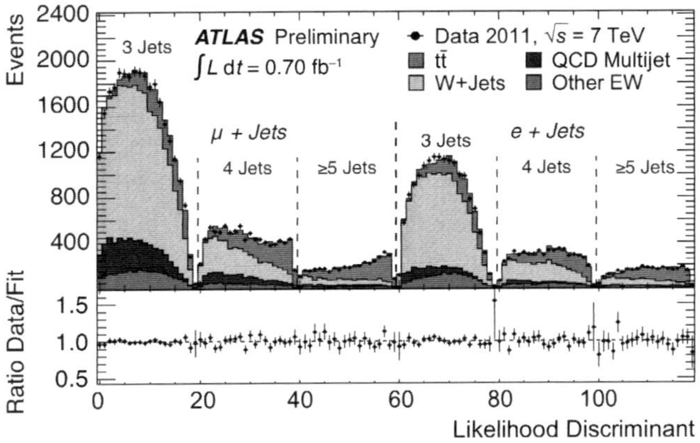

Figure 1. Likelihood discriminant distributions in data compared to simulation for the semileptonic inclusive cross-section measurement at $\sqrt{s} = 7$ TeV. Error bars shown in data refer to the statistical uncertainties [4].

Measurements of the top-quark production in ATLAS are shown in this document as a function of a set of observables, in different channels.

2. Top pair production

In the semileptonic final state of the $t\bar{t}$ system, the inclusive cross-section has been measured by building a likelihood-ratio discriminant D_i for the signal and each background [4]. The likelihood functions were calculated using the lepton η, $\exp(-8 \times \mathcal{A})$ (where $\mathcal{A}$ is the aplanarity), the p_{T} of the leading jet and the $H_{T,3p}$ (a ratio of transverse to longitudinal momenta). This was carried out for the electron and muon channels. The likelihood discriminant for data and simulation is given in Figure 1, for the $\sqrt{s} = 7$ TeV analysis. To enhance heavy flavour content, at least one b-jet was required, using multivariate techniques for the b-tagging algorithm. In this analysis information from the 3, 4 and ≥ 5 jet multiplicities was used and a cut on the transverse mass of the missing transverse energy and the lepton transverse momentum was applied to reduce the QCD-multijet background.

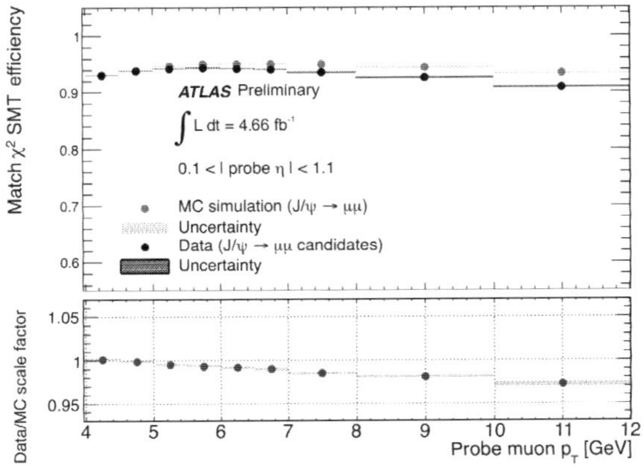

Figure 2. Soft Muon Tagger efficiency in the 7 TeV lepton + jets analysis with semileptonic b decays [6].

The $t\bar{t}$ cross-section was extracted through a maximum-likelihood fit using templates for the discriminant D_i, which resulted in a measurement of $\sigma_{t\bar{t}} = 179.0 \pm 3.9(\mathrm{stat.}) \pm 9.0(\mathrm{syst.}) \pm 6.6(\mathrm{lumi.})$ pb. Systematic uncertainties were found to be mainly due to the generator variations (3%), followed by the jet-energy scale (2.4%).

A similar analysis was carried out using the $\sqrt{s} = 8$ TeV data [5], but building a likelihood discriminator using only the lepton η and the transformed aplanarity. Furthermore, harder cuts on leptons ($p_{\mathrm{T}} > 40$ GeV) were used to reduce the fake leptons contributions. With this setup, a likelihood fit was done to the likelihood discriminant resulting in an inclusive cross-section measurement of $\sigma_{t\bar{t}} = 241 \pm 2(\mathrm{stat.}) \pm 31(\mathrm{syst.}) \pm 9(\mathrm{lumi.})$ pb. The measurement is in good agreement with the approximate NNLO calculations from HATHOR, $\sigma_{t\bar{t}}^{\mathrm{theory}} = 238^{+22}_{-24}$ pb for a top-quark mass of $m_t = 172.5$ GeV.

An alternative b-tagging algorithm using leptonically decaying b-jets has been used to perform a single-lepton channel analysis [6]. It takes advantage of the smaller systematic uncertainty associated with this b-tagging algorithm, which demands a match between a muon and a b-jet candidate. The muon must satisfy quality cuts, including $p_T > 4$ GeV, a match criteria based on a χ^2 between the Muon Spectrometer and the Inner Detector, that it has $\Delta R > 0.01$ from the μ coming from the W-boson decay and that it has $\Delta R < 0.5$ to the b-jet candidate decay. The efficiency of this b-jet selection is shown in Figure 2. The final measured cross-section was extracted by subtracting the background estimate and correcting for the selection efficiency, which results in a cross-section of $\sigma_{t\bar{t}} = 165 \pm 2(\text{stat.}) \pm 17(\text{syst.}) \pm 3(\text{lumi.})$ pb at $\sqrt{s} = 7$ TeV.

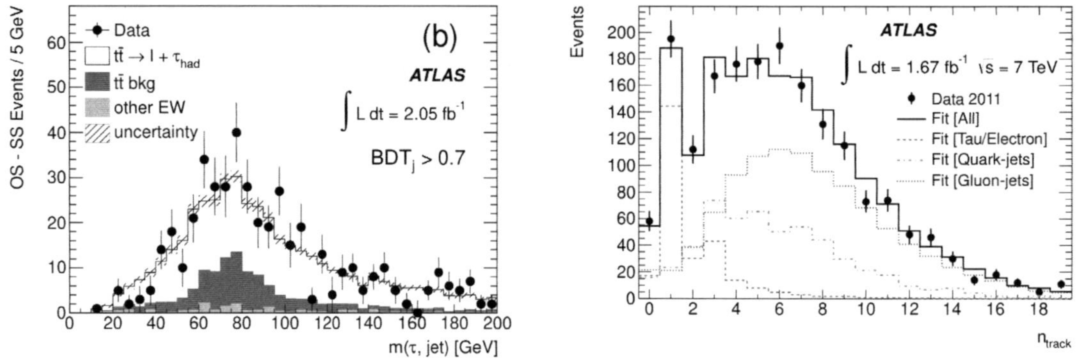

(a) Mass of the tagged τ and jet for selected events in the τ + e/μ + jets analysis.

(b) Track multiplicity for fitted backgrounds in the τ + jets analysis at 7 TeV.

Figure 3. Mass of the tagged τ and jet for the τ + e/μ analysis and fitted track multiplicity for the τ+ jets analysis. Statistical uncertainties are shown for data and systematic uncertainties are shown on the left, for simulation. The solid circles indicate data points and the histograms represent the simulation expectation [7] [8].

A special treatment is given to final states including the τ-lepton, for the τ hadronic decay. For the $\tau + e/\mu+$ jets final state [7] at $\sqrt{s} = 7$ TeV, a set of Boosted Decision Trees are used to identify the τ lepton, separate it from electrons (BDT_e) and separate it from other jets (BDT_j). Since some backgrounds are charge symmetric, it is possible to cancel them by subtracting a set of Opposite Sign (OS) selected sample from a Same Sign (SS) sample. The final cross-section is then extracted from the $(OS - SS)$ yields after a χ^2 fit to the BDT_j output. The mass of the tagged τ + jet system is shown for selected events in Figure 3(a). The final measured cross-section is $\sigma_{t\bar{t}} = 186{\pm}13(\text{stat.})20(\text{syst.}){\pm}7(\text{lumi.})$ pb.

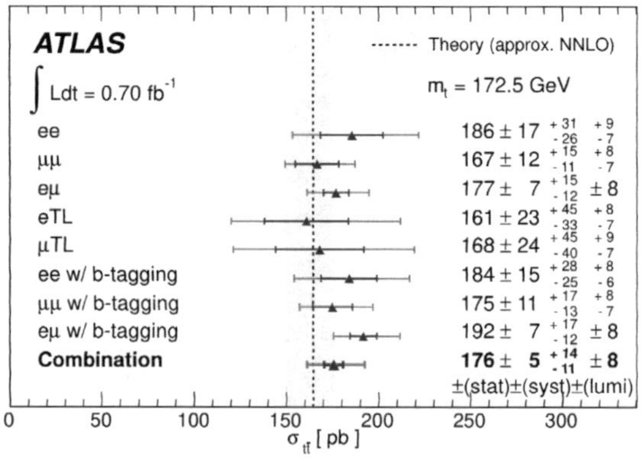

Figure 4. Summary of the $t\bar{t}$ cross-section measurements in the dilepton (e/μ) channel at 7 TeV for each channel and the combined measurement [9].

A τ + jet analysis [8] was also performed at $\sqrt{s} = 7$ TeV from hadronic τ decays, using a one dimensional fit to the number of tracks associated with the τ candidate, taking advantage of the fact that a τ decays preferentially into one or three charged particles. An extended binned likelihood fit was applied to fit the observed events to the number of $\tau + e$. Figure 3(b) shows the track multiplicity distribution and associated fit for signal and backgrounds. The number of $\tau + e$ events were then scaled to the fraction of τ events within this selection, to extract the number of τ + jets events. At least two b-tagged particles, five jets and a veto on electrons or muons were applied to suppress the backgrounds. The final cross-section was measured by correcting for the efficiency of the selection, which results in $\sigma_{t\bar{t}} = 194 \pm 18(\text{stat.}) \pm 46(\text{syst.})$ pb.

A cross-section measurement has been performed using the dilepton final state [9], by selecting electrons and muons using the $\sqrt{s} = 7$ TeV data. It requires two oppositely-charged lepton candidates, two high-p_{T} central jets, and applies constraints on the mass of the two lepton system to reduce the background contribution. A profile likelihood fit was used to estimate the cross-section in different channels and combine the results. A missing transverse energy cut was applied in the ee and $\mu\mu$ channels, while an H_T cut was used in the $e\mu$ channel to suppress the $Z/\gamma^* + jets$ contribution[1]. A summary of cross-section measurements in the dilepton final state is shown in Figure 4.

3. Relative $t\bar{t}$ differential cross-section measurements

Besides inclusive cross-section measurements, which have been shown before, differential cross-section measurements [10] were also done at centre-of-mass energy of $\sqrt{s} = 7$ TeV. In this context, the measurement is normalised by the inclusive cross-section and showed as a function of X using $1/\sigma_{t\bar{t}} d\sigma_{t\bar{t}}/dX$, where X is the mass $m_{t\bar{t}}$, the transverse momentum p_{T} or the rapidity y of the $t\bar{t}$ system.

The analyses were performed in the semileptonic final state, which includes one e/μ lepton, a neutrino and at least four jets. Accordingly, the selection requires at least four jets and large missing transverse energy. Furthermore, a likelihood fit of the measured kinematic variables to a lowest order representation of the $t\bar{t}$ decay was used to reconstruct the $t\bar{t}$ system with the W-boson mass and the top-quark mass contraints. An extra requirement was applied in the likelihood to select events which are consistent with the $t\bar{t}$ decay hypothesis (on the final state of interest).

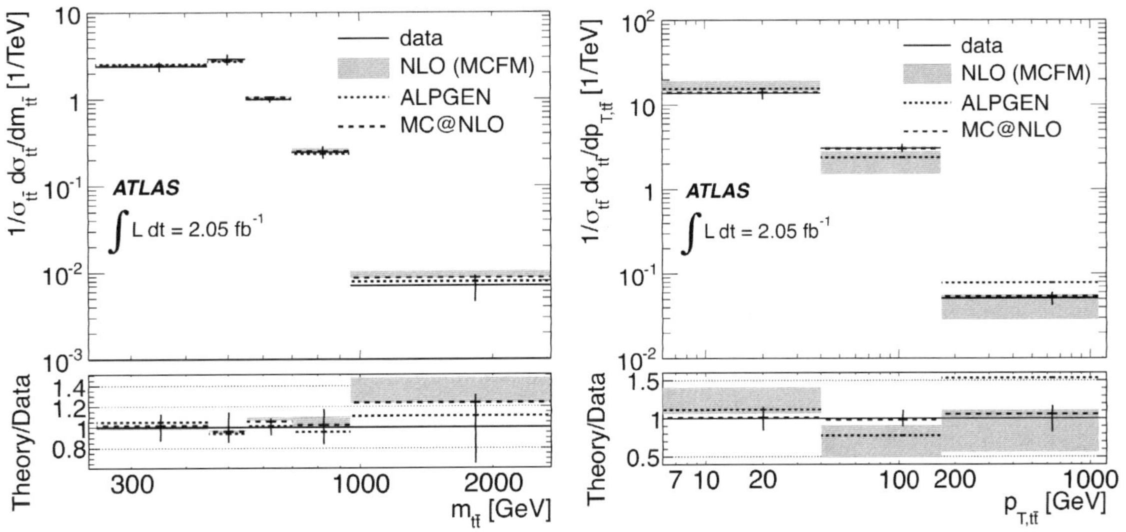

(a) Unfolded relative cross-section binned in $m_{t\bar{t}}$.

(b) Unfolded relative $t\bar{t}$ cross-section binned in p_T.

Figure 5. Unfolded distributions for the relative $t\bar{t}$ cross-section, comparing data and simulation predictions. The shaded band represents the systematic uncertainties on the simulation [10].

[1] H_T is defined as the scalar sum of all objects' transverse momentum.

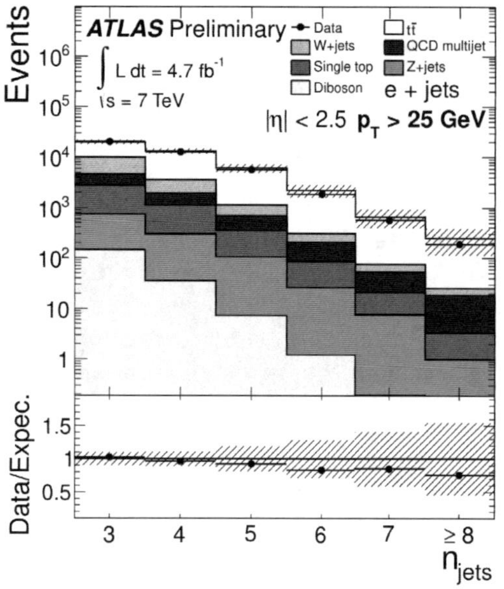

(a) Data/MC comparison for the $t\bar{t}$ jet multiplicity in the e + jets channel.

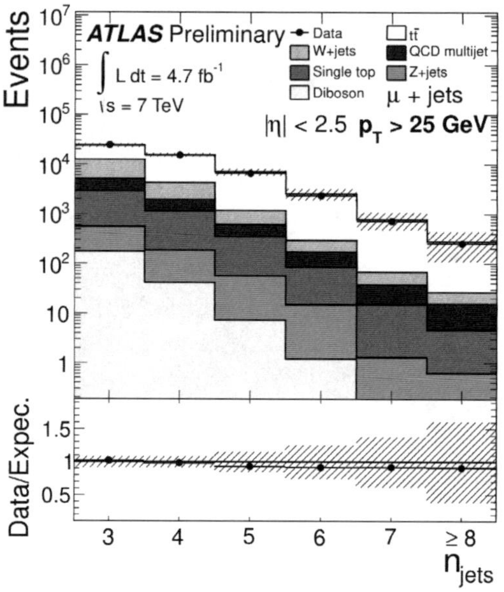

(b) Data/MC comparison for $t\bar{t}$ jet multiplicity in the μ + jets channel.

Figure 6. Data/Expectation comparison for $t\bar{t}$ jet multiplicity in the semileptonic final state. Data points include the statistical uncertainties as error bars and the simulation histograms have the systematic uncertainty as the shaded band [11].

The measurement is performed using an unregularised unfolding procedure, subtracting the estimated background B_i, correcting for the acceptance A_j and migration between bins $(M^{-1})_{ji}$, and using the luminosity L, according to Equation 1. The effect of the systematic uncertainties from the differential cross-section is reduced by the normalisation to the inclusive cross-section, but the final result is still dominated by systematic uncertainties, as it can be seen in Figures 5(a) and 5(b).

$$\sigma_j = \frac{\sum_i (M^{-1})_{ji}(N_i - B_i)}{A_j L} \tag{1}$$

4. $t\bar{t}$ jet multiplicity and jet veto gap fraction

Analyses that work as a test of QCD are very important to study final state radiation. One such analysis is the measurement of the $t\bar{t}$ jet multiplicity in the semileptonic final state [11], which is particularly important since this final state is a significant background for $t\bar{t}H$, $b'\bar{b}'$ and other resonance searches. As a first step, at least three jets, at least one b-tagged jet, and one electron or muon with $p_T > 25$ GeV were required for a first data to simulation comparison, as shown in Figures 6(a) and 6(b) for the electron and muon channels, respectively. Missing transverse energy greater than 30 GeV and transverse mass between[2] greater than 35 GeV were also required to reduce the backgrounds contribution. A veto was applied in the second lepton with $p_T > 20$ GeV to reduce the $t\bar{t}$ dilepton background contribution. The electrons were required to be in the region given by $|\eta| < 2.47$, excluding the $1.37 < |\eta| < 1.52$ region, while muons used had the requirement $|\eta| < 2.5$. Isolation requirements were also applied to the leptons.

[2] In this context the transverse mass is defined using the lepton transverse momentum and the missing transverse energy.

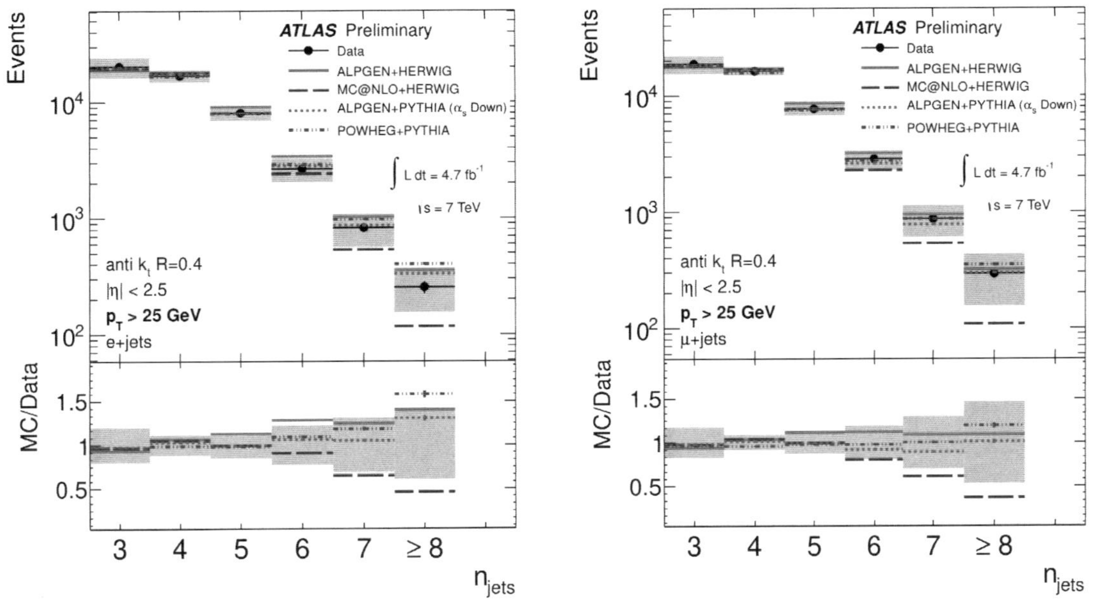

(a) Unfolded $t\bar{t}$ jet multiplicity in the e + jets channel. (b) Unfolded $t\bar{t}$ jet multiplicity in the μ + jets channel.

Figure 7. Unfolded $t\bar{t}$ jet multiplicity for the semileptonic $t\bar{t}$ decay. The shaded band on the data points represents the systematic uncertainty propagated through the unfolding procedure [11].

The result is unfolded to account for detector effects, by subtracting the background estimate ($\vec{f}_{\rm bngd}$) and correcting for the acceptance difference in the reconstruction-level and particle-level simulations, except for the jet multiplicity requirement ($\vec{f}_{\rm accpt}$). Corrections are also applied for events that pass the jet multiplicity requirements at reconstruction-level but not at particle-level ($\vec{f}_{\rm reco!part}$) and events that satisfy the particle-level jet multiplicity requirement, but fail the reconstruction-level requirement. The migration between bins ($\mathbf{M}_{\rm part}$) is taken into account using an iterative unfolding procedure [13]. The unfolding procedure can be summarised in Equation 2.

$$\vec{N}_{\rm part} = \vec{f}_{\rm part!reco} \cdot \mathbf{M}_{\rm part} \cdot \vec{f}_{\rm reco!part} \cdot \vec{f}_{\rm accpt} \cdot (\vec{N}_{\rm reco} - \vec{f}_{\rm bngd}) \qquad (2)$$

The unfolded result is shown in Figures 7(a) and 7(b). It can be seen that MC@NLO+HERWIG underestimates the data for bins with ≥ 6 jets, while ALPGEN+PYTHIA with the downward α_S variation and POWHEG+PYTHIA describe the data very well.

Another important test of extra radiation in the final state is the measurement of the "jet veto gap fraction", defined as in Equation 3, in which $\sigma(Q_0)$ is the cross-section for the production of $t\bar{t}$ events with no additional jet with $p_T > Q_0$. The final state analysed in this study is the dilepton decay of the $t\bar{t}$ system, which includes neutrinos, two leptons (in this study e or μ) and two b-jets, with extra radiation. This particular final state is used to have a clean event selection. The results profit from a reduced systematic uncertainty because the "jet veto gap fraction" is a ratio.

$$f(Q_0) = \frac{\sigma(Q_0)}{\sigma} \qquad (3)$$

The jet veto gap fraction for jets within $|y| < 0.8$ for different generators is shown in Figure 8(a) and it can be seen that the MC@NLO simulation tends to produce fewer jets. Also, it can be seen from Figure 8(b), that this measurement constrains systematic uncertainties coming from ISR/FSR modelling.

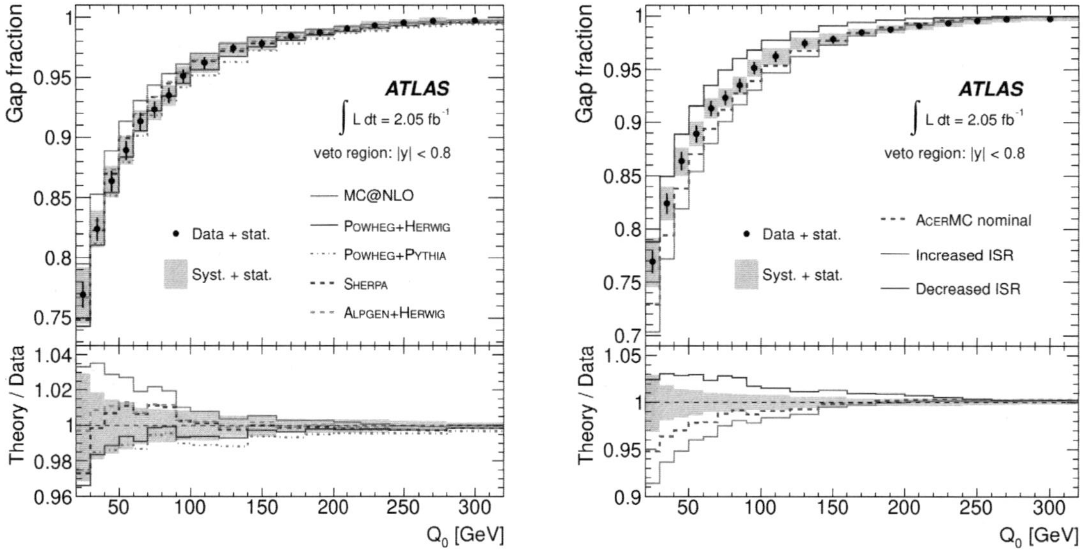

(a) Jet veto gap fraction for jets in $|y| < 0.8$.

(b) Jet veto gap fraction for jets in $|y| < 0.8$ compared to AcerMC with ISR/FSR systematic variations.

Figure 8. Jet veto gap fraction measured in $|y| < 0.8$ rapidity range, including the systematic and statistical uncertainties from the measurement in data in the yellow band [12].

5. Single top-quark production

The single top-quark production cross-section in the LHC is smaller than the dominant strong force mediated production of $t\bar{t}$. It can be produced through a t-channel, Wt-channel and s-channel diagrams, but only the t-channel has been observed so far.

The t-channel production generates a top quark with an extra quark and it has the interesting feature that it allows one to measure the production cross-section for top events and antitop events. Furthermore, the ratio between these cross-sections R_t is sensitive to the extra quark's (u or d) Parton Distribution function. This measurement was carried out at $\sqrt{s} = 7$ TeV, finding $R_t = 1.81 \pm 0.10(\text{stat.}) {}^{+0.21}_{-0.20}(\text{syst.})$ [14]. The measurement of the single top t-channel at $\sqrt{s} = 8$ TeV was also done without the top/antitop separation, finding a final cross-section of $\sigma_t = 95.1 \pm 2.4(\text{stat.}) \pm 18.0(\text{syst.})$ pb [15], in agreement with the $87.2^{+2.8}_{-1.0}{}^{+2.0}_{-2.2}$ pb approximate NNLO theoretical calculations [16].

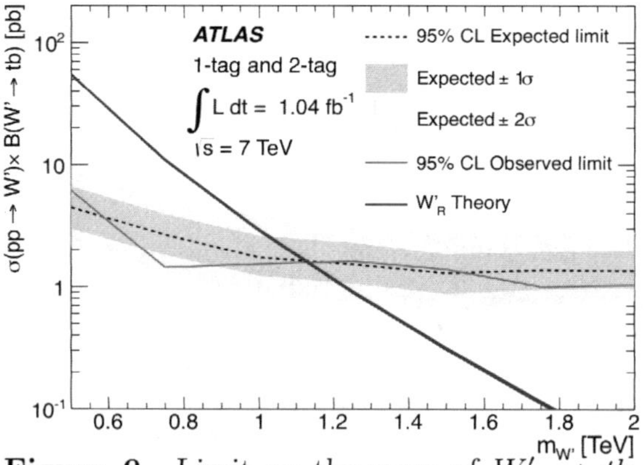

Figure 9. Limit on the mass of $W'_R \to tb$. The systematic uncertainty variations for the expected limits for 1σ and 2σ are shown as the green and yellow bands [19].

The analysis in the Wt-channel at $\sqrt{s} = 7$ TeV was also performed [17] and the background-only hypothesis was excluded at 3.3σ, with a cross-section measurement obtained through a maximum likelihood estimate of $\sigma_{Wt} = 16.8 \pm 2.9(\text{stat.}) \pm 4.9(\text{syst.})$ pb, in agreement with the approximate NNLO theoretical calculations of $15.6 \pm 0.4 \pm 1.1$ pb [16]. This allows for an estimate of the $|V_{tb}|$ CKM matrix element of $|V_{tb}| = 1.03^{+0.16}_{-0.19}$.

The s-channel search [18] was performed at $\sqrt{s} = 7$ TeV and it leads to an observed cross-section upper limit of 26.5 pb, while the predicted Standard Model cross-section is $4.56 \pm 0.07^{+0.18}_{-0.17}$ pb [16]. Using this result a tb-resonance search [19] was also done for a model of a right-handed W'_R with Standard Model-like couplings. Limits are set for this exotic particle as shown in the Figure 9, for $m_{W'_R} > 1.13$ TeV at 95% Confidence Level.

6. Summary

The ATLAS Collaboration has used different techniques to estimate the top-quark production cross-section in different scenarios. The cross-section of the dominant $t\bar{t}$ production process has been measured for dileptonic and semileptonic final states, which also included a cross-check study, using an uncorrelated tagger (Soft Muon Tagger).

Extra radiation in the semileptonic final state and the dileptonic final states are studied in the $t\bar{t}+$ jets analysis and the jet veto gap fraction study. The jet veto gap fraction analysis was used to constrain the AcerMC variations. Smaller variations were used to compare with the $t\bar{t}+$ jets analysis. The results of the $t\bar{t}+$ jets analysis are consistent with the jet veto gap fraction analysis. The former is also important to understand the main background for many searches, while the latter constrains the ISR/FSR systematic uncertainty in the generators by $\sim 50\%$.

The single top-quark production has also been studied in the t-, Wt- and s-channels, which were used to estimate the $|V_{tb}|$ CKM matrix element, in agreement to the world average of $|V_{tb}| = 0.999146^{+0.000021}_{-0.000046}$ [20], and set limits for the s-channel cross-section and for a model for a right-handed $W'_R \to tb$. The cross-section in the t-channel was measured and evidence for the Wt-channel was found.

Other recent results from ATLAS, complementary to the ones shown, include the measurement of the $t\bar{t}$ all hadronic cross section [21] and a search for single top-quark Flavour Changing Neutral Currents [22].

References

[1] J. R. Incandela et al., Prog. Part. Nucl. Phys. 63 (2009) 239-292.
[2] O. S. Bruning, (Ed.), P. Collier, (Ed.), P. Lebrun, (Ed.), S. Myers, (Ed.), R. Ostojic, (Ed.), J. Poole, (Ed.) and P. Proudlock, (Ed.), CERN-2004-003-V-1.
[3] The ATLAS Collaboration, *ATLAS Detector Status and Physics Startup Plans*, JINST 3 S08003 (2008).
[4] The ATLAS Collaboration, ATLAS-CONF-2011-121.
[5] The ATLAS Collaboration, ATLAS-CONF-2012-149.
[6] The ATLAS Collaboration, ATLAS-CONF-2012-131.
[7] The ATLAS Collaboration, Phys. Lett. B 717 (2012) 89-108.
[8] The ATLAS Collaboration, arXiv:1211.7205.
[9] The ATLAS Collaboration, JHEP 1205 (2012) 059.
[10] The ATLAS Collaboration, arXiv 1207.5644.
[11] The ATLAS Collaboration, ATLAS-CONF-2012-155.
[12] The ATLAS Collaboration, Eur. Phys. J. C72 (2012) 2043.
[13] G. D'Agostini, Nucl. Instr. and Meth. A 362 (1995) 487.
[14] The ATLAS Collaboration, ATLAS-CONF-2012-132.
[15] The ATLAS Collaboration, ATLAS-CONF-2012-056.
[16] N. Kidonakis, arXiv: 1210.7813.
[17] The ATLAS Collaboration, Phys. Lett. B 716 (2012) 142-159.
[18] The ATLAS Collaboration, ATLAS-CONF-2011-118.
[19] The ATLAS Collaboration, Phys. Rev. Lett. 109 (2012) 081801.
[20] J. Beringer et al. (PDG), Phys. Rev. D86, 010001 (2012).
[21] The ATLAS Collaboration, ATLAS-CONF-2012-031.
[22] The ATLAS Collaboration, arXiv 1203.0529.

Journal of Physics: Conference Series **455** (2013) 012015 doi:10.1088/1742-6596/455/1/012015

Search for the Standard Model Higgs boson in the $H \rightarrow b\bar{b}$ and $H \rightarrow \tau^+\tau^-$ decay modes with the ATLAS detector

A. Farilla on behalf of the ATLAS Collaboration

INFN Sezione di Roma Tre, Via Della Vasca Navale 84, 00146, Roma, IT

E-mail: ada.farilla@roma3.infn.it

Abstract. We present an updated search for the Standard Model Higgs boson in the $H \rightarrow b\bar{b}$ and $H \rightarrow \tau^+\tau^-$ decay modes performed by the ATLAS experiment using 4.7 fb^{-1}of LHC data at $\sqrt{s}$ = 7 TeV and 13.0 fb^{-1}at $\sqrt{s}$ = 8 TeV delivered in 2011 and 2012 respectively. No significant excess is observed. For m_H = 125 GeV the observed (expected) upper limit at 95% confidence level on the cross section times the branching ratio is found to be 1.8 (1.9) times the Standard Model prediction for $H \rightarrow b\bar{b}$ and 1.9 (1.2) for $H \rightarrow \tau^+\tau^-$.

1. Introduction

The search for the Standard Model (SM) Higgs boson is one of the most important endeavours of the Large Hadron Collider (LHC) [1]. The discovery of a Higgs-like boson decaying to a pair of photons or heavy vector bosons has recently been reported by the ATLAS [2] and CMS [3] collaborations. The precise determination of the nature of this boson is crucial for establishing if it is indeed the SM Higgs boson. The fermionic decays $H \rightarrow b\bar{b}$ and $H \rightarrow \tau^+\tau^-$ will play an important role in such determination. This note presents an update of the results reported in Ref. [4] and Ref. [5], which were based on 4.7 fb^{-1}of LHC data at $\sqrt{s}$ = 7 TeV. A detailed description of the ATLAS detector can be found in [6].

2. $H \rightarrow b\bar{b}$

The decay to b-quarks, for a SM Higgs boson with a mass of m_H = 125 GeV, is expected to be the dominant decay mode (BR($H \rightarrow b\bar{b}$)$\simeq$ 58%). Therefore, an observation in this channel is crucial in order to provide a direct constraint on the dominant decay mode. The three processes with the largest SM Higgs boson production cross-section at the LHC are gluon-gluon fusion (ggF), W or Z boson fusion (VBF) and associated production with a W or a Z boson (VH). The cross-sections for each process at 7 TeV and 8 TeV, for m_H = 125 GeV, are summarized in Table 1.

To reduce the huge background from multijet production and to have a clean signature for triggering, the search for $H \rightarrow b\bar{b}$ has been performed with the Higgs boson produced in association with a W or Z boson in the decay channels $ZH \rightarrow \nu\bar{\nu}b\bar{b}$ (0-lepton channel), $WH \rightarrow l\nu b\bar{b}$ (1-lepton channel) and $ZH \rightarrow l^+l^-b\bar{b}$ (2-lepton channel), where l refers to either an electron or a muon. More details on the analysis can be found in Ref. [7]. Monte Carlo (MC) event samples produced with the full ATLAS detector simulation [8] based on the GEANT4 [9]

Table 1. Cross sections for Higgs production processes for pp collisions at $\sqrt{s}$=7 TeV and $\sqrt{s}$=8 TeV at LHC, for m_H = 125 GeV [10], [11].

$\sqrt{s}$	ggF	VBF	VH
7 TeV	15.3 pb	1.22 pb	0.89 pb
8 TeV	19.5 pb	1.57 pb	1.09 pb

program, corrected for all known detector effects, are used to model the Higgs boson signal and most backgrounds. The total production cross sections and associated uncertainties, computed at next-to-leading order (NLO), are taken from Ref. [10] with NLO corrections as a function of the transverse momentum of the vector boson, p_{T}^{V} [12], [13]. The decay branching ratios are calculated with HDECAY [14].

2.1. Event selection

Three categories of leptons, as defined in lepton reconstruction, are used and are denoted in increasing order of purity as loose, medium and tight leptons. Events containing no loose leptons are assigned to the 0-lepton channel, those with one tight and no additional loose leptons are assigned to the 1-lepton channel. Events with one medium and one additional loose lepton of the same flavour but opposite charge are assigned to the 2-lepton channel. Jets are reconstructed using the anti-k_t algorithm [15] with a radius parameter R = 0.4 and are required to have p_{T}> 20 GeV and $|\eta| < 4.5$. The two jets used to reconstruct the Higgs boson candidate are required to satisfy $|\eta| < 2.5$ and the leading jet is required to have p_{T}> 45 GeV.

b-quark jets are identified using the MV1 b-tagging algorithm [16], [17], tuned to provide an approximate efficiency of 70% for b-jets and rejection factors of 150 and 5 for light and c-quark jets respectively. Selected events are required to have exactly two b-tagged jets, which are used to reconstruct the mass of the Higgs boson candidate, $m_{b\bar{b}}$. In 3-jets events, the leading two jets are required to be b-tagged.

The missing tranverse energy, E_{T}^{miss} , is measured as the negative vector sum of the transverse momenta associated with cluster energies in the calorimeters with $|\eta| < 4.9$. A track-based missing transverse momentum, p_{T}^{miss}, is calculated as the negative vector sum of the transverse momenta of tracks associated to the primary vertex. The transverse momentum of the vector boson, p_{T}^{V}, where V denotes the W or the Z boson, is reconstructed as the E_{T}^{miss} in the 0-lepton channel, as the vector sum of the lepton and the E_{T}^{miss} in the 1-lepton channel and the vector sum of the two leptons in the 2-lepton channel. In order to increase the analysis sensitivity, the events are split in different categories, for a total of 16 categories. Events in the 0-lepton channel are split into six categories depending on whether they contain two or three jets and on the amount of missing transverse energy. Events with one or two leptons are divided into five categories of p_{T}^{V}. Topological cuts, optimized for each category, are used to enhance the signal and reject background.

The multijet background in the 0-lepton channel is suppressed by cuts on E_{T}^{miss}, p_{T}^{miss} and their directions with respect to each other and the reconstructed jets. In the 1-lepton channel, requirements are made on E_{T}^{miss} and the transverse mass on the W boson, m_W, to select events consistent with the presence of a W boson. The background from top production is reduced in the 2-lepton channel by cuts on the di-lepton invariant mass, $83 < m_{ll} < 99$ GeV, and E_{T}^{miss}< 60 GeV.

2.2. Background Estimation

The dominant backgrounds depend on the event category: the 2-lepton channel is dominated by Z+jets events, the 1-lepton channel is dominated by W+jets and top (both $t\bar{t}$ and single top) while all three types of backgrounds (Z+jets, W+jets and top) contribute to the 0-lepton channel. Background from multijet and diboson production (WW, ZZ, WZ) is also present. Most background shapes are taken from simulation and normalised using control regions in data except for multijet, which is estimated entirely from data, and the diboson production for which both normalisation and shape are taken from simulation.

Control regions are used to normalise the top background from data in the 1- and 2-lepton channels. In the 1-lepton channel the control region is defined by requiring an additional jet and for the 2-lepton channel by inverting the requirements on m_{ll} and E_{T}^{miss}. Control distributions for the V+jets background are categorised by the number of b-tagged jets. The 0-tag and 1-tag control regions contain exactly 0 or 1 b-tagged jets.

The normalisations of the V + c-jets and V + light-jet backgrounds are determined by a maximum likelihood fit to control and signal regions in the 1- and 2-lepton channels. Events containing 0,1 and 2 b-tagged jet and the top control regions of each channel are used. This flavour fit exploits the fact that the b-tagging algorithm has very different efficiencies for b, c and light jets, such that the flavour composition varies significantly with the number of b-tagged jets. The normalisations of V+light, V+c, V+b and top are allowed to float in the fit. The distributions of the signal regions for each of the three channels, after having been normalised by the result of the flavour fit, are shown in Figure 1 for the $\sqrt{s} = 8$ TeV data.

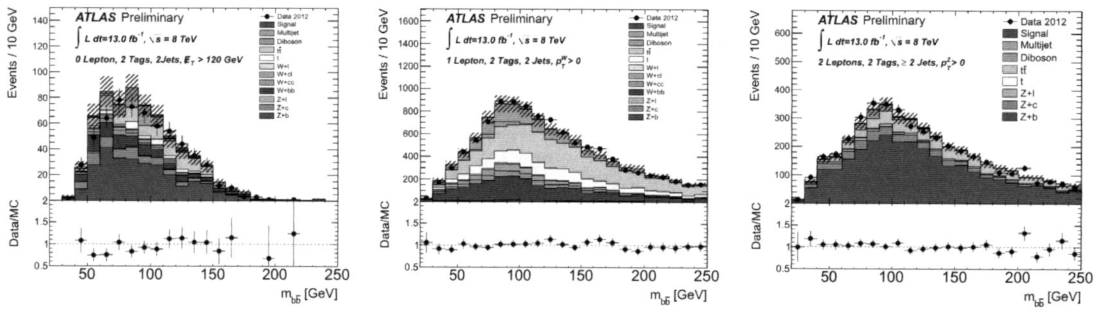

Figure 1. The $m_{b\bar{b}}$ signal distribution integrated over bins of p_{T}^{V}: (left) 0-lepton channel, (middle) 1-lepton channel, (right) 2-lepton channel. The error bands indicate the size of the combined statistical and systematic uncertainties before the profile likelihood fit [7].

Systematic uncertainties are determined on all experimental and theoretical sources. The most significant are on the b-tagging efficiency ($\simeq 7\%$), the calibration of the jet energy ($7-20\%$) and the statistical uncertainty on the MC ($4-8\%$).

2.3. Results

The statistical analysis of the data employs a binned profile likelihood function constructed as a product of the likelihood terms for each category. The likelihood in each category is a product over bins in the distribution of $m_{b\bar{b}}$. A detailed description of the statistical analysis of LHC data can be found in Ref. [18], [19].

The event categories that enter the profile likelihood fit are the 16 individual 2 b-tag signal regions and the top control regions for the 1- and 2-lepton channels. The signal strenght parameter, μ, multiplying the expected signal yield in each bin, is the parameter of interest in the fit procedure. The μ parameter is defined as the ratio of the measured cross section normalised to the Standard Model Higgs boson production cross section times the branching ratio for $H \rightarrow b\bar{b}$. The dependence of the signal and background predictions on the systematic uncertainties is described by nuisance parameters. The normalisation of the top, Z + b-jet and W + b-jet backgrounds floats without constraint in the fit, while the other backgrounds have their uncertainties applied as a constraint.

Diboson production (WZ or ZZ) with a Z boson decaying to a pair of b-quarks, has a signature very similar to the Higgs associated production but with lower p_{T}^{Z} and $m_{b\bar{b}}$ and a cross section 5 times larger. Therefore a separate fit was made to validate the analysis procedure. In the fit the normalisation of the diboson contribution is allowed to vary with a multiplicative scale μ_{D} with respect to the SM expectation.

Figure 2 (top) shows the distribution in data after subtracting all backgrounds except the diboson and SM Higgs contribution. An excess in the data compared to the background is observed at the expected mass for the diboson signal. We measure $\mu_{\mathrm{D}} = 1.05 \pm 0.32$ corresponding to a significance of 4.0 σ and agrees with the SM expectation of $\mu_{\mathrm{D}} = 1$.

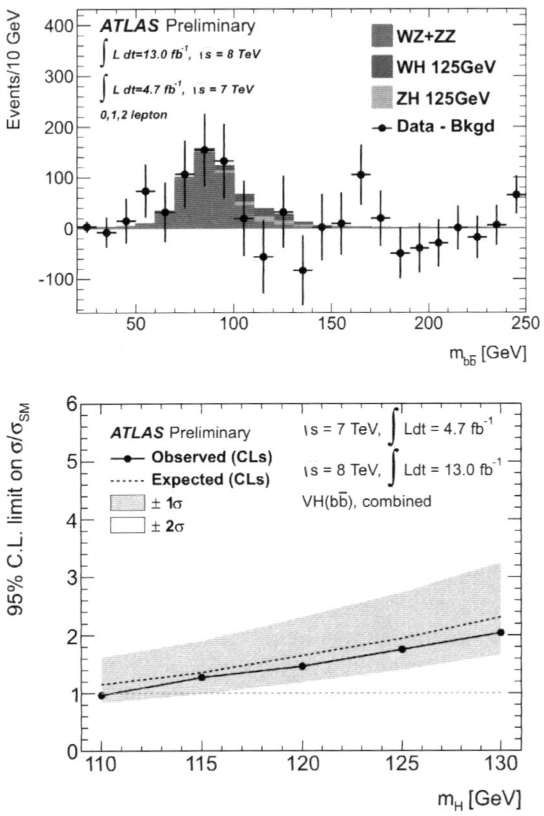

Figure 2. Top plot: the $m_{b\bar{b}}$ distribution after subtraction of all backgrounds except diboson processes (WZ, ZZ) and WH, ZH production. The MC backgrounds are normalised according to the results of the global fit. Only statistical uncertainties are shown. Bottom plot: expected (dashed) and observed (solid) 95% CL on the normalised signal strenght as a function of m_H for the combination of the $\sqrt{s} = 7$ TeV and $\sqrt{s} = 8$ TeV dataset [7].

The profile likelihood fit with the Higgs boson signal strenght floating is performed with the diboson contribution fixed to its SM expectation. Figure 2 (bottom) shows the results for the 95% confidence level (CL) exclusion limits, evaluated using the CL_S method [18], on the Higgs boson production cross section in the mass range 110-130 GeV. The observed limit for the combination of the $\sqrt{s} = 7$ TeV and $\sqrt{s} = 8$ TeV datasets is 1.8 times the SM expectation with an expected limit of 1.9 times the SM. The probability, p_0, that a fluctuation in the background can mimic the presence of a SM Higgs boson is studied by calculating the local p_0 value with respect to the background-only hypothesis. For $m_H = 125$ GeV the combination of both datasets yields an observed (expected) p_0 value of 0.64 (0.15). We measure $\mu = -2.7 \pm 1.1(stat) \pm 1.1(syst)$, $\mu = 1.0 \pm 0.9(stat) \pm 1.1(syst)$ and $\mu = -0.4 \pm 0.7(stat) \pm 0.8(syst)$ in the $\sqrt{s} = 7$ TeV, $\sqrt{s} = 8$ TeV and combined datasets, respectively.

3. $H \to \tau^+\tau^-$

The coupling of the new Higgs-like boson to leptons has not yet been established. In the SM, the Higgs coupling to tau leptons is significant (BR($H \to \tau^+\tau^-$)$\simeq$ 6.3%), due to a relatively high mass of the tau. The search for $H \to \tau^+\tau^-$ is performed across the three main Higgs production processes, shown in Table 1, and this will ultimately lead to the measurement of a number of important Higgs couplings. Both leptonic $\tau_{\rm lep}$ and hadronic $\tau_{\rm had}$ decays of the tau leptons have been exploited, therefore the search has been performed in the following three Higgs decay channels: $H \to \tau_{\rm lep}\tau_{\rm lep}$, $H \to \tau_{\rm lep}\tau_{\rm had}$ and $H \to \tau_{\rm had}\tau_{\rm had}$. More details on the analysis can be found in Ref. [20].

3.1. Event selection

Signal events are selected by requiring exactly two opposite-charge tau decays products which are two light leptons l (l=electron or muon) for the $H \to \tau_{\rm lep}\tau_{\rm lep}$ candidate, one light lepton and a $\tau_{\rm had-vis}$ for $H \to \tau_{\rm lep}\tau_{\rm had}$ candidate and two $\tau_{\rm had-vis}$'s for $H \to \tau_{\rm had}\tau_{\rm had}$ candidate, where $\tau_{\rm had-vis}$ denotes the visible component of hadronically decaying tau lepton. To suppress backgrounds produced via purely QCD processes, signal events are selected by requiring large $E_{\rm T}^{miss}$ to account for the neutrinos produced in tau decays. The invariant mass of the two taus is the final discriminating observable in all the three channels and is reconstructed by means of the Missing Mass Calculator (MMC) [21]. This tecnique provides a reconstruction of event kinematics in the $\tau\tau$ final state with efficiency > 99% and 13 − 20% resolution in $m_{\tau\tau}$, depending on the event topology and final state (better resolution is obtained for events with high-$p_{\rm T}$ jets and boosted topology).

All three channels analyzed exploit both the VBF and Boosted categories. Events in the VBF category are selected requiring two jets of high invariant mass and high angular separation in η. Events in the Boosted category are selected requiring large transverse momenta of the l+l+$E_{\rm T}^{miss}$ system (for the $H \to \tau_{\rm lep}\tau_{\rm lep}$ case), of l+$\tau_{\rm had-vis}$+$E_{\rm T}^{miss}$ system (for the $H \to \tau_{\rm lep}\tau_{\rm had}$ case) or a leading jet in the event (for $H \to \tau_{\rm had}\tau_{\rm had}$ case). The following selections and categorizations are applied for each decay channel to reduce backgrounds and to focus on a particular production process.

$H \to \tau_{\rm lep}\tau_{\rm lep}$: To search for a Higgs within the range 100 GeV $< m_H <$ 150 GeV the di-lepton invariant mass is required to be in the range 30 GeV $< m_{ll} <$ 100 GeV for the $e\mu$ channel, whereas for the ee and $\mu\mu$ channel 30 GeV $< m_{ll} <$ 75 GeV is required to reduce the contamination from $Z/\gamma^* \to ll$. Events are rejected if there is any jet, with $p_{\rm T} >$ 25 GeV and $|\eta| < 2.5$, identified as a b-quark jet. Events are also required to have $H_{\rm T}^{\rm miss} >$ 40 GeV, which is the missing transverse momentum defined by high-$p_{\rm T}$objects, the di-lepton azimutal opening angle $0.5 < \Delta\phi_{ll} < 2.5$ and $0.1 < x_1,x_2 < 1.0$, where x_1 and x_2 denote the visible momentum fraction of the tau leptons carried by the leading and sub-leading light leptons, respectively. For the event selection the momentum of the tau lepton is estimated using the collinear approximation [22]. Five mutually exclusive categories are then defined, depending on the jet multiplicity and kinematics: 2-jet VBF, Boosted, 2-jet VH, 1-jet and 0-jet.

$H \to \tau_{\rm lep}\tau_{\rm had}$: To suppress W+jets and non-resonant backgrounds, the transverse mass, $m_{\rm T}$, of lepton and $E_{\rm T}^{\rm miss}$ system, the $\Sigma\Delta\phi$ variable, defined as $\Delta\phi_{l,E_{\rm T}^{\rm miss}}+\Delta\phi_{\tau,E_{\rm T}^{\rm miss}}$ and the $\Delta(\Delta R)$ variable, defined as the difference between the angular separation $\Delta R_{l\tau_{\rm had-vis}}$ and its expectation value for the Higgs decay, are used. Six exclusive categories are defined for this channel: VBF, Boosted, 1-jet ($e\tau_{\rm had-vis}$, $\mu\tau_{\rm had-vis}$) and 0-jet ($e\tau_{\rm had-vis}$, $\mu\tau_{\rm had-vis}$).

$H \to \tau_{\rm had}\tau_{\rm had}$: Two $\tau_{\rm had-vis}$ candidates are required with $0.8 < \Delta R < 2.8$ and $\Delta\eta < 1.5$. Two exclusive categories are defined for this channel: VBF and Boosted.

3.2. Background Estimation

The background composition and normalisation have been determined using both data-driven and simulation based methods. The main background for this search is due to the largely irreducible $Z/\gamma^* \to \tau\tau$. Other background processes are multijet, W+jets, Z+jets, top (both $t\bar{t}$ and single top) and diboson production (WW, ZZ, WZ). Since it is not possible to select a signal-free $Z/\gamma^* \to \tau\tau$ sample directly from data, this background has been modeled in a partly data-drive way by selecting a control sample of $Z/\gamma^* \to \mu\mu$ data, where the expected signal contamination is negligible. In this sample the muon tracks and associated calorimeter cells are replaced by simulated tau leptons (embedding method). In this way only the tau decays and the corresponding detector response are taken from the simulation, while jets, underlying event and pile-up effects are obtained from data, thus resulting in a very good agreement between the MMC mass distribution obtained with the embedded method and the one obtained with the simulation [20].

The contributions of multijet and W+jets backgrounds are estimated using data-driven methods, either using template fitting of kinematics distributions or derived from a control region defined by inverting the opposite charge requirement. For the top background the normalisation is obtained from control regions, defined by inverting the $m_{\rm T}$ cut and requiring at least one b-tagged jet, whereas kinematic shapes are obtained by simulation. A similar approach, based on selected control regions, is used for the Z+jets and diboson backgrounds.

Systematic uncertainties are determined on all experimental and theoretical sources. The dominant systematic uncertainties are those on the theoretical predictions for the signal production cross-sections and decay branching ratios (3-28 %), on tau energy scale (2-15%), on the tau identification (1-10%), as well as those on the jet energy scale (1-9%) [20] and on the measurement of integrated luminosity (3.6-3.9%).

3.3. Results

The most sensitive categories are the Boosted and VBF categories for all three channels. Figure 3 and Figure 4 show distributions for the final discriminant variables MMC $m_{\tau\tau}$ of the selected events. No significant excess is observed in the data compared to the SM background-only expectation in any of the channels. The statistical analysis of the data employs a binned likelihod function constructed as a product of the likelihood terms for each category. The likelihood in each category is a product over bins in the distributions of $m_{\tau\tau}$.

Figure 5 (left) shows the expected and observed cross section limits for a combination of all three channels as a function of the Higgs boson mass at the 95% CL. The combined expected limit varies between 1.2 and 3.4 times the SM cross section times branching ratio for the mass range 100-150 GeV. The corresponding observed limits are in the range between 1.9 and 3.3 times the SM prediction. For m_H = 125 GeV specifically, the expected limit is 1.2 and the observed 1.9. The best fit value for the signal strenght is $\mu = 0.7 \pm 0.7$. The combination of both datasets yields an observed (expected) p_0 value of 0.135 (0.047) for m_H = 125 GeV. The contributions from the three different production modes in the $H \to \tau^+\tau^-$ channel have

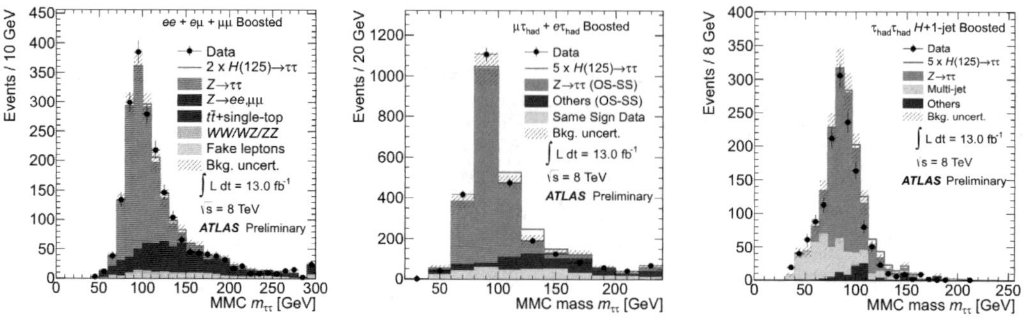

Figure 3. Reconstructed $m_{\tau\tau}$ of the selected events in the 8 TeV sample for the Boosted category: (left)$\tau_{\text{lep}}\tau_{\text{lep}}$, (middle) $\tau_{\text{lep}}\tau_{\text{had}}$, (right) $\tau_{\text{had}}\tau_{\text{had}}$ [20].

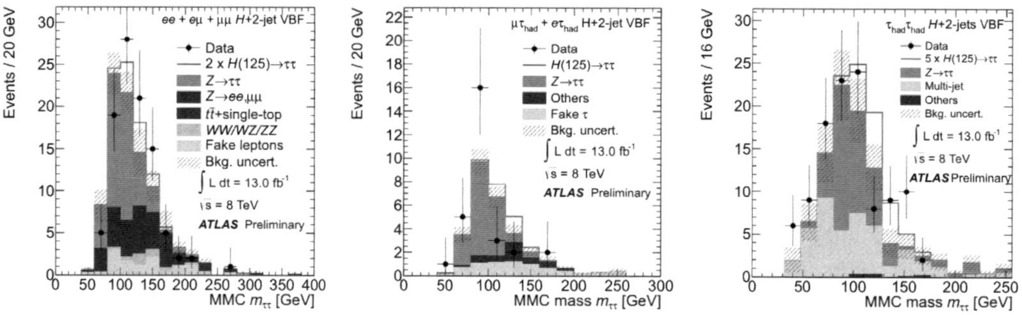

Figure 4. Reconstructed $m_{\tau\tau}$ of the selected events in the 8 TeV sample for the VBF category: (left)$\tau_{\text{lep}}\tau_{\text{lep}}$, (middle) $\tau_{\text{lep}}\tau_{\text{had}}$, (right) $\tau_{\text{had}}\tau_{\text{had}}$ [20].

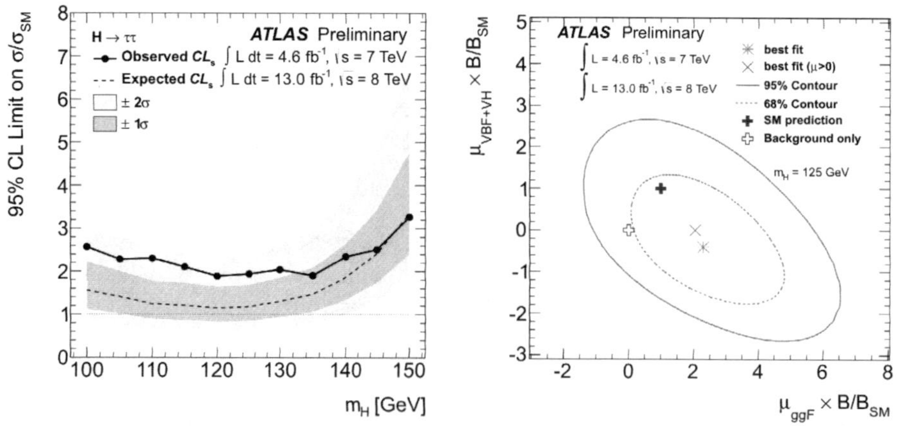

Figure 5. Left plot: expected (dashed) and observed (solid) 95% CL on the normalised signal strenght as a function of $m_H$for the combination of the $\sqrt{s} = 7$ TeV and $\sqrt{s} = 8$ TeV dataset. Right plot: likelihood contours for $H \to \tau^+\tau^-$ in the $\mu_{\text{ggF}} \times B/B_{\text{SM}}$, $\mu_{\text{VBF+VH}} \times B/B_{\text{SM}}$ plane are shown for the 68% and 95% CL by dashed and solid lines, respectively. The best fit to the data is shown for the case when both parameters have been constrained to be non-negative and for the unconstrained case, as explained in the legend of the plot [20].

been studied in order to assess any tension between data and the signal production cross section as predicted by the SM. The VBF and VH production modes have been grouped together as they both scale with the WWH or ZZH coupling in the SM, denoted as $\mu_{\mathrm{VBF+VH}}$ and the strenght parameter for the remaining process is denoted as μ_{ggF}. Two-dimensional contours in the $\mu_{\mathrm{VBF+VH}} \times B/B_{\mathrm{SM}}$ and $\mu_{\mathrm{ggF}} \times B/B_{\mathrm{SM}}$ plane are shown in Figure 5 (right), where B and B_{SM} are the measured and SM branching ratio for $H \to \tau^+\tau^-$. The best fit values are $\mu_{\mathrm{VBF+VH}} \times B/B_{\mathrm{SM}}$=-0.4 and $\mu_{\mathrm{ggF}} \times B/B_{\mathrm{SM}} = 2.4$ and the result is consistent with both the SM Higgs and SM background-only hypothesis within the 95% CL.

3.4. Conclusions

A search for the Standard Model Higgs boson decaying in the $H \to b\bar{b}$ and $H \to \tau^+\tau^-$ channels has been performed by the ATLAS experiment at the LHC using 4.7fb^{-1}of pp collision data recorded at $\sqrt{s} = 7$ TeV and 13.0 fb^{-1}recorded at $\sqrt{s} = 8$ TeV. No significant excess is observed. For $m_H = 125$ GeV the observed (expected) upper limit at 95% CL on the cross section times the branching ratio is found to be 1.8 (1.9) times the Standard Model prediction for $H \to b\bar{b}$ and 1.9 (1.2) for $H \to \tau^+\tau^-$ for the combination of the $\sqrt{s} = 7$ TeV and $\sqrt{s} = 8$ TeV datasets.

References

[1] L. Evans and P. Bryant (Editors), JINST **3** (2008) S08001
[2] ATLAS Collaboration, Phys. Lett. B **716** (2012) 1
[3] CMS Collaboration, Phys. Lett. B **716** (2012) 30
[4] ATLAS Collaboration, Phys. Lett. B **718** (2012) 369
[5] ATLAS Collaboration, JHEP **09** (2012) 070
[6] ATLAS Collaboration, JINST **3** (2008) S08003
[7] ATLAS Collaboration, ATLAS-CONF-2012-161 (2012), https://cds.cern.ch/record/1493625
[8] ATLAS Collaboration, Eur. Phys. J. **C70**, 823 (2010)
[9] S. Agostinelli et al.(GEANT4), Nucl. Instr. and Meth. **A506** (2003) 250
[10] S. Dittmaier et al., LHC Higgs Cross Section Working Group, CERN-2011-002 (CERN, Geneva, 2011)
[11] LHC Higgs Cross Section Working Group, https://twiki.cern.ch/twiki/bin/view/LHCPhysics/CrossSections
[12] S. Dittmaier et al., LHC Higgs Cross Section Working Group, CERN-2012-002 (CERN, Geneva, 2012)
[13] G. Ferrera, M. Grazzini and F. Tramontano, Phys. Rev. Lett. **107** (2011) 152003
[14] A. Djouadi, J. Kalinowski and M. Spira, Comp. Phys. Commun. **108** (1998) 56
[15] M. Cacciari, G.P. Salam and G. Soyez. JHEP **04** (2008) 063
[16] ATLAS Collaboration, ATLAS-CONF-2012-040 (2012), https://cds.cern.ch/record/1435194
[17] ATLAS Collaboration, ATLAS-CONF-2012-043 (2012), https://cds.cern.ch/record/1435197
[18] A. L. Read, J. Phys. G **28** (2002) 2693
[19] G. Cowan, K. Cranmer, E. Gross and O. Vitells, Eur. Phys. J. C **71** (2011) 1554
[20] ATLAS Collaboration, ATLAS-CONF-2012-160 (2012), https://cds.cern.ch/record/1493624
[21] A. Elagin, P. Murat, A. Pranko and A. Safonov, Nucl. Instr. and Meth. **A654** (2011) 481
[22] R. K. Ellis, I. Hinchliffe, M. Soldate and J. J. Van der Bij, Nucl. Phys. B **297** (1988) 221

International Workshop on Discovery Physics at the LHC (Kruger2012) IOP Publishing
Journal of Physics: Conference Series **455** (2013) 012016 doi:10.1088/1742-6596/455/1/012016

Searches for neutral and charged Higgs bosons in the ATLAS experiment as a test of physics beyond the Standard Model

Arnaud Ferrari, on behalf on the ATLAS collaboration
Department of Physics and Astronomy, Uppsala University, Sweden
E-mail: arnaud.ferrari@physics.uu.se

Abstract. In several theories beyond the Standard Model, the Higgs sector consists of more than one complex scalar doublet. For instance, supersymmetric models such as the MSSM have five physical Higgs states: two charged and three neutral. In this paper, a review of the searches for new neutral and charged Higgs bosons at the Large Hadron Collider, with the ATLAS experiment, is presented.

1. Introduction

In the Standard Model (SM), only one doublet of Higgs complex scalars is responsible for the electroweak symmetry breaking. Since three degrees of freedom are used in order to give mass to the Z, W^+ and W^- gauge bosons, there can be only one Higgs physical state, h^0. Recently, a new neutral boson with Higgs-like properties has been observed by both the ATLAS and CMS experiments at the Large Hadron Collider (LHC), with a mass of 125–126 GeV [1, 2]. The remaining mass range is meanwhile excluded with a high confidence level. Detailed investigations are now conducted in order to determine whether this new particle is the SM Higgs boson, or *a* Higgs boson. Indeed, additional Higgs bosons are predicted by several non-minimal Higgs scenarios, such as Two Higgs Doublet Models (2HDM) [3]. In a type-II 2HDM, such as the Higgs sector of the Minimal Supersymmetric extension of the Standard Model (MSSM), five physical Higgs states are predicted [4]: the neutral CP-even h^0 and H^0, the neutral CP-odd A^0, and the charged $H^\pm$ Higgs bosons. At tree level, the MSSM Higgs sector is fully defined by the mass of one of the Higgs bosons and $\tan\beta$, the ratio of the two Higgs vacuum expectation values. In this paper, the searches for neutral and charged Higgs bosons using the LHC proton-proton collision data recorded by the ATLAS [5] experiment in 2011 are reviewed.

2. The ATLAS experiment

The ATLAS detector consists of an inner tracker with coverage in pseudorapidity up to $|\eta| = 2.5$, surrounded by a thin 2 T superconducting solenoid, a calorimeter system up to $|\eta| = 4.9$ for the detection of electrons, photons and hadronic jets, as well as a large muon spectrometer extending up to $|\eta| = 2.7$, which measures the deflection of muon tracks in the field of three superconducting toroid magnets. A three-level trigger system is used, which reduces the recorded event rate to about 300 Hz.

Electrons are reconstructed by matching energy deposits in the electromagnetic calorimeter (except in the pseudorapidity range $1.37 < |\eta| < 1.52$) to tracks reconstructed in the inner detector [6]. Muons are required to contain matching inner detector and muon spectrometer tracks [7]. The combination of all sub-systems provides precise lepton measurements within $|\eta| < 2.5$. The jets, as well as the magnitude of the missing transverse momentum $E_{\mathrm{T}}^{\mathrm{miss}}$, are reconstructed using energy deposits over the full coverage of the calorimeters, i.e. up to $|\eta| = 4.9$. The anti-k_t algorithm [8] with a radius parameter value of $R = 0.4$ is used for the jet reconstruction, taking three-dimensional noise-suppressed clusters of calorimeter-cell energy deposits as input [9]. In order to select jets originating from the hard-scatter interaction, at least 75% of the tracks associated to a jet (weighted by their transverse momenta p_{T}) must point to the primary vertex. A multivariate algorithm combining information from the impact parameter of associated tracks and the reconstruction of b- and c-hadron decay vertices is used in order to identify jets initiated by b-quarks, within $|\eta| < 2.5$. The b-jet identification has an efficiency of about 70% in $t\bar{t}$ events. In order to reconstruct hadronically decaying τ leptons, anti-k_t jets with either one or three associated tracks are considered as τ candidates, however, with a different calibration than hadronic jets. The τ candidates are required to have a visible $p_{\mathrm{T}}^{\tau} > 20$ GeV and to be within $|\eta| < 2.5$. Dedicated algorithms are used to reject electrons and muons. The hadronic τ decays are identified using a likelihood criterion or a boosted-decision-tree (BDT) discriminant, with "loose", "medium" and "tight" selections depending on the efficiency for hadronically decaying τ leptons and on the rejection factor against quark- and gluon-initiated jets [11]. Selected τ candidates fulfilling the identification criteria are referred to as "τ jets".

3. Search for new neutral Higgs bosons in ATLAS

In the MSSM, the couplings of neutral Higgs bosons to τ leptons and b-quarks are strongly enhanced for a large part of the parameter space, especially at large $\tan\beta$. The most common production mechanism for MSSM neutral Higgs bosons at the LHC are the b-quark associated production and the gg fusion (primarily through a b-quark loop), with cross sections increasing with $\tan\beta$. In this paper, searches for new neutral Higgs bosons in the $A^0/H^0/h^0 \to \tau\tau$ mode using 4.7 fb^{-1} of pp collisions at $\sqrt{s} = 7$ TeV are presented. This review is divided into separate search channels, according to the τ lepton decay modes. More details can be found in Ref. [12]. The direct decay into a $\mu\mu$ pair occurs rarely (its branching ratio is 0.04%), however it offers a clean signature. It was also considered in Ref. [12], but is not detailed in the present review.

3.1. Common aspects

Missing Mass Calculator: The full reconstruction of the $A^0/H^0/h^0 \to \tau\tau$ events requires solving equations with more unknowns than constraints. Hence, a technique known as the Missing Mass Calculator (MMC) is used to reconstruct the neutral Higgs boson candidate mass [13]. Assuming that the missing transverse momentum is entirely due to the neutrinos, the algorithm scans over the angles between the neutrinos and the visible τ lepton decay products. At each point, the $\tau\tau$ invariant mass is calculated, and the most likely value is chosen by weighting each solution according to probability density functions derived from simulated τ lepton decays. This method finds a solution in 99% of the cases, with a 13–20% mass resolution.

τ-embedding for the $Z/\gamma^* \to \tau\tau$ background estimation from data: $Z/\gamma^* \to \tau\tau$ events are a largely irreducible background in the search for neutral Higgs bosons $A^0/H^0/h^0 \to \tau\tau$, in all final states. In order to estimate this background from data, the τ-embedding technique is used. It consists in selecting $Z/\gamma^* \to \mu\mu$ events in data and in replacing the detected muon signature (muon track hits and associated calorimeter cells) by the response of a simulated τ lepton decay. After selection of the τ decay products, the embedded sample is normalised to the event yield obtained with simulated $Z/\gamma^* \to \tau\tau$ events.

3.2. $A^0/H^0/h^0 \to \tau\tau \to e\mu$

Events in this channel must fire a single-lepton or electron+muon trigger and they must contain exactly one electron and one muon, with $E_{\mathrm{T}}^{e} > 24$ GeV and $p_{\mathrm{T}}^{\mu} > 20$ GeV for events selected by a single-lepton trigger, or $E_{\mathrm{T}}^{e} > 10$ GeV and $p_{\mathrm{T}}^{\mu} > 6$ GeV for events selected by the electron+muon trigger. The invariant mass of the electron and the muon must satisfy $m_{e\mu} > 30$ GeV. The selected sample is then split according to the flavor content:

- b-tagged: exactly one b-jet, $H_{\mathrm{T}} = \sum_j E_{\mathrm{T}}(j) < 100$ GeV (using only jets with $E_{\mathrm{T}} > 20$ GeV and $|\eta| < 4.5$), $E_{\mathrm{T}}^{\mathrm{miss}} + E_{\mathrm{T}}^{e} + p_{\mathrm{T}}^{\mu} < 125$ GeV, $\Delta\phi_{e\mu} > 2.0$, $\sum_{e,\mu} \cos\Delta\phi_{\ell,E_{\mathrm{T}}^{\mathrm{miss}}} > -0.2$;
- b-vetoed: no b-jet, $E_{\mathrm{T}}^{\mathrm{miss}} + E_{\mathrm{T}}^{e} + p_{\mathrm{T}}^{\mu} < 150$ GeV, $\Delta\phi_{e\mu} > 1.6$, $\sum_{e,\mu} \cos\Delta\phi_{\ell,E_{\mathrm{T}}^{\mathrm{miss}}} > -0.4$,

where ϕ stands for the azimuthal angle. The selections above allow to reduce the top quark, diboson and $W \to \ell\nu$ backgrounds, as well as processes with a higher jet multiplicity or containing jets with higher transverse energies than the signal. The irreducible $Z/\gamma^* \to \tau\tau$ background is estimated with the τ-embedding technique. The contribution from $t\bar{t}$ events is extrapolated from control regions, which have the same selection criteria as for the signal, except that two b-jets are required and $H_{\mathrm{T}} < 100$ GeV is not applied. Finally, the multi-jet background is determined with a so-called ABCD method by splitting the event sample into four regions according to the charge product of the $e\mu$ pair and the isolation requirements on the electron and muon. The signal region is labelled A, and the other regions (B, C, D) are dominated by the background from multi-jet processes. One variable separates A and B from C and D. After correcting the event yields n_B, n_C and n_D in, respectively, the regions B, C and D by subtracting (simulated) processes other than the multi-jet background, an estimate of the multi-jet background in the signal region, n_A, is:

$$n_A = n_B \times (n_C/n_D).$$

The observed event yields in the 2011 data, as well as the MMC distributions of the selected events, are compatible with the SM predictions within uncertainties, see Figure 1.

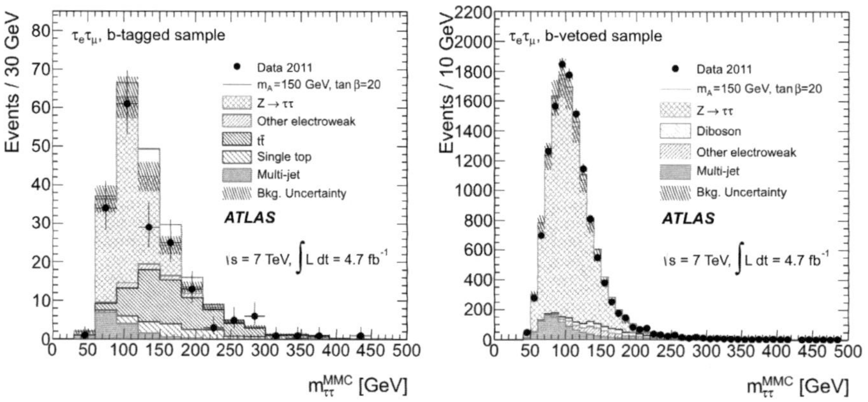

Figure 1. MMC mass distributions for the $A^0/H^0/h^0 \to \tau\tau \to e\mu$ final state, shown for the b-tagged (left) and b-vetoed (right) samples, extracted from Ref. [12]. The data are compared to the SM background expectation and an added hypothetical MSSM signal ($m_A = 150$ GeV and $\tan\beta = 20$). The contributions of $Z/\gamma^* \to ee, \mu\mu$ and W+jets are labelled as "Other electroweak". For the b-tagged sample, the diboson background is included as well. Background contributions from top quarks are included in "Other electroweak" for the b-vetoed sample.

3.3. $A^0/H^0/h^0 \to \tau\tau \to \ell\tau_{\mathrm{had}}$ *with* $\ell = e, \mu$

Events in this channel must fire a single-lepton trigger and they must contain one isolated electron with $E_{\mathrm{T}}^{e} > 25$ GeV or one isolated muon with $p_{\mathrm{T}}^{\mu} > 20$ GeV. Events containing additional electrons or muons with transverse energies or momenta greater than 15 GeV or 10 GeV, respectively, are rejected. One τ jet, with $p_{\mathrm{T}}^{\tau} > 20$ GeV, $|\eta| < 2.5$, a medium BDT identification and a charge of opposite sign to the selected electron or muon is required. A cut on the transverse mass of the lepton ℓ and the missing transverse momentum is then applied to reduce the W+jets and $t\bar{t}$ backgrounds:

$$m_{\mathrm{T}} = \sqrt{2p_{\mathrm{T}}^{\ell}E_{\mathrm{T}}^{\mathrm{miss}}(1 - \cos\Delta\phi_{\ell,E_{\mathrm{T}}^{\mathrm{miss}}})} < 30 \text{ GeV}.$$

At this stage, events are included in the b-tagged sample if the leading jet within $|\eta| < 2.5$ is identified as a b-jet and if its E_{T} is in the range of 20–50 GeV. Events are included in the b-vetoed sample if the leading jet fails the b-jet identification criterion and the event has $E_{\mathrm{T}}^{\mathrm{miss}} > 20$ GeV. After the event selection up to the m_{T} requirement, the W+jets background consists primarily of events in which a jet is misidentified as a τ jet. The jet $\to \tau_{\mathrm{had}}$ misidentification rates are corrected in simulated W+jets events, separately for $e\tau_{\mathrm{had}}$ and $\mu\tau_{\mathrm{had}}$, using control regions enriched with W+jets events in data, selected by requiring 70 GeV $< m_{\mathrm{T}} <$ 110 GeV. The $Z/\gamma^* \to \tau\tau$ background is estimated with the τ-embedding technique. The contribution from $t\bar{t}$ events is extrapolated from control regions, which have the same selection criteria as for the signal, except that there is no requirement on m_{T}, the leading jet must be b-tagged, with E_{T} in the range 50–150 GeV, and a second highest-E_{T} jet must satisfy the same b-jet identification requirement. Finally, the ABCD method is used in order to estimate the multi-jet background. The four regions are defined according to whether the charges of the τ jet and the lepton have opposite or same sign, and whether the selected lepton passes or fails isolation criteria.
The observed event yields in the 2011 data, as well as the MMC distributions of the selected events, are compatible with the SM predictions within uncertainties, see Figure 2.

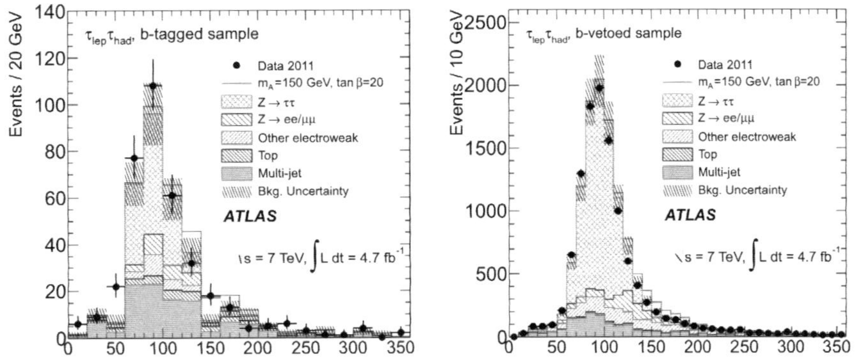

Figure 2. MMC mass distributions for the $A^0/H^0/h^0 \to \tau\tau \to \ell\tau_{\mathrm{had}}$ final state, shown for the b-tagged (left) and b-vetoed (right) samples, extracted from Ref. [12]. The data are compared to the SM background expectation and an added hypothetical MSSM signal ($m_A = 150$ GeV and $\tan\beta = 20$). The diboson and W+jets backgrounds are combined and labelled as "Other electroweak".

3.4. $A^0/H^0/h^0 \to \tau\tau \to \tau_{\text{had}}\tau_{\text{had}}$

Events in this channel must fire a di-τ_{had} trigger and contain two τ_{had} candidates matched to the trigger objects (one passing the tight BDT identification requirement and the second one passing the medium criteria), with opposite-sign charges and with transverse momenta larger than 45 GeV and 30 GeV. Corrections for the τ_{had} trigger and identification efficiencies, as well as for the jet $\to \tau_{\text{had}}$ misidentification rate, are derived from data and applied in the simulation. Events containing identified electrons or muons with $E_T^e > 15$ GeV or $p_T^\mu > 10$ GeV, respectively, are vetoed. The missing transverse momentum is required to be larger than 25 GeV to account for the presence of neutrinos originating from the τ decays and to suppress the multi-jet background. At this stage, events are included in the b-tagged sample if the leading jet within $|\eta| < 2.5$ is identified as a b-jet and if its E_T is in the range of 20–50 GeV. Events are included in the b-vetoed sample if no jet is found or if the leading jet fails the b-jet identification criterion. Because of higher background levels in this sample, the leading τ_{had} must have $p_T^\tau > 60$ GeV. Since the di-τ_{had} trigger cannot be modelled in embedded $Z/\gamma^* \to \tau\tau$ events, this background is estimated with simulation, and validated using a τ-embedded $Z/\gamma^* \to \mu\mu$ sample. Similarly, simulation is used to estimate the $W \to \tau\nu$ background and a τ-embedded $W \to \mu\nu$ sample from data is used for validation purposes. For both backgrounds, correction factors for the efficiency of the b-jet identification requirement on the leading jet are applied. The ABCD method is used in order to estimate the multi-jet background. The four regions are defined according to the charge product of the two leading τ_{had} candidates, and whether their nominal identification requirements are met.

The observed event yields in the 2011 data, as well as the MMC distributions of the selected events, are compatible with the SM predictions within uncertainties, see Figure 3.

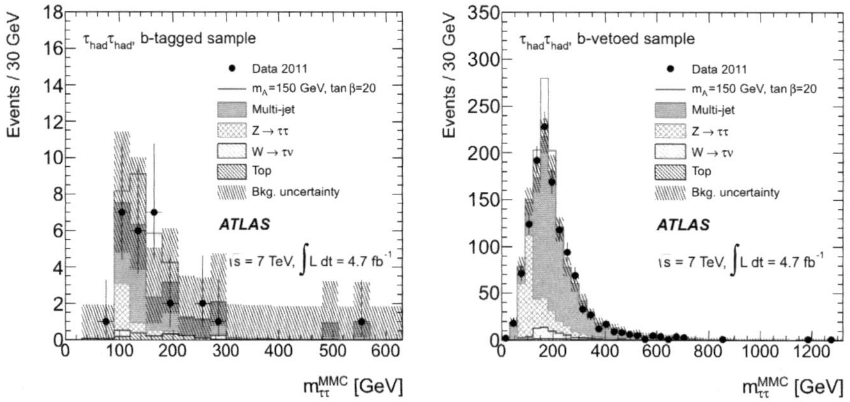

Figure 3. MMC mass distributions for the $A^0/H^0/h^0 \to \tau\tau \to \tau_{\text{had}}\tau_{\text{had}}$ final state, shown for the b-tagged (left) and b-vetoed (right) samples, extracted from Ref. [12]. The data are compared to the SM background expectation and an added hypothetical MSSM signal ($m_A = 150$ GeV and $\tan\beta = 20$).

3.5. Exclusion limits

The statistical analysis of the data employs a binned likelihood function. Each one of the $e\mu$, $e\tau_{\text{had}}$, $\mu\tau_{\text{had}}$ and $\tau_{\text{had}}\tau_{\text{had}}$ final states is split into b-tagged and b-vetoed samples. In each category, the likelihood is a product over bins of the MMC distributions in the signal and control regions. Signal and background predictions depend on systematic uncertainties that are listed in Ref. [12] and parameterised by nuisance parameters. Correlations of the systematic uncertainties across categories are taken into account. Since no significant excess of events above the background-only expectation is observed in the considered channels, using 4.7 fb^{-1} of 7 TeV LHC data, 95% confidence level (CL) limits are set on the production cross section for a single scalar boson ϕ, produced in either the gluon fusion or b-associated production mode, times the branching ratio into $\tau\tau$ pairs, see the left-hand plot of Figure 4. In the $m_h^{\max}$ scenario with $\mu > 0$ [14], 95% CL upper limits can be placed on $\tan\beta$ for each m_A point, see the right-hand plot of Figure 4. In all plots, limits are also shown for $A^0/H^0/h^0 \to \mu\mu$.

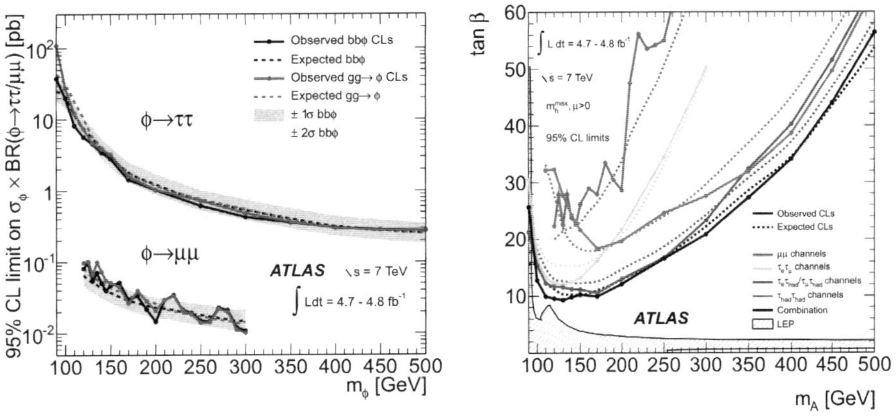

Figure 4. Left: expected (dashed line) and observed (solid line) 95% CL limits on the cross section for gluon fusion and b-associated Higgs boson production times the branching ratio into $\tau\tau$ and $\mu\mu$, along with the 1σ (green) and 2σ (yellow) bands for the expected limit. Right: expected (dashed lines) and observed (solid lines) 95% CL limits on $\tan\beta$ as a function of m_A for each of the $\mu\mu$, $e\mu$, $e\tau_{\rm had}$, $\mu\tau_{\rm had}$ and $\tau_{\rm had}\tau_{\rm had}$ channels and their statistical combination (the 95% CL exclusion region from neutral MSSM Higgs boson searches performed at LEP [15] is shown in a hatched style). Both figures are extracted from Ref. [12].

4. Search for light charged Higgs bosons in ATLAS

The main production mode for charged Higgs bosons (referred to as H^+ in the following) at the LHC is through top quark decays $t \to bH^+$, for charged Higgs boson masses smaller than the top quark mass ($m_{H^+} < m_{\rm top}$). The dominant source of top quarks at the LHC is through $t\bar{t}$ production. For $\tan\beta > 2$, the charged Higgs boson decay via $H^+ \to \tau\nu$ is dominant. At lower $\tan\beta$ values, the decay into a $c\bar{s}$ pair also becomes sizeable. In this paper, searches for charged Higgs bosons using 4.6 fb^{-1} of pp collisions at $\sqrt{s} = 7$ TeV are presented. With the assumption that the branching ratio Br($H^+ \to \tau\nu$) is 100%, H^+ is searched for in $t\bar{t}$ events with a τ lepton in the final state. More details can be found in Refs. [16] and [17]. On the other hand, if the branching ratio Br($H^+ \to c\bar{s}$) is assumed to be 100%, H^+ is searched for in semi-leptonic decays of $t\bar{t}$ pairs [18].

4.1. Search for charged Higgs bosons in the $\tau_{\rm had}$+jets final state

Using 4.6 fb^{-1} of pp collision data at $\sqrt{s} = 7$ TeV, three final states were analysed by the ATLAS collaboration in Ref. [16], in order to search for charged Higgs bosons in the $H^+ \to \tau\nu$ decay mode. The most sensitive channel, $t\bar{t} \to b\bar{b}WH^+ \to b\bar{b}(qq')(\tau_{\rm had}\nu)$, is reviewed here. This analysis uses events which pass a $\tau + E_{\rm T}^{\rm miss}$ trigger and which contain four or more jets with $p_{\rm T} > 20$ GeV, of which at least one is b-tagged, as well as one τ jet with $p_{\rm T}^{\tau} > 40$ GeV and

$|\eta| < 2.3$ (the tight likelihood criterion is used for τ identification), matched to the corresponding trigger object. Events with identified electrons and muons are vetoed. With $p_{\mathrm{T}}^{\mathrm{PVtrk}}$ being the transverse momentum of a track originating from the primary vertex, the following requirements are then applied:

$$E_{\mathrm{T}}^{\mathrm{miss}} > 65 \text{ GeV and } \frac{E_{\mathrm{T}}^{\mathrm{miss}}}{0.5 \cdot \sqrt{\sum p_{\mathrm{T}}^{\mathrm{PVtrk}}}} > 13 \text{ GeV}^{1/2}.$$

Finally, in order to enrich the sample with $t\bar{t}$ events, the jjb candidate with the highest p_{T}^{jjb} value is requested to satisfy $m(jjb) \in [120, 240]$ GeV. The transverse mass of the τ jet and the missing transverse momentum is used as a discriminating variable:

$$m_{\mathrm{T}} = \sqrt{2p_{\mathrm{T}}^{\tau} E_{\mathrm{T}}^{\mathrm{miss}}(1 - \cos\Delta\phi_{\tau,E_{\mathrm{T}}^{\mathrm{miss}}})}.$$

In order to estimate the contribution of the multi-jet background in terms of event yield, a fit of the $E_{\mathrm{T}}^{\mathrm{miss}}$ distribution in the signal region is performed, using a multi-jet $E_{\mathrm{T}}^{\mathrm{miss}}$ template from a control region in data and simulated events for all other processes. The m_{T} shape for the multi-jet background is also measured in data, in the same control region as the $E_{\mathrm{T}}^{\mathrm{miss}}$ template, and is then normalised using the result of the $E_{\mathrm{T}}^{\mathrm{miss}}$ fit. The electroweak backgrounds with misidentified τ jets are first estimated using simulations and later adjusted using data-driven correction factors. In the case of an electron misidentified as τ_{had}, scale factors are derived with a tag-and-probe method in $Z \to ee$ events from data, and they are then applied to all simulated events. Jet $\to \tau_{\mathrm{had}}$ misidentification rates are measured in a data sample enriched with W+jets events, and then applied in the simulation. Finally, the electroweak backgrounds with true τ jets is estimated using a τ-embedded sample of $t\bar{t}$-like μ+jets events from the data, which is then normalised using simulation.
The m_{T} distribution for the τ_{had}+jets channel, after all selection cuts are applied, is shown in Figure 5. The data are found to be consistent with the SM background expectation.

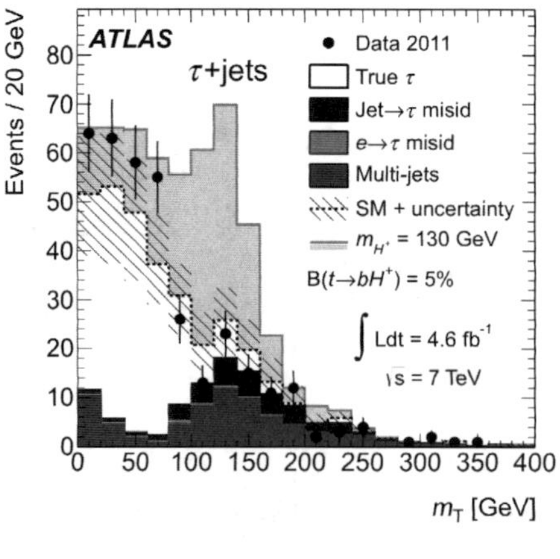

Figure 5. Distribution of m_{T} after all selection cuts in the τ_{had}+jets channel, extracted from Ref. [16]. The dashed line corresponds to the SM-only hypothesis and the hatched area around it shows the total uncertainty. The solid line is the predicted contribution of signal+background with a charged Higgs boson with $m_{H^+} = 130$ GeV, assuming that $\mathrm{Br}(t \to bH^+) = 5\%$ and $\mathrm{Br}(H^+ \to \tau\nu) = 100\%$. The contributions of $t\bar{t} \to b\bar{b}W^+W^-$ events in the backgrounds with misidentified objects are scaled down accordingly.

Journal of Physics: Conference Series **455** (2013) 012016 doi:10.1088/1742-6596/455/1/012016

4.2. Search for charged Higgs bosons via the violation of lepton universality in $t\bar{t}$ events

While W bosons decay equally to all lepton families, H^+ mostly decays to τ leptons, which would lead to an excess of $t\bar{t}$ events in $\ell + \tau_{\text{had}}$ final states ($\ell = e, \mu$). In order to cancel several systematic uncertainties in the analysis presented here, and detailed in Ref. [17], event yield ratios are computed between $\ell + \tau_{\text{had}}$ and dilepton ($e\mu$) final states, again using 4.6 fb^{-1} of pp collision data at $\sqrt{s} = 7$ TeV. Events in this channel must fire a single-lepton trigger and contain one isolated lepton with $E_{\text{T}}^{e} > 25$ GeV or $p_{\text{T}}^{\mu} > 25$ GeV, matched to the corresponding trigger object. In addition, at least two jets with $p_{\text{T}} > 20$ GeV and $|\eta| < 2.4$ must be found, with two b-tags. Then, events are split into two categories ($\ell + \tau_{\text{had}}$ and $\ell\ell'$), depending on whether they contain either exactly one τ jet with $p_{\text{T}} > 25$ GeV and $|\eta| < 2.3$, as well as no other lepton, or exactly one other lepton with E_{T} or p_{T} above 25 GeV and a different flavor than the trigger-lepton. Finally, all events are required to have $E_{\text{T}}^{\text{miss}} > 40$ GeV.

The backgrounds with misidentified leptons are estimated using a data-driven method based on events containing tight or loose leptons (those differ only in lepton identification criteria). The efficiencies p_r and p_m for a real or misidentified lepton, respectively, to be detected as a tight lepton, are determined from data. Based on these efficiencies, the number of misidentified leptons passing the final requirements can be calculated by weighting events in the data sample with one loose lepton, according to the following per-lepton weights: for a loose but not tight lepton $w_{\ell\text{L}} = p_m p_r/(p_r - p_m)$, while for a tight lepton $w_{\ell\text{T}} = p_m(p_r - 1)/(p_r - p_m)$. About 51% of the simulated $t\bar{t}$ events in the $\ell + \tau_{\text{had}}$ final state contain a τ jet matched to a true hadronically decaying τ lepton. For the small background where an electron is misidentified as τ_{had}, scale factors are used, as in the analysis of the τ_{had}+jets channel. For the background where a hadronic jet is misidentified as τ_{had}, it was noticed that giving a negative weight to same-sign (SS) events cancels the gluon and b-quark jet contributions of opposite-sign (OS) events, leaving only light quark jets misidentified as τ objects. The misidentification rate of light quark jets to τ_{had} is derived from a control region in data enriched with OS-SS W + >2 jets events. It is then applied to simulated events, prior to the OS-SS subtraction. Having computed OS-SS event yields $\mathcal{N}$ in the $\ell + \tau_{\text{had}}$ and $\ell\ell'$ final states, ratios are determined as follows:

$$R_\ell = \frac{\mathcal{N}(bb + \ell\tau_{\text{had}} + E_{\text{T}}^{\text{miss}})}{\mathcal{N}(bb + \ell\ell' + E_{\text{T}}^{\text{miss}})}.$$

As shown in Table 1, the measured event yield ratios R_e and R_μ are compatible with those predicted in the SM-only hypothesis, within uncertainties.

Table 1. Predicted and measured ratios of event yields in the ℓ+τ_{had} and $\ell\ell'$ final states [17]. For the values of the ratios predicted using simulation, the statistical and systematic uncertainties are combined.

Event yield ratio	SM-prediction	Data 2011
R_e	0.105 ± 0.012	0.115 ± 0.010 (stat)
R_μ	0.166 ± 0.017	0.165 ± 0.015 (stat)

4.3. Exclusion limits for $H^+ \to \tau\nu$

In the two searches reviewed above, the data are found to be consistent with the SM prediction. Assuming Br($H^+ \to \tau\nu$) = 100%, 95% CL upper limits are placed on Br($t \to bH^+$), see the left-hand plot of Figure 6. The combined limit on the product Br($H^+ \to \tau\nu$) $\times$ Br($t \to bH^+$) is then interpreted in the m_h^{max} scenario, see the right-hand plot of Figure 6.

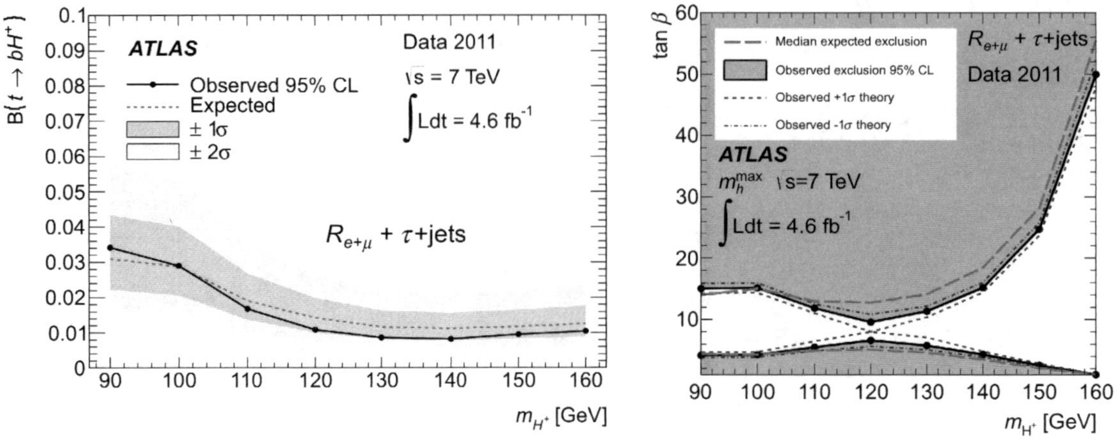

Figure 6. Left: 95% CL upper limits on Br($t \to bH^+$) derived from the transverse mass distribution of $\tau_{\rm had}$+jets events and the combined event yield ratio $R_{e+\mu}$, as a function of the charged Higgs boson mass, obtained for an integrated luminosity of 4.6 fb^{-1}, assuming Br($H^+ \to \tau\nu$) = 1. Right: limits for charged Higgs boson production from top quark decays in the m_{H^+}-$\tan\beta$ plane, in the context of the $m_h^{\rm max}$ scenario of the MSSM. The 1σ band around the observed limit (dashed lines) shows the theoretical uncertainties. Values below $\tan\beta = 1$, where the calculations in the MSSM become non-perturbative, are not considered, as the results become unphysical. Both figures are extracted from Ref. [17].

4.4. Search for charged Higgs bosons in the $c\bar{s}$ decay mode

The decay mode $H^+ \to c\bar{s}$ may have a significant branching ratio at low $\tan\beta$ values. It was searched for in ATLAS, using 4.6 fb^{-1} of pp collision data [18]. In order to select a sample enriched with $pp \to t\bar{t} \to b\bar{b}WH^+$ events, with $W \to \ell\nu$ and $H^+ \to c\bar{s}$, a single-lepton trigger is required. The events used for this anlysis must contain one trigger-matched electron or muon with $E_{\rm T}^{e} > 25$ GeV or $p_{\rm T}^{\mu} > 20$ GeV, respectively, and at least four jets with $p_{\rm T} > 25$ GeV and $|\eta| < 2.5$ (including two b-tags). After some requirements on both $E_{\rm T}^{\rm miss}$ and the transverse mass of the lepton and the missing transverse momentum, $t\bar{t}$ is the dominant background. At this stage, a kinematic fit is applied on the top quark decay products. Among all jet combinations, the one with the smallest χ^2 value is selected as the best assignment:

$$\begin{aligned}\chi^2 &= \sum_{i=\ell,4\rm jets} \frac{(p_{\rm T}^{i,\rm fit} - p_{\rm T}^{i,\rm meas})^2}{\sigma_i^2} \\ &+ \sum_{j=x,y} \frac{(p_j^{\rm SEJ,fit} - p_j^{\rm SEJ,meas})^2}{\sigma_{\rm SEJ}^2} \\ &+ \sum_{k=jjb,b\ell\nu} \frac{(M_k - m_t)^2}{\Gamma_t^2} \quad ,\end{aligned}$$

where SEJ is the vector sum of the energy of the remaining jets in the event. The fitter constrains the invariant mass of the two systems ($b\ell\nu$, bjj) to be within $\Gamma_t = 1.5$ GeV of the top quark mass 172.5 GeV. With the requirement $\chi^2_{\rm min} < 10$, the fit results in a 20–30% improvement in the resolution of the dijet mass, which is the discriminating variable for this analysis.
The left-hand plot of Figure 7 shows the dijet mass distribution in data and simulated events. The data are found to be in good agreement with the SM-only hypothesis. Hence, with the assumption Br($H^+ \to c\bar{s}$) = 100%, 95% CL upper limits are placed on Br($t \to bH^+$), see the right-hand plot of Figure 7.

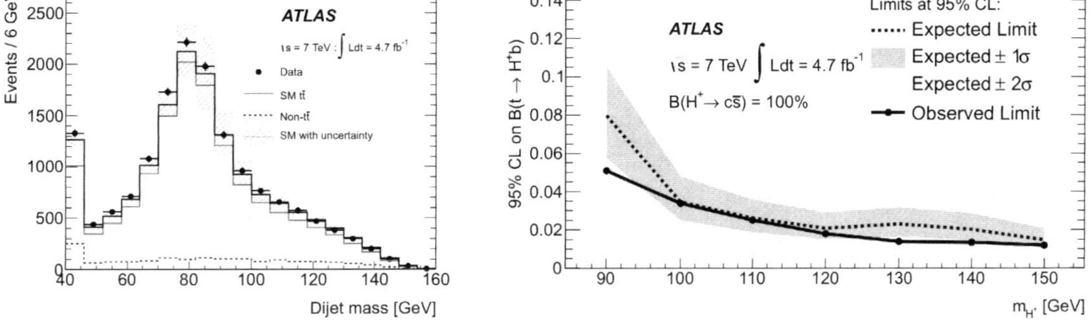

Figure 7. Left: Dijet mass distribution from data and expectation from the SM, with the corresponding uncertainties (the first and last bins contain the underflow and overflow events, respectively). Right: 95% CL upper limits on $\mathrm{Br}(t \to bH^+)$ for charged Higgs boson masses ranging from 90 GeV to 150 GeV, with the assumption $\mathrm{Br}(H^+ \to c\bar{s}) = 100\%$. Both figures are extracted from Ref. [18].

5. Conclusion

Using LHC proton-proton collision data recorded in 2011, the ATLAS collaboration has searched for new neutral and charged Higgs bosons, as a test of physics beyond the Standard Model. This paper reports on the search for $A^0/H^0/h^0 \to \tau\tau$ events, as well as $t \to bH^+$ decays of the top quark in $t\bar{t}$ events, with subsequent $H^+ \to \tau\nu$ or $H^+ \to c\bar{s}$ decays. In all analyses, the data are found to agree with the SM-only hypothesis, hence 95% CL exclusion limits are set on the production for new neutral and charged Higgs bosons.

References

[1] ATLAS Collaboration, *Phys. Lett.* B **716** (2012) 1.
[2] CMS Collaboration, *Phys. Lett.* B **716** (2012) 30.
[3] Lee T D, *Phys. Rev.* D **8** (1973) 1226.
[4] Djouadi A, *Phys. Rept.* **459** (2008) 1.
[5] ATLAS Collaboration, *JINST* **3** (2008) S08003.
[6] ATLAS Collaboration, *Eur. Phys. J.* C **72** (2012) 1909.
[7] ATLAS Collaboration, ATLAS-CONF-2011-063, available at http://cdsweb.cern.ch/record/1345743.
[8] Cacciari M, Salam G P and Soyez G, *JHEP* **0804** (2008) 063.
[9] Lampl W et al., ATL-LARG-PUB-2008-002.
[10] ATLAS Collaboration, ATLAS-CONF-2012-043, available at http://cdsweb.cern.ch/record/1435197.
[11] ATLAS Collaboration, ATLAS-CONF-2012-142, available at http://cdsweb.cern.ch/record/1485531.
[12] ATLAS Collaboration, *JHEP* **02** (2013) 095.
[13] Elagin A, Murat P, Pranko A and Safonov A, *Nucl. Instrum. Meth.* A **654** (2011) 481.
[14] Carena M, Heinemeyer S, Wagner C E M and Weiglein G, *Eur. Phys. J.* C **26** (2003) 601.
[15] ALEPH, DELPHI, L3 and OPAL Collaborations, *Eur. Phys. J.* C **47** (2006) 547.
[16] ATLAS Collaboration, *JHEP* **1206** (2012) 039.
[17] ATLAS Collaboration, arXiv:1212.3572.
[18] ATLAS Collaboration, arXiv:1302.3694.

Searches for Natural Supersymmetry with the ATLAS Detector

Mark Hodgkinson on behalf of the ATLAS Collaboration
University of Sheffield
E-mail: m.hodgkinson@sheffield.ac.uk

Abstract. The ATLAS collaboration has performed searches for signals of natural Suypersymmetry. These signals include the production of charginos and neutralinos, third generation squarks and sleptons and cover an extensive list of scenarios. The direct production of Supersymmetric particles and the indirect production via a gluino decay has been investigated. No excess has been observed in any scenario and limits are placed on the masses of squarks and gluinos in a number of Supersymmetry based models using both 8 TeV and 7 TeV proton-proton collision data recorded with the ATLAS detector at the LHC.

1. Introduction

A typical natural Supersymmetry (SUSY) spectrum contains light third generation squarks (stops and sbottoms), charginos and neutralinos. It is motivated by the fact that the dominant radiative corrections to the Higgs boson mass arise from the top quark in the Standard Model (SM). Thus a relatively light stop can cancel out these corrections. Furthermore the SM Z boson mass, the neutralino and chargino masses are related to the μ parameter in SUSY. Therefore one can argue it is possible the gaugino masses will be as low as the electroweak scale. The ATLAS collaboration has performed searches for stops, sbottoms, charginos, neutralinos and sleptons using 7 TeV and 8 TeV proton-proton collision data recorded at the LHC in the ATLAS detector [1]. Stops and sbottoms are searched for with analyses sensitive to both direct production and indirect production via a gluino decay. Charginos, neutralinos and sleptons are searched for with analyses sensitive to direct production of these sparticles.

2. Search for gluino mediated stop or sbottom production with 3 b-tagged jets and Missing Transverse Energy in 8 TeV data

This analysis [2] is sensitive to pair production of gluinos where each gluino can decay via $\tilde{g} \to \tilde{t}_1 t \to t\bar{t}\tilde{\chi}_0^1$ or $\tilde{g} \to \tilde{b}_1 b \to b\bar{b}\tilde{\chi}_0^1$. Therefore the final state contains at least 4 bottom quark jets and Missing Transverse Energy ($E_{\mathrm{T}}^{\mathrm{miss}}$) from the neutralinos. Both a 4 jet and a 6 jet signal region are defined, each requiring 3 b-tagged jets with transverse momentum (p_{T}) > 50 GeV and $E_{\mathrm{T}}^{\mathrm{miss}} > 200$ GeV. $E_{\mathrm{T}}^{\mathrm{miss}}$ is calculated, in a object based way, using all calorimeter clusters in the objects (electrons, muons, jets, etc.) in the analysis and in addition clusters not contained in such objects. To suppress the multi jet background none of the signal jets are allowed to be within 0.4 in the azimuthal (ϕ) direction of the $E_{\mathrm{T}}^{\mathrm{miss}}$ vector ($\Delta\phi < 0.4$). Furthermore the ratio of E_T^{miss} to m_{eff} (the Effective Mass defined as the scalar sum of the signal jet p_T and E_T^{miss}) must be larger than 0.2 to enhance the signal. The main background comes from top pair production after this data selection. A control region in data is defined which requires exactly 2 b-tagged jets and relaxes both the E_T^{miss} and E_T^{miss}/m_{eff} cuts to increase statistics. This control region, which is dominated by top pair production, is used to normalise the top pair Monte Carlo (MC). Good agreement between data and MC is observed.

No excess is observed in either signal region. Limits are set on the gluino mass in two simplified models where the gluino is lighter than the squarks and either decays exclusively into a bottom-antibottom or top-antitop pair, in both cases with an accompanying neutralino. Figure 1 shows that gluino masses below 1320 GeV in the bottom-antibottom scenario are excluded, and below 1240 GeV in the top-antitop scenario when the neutralino is massless.

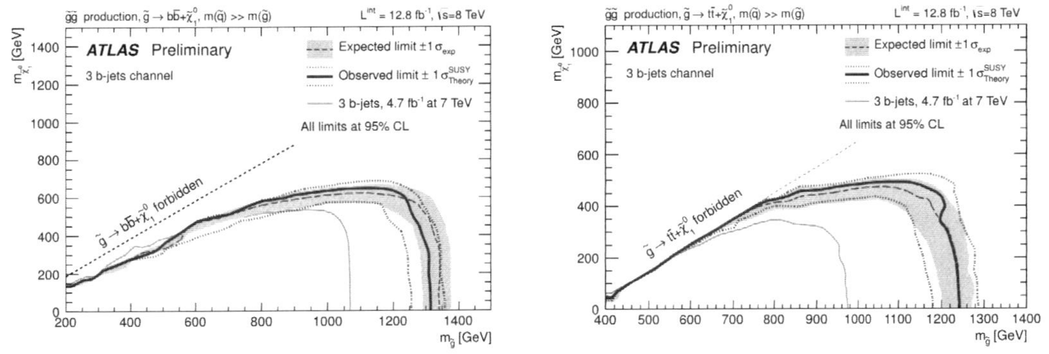

(a) (b)

Figure 1. Limits on the gluino mass, where the final state contains a bottom-antibottom pair (a) and a top-antitop pair (b) [2].

3. Search for gluino mediated stop production, and direct sbottom production with leptons and jets and E_T^{miss} in 8 TeV data

Gluino mediated stop production can also result in a final state with leptons from the top decays. The final state in this analysis is also sensitive to direct sbottom production because of leptons from both W boson and top quark decays: $\tilde{b}_1 \to \tilde{\chi}_1^- t \to \tilde{\chi}_1^0 W^- t \to \tilde{\chi}_1^0 l^- \bar{\nu}_l t$

The analysis [3] defines 6 signal regions based on the possible electron and muon combinations. All of these require at least 4 jets, $E_T^{miss} > 50$ GeV and a same-flavor-opposite-sign (SFOS) lepton pair with an invariant mass outside the range 81 to 101 GeV. Three control regions are used to estimate the main backgrounds in the signal region. A Z boson control region with the SFOS invariant mass in the range 81 to 101 GeV, a top control region with less than 4 jets and a E_T^{miss} control region with $E_T^{miss} < 50$ GeV are used. These control regions are further subdivided into regions sensitive to different fake rates (light or heavy flavour jets faking electrons or muons). A likelihood fit is performed in these regions to correct the MC fake rates. Good agreement between data and MC is seen in all three control regions, and no excess in data is found in any of the signal regions.

The results are interpreted in terms of two simplified models. In the first model the gluino always decays into a top-antitop pair and a neutralino. The second model requires the directly produced sbottoms to decay to a final state with a W boson, neutralino and top quark. In the first model gluinos with masses below 860 GeV are excluded at 95% Confidence Limit (CL) and in the second model sbottoms are excluded up to 430 GeV at 95% CL. The limits are illustrated in the respective mass planes in Figure 2.

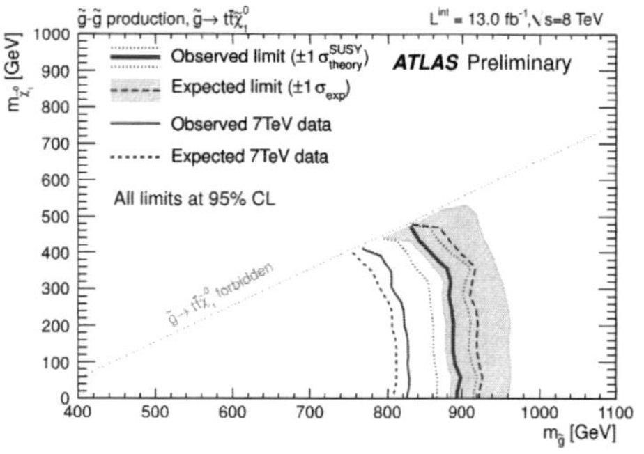

(a)

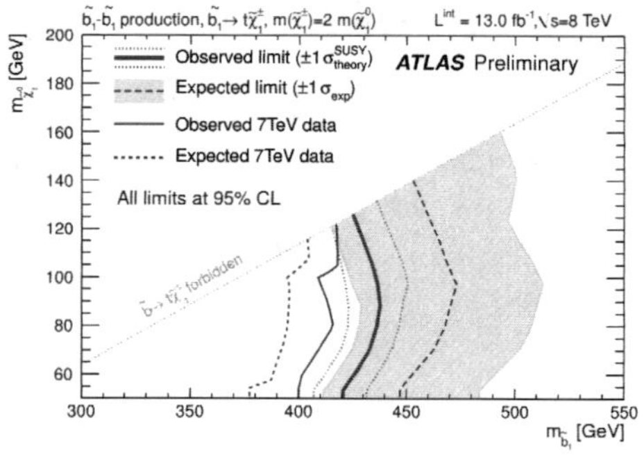

(b)

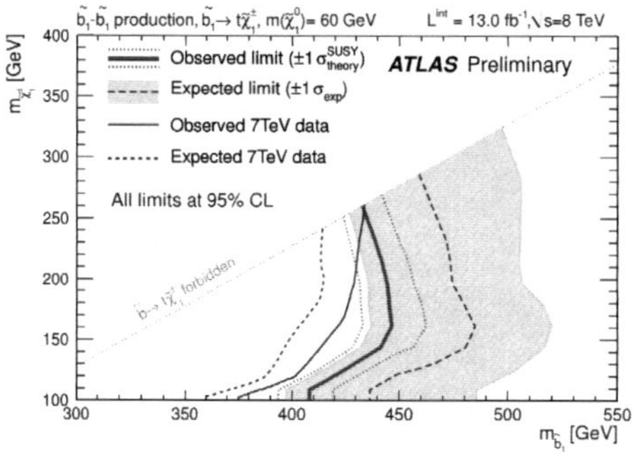

(c)

Figure 2. Limit in neutralino-gluino plane for gluino mediated stop production (a), limit in neutralino-sbottom plane for direct sbottom production, where $m(\tilde{\chi}_1^{\pm}) = 2m(\tilde{\chi}_1^0)$ (b) and limit in neutralino-sbottom plane for direct sbottom production, where $m(\tilde{\chi}_1^{\pm}) = 60$ GeV (c) [3].

4. Search for direct sbottom production with 2 b-jets and $E_{\mathrm{T}}^{\mathrm{miss}}$ in 7 TeV data

The sbottom can decay into a final state with a bottom quark and a neutralino: $\tilde{b} \to b\tilde{\chi}_1^0$. The mass difference between the sbottom and neutralino affects the kinematics of the decay. For large mass differences we expect two hard b-tagged jets, and make use of the contransverse mass (m_{CT}) [4] variable shown in equation 1, defined for two particles ν_1 and ν_2 in the final state, to separate signal and background.

$$m_{\mathrm{CT}}^2(\nu_1, \nu_2) = [E_{\mathrm{T}}(\nu_1) + E_{\mathrm{T}}(\nu_2)]^2 - [p_{\mathrm{T}}(\nu_1) + p_{\mathrm{T}}(\nu_2)]^2 \qquad (1)$$

When the mass difference is small we make use of a third jet from initial state radiation (ISR) being present in each event, which recoils against the two b-tagged jets thus boosting them above the minimal jet p_{T} thresholds used in ATLAS. A 1-lepton control region is used for top pair production and W boson production in association with heavy flavour jets. The 1-lepton selection ensures no signal events of interest are in this control region. A 2-lepton control region is used for Z bosons decaying to neutrinos in association with heavy flavour jets and top pair production. The 2 leptons can be added vectorially to the $E_{\mathrm{T}}^{\mathrm{miss}}$ to mimic neutrinos, and ensure no signal of interest is present in the control region. The control regions are used to normalise the MC and no excess in data was found in the signal regions. Limits are set in a SUSY model where the sbottom decays exclusively to a bottom quark and neutralino. The sbottom mass below 490 GeV is excluded for massless neutralinos, as shown in Figure 3 [5].

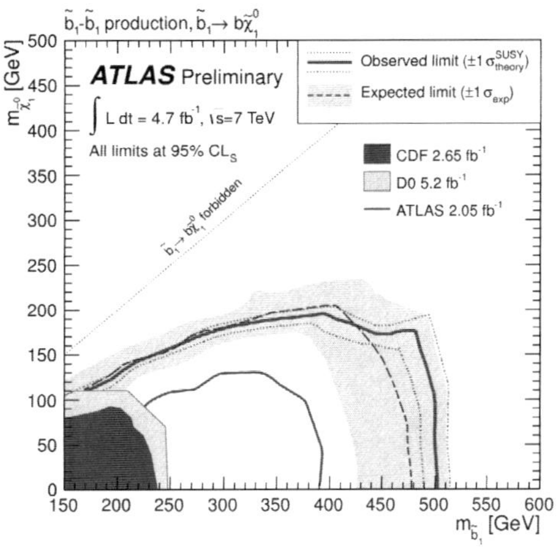

Figure 3. Limit in neutralino-sbottom plane for direct sbottom production [5].

5. Search for direct stop production in 7 TeV data

If the stop is heavier than the top quark, then the stop can decay to a top and a neutralino. We search for this process using final states with 0,1 or 2 leptons in association with jets and $E_{\mathrm{T}}^{\mathrm{miss}}$. When the stop is similar in mass to the top quark it is more likely to decay into a bottom quark and chargino: $\tilde{t}_1 \to \tilde{\chi}_1^- b \to W^- \tilde{\chi}_1^0 b$. We search for this process using a final state with 1 or 2 leptons in association with b-tagged jets and $E_{\mathrm{T}}^{\mathrm{miss}}$. If the stop is extremely light then the leptons will be very soft and a dedicated analysis is performed looking for 1 or 2 soft leptons in association with jets and $E_{\mathrm{T}}^{\mathrm{miss}}$ [6–10].

In the heavy stop scenario in the 0-lepton search we require at least 5 jets and no leptons or tau-like jets. Furthermore the invariant mass of the hadronic top candidate should be between 80 and 270 GeV and the transverse mass constructed from the b-tagged jet and $E_{\mathrm{T}}^{\mathrm{miss}}$ larger than 175 GeV. Two signal regions are then defined with $E_{\mathrm{T}}^{\mathrm{miss}} > 150$ or 260 GeV along with a control region for top pair production with a tau faking a jet - here we require an isolated non-tau lepton with transverse mass between 40 and 120 GeV. In the 1-lepton analysis we construct 5 signal regions based on $E_{\mathrm{T}}^{\mathrm{miss}}$ and $E_{\mathrm{T}}^{\mathrm{miss}}/\sqrt{H_{\mathrm{T}}}$ where H_{T} is the scalar sum of jet p_{T} and the transverse mass. Furthermore three control regions are defined for the top and W boson backgrounds. In the 2-lepton analysis we use m_{T2} [11], defined in equation 2 where $p_{\mathrm{T}}^{\mathrm{miss}}$ is decomposed into two hypotheses for the 2-vectors of invisible particles q_{T} and r_{T}, to separate the signal and background. A same-flavour and different-flavour signal region can then be defined alongside a control region for top and Z boson production.

$$m_{\mathrm{T2}}(p_{\mathrm{T}}^{\mathrm{l1}}, p_{\mathrm{T}}^{\mathrm{l2}}, p_{\mathrm{T}}^{\mathrm{Miss}}) = min[max[m_{\mathrm{T}}(p_{\mathrm{T}}^{\mathrm{l1}}, q_{\mathrm{T}}), m_{\mathrm{T}}(p_{\mathrm{T}}^{\mathrm{l2}}, r_{\mathrm{T}})]] \quad (2)$$

When the stop is lighter, around the top mass, we require 1 or 2 leptons along with b-tagged jets and $E_{\mathrm{T}}^{\mathrm{miss}}$. Then $\sqrt{S_{\mathrm{min}}^{\mathrm{sub}}}$ [12], which can be calculated for a subsystem of final state particles defined in equation 3, is used to separate signal and background. Three control regions are used for the top and W/Z boson backgrounds.

$$\sqrt{S_{\text{min}}^{\text{sub}}} = \sqrt{(\sqrt{m_{\text{sub}}^2 + \bar{p}_{\text{T,sub}}^2} + \sqrt{m_{\text{miss}}^2 + E_{\text{T}}^{\text{miss}}})^2 - (\bar{p}_{\text{T}}^{\text{sub}} + \bar{p}_{\text{T}}^{\text{Miss}})^2} \quad (3)$$

No excess in data is seen in any of the signal regions defined in the five direct stop analyses. Hence limits can be set in the neutralino-stop mass plane. Depending on the assumptions made we exclude a stop mass up to 500 GeV, though as can be seen from Figure 4 we cannot exclude a stop mass from just above 150 GeV to just below 225 GeV using 7 TeV data.

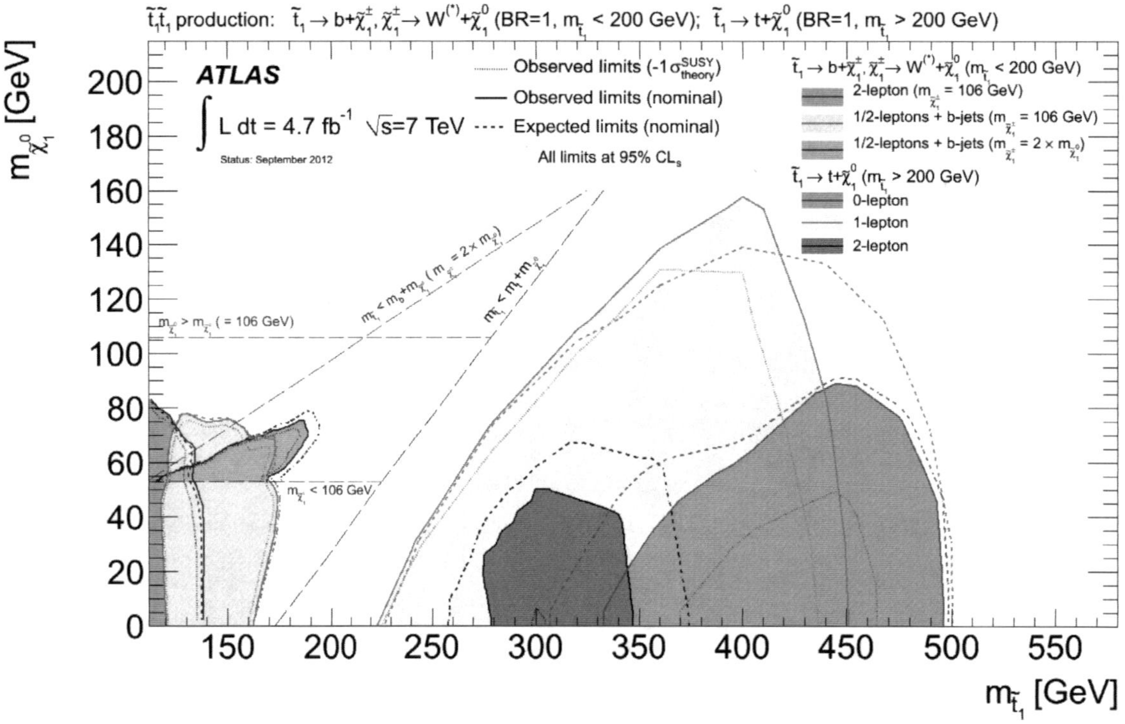

Figure 4. Limits on direct stop production with 7 TeV data [13].

6. Search for direct slepton and gaugino pair production in 7 TeV data with 2 leptons and $E_{\text{T}}^{\text{miss}}$

Direct production of sleptons is possible if the sleptons are light enough, and for masses in the range 70 to 190 GeV cross-sections of 0.05 to 3.9 pb occur in the models considered. The cross-sections are independent of the neutralino mass. This leads to a final state containing 2 leptons and 2 neutralinos via the decay: $\tilde{l}^- \to l\chi_1^0$. Light sleptons can also be produced via chargino pair production, where each chargino decays via $\tilde{\chi}_1^\mp \to \tilde{l}\nu \to \nu l\tilde{\chi}_1^0$ or $\tilde{\chi}_1^\mp \to \tilde{\nu}l \to l\nu\tilde{\chi}_0^1$. This analysis [14] is also sensitive to 3 lepton final states where one of the leptons is not reconstructed or impinges on non-instrumented regions of the detector. In this scenario both a chargino and neutralino are produced, with the latter decaying via $\tilde{\chi}_2^0 \to \tilde{l}l \to ll\tilde{\chi}_1^0$. If the sleptons are heavy then the decays of the chargino and neutralino instead proceed via W and Z bosons. In this 2 lepton analysis the first signal region requires that the opposite sign (OS) di-lepton invariant mass is inconsistent with the Z boson mass, no signal like jets (defined to have $p_{\mathrm{T}} > 30$ GeV and $|\eta| < 2.5$ with a jet vertex fraction larger than 0.75) , $E_{\mathrm{T}}^{\mathrm{miss}}$ larger than 40 GeV and m_{T2} larger than 90 GeV. The second signal region removes the m_{T2} cut and tightens the $E_{\mathrm{T}}^{\mathrm{miss}}$ cut to 100 GeV. A third signal region with same sign (SS) lepton pairs requires no signal like jets and $E_{\mathrm{T}}^{\mathrm{miss}}$ larger than 100 GeV. Finally a fourth signal region with SFOS lepton pairs with an invariant mass inconsistent with the Z boson mass, at least 2 jets, no b-tagged jets, $E_{\mathrm{T}}^{\mathrm{miss}}$ greater than 50 GeV and a m_{CT} veto is defined. Control regions are defined for top, di-boson and Z boson + jets production. No excess of data is seen in the signal regions. Limits are placed in the chargino-neutralino plane and neutralino-slepton plane. For 10 GeV neutralinos we exclude a chargino with mass between 110 and 340 GeV, and for a 20 GeV neutralino we exclude sleptons between 85 and 195 GeV.

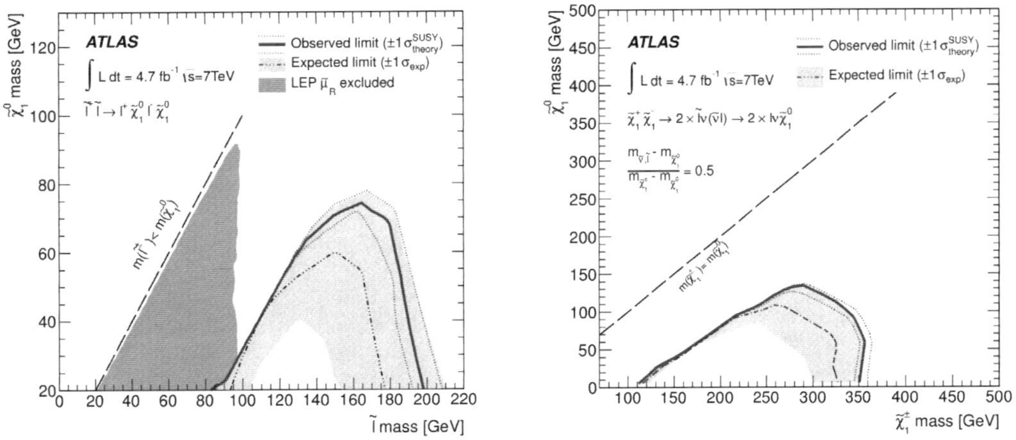

(a) (b)

Figure 5. Limits in the neutralino-slepton plane (a) and the chargino-neutralino plane (b) [14]
.

7. Search for gaugino pair production in 8 TeV data with 3 leptons and $E_{\mathrm{T}}^{\mathrm{miss}}$

This analysis [15], sensitive to the production of a chargino and neutralino as discussed in the previous section, requires 3 isolated leptons in each signal region. The first signal region, targeting the neutralino decay via a slepton, requires the SFOS lepton pair mass to be more than 10 GeV from the Z boson mass, no b-tagged jets, $E_{\mathrm{T}}^{\mathrm{miss}} > 75$ GeV and the lepton p_{T} to be above 10 GeV. The second signal region targeting the neutralino decay via an off-shell Z boson in addition asks for the transverse mass of the lepton and $E_{\mathrm{T}}^{\mathrm{miss}}$ to be larger than 100 GeV and the lepton p_{T} to be greater than 30 GeV. The final signal region, targeting neutralino decay via an on-shell Z boson, requires the SFOS lepton pair invariant mass to be within 10 GeV of the Z boson mass, $E_{\mathrm{T}}^{\mathrm{miss}}$ larger than 120 GeV, the transverse mass of the lepton and $E_{\mathrm{T}}^{\mathrm{miss}}$ to be larger than 100 GeV and the lepton p_{T} to be greater than 10 GeV. A control region with $E_{\mathrm{T}}^{\mathrm{miss}}$ between 50 and 75 GeV and transverse mass between 50 and 100 GeV is defined for WZ boson production. No excess is observed in data in any of the signal regions. We exclude charginos up to 580 GeV in simplified models in the presence of light sleptons, and 150 to 300 GeV if the sleptons are heavy and neutralinos light. This is shown in Figure 6. We also interpret these results in the framework of the phenomenological Minimal Supersymetric Model (pMSSM), which is shown in Figure 7, in terms of the gaugino mass parameters M_1 and M_2, as well as the Higgs mass parameter μ. For all values of M_1, regions low M_2 and μ are excluded.

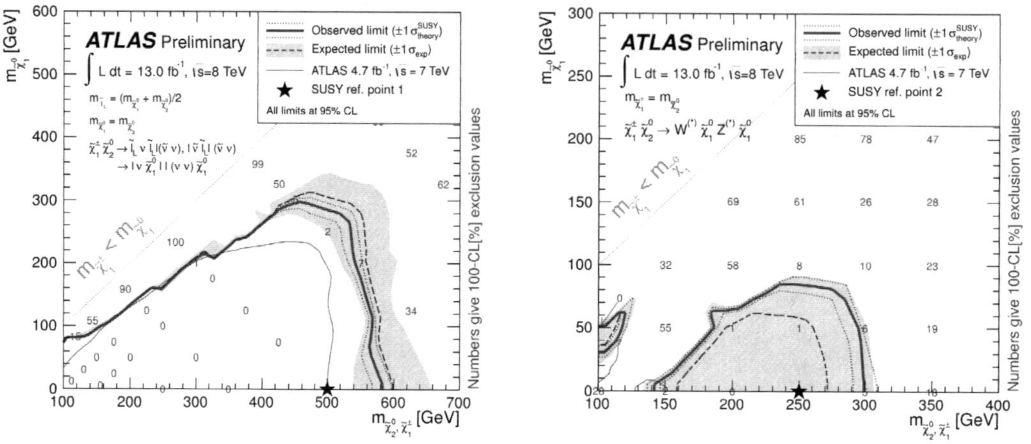

(a) (b)

Figure 6. Limits in the chargino-neutralino1 plane and neutralino2-neutralino1 plane, for decays via a slepton (a). Limits in the chargino-neutralino1 plane and neutralino2-neutralino1 plane, for decays via a W boson (b) [15].

8. Conclusion

ATLAS has performed careful searches for production of third generation squarks and gauginos. An extensive range of scenarios have been covered using leptons, jets and $E_{\mathrm{T}}^{\mathrm{miss}}$ - no analysis observed any excess of data. Natural SUSY Searches in ATLAS exclude a gluino mass below 1320 GeV and stop masses are excluded up to masses of 500 GeV. 8 TeV searches have already extended the 7 TeV limits on SUSY models, whilst other searches are in the process of being updated with 8 TeV data. After the Long Shutdown 1 (LS1) the LHC will resume data taking at a far higher centre of mass energy, and therefore future searches for natural SUSY in ATLAS are expected to be even more sensitive.

[1] The ATLAS Collaboration 2008 *JINST 3* **S08003**
[2] The ATLAS Collaboration 2012 *ATLAS-CONF-2012-145, http://cds.cern.ch/record/1493484*
[3] The ATLAS Collaboration 2012 *ATLAS-CONF-2012-151, http://cds.cern.ch/record/1493490*
[4] Tovey D 2008 *JHEP* **04** 034
[5] The ATLAS Collaboration 2012 *ATLAS-CONF-2012-106, http://cds.cern.ch/record/1472685*
[6] The ATLAS Collaboration 2012 *EPJC* **72** 2237
[7] The ATLAS Collaboration 2012 *arxiv:1209.102*
[8] The ATLAS Collaboration 2012 *PRL* **109** 211802
[9] The ATLAS Collaboration 2012 *PRL* **109** 211803
[10] The ATLAS Collaboration 2012 *JHEP* **11** 094
[11] Lester C G and Summers D J 1999 *Phys.Lett. B* **463** 99–103
[12] Konar P *et al.* 2011 *JHEP* **041** 1106
[13] The ATLAS Collaboration 2012 `https://twiki.cern.ch/twiki/pub/AtlasPublic/CombinedSummaryPlots/ATLAS_directstop_all_sep12.eps`

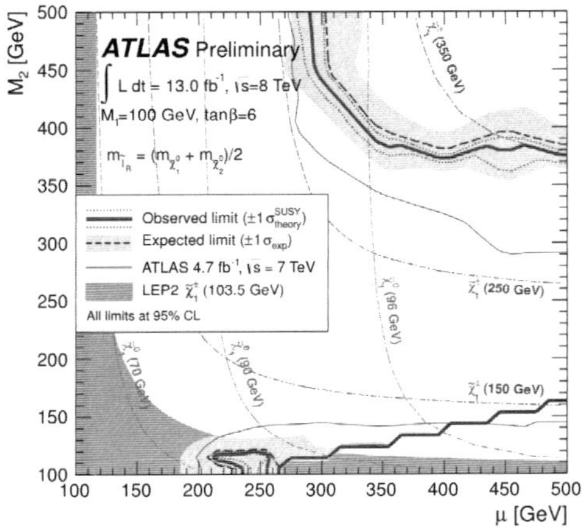

(a)

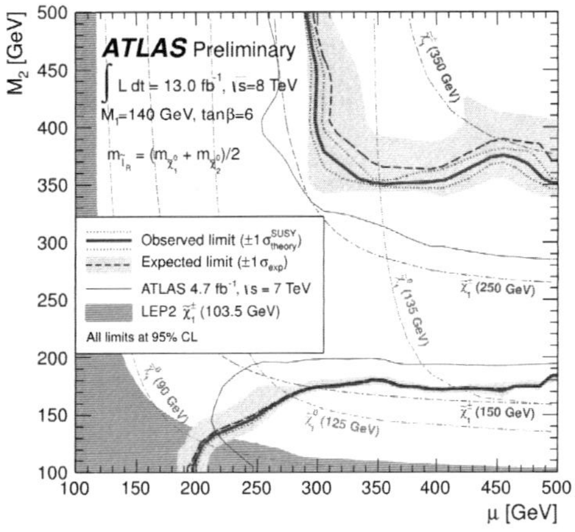

(b)

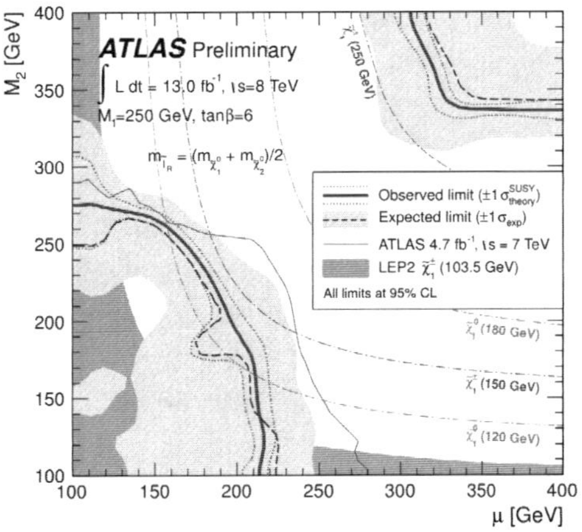

(c)

Figure 7. Limits in the M_2-μ plane, for M_1 = 100 GeV (a), M_1 = 140 GeV (b) and M_1 = 250 GeV (c) [15].

[14] The ATLAS Collaboration 2013 *PLB* **718** 879
[15] The ATLAS Collaboration 2012 *ATLAS-CONF-2012-154*, *http://cds.cern.ch/record/1493493*

Journal of Physics: Conference Series **455** (2013) 012018 doi:10.1088/1742-6596/455/1/012018

Production of beauty and quarkonia in ATLAS

T Jakoubek on behalf of the ATLAS Collaboration

Faculty of Nuclear Sciences and Physical Engineering, Czech Technical University in Prague, Brehova 7, Prague, Czech Republic

E-mail: tomas.jakoubek@cern.ch

Abstract. Quarkonia and b-hadron production cross sections were measured using the ATLAS detector in proton-proton collisions at the Large Hadron Collider at $\sqrt{s} = 7$ TeV using data collected in 2010 and 2011 runs. The new bottomonium state $\chi_b(3P)$ has been observed.

1. Introduction

Quarkonia (bound states of a quark and its anti-quark) and other particles containing heavy quarks provide a good ground for testing the validity of quantum chromodynamics (QCD). Quarkonia production mechanisms are still not well understood. Several models (Colour Singlet and Colour Octet production mechanisms) describe the transverse momentum (p_{T}) spectra and production cross sections observed at the Tevatron, but polarization predictions are not in agreement with measurements. The b-hadron production cross section has been predicted with next-to-leading-order (NLO) accuracy and several measurements were performed at the Sp$\bar{\text{p}}$S and the Tevatron colliders. Theory predictions and measurements are in a good agreement, but they still have large uncertainties. The Large Hadron Collider (LHC) opens a new region of energies for all these studies.

2. The ATLAS detector

The ATLAS experiment [1] is a multipurpose particle physics detector at the LHC, optimized for high p_{T} physics and instantaneous luminosity up to 10^{34} cm^{-2}·s^{-1}. The detector has forward-backward symmetric cylindrical geometry with almost 4π coverage. Precise tracking is provided by its innermost part, the Inner Detector (ID), which consists of a silicon pixel detector, a silicon microstrip detector and a transition radiation tracker, all immersed in a 2 T axial magnetic field. The ID covers the pseudorapidity region of $|\eta| < 2.5$ and is enclosed by the electromagnetic and hadronic calorimeters. The outermost part of the ATLAS detector is the Muon Spectrometer (MS). It covers the pseudorapidity in the interval of $|\eta| < 2.7$ and is located within the magnetic field produced by three large superconducting air-core toroid systems.

The ATLAS detector uses a three-level trigger system (hardware-based Level 1 and software-based Level 2 and Event Filter), which reduces the event rate from 20 MHz to several hundred Hz. The triggers used in the analyses presented rely on the Minimum Bias Trigger Scintillators (MBTS) and the muon trigger chambers. During the later periods of 2010 data taking with higher instantaneous luminosity, selective muon triggers with high p_{T} threshold (4, 6 or 10 GeV/c) were required.

Journal of Physics: Conference Series **455** (2013) 012018 doi:10.1088/1742-6596/455/1/012018

3. Measurement of the J/ψ production

The J/ψ meson is perhaps the best known state of charmonium, the bound state of $c\bar{c}$ quarks. This meson was discovered in the 1970s and since then its properties have been deeply studied. Thus it can be used not only for heavy quark studies, but also for calibration of the high-energy physics detectors.

The J/ψ double differential production cross-section was measured [2] in bins of p_{T} and rapidity y on the ATLAS detector using data from 2010 proton-proton 7 TeV collisions with integrated luminosity of 2.2 pb^{-1}. The differential cross-section was measured as

$$\frac{\mathrm{d}^2\sigma(J/\psi)}{\mathrm{d}p_{\mathrm{T}}\mathrm{d}y}\mathrm{Br}(J/\psi \to \mu^-\mu^+) = \frac{N^{J/\psi}_{\mathrm{corr}}}{\mathcal{L}\cdot\Delta p_{\mathrm{T}}\Delta y}, \tag{1}$$

where $N^{J/\psi}_{\mathrm{corr}}$ is the J/ψ yield in a given $p_{\mathrm{T}} - y$ bin after continuum background subtraction and correction for detector efficiency, bin migration and acceptance effects, $\mathcal{L}$ is the integrated luminosity of the data sample and Δp_{T} and Δy are the p_{T} and rapidity bin widths respectively. In order to obtain the true number of J/ψ s every candidate was weighted by:

$$w^{-1} = \mathcal{A}\cdot\mathcal{M}\cdot\epsilon^2_{\mathrm{trk}}\cdot\epsilon^-_\mu(p^-_{\mathrm{T}},\eta^-)\cdot\epsilon^+_\mu(p^+_{\mathrm{T}},\eta^+)\cdot\epsilon_{\mathrm{trig}}, \tag{2}$$

where $\mathcal{A}$ is the kinematic acceptance, describing the probability of the muons from a J/ψ falling into the fiducial volume of the detector. $\mathcal{M}$ is the correction factor for bin migrations due to the finite detector resolution, ϵ_{trk} is the ID tracking efficiency, $\epsilon^\pm_\mu$ is the single-muon offline reconstruction efficiency and ϵ_{trig} is the trigger efficiency. The acceptance $\mathcal{A}$ depends on the p_{T} and y of the J/ψ candidate and also on its spin-alignment, which is not yet known for the LHC conditions. Five extreme cases can be identified in the general parametrization of the angular distribution and these were used to calculate the maximal variation of the acceptance. The central values were obtained using the hypothesis with flat polarization. The kinematic acceptance was calculated using generator-level Monte Carlo.

The total inclusive cross-section for a J/ψ produced within $|y| < 2.4$ and $p_{\mathrm{T}} > 7$ GeV, multiplied by the branching fraction into muons, has been measured to be:

$$Br(J/\psi \to \mu^+\mu^-)\sigma(pp \to J/\psi X) = 81 \pm 1\,(\mathrm{stat.}) \pm 10\,(\mathrm{syst.}) \pm^{25}_{20}\,(\mathrm{spin}) \pm 3\,(\mathrm{lumi.})\ \mathrm{nb}. \tag{3}$$

3.1. Measurement of the non-prompt J/ψ fraction

The J/ψ mesons can be produced directly in pp collision, moreover they can originate from decays of heavier charmonium states or from the decay of a B-hadron (non-prompt production). As B-hadrons are long-lived, the pseudo-proper lifetime can be used as a discriminating factor between prompt and non-prompt production. It is defined as:

$$\tau = \frac{L_{xy}\cdot m^{J/\psi}_{\mathrm{PDG}}}{p^{J/\psi}_{\mathrm{T}}} \tag{4}$$

where L_{xy} is the vertex distance projected onto the transverse momentum direction, $m^{J/\psi}_{\mathrm{PDG}}$ is the world average value of the J/ψ mass and $p^{J/\psi}_{\mathrm{T}}$ is the transverse momentum of the J/ψ candidate. The fraction of prompt and non-prompt J/ψ production as a function of p_{T} was determined using a simultaneous fit of mass and pseudo-proper lifetime in four y bins. The results, together with CDF and CMS measurements for comparison, are shown in Fig. 1.

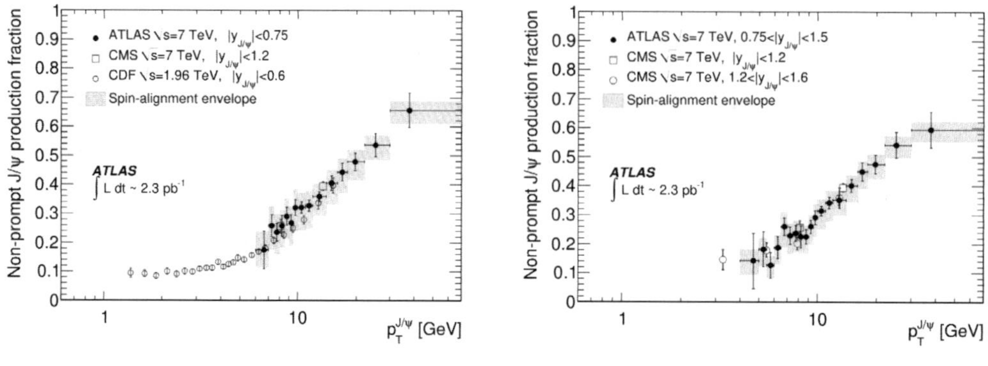

Figure 1. The non-prompt to prompt J/ψ fraction as the function of p_{T} in the first two rapidity bins. The yellow band represents the variation of the result under various spin-alignment scenarios. The equivalent results from CMS and CDF are included. [2]

3.2. The prompt and non-prompt differential production cross-section

The prompt J/ψ fraction was also used to derive the prompt and non-prompt J/ψ production cross-sections from the inclusive production cross-section. The total production cross-section for non-prompt J/ψ within $|y| < 2.4$ and $p_{\mathrm{T}} > 7$ GeV, multiplied by the branching fraction into muons, has been measured to be:

$$Br(J/\psi \to \mu^+\mu^-)\sigma(pp \to B + X \to J/\psi + X') = \\ = 23.0 \pm 0.6(\text{stat.}) \pm 2.8(\text{syst.}) \pm 0.2(\text{spin}) \pm 0.8(\text{lumi.})\ \text{nb}. \quad (5)$$

Results were compared to Fixed Order Next-to-Leading Logarithm (FONLL) calculations [6]. This comparison is shown in Fig. 2, where a good agreement between the experimental data and the theoretical prediction can be seen. The total integrated prompt J/ψ production cross-section within $|y| < 2.4$ and $p_{\mathrm{T}} > 7$ GeV, multiplied by the branching fraction into muons, has been measured to be:

$$Br(J/\psi \to \mu^+\mu^-)\sigma(pp \to \text{prompt } J/\psi X) = 59{\pm}1\,(\text{stat.}){\pm}8\,(\text{syst.}){\pm}^{9}_{6}\,(\text{spin}){\pm}2\,(\text{lumi.})\ \text{nb}. \quad (6)$$

Comparison with the predictions of the Colour Evaporation Model (CEM) [7, 8, 14] and the Colour Singlet Model (CSM) [9, 10] at next-to-leading order (NLO) and partial next-to-next-leading order (NNLO*) calculations is shown in Fig. 3. One can see that contributions from CSM NNLO* show better agreement than the ones from NLO. The shape of the CEM predictions does not provide a good description.

4. Measurement of the Υ production

Production cross-section measurement of the bottomonium states Υ(1S, 2S, 3S) in the di-muon decay mode has been done similarly as the cross-section of the J/ψ in bins of p_{T} and y [3]. To remove the uncertainty due to spin-alignment, this measurement was performed in the ATLAS fiducial region with $p_{\mathrm{T}}^{\mu} > 4$ GeV and $|\eta^{\mu}| < 2.3$. Data from 2011 proton-proton 7 TeV LHC collisions were used with integrated luminosity of 1.8 fb^{-1}. The mass difference was used to separate the three bottomonium states. The number of Υs was then determined using an unbinned maximum likelihood fit to the di-muon invariant mass distribution. Corrected cross-sections integrated over p_{T} and y bins are shown in table 1. Differential cross-sections multiplied by the di-muon branching fraction of Υ(1S) production were extrapolated to the full phase space

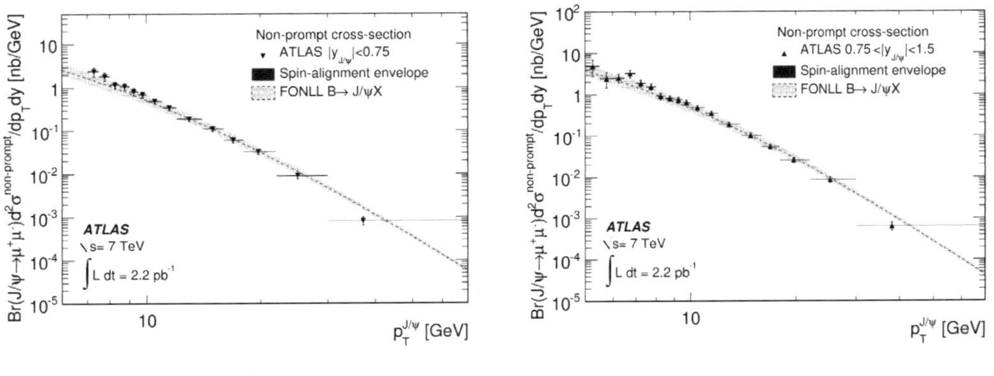

Figure 2. The non-prompt J/ψ production cross-section as the function of p_{T} in the first two rapidity bins, compared with the predictions of FONLL theory. The blue band represents spin-alignment uncertainty. [2]

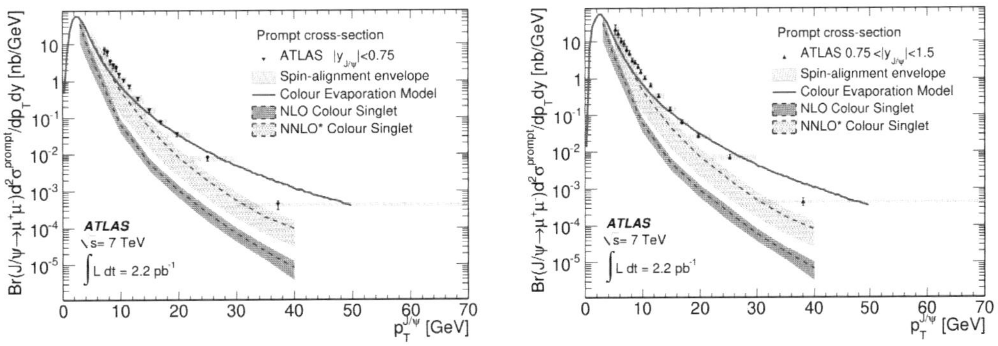

Figure 3. The prompt J/ψ production cross-section as the function of p_{T} in the first two rapidity bins. The yellow band represents spin-alignment uncertainty. Predictions from NLO and NNLO* calculations, and the Colour Evaporation Model are also shown. [2]

Table 1. Corrected cross-section measurements in the range $p_{\mathrm{T}}^{\Upsilon} < 70$ GeV and $|y^{\Upsilon}| < 2.25$. [3]

State	$\sigma(pp \to \Upsilon)Br(\Upsilon \to \mu^+\mu^-)$
Υ (1S)	8.01 $\pm$ 0.02 (stat.) $\pm$ 0.36 (syst.) $\pm$ 0.31 (lumi.) nb
Υ (2S)	2.05 $\pm$ 0.01 (stat.) $\pm$ 0.12 (syst.) $\pm$ 0.08 (lumi.) nb
Υ (3S)	0.92 $\pm$ 0.01 (stat.) $\pm$ 0.07 (syst.) $\pm$ 0.04 (lumi.) nb

of produced Υ mesons and compared to the NNLO* CSM [10–12] and CEM [7,13,14] predictions. These comparisons are shown in Fig. 4. CSM underestimates the data in bins of higher p_{T}, while CEM predictions show a better match with the data at higher p_{T}, but have difficulties with modelling the shape of the spectrum at lower p_{T}. The ratio of Υ(2S) and Υ(3S) cross-sections over Υ(1S) cross-section was also determined. It is flat across the full rapidity interval, as can be seen in Fig. 5 (left). In p_{T} it steadily rises and saturates above the value in the range

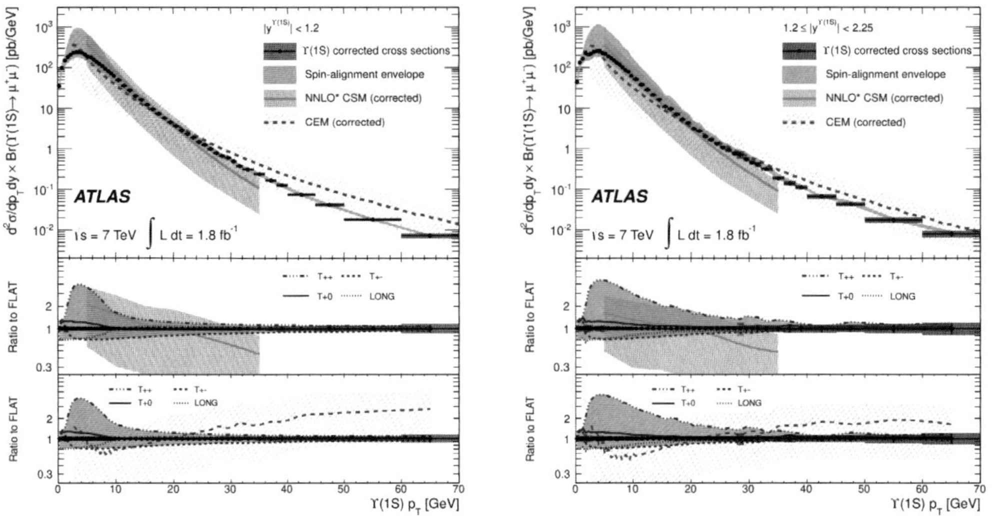

Figure 4. The differential $\Upsilon(1S)$ cross-sections multiplied by corresponding the di-muon branching fraction extrapolated to the full phase space in two rapidity bins. The blue band represents the spin-alignment uncertainty. Predictions of the NNLO* CSM and CEM are also shown. The ratios of these predictions to the measured cross-sections are in the lower panes for CEM (middle) and CSM (bottom). [3]

30-40 GeV/c, where the direct production dominates (see Fig. 5 middle and right). However, the measured values are higher than theory predictions.

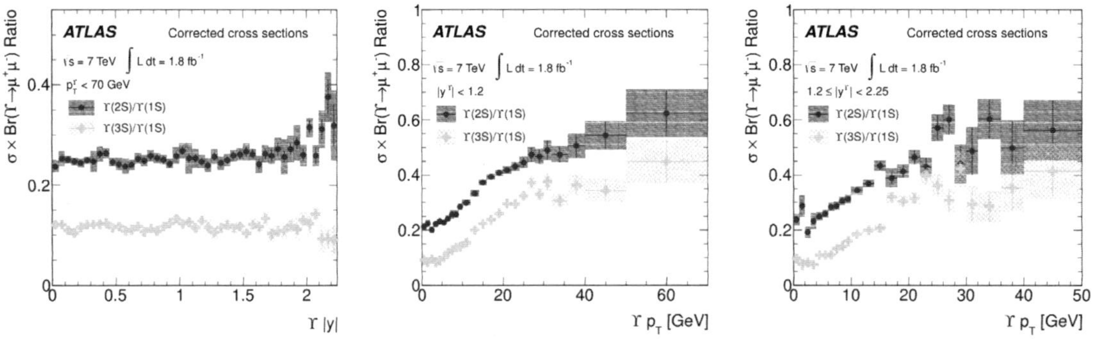

Figure 5. The ratios of the $\Upsilon(2S)$ /$\Upsilon(1S)$ and $\Upsilon(3S)$ /$\Upsilon(1S)$ cross-section as a function of di-muon rapidity (left) and di-muon p_T (middle and right). Error bars do not include spin-alignment effects. [3]

5. Observation of the new $\chi_b(3P)$ states

A new quarkonium state has been observed [4] in radiative transitions to the $\Upsilon(1S)$ and $\Upsilon(2S)$ in the ATLAS measurements in 2011 proton-proton 7 TeV collisions with integrated luminosity of 4.4 fb^{-1}. Υ(1S, 2S) then decays into a $\mu^+\mu^-$ pair and the photon is reconstructed either through conversion to e^+e^- pair (converted photons) or by the direct calorimetric measurement (unconverted photons). States $\chi_b(1P)$ and $\chi_b(2P)$ have already been observed, but the existence of the third bottomonium state $\chi_b(3P)$ was only predicted. The invariant mass distribution of di-muon plus photon candidates is shown in Fig. 6. The $\Upsilon(1S)$ mass window was set to be $9.25 < m(\mu\mu) < 9.65$ GeV and for $\Upsilon(2S)$ to be $9.80 < m(\mu\mu) < 10.10$ GeV. An unbinned maximum likelihood fit was performed in order to obtain the three mean values $m_{n=1,2,3}$, where m_3 is an estimate of the mass barycenter of the observed $\chi_b(3P)$ signal multiplet:

$$m_3(\text{unconverted photons}) = 10.541 \pm 0.011(\text{stat.}) \pm 0.030(\text{syst.})\ \text{GeV}, \tag{7}$$

$$m_3(\text{converted photons}) = 10.530 \pm 0.005(\text{stat.}) \pm 0.009(\text{syst.})\ \text{GeV}. \tag{8}$$

Due to the substantially smaller systematic uncertainties in the conversion measurement, the value (8) has been chosen as the final result of the new $\chi_b(3P)$ state mass.

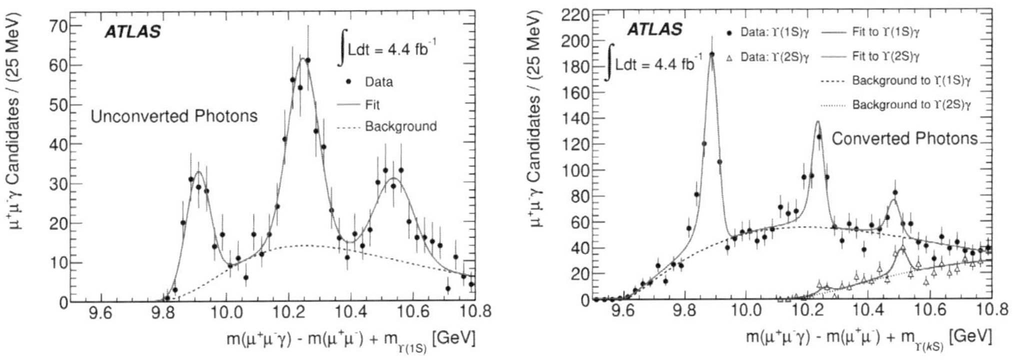

Figure 6. The invariant mass distribution of the $\chi_b \to \Upsilon + \gamma$ candidates for unconverted (left) and converted (right) photons. [4]

6. Measurement of the b-hadron production cross section

The b-hadron (H_b) production cross-section has been measured [5] using partially reconstructed $D^{*\pm}\mu^{\mp}X$ final states. The muon in this case acts as a trigger and the decay of the D meson can be fully reconstructed in the ATLAS detector ($D^{*\pm} \to D^0\pi^{\pm}$; $D^0 \to \pi K$). For this purpose, 3.3 pb^{-1} of integrated luminosity of 2010 proton-proton 7 TeV ATLAS run have been used. The differential cross-section as a function of p_T and pseudorapidity together with the theoretical predictions is shown in Fig. 7. NLO QCD [15–18] calculations underestimate the measured data, but are still consistent within the experimental and theoretical uncertainties. The integrated production cross section for $p_T > 9$ GeV and $|\eta| < 2.5$ was measured to be

$$\sigma(pp \to H_b X) = 32.7 \pm 0.8\,(\text{stat.}) \pm 3.1\,(\text{syst.})^{+2.1}_{-5.6}\,(\alpha) \pm 2.3\,(Br) \pm 1.1\,(\text{lumi.})\ \mu\text{b}, \tag{9}$$

where α stands for the acceptance contribution.

7. Conclusions

The ATLAS experiment has produced impressive and competitive results in Onia and b-hadron measurements. This includes measurement of production cross-sections of prompt and non-prompt J/ψs and b-hadrons. Especially in the case of Υ(1S, 2S, 3S) also production ratios between them have been estimated. In radiative transitions to bottomonium states a new $\chi_b(3P)$ state has been observed. All these results will help to improve our understanding of quarkonium production and dynamics.

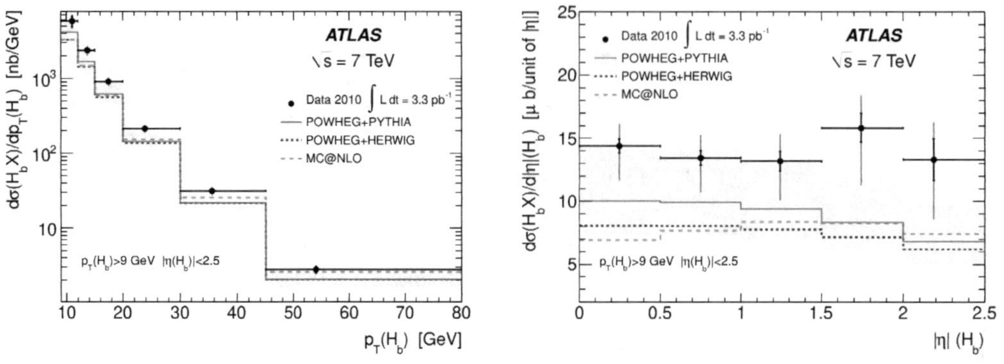

Figure 7. The differential cross section for b-hadron production as a function of p_{T} (left) and $|\eta|$ (right). The comparison with the theoretical predictions is also shown. The inner error bars of the data points represent statistical uncertainties, the outer statistical+total systematic uncertainties. [5]

References

[1] ATLAS Collaboration 2008 *JINST* **3** S08003
[2] ATLAS Collaboration 2011 *Nucl. Phys.* B **850** 387
[3] ATLAS Collaboration 2013 *Phys. Rev.* D **87** 052004
[4] ATLAS Collaboration 2012 *Phys. Rev. Lett.* **108** 152001
[5] ATLAS Collaboration 2012 *Nucl. Phys.* B **864** 341
[6] Cacciari M, Greco M and Nason P 1998 *JHEP* **9805** 007
Cacciari M, Greco M and Nason P 2001 *JHEP* **0103** 006
[7] Ullrich T, Frawley A D and Vogt R 2008 *Phys. Rept.* **462** 125
[8] Barger V D, Keung W Y and Phillips R J N 1980 *Z. Phys.* C **6** 169
[9] Lansberg J P 2010 Total J/ψ production cross section at the LHC (*Preprint* hep-ph/1006.2750)
Brodsky S J and Lansberg J P 2010 *Phys. Rev.* D **81** 051502(R)
[10] Lansberg J P 2009 *Eur. Phys. J.* C **61** 693
[11] Owens J F, Reya E and Gluck M 1978 *Phys. Rev.* D **18** 1501
Kartvelishvili V G, Likhoded A K and Slabospitsky S R 1978 *Sov. J. Nucl. Phys.* **28** 678
Chang C H 1980 *Nucl. Phys.* B **172** 425
Berger E L and Jones D L 1981 *Phys. Rev.* D **23** 1521
Baier R and Ruckl R 1982 *Nucl. Phys.* B **201** 1
Baier R and Ruckl R 1983 *Z. Phys.* C **19** 251
[12] Artoisenet P, Campbell J M, Lansberg J P, Maltoni F and Tramontano F 2008 *Phys. Rev. Lett.* **101** 152001
[13] Fritzsch H 1977 *Phys. Lett.* B **67** 217
Halzen F 1977 *Phys. Lett.* B **69** 105
Gluck M, Owens J F and Reya E 1978 *Phys. Rev.* D **17** 2324
Amundson J F, Eboli O J P, Gregores E M and Halzen F 1996 *Phys. Lett.* B **372** 127
Amundson J F, Eboli O J P, Gregores E M and Halzen F 1997 *Phys. Lett.* B **390** 323
[14] Barger V D, Keung W Y and Phillips R J N 1980 *Phys. Lett.* B **91** 253
[15] Frixione S and Webber B R 2002 *JHEP* **0206** 029
[16] Frixione S, Nason P and Webber B R 2003 *JHEP* **0308** 007
[17] Nason P 2004 *JHEP* **0411** 040
[18] Frixione S, Nason P and Ridolfi G 2007 *JHEP* **0709** 126

Topics in the Relation of the LHeC and the LHC

Bruce Mellado for the LHeC Study Group

University of the Witwatersrand, School of Physics, Private Bag 3, Wits 2050, Johannesburg, South Africa

E-mail: Bruce.Mellado@wits.ac.za

Abstract. The LHeC is envisioned to collider electrons and protons concurrently with proton-proton collisions at the LHC. The present report is based on the recently published conceptual design report of the LHeC and on contributions to the European strategy debate in emphasizing the role of the LHeC to complement and complete the high luminosity LHC programme. The report focusses on the importance of high precision PDF and α_s determinations for the physics of the Higgs boson and beyond the Standard Model.

1. Introduction

Deep inelastic lepton-hadron scattering is the cleanest and most precise probe of parton dynamics in protons and nuclei. The LHeC is the only current proposal for TeV-scale lepton-hadron scattering and the only medium-term potential complement to the LHC pp, AA and pA programme at the energy frontier. As such, it has a rich and diverse physics programme of its own, as documented extensively in the recent conceptual design report (CDR) [1] and summarised in an initial submission by the LHeC Study Group to the European Stratey of Particle Physics (ESPP) discussion prior to the Cracow Symposium [2]. The focus of this second submission to the ESPP is a further exploration of the relationship between the LHeC and the LHC. Specifically, by improving the understanding of the LHC initial state through tighter constraints on parton densities and providing information on many other aspects of strong interactions, the LHeC extends the capabilities of the LHC programme substantially.

The LHeC offers the prospect of synchronous operation of a new ep (and eA) facility with the GPDs in the high luminosity phase of the LHC (HL-LHC). The physics programme of the GPD experiments on this timescale, as illustrated in the ATLAS and CMS contributions to the ESPP process, is centered around two major objectives: i) the most precise and complete exploration possible of the newly discovered Higgs-like particle and ii) maximising the sensitivity of the continuing search for new particles and symmetries or extra dimensions in the few TeV range of mass. This document investigates the potential impact of precision LHeC results on these objectives, as we understand them at present, recognising that the situation will evolve with time and deserves continuous further study. It also recalls the importance of the deep inelastic electron-ion scattering for the completion of the LHC heavy ion programme. The mutual relations between the LHeC and the LHC are of course much deeper than can be covered in a brief communication such as this.

As documented in detail in [1], the parton density (PDF) determinations offered by the LHeC are substantially superior to the possibilities using LHC data alone and, for the first time, provide a full flavour decomposition essentially free of assumptions. The LHeC also promises a broad and unique programme of further strong interaction physics, such as the exploration of a newly accessed high density, low coupling regime at low x and a new level of precision and hugely extended kinematic coverage on the partonic structure of nuclei. Combined with competitive sensitivity to new physics in channels where initial state lepton quantum numbers are an advantage, the LHeC represents a cost effective means of fully exploiting the LHC and substantially extending its physics programme.

Following the discovery by ATLAS [3] and CMS [4] of a new boson with a mass of about 126 GeV, the future focus of the field will be firstly on determining whether this particle is the Standard Model Higgs boson or something more exotic and secondly on extending the sensitive range for discovery of other new particles as far as possible. ATLAS and CMS have reported exclusion limits for a wide range of massive new particles in the $1-2$ TeV range. With no strong evidence for new effects so far, the need to further extend such searches to the largest masses possible is paramount. The increased beam energy following LS1 will provide a first major step in this direction. Beyond that, further progress is limited by luminosity, due to the fast-falling cross sections as higher and higher x PDFs are involved (especially for gluon initiated processes), and by the uncertainities on those PDFs. To fully exploit the new particle discovery range of the LHC, both a luminosity upgrade and tighter external PDF constraints are therefore required.

The future exploration of the Higgs boson sector at the LHC, for example by measuring relative couplings and testing the CP structure, may similarly become limited by theory uncertainties derived from PDF measurements once the very high luminosities possible at HL-LHC have been accumulated [5]. This is particularly true for the 'working horse' channels in which the Higgs boson decays to $\gamma\gamma$ or four charged leptons. Whilst LHC inclusive W and Z production data will somewhat improve constraints on the quark densities of the proton at the electroweak scale, they will have a limited impact on the gluon density, which is more pertinent to Higgs boson physics, given the dominant gg production mechanism.

2. The LHeC Linac-Ring Collider

The LHeC aims at colliding the high-energy protons and heavy ions circulating in the LHC with 60-GeV polarized electrons and possibly also positrons. The LHeC is realized by adding to the LHC a separate 9-km racetrack-shaped recirculating superconducting (SC) energy-recovery linac (ERL). The key components of the LHeC are the two 1-km 10-GeV SC linacs of the ERL, comparable in scale to the 17.5-GeV SC linac of the European XFEL presently under construction. The LHeC ERL provides a design lepton beam current of 6.4 mA at the ep collision point, which is taken to be at IP2 of the LHC. Aside from the IP2 interaction region, the LHeC underground infrastructure is fully decoupled from the existing LHC tunnel. Two of the access shafts could be located on the CERN Prevessin site.

The LHeC is designed to operate with simultaneous LHC pp (or AA) collisions. LHeC operation is fully transparent to the other LHC experiments thanks to the low lepton bunch charge and resulting minuscule beam-beam tune shift experienced by the protons, together with the choice of the LHeC circumference to be equal to a third of the LHC's in order to allow for ion-clearing gaps in the ERL without perturbing LHC steady-state operation.

LHeC has been designed under the constraint that the total electrical power for the LHeC lepton branch should not exceed 100 MW (about half the present maximum CERN site power). The LHeC electrical power budget is dominated by the RF and by the cryo power for the two 1-km long SC linacs. The cryo power required and, therefore, also the size of the cryoplants (as well as the maximum lepton current) are directly linked to the unloaded quality factor of the cavities, Q_0. With a Q_0 of 2.5×10^{10}, the total main-linac cryopower amounts to 23 MW. The RF power needed for RF microphonics control is about 24 MW, and the extra-RF power needed for compensating SR losses at 6.4-mA current also to 23 MW. The remaining components, like injectors or arc magnets, require a few MW each.

parameter [unit]	LHeC	
species	e	p, $^{208}Pb^{82+}$
beam energy (/nucleon) [GeV]	60	7000, 2760
bunch spacing [ns]	25, 100	25, 100
bunch intensity (nucleon) [10^{10}]	0.1 (0.2), 0.4	17 (22), 2.5
beam current [mA]	6.4 (12.8)	860 (1110), 6
rms bunch length [mm]	0.6	75.5
polarization [%]	90 (e^+ none)	none, none
normalized rms emittance [μm]	50	3.75 (2.0), 1.5
geometric rms emittance [nm]	0.43	0.50 (0.31)
IP beta function $\beta^*_{x,y}$ [m]	0.12 (0.032)	0.1 (0.05)
IP spot size [μm]	7.2 (3.7)	7.2 (3.7)
synchrotron tune Q_s	—	1.9×10^{-3}
hadron beam-beam parameter	0.0001 (0.0002)	
lepton disruption parameter D	6 (30)	
crossing angle	0 (detector-integrated dipole)	
hourglass reduction factor H_{hg}	0.91 (0.67)	
pinch enhancement factor H_D	1.35 (0.3 for e^+)	
CM energy [TeV]	1.3, 0.81	
luminosity / nucleon [10^{33} cm^{-2}s^{-1}]	1 (10), 0.2	

Table 1. LHeC parameters. The numbers give the default values with optimum values for maximum *ep* luminosity in paranthesis and values for the *ePb* configuration separated by a komma.

Together with rather conservative assumptions for most parameters, the 100 MW power limit yields the LHeC *ep* target luminosity of 10^{33} cm^{-2}s^{-1}. However, extensions to significantly higher luminosity, e.g., 10^{34} cm^{-2}s^{-1}, are possible by a combination of improvements, namely (1) by considering a normalized proton beam emittances of 2 μm (as achieved in 2011/12 LHC operation) instead of 3.75 μm; (2) by a further reduction of the proton IP beta function from 0.1 m down to 0.05 m, which should be possible by using a variant of the so called ATS optics; (3) by increasing the proton bunch intensity from 1.7×10^{11} to the HL-LHC 25 ns target value of 2.2×10^{11} [for the 50-ns HL-LHC scenario it would be even 3.3×10^{11} with a possible further factor 2.5 increase of luminosity, to more than 2×10^{34} cm^{-2}s^{-1}]; and (4) by doubling the lepton beam current, which should be possible without exceeding the 100-MW power limit if the unloaded Q_0 value of the SC RF cavities can be raised to 4×10^{10} (as it is assumed for the similar eRHIC design). Table 1 shows LHeC parameters, including, in parentheses, values for a higher-luminosity variant.

The LHeC represents an interesting possibility for further efficient exploitation of the LHC infrastructure investment. The development of a CW SC recirculating energy-recovery linac for LHeC would prepare for many possible future projects, e.g., for an International Linear Collider, for a neutrino factory, for a proton-driven plasma wake field accelerator, or for a muon collider. With some additional arcs, using 4 instead of 3 passes through the linacs, a machine like the LHeC ERL (without energy recovery) could also operate as a Higgs boson factory $\gamma\gamma$ collider (SAPPHiRE).

Journal of Physics: Conference Series **455** (2013) 012019 doi:10.1088/1742-6596/455/1/012019

3. Parton Distributions

3.1. Valence and Sea Quarks

The LHeC is in a unique position to unravel all quark densities in the proton with a complete quark flavour separation for the first time and with unprecented precision. The huge phase space covered matches the needs of the LHC and includes the extreme values of Bjorken x, lowest, $x \simeq 10^{-5}$, where saturation may set in and largest, near to 1, which determine the multi-TeV BSM cross sections at the LHC. The detailed shape measurements of the various parton distributions, for example of the strange density versus Q^2 and x, imply that the nowadays large uncertainties due to PDF parameterisations in pQCD fits will be drastically reduced. A complete base of PDFs as the LHeC promises to deliver will at many places lead to possibly sizeable deviations of the now canonical PDF pattern. This is a necessary input for future LHC measurements, as precision Higgs boson coupling and cross section determinations. There are also expectations that discoveries are made in the conventional PDF pattern, as by possible observations of anti-quarks to be different from their sea quarks or an intrinsic heavy flavour part. Since the momentum is conserved, and shared between quarks and gluons, any deviation affects the overall pattern, which reflects on other parts of physics. Moreover, a crucial variety of non-canonical PDFs will be accessed: generalised, unintegrated, diffractive, neutron, photon and nuclear parton distributions.

The basis of LHC physics and discoveries BSM is QCD at high orders and the accurate knowledge of the classic PDFs. In the following, based on [1], some brief remarks are made on the unique potential of the LHeC in the determination of the complete set of quark densities, while the mapping of the gluon density is described subsequently below.

- **Valence quarks:** The knowledge of the valence quark distributions, both at large and at low Bjorken x, as derived in the current world data QCD fit analyses is amazingly limited. An impressive improvement is expected from the LHeC. A NLO QCD fit to simulated inclusive neutral and charged current LHeC data (see [1]) shows that the uncertainty of the down valence quark distribution at, for example, $x = 0.7$ can be reduced from a level of $50 - 100\,\%$ to about $5\,\%$. This will be crucial for searches of new physics at the LHC at the high energy frontier, in order to verify any excess (or deficiency) compared to the SM prediction. Direct access to valence quarks down to low $x \sim 0.001$ can be obtained at LHeC from the NC, Z exchange related $e^{\pm}p$ cross section difference, which can resolve possible sea-antiquark differences.
- **Light sea quarks:** The measurement of the structure functions $F_2 \propto 4U + D$, in ep and $F_2 \propto U + D$, in eD is the basis for determining the light sea quark densities in the nucleon. LHeC will extend greatly the HERA kinematic coverage to much lower x and to higher scales Q^2. From NC and CC measurements and comparing ep with eD data, the up and down sea quark densities will be unfolded, which nowadays are assumed to be equal at $x < 0.01$.
- **Strange:** Several long-standing questions are related to the strange quark density in the proton: how much is it suppressed with respect to the other two light quarks? Is there an asymmetry between the strange and anti-strange density? The knowledge of the strange-quark density itself is important for many processes, for instance for the precision measurement of the W boson mass. Information on the strange quark density is available from several experiments, in particular from previous Neutrino DIS experiments, but overall there is no real understanding of the strange quark distribution. The strange quark distribution is accessible at LHeC in charged current scattering through the subprocesses $W^+s \to c$ (for positron beams) and $W^-\bar{s} \to \bar{c}$ (for electron beams), with charm tagging in the final state. The LHeC simulation studies show that for the first time accurate measurements of the s and $\bar{s}$ densities can be performed over a large kinematical phase space in x and Q^2.

- **Charm:** Information on the charm content in the proton can be accessed at LHeC by measuring the inclusive charm production cross section in neutral current DIS. At low scales $Q^2 \sim m_c^2$ (with m_c being the charm quark mass) one has to treat the charm production fully massive, i.e. the charm quarks can be only dynamically produced in the reaction $\gamma g \to c\bar{c}$ and thus are themselves no active flavours in the proton. However, at large scales $Q^2 \gg m_c^2$ one can treat the charm quarks as massless partons, which are contributing to the sea. The charm quark mass m_c is a crucial parameter: it regulates the ratio of charm and light quarks in the sea and thus affects predictions for almost any quark driven process at the LHC. At LHeC one expects much more precise and kinematically extended measurements of inclusive charm production compared to HERA. This will allow to map for the first time the transition from the massive to the massless regime. Simulations show that one can use the data for a m_c determination at a precision of two permille. With very good forward charm tagging one can also test the hypothesis of an intrinsic charm component in the proton wave function, which could appear at high $x \simeq 0.2$.
- **Beauty:** Simulation studies show that one expects at LHeC precise measurements of inclusive beauty production in DIS. For large squared momentum transfer $Q^2 \gg m_b^2$ (with m_b being the beauty quark mass) these measurements can be directly translated into an effective beauty quark density in the proton. There is a huge interest in these densities, since in many scenarios of new physics the beauty quarks are the original particles. For instance in the minimal supersymmetric extension of the standard model the production of the neutral Higgs boson A is driven by $b\bar{b} \to A$. While at HERA the inclusive beauty production results were statistically limited to about 20% precision, very accurate results can be expected at the LHeC.
- **Top:** The production of top quarks can be studied at the LHeC for the first time in DIS experiments. The dominant process is single top (or anti-top) production in Wb to t fusion. The unique top physics program that can be performed at the LHeC includes possibly the consideration of a quark density for the top, from NC, a high precision measurement of the top mass from its decay and cross section. Top physics at the LHeC is a promising subject for further study.

In summary, while the LHC data can add information to certain aspects of the quark densities in the proton using specific reactions (e.g. Drell Yan), it remains reserved for the LHeC to uniquely resolve the complete quark and antiquark structure of the proton, for all quark flavours, over the largest kinematic range ever and last but not least with the best theoretical understanding.

3.2. Gluon Distribution

As has been summarised in the CDR, there are many fundamental reasons for the necessity to understand the gluon distribution and the gluon-parton interactions deeper than hitherto. Half of proton's momentum is carried by gluons. Gluon self-interaction is responsible for the creation of baryonic mass. In pp scattering at the LHC, the Higgs particle is predominantly produced by gluon-gluon interactions. Gluino pair production, predominantly proceeds via $gg \to g \to \tilde{g}\tilde{g}$ production and is hugely uncertain at high masses. On the other hand the rise of the gluon density towards low Bjorken $x < 10^{-5}$ is expected to be tamed and a new phase of hadronic matter to be discovered, in which gluons interact non-linearly while α_s is smaller than 1.

From the simulations as described one derives a typical uncertainty for the gluon density to be reduced to about 3, 1, 5 % at $x = 5 \cdot 10^{-6}$, 0.005, 0.5, respectively. These values of Bjorken x mark the low value for saturation to be discovered, the central rapidity value for Higgs boson production and about the high mass limit for gluino pair production. Obviously, see Figure 1, the potential of the LHC for the determination of the gluon density over $5-6$ orders of magnitude in x, simultaneously with all quark PDFs and α_s, is striking. It has to be compared with the

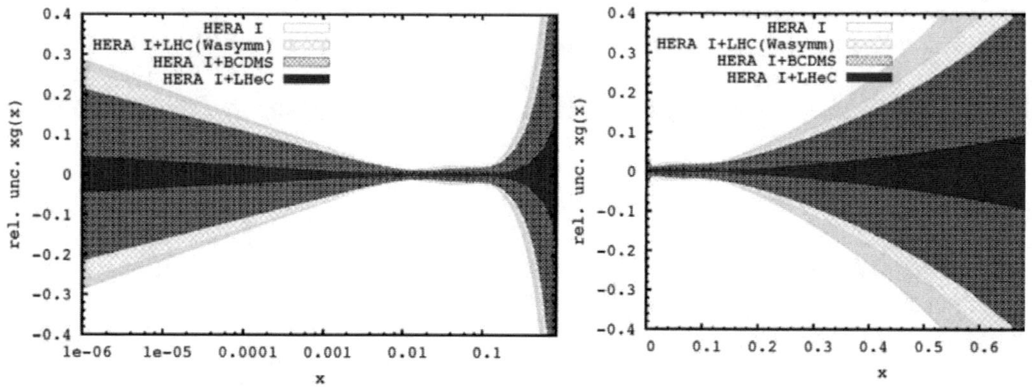

Figure 1. Relative uncertainty of the gluon distribution at $Q^2 = 1.9\,\text{GeV}^2$, as resulting from an NLO QCD fit to HERA (I) alone (green, outer), HERA and BCDMS (crossed), HERA and LHC (light blue, crossed) and the LHeC added (blue, dark). Left: logarithmic x scale, right: linear x scale.

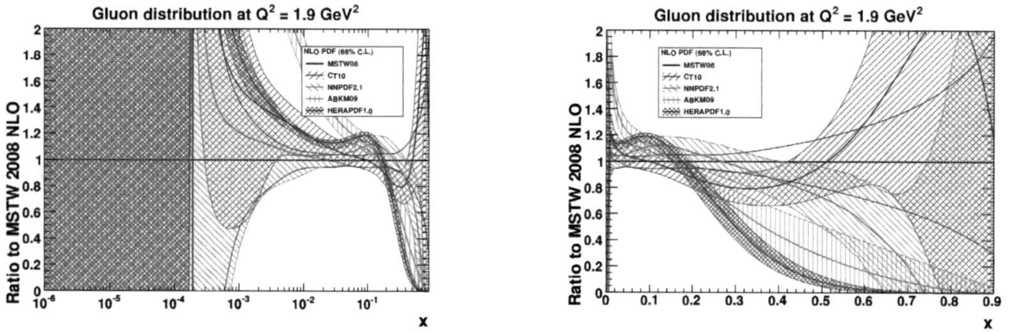

Figure 2. Ratios to MSTW08 of gluon distribution and uncertainty bands, at $Q^2 = 1.9\,\text{GeV}^2$, for most of the available recent PDF determinations, taken from [1]. Left: logarithmic x scale, right: linear x scale.

current status of huge uncertainty on xg at low and high x as is illustrated in Figure 2.

4. Higgs Boson Measurements

In the Standard Model, the breaking of the electroweak $SU(2)_L \times U(1)_Y$ symmetry gives mass to the electroweak gauge bosons via the Brout-Englert-Higgs mechanism, while the fermions obtain their mass via Yukawa couplings with a scalar Higgs field. With the observation of a Higgs-like boson by the ATLAS [3] and CMS [4] collaborations with a mass around 126 GeV, a new research field has opened in particle physics. The measurement of the couplings of the newly found boson to the known fundamental particles will be a crucial test of the SM and a window of opportunity to establish physics beyond the SM.

At the LHeC, a light Higgs boson could be uniquely produced and cleanly reconstructed either via HZZ coupling in neutral current DIS or via HWW coupling in charged current DIS. Those vector boson fusion processes have sizeable cross sections, O(100) fb for 126 GeV mass, and they can be easily distinguished, which is a unique advantage in comparison to the VBF

Higgs production in pp scattering. The observability of the Higgs boson signal at the LHeC was investigated in the CDR [1] using initially the dominant production and decay mode, i.e. the CC reaction $e^-p \to H(\to b\bar{b}) + \nu + X$, for the nominal 7 TeV LHC proton beam and electron beam energies of 60 and 150 GeV. Simple and robust cuts are identified and found to reject effectively e.g. the dominant single-top background, providing an excellent S/B ratio of about 1 at the LHeC, which may be further refined using sophisticated neural network techniques. At the default electron beam energy of 60 GeV, for 80 % e^- polarisation and an integrated luminosity of 100 fb^{-1}, the Hbb coupling is estimated to be measurable with a statistical precision of about 4 %, which is not far from the current theoretical uncertainty. Typical coupling measurements, as for $\gamma\gamma$ or $4l$, are of about 10 % precision with the HL-LHC, while the specific $b\bar{b}$ coupling will be particularly difficult to measure due to high combinatorial backgrounds in pp.

The LHC is said to be inferior to a linear collider in its coupling measurement prospects. Part of this statement comes from large uncertainties, which are related to the imperfect knowledge of the PDFs and theory parameters. The LHeC, with its high precision PDF and QCD programme, will render many of these uncertainties unimportant. Currently, for example, an uncertainty of the $H \to \gamma\gamma$ cross section due to PDFs is quoted of nearly 10 % [6], based on the variation of the cross section predictions from different PDFs. This will be very much reduced with the LHeC: an iHix calculation of the NLO Higgs cross section for MSTW08, NNPDF2.3 and HERAPDF1.5 leads to intrinsic uncertainties of 1.7, 1.2 and 2.2 %, respectively, with a maximum deviation of 6.9 %. The full experimental LHeC uncertainty, however, is 0.2 %. The main advantage will be that the precision LHeC data, possibly combined with HERA, will replace essentially all previous data sets and thus lead to a much better agreement between various PDF determinations, besides the huge reduction in uncertainty with the LHeC.

A sizeable uncertainty is also related to the strong coupling constant, a difference in α_s of ± 0.005 corresponding to an about 10 % cross section uncertainty, see e.g. [7] or [8]. Obviously, the large improvement in the determination of α_s with the LHeC will reduce this uncertainty strongly too. Essentially with such a high quality data set as the LHeC, one will simultaneously determine the coupling and the PDFs, and control their correlations at a very high level. In [9] a systematic evaluation has been presented of the effect of the heavy quark masses and of α_s on the uncertainties of the Higgs boson branching fractions in various channels. One finds partially large effects as 6 % from M_c on the $H \to c\bar{c}$ branching or 5.6 % from α_s on $H \to gg$. These will certainly be much reduced. It is for future studies to more systematically analyse the striking potential of the LHeC to remove or reduce the QCD uncertainties of the Higgs boson cross sections and couplings. There will also be improvements related to QCD measurements at the LHC. Their level of precision, however, especially for the gluon, α_s and heavy quark QCD cannot compete with the genuine DIS process.

It has also been observed [1], that the LHeC can specifically explore well the CP structure of the HWW coupling by separating it from the HZZ coupling and the other signal production mechanisms. Any determination of an anomalous HWW vertex will thus be free from possible contaminations of these. Compared to the pp situation, ep lacks the complications due to underlying event and pile-up driven backgrounds. In Ref. [10] it has been recently pointed out that the kinematics of the scattered quarks depend strongly on the CP structure of the HWW. This not only affects the azimuthal angle difference between the scattered ququarks but also their transverse momenta, separation in rapidity and other relevant quantities. The analysis performed in [1] needs to be re-assessed to take this into account.

The few initial studies performed so far will be pursued further given that the Higgs boson is now likely to indeed exist. For the projected analyses, this first concerns using the appropriate LHeC detector simulation, and optimising further its design. For the accelerator design it is obvious that a luminosity in excess of 10^{33} cm^{-2}s^{-1} is very desirable, see the discussion on machine parameters in Section 2. The then possible precision measurements of rarer (τ, Z,

W, perhaps photon) decay channels, of the CP angular distributions, for both the HWW and HZZ couplings, and NC initiated production would make the LHeC, by design a QCD machine, a collider to study the Higgs boson and mechanism of very remarkable and complementary potential.

5. Summary

The LHeC is a new *ep* collider of unprecedent kinematic range, luminosity and precision in deep inelastic scattering. This leads to the first ever complete measurement of PDFs, including, for example, the strange density. It furthermore extends to very high Bjorken x in *ep* and to so large Q^2 that no nuclear or higher twist corrections affect the high x PDF determinations. A much deeper understanding of quark-gluon dynamics is needed and in sight, exemplified by the potential of the LHeC to measure $\alpha_s(M_Z^2)$ to per mille accuracy.

The LHeC is designed to operate synchronously with the HL-LHC. Preparations have begun, following the detailed design concept report and a corresponding mandate from CERN, supported by ECFA and NuPECC, to prototype critical components and to carry out more detailed studies, as of the interaction region and high luminosity optics, in major international collaborations with CERN. As has been indicated here, there is a potential for the LHeC to achieve luminosities in excess of $10^{33}\,\mathrm{cm}^{-2}\mathrm{s}^{-1}$.

There are two major questions for the HL-LHC to investigate, the properties of the 126 GeV boson, which likely is the Higgs particle, and the search for maximum mass particles, of several TeV in direct production mode. It has been argued here that the LHeC can assist the transformation of the LHC into a precision Higgs facility, with its own coupling and CP measurements based on clean WW fusion in *ep*, and with its ultra-precise PDF, heavy quark and α_s determinations, which will reduce the theoretical uncertainties of Higgs measurements in *pp* to a negligible level. It is similarly apparent that the discovery potential of new particles at high masses at the HL-LHC is severely limited by the deviations and uncertainties of the predictions based on currently available PDF sets and as well the uncertainties of input parameters such as the heavy quark masses and α_s. For the Higgs the prominent decay $H \to b\bar{b}$ and for SUSY gluino pair production have been used as first representative examples.

References

[1] J. L. Abelleira Fernandez *et al.* [LHeC Study Group], “A Large Hadron Electron Collider at CERN” J.Phys.G. **39**(2012)075001, arXiv:1206.2913.
[2] LHeC Study Group, Contribution (No 147) to the Cracow meeting on the ESPP, "A Large Hadron Electron Collider at CERN", LHeC-Note-2012-004 GEN, CERN, August 2012.
[3] G. Aad *et al.* [ATLAS Collaboration], Phys. Lett. B **716** (2012) 1 [arXiv:1207.7214 [hep-ex]].
[4] S. Chatrchyan *et al.* [CMS Collaboration], Phys. Lett. B **716** (2012) 30 [arXiv:1207.7235 [hep-ex]].
[5] ATLAS Collaboration, Contributions to the ESPP, "Physics at High Luminosity LHC with ATLAS", CERN, August 2012 and Update October 2012.
[6] J. Baglio, A. Djouadi and R. M. Godbole, Phys. Lett. B **716** (2012) 203 [arXiv:1207.1451 [hep-ph]].
[7] J. Rojo, Talk at the PDF4LHC Workshop, 8th of October, 2012, CERN.
[8] J. Baglio and A. Djouadi, JHEP **1103** (2011) 055 [arXiv:1012.0530 [hep-ph]].
[9] A. Denner, S. Heinemeyer, I. Puljak, D. Rebuzzi and M. Spira, Eur. Phys. J. C **71** (2011) 1753 [arXiv:1107.5909 [hep-ph]].
[10] A. Djouadi, R. M. Godbole, B. Mellado and K. Mohan, arXiv:1301.4965 [hep-ph].

Journal of Physics: Conference Series **455** (2013) 012020 doi:10.1088/1742-6596/455/1/012020

Searches for Physics Beyond the Standard Model in Monojets- and Monophoton-Events with the ATLAS Detector

Ruth Pöttgen, on behalf of the ATLAS Collaboration

CERN, CH-1211 Geneva 23, Switzerland,
Johannes Gutenberg - University, DE-55122 Mainz, Germany

E-mail: ruth.poettgen@cern.ch

Abstract. Various models for new physics not described by the Standard Model predict event topologies with large missing transverse energy due to invisible new particles. Such events can be identified in the detector if they are accompanied by an energetic photon or a jet with high transverse energy. This article presents results from searches for new physics with both signatures in proton-proton collision data with the ATLAS detector at the LHC. The focus is on the analyses using the full 2011 data set with an integrated luminosity of $\mathcal{L} = 4.7\,\mathrm{fb}^{-1}$ at a center of mass energy of $\sqrt{s} = 7\,\mathrm{TeV}$ but the article also includes an update of the monojet analysis with $\mathcal{L} = 10.5\,\mathrm{fb}^{-1}$ of data recorded in 2012 at $\sqrt{s} = 8\,\mathrm{TeV}$. The results are translated into exclusion limits on parameters of different theoretical models.

1. Introduction

A variety of models for physics beyond the Standard Model (BSM) include the production of new particles that do not interact in the detector. Such events can be tagged, however, if in addition - for example - a jet or a photon with high transverse momentum p_{T} is produced, recoiling against the invisible particles. This leads to event topologies with large missing transverse energy $E_{\mathrm{T}}^{\mathrm{miss}}$. Among the theories predicting such event signatures are models for large extra dimensions (LED) [1, 2], pair production of weakly interacting massive particles (WIMP), candidates for dark matter [3], and supersymmetry (SUSY) scenarios [4, 5, 6].
Analyses with both final states have been performed with proton-proton collision data at the ATLAS experiment [8] at the LHC and since no significant deviation from the Standard Model (SM) prediction is observed, exclusion limits are set on the respective model parameters: the Planck scale of the Arkani-Hamed, Dimopoulos, Dvali (ADD) model for LED [2], the suppression scale of an effective theory used to describe the WIMP pair production and on the mass of the gravitino which is produced in association with a squark or a gluino in gauge-mediated SUSY breaking (GMSB) scenarios [7].

2. Mono-jet Analysis

The monojet analysis was performed with $\mathcal{L} = 4.7\,\mathrm{fb}^{-1}$ of data recorded at $\sqrt{s} = 7\,\mathrm{TeV}$ in 2011 [9] and updated with $\mathcal{L} = 10.5\,\mathrm{fb}^{-1}$ of $\sqrt{s} = 8\,\mathrm{TeV}$ data recorded in 2012 [10] . This update includes, for the first time, the GMSB scenario as a new interpretation in addition to the WIMP and ADD model. In the following, the event selection as well as the determination of background contributions are outlined and the results are presented.

2.1. Event Selection

At first, a basic preselection of good quality data with potentially interesting events is applied: All data taken with a completely functional detector are being used, events have to have at least one reconstructed primary vertex and have to be selected by a trigger which accepts events with a missing transverse energy of more than 80 GeV. This trigger has an efficiency above 95% for an offline reconstructed $E_{\mathrm{T}}^{\mathrm{miss}}$ higher than 120 GeV. Events are required to have an $E_{\mathrm{T}}^{\mathrm{miss}}$ above this threshold and at least one jet within $|\eta| < 2.0$ with $p_T > 120$ GeV as well. Jets are reconstructed with the anti-Kt algorithm with a distance parameter of $R = 0.4$. Events with more than one additional jet with $p_T > 30$ GeV and $|\eta| < 4.5$ are rejected.
Mulitjet events can contribute to the background if a mis-measurement of one of the jets gives rise to fake $E_{\mathrm{T}}^{\mathrm{miss}}$, which then typically is aligned with the direction of the jet. To suppress such events, sub-leading jets which survive the jet veto have to be well separated from the direction of $E_{\mathrm{T}}^{\mathrm{miss}}$ - vector, which is ensured by requiring that their azimuthal distance be larger than 0.5: $|\Delta\phi(E_{\mathrm{T}}^{\mathrm{miss}}, jet_2)| > 0.5$. Contributions from the production of Z or W bosons in association with jets are reduced by rejecting any event containing a reconstructed electron or muon. The search is performed in 4 signal regions (SR), defined by symmetric lower bounds on the leading jet p_{T} and the $E_{\mathrm{T}}^{\mathrm{miss}}$ with values of 120 GeV, 220 GeV, 350 GeV, and 500 GeV, labelled SR1 - SR4, respectively.

2.2. Estimation of Background Contributions

The main (and irreducible) background contribution is the production of Z bosons together with jets where the boson decays to two neutrinos, $Z(\nu\bar{\nu})$+jets. Other significantly contributing processes are $W(\ell\nu)$+jets ($\ell = e, \mu, \tau$) with the decay leptons escaping detection or - in case of $W(\tau\nu)$+jets - the τ decaying hadronically. $Z/\gamma^{\star}(\ell^+\ell^-)$+jets ($\ell = e, \mu, \tau$) contribute when both leptons are not identified or are outside the detector acceptance. Together, those make up for about 97% of the total background. The remaining contribution are QCD multijet events, non-collision backgrounds such as muons from cosmic rays or beam halo, as well as top quark and diboson production ($WW/ZZ/WZ$). The top and diboson backgrounds are taken directly from the simulation while the other contribution are estimated in a data driven way. For the determination of the W/Z+jets background contributions, control regions (CR) orthogonal to the signal regions are defined by requiring good quality reconstructed electrons or muons in the event and otherwise applying the same cuts on jets and $E_{\mathrm{T}}^{\mathrm{miss}}$ as in the signal selection. This means that, for each control region process, there are 4 control regions, corresponding to the 4 signal regions.
A $W(\mu\nu)$+jet control region is defined by requiring exactly one good muon with a p_{T} greater than 20 GeV and rejecting additional electrons or muons. In addition, the transverse mass - defined as $m_T = \sqrt{2p_T E_T^{miss,\mu}(1 - cos\Delta\phi(\ell, E_T^{miss,\mu}))}$, with p_{T} of the lepton - has to fulfill 40 GeV $< m_T <$ 100 GeV and the $E_{\mathrm{T}}^{\mathrm{miss}}$ corrected for the muon contribution, $E_{\mathrm{T}}^{\mathrm{miss},\mu}$, should be larger than 25 GeV.
For the $Z(\mu^+\mu^-)$+jets CR exactly two oppositely charged muons are selected, restricting the invariant mass of the muon pair around the Z-mass peak, 66 GeV $< m_{\mu^+\mu^-} <$ 116 GeV.
Similarly, control regions in the electron channel are defined: The $W(e\nu)$+jets CR uses events which have been selected by an electron trigger and which contain exactly one good electron with $p_T >$ 25 GeV. An overlap removal between jets and the electron in the event is performed and the same invariant mass cut as for the muon channel is applied. The $E_{\mathrm{T}}^{\mathrm{miss}}$ in the event has to be larger than 25 GeV.

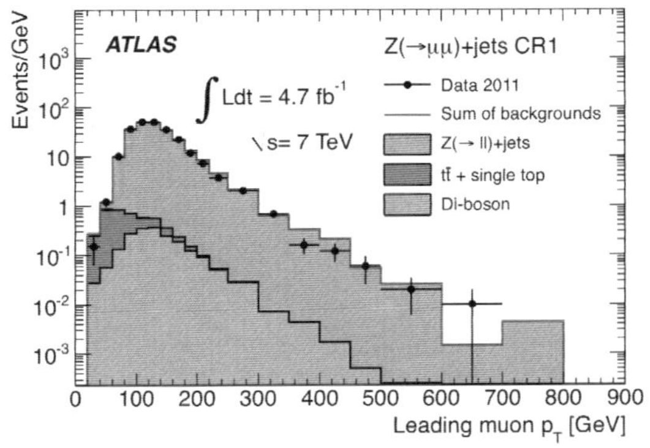

Figure 1. p_{T} distribution of the leading muon in the $Z(\mu^+\mu^-)$+jets control region corresponding to SR1. The background prediction is shown in comparison to the data [9].

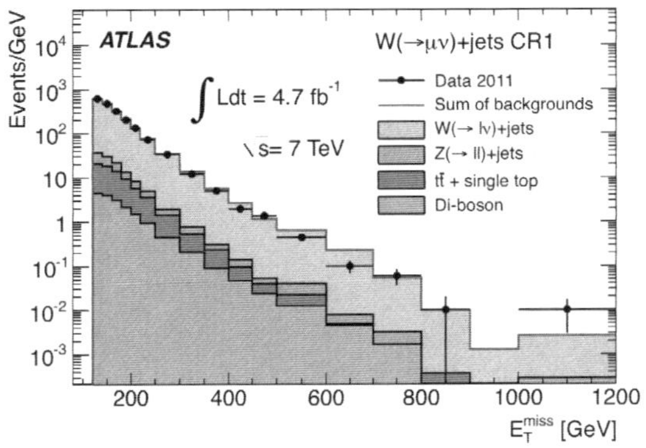

Figure 2. $E_{\mathrm{T}}^{\mathrm{miss}}$ distribution in the $W(\mu\nu)$+jets control region corresponding to SR1. The background prediction is shown in comparison to the data [9].

Finally, there is a $Z(e^+e^-)$+jets control region of events selected by an electron trigger and containing exactly two oppositely charged electrons and no muons. The electrons have to have p_T >25 and 20 GeV, respectively. Again, a jet-electron overlap removal is performed and the cut on the invariant mass of the lepton pair is applied as for the muon channel.
Fig. 1 shows the distribution of the leading muon p_{T} in the $Z(\mu^+\mu^-)$+jets CR, while in Fig. 2 the $E_{\mathrm{T}}^{\mathrm{miss}}$ distribution in the $W(\mu\nu)$+jets CR is displayed. Good agreement between data and simulation is observed.
The selected data in a control region has to be corrected for background contributions to this control region (from multijet events or other electroweak processes). By applying transfer factors to this corrected number, the contribution of a given background process in the signal region can be estimated. These transfer factors are based on ratios of event numbers predicted by simulation which allows for a significant reduction of systematic uncertainties compared to using simulation only. For example, the dominating $Z(\nu\bar{\nu})$+jets background contribution in the signal region is estimated from the $W(\mu\nu)$+jets control region as

$$N(Z(\nu\bar{\nu})\text{+jets})_{signal} = (N_{CR}^{data} - N_{CR}^{bkg}) \cdot (1 - f_{EW}) \cdot C \cdot \frac{N(Z(\nu\bar{\nu})\text{+jets})_{signal}^{sim}}{N(W(\mu\nu)\text{+jets})_{CR,jet,E_T^{miss}}^{sim}} \quad (1)$$

where N_{CR}^{data} is the number of data events observed in the CR, N_{CR}^{bkg} is the number of non-electroweak background events in the CR and f_{EW} the remaining fraction of events from electroweak processes other than $W(\mu\nu)$+jets. The predicted number of events in the SR is $N(Z(\nu\bar{\nu})\text{+jets})_{signal}^{sim}$, while $N(W(\mu\nu)\text{+jets})_{CR,jet,E_T^{miss}}^{sim}$ is the number of $W(\mu\nu)$+jets events in the CR from simulation after applying only the SR selection of jets and $E_{\mathrm{T}}^{\mathrm{miss}}$. The ratio of the two is the transfer factor. The factor C accounts for differences in the lepton selection between signal and control region.
Di-jet or tri-jet events can enter the signal region if the energy of one of the sub-leading jets is measured badly so that its p_{T} falls below the threshold of 30 GeV. To estimate this contribution from data, two control regions are defined: For the first region, the $\Delta\phi$-cut between the second jet and the $E_{\mathrm{T}}^{\mathrm{miss}}$ is inverted, for the second region a third jet above 30 GeV is required which again is aligned with the missing E_T: $\Delta\phi(jet_3, \mathrm{E}_T^{miss}) < 0.5$. The p_{T} distributions of these

Journal of Physics: Conference Series **455** (2013) 012020 doi:10.1088/1742-6596/455/1/012020

sub-leading jets are extrapolated below the 30 GeV threshold to get an estimate of the multijet contribution in the signal region.

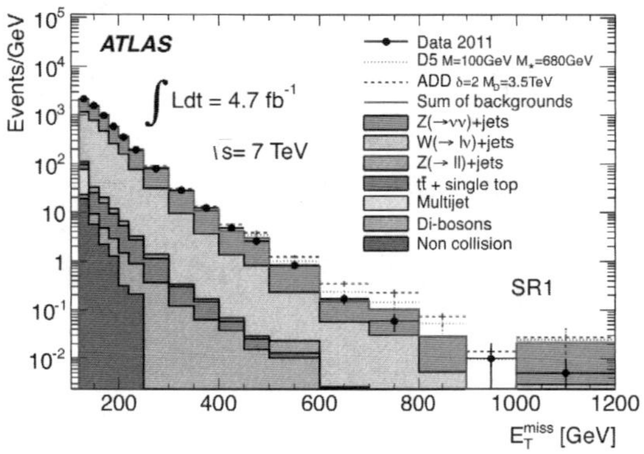

Figure 3. $E_{\mathrm{T}}^{\mathrm{miss}}$ distribution in SR1. The background prediction is shown in comparison to the data. For illustration purposes shapes from ADD and WIMP signal simulations are shown as well (dashed lines) [9].

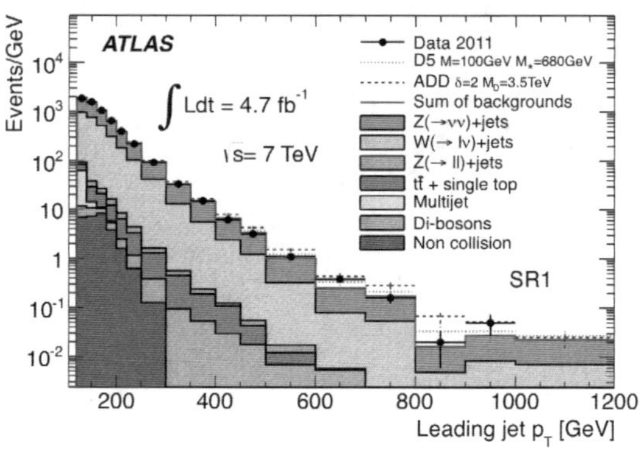

Figure 4. p_{T} distribution in SR1. The background prediction is shown in comparison to the data. For illustration purposes shapes from ADD and WIMP signal simulations are shown as well (dashed lines) [9].

2.3. Results

As can be seen from the $E_{\mathrm{T}}^{\mathrm{miss}}$ and p_{T} distributions (Figs 3 and 4, respectively) good agreement between data and prediction from the SM is observed. Therefore, limits on model parameters for the different theoretical interpretations are derived using the CLs method [11].

Fig. 5 shows the 95%CL limits from SR4 on the fundamental Planck scale in 4+n dimensions, M_D, as a function of the number of extra dimensions, n. The expected limit is shown as a black dashed line with the $\pm 1\sigma$ error band in gray, while the red solid line marks the observed limit. The theoretical uncertainties have not been included as nuisance parameters in the limit calculation. Instead, their impact is illustrated by the red dotted lines which show the observed limit recomputed after varying the signal cross section up and down within the theoretical uncertainties. The comparison with the violet line, which indicates the limits obtained from the monojet analysis performed with 2010 data [12], shows an increase in sensitivity.

The WIMP interpretation is based on the assumption that the interaction between dark matter and SM particles is mediated by a new particle, too heavy to be directly produced at the LHC. This allows for an effective field theory approach with 14 different operators for the contact interaction, assuming WIMPs are Dirac fermions. One characteristic operator is D5 (following the naming scheme from [13]), which describes a vector interaction with a quark-anitquark pair in the initial state. The 90%CL lower limit on the suppression scale of the effective theory, $M_\star$, for D5 is shown in Fig. 6 as a function of the WIMP mass. For these limits the results from SR3 are used. The colour scheme is the same as described for Fig. 5, the gray shaded area (in the bottom right corner) marks the region where the effective theory is not valid any more. In addition, a green line indicates the thermal relic limit, i.e. those values of $M_\star$ that for a given WIMP mass m_χ result in the correct cross section to yield the total relic density of dark matter in the universe as measured by the WMAP satellite [14]. This means that above this green line, annihilation channels other than the one considered by the given operator must also be open.
The lower limits on $M_\star$ can be translated into upper limits on WIMP-nucleon scattering [13] and these can be compared to direct detection experiments which are sensitive to spin-dependent and

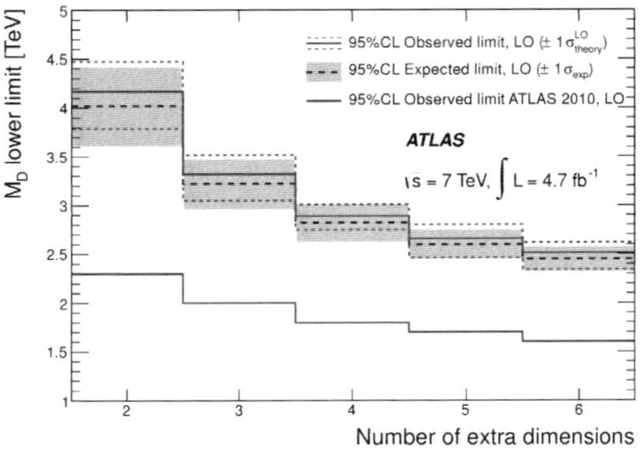

Figure 5. 95%CL lower limits from SR4 on the fundamental Planck scale M_D as function of the number of extra dimensions. The expected limit is shown as a black dashed line with a gray $\pm 1\sigma$ error band. The red solid line shows the observed limit with the impact of theoretical uncertainties illustrated by the red dotted lines. For comparison the observed limit from the 2010 analysis is shown [9].

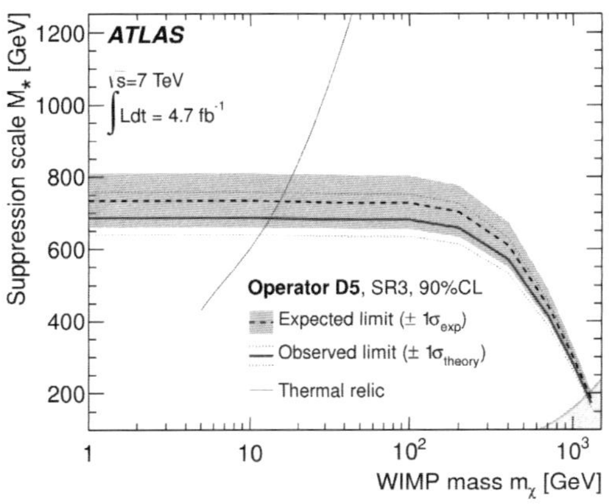

Figure 6. 90%CL lower limits from SR3 on the suppression scale $M_\star$. The colour coding is the same as in fig. 5. The gray shaded area marks the region where the effective theory is not valid. The green line indicates those values of $M_\star$ that for a given value of m_χ result in the measured relic density [9].

to spin-independent interactions. In Fig. 7 the limits for operators describing spin-independent interactions are compared to other collider limits as well as limits from direct searches, again as a function of the WIMP mass. It can be seen that the colliders have a higher sensitivity at small WIMP masses (of the order of MeV). For spin-dependent interactions on the other hand, colliders are competitive over a large range of WIMP masses, as Fig. 8 illustrates.
The monojet analysis was updated with $10.5\,\text{fb}^{-1}$ of 8 TeV data with only minor changes to the analyses strategy and cuts. Again, no significant excess over the SM prediction was observed as can be seen from the E_T^{miss} distribution in the first signal region (Fig. 9). A new theoretical interpretation has been included in the analysis: the associated production of a gravitino together with a squark or a gluino, assuming a GMSB scenario. Lower limits on the gravitino mass for different squark/gluino mass configurations can be derived. Fig. 10 shows the limit for the degenerate case as a function of the squark/gluino mass. The dashed-dotted line defines the validity of the narrow-width approximation (NWA) used to derive the decay rate of a squark/gluino to a gravitino and a parton. For comparison, the current limit from LEP on the gravitino mass assuming very heavy squarks/gluino is shown (solid red line).

3. Monophoton Analysis

The monophoton analysis has been performed at the ATLAS experiment with $4.7\,\text{fb}^{-1}$ of 7 TeV data [15]. In the following the event selection, background determination and results are presented.

3.1. Event Selection

The $4.7\,\text{fb}^{-1}$ remain after applying quality criteria on the data. Out of these, events are selected using an E_T^{miss} trigger with a threshold of 70 GeV and requiring at least one primary vertex. The photon is required to be within $|\eta| < 2.37$ where the transition region between barrel and endcap ($1.35 < |\eta| < 1.52$) is excluded. At most one jet with $p_T > 30$ GeV and $|\eta| < 4.5$ is allowed in

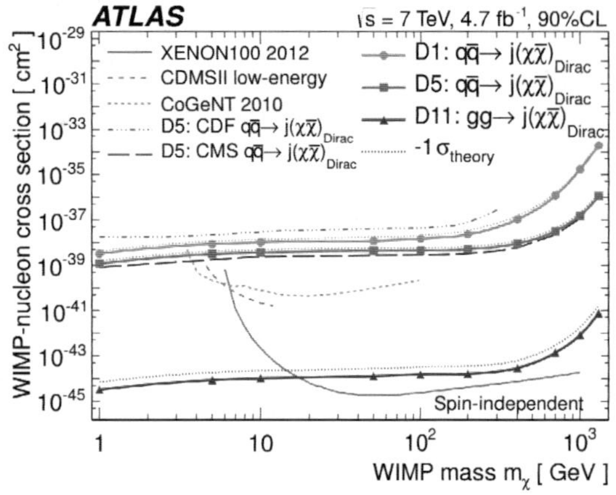

Figure 7. 90%CL upper limits on the spin-independent WIMP-nucleon scattering cross section for different collider and direct searches as a function of the WIMP mass. Solid lines are the observed limits without theory uncertainties, the dotted lines show the impact of the theoretical uncertainties. The ATLAS limits are calculated considering the four light quark flavours assuming equal coupling strength to WIMPs for all flavours [9].

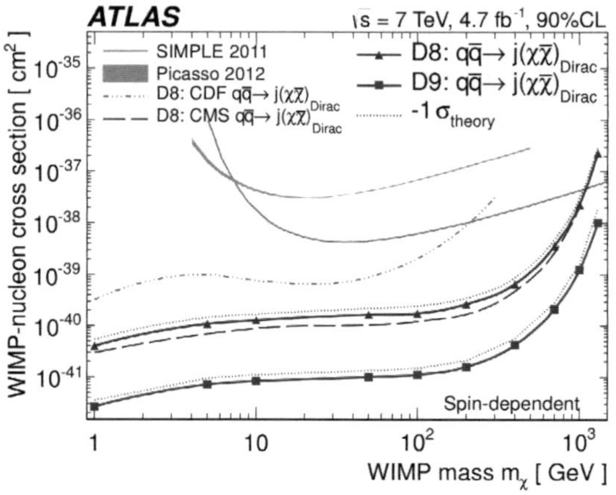

Figure 8. 90%CL upper limits on the spin-dependent WIMP-nucleon scattering cross section for different collider and direct searches as a function of the WIMP mass. Solid lines are the observed limits without theory uncertainties, the dotted lines show the impact of the theoretical uncertainties. The ATLAS limits are calculated considering the four light quark flavours assuming equal coupling strength to WIMPs for all flavours [9].

the event. The signal region is defined by requiring the $E_{\mathrm{T}}^{\mathrm{miss}}$ and the p_{T} of the photon to be larger than 150 GeV. The photon, $E_{\mathrm{T}}^{\mathrm{miss}}$ and jet (if present) have to be well separated, which is ensured by applying the following cuts: $|\Delta\phi(E_{\mathrm{T}}^{\mathrm{miss}},\gamma)| > 0.4$, $|\Delta\phi(E_{\mathrm{T}}^{\mathrm{miss}},jet)| > 0.4$ and $|\Delta R(\gamma, jet)| > 0.4$. As in the monojet analysis a veto on any reconstructed electron or muon is applied.

3.2. Background Estimation

The dominant background contribution is events with a photon and a Z boson where the latter decays into two neutrinos. Further contributions are $W/Z+\gamma$ events with lost electrons, muons or hadronically decaying taus, and W/Z+jet events where an electron or jet is mis-identified as a photon. The remaining backgrounds consist of top, diboson ($WW/ZZ/WZ$), $\gamma\gamma$, $\gamma + jet$ and multijet processes. Non-collision backgrounds are found to be negligible. The contributions from top, diboson and $\gamma\gamma$ are taken from the simulation while for the W/Z, γ+jet and multijet backgrounds the same data-driven techniques as in the monojet analysis are used.

3.3. Results

As in the monojet search no significant deviation from the SM prediction is observed as illustrated by Fig. 11. Hence, lower limits on M_D are derived for the ADD model. These are shown in Fig. 12 as a function of the number of extra dimension in comparison to results from other experiments. Upper limits on the WIMP-nucleon scattering cross section are also derived as was done for the monojet analysis. The 90%CL bounds are shown in Fig. 13 as a function of the WIMP mass in comparison to other collider and direct search experiments both for spin-independent and spin-dependent interactions.

Journal of Physics: Conference Series **455** (2013) 012020 doi:10.1088/1742-6596/455/1/012020

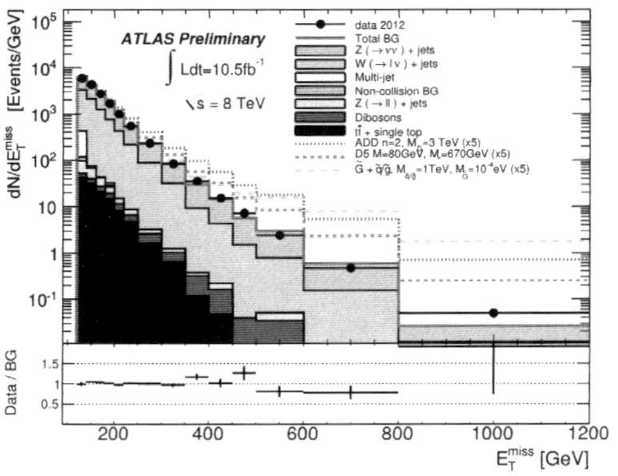

Figure 9. Distribution of $E_{\mathrm{T}}^{\mathrm{miss}}$ in SR1 for the 8 TeV monojet analysis. The data are shown in comparison to the SM prediction. For illustration purposes signal simulation for all three theoretical interpretations are also shown as dashed lines [10].

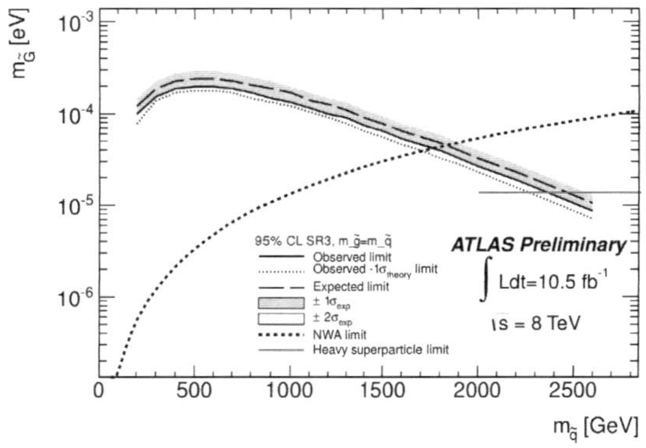

Figure 10. 95% CL lower limits on the gravitino mass as a function of the squark mass for degenerate squark/gluino masses. The dotted line indicates the impact of the theoretical uncertainty on the observed limit (solid line). The expected limit (dashed line) is shown with its $\pm 1\sigma$ and $\pm 2\sigma$ error bands. The solid red line denotes the current limit from LEP [10].

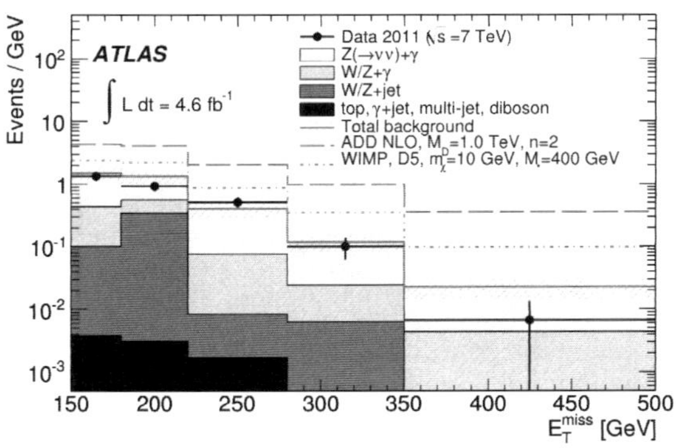

Figure 11. Distribution of $E_{\mathrm{T}}^{\mathrm{miss}}$ in the signal region for the monophoton analysis. The data are shown in comparison to the SM prediction. The band around the total background prediction includes uncertainties on the data-driven background estimates and statistical uncertainties on the MC samples. For illustration purposes signal simulations for the theoretical interpretations are also shown as dashed lines [15].

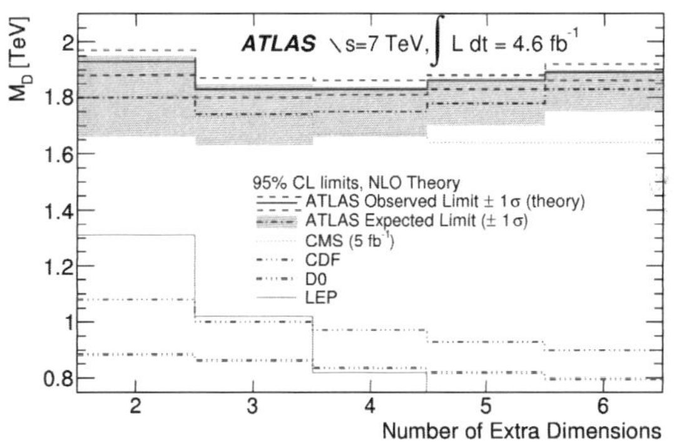

Figure 12. Observed (solid lines) and expected (dashed-dotted lines) 95% CL limits on M_D as a function of the number of extra dimensions in the ADD model. The dashed lines around the observed limit indicate the impact of the $\pm 1\sigma$ NLO theoretical uncertainty on the limit computation. The shaded band indicates the expected $\pm 1\sigma$ range of limits in the absence of a signal. The results are compared with previous results [15].

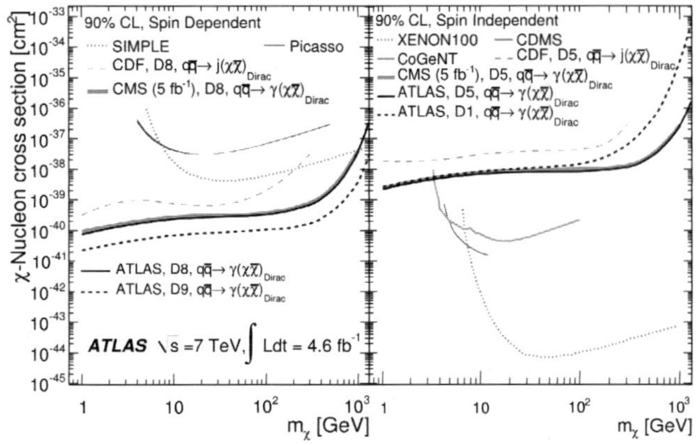

Figure 13. 90% CL upper limits on the nucleon-WIMP cross section as a function of m_χ for spin-dependent (left) and spin-independent (right) interactions. Collider bounds are compared to results from direct search experiments [15].

4. Conclusion

The ATLAS experiment has performed analyses with monojet and monophoton signatures with proton-proton collision data at center of mass energies of 7 TeV and 8 TeV. Data-driven techniques were used to determine the largest background contributions from Standard Model processes. No excess over the SM predictions is observed and the results are used to set limits on model parameters for different BSM scenarios: the ADD model of large extra dimensions, associated squark(gluino)+gravitino production in gauge-mediated SUSY breaking and WIMP pair production.

References

[1] Giudice G F, Rattazzi R and Wells D 1999 Quantum gravity and extra dimensions at high-energy colliders *Nucl. Phys.* B **544** 3-38

[2] Arkani-Hamed N, Dimopoulos S and Dvali G R 1998 The Hierarchy problem and new dimensions at a millimeter *Phys. Lett.* B **429** 263-72

[3] Fox P J, Harnik R, Kopp J and Tsai Y 2012 Missing energy signatures of dark matter at the LHC *Phys. Rev.* D **85** 056011

[4] Giudice G F and Rattazzi R 1999 Theories with gauge mediated supersymmetry breaking *Phys. Rep.* **322** 419

[5] Fayet P 1977 Mixing between gravitational and weak interactions through the massive gravitino *Phys. Lett.* B **70** 461

[6] Casalbuoni R, De Curtis S, Dominici D, Feruglio F and Gatto R 1988 A gravitino - goldstino high-energy equivalence theorem *Phys. Lett.* B **215** 313

[7] Klasen M, Pignol G 2007 New Results for Light Gravitinos at Hadron Colliders: Tevatron Limits and LHC Perspectives *Phys. Rev.* D **75** 115003

[8] ATLAS Collaboration 2008 The ATLAS experiment at the CERN Large Hadron Collider *J. Inst.* **3** S08003

[9] ATLAS Collaboration 2012 Search for dark matter candidates and large extra dimensions in events with a jet and missing transverse momentum with the ATLAS detector *Preprint* arXiv:1210.4491 [**hep-ex**]

[10] ATLAS Collaboration 2012 Search for New Phenomena in Monojet plus Missing Transverse Momentum Final States using 10 fb^{-1} of pp Collisions at $\sqrt{s} = 8$ TeV with the ATLAS detector at the LHC, ATLAS-CONF-2012-147, https://cdsweb.cern.ch/record/1493486

[11] A. L. Read A L 2002 Presentation of search results : The CLs technique *J.Phys* G **28** 26932704

[12] ATLAS Collaboration 2011 Search for new phenomena with the monojet and missing transverse momentum signature using the ATLAS detector in $\sqrt{s} = 7$ TeV proton-proton collisions *Phys. Lett.* B **705** 294-312

[13] Goodman J, Ibe M, Rajaraman A, Shepherd W, Tait T M P and Yu H-B 2010 Constraints on Dark Matter from Colliders *Phys. Rev.* D **82** 116010

[14] WMAP Collaboration 2011 Seven-year Wilkinson Microwave Anisotropy Probe (WMAP) observations: Cosmological interpretation *Astrophys. J. Suppl.* **192** 18

[15] ATLAS Collaboration 2012 Search for dark matter candidates and large extra dimensions in events with a photon and missing transverse momentum in pp collision data at $\sqrt{s} = 7$ TeV with the ATLAS detector *Preprint* arXiv:1209.4625 [**hep-ex**]

Journal of Physics: Conference Series **455** (2013) 012021 doi:10.1088/1742-6596/455/1/012021

Searches for Exotics: Heavy resonances with the ATLAS detector

Simon Viel, on behalf of the ATLAS Collaboration
University of British Columbia - TRIUMF, Vancouver, Canada
E-mail: sviel@cern.ch

Abstract. Many theories that go beyond the Standard Model predict the existence of new heavy resonances decaying into pairs of particles. This review summarizes a wide collection of recent results from the ATLAS experiment at the Large Hadron Collider on searches for resonances decaying into various combinations of charged leptons, neutrinos, jets from gluons or light quarks, top quarks, photons and heavy gauge bosons. Limits are set on a variety of theories beyond the Standard Model used as benchmarks, among them Kaluza-Klein, Randall-Sundrum and ADD models with extra dimensions, as well as Grand Unified Theories and Technicolour.

1. Introduction

Resonances decaying into pairs of particles have played a very significant role in the development of particle physics, and are an important place to look for phenomena beyond the Standard Model (SM). This review summarizes several searches for such new particles, conducted using proton-proton collision data at a centre-of-mass energy of 7 TeV or 8 TeV recorded with the ATLAS detector [1] at the Large Hadron Collider (LHC). Since no significant excess is observed over expectations from SM processes, limits are set at the 95% confidence level (CL) on resonances decaying into jets from light quarks or gluons, charged leptons (including taus), neutrinos, top quarks, photons or heavy gauge bosons.

The searches are signature-based, and interpreted in terms of theories beyond the Standard Model, used as benchmarks.[1] The Sequential Standard Model (SSM), where new gauge bosons W'_{SSM} and Z'_{SSM} only differ from the SM W and Z by their mass, is commonly used for this purpose. Better theoretically motivated are the additional gauge bosons W' and Z' which arise from Grand Unified Theories (GUT), for example with E_6 as the unifying gauge group, or in parametrizations such as the Minimal Z' models. Another class of models postulates extra dimensions of space in an attempt to solve the hierarchy problem, and these theories predict new resonances such as Kaluza-Klein (KK) modes of the graviton in ADD models, excited gravitons (G^*) or KK gluon modes (g_{KK}) in Randall-Sundrum (RS) models, or KK modes of other gauge bosons (γ_{KK}, Z_{KK}) in TeV^{-1} models. Other possible solutions to the hierarchy problem, including Technicolour and theories with chiral bosons (W^*, Z^*), predict resonances that may be observed at the LHC, and likewise other theories beyond the SM predict excited fermions, colour-octet scalars, string resonances, torsion resonances, quantum black holes, etc. Some searches can also be interpreted in terms of effective four-fermion contact interactions.

[1] Theoretical references can be found in the respective experimental works cited.

Journal of Physics: Conference Series **455** (2013) 012021 doi:10.1088/1742-6596/455/1/012021

2. The ATLAS detector

The analyses described in this review collectively make use of all the detector systems that constitute the ATLAS detector. The inner detector (ID), used to reconstruct charged particle tracks and vertices, consists of silicon pixels, silicon strips and transition radiation detectors, covering $|\eta| < 2.5$ in pseudorapidity[2]. It is immersed in a homogeneous solenoidal magnetic field, which allows to measure the momentum of charged particles using track curvature. The calorimeters, responsible for the reconstruction of particle showers, are made of liquid argon and lead in the electromagnetic (EM) part, while the hadronic part is composed of scintillating tiles and iron in the central region and liquid argon, copper and tungsten in the forward region. Outside the calorimeter, toroid magnets provide the field for the muon spectrometer (MS), which consists of resistive-plate and thin-gap trigger chambers, and three sets of drift tubes and cathode strip chambers allowing for the reconstruction of muon tracks with high precision.

3. Dijet resonances

Collisions at the LHC often result in a pair of jets of particles with high transverse momentum (p_{T}). and such events can reach the highest energies accessible at the LHC. Jets in ATLAS are reconstructed using the anti-k_t clustering algorithm [2] with the distance parameter $R = 0.6$, and their energy is measured in the calorimeters. The dijet invariant mass, m_{jj}, is calculated from the four-momenta of the two highest-p_{T} central jets passing the selection criteria, and the resulting spectrum from data is fitted with a smooth function: $f(x) = p_1(1-x)^{p_2}x^{p_3+p_4 \ln x}$, where $x = m_{jj}/\sqrt{s}$ and the p_i are fitting parameters, to obtain the background estimate.

The dijet invariant mass distribution obtained using 13 fb^{-1} of data collected in 2012 at $\sqrt{s} = 8$ TeV [3] is shown along with the corresponding background estimate in Figure 1. Limits are set on excited quarks (q^*), excluding masses above 3.84 TeV at the 95% CL, as well as on the cross section times acceptance ($\sigma \times A$) of Gaussian signal shapes, as shown in Figure 2.

The analysis performed using the full dataset taken in 2011 at $\sqrt{s} = 7$ TeV [4], set limits on a wider range of hypothetical resonances decaying into dijets: colour octet scalars, W'_{SSM} and string resonances, in addition to excited quarks. In addition, an angular analysis was performed and data were shown to be consistent with QCD NLO simulations. Limits were therefore set on the scale of four-quark contact interactions, and on the quantum gravity scale for a quantum black hole signal in six additional space-time dimensions.

4. Dilepton resonances

Searches for resonances decaying into pairs of high-energy electrons and muons rely on accurately measuring the final-state leptons. Electron candidates are identified using clusters in the electromagnetic calorimeter associated with inner detector tracks, while muon candidates are combined tracks from the independent inner detector and muon spectrometer measurements.

The main experimental challenge for this analysis in the muon channel is to ensure a good momentum measurement. Muon momenta are measured using the curvature of the associated tracks due to the magnetic field. At high momentum, tracks are very straight, and the resolution is dominated by the intrinsic curvature resolution. Therefore, only the best-aligned chambers of the muon spectrometer can be used for this analysis, and muons are required to pass stringent hit requirements in both the ID and the MS. Impact parameter and track isolation requirements suppress muons from cosmic rays and QCD jets, then evaluated to be negligible in this channel.

In the electron channel, the main challenge is particle identification: quality cuts on the shape of the energy deposit, inner detector hits and calorimeter isolation are needed to distinguish genuine high-energy electrons from QCD jets and converted photons. Data-driven strategies are employed to estimate the remaining background due to these sources.

[2] Pseudorapidity is defined in terms of the polar angle θ as $\eta = -\ln\tan(\theta/2)$, while φ is the azimuthal angle.

Journal of Physics: Conference Series **455** (2013) 012021 doi:10.1088/1742-6596/455/1/012021

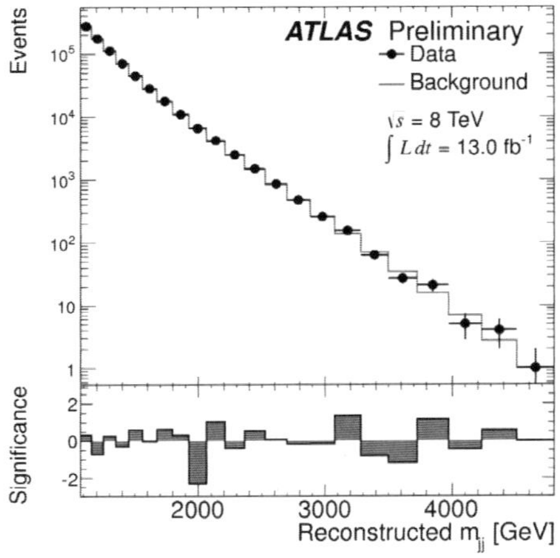

Figure 1. Reconstructed dijet mass distribution with statistical uncertainties, fitted with a smooth functional form. The bin-by-bin significance of the data-fit difference is shown in the lower panel. [3]

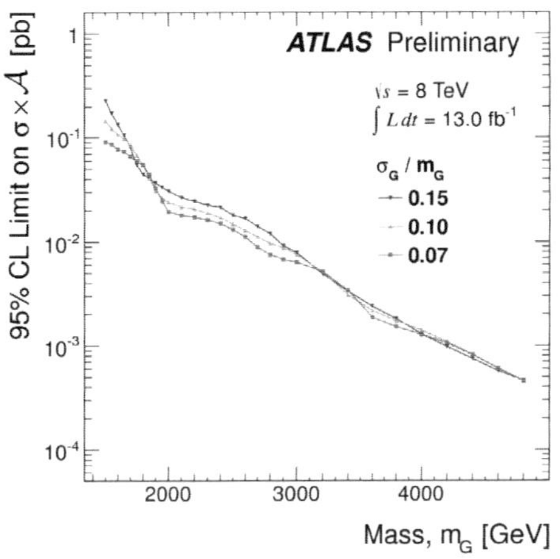

Figure 2. The 95% CL upper limits on $\sigma \times A$ for Gaussian signal shapes as a function of the mean mass m_G, for three values of the relative standard deviation σ_G/m_G. [3]

The dilepton invariant mass spectra obtained using 6 fb^{-1} of data taken in 2012 at $\sqrt{s} = 8$ TeV [5] are shown in Figure 3. The dominant background to this search is the SM Z/γ^* production and decay into dileptons, followed by contributions from diboson processes and top quark pairs. Background from these sources is estimated using Monte Carlo (MC) simulations. The sum of all background contributions is normalized to the data on the Z peak.

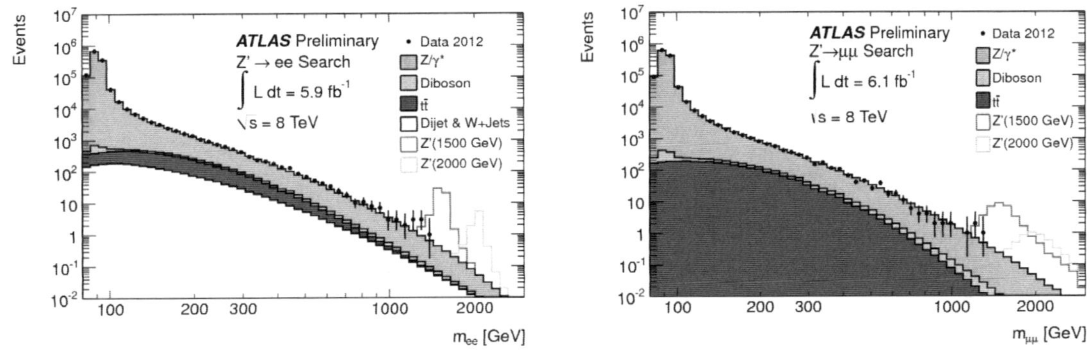

Figure 3. Dielectron (left) and dimuon (right) invariant mass distributions after final selection, compared to the stacked sum of all expected backgrounds, with three example Z'_{SSM} signals overlaid. The bin width is constant in log $m_{\ell^+\ell^-}$. [5]

This analysis sets limits on Z' from the SSM and E_6 GUT, excluding Z'_{SSM} with masses below 2.49 TeV at the 95% CL. Limits on further models were set using data collected in 2011 at $\sqrt{s} = 7$ TeV [6]: RS G^*, Z_{KK}, Z^*, Technicolour, torsion resonances and Minimal Z' models. A non-resonant search was also performed using the integral of the dilepton spectrum [7], setting limits on the scale of $\ell\ell qq$ contact interactions, as well as on the string scale in the ADD model.

5. Diphoton resonances

Photons in ATLAS are measured primarily using the EM calorimeter, but the inner detector helps to identify photon conversions, by determining whether charged particle tracks come directly from the primary collision vertex or from a subsequent decay. Requirements on the shower width and energy distribution, as well as a tight calorimeter isolation requirement help distinguish photons from jets. As well, an ambient energy correction is applied to remove contributions to the photon energy measurement from the underlying event and pile-up.

The invariant mass distribution of diphoton events is shown in Figure 4, for the analysis using data taken at $\sqrt{s} = 7$ TeV in 2011 [8]. The dominant, irreducible background in this search is $\gamma\gamma$ production from the SM, estimated from MC simulations. Reducible backgrounds due to events with one or more jets mis-identified as a photon are estimated using a data-driven technique.

Limits at the 95% CL are set on the RS G^* in combination with the results in the dilepton channel at $\sqrt{s} = 7$ TeV, in the plane of $k/\overline{M}_{\mathrm{Pl}}$ versus M_{G^*}, are presented in Figure 5.

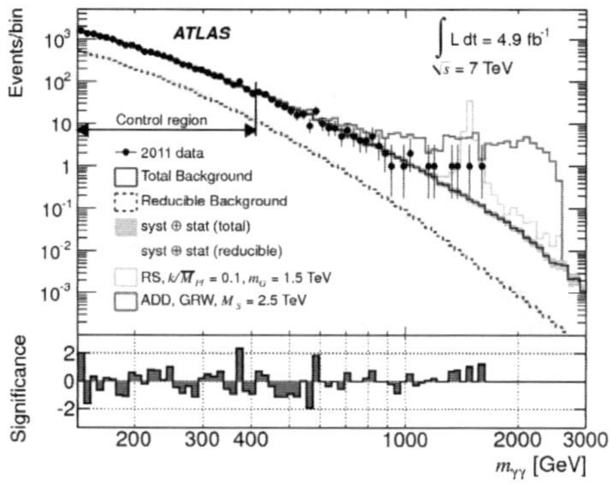

Figure 4. Observed invariant mass distribution of diphoton events. The SM background expectation and two expected signal examples are superimposed. The bin-by-bin significance of the difference between data and predicted background is shown in the lower panel. [8]

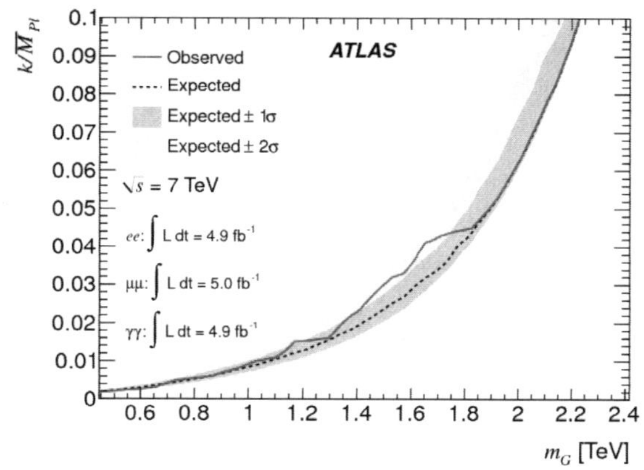

Figure 5. Expected and observed limits from the combination of $G^* \to \gamma\gamma/ee/\mu\mu$ channels, interpreted in the plane of $k/\overline{M}_{\mathrm{Pl}}$ versus graviton mass. The region above the curve is excluded at the 95% CL. [8]

6. Resonances decaying into a lepton and a neutrino

Neutrinos leave ATLAS undetected, but their production can be inferred via the presence of missing transverse momentum ($E_{\mathrm{T}}^{\mathrm{miss}}$). In the search for resonances decaying into a neutrino and exactly one high-p_{T} muon or electron, $E_{\mathrm{T}}^{\mathrm{miss}}$ is combined with the lepton momentum to form the transverse mass (m_{T}):

$$m_{\mathrm{T}} = \sqrt{2 p_{\mathrm{T}} E_{\mathrm{T}}^{\mathrm{miss}} (1 - \cos\varphi_{\ell\nu})}, \qquad (1)$$

where $\varphi_{\ell\nu}$ is the angle between the p_{T} and $E_{\mathrm{T}}^{\mathrm{miss}}$. The lepton measurement is especially important in this channel, because any mis-measurement also biases the $E_{\mathrm{T}}^{\mathrm{miss}}$, and therefore has a double impact on m_{T}. The selected leptons satisfy stringent identification and quality requirements.

The resulting m_{T} spectra with the complete dataset taken in 2011 at $\sqrt{s} = 7$ TeV [9] are shown in Figure 6. The main background to this search is SM W production and decay in the leptonic channel, and also Z/γ^* where one of the two final state leptons is not reconstructed, as well as diboson processes, top quark pairs and single top. All of the above are estimated from MC simulation, and contributions from mis-identified QCD jets are estimated from the data.

Limits at the 95% CL are set on $W'_{\rm SSM}$ excluding masses below 2.55 TeV, as well as on W^* excluding masses below 2.42 TeV.

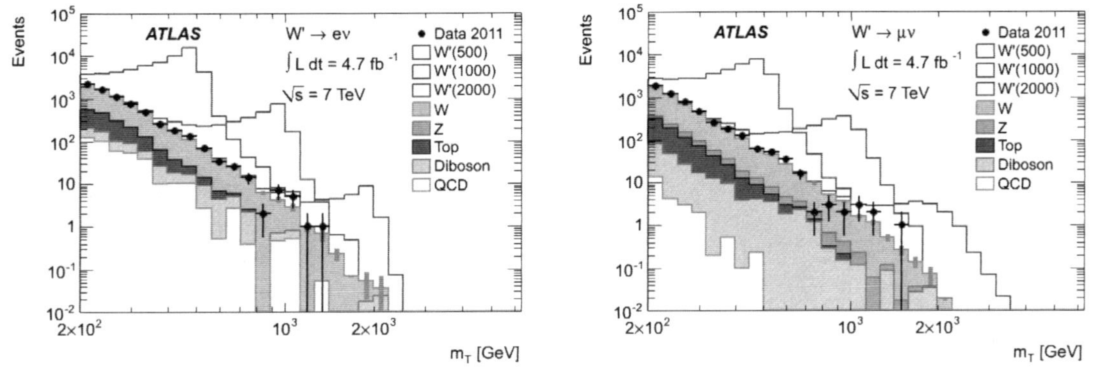

Figure 6. Spectra of $m_{\rm T}$ for the electron (left) and muon (right) channels, with three example $W'_{\rm SSM}$ signals overlaid. The bin width is constant in log $m_{\ell^+\ell^-}$. [9]

7. Resonances decaying into a pair of weak bosons

7.1. $WW \to \ell\nu\ell\nu$

Searches looking for exactly two oppositely-charged high-$p_{\rm T}$ electrons or muons and large $E_{\rm T}^{\rm miss}$ are sensitive to the production of RS G^* decaying into W boson pairs. In addition to the original RS model, one of its extensions, the bulk RS model, is considered in this search because there the decay probability of the new boson G^*_{bulk} in light fermion and photon channels is small and the coupling to heavy particles such as W, Z, Higgs bosons and top quarks is strongly enhanced.

The observable in this search is defined as follows, with $p^{\ell i}$ the momentum of the $i^{\rm th}$ lepton:

$$m_{\rm T}^{\rm WW} = \sqrt{(\sum_{i=1}^{2} p_{\rm T}^{\ell i} + E_{\rm T}^{\rm miss})^2 - (\sum_{i=1}^{2} p_x^{\ell i} + E_{\rm T}^{\rm miss})^2 - (\sum_{i=1}^{2} p_y^{\ell i} + E_{\rm T}^{\rm miss})^2} \tag{2}$$

and the resulting distributions with data collected in 2011 at $\sqrt{s}$ = 7 TeV [10] show good agreement with expectations from the SM. Reducible backgrounds are suppressed using dedicated cuts: requiring that the invariant mass of the two leptons in the event be higher than 106 GeV reduces the contribution from Z/γ^*, and a b-tag veto helps reject top quark events. These two sources of background are estimated from simulation, and the estimates are validated using dedicated control regions. For $m_{\rm T}^{\rm WW} > 300$ GeV, the predicted backgrounds are obtain from fits to the MC samples. This analysis sets limits excluding, both for $k/\overline{M}_{\rm Pl} = 0.1$, G^* masses below 1.23 TeV and G^*_{bulk} masses below 0.84 TeV, at the 95% CL.

7.2. ZW or $ZZ \to \ell\ell qq$

In the $\ell\ell qq$ decay channel, 7.2 fb^{-1} of data taken at $\sqrt{s} = 8$ TeV in 2012 are analyzed to look for a resonant excess in the diboson mass spectrum [11]. Two final states are considered: while the jets from many W and Z hadronic decays can be resolved, in events with high invariant mass the gauge bosons are highly boosted and hadronic decays merge into a single large jet.

To obtain the background prediction, data in each of these two event selections are fitted using the same function as in the dijet analysis: $f(x) = p_1(1-x)^{p_2}x^{p_3+p_4 \ln x}$, with $x = m_{lljj}/\sqrt{s}$ or $x = m_{llj}/\sqrt{s}$ depending on the selection, and the results are shown in Figure 7. Limits are set excluding G^*_{bulk} for $k/\overline{M}_{\rm Pl} = 1.0$ for masses below 0.88 TeV, at the 95% CL.

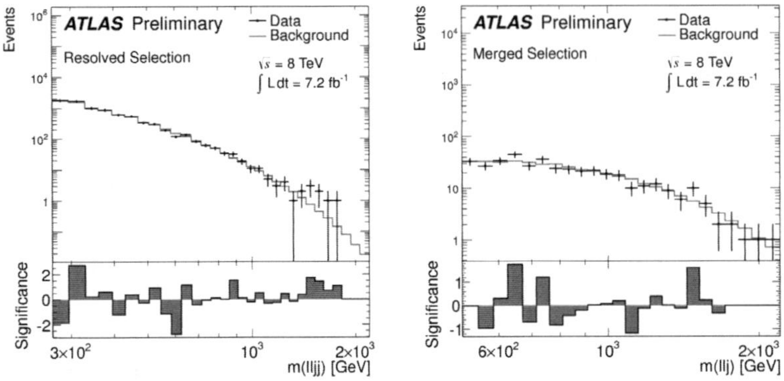

Figure 7. Distributions of reconstructed m_{lljj} (left) and m_{llj} (right) for the combined electron and muon channels. The local deviation of the data from the background expectation is also shown and takes into account statistical uncertainties only. [11]

8. Third-generation searches

Analyses looking for new processes decaying into third-generation particles are complementary to their light-flavour counterparts. For example, some extensions of the standard model proposed to explain the high mass of the top quark predict new gauge bosons that couple preferentially to third-generation fermions. All results in this section were obtained using the complete dataset collected in 2011 at $\sqrt{s} = 7$ TeV [12, 13, 14, 15].

8.1. Ditau resonances

The identification of hadronic tau decays is performed using Boosted Decision Trees, trained to discriminate against QCD jets using shower shape and tracking identification variables.

The search for ditau resonances has the same backgrounds as the dilepton analysis, and is carried out by counting events above thresholds in $m_{\mathrm{T}}^{\mathrm{tot}}$, defined as

$$m_{\mathrm{T}}^{\mathrm{tot}} = \sqrt{2p_{\mathrm{T}1}p_{\mathrm{T}2}C + 2E_{\mathrm{T}}^{\mathrm{miss}}p_{\mathrm{T}1}C_1 + 2E_{\mathrm{T}}^{\mathrm{miss}}p_{\mathrm{T}2}C_2} \quad (3)$$

where $C = 1 - \cos(\Delta\varphi)$, with $\Delta\varphi$ the angle between the two visible tau decay products, and C_1 and C_2 are defined analogously for the angles between the respective decay products and $E_{\mathrm{T}}^{\mathrm{miss}}$.

Four decay channels are considered, depending on the combinations of hadronic (τ_{had}) or leptonic (τ_e or τ_μ) tau decays. The $m_{\mathrm{T}}^{\mathrm{tot}}$ spectra obtained are shown in Figure 8 [12]. This analysis sets limits on the cross-section of a new resonance times its branching fraction to tau lepton pairs, with Z'_{SSM}, used as a benchmark, excluded for masses below 1.4 TeV.

8.2. Ditop resonances in the fully hadronic channel

Forty-six percent of top quark pair ($t\bar{t}$) events have both top quarks decaying in the hadronic channel, and novel hadronic top-tagging techniques allow a search for resonances decaying into $t\bar{t}$. Two complementary algorithms are used: the HEPTopTagger tests the substructure of jets with a large distance parameter $R = 1.5$ and is effective in identifying top quarks with $p_{\mathrm{T}} > 200$ GeV, while the Top Template Tagger matches the observed energy deposits of jets with $R = 1.0$ to a large set of possible patterns of energy deposits from simulated hadronic top quark decays, in order to reject jets from light quarks and gluons, and is optimized for top quarks with $p_{\mathrm{T}} > 450$ GeV.

In addition to $t\bar{t}$ production from the SM, there is a background contribution from QCD multijets, which is evaluated from data. Figure 9 shows the resulting $t\bar{t}$ invariant mass distributions [13]. Limits at the 95% CL are set excluding g_{KK} for masses between 0.70 and 1.62 TeV, as well as a leptophobic Z' boson for masses between 0.70 and 1.00 TeV and between 1.28 and 1.32 TeV.

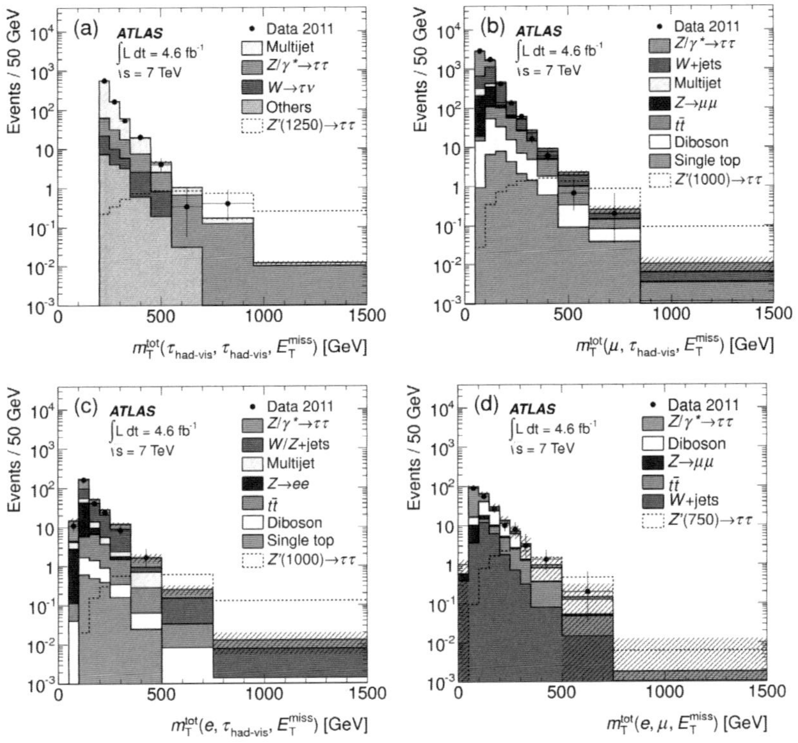

Figure 8. Spectra of m_T^{tot} for each channel: (a) $\tau_{had}\tau_{had}$, (b) $\tau_\mu\tau_{had}$, (c) $\tau_e\tau_{had}$ and (d) $\tau_e\tau_\mu$, with an example Z'_{SSM} signal overlaid. The signal mass point closest to the Z'_{SSM} exclusion limit in each channel is chosen and is indicated in parentheses in the legend in units of GeV. The uncertainty on the total estimated background (hatched) includes only the statistical uncertainty from the simulated samples. The visible decay products of hadronically decaying taus are denoted by $\tau_{had-vis}$. [12]

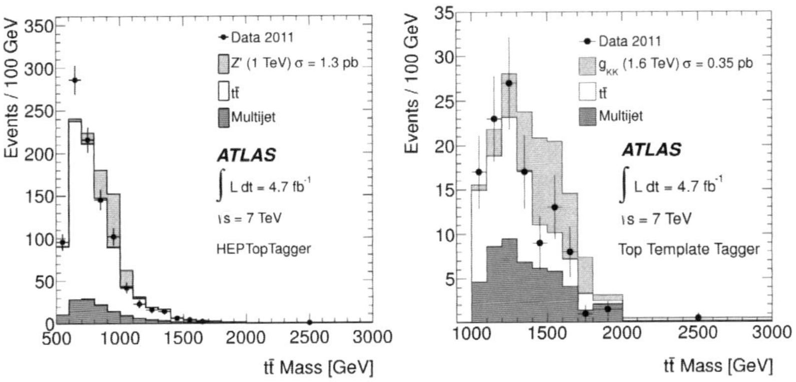

Figure 9. Distributions of the $t\bar{t}$ invariant mass in the fully hadronic channel, for the HEPTopTagger with a Z' signal (left) and for the Top Template Tagger with a g_{KK} signal (right). [13]

8.3. Ditop resonances in the semi-leptonic channel

The analysis looking for resonances decaying into $t\bar{t} \rightarrow \ell\nu b + jjb$, where $\ell = e$ or μ, makes use of two event selections. In the "resolved" selection, jets in the event are assigned to either top quark using a χ^2 algorithm, while the "boosted" selection requires one jet with $R = 1.0$, interpreted as a hadronic top decay, and a b-tagged jet close to the lepton ($\sqrt{\Delta\eta^2 + \Delta\varphi^2} < 1.5$). Additional requirements are applied on m_T and E_T^{miss} to reduce the QCD multijet background.

The $t\bar{t}$ invariant mass spectrum observed in this analysis [14] is in good agreement with the SM, hence 95% CL limits are set excluding g_{KK} for masses below 1.7 TeV and a leptophobic Z' for masses below 1.9 TeV.

Journal of Physics: Conference Series **455** (2013) 012021 doi:10.1088/1742-6596/455/1/012021

8.4. Resonances decaying into a top quark and a light quark

Finally, a search for resonances decaying into a light quark and a top quark decaying leptonically is carried out to probe models proposed to explain the top quark forward-backward asymmetry (A_{FB}) anomaly from the Tevatron experiments. This analysis makes use of the semi-leptonic $t\bar{t}$ selection and requires at least one additional jet, for a total of one lepton and at least 5 jets. A likelihood test is performed to assign jets to one of the top quarks in the event, and the additional jets forming the highest jet-top mass (m_{tj}) and jet-anti-top mass ($m_{\bar{t}j}$) are considered.

The m_{tj} and $m_{\bar{t}j}$ spectra [15] are consistent with SM expectations and limits on new colour-singlets (W') or colour-triplets (ϕ) are shown in Figure 10. The parameter space favoured by the Tevatron A_{FB} and $\sigma_{t\bar{t}}$ measurements is notably almost completely ruled out by this result.

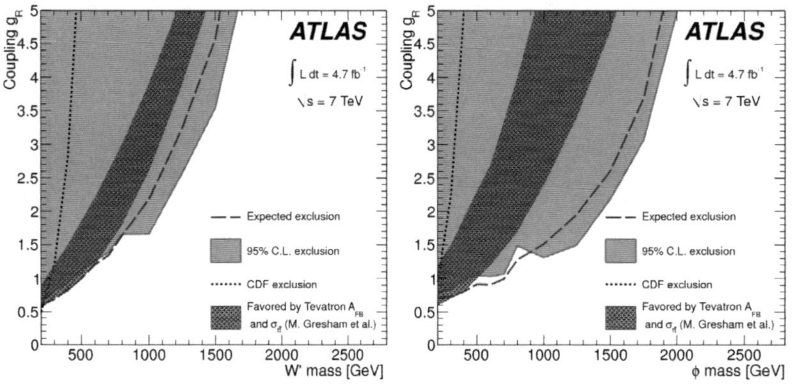

Figure 10. Expected and observed 95% CL upper limits on the colour-singlet (W', left) and colour-triplet (ϕ, right) model cross-sections. The region favoured by the Tevatron A_{FB} and $\sigma_{t\bar{t}}$ measurements is shown as the dark band. [15]

9. Conclusion

To summarize, no significant excess above expectations from the Standard Model has been observed in searches for heavy resonances carried out by the ATLAS collaboration at the LHC. Limits are set on a variety of resonances decaying into combinations of charged leptons, neutrinos, gauge bosons, top quarks and jets, and these limits are interpreted in the context of theoretical models used as benchmarks such as Z', W', Randall-Sundrum G^* and many others.

Acknowledgments

The author acknowledges support from the Vanier Canada Graduate Scholarship program, and the Natural Sciences and Engineering Research Council of Canada.

References

[1] ATLAS Collaboration 2008 *JINST* **3** S08003
[2] Cacciari M, Salam G P and Soyez G 2008 *Journal of High Energy Physics* **04** 063 (*Preprint* hep-ph/0802.1189)
[3] ATLAS Collaboration 2012 *ATLAS-CONF-2012-148,* http://cdsweb.cern.ch/record/1493487
[4] ATLAS Collaboration 2013 *Journal of High Energy Physics* **1301** 029 (*Preprint* hep-ex/1210.1718)
[5] ATLAS Collaboration 2012 *ATLAS-CONF-2012-129,* http://cdsweb.cern.ch/record/1477926
[6] ATLAS Collaboration 2012 *Journal of High Energy Physics* **1211** 138 (*Preprint* hep-ex/1209.2535)
[7] ATLAS Collaboration 2013 *Phys. Rev.* D **87** 015010 (*Preprint* hep-ex/1211.1150)
[8] ATLAS Collaboration 2012 *Submitted to Phys. Lett.* B (*Preprint* hep-ex/1210.8389)
[9] ATLAS Collaboration 2012 *Eur. Phys. J.* C **72** 2241 (*Preprint* hep-ex/1209.4446)
[10] ATLAS Collaboration 2013 *Phys. Lett.* B **718** 860 (*Preprint* hep-ex/1208.2880)
[11] ATLAS Collaboration 2012 *ATLAS-CONF-2012-150,* http://cdsweb.cern.ch/record/1493489
[12] ATLAS Collaboration 2013 *Phys. Lett.* B **719** 242 (*Preprint* hep-ex/1210.6604)
[13] ATLAS Collaboration 2013 *Journal of High Energy Physics* **1301** 116 (*Preprint* hep-ex/1211.2202)
[14] ATLAS Collaboration 2012 *ATLAS-CONF-2012-136,* http://cdsweb.cern.ch/record/1478974
[15] ATLAS Collaboration 2012 *Phys. Rev.* D *Rapid Communications* **86** 091103 (*Preprint* hep-ex/1209.6593)

International Workshop on Discovery Physics at the LHC (Kruger2012) IOP Publishing
Journal of Physics: Conference Series **455** (2013) 012022 doi:10.1088/1742-6596/455/1/012022

Searches for supersymmetry with the CMS detector at the LHC

Andrew Askew
Assistant Professor, Florida State University, 77 Chieftan Way, Tallahassee FL 32306, USA
E-mail: askew@hep.fsu.edu

Abstract. We present the results of searches for supersymmetry performed using the datasets collected in 2011-2012 by the CMS experiment at the LHC in pp-collisions at center-of-mass energy of 7 and 8 TeV. The results are interpreted in the light of simplified models.

1. Introduction

The standard model (SM) of particle physics is a very successful theory which has been confirmed by an enormous amount of experimental data. However, the consensus view is that this model is incomplete. As yet, there is no candidate particle for dark matter, and the loop corrections to the mass of the Higgs boson require many orders of magnitude of tuning to stabilize. Supersymmetry (SUSY) is an attractive alternative which would augment the SM, providing new particles which would stabilize the Higgs mass, provide for gauge coupling unification, and provide a dark matter candidate. SUSY is not simply a single theory with a single phenomenology, it is an entire family of theories each with their own free parameters and unique phenomenologies.
Many of the first analyses performed using LHC data were constructed within the confines of particular models of SUSY, such as the minimal supersymmetric standard model. As yet, no excesses characteristic of these models have been observed. This has largely prompted a new strategy in which simplified models which broadly encompass large portions of SUSY phenomenology are used to characterize the observations made in the signature based searches that are performed.
The analyses discussed here are based on the 7 TeV and 8 TeV datasets, and utilize between 5 and 11.7 fb^{-1} of integrated luminosity using the CMS detector. The CMS experiment is described in detail elsewhere [1], but provides for precise reconstruction and identification of electrons, photons and muons. Taus, and hadronic jets, and missing transverse energy are identfied using particle flow techniques which allow for better efficiency and precision for these objects [2].

2. Gluino production

2.1. All hadronic search using α_T

Searches in all hadronic final states are sensitive to a wide variety of SUSY processes, and also tend to have the largest cross sections. These searches also have enormous backgrounds due to inclusive QCD jet production, which requires both a precise measurement of the missing transverse energy and all other event information that can discriminate between jet production and the production of SUSY particles.

International Workshop on Discovery Physics at the LHC (Kruger2012) IOP Publishing
Journal of Physics: Conference Series **455** (2013) 012022 doi:10.1088/1742-6596/455/1/012022

The variable α_T is created in order to separate QCD events from events with true missing transverse energy from the escape of the lightest SUSY particle (LSP). In the simplest case of a QCD dijet event, α_T reduces to Equation 1, where

$$\alpha_T = \sqrt{\frac{\frac{p_{T,j2}}{p_{T,j1}}}{2(1 - \cos\Delta\phi)}} \tag{1}$$

$p_{T,j1}$ is the leading jet transverse momentum (p_T), $p_{T,j2}$ is the trailing jet p_T, and $\Delta\phi$ is the difference in azimuthal angle between the jets. If one assumes that these two jets have exactly the same measured transverse momentum, and are exactly back to back, one would calculate 0.5 for α_T. For any imbalance between the jets, this value would have to be smaller (assuming the jets are still back to back, the ratio between the trailing jet and the leading jet would be less than one). Only in the case where the angular difference indicates that the jets are not back to back, and the transverse momenta are almost equal, this value is larger than 0.5 which indicates that this is a topology more likely to originate from an event with true missing tranverse energy, and not simply the result of a mismeasurement. The generalized form of this equation (for any number of jets) localizes most backgrounds with no true missing transverse energy at low values of α_T. The remaining electroweak backgrounds, those with neutrinos and true missing transverse energy are estimated from a combination of data control samples and Monte Carlo translation factors.

No significant excess is observed, and limits on the cross section as a function of the gluino mass and the mass of the LSP are set in a number of different simplified scenarios [3]. Cross section limits are set by performing a maximum likelihood fit across all bins of the sum of hadronic transverse energy (H_T), the number of jets, and number of jets tagged as being associated with the decay of a b-quark (or b-jets), to maximize sensitivity to those models with many jets and associated heavy flavor production. Cross section limits on one particular model, in which gluinos are pair produced, are shown in Figure 1. This analysis makes use of 11.7 fb^{-1} of 8 TeV data.

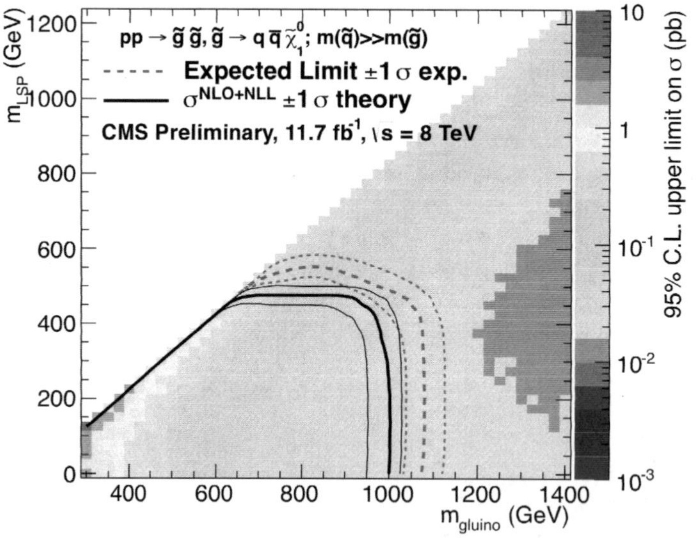

Figure 1. Cross section limits in one particular simplified model using the α_T variable as a function of gluino mass and the LSP mass.

Journal of Physics: Conference Series **455** (2013) 012022 doi:10.1088/1742-6596/455/1/012022

2.2. Seaches with leptons and hadronic jets
As an alternative to searching inclusively only requiring large amounts of hadronic energy, searches with multiple leptons have far smaller cross sections for SUSY production, but also smaller backgrounds from SM processes. One particular simplified model gives a large number of potential leptons in the final state: gluinos are pair produced, and then each gluino decays into a pair of top-quarks and the LSP. Two particular searches are highlighted here, one in which two same sign leptons are required with hadronic jets, and one in which at least three leptons are required along with additional hadronic actvity.
The dilepton search required at least two leptons (e or μ) p_T greater than 20 GeV, which also have the same charge, and two hadronic jets with p_T >40 GeV. The significant backgrounds to this final state originate from fake leptons (which have random charges), real leptons in which the charge is mismeasured (a more significant problem for e than for μ), and rare SM processes such as $t\bar{t}W$ and $t\bar{t}Z$. Events are separated into bins of jet multiplicity and how many jets had associated heavy flavor tags. The expected number of background events model the observed number of candidate events well, bin by bin, and cross section limits are again obtained, see Figure 2 as a representative example [4]. This analysis makes use of 10.5 fb^{-1} of 8 TeV data.

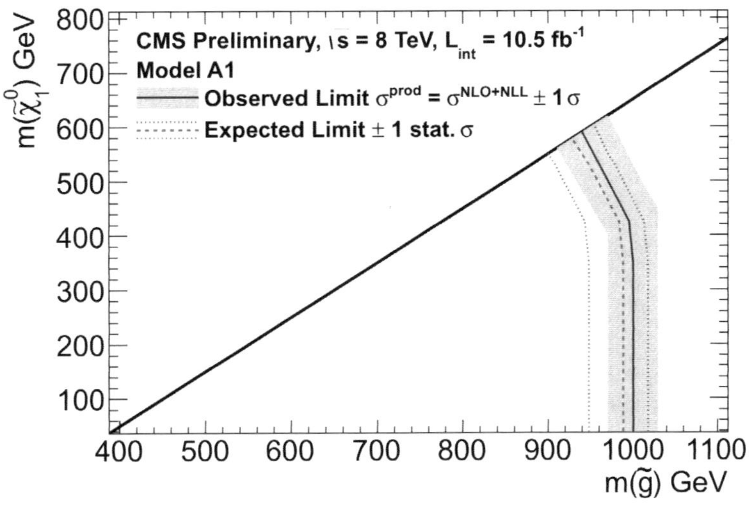

Figure 2. Cross section limit in simplified model where same sign leptons and jets are required.

Requiring at least three leptons in the final state substantially further reduces backgrounds from SM sources. In this case, three leptons are required, where these leptons can be e, μ, or hadronically decaying τ. Topologies are then split into categories by flavor, presence of a τ, missing transverse energy, additional leptons, hadronic activity, and heavy flavor jets. Over all of these categories, there are no significant excesses observed, and limits are set on the cross section in simplified models, an example of which is shown in Figure 3 [5]. This analysis makes use of 9.2 fb^{-1} of 8 TeV data.

3. Stop production
Given the large Yukawa coupling of the top-quark to the Higgs boson, the stop (supersymmetric partner to the top) is expected to play a significant role in naturalizing the Higgs mass (providing corrections that reduce the amount of tuning required in the model). In two simplified scenarios considered here, the pair produced stop decays either to the top quark and the LSP,

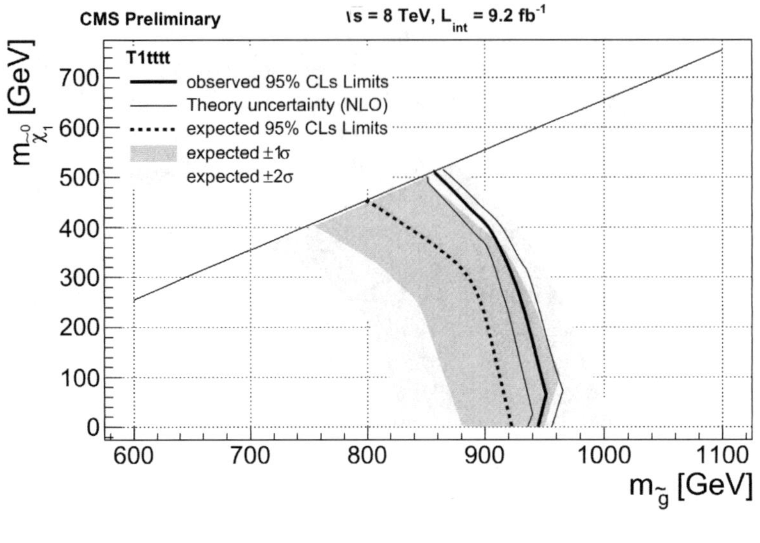

Figure 3. Cross section limits in simplified model using multilepton final states.

or decays to a b-quark and a chargino, which subsequently decays to a W boson and the LSP. Two separate analyses (disjoint at the selection level) are shown here, one in which the decay products are all hadronic (no leptons allowed), and a second in which at least one lepton is required and the transverse mass of the lepton and missing transverse energy is used to search for the presence of the stop.
The stop search in the all hadronic mode attempts to maintain sensitivity to both of the previously described stop decay scenarios. At least five hadronic jets are required (with at least one b-tagged), and isolated leptons are vetoed. The jets are required to be separated from the missing transverse energy. Three separate selections are examined beyond the baseline requirements, with tighter cuts on the missing transverse energy, the number of jets and the separation between the b-tagged jet and the missing tranverse energy. In all three cases the number of observed events is consistent with the background expectations, but the uncertainties on the backgrounds in the end gave the best expected limits with the simpler baseline selection, and thus these selection requirements are chosen for the result. One cross section limit is shown in Figure 4, and the applicable simplified model limits may be found in [6]. This analysis makes use of 4.96 fb^{-1} of 7 TeV data.

In the second stop search analysis at least one lepton is explicitly required. The transverse mass (M_T)of the lepton and missing transverse energy is then formed. The transverse mass in this case can be larger than that of the W-mass, which minimizes backgrounds from inclusive top production. The candidate sample is then split into control regions by number of leptons and number of b-tagged jets to validate the Monte Carlo description of the backgrounds, where the signal region is with one and only one lepton and at least one b-tagged jet. A number of different mimimum missing transverse energy (from 100-400 GeV) and tranvserse mass selections (from 120 to 150 GeV) were used to optimize the expected limit in multiple simplified model scenarios. No excess above the expected number of events is observed and the limits from one of the simplified models are shown in Figure 5 [7]. This analysis uses 9.7 fb^{-1} of 8 TeV data.

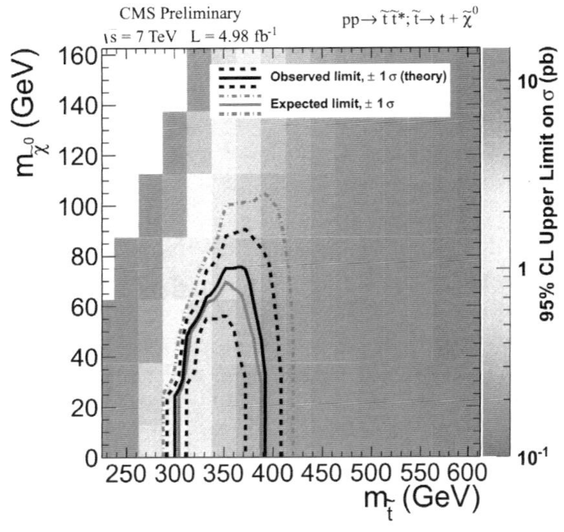

Figure 4. Cross section limits from all hadronic stop search as a function of stop mass and LSP mass, in the scenario where the stop decays to the top quark and LSP.

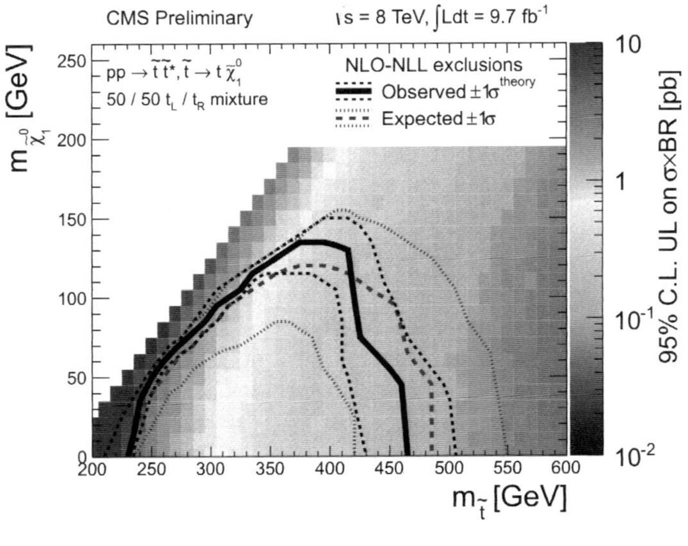

Figure 5. Limits on stop mass and LSP mass from M_T based stop search.

4. Electroweak production

All of the previously discussed analyses made use of lepton multiplicity, hadronic energy and missing transverse energy in searching for gluino and stop production. If instead we search for production of charginos and neutralinos, the presence of additional hadronic activity is no longer assured (except for the cases where that hadronic activity originates from the decay of an electroweak boson). The results of searches mentioned here span multiple final states, with three isolated leptons, two same sign leptons, four isolated leptons, two isolated leptons and two jets, and two non-resonant leptons. In each of these cases heavy flavor jets are explicitly vetoed.

These analyses are all performed on 9.2 fb^{-1} of 8 TeV data.
In the trilepton search, events which have at least three leptons (e, μ or τ) are selected, where at least two of the leptons must be electrons or muons, and the lead electron or muon must have a p_T >20 GeV. All other leptons are required to have p_T >10 GeV, except for taus which must also be p_T >20. Missing tranvserse energy is required to be greater than 50 GeV. The major SM backgrounds to this final state are inclusive WZ production and $t\bar{t}$.
As a complement to the trilepton final states, two same sign leptons can be required to regain some of the acceptance that is lost from having required three well reconstructed leptons, though the p_T requirements and isolation are tightened in order to reduce hadronic backgrounds.
Since in both channels the observed number of events are consistent with the expectation, cross section limits are set on chargino and neutralino production in simplified models. The observed three-lepton and same-sign lepton limits for a particular choice of simplified model are shown in Figure 6 [8].

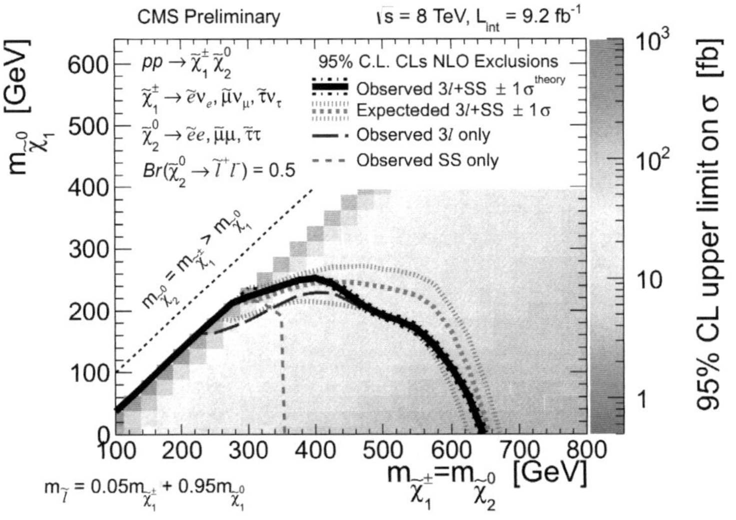

Figure 6. Chargino and neutralino cross section limits from the three-lepton and same-sign lepton analyses, along with the combined limits.

In a similar attempt to mitigate the acceptance cost of requiring an three leptons, an analysis in which two (and only two) leptons and at least two jets are required. The missing transverse energy is then used to search for an excess above the SM WZ/ZZ processes. No excess is observed, and the cross section limits, along with the three-lepton search cross section limits are shown in Figure 7. In an additional dilepton search, events in which the dilepton invariant mass was inconsistent with that of the Z boson, and at least 60 GeV of missing transverse energy are used along with a topological variable (the transverse contransverse mass defined in [9]). Production of charginos would lead to a sensitive to mass endpoint in the distribution at the mass of the chargino. Template shapes in this variable are defined and then normalized to the region of the topological variable where the background dominates. No excess in the expected signal region was observed, and cross section limits on direct chargino pair production were obtained, see Figure 8.

Four lepton events are used to search for pair-produced neutralinos in which each neutralino

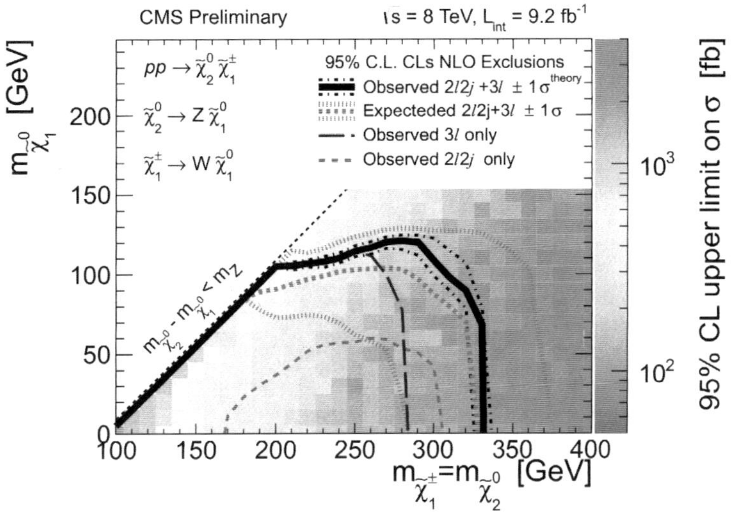

Figure 7. Chargino and neutralino cross section limits from the three-lepton and two-lepton plus two jets search, along with the combined limits.

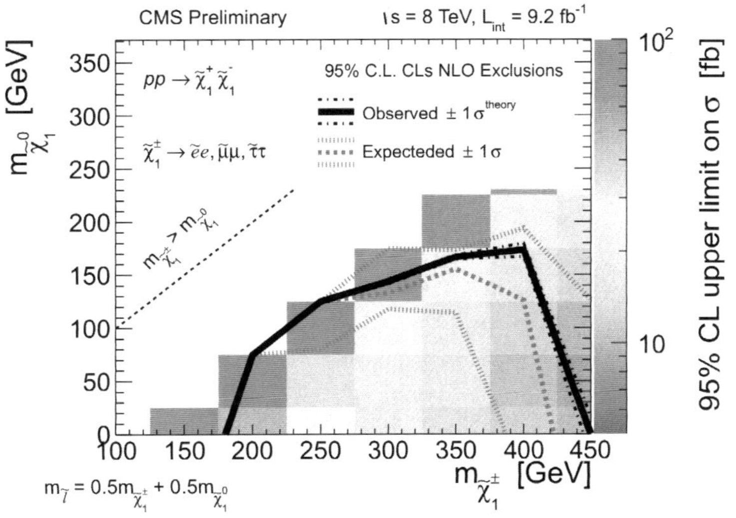

Figure 8. Chargino cross section limits using non-resonant dilepton events and topological variable.

decays to an LSP and a Z-boson which subsequently decays to four leptons, of which at most one can be a hadronic tau. Each event must have exactly four leptons, and then candidate events are binned in missing transverse energy. The major SM background process is continuum ZZ production. No excess is observed and limits are set both in the simplified model scenarios and in a higgino enhanced gauge mediated supersymmetry breaking model (GMSB). The cross section limits for this GMSB scenario are shown in Figure 9.

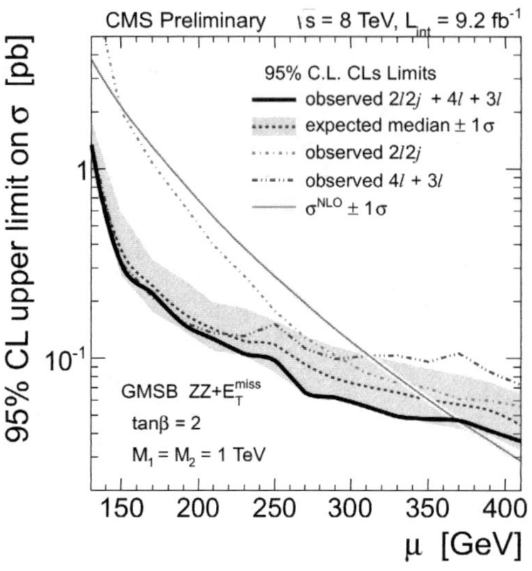

Figure 9. Four lepton cross section limit in the higgino enhanced GMSB model.

5. Stealth SUSY

In a final search topology, we consider the case where there is neither an abundance of leptons, nor is there significant missing transverse energy. Such models arise when SUSY is augmented with a hidden sector wherein there is a singlet state which is almost mass degenerate with its SM counterpart. One can therefore have pair produced squarks which then decay to a neutralino via the emission of a quark. In a GMSB like scenario, the neutralino then decay to the singlino via the emission of a photon, and then the singlino state can decay to the SM singlet state via the emission of a gravitino (the LSP), which carries away very little energy due to the small mass difference, leading to a final state with very little missing transverse energy (see Figure 10). The end result is an event with two photons (given the decay of both the squarks), many jets and low missing transverse energy. This kind of final state would evade most constraints from the previously shown analyses, and calls for a different strategy to maintain sensitivity.

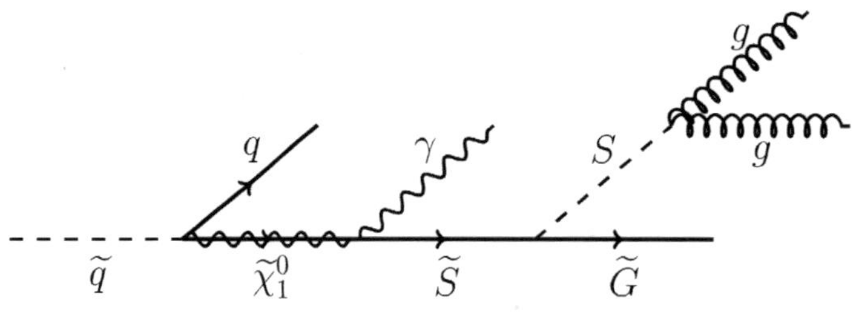

Figure 10. An example Stealth SUSY decay chain.

To search for these final states, the following assumptions are made: that the SUSY events will populate a region of high total energy (due to the high mass of the squarks) and high object multiplicity, and also that the total scalar sum of the object transverse energy (S_T) from the QCD background will be the same regardless of the object multiplicity (this is verified in both data and Monte Carlo). The shape of the S_T distribution can therefore be taken from the low object multiplicity sideband, and then normalized to the low S_T region of the high multiplicity signal region (allowing extrapolation to where the SUSY signal should live). This is performed both in the 4 and 5 (or more) bins of jet multiplicity, in order to optimize sensitivity. No excess is observed in the data, and limits are placed both on the Stealth SUSY model, but also as a function of the the S_T variable itself (generalizing to any new physics that would manifest itself in high energy and object multiplicity)[10], see Figure 11. This analysis utilizes 4.96 fb^{-1} of 7 TeV data.

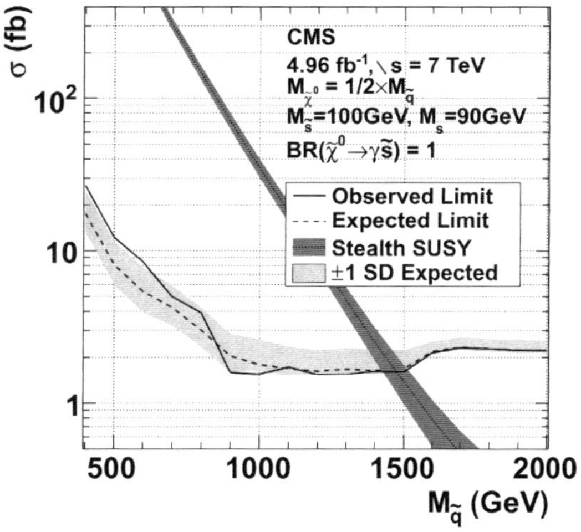

Figure 11. Limit on the squark mass in the Stealth SUSY model.

6. Conclusions

We have presented an array of searches for supersymmetry at the LHC, presented in a more generalized formalism of simplified models. No excesses are observed. These models provide a more useful set of constraints across a variety of SUSY models than previous more constrained analyses.

References

[1] Chatrchyan S et al. [CMS Collaboration], *Journ. of Instrumentation* **3** S08004, 2008.
[2] Chatrchyan S et al. [CMS Collaboration], *"Commissioning of the Particle-Flow Reconstruction in Minimum-Bias and Jet Events from pp Collisions at 7 TeV"*, CMS Physics Analysis Summary CMS-PAS-PFT-10-002, (2010).
[3] Chatrchyan S et al. [CMS Collaboration], *"Search for supersymmetry in final states with missing transverse energy and 0, 1, 2, 3, or at least 4 b-quark jets in 8 TeV pp collisions using the variable AlphaT"*, CMS Physics Analysis Summary CMS-PAS-SUS-12-028, (2012).
[4] Chatrchyan S et al. [CMS Collaboration], *"Search for new physics in events with same-sign dileptons and b jets in pp collisions at sqrt(s) = 8 TeV"*, *Preprint* hep-ex/1212.6194.
[5] Chatrchyan S et al. [CMS Collaboration], *"A Search for Anomalous Production of Events with three or more leptons using 9.2 fb^{-1}"*, CMS Physics Analysis Summary CMS-PAS-SUS-12-026, (2012).
[6] Chatrchyan S et al. [CMS Collaboration], *"Scalar Top Quark Search with Jets and Missing Momentum in pp Collisions at sqrt(s) = 7 TeV"*, CMS Physics Analysis Summary CMS-PAS-SUS-11-030, (2012).

[7] Chatrchyan S et al. [CMS Collaboration], *"Search for direct top squark pair production in events with a single isolated lepton, jets and missing transverse energy at $\sqrt{s} = 8$ TeV"*, CMS Physics Analysis Summary CMS-PAS-SUS-12-023, (2012).
[8] Chatrchyan S et al. [CMS Collaboration], *"Search for direct EWK production of SUSY particles in multilepton modes with 8 TeV data"*, CMS Physics Analysis Summary CMS-PAS-SUS-12-022 (2012).
[9] Matchev K and Park M. *"A general method for determining the masses of semi-invisibly decaying particles at hadron colliders"*, *Preprint* hep-ex/0910.1584.
[10] Chatrchyan S et al. [CMS Collaboration], *"Search for supersymmetry in events with photons and low missing transverse energy in pp collisions at 7 TeV"*, *Phys. Lett. B* **719** 42, (2013).

Heavy ion physics from the CMS collaboration

P Sarin[1] **for the CMS collaboration**
Indian Institute of Technology Bombay, Mumbai, India 400076

E-mail: pradeepsarin@iitb.ac.in

Abstract. We present an overview of results obtained by analyses of data from heavy ion (PbPb) collisions in the CMS experiment at LHC. The heavy ion physics program of the CMS collaboration spans a gamut of interesting measurements on high density multi-particle systems created in PbPb collisions at the TeV scale. We highlight key results from the inclusive production of charged particles at high p_T, the effect of QCD dynamics on jets and heavy quark bound states. Intriguing new results on long range angular two-particle correlations observed in PbPb and pPb collisions are also presented.

1. Overview of the CMS physics program and the context for heavy ion collision studies

The CMS experiment[1] has excellent capabilities for particle tracking and energy measurement. The CMS superconducting solenoid, 12.5 m long with an internal diameter of 6m, provides a uniform magnetic field of 3.8 T.

The inner tracking system is composed of a pixel detector with three barrel layers at radii between 4.4 and 10.2 cm and a silicon strip tracker with 10 barrel detection layers extending out to a radius of 1.1 m. This is complemented by two endcaps, extending the acceptance up to $|\eta|=2.5$. The momentum resolution for reconstructed tracks in the central region is about 1% at p_T = 100 GeV/c.

The calorimeters inside the magnetic coil consist of a lead-tungstate crystal electromagnetic calorimeter (ECAL) and a brass-scintillator hadron calorimeter (HCAL) with coverage up to $|\eta|=3$. Quartz/steel forward hadron calorimeters extend the calorimetry coverage up to $|\eta|=5.2$. Muons are measured in gas-ionization detectors embedded in the steel return yoke of the magnet. The calorimeter cells are grouped in projective towers of granularity $\Delta\eta \times \Delta\varphi=0.087 \times 0.087$ for central rapidity.

Fig 1 shows a schematic view of these numbers for tracking and calorimetric coverage. The azimuthal φ coverage is complete.

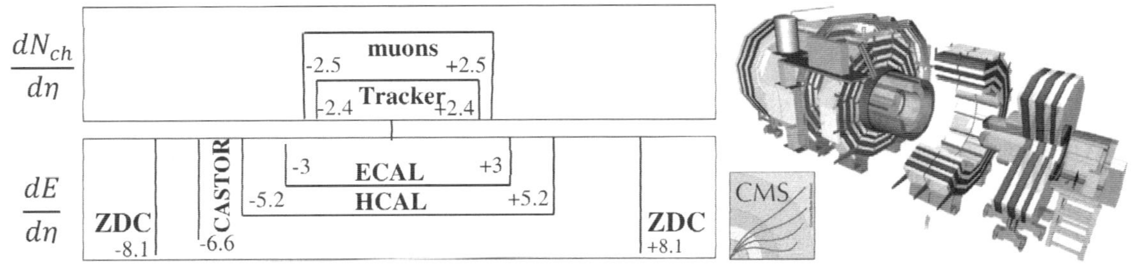

Fig 1: η coverage of the tracking and calorimeter detectors (left), and CMS (cutaway view right). Beam pseudorapidity for PbPb collisions at $\sqrt{s_{NN}}$ = 2.76 TeV is 8.

[1] To whom any correspondence should be addressed.

For the high multiplicity of particles generated in PbPb collisions, this excellent (η,φ) coverage and resolution are invaluable, though tracking does present a formidable challenge. Special high level triggers are used to record rare events with signals of interest like events with high p_T jets and quarkonia. The details of the following analyses are described in the literature[2,7,8,13,14,15,16,17] where the results are published and not elaborated here.

A major advantage of the CMS detector setup is that it is used to perform a variety of physics measurements in pp collisions. By making our observations at the same centre of mass energy scaled by the number of binary nucleon-nucleon collisions in the same apparatus, many uncertainties are minimized that may creep in when comparing to baseline pp measurements in different setups.

The results presented in the following sections are derived from data acquired during the month long PbPb running periods in 2010 and 2011 at $(s_{NN})^{1/2}$=2.76 TeV. Each period corresponds to an integrated luminosity of 8 μb^{-1} and 150 μb^{-1} respectively. Comparisons to pp measurements are made using data from pp collisions at the *same* centre of mass energy in CMS. A short pPb test run lasting approximately 4 hours was performed at the end of 2012, in preparation of the pPb run planned for spring 2013. This run yielded about two million pPb collisions which show very interesting results – these results are also presented in this paper.

The rest of this paper is organized in four sections – each section deals with a key result from the CMS heavy ion program:

Section 2: Study of high p_T charged particle suppression
Section 3: Measurements of jet quenching
Section 4: Observation of sequential heavy quark bound state suppression in PbPb collisions
Section 5: Measurements of long-range near-side angular correlations in PbPb and pPb collisions

A complete, up-to-date list of all the physics results published by the heavy ion physics group in CMS can be found at: https://twiki.cern.ch/twiki/bin/view/CMSPublic/PhysicsResultsHIN

2. Suppression of high p_T charged particles

The charged particle spectrum at large transverse momentum p_T is dominated by hadrons originating from parton fragmentation. A large density of energy is created in the centre of mass of two Pb ions colliding at TeV scale which is expected to produce a many body system with dynamics governed by many-body QCD interactions. The dynamics of such a system are vital to the understanding of many-body QCD in asymptotically large momentum exchange (and hence small distance) regime. The study of the modifications of the p_T spectrum of charged particles in PbPb compared to pp collisions at the same collision energy can shed light on the mechanism by which hard partons lose energy traversing the medium

The results of the invariant charged particle differential yield Ed^3N_{ch}/dp^3 [2] as a function of p_T for mid-rapidity |η|<1 are compared between pp and PbPb collisions in Fig 2. The left panel indicates that the yield in pp collisions is consistent with a global power law scaling prediction from NLO QCD[3]. The right panel shows the yield in PbPb collisions for six centrality bins. The collision centrality is determined from the total event-by-event energy deposited in both HF calorimeters[2].

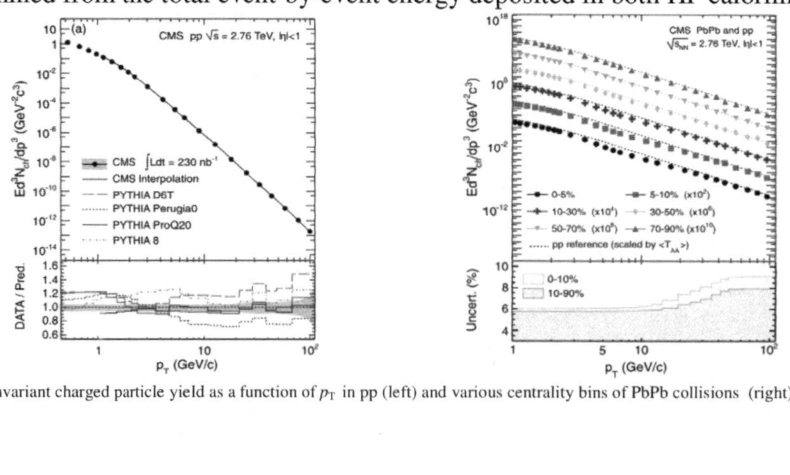

Fig 2: Invariant charged particle yield as a function of p_T in pp (left) and various centrality bins of PbPb collisions (right)

By comparing the PbPb measurements to the dashed lines representing the scaled pp reference spectrum, it is clear that the charged particle spectrum is strongly suppressed in central PbPb events compared to pp, with the most pronounced suppression at around 5–10 GeV/c.

R_{AA} quantifies the amount of this suppression as a function of p_T:

$$R_{AA}(p_T) = \frac{1}{\langle T_{AA}\rangle} \frac{d^2N_{ch}^{AA}/dp_T d\eta}{d^2\sigma_{ch}^{pp}/dp_T d\eta}$$

where N_{ch}^{AA} and σ_{ch}^{pp} represent the charged particle yield per event in AA collisions and the charged particle cross section in pp collisions, respectively. In order to compare the yield of high-p_T charged particles produced in PbPb and pp collisions, a scaling factor, the nuclear overlap function T_{AA}, calculated from the Glauber model[5,6] is used to provide a proper normalization as a function of centrality. The suppression $R_{AA}(p_T)$ is shown in Fig 3.

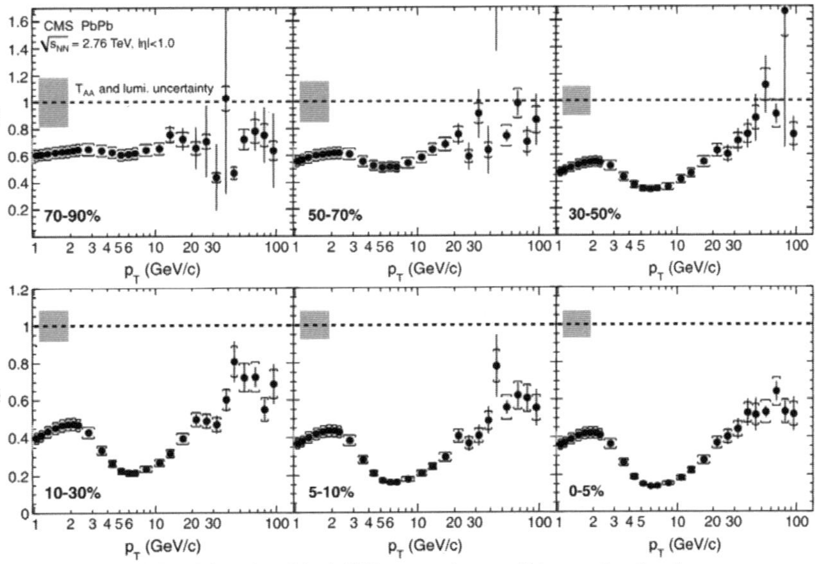

Fig 3: Relative yield R_{AA} of charged particles in PbPb compared to pp collisions as a function of p_T.

In the most peripheral events 70–90 % (top left in Fig 3), a moderate suppression R_{AA}≈0.6 is observed at low p_T. The suppression becomes more pronounced in central collisions (bottom right in Fig 3) as is expected from the increasingly dense final-state system and longer average path-length of hard-scattered partons before fragmenting into final hadrons.

In stark contrast to charged particles, isolated photon yield measured in CMS[7] shown in Fig 4 indicates an R_{AA} consistent with 1 even for very high momentum photons in the most central (0-10%) collisions.

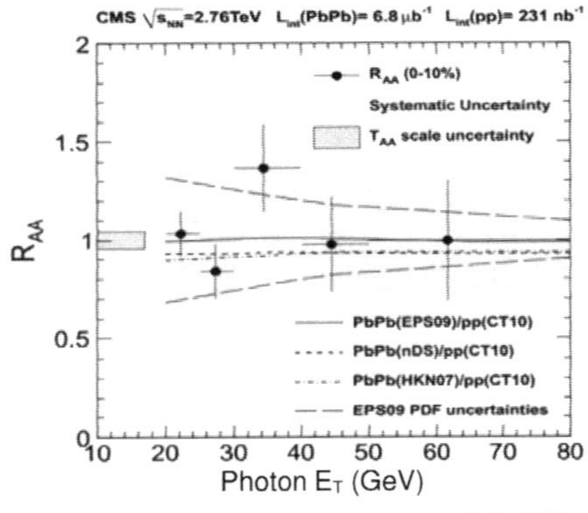

Fig 4: R_{AA} for isolated photons measured in CMS for most central (0-10%) collisions

3. Measurements of jet quenching

The suppression of inclusive charged yield described in the previous section provides a hint of energy loss by fast partons traversing through a dense medium. Since these fast partons fragment into jets, further quantitative information on the transport properties of the medium can be gained by looking at the momentum imbalance between fully reconstructed jets[8].

To characterize the momentum imbalance between back-to-back jets produced in PbPb collisions, we define the asymmetry ratio:

$$A_J = \frac{p_{T,1} - p_{T,2}}{p_{T,1} + p_{T,2}}$$

where $p_{T,1}$ is the total transverse momentum of the 'leading' jet and $p_{T,2}$ is the total transverse momentum of the azimuthally opposed 'sub-leading' jet. In the absence of any effect caused by the dense many-body system that the jets have to traverse, $p_{T,1}= p_{T,2}$ and $A_J=0$ by construction. Fig 5 shows the measured values of A_J in various bins of centrality. A substantial degree of asymmetry is observed for the most central events.

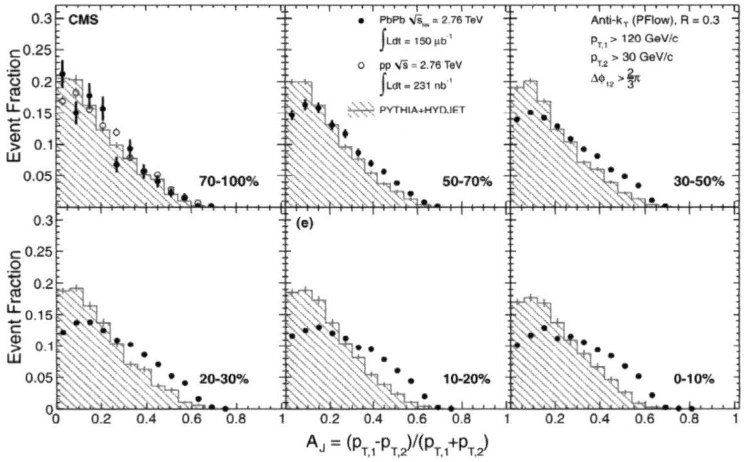

Fig 5: Dijet asymmetry ratio, A_J for leading jets of $p_{T,1}$ > 120 GeV/c and subleading jets of $p_{T,2}$ > 30 GeV/c with a selection of $\Delta\varphi_{1,2} > 2\pi/3$ between the two jets[8]. Results are shown for six bins of collision centrality, corresponding to selections of 70–100% to 0–10% of the total inelastic cross section. Results from data are shown as points, while the histogram shows the results for PYTHIA dijets embedded into HYDJET PbPb simulated events. Data from pp collisions at 2.76 TeV are shown as open points in comparison to PbPb results of 70–100% centrality. The error bars represent the statistical uncertainties

The distributions of the mean ratio of sub-leading to leading jet momentum $\langle p_{T,2}/p_{T,1}\rangle$, shown in Fig 6, provide a more intuitive way of quantifying the energy loss[8]. The PbPb data and results of simulation of jet production and quenching performed using PYTHIA+HYDJET event generator reveal an increasing trend for the mean value of this ratio, as a function of the leading jet $p_{T,1}$. This can be understood as the reduction in the effects of jet splitting and energy resolution as one goes to higher jet momenta. However, the central PbPb data points consistently fall below the predictions of PYTHIA+HYDJET.

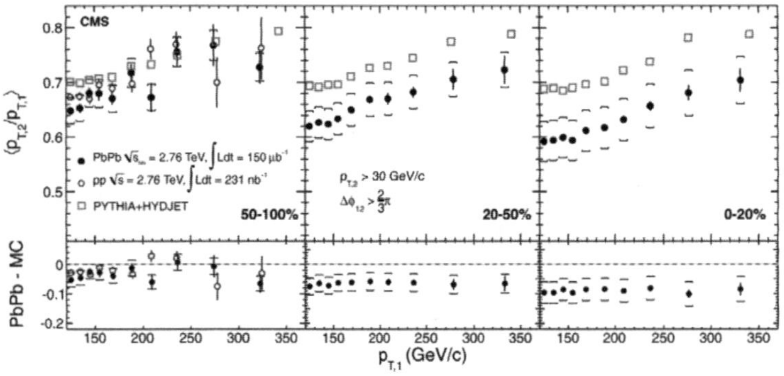

Fig 6: Average dijet momentum ratio $p_{T,2}/p_{T,1}$ as a function of leading jet p_T for three bins of centrality, from peripheral to central collisions[8]. Results for PbPb data are shown as points with vertical bars and brackets indicating the statistical and systematic uncertainties respectively. Results for PYTHIA+HYDJET are shown as squares. In the 50–100% centrality bin, results are also compared with pp data, which are shown as open circles.

4. Sequential suppression of heavy quark bound states in PbPb collisions

A dense and strongly interacting many particle QCD system, like the one formed in heavy-ion collisions, is expected to screen the confining potential of heavy quark-antiquark pairs[9], leading to the melting of charmonia (J/ψ, ψ, χ_c ...) and bottomonia (Y(1S), Y(2S), Y(3S)). The melting temperature depends on the binding energy of the quarkonium state. The ground states J/ψ, Y(1S) are expected to dissolve at significantly higher temperatures than the more loosely bound excited states. Quenched lattice QCD calculations [10,11] predict that the Y(nS) states melt at $1.2T_c$ (3S), $1.6T_c$ (2S), and above $4T_c$(1S). This sequential melting pattern is generally considered a 'smoking-gun' signature of the QCD deconfinement transition. See the excellent review by H. Satz in these proceedings[12]

The muon momentum resolution of the CMS detector gives well-resolved Y peaks in the dimuon mass spectrum as shown in Fig 7. Y states are identified through their dimuon decay[13]

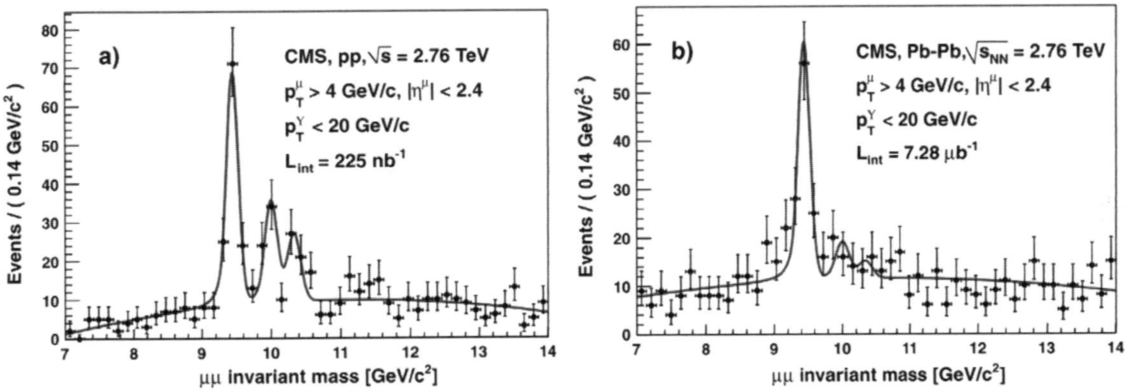

Fig 7: Dimuon invariant-mass distributions from the pp(a) and PbPb(b) data at $\sqrt{s_{NN}}$=2.76 *TeV*. The same reconstruction algorithm and analysis criteria are applied to both data sets, including a transverse momentum requirement on single muons of p_T > 4 GeV/c.

The yields of the Y(nS) states as a function of centrality[14]. Absolute suppression of the individual Y states and their dependence on the collision centrality are studied using the nuclear modification factor, R_{AA}, defined as the yield per nucleon-nucleon collision in PbPb relative to that in pp . The R_{AA} observable:

$$R_{AA} = \frac{L_{pp}}{T_{AA}N_{MB}} \frac{Y(nS)|_{PbPb}}{Y(nS)|_{pp}} \frac{\varepsilon_{pp}}{\varepsilon_{PbPb}}$$

is evaluated from the ratio of total Y(nS) yields in PbPb and pp collisions corrected for the ratio of efficiencies $\varepsilon_{pp}/\ \varepsilon_{PbPb}$, the average nuclear overlap function T_{AA}, number of minimum bias (MB) events sampled by the event selection N_{MB}, and integrated luminosity of the pp data set $\mathcal{L}_{pp}$ accounting for normalization. Fig 8 shows the obtained values of R_{AA} for Y(1S) and Y(2S) as a function of centrality.

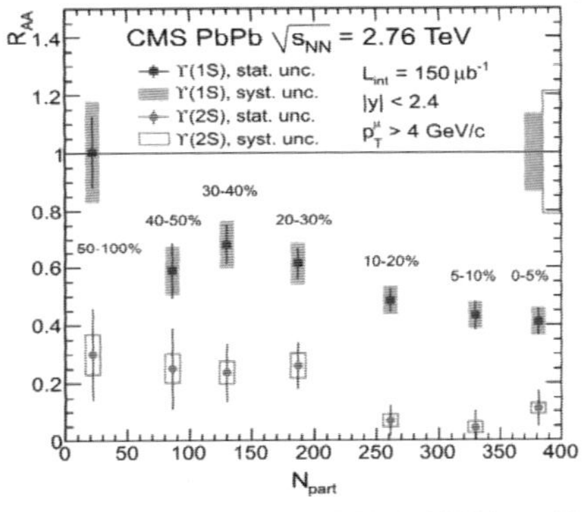

Fig 8: The nuclear modification factor R_{AA} for the yield of Y(1S) and Y(2S) in PbPb/pp collisions at $\sqrt{s_{NN}}$=2.76 *TeV*. The ratio benefits from cancellation of the efficiencies $\varepsilon_{pp}/\ \varepsilon_{PbPb}$ since the same muon reconstruction algorithms are used for both analyses.

The centrality integrated (0-100%) R_{AA} values for the individual Y states are:

$R_{AA}(Y(1S)) = 0.56 \pm 0.08(stat) \pm 0.07(sys)$

$R_{AA}(Y(2S)) = 0.12 \pm 0.04(stat) \pm 0.02(sys)$

$R_{AA}(Y(3S)) = 0.03 \pm 0.04(stat) \pm 0.01(sys) < 0.1\ (95\%\ CL)$

The measurement of the double ratio: 'ratio of Y(nS)/Y(1S) ratios in PbPb to pp collisions' is much more interesting from a physics perspective[12]. It is expected that it corresponds to the lowered melting temperature of the higher (loosely) bound quarkonium states. The double ratio benefits from an almost complete cancellation of possible acceptance or efficiency differences among the reconstructed resonances. A simultaneous fit to the PbPb and pp mass spectra gives the results:

$$\frac{Y(2S)/Y(1S)|_{PbPb}}{Y(2S)/Y(1S)|_{pp}} = 0.21 \pm 0.07(stat) \pm 0.02(sys) \quad 5\sigma\ result$$

$$\frac{Y(3S)/Y(1S)|_{PbPb}}{Y(3S)/Y(1S)|_{pp}} = 0.06 \pm 0.06(stat) \pm 0.06(sys) < 0.17\ (95\%\ CL)$$

The double ratios are expected to be compatible with unity in the absence of suppression of the excited states relative to the Y(1S) state. The measured values are, instead, considerably smaller than unity. The significance of the observed suppression exceeds 5σ and does not show a significant dependence on the collision centrality[9].

5. Observation of long range, near-side angular correlations in PbPb and pPb collisions

Most of the results presented in sections 2 to 4 above shed light on the properties of the strongly interacting multi-particle system created in high energy PbPb collisions. Dihadron correlations provide information on how the medium affects the properties of the created particles. CMS has studied dihadron correlations over a range of relative azimuthal angles ($|\Delta\varphi|$) and pseudorapidity ($|\Delta\eta|$) for all collision centralities and a broad range of p_T.

The correlations are performed with respect to a randomly chosen 'trigger' particle originating from the primary vertex and detected within the acceptance $|\eta|<2.4$ of the detector. A variety of bins of trigger transverse momentum, denoted by p_T^{trig} are considered. There can be more than one such trigger particle in a single event and their total multiplicity in a particular data sample is denoted by N_{trig}. Within each event, every trigger particle is paired with all of the remaining particles (again within $|\eta| < 2.4$). As for the trigger particles, these associated particles are binned in transverse momentum (p_T^{assoc}). The differential yield of associated particles per trigger particle is:

$$\frac{1}{N_{trig}}\frac{d^2N_{pair}}{d\Delta\eta\, d\Delta\phi} = B(0,0) \times \frac{S(\Delta\eta,\Delta\phi)}{B(\Delta\eta,\Delta\phi)}$$

The signal distribution is the per-trigger-particle yield of pairs found in the same event:

$$S(\Delta\eta,\Delta\phi) = \frac{1}{N_{trig}}\frac{d^2N_{same}}{d\Delta\eta\, d\Delta\phi}$$

The background distribution is obtained by mixing the trigger particle from one event with 10 randomly chosen other events:

$$B(\Delta\eta,\Delta\phi) = \frac{1}{N_{trig}}\frac{d^2N_{mix}}{d\Delta\eta\, d\Delta\phi}$$

The background distribution gives correlation if the only effects present were random combinatorics and pair acceptance. $B(\Delta\eta,\Delta\varphi)$ at $\Delta\eta=0$ and $\Delta\varphi=0$ (with a bin width of 0.3 in $\Delta\eta$ and $\pi/16$ in $\Delta\varphi$) is used to find the normalisation factor B(0,0).

Fig 9 shows the two-dimensional per-trigger-particle associated yield distribution of charged hadrons as a function of $|\Delta\eta|$ and $|\Delta\varphi|$ interval for a sample $3<p_T^{trig}<3.5$ GeV/c and $1<p_T^{assoc}<1.5$ GeV/c in different centrality classes of PbPb collisions. The 2D correlations are rich in structure, and evolve with centrality. For the most central PbPb collisions, a clear and significant ridge structure mostly flat in $\Delta\eta$, and extending to the limit of $|\Delta\eta| = 4$, is observed at $\Delta\varphi \approx 0$. At mid-peripheral events, a pronounced $\cos(2\Delta\varphi)$ component emerges, originating predominantly from elliptic flow[18]. In the

Journal of Physics: Conference Series **455** (2013) 012023 doi:10.1088/1742-6596/455/1/012023

most peripheral collisions, the near-side ridge structure has largely diminished, while the away-side back-to-back jet correlations can be clearly seen at $\Delta\varphi \approx \pi$, but spread out in $\Delta\eta$.

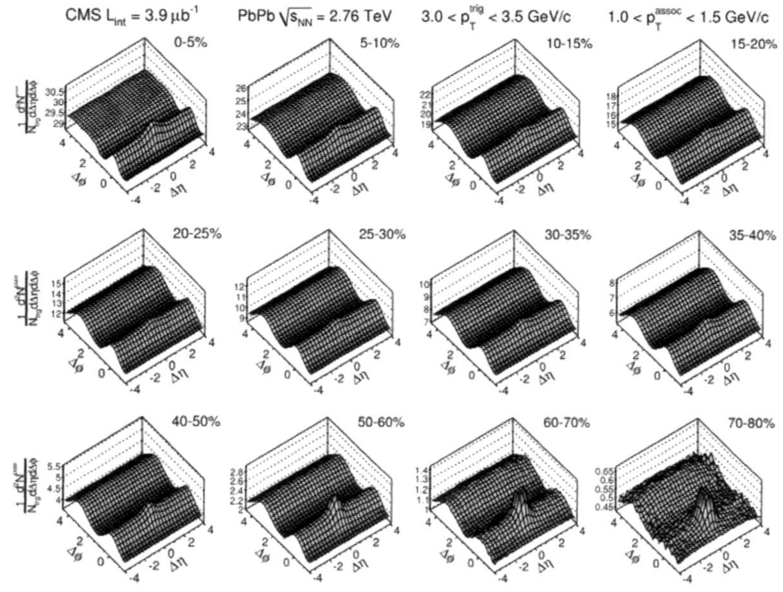

Fig 9: Two-dimensional (2D) per-trigger-particle associated yield of charged hadrons as a function of $|\Delta\eta|$ and $|\Delta\varphi|$ for $3 < p_T^{trig} < 3.5$ GeV/c and $1 < p_T^{assoc} < 1.5$ GeV/c, for twelve centrality ranges of PbPb collisions at $\sqrt{s_{NN}}=2.76$ *TeV*. The near-side peak height is truncated in the two most peripheral distributions to better display the surrounding structure.

In September 2012, LHC performed a short 8 hour pilot run colliding protons on Pb, in preparation for the dedicated pPb run planned for spring 2013. A total of 2 million events passed all selection criteria, corresponding to an integrated luminosity of ~ 1 μb^{-1} at $\sqrt{s_{NN}}=5.02$ *TeV* assuming a pPb interaction cross section of 2.1 barns. In the absence of a concept of 'centrality' in pPb collisions, the two particle correlations are obtained in classes of the number of reconstructed tracks $N_{trk}^{offline}$.

Fig 10 compares 2-D two-particle correlation functions for events with low(a) and high(b) multiplicity, for pairs of charged particles with $1 < p_T < 3$ GeV/c. For the low-multiplicity selection ($N_{trk}^{offline} < 35$), the dominant features are the correlation peak near $(\Delta\eta,\Delta\varphi)=(0,0)$ for pairs of particles originating from the same jet and the elongated structure at $\Delta\varphi \approx \pi$ for pairs of particles from back-to-back jets. High-multiplicity events ($N_{trk}^{offline} \geq 110$) also show the same-side jet peak and back-to-back correlation structures. However, in addition, a pronounced "ridge"-like structure emerges at $\Delta\varphi \approx 0$ extending to $|\Delta\eta|$ of at least 4 units.

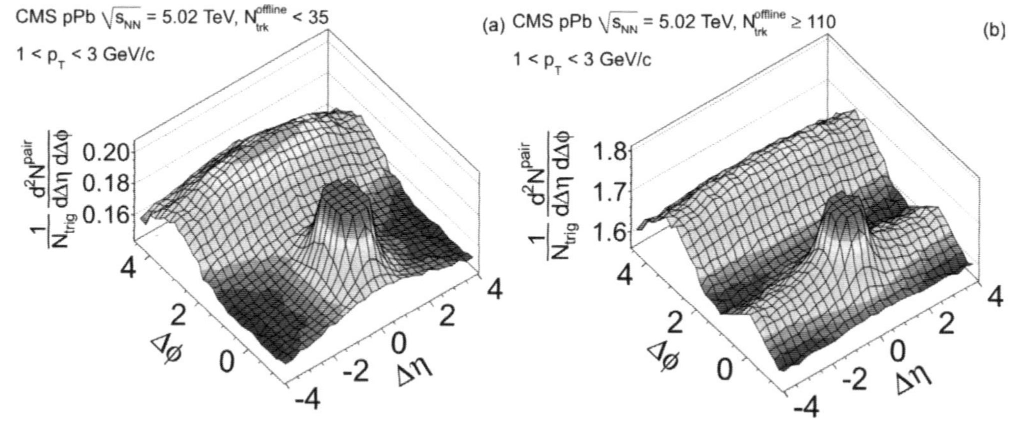

Fig 10: 2D two-particle correlation function for 5.02 TeV pPb collisions for pairs of charged particles with 1<pT<3 GeV/c for (a) $N_{trk}^{offline} < 35$ and (b) for $N_{trk}^{offline} \geq 110$

When interpreting the similarities and differences in the correlation structure between the PbPb and pPb collisions as shown in Fig 9 and 10, it is important to consider the relative contributions of different particle production mechanisms to the observed particle yields. High-multiplicity pPb events should mostly result from particle production in multiple soft proton–nucleon scatterings. The ridge structure in PbPb collisions is expected from collective flow, but its origin in the smaller pPb system is unclear. A simultaneous description of the measurements in pPb and PbPb should provide significant constraints on models of the underlying physics processes.

Summary

An overview of results from the first two years of LHC running has been presented here. CMS has collected data on PbPb collisions with an integrated luminosity of 8 μb^{-1} (2010) and 150 μb^{-1} (2011) at $\sqrt{s_{NN}}$ = 2.76TeV. The results provide a fascinating insight into the dynamics of a many body system of partons created in the collisions. We have also highlighted how the system affects the properties of particles emerging from the collisions. All data presented in this review is published in 19 peer-reviewed journals.

The key results presented here are:

1. Charged particles detected at high p_T in PbPb collisions scaled by the number of binary nucleon-nucleon collisions are significantly lesser than in pp collisions at the same centre of mass energy. The effect becomes more pronounced in central collisions relative to peripheral collisions and has a pronounced p_T dependence (Fig 3)
2. A substantial imbalance characterized by the asymmetry parameter A_J is observed between the momentum of back-to-back jets. The imbalance is most pronounced for central events (Fig 5). The p_T dependence of the mean sub-leading to leading jet p_T ratio shows an increasing trend with the collision centrality, indicating an increased parton energy loss. The magnitude of the trend is however over-predicted a PYTHIA+HYDJET simulation (Fig 6).
3. The yield of Υ(nS) heavy quark bound states in PbPb collisions is suppressed compared to that in pp collisions. Specifically, we find that the *relative* yield of Υ(2S) to Υ(1S) is suppressed by a factor of 0.21 at the 5σ confidence level, and we place a 95% CL upper limit of 0.17 on the relative yield of Υ(3S) to Υ(1S).
4. We observe long-range near-side angular correlations of the produced charged particle tracks in *both* PbPb collisions (increasing with centrality) and high multiplicity pPb collisions.

A lot more data from the 2012 run period is being analysed and a rich new set of results are expected from the dedicated pPb LHC run in the spring of 2013.

Acknowledgements

We congratulate our colleagues in the CERN accelerator departments for the excellent performance of the LHC and thank the technical and administrative staffs at CERN and at other CMS institutes for their contributions to the success of the CMS effort. In addition, we gratefully acknowledge the computing centres and personnel of the Worldwide LHC Computing Grid for effectively delivering the computing infrastructure essential to our analyses. Finally, we acknowledge the enduring support for the construction and operation of the LHC and the CMS detector provided by the following funding agencies: BMWF and FWF (Austria); FNRS and FWO (Belgium); CNPq, CAPES, FAPERJ, and FAPESP (Brazil); MEYS(Bulgaria); CERN; CAS, MoST, and NSFC (China); COLCIENCIAS(Colombia); MSES (Croatia); RPF (Cyprus); MoER, SF0690030s09 and ERDF (Estonia); Academy of Finland, MEC, and HIP (Finland); CEA and CNRS/IN2P3 (France); BMBF, DFG, and HGF (Germany); GSRT (Greece); OTKA and NKTH (Hungary); DAE and DST (India); IPM (Iran); SFI (Ireland); INFN (Italy); NRF and WCU (Republic of Korea); LAS (Lithuania); CINVESTAV, CONACYT, SEP, and UASLP-FAI (Mexico); MSI (New Zealand); PAEC (Pakistan); MSHE and NSC(Poland); FCT (Portugal); JINR (Armenia, Belarus, Georgia, Ukraine, Uzbekistan); MON, RosAtom, RAS and RFBR (Russia); MSTD (Serbia); SEIDI and CPAN (Spain); Swiss Funding Agencies (Switzerland); NSC (Taipei); ThEP, IPST and NECTEC (Thailand); TUBITAK and TAEK (Turkey); NASU (Ukraine); STFC (United Kingdom); DOE and NSF (USA).

References

[1] CMS collaboration, S Chatrchyan et al, *The CMS experiment at the CERN LHC,* 2008 *JINST* **3** S08004

[2] CMS collaboration, *Study of high-*p_T *charged particle suppression in PbPb compared to pp collisions at* $\sqrt{s_{NN}}$*=2.76 TeV* 2012 *Eur. Phys. J.* C**72** 1945

[3] F. Arleo, D. d'Enterria, A.S. Yoon, *Single-inclusive production of large-*p_T *charged particles in hadronic collisions at TeV energies and perturbative QCD predictions*, 2010, *J. High Energy Phys.*, **06**, 035

[4] CMS Collaboration, *Charged particle transverse momentum spectra in pp collisions at* $\sqrt{s_{NN}}$*=0.9 and 7 TeV*, 2011 *J. High Energy Phys.* **08**, 086

[5] B. Alver et al., *Importance of correlations and fluctuations on the initial source eccentricity in high energy nucleus-nucleus collisions*, 2008, *Phys. Rev.* C **77** 014906

[6] Particle Data Group Collaboration, *Review of particle physics*, 2010, *J. Phys.* G **37** 075021

[7] CMS collaboration, *Measurement of isolated photon production in pp and PbPb collisions at* $\sqrt{s_{NN}}$*=2.76 TeV,* 2012, *Phys. Lett* B**710** 256-277

[8] CMS collaboration, *Jet momentum dependence of jet quenching in PbPb collisions at* $\sqrt{s_{NN}}$*=2.76 TeV,* 2012, *Phys. Lett* B**712** 176-197

[9] T. Matsui and H. Satz, *J/ψ suppression by quark-gluon plasma formation,* 1986, *Phys. Lett.* B**178** 416

[10] H. Satz, *Color deconfinement and quarkonium binding*, 2006, *J. Phys.* G**32** R25

[11] C. Y. Wong, *Heavy quarkonia in quark gluon plasma,* 2005, *Phys. Rev.* C**72** 034906

[12] H. Satz, 2012 *Proceedings of Kruger2012 workshop*

[13] CMS collaboration, *Indications of Suppression of Excited Υ States in PbPb Collisions at* $\sqrt{s_{NN}}$*=2.76 TeV*, 2011, *Phys. Rev. Lett.* **107** 052302

[14] CMS collaboration, *Observation of sequential suppression of Υ states in PbPb collisions,* 2012 *Phys. Rev. Lett.* **109** 222301

[15] CMS collaboration, *Long-range and short-range dihedron angular correlations in central PbPb collisions at* $\sqrt{s_{NN}}$*=2.76 TeV* 2011 *J. High Energy Phy.* **07** 076

[16] CMS collaboration, *Centrality dependence of dihadron correlations and azimuthal anisotropy harmonics in PbPb collisions at* $\sqrt{s_{NN}}$*=2.76 TeV*, 2012 *Eur. Phys. J.* C**72** 2012

[17] CMS collaboration, *Observation of long-range near side angular correlations in pPb collisions at the LHC,* 2013, *Phys. Lett. B* **718** 795-814

[18] PHOBOS Collaboration, *System size dependence of cluster properties from two-particle angular correlations in Cu + Cu and Au + Au collisions at* $\sqrt{s_{NN}}$ *= 200 GeV*, 2010, Phys. Rev. C **81** 024904

Journal of Physics: Conference Series **455** (2013) 012024 doi:10.1088/1742-6596/455/1/012024

Search for the neutral Supersymmetric Higgs boson with the CMS experiment at LHC

Antonio Branca for the CMS Collaboration

University of Padova - Department of Physics and Astronomy and INFN, Via Marzolo 8, 35131 Padova - Italy

E-mail: antonio.branca@pd.infn.it

Abstract. The search of neutral Higgs boson in two different decay channels, a pair of τ leptons and a pair of b-quarks, is presented. The search is performed in the context of the Minimal Supersymmetric Standard Model (MSSM), using data collected by the CMS experiment at the LHC during 2011-12 at a center-of-mass energy of 7-8 TeV. In the τ decay four independent pair final states, where one or both τs decay leptonically, are studied. In the $b\bar{b}$ channel we look for final states with two high energetic jets plus a third jet, exploiting the production in association with b-quarks to enhance the signal purity in the selections of events. A description of the analyses strategies adopted for the two channels is provided. The results are presented and interpreted in the m_h^{max} scenario of the MSSM framework.

1. Instability of the Standard Model Higgs mass

The Standard Model (SM) Higgs boson suffers a theoretical problem, known as the *Hierarchy problem*, causing instability of its mass [1]. The Higgs boson mass receives enormous quantum corrections from the virtual effects of every particle that couples, directly or indirectly, to the Higgs field; drawing the value of the mass away from the experimental results. Direct searches at LEP e^+e^- and Tevatron $p\bar{p}$ colliders have led, respectively, to a lower-mass bound of $m_h > 114.4\ GeV$ [2], and to an exclusion in the range $162 - 166\ GeV$ at 95% of confidence level (CL) [3]. Indirect constraints from precision measurements favour the mass range $m_h < 158\ GeV$ at 95% of CL [4]. More recently, searches performed at the LHC [5] put more stringent limits on the SM Higgs particle mass [6] and eventually a boson with the signatures compatible with a SM Higgs boson has been observed with a mass around 125 GeV [7, 8].

The Minimal Supersymmetric Standard Model (MSSM) [9] provide a solution to the hierarchy problem. Because of the new symmetry between fermions and bosons, the supersymmetry, the number of particles is doubled in the MSSM with respect to the SM. Also the new particles gives enormous quantum corrections to the Higgs mass, but they have sign such that cancel the SM divergencies.

2. MSSM Higgs boson sector

The Higgs boson sector of the MSSM is more complex than the SM one. Electroweak (EWK) anomaly cancellation requires two doublets of complex scalars, corresponding to eight degree of freedom. These turn into five Higgs boson after EWK symmetry breaking through the Higgs

mechanism: 2 neutral CP even, h and H, 1 neutral CP odd, A, 2 charged, $H^{\pm}$. Their masses depend only by two parameters of the MSSM model, the A mass M_A and $tan\beta$, at tree-level. Important radiative corrections, depending on other MSSM parameters, affect the Higgs masses. In the experimental searches these parameters are fixed defining the so called benchmark scenarios. The benchmark scenario adopted for the searches presented here is the m_h^{max} [10].
It can be shown that if M_A is not too large and $tan\beta$ is large, the couplings of the neutral Higgs bosons to down-type fermions are enhanced. Grand unification theories (GUTs) prefer large values of $tan\beta$. Moreover, two out of the three neutral Higgs boson states are degenerate in mass for large values of $tan\beta$. This contributes to enhance a possible experimental signal. Thus it is experimentally important to search for signals produced by channels where the Higgs boson decays into a τ-leptons or b-quarks pair.

3. MSSM neutral Higgs boson production at LHC and branching ratios

The main production mechanisms of the neutral Higgs bosons at LHC are the gluonn-gluon fusion and the production in association with b-quarks. The cross section for these mechanisms is shows in Figure 1 for pp collisions at a center of mass energy of 7 TeV. The distributions

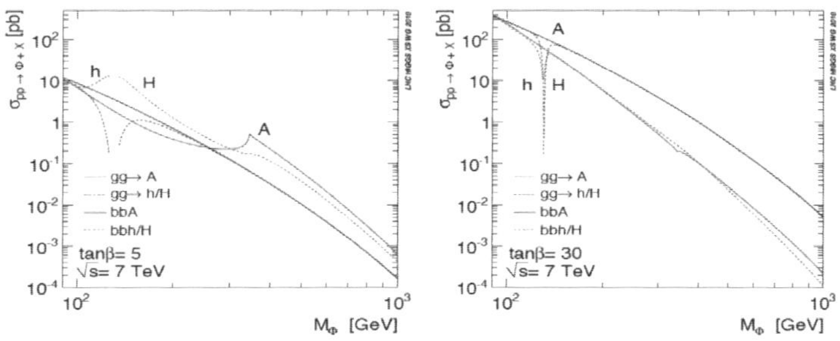

Figure 1. Neutral MSSM Higgs production cross sections at the LHC at the center of mass energy of 7 TeV for gluon fusion and the associated production with bottom quarks, including QCD corrections. The values are shown as a function of the h, H, A masses and for $tan\beta = 5$ (left) and $tan\beta = 30$ (right). The m_h^{max} scenario is considered.

show that the production in association with b-quarks is dominant for large values of $tan\beta$. Specifically, cross sections between $O(200) - O(0.01)$ pb are obtained at $tan\beta = 30$. Figure 2 shows the branching ratios of the neutral Higgs bosons into SM particles for $tan\beta = 30$. The branching ratio values are around 90% for the b-quarks pair decay and around 10% for the τ pair decay.
The decay channels in τ-leptons and b-quarks pairs of the MSSM Higgs boson have been explored by the analyses performed by the Collaboration of the CMS experiment [11] at the LHC.

4. $\tau\tau$ channel search

In this analysis [12] the channel where a neutral Higgs boson decays into a τ pair, producing two energetic opposite charge leptons in the final state, is considered. The analyzed data sample is made of 4.9 fb^{-1} and 12.1 fb^{-1} at pp center of mass energy of 7 TeV and 8 TeV, respectively, recorded by the CMS experiment. Specific High Level Trigger (HLT) paths, requiring a combination of electron, muon and tau particles, are implemented to select the final states from the LHC pp collisions compatible with the expected final state. Four final states of the $\tau\tau$

Journal of Physics: Conference Series **455** (2013) 012024 doi:10.1088/1742-6596/455/1/012024

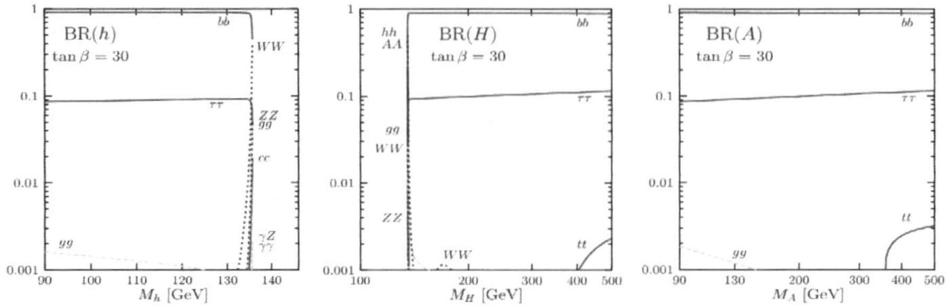

(a) h branching ratios. (b) H branching ratios. (c) A branching ratios.

Figure 2. The decay branching ratios of the neutral Higgs bosons as a function of their masses for $tan\beta = 30$. The values are obtained in the m_h^{max} scenario.

pair are considered for the analysis: $e\tau_h$, $\mu\tau_h$, $e\mu$ and $\mu\mu$. Here τ_h stands for the hadronic decay of the τ lepton. The recorded data sample is further reduced by selections requiring the presence of two well reconstructed high p_T opposite charge leptons, matching one of the four final states, in the event. The data sample is split into two categories: a sample where at least one jet is identified as coming from b-quark (*b-tagged* jet) with transverse momentum of $p_T > 20\ GeV$ and no more than one jet with $p_T > 30\ GeV$ is present (*B-Tag* category); a sample where are not present b-tagged jets with $p_T > 20\ GeV$ (*No B-Tag* category). This separation is done in order to enhance the sensitivity to the production in association with b-quarks (*B-Tag* category) and to the production through gluon-gluon fusion (*No B-Tag* category).
The largest background comes from $Z \to \tau\tau$ events, which is irreducible, followed by the reducible QCD multi-jets. Other reducible backgrounds come from EWK processes: $W + jets$, $t\bar{t}$, WW, ZZ, WZ and $Z \to e^+e^-/\mu^+\mu^-$. The reduction of the reducible backgrounds is achieved by different cuts. The QCD multi-jet is reduced by requiring low charged/neutral particle activity around the selected leptons (lepton isolation). For $e\tau_h$ and $\mu\tau_h$ the $W + jets$ is reduced requiring the transverse mass between the p_T of the electron or muon and the missing transverse energy to be less than 40 GeV. For $e\mu$ and $\mu\mu$ a topological requirement is applied to reduce the $W + jets$ and $t\bar{t}$ backgrounds. In the τ decay the visible (from the detected particles) and the invisible (from neutrinos) momenta have about the same direction. This is not true, in average, for fake τs. This observation leads to the construction of a topological variable sensible to the visible and invisible momenta directions in the $\tau\tau$ final states. This is used as discrimination variable for signal and background events. To reduce $Z \to e^+e^-/\mu^+\mu^-$, events with more than one electron or muon with $p_T > 25\ GeV$ are excluded for the $e\tau_h$ and $\mu\tau_h$ final states; for $e\tau_h$ events an additional requirement on transverse missing energy is performed. Different techniques are exploited to estimate the backgrounds that pass all the selections. The irreducible $Z \to \tau\tau$ background is estimated in shape using a $Z \to \mu\mu$ data sample, where the muon is interpreted as tau lepton (*embedding* technique). This way, a better estimation of the event migrations between the two data sample categories is achieved. Also the normalization is obtained from $Z \to \mu\mu$ events. For the QCD multi-jet in the $e\mu$ final state, the shape is obtained from same-sign events, whereas the normalization is given by $e\mu$ events with fake-electron selections weighted by the efficiency of jet misidentification as electron (ϵ_{fake}). For the $e\tau_h$ and $\mu\tau_h$ final states, QCD multi-jet background come from random combination of a reconstructed electron or muon and a τ. Thus the ratio of same-sign and opposite-sign events is close to one. Same-sign events give the background prediction for both shape and normalization. Monte Carlo (MC) simulated W events are used to predict the shape for the $W + jets$ background. The normalization is given by a W enriched control sample from data.

MC simulation is also used for $t\bar{t}$ shape prediction. The measured cross section for $t\bar{t}$ events is used for the normalization. Finally, the $Z \rightarrow e^+e^-/\mu^+\mu^-$ backgrounds are predicted using data, where the fraction of leptons misidentified as τ_h is measured from the same events exploiting the Tag and Probe technique [13].
The signal extraction is performed by a binned maximum likelihood fit to the full reconstructed invariant mass of the $\tau\tau$ final states. A $\tau\tau$ invariant mass resolution of about 20% is achieved. In Figure 3 is shown the $\tau\tau$ invariant mass for the $e\tau_h$ final state. Since no evidence of a Higgs

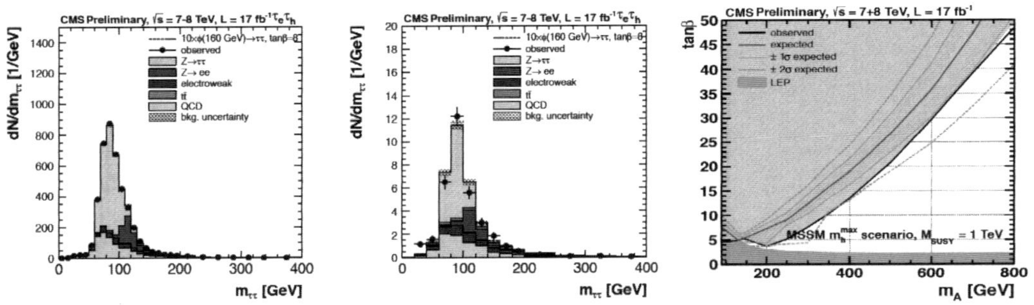

(a) No B-tag sample fit. (b) B-Tag sample fit. (c) 95% CL limits.

Figure 3. (a-b) $e\tau_h$ invariant mass for the *No B-Tag* and *B-Tag* categories. The data (black dots) are fitted with the background predictions (full histograms) within the systematic uncertainties (grey dashed band). (c) 95% CL exclusion contours in the $(M_A, tan\beta)$ parameter space of the MSSM model in the m_h^{max} benchmark scenario. The black line is the observed limit in data. The grey line is the median expected limit for the background only hypothesis. The dashed grey band (dotted grey line band) is the range that is expected to contain 68% (95%) of all the observed limit excursions from the median. The [CELESTE] area is the excluded area corresponding to the observed limit. The green area is the excluded area by the LEP experiment.

boson signal is found, 95% CL limits on the cross section times the branching ratio in τ pair are calculated with the CL_s method [14]. These limits are converted in exclusion plots in the $(M_A, tan\beta)$ parameter space, shown in Figure 3. The analysis excludes values of $tan\beta$ as low as 4 at $M_A = 200\ GeV$. The limits calculation take into account the statistics and systematic uncertainties, that contribute to worsen the sensitivity of the analysis. Different systematic uncertainties have been identified. They affect the event yield, the largest contribution coming from b-tagging efficiency (10%) and background normalization (10 – 30%), and the $\tau\tau$ mass shape, where contributions come from τ_h (33%), muon (1%) and electron (1 – 2.5%) energy scale and resolution. Theoretical systematic uncertainties range from 20% to 25%.

5. $b\bar{b}$ channel search: all hadronic

In this analysis [15] the search is performed in the channel where the Higgs boson is produced in association with b-quarks and decays into a b-quarks pair, producing up to four jets in the final state. In average, the two jets from the Higgs decay are high p_T jets, while the associated jets are softer. The analyzed data sample is made of a total of 6.7 fb^{-1} at pp center of mass energy of 7 TeV, selected with specific HLT paths requiring 2 or 3 jets, with 2 of them b-tagged. The final analyzed data sample is filtered requiring 3 well reconstructed high p_T b-tagged jets in the tracker region of the CMS detector. The final state contains only jets, and it is said all hadronic. The analysis is performed in two different Higgs mass regions: *low Higgs mass* ($M_\Phi < 180\ GeV$), where data are collected with HLT paths with low p_T thresholds, and

high Higgs mass ($M_\Phi \geq 180\ GeV$), where data are collected with HLT paths with higher p_T thresholds.
The dominant background comes from heavy-flavour multi-jet QCD, made of 3 real b-jets or 2 real b-jets plus 1 jet originating from a light parton or c-quark. Other backgrounds are EWK $t\bar{t}$ and $Z + jets$, but they are negligible as shown by MC studies.
The Higgs boson search is performed in the two dimensional distribution of the two leading b-tagged jets invariant mass (M_{12}), and an event b-tagging variable condensing the b-jet properties of the 3 b-tagged jets in the event. Data driven background templates are built for the background prediction. Since about the 98% of the events in the data sample of 3 b-tagged jets is made of two real b-jets (from MC studies), the data sample used for the background modeling is made of 2 b-tagged jets (b) plus 1 jet (x) where the b-tagging requirement is not applied (*untagged* jet). Three categories are defined depending on the rank of the x jet p_T in this sample: bbx, bxb and xbb. The untagged jet x can originate from partons with three possible flavours: light ($u\ d\ s\ g$), charm (c) and beauty (b). Thus, a total of 9 templates have to be built for the background modeling. These are reduced to 5, since there are templates with similar shapes. The templates are obtained by weighting the x jet with the probability for the jet to be b-tagged (from MC simulation), to model the M_{12} variable, and the probability to have different event b-tagging variable values (from $t\bar{t}$ MC simulated events). The 2 b-tagged data sample used for background modeling contains $2-3\%$ of the events where one or both the b-tagged jets are not originated by b-quarks (from MC simulation). This can have slight effects on the M_{12} shape of the templates. The contribution is evaluated from non-bb events from mistag rate measurements [16]. The effect is small and is subtracted from the three categories.
The signal extraction is done by performing a binned least-squares fit in the M_{12} and event b-tagging variable space of the 3 b-tagged data sample. The linear combination of the background templates and signal template, from MC simulation, is used for the fit. Only the background template shapes are predicted by the data-driven method, the normalization is obtained from the fit. Figure 4 shows the projection of the fit in the M_{12} and event b-tagging variables. Since

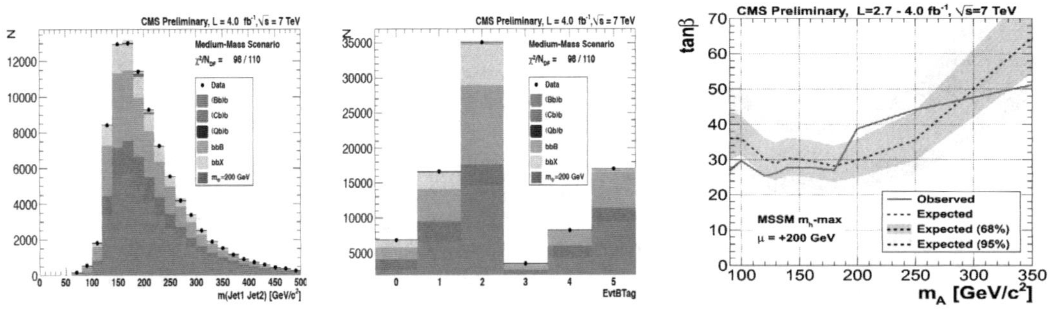

(a) M_{12} fit projection. (b) Event b-tagging variable fit projection. (c) 95% CL limits.

Figure 4. (a-b) Results of the fit in data (black dots) using the background and signal ($M_\Phi = 200\ GeV$) templates (full distributions). The projection of the fit in each variable is shown in the plots. (c) 95% CL exclusion limits in the ($M_A, tan\beta$) parameter space of the MSSM model in the m_h^{max} benchmark scenario. The red line is the observed limit in data. The blue dotted line is the median expected limit for the background only hypothesis. The green (yellow) band is the range that is expected to contain 68% (95%) of all the observed limit excursions from the median.

no evidence of a Higgs boson signal is found, 95% CL limits on the cross section times the branching ratio in b pair are calculated with the CL_s method. These limits are converted in exclusion plots in the $(M_A, tan\beta)$ parameter space, shown in Figure 4. The analysis excludes values of $tan\beta$ as low as 25 at $M_A = 120\ GeV$. The limits calculation take into account the statistics and systematic uncertainties, that contribute to worsen the sensitivity of the analysis. Different systematic uncertainties have been identified. They affect the signal yield and shape, the largest contribution coming from b-tagging efficiency $(10 - 13\%)$, and signal yield only, the largest contribution coming from the online b-tagging efficiency (32%). Theoretical systematic uncertainties range from 3% to 28%.

6. $b\bar{b}$ channel search: semi leptonic

This analysis [17] exploits the same channel used in the all hadronic analysis to perform the Higgs boson search. A data sample of a total of 4.7 fb^{-1} at pp center of mass energy of 7 TeV are analyzed, selected by HLT paths requiring one muon, one or two jets with one or two of them b-tagged. The final analyzed data sample is filtered requiring 3 well reconstructed high p_T b-tagged jets in the tracker region of the CMS detector and a muon inside one of the two leading jets. The semi leptonic decay of the b-quark is exploited in this analysis.
The semi leptonic analysis is affected by the same backgrounds of the all hadronic analysis, with the QCD multi-jet the dominant source and the $t\bar{t}$ and $Z + jets$ negligible.
The Higgs boson search is performed in the invariant mass of the two leading b-tagged jets M_{12}. Two data-driven techniques are used to estimate shape and normalization of the QCD background: *B-Tagging matrix* and *nearest-neighbour* methods. The *B-Tagging matrix* uses a data sample of 2 b-tagged jets plus a third jet where the b-tagging requirement is not applied (bbj sample). The prediction for the distribution of the relevant kinematic variables in the 3 b-tagged jets sample (bbb sample), is obtained by weighting each bbj event with the probability for the third jet j to be b-tagged. This probability is built with the efficiencies to b-tag a jet originating from a x-flavour parton (from MC simulation) and the fractions of jets originated by that x-flavour parton (from data). To avoid a bias in the measure of the fraction of jets originated by b-quarks due to a possible Higgs boson signal, control and signal regions are defined from a discriminator variable built using QCD and signal simulated events. The discriminator is defined for two different Higgs mass regions: *low Higgs mass* ($M_\Phi < 200\ GeV$) and *high Higgs mass* ($M_\Phi \geq 200\ GeV$). The *nearest-neighbour* method use a data samle of 1 b-tagged jet plus a second and a third jets where the b-tagging requirement is not applied (bjj sample). A sample of similar events to those in the signal region are selected from a large training sample obtained in the control region. The probability for a bjj event to have 3 b-tagged jets is obtained from the ration of bbb and bjj events in the sample of similar events. This probability is used to predict the relevant kinematic distributions in the bbb sample. The two data-driven methods are independent, thus the two background templates are combined.
The Higgs boson signal search is done performing a binned likelihood fit of the M_{12} variable in the 3 b-tagged data sample, shown in Figure 5. The combined background template from the two data-driven methods and the signal templates from MC simulation are used in the fit. Since no evidence of a Higgs boson signal is found, 95% CL limits on the cross section times the branching ratio in b pair are calculated with the CL_s method. These limits are converted in exclusion plots in the $(M_A, tan\beta)$ parameter space, shown in Figure 5. The analysis excludes values of $tan\beta$ as low as 20 at $M_A = 200\ GeV$. The limits calculation take into account the statistics and systematic uncertainties, that contribute to worsen the sensitivity of the analysis. Different systematic uncertainties have been identified. They affect the signal yield, the largest contribution coming from b-tagging efficiency (12%), and the background normalization (5%). Theoretical systematic uncertainties range from 3% to 28%.

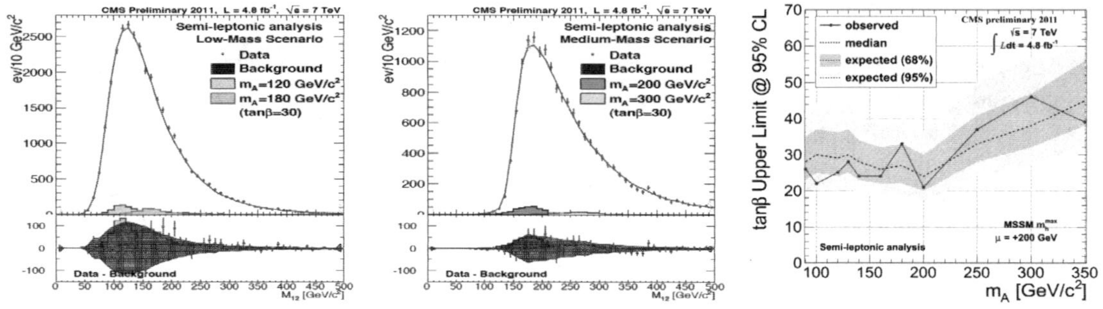

(a) Low Higgs mass region. (b) High Higgs mass region. (c) 95% CL limits.

Figure 5. (a-b) Results of the fit in data (red dots) using the combined background from the two data-driven methods (blue line) and signal templates (full distributions) for the low and high Higgs mass regions. In the lower section of the distributions is shown the ratio of data and background prediction (red dots) compared to the uncertainties (blue band). (c) 95% CL exclusion limits in the $(M_A, tan\beta)$ parameter space of the MSSM model in the m_h^{max} benchmark scenario. The red line is the observed limit in data. The black dotted line is the median expected limit for the background only hypothesis. The green (yellow) band is the range that is expected to contain 68% (95%) of all the observed limit excursions from the median.

7. All hadronic and semi leptonic combination

The same Higgs boson production mechanism and decay channels are exploited by the all hadronic and semi leptonic analyses. The sensitivities reached by the two analyses are similar and a combination of the results is therefore performed to gain sensitivity [18]. All the systematic uncertainties have been considered for the calculation. The background systematic uncertainties are all uncorrelated, the common signal systematic uncertainties are 100% correlated and the analyses specific signal systematic uncertainties are uncorrelated. The small overlap of about 2.5% between the data samples of the two analyses has been removed for the combination. Figure 6 shows the result of the combined limits. As it can be seen the overall sensitivity is

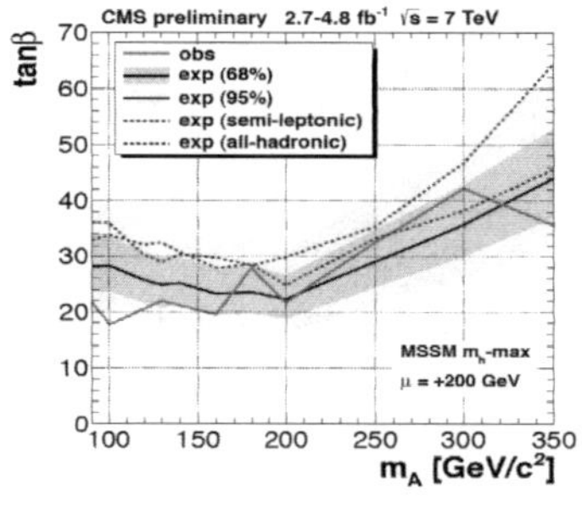

Figure 6. 95% CL combined exclusion limits in the $(M_A, tan\beta)$ parameter space of the MSSM model in the m_h^{max} benchmark scenario. The red line is the observed limit in data. The black line is the median expected limit for the background only hypothesis. The green (yellow) band is the range that is expected to contain 68% (95%) of all the observed limit excursions from the median. The limits are compared to the expected limits from the all hadronic (magenta dotted line) and semi leptonic (blue dotted line) analyses.

improved with respect to the sensitivity of the single analyses.

8. Conclusions

The search for a neutral MSSM Higgs boson with the CMS experiment at the LHC has been reported for two decay channels: $\tau\tau$ and $b\bar{b}$. The data show no significant excess with respect to the expected background. Limits at 95% of CL are then calculated and interpreted as exclusion limits on $tan\beta$ as a function of M_A in the m_h^{max} of the MSSM theory. In the $\tau\tau$ final state the analysis excluded $tan\beta$ as low as 4 for $M_A = 200\ GeV$. In the $b\bar{b}$ final state the analyses excluded $tan\beta \sim 20 - 40$ in the full M_A range up to $M_A = 350\ GeV$. These results extend the area in the $(M_A, tan\beta)$ parameter space excluded by the Tevatron experiments [19, 20, 21]. Both the $\tau\tau$ and $b\bar{b}$ analyses are important since they test couplings to different particles of the Higgs boson.

References

[1] Steven Weinberg 1976 Implications of dynamical symmetry breaking *Phys. Rev.* D **13** pp. 974-996.
[2] R. Barate et al. 2003 Search for the standard model Higgs boson at LEP *Phys. Lett.* B **565** pp. 61-75.
[3] T. Aaltonen et al. 2010 Combination of Tevatron searches for the standard model Higgs boson in the W+W- decay mode *Phys. Rev. Lett.* **104** p. 061802.
[4] S. Schael et al. 2006 Precision electroweak measurements on the Z resonance *Phys. Rept.* **427** pp. 257-454.
[5] (ed.) Evans Lyndon and (ed.) Bryant Philip 2008 LHC Machine *JINST* **3** S08001.
[6] CMS Collaboration 2012 Combination of SM, SM4, FP Higgs boson searches *CMS-PAS-HIG-12-008*.
[7] CMS Collaboration 2012 Observation of a new boson at a mass of 125 GeV with the CMS experiment at the LHC *Phys. Lett.* B **716** pp. 30-61.
[8] ATLAS Collaboration 2012 Observation of a new particle in the search for the Standard Model Higgs boson with the ATLAS detector at the LHC *Phys. Lett.* B **716** pp. 1-29.
[9] I. J. R. Aitchison 2005 Supersymmetry and the MSSM: An Elementary introduction *hep-ph/0505105*.
[10] Marcela S. Carena et al. 2003 Suggestions for benchmark scenarios for MSSM Higgs boson searches at hadron colliders *Eur. Phys. J.* C **26** pp. 601-607.
[11] CMS Collaboration 2008 The CMS experiment at the CERN LHC *JINST* **3** S08004.
[12] CMS Collaboration 2012 Search for neutral Higgs bosons decaying to τ pairs in pp collisions at $\sqrt{s} = 7$ TeV *Phys.Lett.* B **713**, 68 [arXiv:1202.4083 [hep-ex]].
[13] CMS Collaboration 2011 Measurement of Inclusive W and Z Cross Sections in pp Collisions $\sqrt{s} = 7\ TeV$ *JHEP* **1110** 132.
[14] A. L. Read 2000 Modified frequentist analysis of search results (the CLs method) *CERN-OPEN-2000-205*.
[15] CMS Collaboration 2012 MSSM Higgs production in association with b quarks - all hadronic *CMS-PAS-HIG-12-026*.
[16] CMS Collaboration 2011 b-Jet Identification in the CMS Experiment *CMS-PAS-BTV-11-004*.
[17] CMS Collaboration 2012 Search for SuperSymmetric Higgs boson states decaying into $b\bar{b}$ and produced in association with b-quarks in events collected by semi-leptonic triggers in pp collisions at $\sqrt{s} = 7\ TeV$ *CMS-PAS-HIG-12-027*.
[18] CMS Collaboration 2013 Search for a Higgs boson decaying into a b-quark pair and produced in association with b quarks in proton-proton collisions at 7 TeV *arXiv:1302.2892* [hep-ex].
[19] Tevatron New Phenomena & Higgs Working Group Collaboration 2010 Combined CDF and D0 Upper Limits on MSSM Higgs Boson Production in tau-tau Final States with up to 2.2 fb-1 *arXiv:1003.3363* [hep-ex].
[20] CDF Collaboration 2012 Search for Higgs Bosons Produced in Association with *b*-quarks *Phys. Rev.* D **85**, 032005 [arXiv:1106.4782 [hep-ex]].
[21] D0 Collaboration 2011 Search for neutral Higgs bosons in the multi-*b*-jet topology in 5.2fb^{-1} of $p\bar{p}$ collisions at $\sqrt{s} = 1.96$ TeV *Phys. Lett.* B **698**, 97 [arXiv:1011.1931 [hep-ex]].

International Workshop on Discovery Physics at the LHC (Kruger2012) IOP Publishing
Journal of Physics: Conference Series **455** (2013) 012025 doi:10.1088/1742-6596/455/1/012025

Electroweak results from CMS

V Ciulli[1] on behalf of the CMS Collaboration

[1] Dipartimento di Fisica, Università di Firenze, via G. Sansone 1, 50019 Sesto F. (FI), Italy

E-mail: vitaliano.ciulli@fi.infn.it

Abstract. The latest electroweak results from CMS, based on pp collision data taken at 7 and 8 TeV, are presented. In particular precise measurements of the inclusive and differential cross sections for W and Z boson production, and for multi-boson production are reviewed and compared with theory predictions. Results agree with Standard Model and limits are set on anomalous triple gauge couplings.

1. Introduction

The dominant mechanism for W and Z boson production in pp collisions at the Large Hadron Collider (LHC) is the annihilation of a valence quark from one proton with a sea antiquark from the other. The contribution from sea quark-antiquark scattering is also relevant, as a value of $Q^2 \sim M_Z^2$ is reached in the interactions between partons carrying a proton momentum fraction $10^{-3} < x < 10^{-1}$. Furthermore, the process where a vector boson is produced in association with a jet, depends primarily on the gluon content of the proton. Therefore, the W and Z production allows probing both the electroweak couplings to the quarks and constraining the PDFs.

Similarly the measurement of the diboson production is a direct test of triple gauge couplings (TGC). In the Standard Model only the charged triple gauge couplings, WWZ and WWγ, are allowed, while the neutral, ZZZ, ZZγ and Z$\gamma\gamma$, are forbidden. Any deviation of the latter from a null value would be therefore an evidence of new physics beyond the Standard Model.

Theory calculations and Monte Carlo generators are now available at next-to-leading order (NLO), and even next-to-next-to-leading order (NNLO) predictions for single W and Z cross-section.

In the following the latest measurements of single and diboson cross-section measurements done by the Compact Muon Solenoid experiment at the LHC are reviewed and compared with expectations from the theory.

2. The CMS detector

The central feature of the Compact Muon Solenoid (CMS) apparatus is a superconducting solenoid of 6 m internal diameter, providing a magnetic field of 3.8 T. Within the superconducting solenoid volume are a silicon pixel and strip tracker, a lead tungstate crystal electromagnetic calorimeter (ECAL), and a brass/scintillator hadron calorimeter (HCAL). Muons are measured in gas-ionization detectors embedded in the steel return yoke outside the solenoid. Extensive forward calorimetry complements the coverage provided by the barrel and endcap detectors. The ECAL energy resolution for electrons with $E_{\perp}\approx$45 GeV from Z$\rightarrow$ ee decays is better than 2%

in the central region of the ECAL barrel ($|\eta| < 0.8$), and is between 2% and 5% elsewhere. The electron energies are measured by the ECAL and their directions are measured by the tracker. The energy resolution for photons with $E_{\perp} \approx 60$ GeV varies between 1.1% and 2.5% over the solid angle of the ECAL barrel, and from 2.2% to 5% in the endcaps. Muons are measured in the pseudorapidity range $|\eta| < 2.4$, with detection planes made using three technologies: drift tubes, cathode strip chambers, and resistive plate chambers. Matching muons to tracks measured in the silicon tracker results in a transverse momentum resolution between 1 and 5%, for $p_{\perp}$ values up to 1 TeV. A more detailed description can be found in Ref. [1].

3. Single W and Z cross-section measurements

A precision measurement of the inclusive W and Z boson cross sections at $\sqrt{s} = 8$ TeV [2] has been performed with the same method used at 7 TeV [3]. The instantaneous luminosity was reduced to accumulate a dataset with an average low number of proton-proton interactions (pileup) and low transverse momentum trigger thresholds. The beam separation was periodically adjusted to compensate the loss of instantaneous luminosity, which therefore was maintained in the range 3×10^{32} to 6×10^{32} cm^{-2} s^{-1} ("luminosity leveling"). A data sample corresponding to an integrated luminosity of 18.7 ± 0.9 pb^{-1} was collected, where both electron and muon final states were analyzed. The measured values are shown in Fig. 1 and agree with NNLO cross section calculations obtained with FEWZ [4] and the MSTW2008 PDF set.

In addition to the analysis of 8 TeV data, several analyses have been updated with the full dataset at 7 TeV. A measurement of the Drell-Yan differential cross section $d\sigma/dM$ in the dielectron and dimuon channels and the first measurement of the double-differential cross section $d^2\sigma/dMdY$ in the dimuon channel, have been been performed with an integrated luminosity of 4.5 fb^{-1} [5]. Distributions are unfolded for the effects of resolution and final state QED radiation using POWHEG Monte Carlo [6], and an additional acceptance correction is applied to take the NNLO effects into account. The $d\sigma/dM$ results are overall in very good agreement with the NNLO theoretical predictions, as computed with FEWZ using the MSTW2008 PDFs. A small deviation is observed for $d^2\sigma/dMdY$ measurement for dimuon mass $M < 45$ GeV.

The forward-backward asymmetry (A_{FB}) of Drell-Yan lepton pairs has been also measured as a function of dilepton mass and rapidity in the dielectron and dimuon channels [7]. The A_{FB} measurement in the dimuon channel and the combination of the two channels are the first such results obtained at a hadron collider. The measured asymmetries are consistent with the Standard Model predictions.

Lepton charge asymmetry have been measured in the $W \to e$ [8] and $W \to \mu$ [9] channels. Electrons (muons) are selected with a $p_{\perp} > 35(25)$ GeV in a sample of proton-proton collisions at 7 TeV corresponding to an integrated luminosity of 840 (234) pb^{-1}. The measured distributions are shown in Fig. 2. In the muon channel the distribution is flatter than expected, while in the electron channel it is difficult to draw any conclusion as the uncertainty increases at high values of the pseudorapidity because of the larger background. Overall the experimental data are in good agreement with the predictions from CT10, NNPDF, and HERAPDF, while the predictions from MSTW are systematically lower than the observed asymmetry in the region $|\eta| < 1.4$.

Using the dataset collected at $\sqrt{s} = 7$ TeV, the Z boson decaying to four leptons (electrons and/or muons) has been observed for the first time in proton-proton collisions [10]. A pronounced resonance peak, with a statistical significance of $9.7\,\sigma$, is present in the distribution of the invariant mass of four leptons with mass and width consistent with expectations for Z boson decays. The measured branching fraction is $\mathcal{B}(\mathrm{Z} \to 4l) = \left(4.2^{+0.9}_{-0.8}\,(\mathrm{stat.}) \pm 0.2\,(\mathrm{syst.})\right) \times 10^{-6}$, in agreement with the Standard Model prediction of 4.45×10^{-6}. The measured cross section times branching fraction is $\sigma(\mathrm{pp} \to \mathrm{Z})\,\mathcal{B}(\mathrm{Z} \to 4l) = 112^{+23}_{-20}\,(\mathrm{stat.})^{+7}_{-5}\,(\mathrm{syst.})^{+3}_{-2}\,(\mathrm{lumi.})$ fb^{-1}, also consistent with the Standard Model prediction of 120 fb. The four-lepton mass peak arising

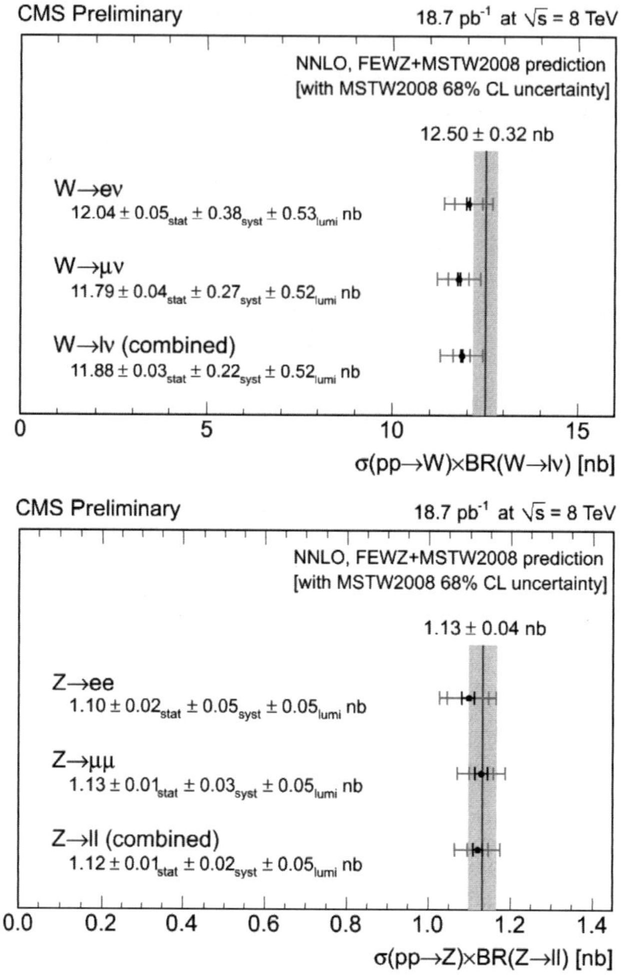

Figure 1. Summary of the W (top) and Z (bottom) production cross section times branching ratio measurements. Measurements in the electron and muon channels, and combined, are compared to the theoretical predictions (yellow band) computed at the NNLO in QCD with FEWZ and the MSTW2008 PDF set. Statistical uncertainties are represented as black error bars, while the red error bars also include systematic uncertainties, and the green error bars also include luminosity uncertainties

from $Z \to 4l$ decays provides a very important calibration channel for the Higgs boson search in the $H \to ZZ \to 4l$ decay mode.

4. Diboson production and limits on TGC couplings

The WW and ZZ production cross sections are measured in proton-proton collisions at $\sqrt{s} =$ 8 TeV in data samples corresponding to an integrated luminosity of up to 5.3 fb^{-1} [11]. The measurements are performed in the leptonic decay modes $WW \to l\nu l'\nu$ and $ZZ \to 2l2l'$, where $l = e$, μ and $l' = e$, μ, τ. The measured cross sections $\sigma(pp \to WW) = 69.9 \pm 2.8(\text{stat.}) \pm 5.6(\text{syst.}) \pm 3.1(\text{lum.})$ pb and $\sigma(pp \to ZZ) = 8.4 \pm 1.0(\text{stat.}) \pm 0.7(\text{syst.}) \pm 0.4(\text{lum.})$ pb, for both Z bosons produced in the mass region $60 < M_Z < 120$ GeV are consistent with Standard Model predictions. Figure 3 shows the dilepton transverse momentum of the WW candidate events.

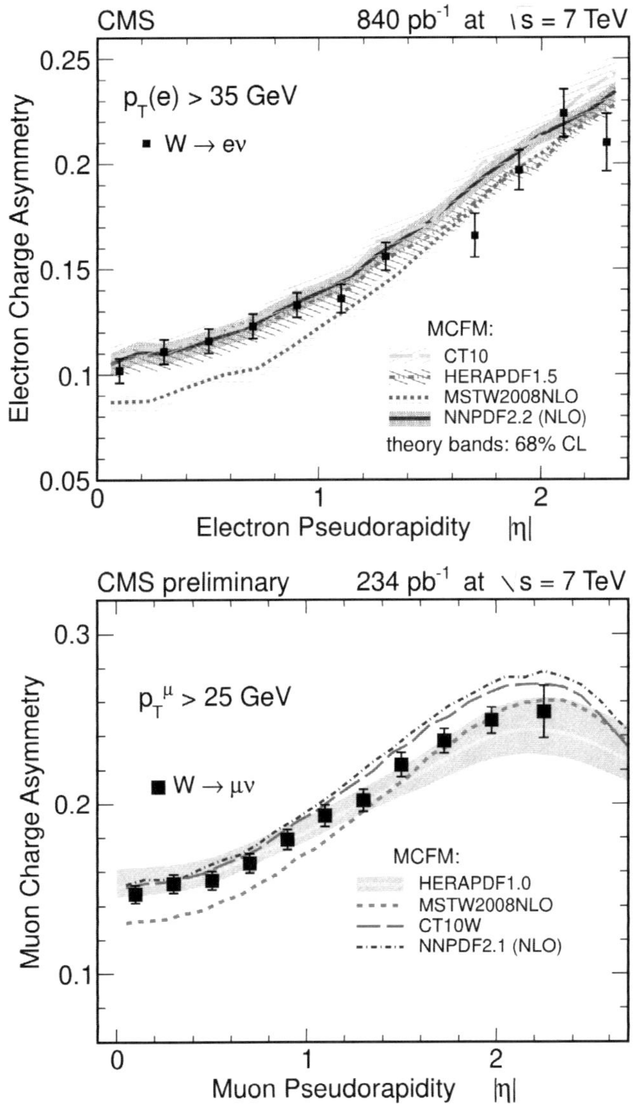

Figure 2. Comparison of the measured electron (top) and muon (bottom) asymmetry to the predictions of different PDF models. The error bars include both statistical and systematic uncertainties. The data points are placed in the center of the $|\eta|$ bins. The PDF uncertainty bands are estimated using the PDF reweighting technique and correspond to 68% confidence level.

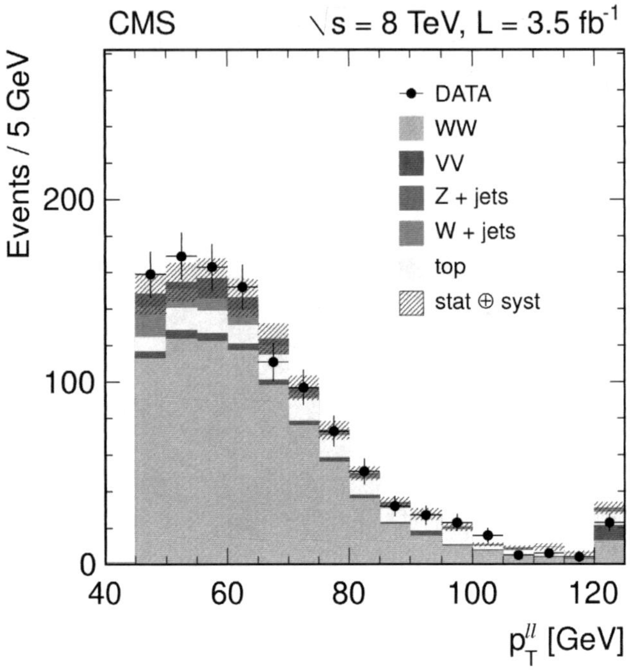

Figure 3. Distributions for WW candidate events of the dilepton transverse momentum. Points represent the data, and shaded histograms represent the WW signal and the background processes. The last bin includes the overflow. The WW signal is scaled to the measured cross section, and the background processes are normalized to the corresponding estimated values

These are the first measurements of the diboson production cross sections at $\sqrt{s} = 8$ TeV.

The full data sample recorded at $\sqrt{s} = 7$ TeV, corresponding to an integrated luminosity of 5 pb^{-1}, has been used to measure Wγ and Zγ production cross sections and set limits on anomalous triple WWγ, ZZγ, and Z$\gamma\gamma$ gauge couplings.

The total cross sections for W$\gamma \to l\nu\gamma$ and Z$\gamma \to ll\gamma$, where $l = e$ or μ, are measured for photon transverse energy $E_\perp > 15$ GeV and angular separation from charged leptons $\Delta R(l, \gamma) > 0.7$. For the Z$\gamma$ final state an additional dilepton invariant mass requirement of $M(ll) > 50$ GeV is imposed. Preliminary results shows no deviation from the Standard Model values.

The first measurement of the Z$\gamma \to \nu\nu\gamma$ cross section in pp collisions at $\sqrt{s} = 7$ TeV has been also done. The measurement is performed normalizing to photons with transverse energy of 145 GeV in the pseudorapidity region $|\eta| < 1.4$. Figure 4 shows the photon $p_\perp$ distribution for the candidates on a data sample corresponding to an integrated luminosity of 5 fb^{-1}. The measured cross is 21.3 ± 4.2(stat.) $\pm$ 4.3(syst.) $\pm$ 0.5(lumi.) fb, which is in good agreement with the theoretical prediction of 21.9 ± 1.1 fb, obtained at NLO accuracy using the WGRAD generator [12]. Combining this measurement with the one of Z$\gamma \to ll\gamma$, the most stringent limits to date have been set on anomalous ZZγ and Z$\gamma\gamma$ couplings.

The sum of WW and WZ diboson production cross section has been also measured at $\sqrt{s} = 7$ TeV, using events containing a leptonically decaying W boson and exactly two jets [13]. The measured value of the total cross section is 68.89 ± 8.71(stat) $\pm$ 9.70(syst) $\pm$ 1.52(lumi) pb, consistent with the Standard Model NLO prediction of 65.6 ± 2.2 pb [14] and no evidence for

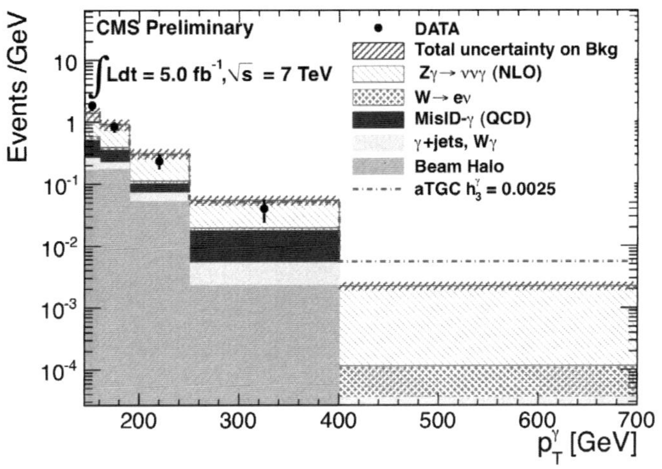

Figure 4. The photon $p_\perp$ distribution for the candidate sample, compared with estimated contribution from SM background, SM signal and aTGC signal. The last bin includes overflows.

anomalous triple gauge couplings is found.

In addition, the invariant mass spectrum of the two jets with highest transverse momentum in W+2-jets and W+3-jets events has been studied [15]. The relative contribution of the known Standard Model processes to the observed dijet invariant mass spectrum has been determined using an unbinned maximum likelihood fit in the range between 40 GeV and 400 GeV. The fit is performed separately for the electron and muon channels and for the 2 jets and 3 jets samples since relative background contributions and composition differ. Figure 5 shows the distribution of the invariant mass spectrum of the leading two jets observed in data, after subtraction of all SM components except the electroweak diboson WW/WZ. No evidence is found for the anomalous structure reported by the CDF Collaboration, and an upper limit of 5.0 pb is established at 95 % confidence level on the production cross section for a generic Gaussian signal with mass near 150 GeV.

5. Conclusions

The CMS experiment at the LHC accomplished a large program of electroweak measurements that allows testing the Standard Model with a precision comparable or even better than available (N)NLO predictions. The results are in agreement with expectations for production cross sections that span over several order of magnitude. The measurements of inclusive and differential cross sections of W and Z bosons provides significant constraints on PDFs. The measurements of diboson production set stringent limits on the triple gauge couplings. No significant deviation from the Standard Model is observed.

[1] S. Chatrchyan et al. The CMS experiment at the CERN LHC. *JINST*, 03:S08004, 2008.
[2] CMS Collaboration. Inclusive W/Z cross section at 8 TeV. CMS Physics Analysis Summary CMS-PAS-SMP-12-011, 2012.
[3] S. Chatrchyan et al. Measurement of the inclusive W and Z production cross sections in pp collisions at $\sqrt{s} = 7$ TeV with the CMS experiment. *Journal of High Energy Physics*, 2011:1–76, 2011.
[4] Ryan Gavin, Ye Li, Frank Petriello, and Seth Quackenbush. FEWZ 2.0: A code for hadronic Z production at next-to-next-to-leading order. *Comput.Phys. Commun.*, 182:2388–2403, 2011.
[5] CMS Collaboration. Measurement of the differential and double-differential Drell-Yan cross section in proton-proton collisions at 7 TeV. CMS Physics Analysis Summary CMS-PAS-EWK-11-007, 2012.

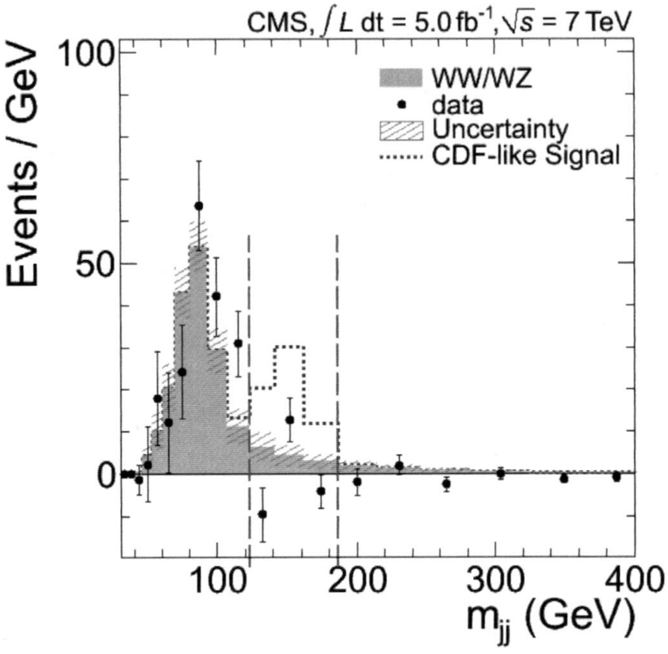

Figure 5. Distribution of the invariant mass spectrum of the leading two jets observed in data, after subtraction of all SM components except the electroweak diboson WW/WZ. The region between the vertical dotted lines is excluded in the fit. Error bars correspond to the statistical uncertainty. The band represents the systematic uncertainty in the sum of the SM components.

[6] Simone Alioli, Paolo Nason, Carlo Oleari, and Emanuele Re. NLO vector-boson production matched with shower in POWHEG. *JHEP*, 07:060, 2008, 0805.4802.

[7] S. Chatrchyan et al. Forward-backward asymmetry of Drell-Yan lepton pairs in pp collisions at $\sqrt{s} = 7$ TeV. *Physics Letters B*, 718(3):752 – 772, 2013.

[8] S. Chatrchyan et al. Measurement of the Electron Charge Asymmetry in Inclusive W Production in pp Collisions at $\sqrt{s} = 7$ TeV. *Phys. Rev. Lett.*, 109:111806, Sep 2012.

[9] CMS Collaboration. Measurement of the muon charge asymmetry in inclusive W production in pp collisions at $\sqrt{s} = 7$ TeV. CMS Physics Analysis Summary CMS-PAS-EWK-11-005, 2012.

[10] S. Chatrchyan et al. Observation of Z decays to four leptons with the CMS detector at the LHC. *Journal of High Energy Physics*, 2012:1–28, 2012.

[11] Serguei Chatrchyan et al. Measurement of W^+W^- and ZZ production cross sections in pp collisions at $\sqrt{s}$=8 TeV. CERN-PH-EP-2012-376, arXiv:1301.4698 [hep-ex], 2013.

[12] U. Baur and Edmond L. Berger. Probing the weak boson sector in $Z\gamma$ production at hadron colliders. *Phys.Rev.*, D47:4889–4904, 1993.

[13] S. Chatrchyan et al. Measurement of the sum of WW and WZ production with W+dijet events in pp collisions at $\sqrt{s} = 7$ TeV. *The European Physical Journal C*, 73:1–18, 2013.

[14] John M. Campbell, R. Keith Ellis, and Ciaran Williams. Vector boson pair production at the LHC. *JHEP*, 1107:018, 2011, 1105.0020.

[15] S. Chatrchyan et al. Study of the Dijet Mass Spectrum in $pp \to W +$jets Events at $\sqrt{s} = 7$ TeV. *Phys. Rev. Lett.*, 109:251801, Dec 2012.

Vector boson plus jet physics at CMS

Tom Cornelis[1] ***on behalf of the CMS collaboration***
[1] University of Antwerp, Belgium
E-mail: tom.cornelis@cern.ch

Abstract. We present recent results on the jet production rates in association with weak vector bosons and their properties using proton-proton collision data collected with the CMS detector. Azimuthal correlations and events shapes in Z + jets events are also shown. A first measurement of the electroweak production cross section of a Z boson with 2 jets is presented, including measurements of the hadronic activity between the two associated jets.

1. Jet production in association with vector bosons

The study of jet production in association with a W or Z boson allows for tests of perturbative QCD calculations. On the other hand the productions of vector bosons with jets is an important background in searches for new physics and for studies of the top quark. We report here results [1] on jet production rates in association with W and Z bosons, using 36 pb^{-1} of proton proton collision data at $\sqrt{s} = 7$ TeV recorded with the CMS detector [2].

Events are selected if they have at least one lepton (electron or muon) with transverse momentum $p_T > 20$ GeV, and satisfying quality and isolation requirements. Electron candidates are required to be in the fiducial region of the electromagnetic calorimeter with pseudorapidity $|\eta| < 2.5$, with the exclusion of $1.44 < |\eta| < 1.57$. Muon candidates are required to have $|\eta| < 2.1$. Z bosons are reconstructed if a second lepton of the same flavour is found, which fullfills looser quality requirements and $p_T > 10$ GeV, and the dilepton invariant mass falls between 60 GeV and 120 GeV. Events which are not assigned to the Z+jets sample, could qualify for the W+jets sample. Due to the undetected neutrino, these events are characterized by a large missing transverse energy $\not{E}_T$. Therefore, the events are assigned to the W+jets sample if they have a transverse mass $M_T = \sqrt{2p_T\not{E}_T(1-\cos\Delta\phi)}$ of at least 20 GeV, where $\Delta\phi$ is the angle in the transverse plane between p_T and $\not{E}_T$. Jets constituents are reconstructed with the particle flow (PF) algorithm and are clustered using the anti-k_T algorithm with size parameter of 0.5. These jets are accepted if they have $p_T > 30$ GeV, $|\eta| < 2.4$ and fullfill standard quality identification criteria.

Figures 1 and 2 show the ratio of $\sigma(\mathrm{V}+ \geq n \text{ jets})/\sigma(\mathrm{V})$ for W and Z bosons respectively, where $\sigma(\mathrm{V})$ is the inclusive cross section. The data agrees with MADGRAPH [3] predictions, both fusing either the Z2 or D6T PYTHIA [4] tunes, while the pure parton model from PYTHIA fails to describe the data in the presence of a large number of jets.

A fit is performed in order to test Berends-Giele scaling:

$$C_n \equiv \frac{\sigma_n}{\sigma_{n+1}} = \alpha + \beta n$$

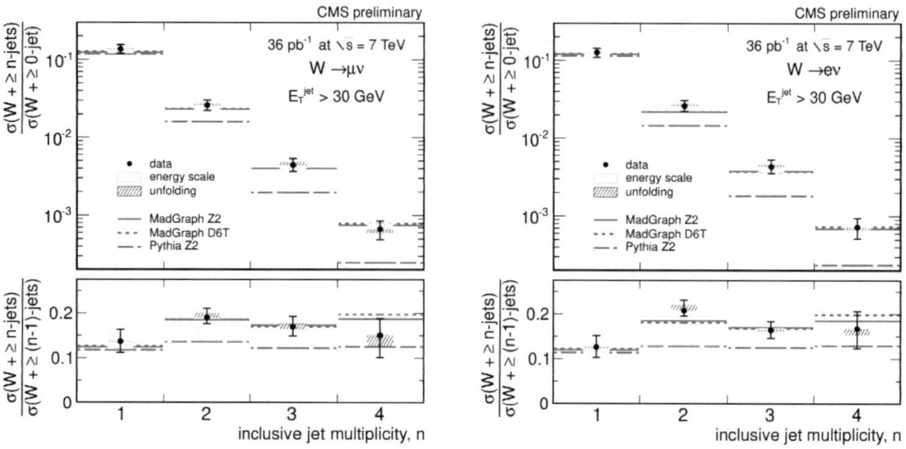

(a) muon channel (b) electron channel

Figure 1. The ratios $\sigma(\mathrm{W}+\geq n \text{ jets})/\sigma(\mathrm{W})$ and $\sigma(\mathrm{W}+\geq n \text{ jets})/\sigma(\mathrm{W}+\geq (n-1) \text{ jets})$. The error bars show the total uncertainty

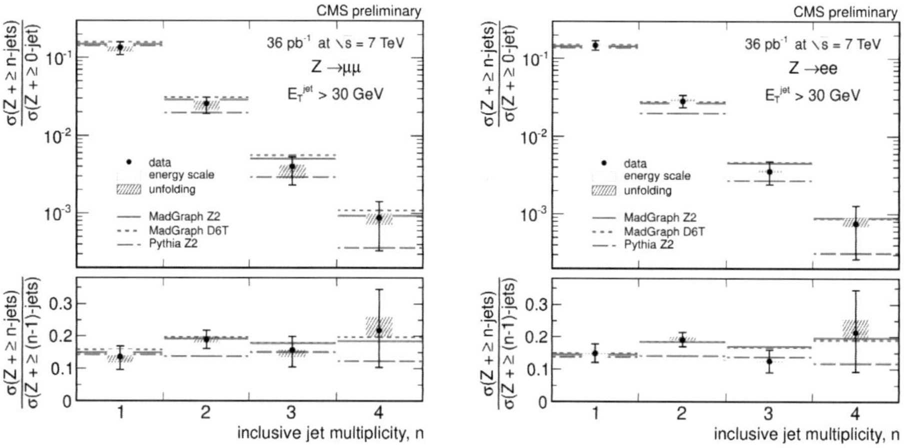

(a) muon channel (b) electron channel

Figure 2. The ratios $\sigma(\mathrm{Z}+\geq n \text{ jets})/\sigma(\mathrm{Z})$ and $\sigma(\mathrm{Z}+\geq n \text{ jets})/\sigma(\mathrm{Z}+\geq (n-1) \text{ jets})$. The error bars show the total uncertainty

where $\sigma_n = \sigma(V+\geq n \text{ jets})$, and α and β are constants. Leading order calculations predict $C_n = \alpha$, where the constant α is inversely proportional to the strong coupling constant. The β parameter is introduced to allow for possible deviations which could appear due to next-to-leading order effects and phase space effects. Figures 3 and 4 show the fit of α and β for both Z + jets and W + jets. The data is in agreement with the MadGraph predictions within one or two standard deviations depending on the channel. The β parameter lies within one standard deviation from zero.

A measurement on the ratio of W + jets and Z + jets cross sections, in which many systematic uncertainties such as the luminosity and jet energy scale cancel, are shown in figure 5. We see again a good agreement between data and simulation.

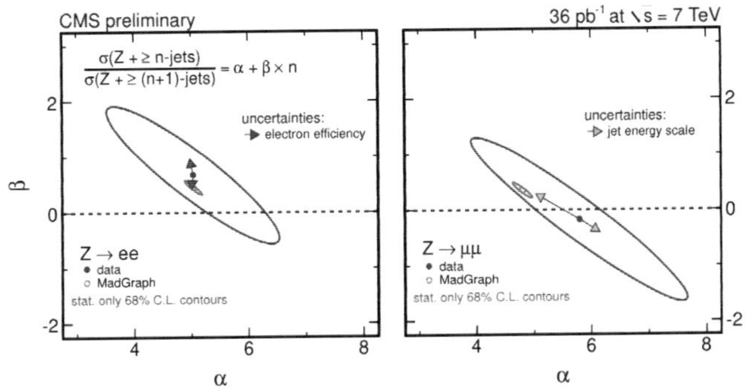

Figure 3. Fit results for the scaling parameters α and β in Z + jets

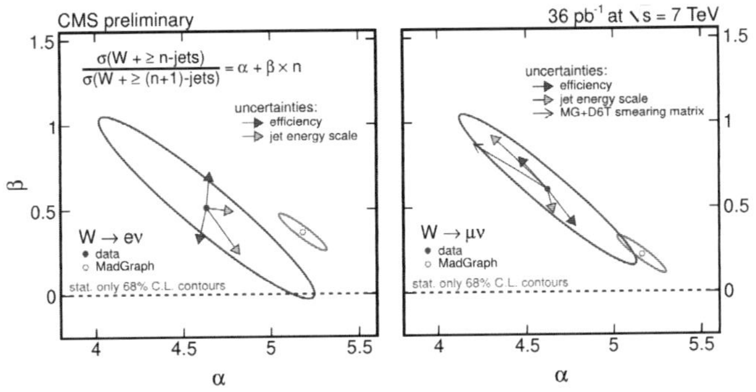

Figure 4. Fit results for the scaling parameters α and β in W + jets

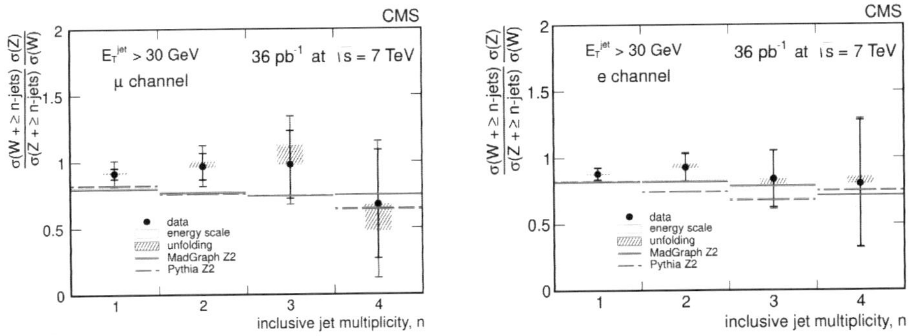

(a) muon channel (b) electron channel

Figure 5. Ratio of the W+jets and Z+jets cross sections normalized to the inclusive W/Z cross section. Error bars on the data are shown for both statistical and total uncertainties.

2. Azimuthal correlations and event shapes in Z plus jet events

A detailed study [5] of azimuthal correlations in Z plus at least one jet events in pp collisions at $\sqrt{s} = 7$ TeV was done using CMS data, corresponding to an integrated luminosity of 5 fb^{-1}. Events with Z bosons were selected by requiring an oppositely charged electron or muon pair with an invariant mass within a mass window of 20 GeV around the Z boson mass. Both leptons satisfy $p_T > 20$ GeV and $|\eta| < 2.4$ and fullfill quality and isolation criteria. PF jets with $p_T > 50$ GeV and $|\eta| < 2.5$ were used, clustered using the anti-k_T algorithm with size parameter of 0.5.

2.1. Azimuthal angles

Figure 6 shows the azimuthal angle between the Z boson and the leading jet, either for all Z bosons or in a subset of events with $p_T^Z > 150$ GeV. The boosted region is of interest because it is critical in searches for new phenomena. The results agree within uncertainties with the MADGRAPH prediction. In $Z+\geq 1$ jet events, the SHERPA [6] prediction is about 10% too low whereas POWHEG [7, 8, 9, 10] overestimates the distribution by about 10%. The Z boson and leading jet are completely correlated for $N_{jets} = 1$, as $\Delta\phi(Z, j_1)$ tends to π. If $\Delta\phi(Z, j_1) < \frac{\pi}{2}$, the leading jet is in the same hemisphere as the Z boson and the Z boson is balanced by subleading jets in the opposite hemisphere. Therefore the $\Delta\phi(Z, j_1)$ becomes more isotropic at larger inclusive jet multiplicities.

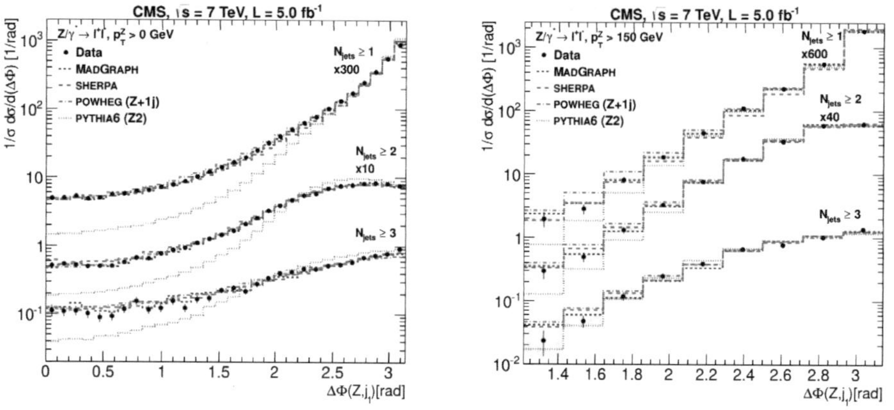

(a) all p_T^Z (b) $p_T^Z > 150$ GeV

Figure 6. Normalized $\Delta\phi(Z, j_1)$ distributions categorized as a function of jet multiplicity. The yellow band shows the sum of statistical and systematic errors on the data.

Figure 7 shows the azimuthal angles among the three leading jets, which show good agreement between MADGRAPH, SHERPA and PYTHIA predictions, and the data. In the boosted regime, the angles between the jets decorrelate.

2.2. Transverse thrust

The properties of Z plus jet events have also been studied using the transverse thrust τ_T, defined as:

$$\tau_T \equiv 1 - \max_{\vec{n}_\tau} \frac{\sum_i |\vec{p}_{T,i} \cdot \vec{n}_\tau|}{\sum_i p_{T,i}}$$

where the sum over i runs over the Z and each accepted jet in the event. The sum is maximized by the unit vector $\vec{n}_\tau$, called the thrust axis. In the limit of a perfectly balanced Z plus one jet event, τ_T tends to zero ($\ln \tau_T \to -\infty$). In the limit of a homogeneously distributed event, the transverse thrust reaches $1-\frac{\pi}{2}$ ($\ln \tau_T \to -1$), its maximum possible value. The $\ln \tau_T$ distribution is shown in figure 8. The data is best described by POWHEG and MADGRAPH, except at large negative values for the transverse thrust where an overestimation by MADGRAPH is seen. PYTHIA and SHERPA have larger discrepancies and predict also too small values for $\ln \tau_T$. The peak at $\ln \tau_T \approx -2$ in the boosted region corresponds with events where the Z is balanced by two or more jets.

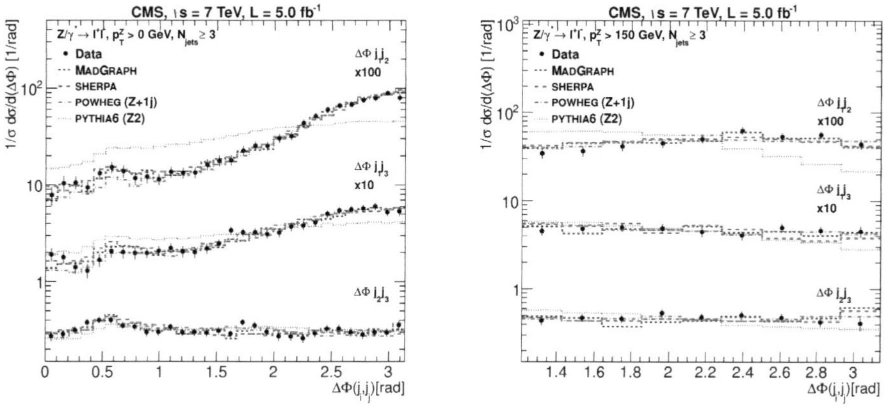

(a) all p_T^Z (b) $p_T^Z > 150$ GeV

Figure 7. Normalized $\Delta\phi(j_i, j_j)$ distributions for the inclusive $N_{jets} > 3$ multiplicity bin. The yellow band shows the sum of statistical and systematic errors on the data.

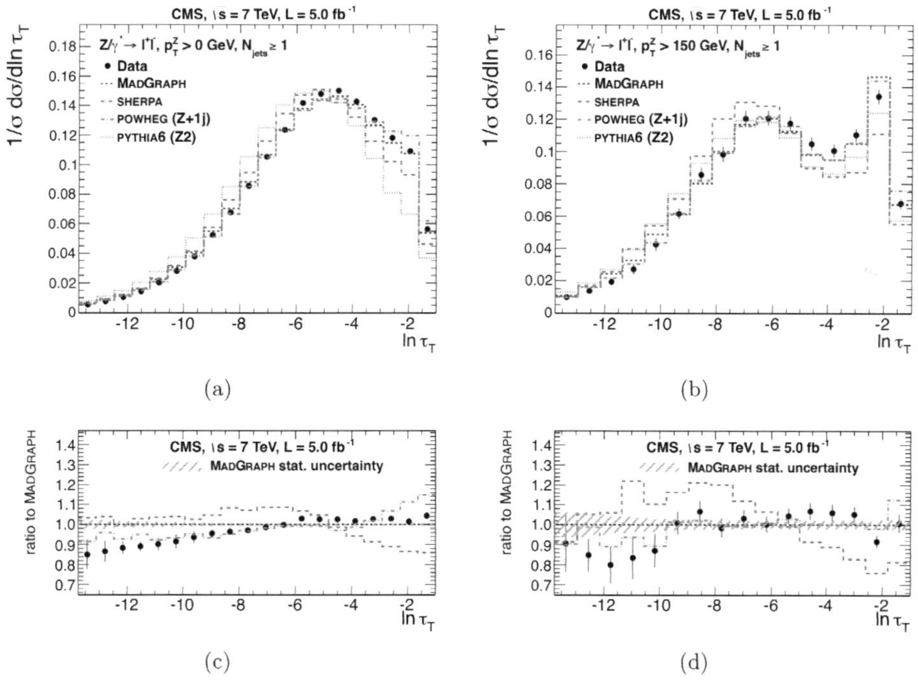

(a) (b) (c) (d)

Figure 8. Normalized distributions in $\ln\tau_T$ for Z plus at least 1 jet events for (a) all p_T^Z (b) $p_T^Z > 150$ GeV. The ratios to MADGRAPH are shown in (c) and (d). The yellow band shows the sum of statistical and systematic errors on the data.

3. Electroweak production of Z+jets

We have measured [11] the pure electroweak cross section of a Z boson in association with two jets in pp collisions at $\sqrt{s} = 7$ TeV, using 5 fb^{-1} of data recorded by the CMS detector. In addition, the hadronic activity in the rapidity interval between the two jets has also been studied. There are three classes of diagrams, shown in figure 9, contributing to the electroweak

Z + 2 jets production: bremsstrahlung, vector boson fusion (VBF) processes and multi-peripheral. The standard model predicts large negative interference effects between the VBF process and the other two categories.

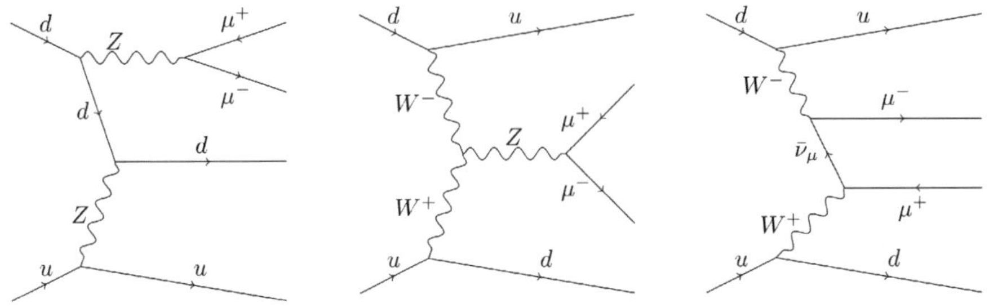

(a) Bremsstrahlung process (b) VBF process (c) Multiperipheral process

Figure 9. Representative diagrams for the EW $\ell\ell$jj production processes.

3.1. Event selection
Events are selected by requiring a muon or electron pair with opposite charge, in which both leptons satisfy a series of quality and isolation criteria, a transverse momentum of at least 20 GeV and $|\eta| < 2.4$. The Z bosons are reconstructed by taking events with a dimuon (dielectron) mass within the 15 GeV (20 GeV) mass interval around the nominal Z boson mass. The two leading PF jets within $|\eta| < 4.7$ are selected as the tagging jets. Further selections were optimized by maximizing the signal significance defined as $N_S/\sqrt{N_B}$ and lead to a requirement of $p_T^{j_1,j_2} > 65, 40$ GeV and $|\eta_j| < 3.6$ for the two tagging jets.

3.2. Cross section measurement
The signal cross section has been extracted from a fit of the signal and background contributions to the output of a boosted decision tree (BDT) applied to the selected events. The BDT was trained with EW $\ell\ell$jj simulated events as signal model and DY $\ell\ell$jj as background model using the following variables:

- the transverse momenta of the tagging jets
- the invariant dijet mass and the pseudorapidity separation between the two tagging jets
- the y^* variable, defined as $y^* = y_{ll} - 0.5(y_{j1} + y_{j2})$ using the rapidities of the two tagging jets and the dilepton system
- the transverse momentum and rapidity of the dilepton system
- the sum of the pseudorapidities of the two tagging jets
- the difference between the azimuthal angles: $(\phi_{ll} - \phi_{j1})$, $(\phi_{ll} - \phi_{j2})$ and $(\phi_{j1} - \phi_{j2})$

In the dielectron channel, a quark-gluon likelihood for both tagging jets is also used as input. The jets in the EW signal process originate only from quarks while in the DY background half of the jets are initiated by gluons. Using this likelihood discriminator, built out of 5 internal jet properties, a decrease of the statistical uncertainty of the measured signal by 5% was achieved.

Figure 10 shows the BDT output for the data, simulated signal and backgrounds. The signal and background contributions in the data, relative to the expected contributions in the simulation are denoted as s and b, and are evaluated from a maximum likelihood fit of the data BDT output distribution to the expected signal and background output shapes from the simulation. The signal cross section measured in the data is then given by

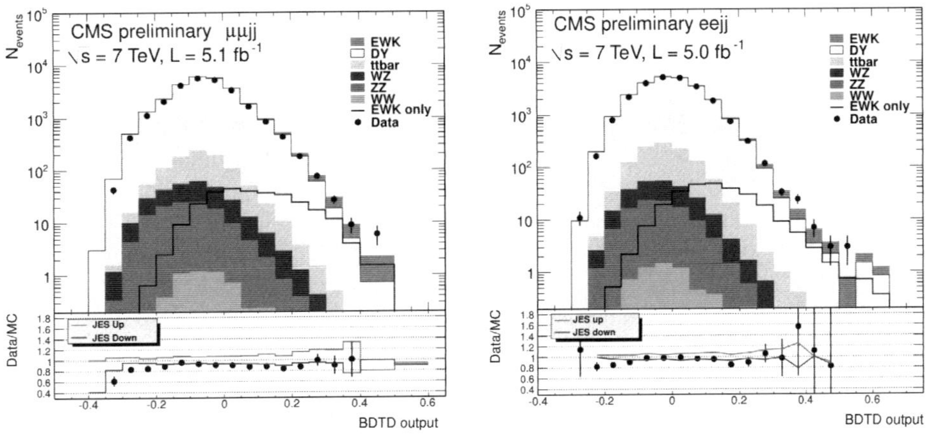

(a) dimuon channel (b) dielectron channel

Figure 10. The BDT output distributions after events selections. The bottom panels show the ratio of data over the expected contribution of signal plus background.

$\sigma_{\text{meas}} = s \times \sigma_{M\text{ADGRAPH}}(\text{EW } \ell\ell\text{jj})$, where $\sigma_{M\text{ADGRAPH}}(\text{EW } \ell\ell\text{jj})$ is obtained from MADGRAPH simulation using PDFs CTEQ6L1 with the following parton-level selections: $m_{ll} > 50$ GeV, $p_T^j > 25$ GeV, $|\eta| < 4.0$ and $m_{jj} > 120$ GeV.

The combined cross section from the $\mu^+\mu^-$ and e^+e^- channels is:

$$\sigma^{EW}_{\text{meas},\mu\mu+ee} = 154 \pm 24 \text{ (stat) } \pm 46 \text{ (syst) } \pm 27 \text{ (theory) } \pm 3 \text{ (lumi) fb}$$

This is in good agreement with a theoretical cross section of 166 fb, calculated with next-to-leading order QCD corrections.

3.3. Central hadronic activity

The hadronic activity between the tagging jets is expected to be small in the case of a purely electroweak production. Although we do not use the hadronic activity to separate the EW $\ell\ell$jj from the backgrounds, studies have been done to test the agreement between data and simulation. For this study, tracks associated with the primary vertex and not associated with either the two leptons or the two tagging jets are selected if they satisfy a high purity requirement and have a transverse momentum exceeding 300 MeV. These tracks are clustered into soft track jets with the anti-k_T algorithm with distance parameter of 0.5. This method gives us a collection of jets with energy as low as a few GeV which are not affected by the pileup.

In order to monitor the hadronic activity in the rapidity gap between the two tagging jets, only track jets with pseudorapidity $\eta^{\text{tag.jet}}_{\text{min}} + 0.5 < \eta < \eta^{\text{tag.jet}}_{\text{max}} - 0.5$ are considered. Figure 11 shows the scalar sum of the three leading soft track jets (H_T). The evolution of the average H_T as a function of the dijet invariant mass and the pseudorapidity separation between the two tagging jets is shown in figure 12 and demonstrates a good agreement between data and simulation up to the most highest regions of dijet invariant mass and pseudorapidity separation.

These results establish an important benchmark in the more general study of VBF processes, of great interest for studies of the Higgs boson, electroweak gauge couplings, and vector boson scattering.

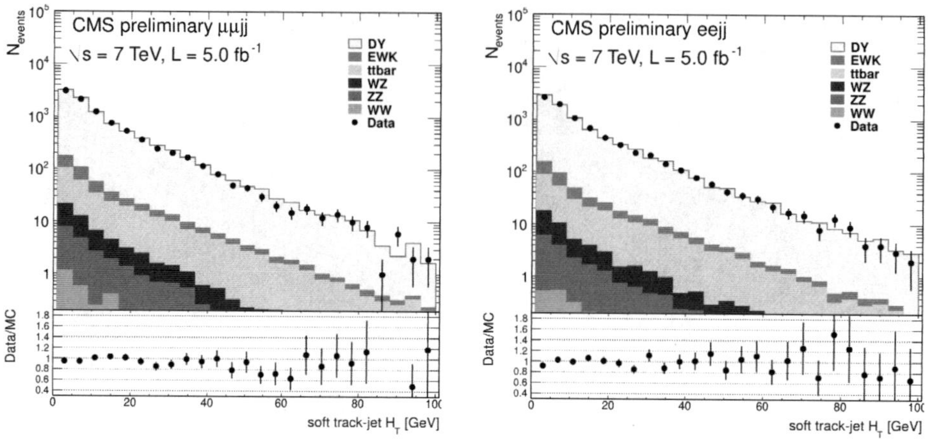

(a) dimuon channel (b) dielectron channel

Figure 11. The $\mathrm{H_T}$ distribution of the three leading soft track jets in the rapidity gap between the two leading jets with $p_T^{j_1,j_2} > 65, 40$ GeV in Z + 2 jets events

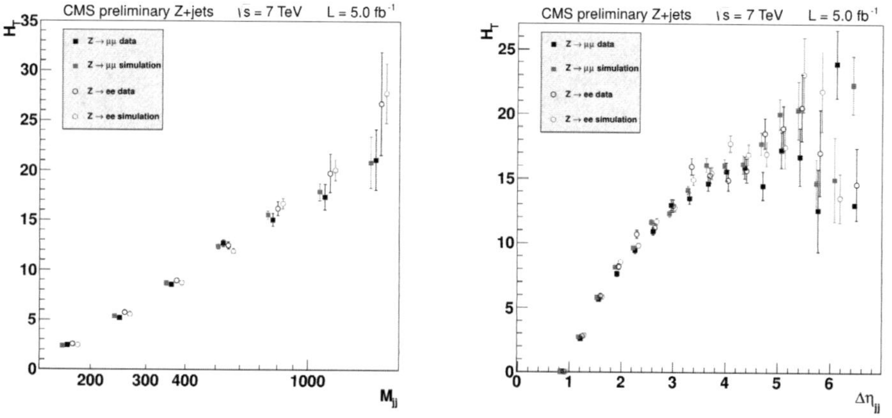

(a) (b)

Figure 12. Average $\mathrm{H_T}$ of the three leading soft track jets in the rapidity gap between the two leading jets with $p_T^{j_1,j_2} > 65, 40$ GeV in Z + 2 jets events versus (a) the dijet invariant mass (b) the pseudorapidity separation between the two tagging jets

References

[1] CMS Collaboration 2012 *JHEP* **01** 010 (*arXiv:* 1110.3226)
[2] CMS Collaboration 2008 *JINST* **3** S08004
[3] J. Alwall et al. 2011 *JHEP* **06** 128 (*arXiv:* 1106.0522)
[4] T. Sjostrand, S. Mrenna, and P.Z. Skands 2006 *JHEP* **05** 026 (*arXiv:* hep-ph/0603175)
[5] CMS Collaboration 2012 Submitted to *Phys. Lett.* B (*arXiv:* 1301.1646)
[6] T. Gleisberg et al. 2009 *JHEP* **02** 007 (*arXiv:* 0811.4622)
[7] P. Nason 2004 *JHEP* **11** 040
[8] S. Frixione, P. Nason and C. Oleari 2007 *JHEP* **11** 070 (*arXiv:* 0709.2092)
[9] S. Alioli et al. 2010 *JHEP* **06** 043 (*arXiv:* 1002.2581)
[10] S. Alioli et al. 2011 *JHEP* **01** 095 (*arXiv:* 1009.5594)
[11] CMS Collaboration 2012 CMS Physics Analysis Summary CMS-PAS-FSQ-12-019

Journal of Physics: Conference Series **455** (2013) 012027 doi:10.1088/1742-6596/455/1/012027

Search for the Standard Model Higgs in $\mathrm{H} \to \mathrm{ZZ} \to 4\mathrm{l}$ channel with the CMS experiment

Nicola De Filippis
on behalf of the CMS Collaboration
Dipartimento Interateneo di Fisica "M. Merlin", Politecnico and INFN Bari, Via Orabona 4, 70126 Bari, IT
E-mail: nicola.defilippis@ba.infn.it

Abstract. Results of the experimental search for the Standard Model Higgs in the $\mathrm{H} \to \mathrm{ZZ} \to 4\mathrm{l}$ (with l=e,μ) decay channel at collision energies of 7 and 8 TeV at the LHC are presented. The analysis uses pp collision data recorded by the CMS detector at the LHC, corresponding to integrated luminosities of 5.05 fb^{-1} at collision energy of 7 TeV and 12.2 fb^{-1} at 8 TeV. A complete description of the $\mathrm{H} \to \mathrm{ZZ} \to 4\mathrm{l}$ analysis is provided by inspecting the most important aspects such as the lepton identification and isolation, the kinematics, the selection steps and the background control from data. The existence of the Higgs boson would be statistically revealed by the presence of a resonance peak in the four-lepton invariant mass distribution. The search covers the overall Higgs boson mass hypotheses in the range $110 < m_H < 1000\,\mathrm{GeV/c^2}$ and a statistical interpretation of the results in that range is provided. The existence of the Standard Model Higgs boson has been excluded in a wide region of mass at 95% of confidence level, while a 4.5σ excess has been observed around a mass of $126\,\mathrm{GeV/c^2}$. A measurement of the mass of the new boson discovered gives $126.2 \pm 0.6(stat) \pm 0.2(syst)$ $\mathrm{GeV/c^2}$. The hypothesis 0^+ of the standard model for the spin J=0 and parity P= +1 quantum numbers is found to be consistent with the observation.

1. Introduction

The Higgs boson is a scalar particle predicted by the Standard Model (SM) of electroweak interactions [1, 2, 3], responsible for the mechanism of spontaneous symmetry breaking [4, 5, 6, 8, 7, 9] that allow fermions and gauge bosons to acquire mass. The mass of the boson is not predicted by the theory and needs to be established experimentally.

The search for the Standard Model (SM) Higgs boson is one of the major goals of the LHC physics program. In July 2012, the CMS and ATLAS experiments announced the discovery of a new boson at a mass around $125\,\mathrm{GeV/c^2}$, with properties compatible with the SM Higgs boson [10, 11]. By using 2011 data at 7 TeV, CMS excluded the SM Higgs boson in the mass range 127–600 $\mathrm{GeV/c^2}$ at 95% confidence level (CL) [12] while ATLAS excluded the ranges 111.4–116.6 $\mathrm{GeV/c^2}$, 119.4–122.1 $\mathrm{GeV/c^2}$, and 129.2– 541 $\mathrm{GeV/c^2}$ at 95% CL [13, 14].

The production of the SM Higgs boson at the LHC proceeds through several mechanisms. The dominant production mechanism is gluon-gluon fusion with a cross section between 0.1 and 50 pb depending on the SM Higgs mass, after including the most recent next-to-next-to-leading-order and next-to-next-leading-log calculations of the cross sections [15]. Other production mechanisms including associated production with the weak vector bosons, the weak vector boson fusion processes and the associated production with heavy top or bottom quarks are less favoured.

In the high mass range the most important decay channels of the SM Higgs particle are $H \rightarrow WW^{(*)}$ and $H \rightarrow ZZ^{(*)}$ which give a clear signature of multi-leptons in the final state. At low and intermediate the values of the Higgs mass the decay to $b\bar{b}$ and $\tau\tau$ fermions are favoured but the high background and the experimental resolution for the searches in those channels make them much less powerful than the $H \rightarrow ZZ$ decay which has a clean experimental signature and the possibility to reconstruct the mass; the the $H \rightarrow \gamma\gamma$ channel suffers from a low branching ratio and is quite sensitive to the background modeling. The product of the cross section times the branching fraction ranges from a few to about 10 fb for the $H \rightarrow ZZ \rightarrow 4l$ final state.

2. CMS detector and physics object reconstruction

The CMS detector [16] consists of a superconducting solenoid providing a uniform magnetic field of 3.8 T in the core, equipped with silicon pixel and strip tracking systems ($|\eta| < 2.5$) surrounded by a lead tungstate crystal electromagnetic calorimeter (ECAL) and a brass-scintillator hadronic calorimeter (HCAL) covering $|\eta| < 3.0$. The steel return yoke outside the solenoid is instrumented with gas ionization detectors used to trigger and identify muons up to $|\eta| < 2.4$.

The reconstruction and the identification of individual particles in the collision event relies on the the Particle Flow algorithm [17, 18] that uses the information from all CMS sub-detectors to provide a list of mutually exclusive categories: charged hadrons, neutral hadrons, photons, muons, and electrons.

Muons are reconstructed within the geometrical acceptance $|\eta| < 2.4$ and with transverse momentum $p_T > 5\,GeV/c$ by using information from both the inner tracker and the muon spectrometer [19]. Minimal requirements on the track components in the muon system and on the compatibility with energy deposits in the calorimeters are applied for the Particle Flow identification.

Electrons are reconstructed within the geometrical acceptance of the detector ($|\eta| < 2.5$) and with $p_T > 7\,GeV/c$ by combining the information from clusters of energy deposits in the ECAL and the trajectory in the inner tracker [20]. The criteria for the electron identification are based on a multivariate technique that combines observables sensitive to the amount of bremsstrahlung along the electron trajectory, the geometrical and momentum matching between the electron trajectory and associated clusters, as well as shower-shape observables.

The isolation of individual e or μ leptons is measured relative to their transverse momentum, by summing over the transverse momenta of the charged hadrons originating from the primary vertex, the neutral hadron and the photon candidates provided by the Particle Flow algorithm. The electrons or muons are considered isolated if the isolation variable is smaller than 0.4, that corresponds to a loose requirement, aimed to preserve the signal efficiency.

The impact parameter of the track associated to the lepton with respect to the primary vertex normalized to the related uncertainty, hereafter referred as the significance of the impact parameter, SIP, is used to select prompt leptons coming from the primary vertex. Only leptons with $|SIP| < 4$ and passing the identification and isolation criteria mentioned above are considered 'good'" leptons for the $H \rightarrow ZZ \rightarrow 4l$ analysis.

The tag-and-probe method [21] is used to measure the muon/electron reconstruction, identification and isolation efficiency from data and compare with expectation from simulation. The method relies upon $Z \rightarrow \mu^+\mu^-/e^+e^-$ decays to provide an unbiased and high purity muon/electron sample from which the efficiency of a particular selection step or cut can be measured by using the Z mass constraint.

3. Event selection

Data collected by the CMS experiment for which all sub-detectors performed properly have been used for this analysis. The data were officially-validated by the CMS collaboration for trigger selection and event reconstruction in 2011 and 2012.

A full set of Monte Carlo (MC) simulated samples are produced using several generators according to the given production process and decay: POWHEG [22], PYTHIA 6.4 [23], Madgraph 5 [24]. The full CMS detector simulation and reconstruction chain is used at a later stage after the event production. The cross sections for all the background processes are re-weighted to next-to-leading-order (NLO) calculations.

Collision events are selected first by the trigger system that requires the presence of a pair of electrons or a pair of muons, or a triplet of electrons, with threshold on the p_T of leptons and additional selection according to the instantaneous luminosity of the LHC, varied over the range $10^{29} - 10^{33} cm^{-2}s^{-1}$ during 2011 and 2012 periods. The trigger efficiency within the acceptance of this analysis is greater than 98% for a Higgs boson signal with $m_H > 120\,\mathrm{GeV/c^2}$.

The event selection is designed to reconstruct two pairs with opposite-sign and same-flavor "good" leptons, preserving the signal efficiency, especially in low Higgs mass region, while rejecting the background from SM processes, mostly $ZZ/Z\gamma^*$ events and $Zb\bar{b}/t\bar{t}$ events including non-prompt leptons from the secondary vertex of the b-hadrons decay. The background sources related to the probability of a jet to be mis-identified as a "good" lepton ("fake" leptons) are referred as "instrumental" background and include Z+light jets and WZ+light jets events with at least two or one 'fake" leptons, respectively.

Between all the pairs of opposite-sign and same-flavor "good" leptons the one with an invariant mass closest to the nominal Z mass is chosen (Z_1) and retained if it satisfies the condition $40 < m_{Z_1} < 120\,\mathrm{GeV/c^2}$. A second pair of opposite-sign and same-flavor 'good" leptons (Z_2) is built with the remaining leptons and chosen if it satisfies the condition: $12 < m_{Z_2} < 120\,\mathrm{GeV/c^2}$. In case of ambiguities for the choice of the Z_2 the one built with the leptons of higher p_T is kept. In order to ensure that the events passing the selection have leptons on the high-efficiency plateau for the trigger, at least one lepton of the Z_1 or the Z_2 candidates is required to have $p_T > 20\,\mathrm{GeV/c}$ and another one to have $p_T > 10\,\mathrm{GeV/c}$.

Any opposite-charge pair of leptons chosen among the four selected leptons (irrespective of flavour) has to satisfy the requirement $m_{ll'} > 4\,\mathrm{GeV/c^2}$ in order to further reject events with leptons originating from hadron decays in jet fragmentation or from the decay of low-mass hadronic resonances.

The phase space for the search of the SM Higgs boson is defined by restricting the mass range to $m_{4l} > 100$ GeV/c^2.

The kinematics of the Higgs or exotic boson decay to ZZ final state can be further described by five angles in its centre-of-mass frame $\vec{\Omega} = (\theta^*, \Phi_1, \theta_1, \theta_2, \Phi)$, defined in Ref. [25], and the invariant masses of the lepton pairs, m_{Z_1} and m_{Z_2}. The distribution of those observables is related to the spin/parity quantum numbers of the decaying resonance. By using a matrix element likelihood analysis (MELA) a kinematic discriminant (K_D) built with that set of observables can be defined for each value of m_{4l}, based on the probability ratio of the signal and background hypotheses, $K_D = \mathcal{P}_{\mathrm{sig}}/(\mathcal{P}_{\mathrm{sig}} + \mathcal{P}_{\mathrm{bkg}})$, as described in Refs. [25, 26]. The use of the K_D provides additional discrimination between the signal and the background (especially the ZZ).

A complete description of the $H \to ZZ \to 4l$ analysis with additional details can be found in Ref.[27].

4. Background control

The ZZ background is evaluated from the Monte Carlo simulation; the shape of the four-lepton mass distribution is taken from the simulated samples while the event yield is normalized to the cross section for ZZ production at NLO calculated with MCFM [28, 29, 30], including the contribution from $q\bar{q}$ annihilation, as well as that from the gluon fusion and the luminosity of the data sample.

To estimate the reducible background ($Zb\bar{b}$, $t\bar{t}$) and the "instrumental" (Z + light jets, WZ + jets) background, a Z+X background control region is defined. The event rates measured in the background control region are then extrapolated to the signal region.

Two approaches are used to defined the background control region both starting from events with a well reconstructed Z_1 and relaxing the isolation and identification criteria for two additional reconstructed lepton objects. For the first method the additional pair of leptons is required to have the same charge (to avoid signal contamination) and same flavour ($e^{\pm}e^{\pm}$, $\mu^{\pm}\mu^{\pm}$), a reconstructed invariant mass $m_{Z_2} > 12\,\mathrm{GeV/c^2}$ and $m_{4l} > 100\,\mathrm{GeV/c^2}$. The second method relies on the control region with two opposite-sign leptons failing the isolation and identification criteria.

The expected number of Z+X background events in the signal region is obtained by taking into account the lepton mis-identification probability for each of the two additional leptons; that probability is measured with a sample $Z + l$, with at least one reconstructed lepton object. The lepton misidentification probability computed as the probability for a reconstructed lepton to pass identification and isolation requirements is compared with the one derived from MC simulation and found compatible.

Since both methods provide comparable background counts in the signal region within uncertainties, an envelope of these is used as the final estimate in Table 1.

5. Systematic uncertainties

Systematic uncertainties on the theoretical calculation of the SM Higgs boson cross section (17 – 20%) and branching fraction (2%) are described in Ref. [15].

The uncertainties on the evaluation of the ZZ background come from the QCD and PDF scales and are computed for each final state following standard prescriptions; they amount to 8% on average.

The uncertainty on the integrated luminosity amounts to 2.2% at 7 TeV, 4.4% at 8 TeV [31] and enters in the normalization of the ZZ background and of the signal.

An uncertainty of approximately 50% comes from the limited statistical precision in the reducible and "instrumental" background control regions as well as from the difference in background composition between the control regions and the sample on which the lepton misidentification probability is derived.

Systematic uncertainties on the selection efficiency coming from trigger (1.5%), and combined lepton reconstruction, identification and isolation efficiencies (varying from 1.2% to 3.8% in the 4μ channel and from 5.5% to 11% in $4e$ channel, depending on the considered mass) are evaluated from data.

6. Final results and statistical interpretation

The reconstructed four-lepton invariant mass distributions are shown in Figure 1 in the full mass range (left) and in the low mass region (right), using 2011 data at 7 TeV and 2012 data at 8 TeV. Data distributions are compared with the expectation from SM background processes. The measured distribution at higher mass is dominated by the irreducible ZZ background while at low mass the contribution of the Z+X estimated from data is visible. A clear peak around $m_{4l} = 126\,\mathrm{GeV/c^2}$ is seen, confirming the previous observation of the a new boson. The peak at around $m_{4l} = 91\,\mathrm{GeV/c^2}$ corresponds to the $Z \rightarrow 4l$ standard candle, as predicted by the SM.

Table 1 reports the number of candidates observed in data, the estimated background and the expected number of signal events for several SM Higgs boson mass hypotheses, in the full mass measurement range, $100 < m_{4l} < 1000\,\mathrm{GeV/c^2}$. The observed event rates for the various channels are compatible with SM background expectation.

Table 1. The number of event candidates observed, compared to the mean expected background and signal rates for each final state in the range $100 < m_{4l} < 1000\,\mathrm{GeV/c^2}$, with 2011 and 2012 data.

Channel	$4e$	4μ	$2e2\mu$
ZZ background	53.0 ± 6.3	82.7 ± 8.9	131.1 ± 14.3
Z + X	$7.6^{+6.9}_{-5.2}$	$2.9^{+2.2}_{-1.6}$	$10.1^{+9.9}_{-6.5}$
All background expected	$60.7^{+9.3}_{-8.2}$	$85.6^{+9.2}_{-9.1}$	$141.3^{+17.3}_{-15.7}$
m_H = 125 GeV/c^2	2.4 ± 0.4	4.6 ± 0.5	6.0 ± 0.7
m_H = 126 GeV/c^2	2.7 ± 0.4	5.1 ± 0.6	6.6 ± 0.8
m_H = 200 GeV/c^2	15.5 ± 1.9	23.1 ± 2.6	38.5 ± 4.3
m_H = 350 GeV/c^2	9.5 ± 1.2	13.6 ± 1.5	23.2 ± 2.7
m_H = 500 GeV/c^2	3.3 ± 0.4	4.7 ± 0.6	8.1 ± 0.9
m_H = 800 GeV/c^2	0.5 ± 0.1	0.6 ± 0.1	1.1 ± 0.1
Observed	59	95	162

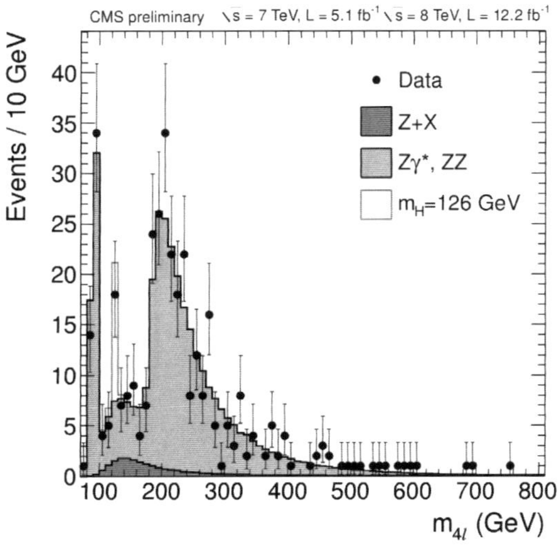

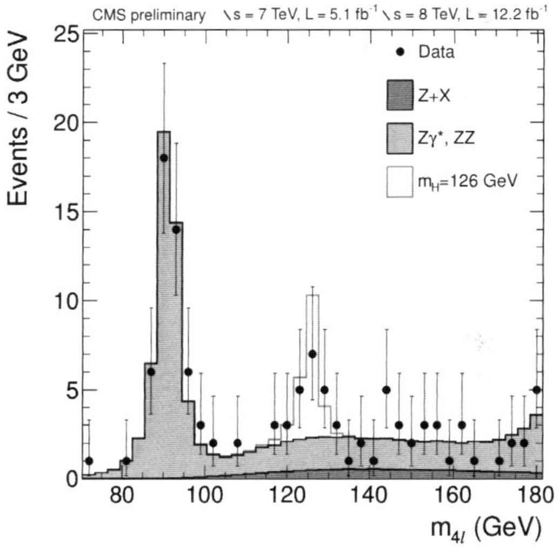

Figure 1. Distribution of the four-lepton reconstructed mass in the full mass range (left) and low mass region (right) for the sum of the 4e, 4μ and $2e2\mu$ channels. Points represent the data, shaded histograms represent the background and unshaded histogram the signal expectations. The expected distributions are presented as stacked histograms. The measurements are presented for the sum of the data collected at $\sqrt{s} = 7$ TeV and $\sqrt{s} = 8$ TeV.

The distributions of the K_D as a function of the four-lepton reconstructed mass m_{4l} for the selected events in data and compared to signal and background expectation are shown in Figure 2; values of K_D close to 1 mean a better compatibility with the hypothesis of the SM Higgs signal.

A statistical approach for testing hypotheses is used to derive both the exclusion limits at 95% CL on the ratio of the production cross section for the SM Higgs boson to the SM expectation and the significance for the discovery of a new particle. The modified frequentist construction $\mathrm{CL_s}$ [32, 33] is used; details and current implementation can be found in Ref. [34]. For each mass

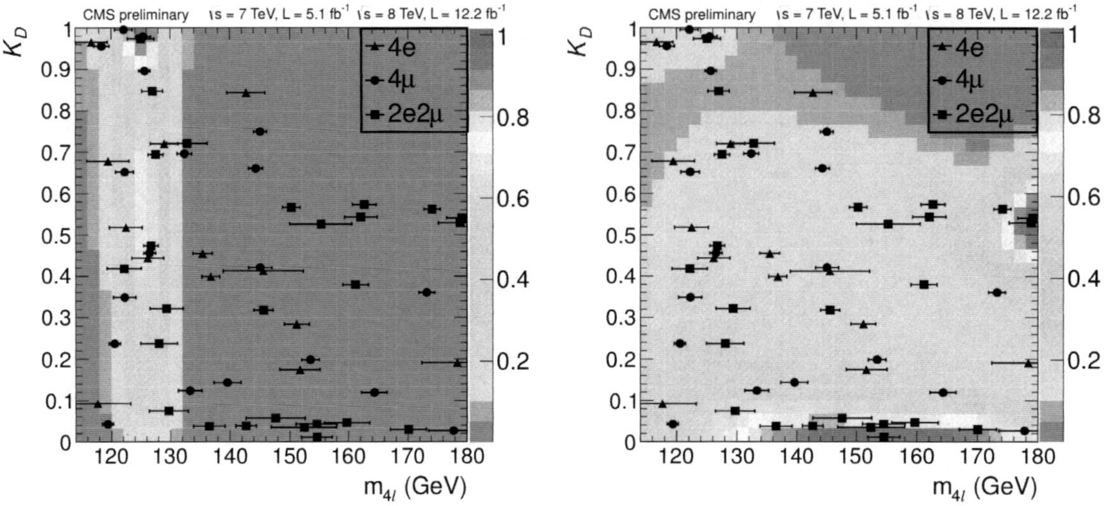

Figure 2. Distribution of the K_D versus the four-lepton reconstructed mass m_{4l} in the low-mass regions for the signal at $m_H = 126\,\mathrm{GeV/c^2}$ (left) and for the background (right). The contours represent the expected relative density of signal (left) or background events (right). The points show data with measured invariant mass uncertainties.

hypothesis, a simultaneous likelihood fit of the the two-dimensional (m_{4l}, K_D) distributions is performed, separately for the three channels and the two data taking periods at 7 and 8 TeV. All systematic uncertainties are included in the likelihood with log-normal distributions.

The upper limit at 95% CL on the ratio of the production cross section for the SM Higgs boson to the SM expectation is shown in Figure 3 (top left) and a zoom at low mass is shown at bottom left. The SM Higgs boson is excluded by the four-lepton channels at 95% CL in the ranges 113–116 GeV/c^2 and 129–720 GeV/c^2.

The local p-values, representing the probability that the background can fluctuate to give an excess of events equal to or larger than the observed number of events, are shown for the full and the low mass range as a function of SM Higgs mass hypothesis in Figure 3 (right). The minimum p-value is reached at mass around $m_{4l} = 125.9\,\mathrm{GeV/c^2}$ and corresponds to a local significance of 4.5σ.

7. Mass and spin/parity measurements

A simultaneous likelihood fit using for each event the three-dimensional distribution (m_{4l}, er_{4l}, K_D) of the four-lepton invariant mass, the associated per-event mass error (er_{4l}), and the kinematic discriminant, is performed to get the mass measurement of the new boson. Per-event errors on the four-lepton invariant mass are calculated from the individual lepton momentum errors by using the full error matrix, as obtained from the muon track fit, in case of muons and by using the estimated momentum error, as obtained from the combination of the ECAL and tracker measurements, in case of electrons. Figure 4 (left) shows the two-dimensional confidence level regions for the ratio of the production cross section for the SM Higgs boson to the SM expectation as a function of the Higgs mass hypothesis. The result of the fit gives $m_H = 126.2 \pm 0.6$ (stat) ± 0.2 (syst) GeV/c^2.

The spin and quantum numbers of the new boson are studied following the MELA methodology, replacing the signal-to-background probability ratio with the probability ratio for two different signal hypotheses as the new discriminant. The pure pseudo-scalar state $J^P = 0^-$

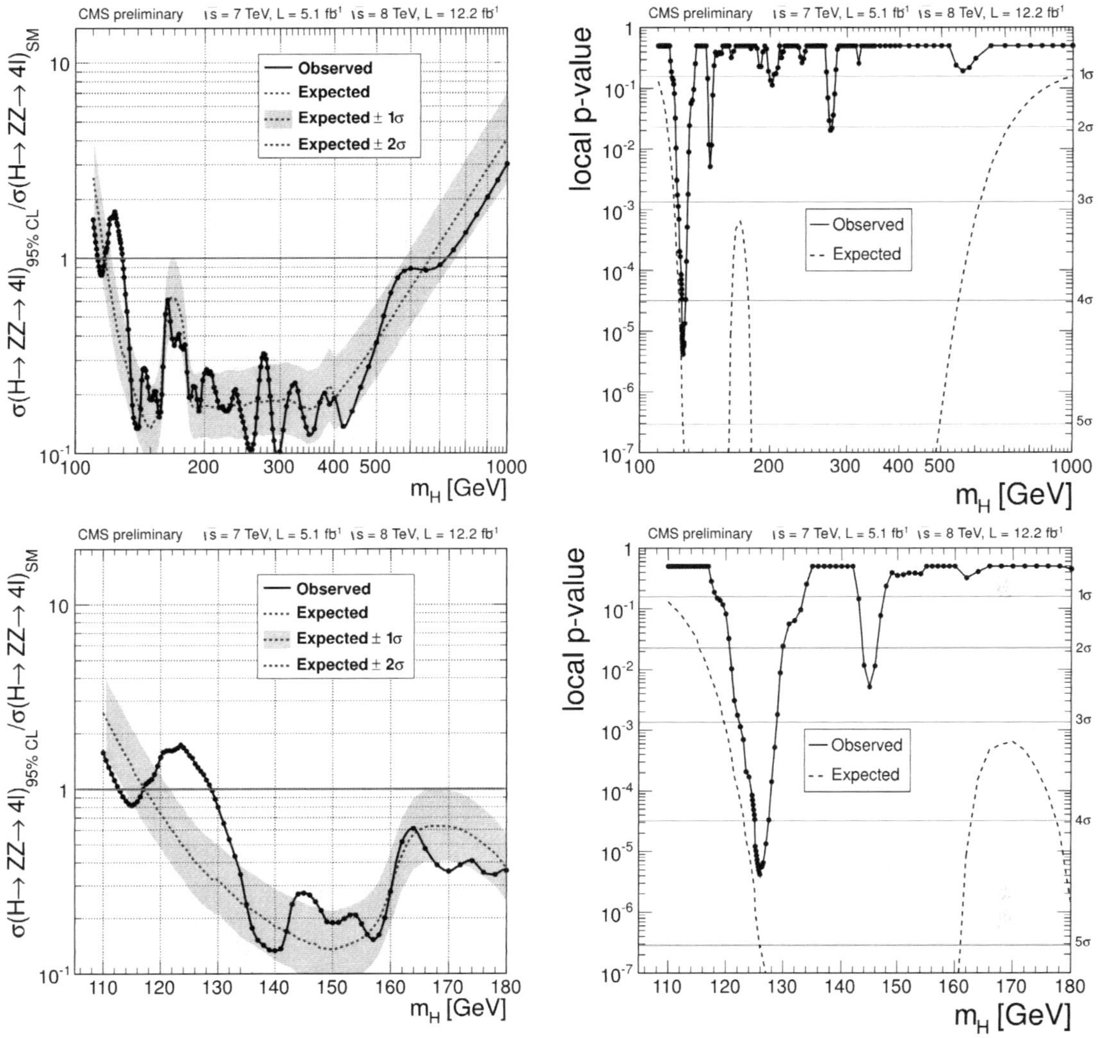

Figure 3. Observed and expected 95% CL upper limit (left) on the ratio of the production cross section to the SM expectation in the full mass range (top) and the low-mass region (bottom). The 68% and 95% ranges of expectation for the background-only model are also shown with green and yellow bands, respectively. Local p-values (right) represent the significance of the local excess with respect to the standard model background expectation as a function of the Higgs boson mass in the full and low mass range.

is considered here as an alternative hypothesis. The same statistical approach previously introduced is used and the likelihoods involved, $\mathcal{L}$, are calculated with the signal rates allowed to float independently for each signal type. The distributions of $q = -2\ln(\mathcal{L}_{0^-}/\mathcal{L}_{0^+})$ obtained with simulated events for both a 0^- resonance and the SM 0^+ Higgs boson with $m_H = 126\,\mathrm{GeV/c^2}$ are shown in Figure 4 (right). The observed value of q, indicated by an arrow in the previous figure, is consistent with expectation within 2.4σ if assuming $J^P = 0^-$ and within 0.5σ if assuming $J^P = 0^+$.

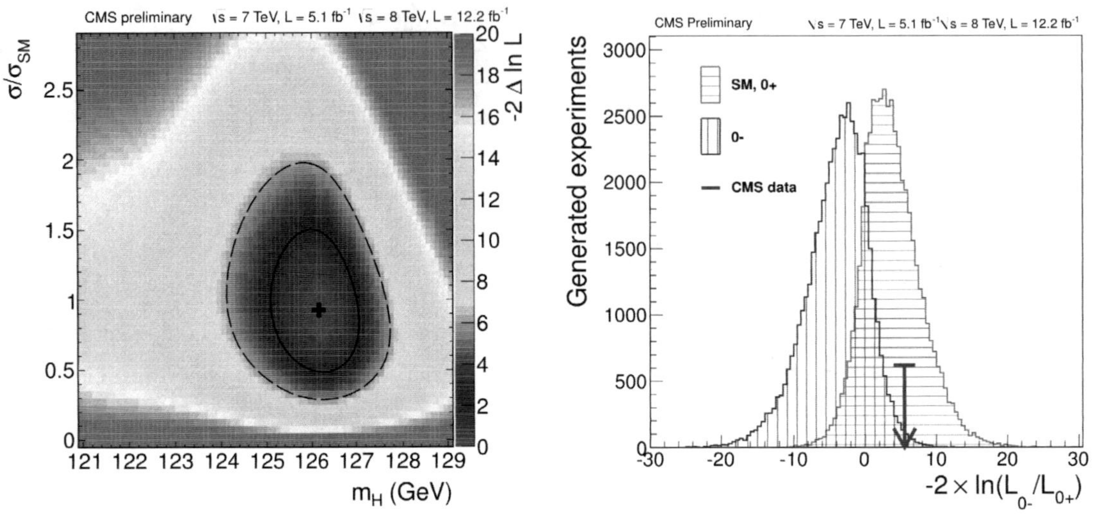

Figure 4. (Left) Distribution of $-2\ln\mathcal{L}$ as a function mass m_H and ratio of the production cross section for the SM Higgs boson to the SM expectation. The central point shows the minimum value of $-2\ln\mathcal{L}$, the solid and dashed contours show 68% and 95% CL contours in two dimensions. (Right) Distribution of $q = -2\ln(\mathcal{L}_{0^-}/\mathcal{L}_{0^+})$ for two signal types (0^+ horizontally hatched histogram, 0^- vertically hatched histogram) for $m_H = 126$ GeV shown with a large number of generated experiments. The arrow indicates the observed value.

8. Conclusions

A search for the SM Higgs boson has been performed in the four-lepton decay modes by using 2011 data at $\sqrt{s} = 7$ TeV and 2012 data at $\sqrt{s}$ 8 TeV. The SM Higgs boson is excluded at 95% CL in the mass ranges 113–116 and 129–720 GeV/c^2. The recently discovered new boson is observed with a local significance of 4.5σ and its mass is measured to be 126.2 ± 0.6 (stat) ± 0.2 (syst) GeV/c^2. The hypothesis 0^+ of the standard model for the spin $J = 0$ and parity $P = +1$ quantum numbers is found to be consistent with the observation.

References

[1] Glashow S L 1961 Partial Symmetries of Weak Interactions *Nucl. Phys.* **22** 579-588
[2] Weinberg S 1967 A Model of Leptons *Phys. Rev. Lett.* **19** 1264-1266
[3] Salam A 1968 Elementary Particle Physics *N. Svartholm ed., Almquvist and Wiksell, Stockholm*
[4] Englert F, Brout R 1964 Broken Symmetry and the Mass of Gauge Vector Mesons *Phys. Rev. Lett.* **13** 321-323
[5] Higgs P W 1964 Broken symmetries, massless particles and gauge fields *Phys. Lett.* **12** 132-133
[6] Higgs P W 1964 Broken Symmetries and the Masses of Gauge Bosons *Phys. Rev. Lett.* **13** 508-509
[7] Higgs P W 1966 Spontaneous Symmetry Breakdown without Massless Bosons *Phys. Rev.* **145** 1156-1163
[8] Guralnik, G S, Hagen C R and Kibble T W B 1964 Global Conservation Laws and Massless Particles *Phys. Rev. Lett.* **13** 585-587
[9] Kibble T W B 1967 Symmetry breaking in non-Abelian gauge theories *Phys. Rev.* **155** 1554-1561
[10] CMS Collaboration 2012 Observation of a new boson at a mass of 125 GeV with the CMS experiment at the LHC *Phys. Lett. B* **716** 30-61 (*Preprint* arXiv:1207.7235 [hep-ex])
[11] ATLAS Collaboration 2012 Observation of a new particle in the search for the Standard Model Higgs boson with the ATLAS detector at the LHC *Phys. Lett. B* **716** 1-29 (*Preprint* arXiv:1207.7214 [hep-ex])
[12] CMS Collaboration 2012 Combined results of searches for the standard model Higgs boson in pp collisions at $\sqrt{s} = 7$ TeV *Phys. Lett. B* **710** 26-48 (*Preprint* arXiv:1202.1488 [hep-ex])
[13] ATLAS Collaboration 2012 Combined search for the Standard Model Higgs boson using up to 4.9 fb^{-1} of

pp collision data at $\sqrt{s} = 7$ TeV with the ATLAS detector at the LHC *Phys. Lett. B* **710** 49-66 (*Preprint* arXiv:1202.1408 [hep-ex])
[14] ATLAS Collaboration 2012 Combined search for the Standard Model Higgs boson in pp collisions at $\sqrt{s} = 7$ TeV with the ATLAS detector *Phys. Rev. D* **86** 032003 (*Preprint* arXiv:1207.0319 [hep-ex])
[15] LHC Higgs Cross Section Working Group 2011 Handbook of LHC Higgs Cross Sections: 1. Inclusive Observables CERN-2011-002 (*Preprint* arXiv:1101.0593 [hep-ph])
[16] CMS Collaboration 2008 The CMS experiment at the CERN LHC *JINST* **3** S08004
[17] CMS Collaboration 2009 Particle–Flow Event Reconstruction in CMS and Performance for Jets, Taus, and MET CMS-PAS-PFT-09-001
[18] CMS Collaboration 2010 Commissioning of the Particle-Flow Reconstruction in Minimum-Bias and Jet Events from pp Collisions at 7 TeV CMS-PAS-PFT-10-002
[19] CMS Collaboration 2010 Performance of muon identification in pp collisions at $\sqrt{s} = 7$ TeV CMS-PAS-MUO-10-002
[20] CMS Collaboration 2010 Electron reconstruction and identification at $\sqrt{s} = 7$ TeV CMS-PAS-EGM-10-004
[21] CMS Collaboration 2011 Measurement of the Inclusive W and Z Production Cross Sections in pp Collisions at $\sqrt{s} = 7$ TeV *JHEP* **10** 132 (*Preprint* arXiv:1107.4789 [hep-ph])
[22] Frixione S, Nason P and Oleari C 2007 Matching NLO QCD computations with Parton Shower simulations: the POWHEG method *JHEP* **11** 070 (*Preprint* arXiv:0709.2092 [hep-ph])
[23] Sjöstrand T, Mrenna S and Skands P Z 2006 PYTHIA 6.4 Physics and Manual *JHEP* **05** 026 (*Preprint* hep-ph/0603175)
[24] Alwall J, Demin P, de Visscher S *et al.* 2007 MadGraph/MadEvent v4: The New Web Generation *JHEP* **09** 028 (*Preprint* arXiv:0706.2334 [hep-ph])
[25] Gao Y, Gritsan A V, Guo Z *et al.* others 2010 Spin determination of single-produced resonances at hadron colliders *Phys. Rev. D* **81** 075022 (*Preprint* arXiv:1001.3396 [hep-ph])
[26] Bolognesi S, Gao Y, Gritsan A V *et al.* 2012 On the spin and parity of a single-produced resonance at the LHC *Preprint* arXiv:1208.4018 [hep-ph]
[27] CMS Collaboration 2013 Study of the Mass and Spin-Parity of the Higgs Boson Candidate via Its Decays to Z Boson Pairs *Phys. Rev. Lett. B* **110** 081803 (*Preprint* arXiv:1212.6639 [hep-ph])
[28] Campbell J M and Ellis R K 2010 MCFM for the Tevatron and the LHC *Nucl. Phys. Proc. Suppl.* **205-206** 10-15 (*Preprint* arXiv:1007.3492 [hep-ph])
[29] Campbell J M and Ellis R K 1999 An update on vector boson pair production at hadron colliders *Phys. Rev. D* **60** 113006 (*Preprint* hep-ex/9905386)
[30] Campbell J M, Ellis R. K. and Williams C 2011 Vector boson pair production at the LHC *JHEP* **07** 018 (*Preprint* arXiv:1105.0020 [hep-ph])
[31] CMS Collaboration 2011 Absolute Calibration of the CMS Luminosity Measurement: Summer 2011 Update CMS-PAS-EWK-11-001
[32] Junk T 1999 Confidence level computation for combining searches with small statistics *Nucl. Instrum. Methods Phys. Res. A* **434** 435 (*Preprint* hep-ex/9902006)
[33] Read A L 2000 Modified frequentist analysis of search results (the CL_s method) CERN-OPEN-2000-205
[34] ATLAS and CMS Collaborations, LHC Higgs Combination Group 2011 Procedure for the LHC Higgs boson search combination in Summer 2011 ATL-PHYS-PUB-2011-011, CMS NOTE-2011/005

International Workshop on Discovery Physics at the LHC (Kruger2012) IOP Publishing
Journal of Physics: Conference Series **455** (2013) 012028 doi:10.1088/1742-6596/455/1/012028

Performance of the CMS electromagnetic calorimeter and its role in the hunt for the Higgs boson in the two-photon channel

Federico De Guio[1] ***on behalf of the CMS collaboration***
[1]University of Milano-Bicocca

Abstract. The electromagnetic calorimeter of CMS (ECAL) is a hermetic, fine grained and homogeneous calorimeter containing 75848 lead-tungstate ($PbWO_4$) crystals, completed by a silicon preshower installed in front of the endcaps. The ECAL sensitivity to decay modes with electromagnetic objects in the final state, such as narrow resonances decaying into two photons, is achieved through its excellent energy and position resolution. The ECAL performance from 2010-2012 is presented in detail and its role in the hunt for the Higgs boson, through the 2-photon decay mode, is discussed.

1. Introduction

The ambitious physics program conducted by the Compact Muon Solenoid (CMS) [1] collaboration ranges from the investigation of electroweak symmetry breaking through the direct search for the Standard Model (SM) Higgs boson to the search for evidence of new physics at the TeV scale. The design of the CMS electromagnetic calorimeter (ECAL) [2] was driven by the requirements imposed by the search for the Higgs boson in the $H \to \gamma\gamma$ channel, in which a narrow peak in the di-photon invariant mass has to be distinguished from a continuous background [3]. Good energy resolution and fine granularity are therefore required: both of them improve the invariant mass resolution of the di-photon system by improving respectively the energy and angle measurement of the two photons. The excellent performance of the ECAL and its stability during the 2011 and 2012 data taking, played a central role in the discovery of a new resonance with a mass of 125 GeV [4, 5]. At such a low mass, in fact, the natural width of the Standard Model Higgs boson is negligible compared to the width due to the instrumental resolution. In this report the instrumental and operational aspects of the CMS ECAL relevant in the 'hunt' for the $H \to \gamma\gamma$ decay channel are discussed.

2. The electromagnetic calorimeter of CMS

The ECAL measures the energy of electrons and photons. In order not to deteriorate its energy resolution, the ECAL is placed inside the solenoid, thus requiring a compact calorimeter. The ECAL is a hermetic, homogeneous calorimeter made of lead tungstate ($PbWO_4$) crystals, 61200 crystals mounted in the central barrel part, and 7324 crystals in each end-cap. Lead tungstate scintillating crystals have a short radiation length (X_0 = 0.89 cm) and a small Molière radius (R_M = 2.2 cm); they also have a fast response time, as the 80% of the scintillation light is emitted within 25 ns and radiation hard. However the relative low light yield (30 photons/MeV)

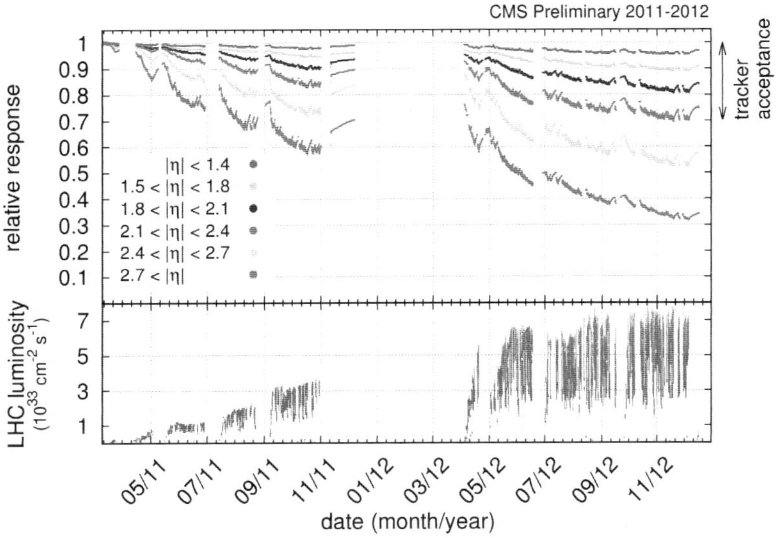

Figure 1. Relative response to laser light (440 nm) measured by the ECAL laser monitoring system [7], averaged over all crystals in bins of pseudorapidity, for the 2011 and 2012 data taking periods.

requires the use of photodetectors with intrinsic gain that can be operated in a magnetic field. In the barrel, silicon avalanche photodiodes (APDs) are used as photodetectors, while vacuum phototriodes (VPTs) have been chosen for the endcaps because of their high radiation hardness. The sensitivity of both the crystal and the APD response to temperature changes requires temperature stability. A water cooling system guarantees a long term stability at the 0.1 °C level [6]. In addition the high stability of the APD high voltage guaranties a negligible contribution to the final energy resolution due to APD gain variation [8, 9].
The performance of the components of the calorimeter has been extensively tested with 120 GeV electron beams: the stochastic and the electronic noise contributions to the energy resolution of ECAL have been shown to match the design requirements; the overall resolution was excellent and well below 1% at high energies, with an irreducible constant term of about 0.3% for particles impinging on the centre of the crystals [10]. In contrast to the test beam setup, in CMS additional contributions to the energy resolution are caused by residual miscalibrations of the channel-to-channel response within ECAL and by channel response changes with time, due to radiation damage to the crystals (see fig. 1) and environmental instability. The energy resolution is further degraded for electrons and photons that interact in the tracker material in front of ECAL, for which specific reconstruction algorithms and additional calibration factors are needed.

3. Reconstruction of the electromagnetic showers and energy calibration

Electromagnetic particles deposit their energy over several ECAL crystals and the energy estimate implies a sum over the corresponding channels. Several factors contribute to the final reconstructed energy of electrons and photons:

$$E_{e,\gamma} = F_{e,\gamma}(E_T, \eta) \cdot \sum_i G(ADC/GeV) \times S_i(T,t) \times c_i \times A_i \qquad (1)$$

where

Journal of Physics: Conference Series **455** (2013) 012028 doi:10.1088/1742-6596/455/1/012028

- A_i: the energy deposited in each ECAL crystal is digitized by a sampling ADC.
- c_i: the inter-calibration coefficient is meant to account for crystal-to-crystal variations in the response by equalizing the different crystal responses and normalizing them to an arbitrary reference value;
- $S_i(T,t)$: a monitoring correction factor to each crystal signal amplitude is needed because of the radiation-induced transparency change T with time t;
- G: the global ECAL energy scale converts ADC counts into GeV. It is determined separately for the ECAL barrel and the endcaps using the the $Z \to e^+e^-$ invariant mass peak;
- $F_{e,\gamma}(E_T,\eta)$: small algorithmic energy corrections are required to correct for residual cluster non-containment effects, particularly shower leakage and bremsstrahlung losses which are dependent on the type of the particle, its momentum, direction and impact point position.

As an illustration of the impact of the inter-calibration and the transparency corrections on the ECAL overall performance, the $Z \to ee$ invariant mass peak has been reconstructed with and without the application of the c_i and $S_i(T,t)$ terms in formula 1 (see fig. 2).

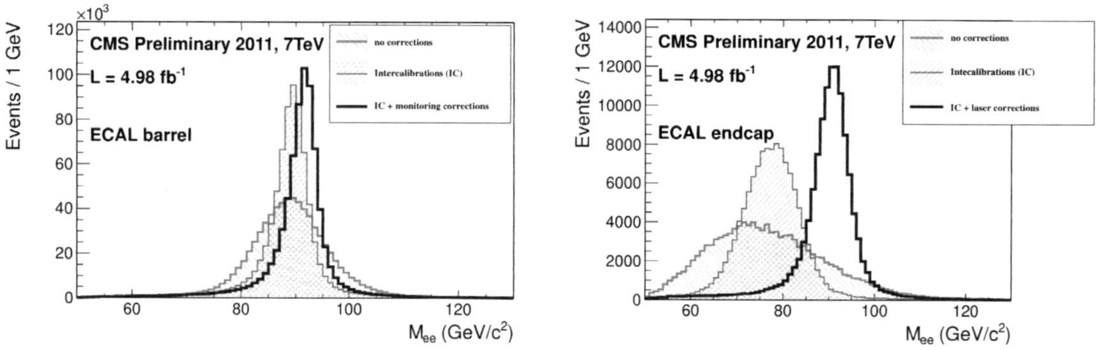

(a) (b)

Figure 2. Effect of single channel inter-calibration and transparency corrections on the $Z \to ee$ invariant mass in the ECAL barrel (a) and endcap (b).

3.1. Response variation with irradiation and monitoring of the performance stability

The energy measured by ECAL for isolated electrons from W and Z decays is used to monitor the stability of the response and resolution.

In fig. 3 top row, the ratio of the supercluster energy of an electron measured by ECAL to the momentum measured in the tracker has been exploited to monitor the stability and uniformity of the ECAL response. The analysis gives an unbiased answer on the electron energy measurement in the assumption that the measurement of the electron momentum is stable and uniform. The overall relative response stability throughout 2011 is at the 0.12% level in EB and 0.45% in EE while the signal loss is 3% and 10% respectively.

The stability of the ECAL resolution in 2011, monitored with the relative instrumental width of the $Z \to e^+e^-$ invariant mass peak (see fig. 3 bottom row), is excellent in EB while a slight degradation is observed in EE partially due to the increased number of collisions per beam crossing (pileup interactions) during the year. The instrumental contribution to the Z width (σ_{CB}) is extracted from a fit to the invariant mass distribution of a Breit-Wigner convoluted with a Crystal-Ball response function [11].

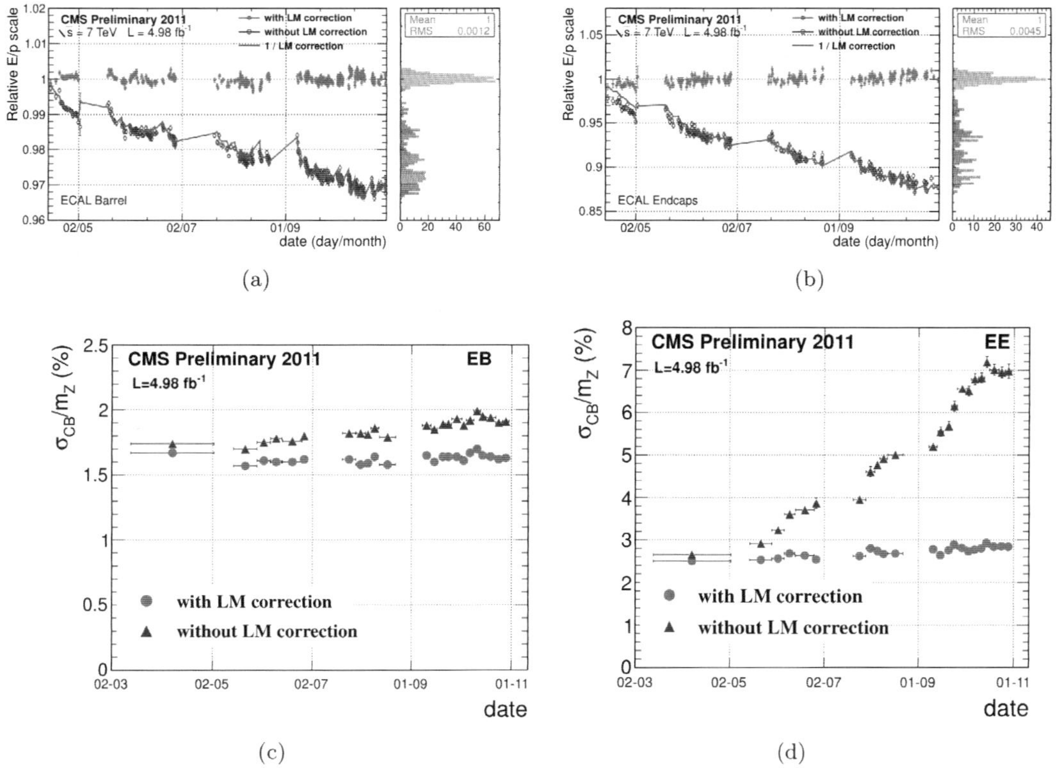

Figure 3. Stability of the ECAL response during 2011 with isolated electrons in the ECAL barrel (a) and endcap (b) and the stability of the ECAL resolution during 2011 with $Z \rightarrow ee$ events in the ECAL barrel (c) and endcap (d). Red (green) points show data upon (after) the application of monitoring corrections.

3.2. Single channel inter-calibration

The main sources of channel-to-channel response variation are the crystal light yield variation in EB, about 15% at construction, and the gain spread of the photodetectors in EE, about 25%. To reduce this spread and provide an acceptable performance at startup, calibration procedures with different levels of accuracy have been adopted during the construction and commissioning phase of ECAL [12]. The inter-calibration precision has been improved with several techniques exploiting the properties of collision events [13]. These include the invariance around the beam axis of the energy flow in minimum bias events, the π^0/η mass constraint on the energy of the two photons from the $\pi^0/\eta \rightarrow \gamma\gamma$ decays, and the momentum constraint on the energy of isolated electrons from W decays. The precision of each method has been estimated through the cross-comparison of the individual results, and cross-checked against precalibration constants. The inter-calibration precision upon combination of the different methods and of the precalibration obtained with the 2010/2011 data are shown in fig. 4 as a function of pseudorapidity for EB (left) and EE (right). The residual miscalibration of the channel response already ensures a contribution to the energy resolution below 0.5% in the central part of the barrel ($\eta < 1$), and below 2% in the endcaps.

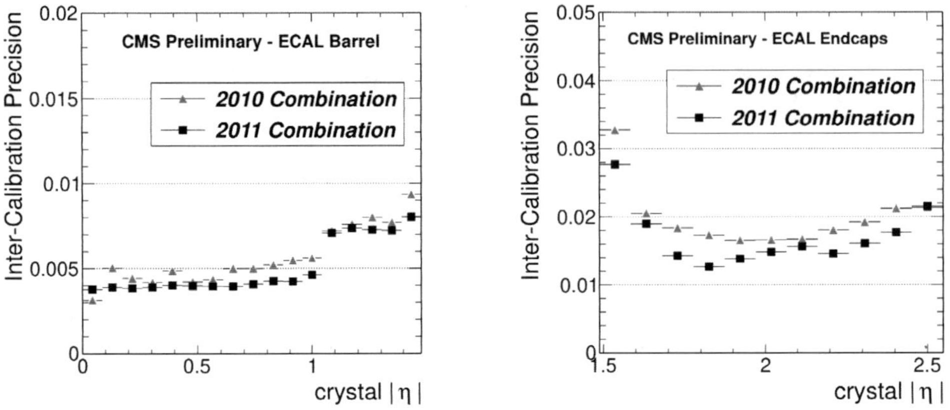

(a) (b)

Figure 4. Precision of the final inter-calibration sets used in 2010 and 2011 in EB (a) and EE (b).

3.3. Corrections to the reconstructed energy

Dynamic clustering algorithms are used in the energy reconstruction to mitigate the impact of the material upstream of ECAL. However not all the energy is clustered due to algorithm inefficiencies and part of the energy, because of the tracker material and the high magnetic field, never reaches ECAL. Thus, algorithmic corrections must be applied. These corrections are particle, energy and position dependent due to the different interaction mechanisms of e/γ upstream of ECAL and to the CMS geometry. Corrections have been optimized separately for electrons and photons on MC events with a multivariate analysis (MVA). Input variables include shower shape information on the azimuthal spread of the energy deposit, shower position in ECAL local and CMS global coordinates, and global event variables sensitive to pileup effects. Corrections are sizable at $1 < |\eta| < 2$, where the tracker material is up to two radiation lengths thick. Local structures are related to the ECAL geometry 5(a). Due to unavoidable imperfections of the MC model, algorithmic corrections must be tested and tuned on collision data. To this end, the stability of the E/p ratio of electrons is studied as a function of the MVA input variables. An identified case of imperfect corrections is shown in Fig. 5(b), displaying the variation of the E/p ratio versus the impact point of the electron on the crystal for four different $|\eta|$ intervals in EB. MC-driven corrections do not fully compensate for the energy leakage in the inter-crystal gaps, yielding a residual response variation of 0.3-0.5% RMS.

4. Measurement of the energy resolution and impact on the $H \to \gamma\gamma$ search

The energy resolution for high energy electrons is measured using $Z \to ee$ events. The electron energies are reconstructed from the ECAL energy deposits with the calibrations and corrections described in the previous sections. An unbinned maximum likelihood fit is used to extract the ECAL energy resolution as a function of the pseudo-rapidity of the final state electrons and it is performed in the di-electron invariant mass range between 89 GeV and 100 GeV. In fig. 6 the results of the fit are reported for both data and MC. The inclusive energy resolution in data varies between 1.5% in the central EB, to 3-4% in the outer EB and 4% in EE. The resolution in the barrel depends on the amount of material in front of the ECAL, and is degraded in the vicinity of the ECAL module boundaries, as indicated by vertical lines in the plots. The resolution in the endcaps shows an η dependence that is also correlated with the amount of

Journal of Physics: Conference Series **455** (2013) 012028 doi:10.1088/1742-6596/455/1/012028

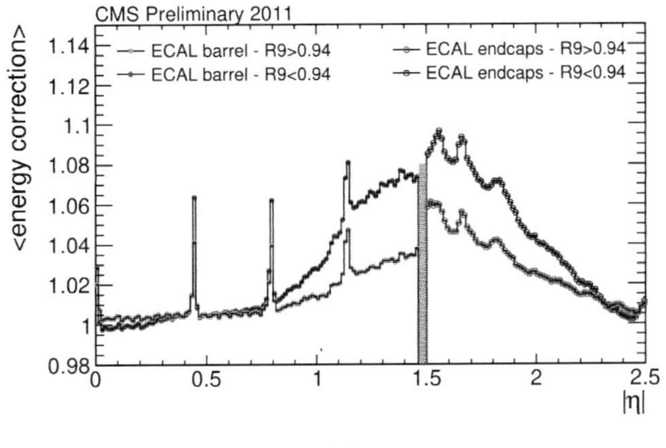

(a)

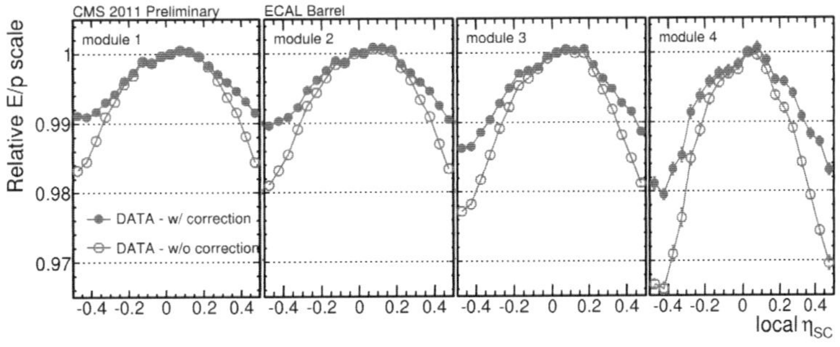

(b)

Figure 5. (a) The average energy correction factors for $W \to e\nu$ electrons with $R9 > 0.94$ and $R9 < 0.94$ are shown. (b) The dependence of the electron E/p on the supercluster centre along η, referring to a local coordinate system where crystal edges are at -0.5 and 0.5. Results for EB crystals are shown in the η ranges: [0, 0.43], [0.43, 0.79], [0.79, 1.10] and [1.10, 1.48] from left to right.

material in front of the ECAL, up to $\eta \sim 2.0$. At larger pseudo-rapidity, a contribution from single-channel response variations, not fully accounted for in simulation, is also contributing to the difference between data and MC.

The search for the Higgs boson in the two-photon decay channel relies on the energy resolution and the position measurement within ECAL of the reconstructed photons. In order to precisely simulate the response to photons, the constant term of the energy resolution in the MC is tuned in different regions of ECAL and for the different electromagnetic shower shapes to match the observed resolution in $Z \to ee$ data, with electrons reconstructed as photons, except for the vertex. After this MC tuning using data, the relative invariant mass width of simulated $H \to \gamma\gamma$ events (expressed as FWHM/2.35) is 1.13%. This result takes into account also the additional contribution to the invariant mass resolution coming from the uncertainty of the measured angle between the two photons. The latter contribution, given the excellent precision of the ECAL shower position reconstruction, is however negligible with respect to the photon energy resolution term in case of a correct assignment of the vertex.

In fig. 7 the gradual improvement of the di-photon invariant mass resolution for a 120 GeV Higgs boson is reported in correspondence with the main public conferences in 2011 and 2012.

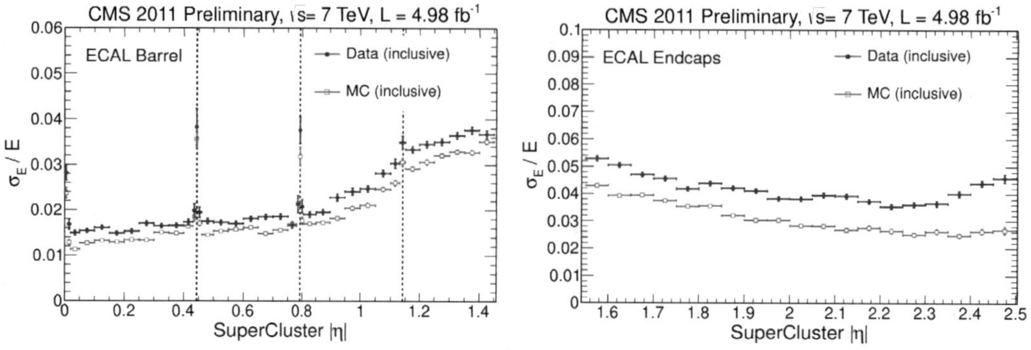

(a) (b)

Figure 6. Relative energy resolution as a function of absolute pseudo-rapidity for $Z \to ee$ electrons from data and MC for EB (a) and EE (b). The vertical dashed lines are the module boundaries in EB.

5. Conclusions

The energy resolution achieved and the stability in time of the ECAL performance played a central role in the search for the Higgs boson in the di-photon final state and in the discovery of a new resonance with a mass of 125 GeV.
Limitations to maintaining the excellent ECAL energy resolution come predominantly from the impact of the material in front of the ECAL and from imperfect algorithmic corrections for which further tuning is ongoing. The ECAL endcaps, with lower acceptance for $H \to \gamma\gamma$ photons, contribute less to this search because of non-optimal resolution. Further progress in the precision of the single channel calibrations is needed.
An exhaustive discussion of the performance of the CMS electromagnetic calorimeter with a complete description about the electron and photon energy calibration at $\sqrt{s} = TeV$ can be found in [14].

6. Acknowledgements

We congratulate our colleagues in the CERN accelerator departments for the excellent performance of the LHC and thank the technical and administrative staffs at CERN and at other CMS institutes for their contributions to the success of the CMS effort. In addition, we gratefully acknowledge the computing centres and personnel of the Worldwide LHC Computing Grid for delivering so effectively the computing infrastructure essential to our analyses. Finally, we acknowledge the enduring support for the construction and operation of the LHC and the CMS detector provided by the following funding agencies: BMWF and FWF (Austria); FNRS and FWO (Belgium); CNPq, CAPES, FAPERJ, and FAPESP (Brazil); MEYS (Bulgaria); CERN; CAS, MoST, and NSFC (China); COLCIENCIAS (Colombia); MSES (Croatia); RPF (Cyprus); MoER, SF0690030s09 and ERDF (Estonia); Academy of Finland, MEC, and HIP (Finland); CEA and CNRS/IN2P3 (France); BMBF, DFG, and HGF (Germany); GSRT (Greece); OTKA and NKTH (Hungary); DAE and DST (India); IPM (Iran); SFI (Ireland); INFN (Italy); NRF and WCU (Republic of Korea); LAS (Lithuania); CINVESTAV, CONACYT, SEP, and UASLP-FAI (Mexico); MSI (New Zealand); PAEC (Pakistan); MSHE and NSC (Poland); FCT (Portugal); JINR (Armenia, Belarus, Georgia, Ukraine, Uzbekistan); MON, RosAtom, RAS

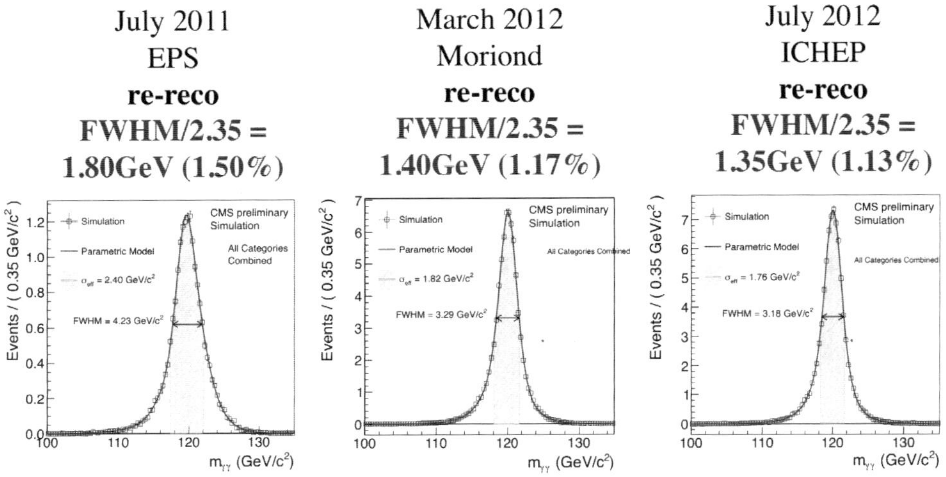

Figure 7. Evolution in time of the invariant mass width for a Higgs of $m_H = 120\ GeV$ decaying into two photons.

and RFBR (Russia); MSTD (Serbia); SEIDI and CPAN (Spain); Swiss Funding Agencies (Switzerland); NSC (Taipei); ThEPCenter, IPST and NSTDA (Thailand); TUBITAK and TAEK (Turkey); NASU (Ukraine); STFC (United Kingdom); DOE and NSF (USA).

References

[1] Chatrchyan S et al., *The CMS experiment at the CERN LHC*, JINST 3 (2008) S08004, http://dx.doi.org/10.1088/1748-0221/3/08/S08004.
[2] The CMS Collaboration, *The Electromagnetic Calorimeter Project: Technical Design Report*, CERN/LHCC 1997, CMS TDR 4.
[3] CMS Collaboration. *Search for a higgs boson decaying into two photons in the cms detector.* CMS Physics Analysis Summary: HIG-11-021, 2011.
[4] Chatrchyan S et al., *Observation of a new boson at a mass of 125 GeV with the CMS experiment at the LHC*, Phys.Lett. B 716 (2012) 30-61.
[5] G. Aad et al., *Observation of a new particle in the search for the Standard Model Higgs boson with the ATLAS detector at the LHC*, Phys.Lett. B716 (2012) 1-29.
[6] Baillon P et al., *Performance of the cooling system of the CMS ECAL*, Proceedings of the 8th conference on Astroparticle, Particle and space physics, detectors and medical applications, Como, 2003,World Scientific Publishing Co., 2003, 203-207.
[7] Anfreville M et al., *Laser monitoring system for the CMS lead tungstate crystal calorimeter*, Nucl. Instr. and Meth. A 594 (2008) 292-320.
[8] Bartoloni A et et al., *High Voltage system for the CMS electromagnetic calorimeter*, Nucl. Instr. Meth. A 582 (2007) 462468.
[9] Chatrchyan S et al., *Performance and Operation of the CMS Electromagnetic Calorimeter*, JINST 5 (2010) T03010.
[10] Adzic P et al., *Energy resolution of the barrel of the CMS Electromagnetic Calorimeter* JINST 2 (2007) P04004.
[11] Skwarnicki T, Ph.D Thesis, DESY F31-86-02(1986), Appendix E, and references therein.
[12] Adzic P et al., *Intercalibration of the barrel electromagnetic calorimeter of the CMS experiment at start-up*, JINST 3 (2008) P10007.
[13] CMS Collaboration, *Electromagnetic calorimeter calibration with 7 TeV data*, CMS PAS-EGM-2010-003 (2010).
[14] CMS Collaboration, *Electron and photon energy calibration and resolution with the CMS ECAL at $\sqrt{s} = TeV$*, CMS PAS-EGM-2011-001 (2011).

International Workshop on Discovery Physics at the LHC (Kruger2012) IOP Publishing
Journal of Physics: Conference Series **455** (2013) 012029 doi:10.1088/1742-6596/455/1/012029

Search for exotic resonances in diboson final states with the CMS detector at the LHC

Flavia A Dias
On behalf of the CMS Collaboration
Instituto de Física Teórica, Universidade Estadual Paulista, São Paulo, Brazil
E-mail: flavia.dias@cern.ch

Abstract. The results of searches for new resonances decaying to a pair of massive vector bosons (WW, WZ, ZZ) are presented. All searches are performed using 5.0 fb^{-1} of proton-proton collisions, at $\sqrt{s} = 7$ TeV of center of mass energy, collected by the Compact Muon Solenoid detector at the Large Hadron Collider. No significant excess compared to the standard model background expectation is observed, and upper limits at 95% confidence level are set on the production cross section times the branching fraction of hypothetical particles decaying to a pair of vector bosons. The results are interpreted in the context of several benchmark models, such as the Randall-Sundrum gravitons, the Sequential Standard Model W′, and Technicolor. Graviton resonances in the Randall-Sundrum model with masses smaller than 940 GeV/c^2, for coupling parameter $k/\overline{M}_{Pl} = 0.05$ are excluded. Bulk (ADPS) Randall-Sundrum gravitons with masses smaller than 610 GeV/c^2 are excluded, for $k/\overline{M}_{Pl} = 0.05$. Sequential Standard Model W′ with masses smaller than 1143 GeV/c^2 are excluded, as well as ρ_{TC} in the 167 – 687 GeV/c^2 mass range, in Low Scale Technicolor models with $M(\pi_{TC}) = 3/4\, M(\rho_{TC}) - 25$ GeV/c^2.

1. Introduction

Searches for exotic resonances decaying to a pair of massive vector bosons are historically connected to electroweak symmetry breaking models, other than the Higgs mechanism. However, a standard model (SM) Higgs boson with mass around $M_H = 126$ GeV/c^2 would not exclude all models being considered, as the Higgs could be accommodated within the theoretical framework. One of the fundamental questions the LHC was designed to elucidate is the electroweak symmetry breaking mechanism, and in many extensions of the SM the generation of mass can be associated with new strong dynamics appearing at the TeV scale. This new dynamics can be due to new interactions arising from extended gauge groups, a composite Higgs boson, or from compact extra dimensions.

Searches for new heavy resonances decaying to a pair of vector bosons VV (V = W,Z) are performed, using 5 fb^{-1} of $\sqrt{s} = 7$ TeV center-of-mass energy pp collisions data collected by the CMS experiment [1] in 2011. Several channels were considered: WZ decaying into three leptons and missing transverse energy [2]; ZZ decaying into a pair of leptons and two jets [3]; WZ or ZZ decaying into a pair of leptons or neutrinos and a single merged jet [4]; and WW, WZ or ZZ decaying into two merged jets [5]. The results were interpreted in several benchmark models: the Sequential Standard Model (SSM) [6], Randall-Sundrum (RS) with SM particles confined in the electroweak brane [7] or allowed to propagate in the bulk (ADPS model) [8], and Low Scale Technicolor (LSTC) [9].

2. WZ $\rightarrow 3\ell$ + MET

A search for new particles decaying via a WZ pair is performed, with W $\rightarrow \ell\nu$ and Z $\rightarrow \ell\ell$ in the final state, where ℓ = e, μ. The decay under study is characterized by a pair of same-flavor, opposite-charge, isolated leptons with high transverse momentum (p_T) and invariant mass consistent with that of the Z boson, along with a third, high-p_T isolated lepton, and missing transverse energy (MET) associated with the escaping neutrino. The SM backgrounds for this search include the irreducible SM WZ background, top pair events, Z+jets, Zγ, and ZZ$\rightarrow 4\ell$.

The background is estimated using simulated samples produced with MadGraph 5.1 [11] and pythia 6.4.22 [12] generators, which were subject to a full simulation of the CMS detector using Geant4 [19]. The cross sections are corrected for higher-order effects using next-to-leading-order (NLO) K-factor corrections, obtained using mcfm 6.1 [17]. The signal is simulated using pythia with mass-dependent next-to-NLO (NNLO) cross sections obtained using fewz 2.0 [18].

Candidate events are required to have three reconstructed leptons within the acceptance of $|\eta| < 2.5$ (2.4) for electrons (muons), where $\eta \equiv -\ln(\tan(\theta/2))$ and θ is the polar angle. Z boson candidates are reconstructed from pairs of opposite-sign, same-flavor leptons with the highest and second-highest p_T greater than 20 and 10 GeV/c, respectively, and with invariant mass between 60 and 120 GeV/c^2. The lepton from the W candidate is required to have tighter identification and isolation requirements, and the events with MET > 30 GeV are selected. The observed invariant mass distribution of WZ candidates (M_{WZ}) is shown in Fig. 1. An additional selection requirement is applied on the scalar sum of the transverse momenta of the charged leptons (L_T), dependent on the mass of the signal hypothesis. Fig. 2 shows the M_{WZ} distribution corresponding to the $W' = 600$ GeV hypothesis, where $L_T > 290$ GeV.

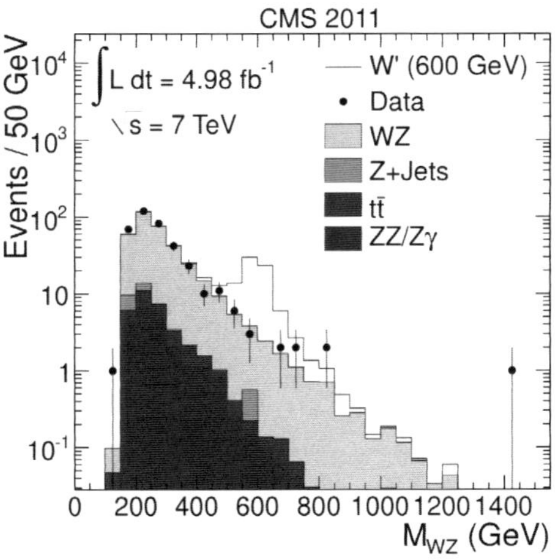

Figure 1. Distribution of the WZ invariant mass for the SM background, W′ mass point at 600 GeV/c^2 and observed data, before application of the L_T requirement.

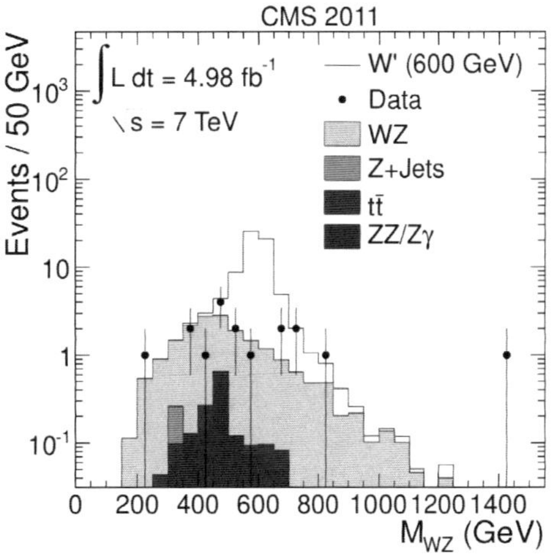

Figure 2. Distribution of the WZ invariant mass for the SM background, W′ mass point at 600 GeV/c^2 and observed data. The selection of $L_T > 290$ GeV/c^2 is applied.

No significant excess above the SM expectation is observed. Exclusion limits on the production cross section σ (pp $\rightarrow W'/\rho_{TC} \rightarrow$WZ) $\times\ \mathcal{B}$ (WZ $\rightarrow 3\ell\nu$) are determined using CL$_S$ [10] statistics (Fig. 3). SSM W′ with masses below 1143 GeV/c^2 are excluded. For LSTC, with the chosen parameters M(π_{TC}) = 3/4 M(ρ_{TC}) – 25 GeV/c^2, ρ_{TC} hadrons with masses between 167 and 687 GeV/c^2 are also excluded. Fig. 4 shows the LSTC limits as a function of the ρ_{TC} and π_{TC} masses.

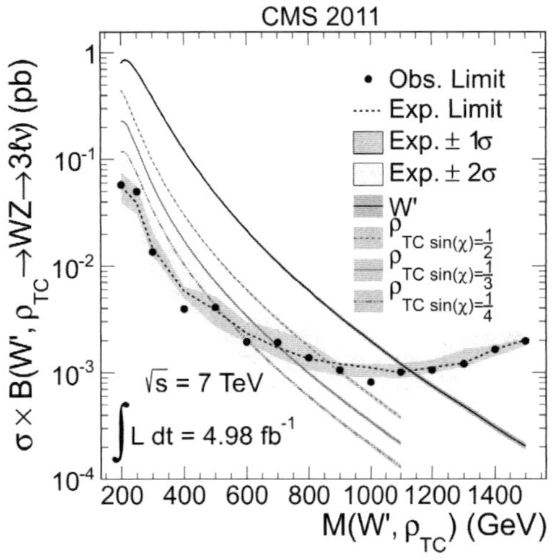

Figure 3. Expected and observed upper limits as a function of the WZ mass for W′ and ρ_{TC} along with the 1 and 2 σ uncertainties.

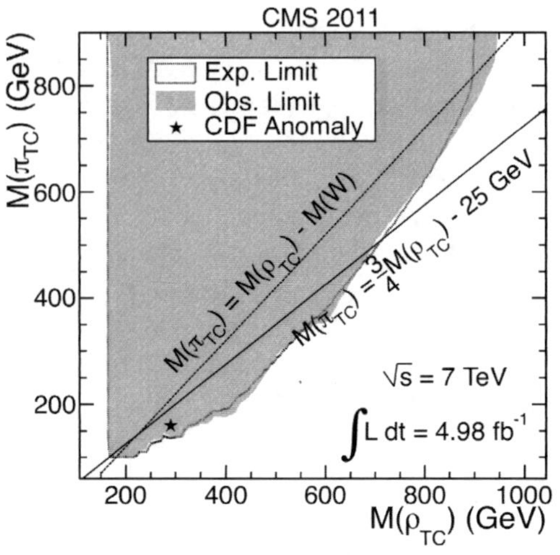

Figure 4. Exclusion region at 95% confidence level (CL) for Low Scale Technicolor as a function of the ρ_{TC} and π_{TC} masses.

3. ZZ $\rightarrow$ 2ℓ + jj

A search for new heavy narrow spin-2 resonances decaying via a ZZ pair is performed, with Z $\rightarrow \ell^+\ell^-$ and Z $\rightarrow q\bar{q}$ in the final state, where ℓ = e, μ. The full decay chain is reconstructed and the invariant mass of the candidate Z boson pair, M_{ZZ}, is investigated in a mass range between 400 and 1200 GeV/c^2. The dominant SM background is Z boson production with associated jets (Z+jets).

Candidate events are required to have two reconstructed leptons within the acceptance of $|\eta| < 2.5$ (2.4) for electrons (muons), and highest and second-highest p_T greater than 40 and 20 GeV/c, respectively. Events are also required to have jet pairs, reconstructed with the anti-k_T [15] algorithm with a distance parameter R = 0.5, within $|\eta| < 2.4$ and with $p_T > 30$ GeV/c. Z boson candidates are reconstructed from pairs of opposite-sign, same-flavor leptons with invariant mass between 70 and 110 GeV/c^2 and from pair of jets with invariant mass between 75 and 105 GeV/c^2. The events are separated in categories of jet flavor, introducing tags for b flavor jets (b-tag), in order to better discriminate signal and background. An angular analysis is also performed, using an angular likelihood discriminant based on the probability ratio of the signal and background hypotheses.

The background is estimated from data, using a sideband in the M_{ZZ} distribution. A cross-check of the estimated background is performed, using simulated samples produced with MadGraph 5, sherpa 1.13 [13] and pythia 6.4.24. The signal is modeled using simulated samples produced with a dedicated generator for spin-2 resonances [20], for both RS and ADPS models. The M_{ZZ} distribution for 1 b-tag category is shown in Figs. 5 and 6, for the electron and muon channels, respectively.

Since no significant excess is observed, exclusion limits at the 95% confidence level (CL) were set on the product of the graviton cross section and branching fraction of $G_{KK} \rightarrow$ ZZ. The upper limits are calculated using the CL$_S$ method and compared to the theoretical values. Fig. 7 shows the upper limits for the RS model and compares them to the LO theoretical expectations for two values of the coupling parameter $k/\overline{M}_{Pl}$, 0.05 and 0.1. Gravitons with masses smaller than 945 GeV/c^2 are excluded, for $k/\overline{M}_{Pl} = 0.1$, and gravitons with masses smaller than 720 GeV/c^2

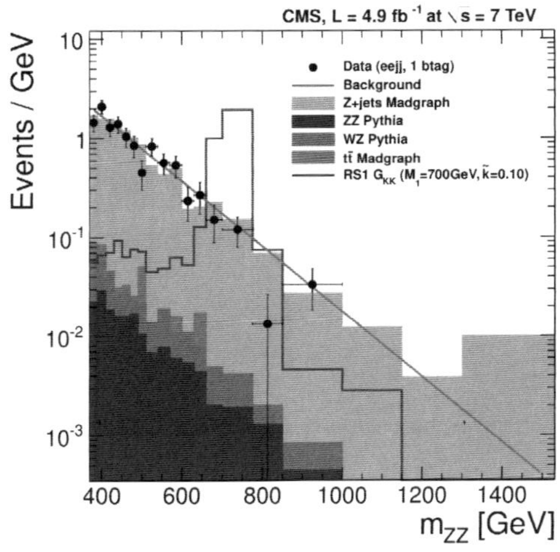

Figure 5. The m_{ZZ} invariant mass distribution for data, simulated background and data-estimated background, after final selection in the 1-btag category, for the electron channel.

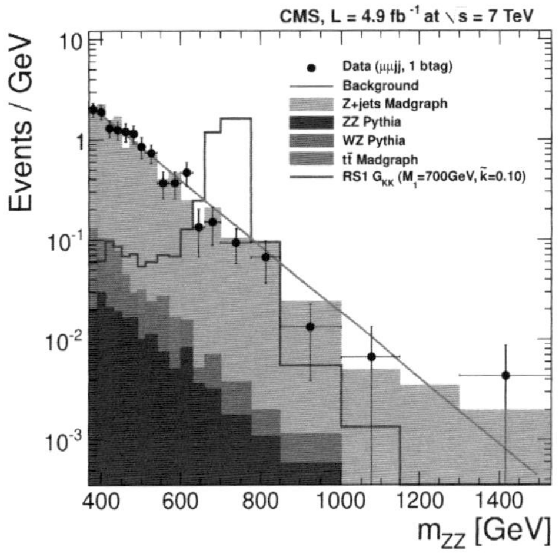

Figure 6. The m_{ZZ} invariant mass distribution for data, simulated background and data-estimated background, after final selection in the 1-btag category, for the muon channel

and in the range 760 – 850 GeV/c^2 are excluded for $k/\overline{M}_{Pl} = 0.05$. Fig. 8 shows the results for ADPS model, excluding graviton masses smaller than 720 GeV/c^2, for $k/\overline{M}_{Pl} = 0.1$, and smaller than 610 GeV/c^2, for $k/\overline{M}_{Pl} = 0.05$.

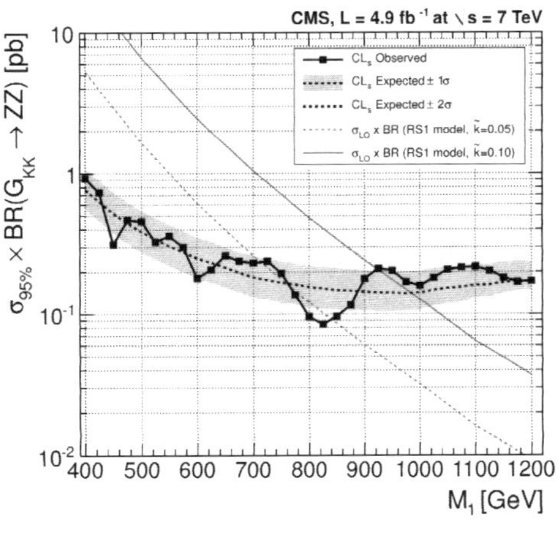

Figure 7. Observed and expected 95% CL upper limits on $\sigma\times$ BR($G_{KK} \to$ ZZ). The graviton signal was generated according to the RS1 model.

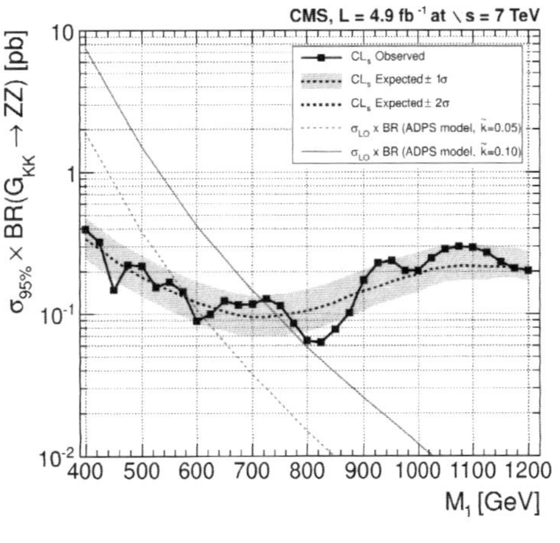

Figure 8. Observed and expected 95% CL upper limits on $\sigma\times$ BR($G_{KK} \to$ ZZ). The graviton signal was generated according to the ADPS model.

4. WZ/ZZ $\to 2\ell$ or 2ν + j

A search for heavy resonances decaying to WZ and ZZ states is performed, where the Z boson is allowed to decay to leptons (Z $\to \ell^+\ell^-$, ℓ = e, μ) or neutrinos (Z $\to \nu\bar{\nu}$) and the second boson decay to hadrons, V $\rightarrow q\bar{q}$. For heavy resonances, its decay produces a highly boosted system, and each the two quarks from the V are emitted very close to each other, producing overlapping jets reconstructed as a single jet with mass close to the V mass, a topology different from the typical quark or gluon jet. This feature is exploited, for resonances heavier than 700 GeV/c^2, complementing the analysis for lower masses in the ZZ $\rightarrow 2\ell$jj channel. The main SM backgrounds are from Z+jets, $t\bar{t}$+jets, W+jets, and di-boson production.

For the leptonic channel, candidate events are required to have two reconstructed leptons within detector acceptance and $p_T > 45$ GeV/c, and one anti-k_T R = 0.7 jet with $p_T > 30$ GeV/c and $|\eta| < 2.4$. Z candidates are reconstructed with invariant mass of leptons in the range $70 < M_Z < 110$ GeV/c^2 and with $p_T > 150$ GeV, and V candidates have mass in the range 65 $< M_j < 120$ GeV/c^2 and $p_T > 250$ GeV/c. The background is determined from data, using the sideband of the V mass, and the signal is generated with PYTHIA 6.4.24. Fig. 9 shows the distribution of the M_{VZ} for the electron channel.

For the MET channel, events are required to have one anti-k_T R = 0.7 jet with $p_T > 300$ GeV/c and $|\eta| < 2.4$, and MET larger than 300 GeV. Events are rejected if they have isolated leptons or more than two jets with $p_T > 30$ GeV. The background is determined from data, using the ABCD method in the variables of the mass of the jet and the transverse mass of the jet-MET system. A correction factor ρ is introduced to account for the correlation between the jet mass and the jet-MET transverse mass. Signal events are generated with PYTHIA 6.4.24, as well as SM background events for cross-check purposes. Fig. 10 shows the distribution of the jet-MET transverse mass for the simulated backgrounds and data in the search region.

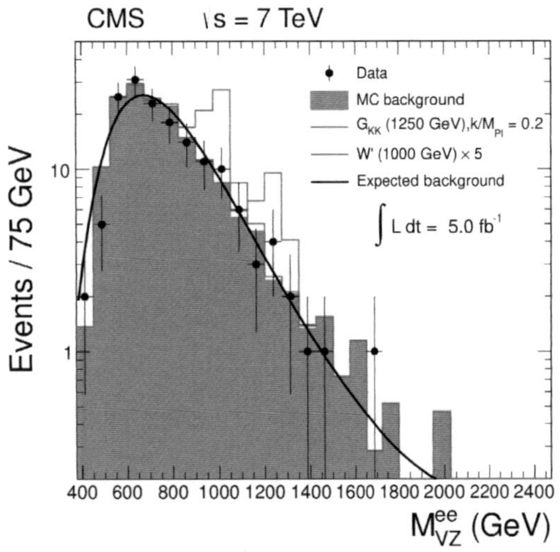

Figure 9. The comparison of the estimated background with the total MC background and the data for M_{VZ} distributions for the electron channel.

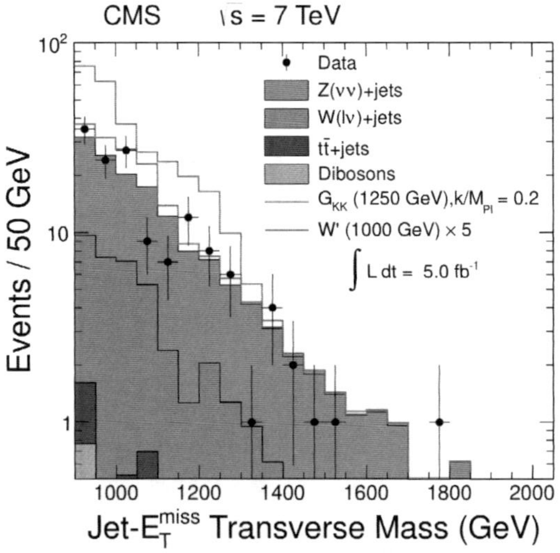

Figure 10. Comparison between ρ-corrected simulated backgrounds and data in the search region for jet-MET transverse mass distribution.

No significant excess is observed over the expected background. The CL$_S$ statistical method is employed to search for exotic VZ resonances. 95% CL exclusion limits on the combined products of the cross section times the branching ratio σ (pp $\rightarrow$ W$'$) $\times$ $\mathcal{B}$(W$' \rightarrow$ WZ) and σ (pp $\rightarrow G_{KK}$) $\times$ $\mathcal{B}(G_{KK} \rightarrow$ ZZ) for the three final states are set, as a function of the mass of the hypothetical resonance. In the Randall-Sundrum model, graviton resonances with masses between 750 – 880 (800) GeV/c^2 at NLO (LO) are excluded for $k/\overline{M}_{Pl} = 0.05$, as seen in Fig. 11. SSM W$'$ masses are excluded between 700 – 940 (890) GeV/c^2 at NNLO (LO), as seen in Fig. 12.

Journal of Physics: Conference Series **455** (2013) 012029 doi:10.1088/1742-6596/455/1/012029

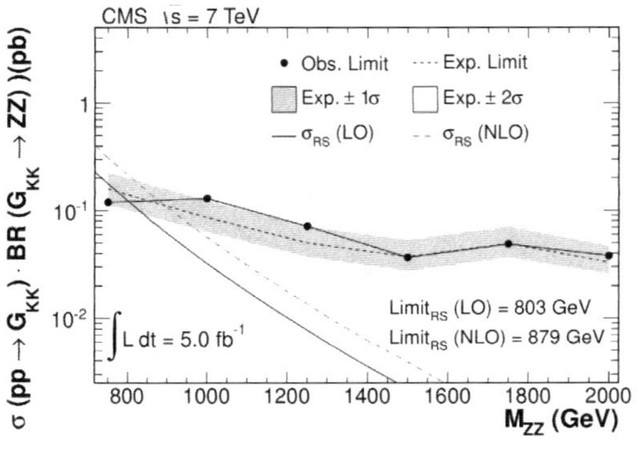

Figure 11. Observed and expected 95% CL upper cross section limits and comparison with the theoretical predictions in RS graviton model with $k/\overline{M}_{Pl} = 0.05$ for the combination of electron, muon, and MET channels.

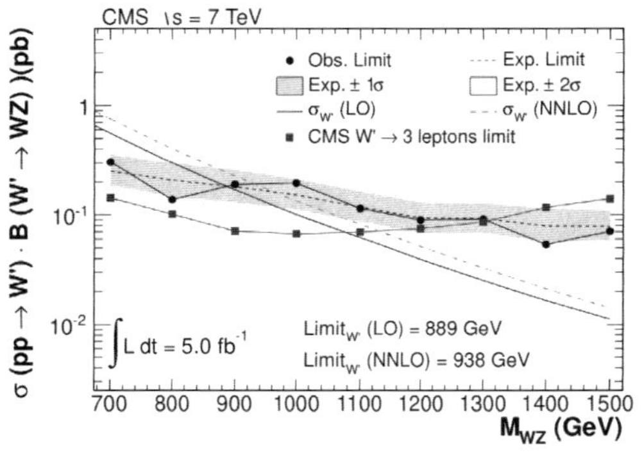

Figure 12. Observed and expected 95% CL upper cross section limits and comparison with the theoretical predictions in W′ model for the combination of electron, muon, and MET channels.

5. WW/WZ/ZZ → jj

A search focused on models predicting new heavy particles with a pair of vector bosons as decay products is performed, with the two bosons each decaying to a single merged jet. The background is substantially suppressed by identifying subjets which correspond to the original daughter jets that have merged, using recent developments in the area of jet substructure. The subjets are used to explicitly reconstruct the W or Z boson. A massive jet which is a candidate to be a W or a Z is called W/Z-tagged.

Jets are reconstructed in the events using two algorithms: anti-k_T with R = 0.5 and Cambridge-Aachen (CA) [16] with R = 0.8. Candidate events must have at least two anti-k_T jets with $p_T > 30$ GeV/c and $|\eta| < 2.5$, and the two leading p_T jets are required to have $|\Delta\eta| < 1.3$ and $M_{jj} > 890$ GeV/c^2. For the W/Z-tag, the jets reconstructed with CA are pruned, in a process where soft and wide-angle particles (relative to the parent in the cluster) are ignored, and events are selected if the number of subjets $N_{sub} = 2$, the pruned jet mass M_{jet} in the range 70 – 100 GeV/c^2 and the mass drop $M_1/M_{jet} < 0.25$, where M_1 is the mass of the hardest subjet. The discrimination power of the mass of the pruned jet selection is shown in Fig. 13.

The QCD background shape is modeled in the dijet spectrum using a simple parametrization which has been successfully deployed in previous searches in the dijet mass spectrum. In the limit setting, the background fit parameters used are obtained by best signal plus background fit to the data points for each signal hypothesis. The signal shapes are simulated with PYTHIA 6.4.24 and HERWIG++ [14]. The double W/Z-tagged M_{jj} distribution in data fitted with the background parametrization is shown in Fig. 14, along with the signal shape distribution for RS graviton decaying to WW with arbitrary cross sections.

There is no significant evidence for new particle production in the W/Z-tagged dijet spectrum. 95% CL upper limits on the cross section for double W/Z-tagged dijet events are set, as shown in Fig. 15, for WW resonances, and Fig. 16, for WZ resonances. The sensitivity of the measurement with the present dataset is not sufficient to extract meaningful mass limits on the RS graviton with $k/\overline{M}_{Pl} = 0.1$ and SSM W′, but the cross section limits are the most stringent in this final state to date.

Journal of Physics: Conference Series **455** (2013) 012029 doi:10.1088/1742-6596/455/1/012029

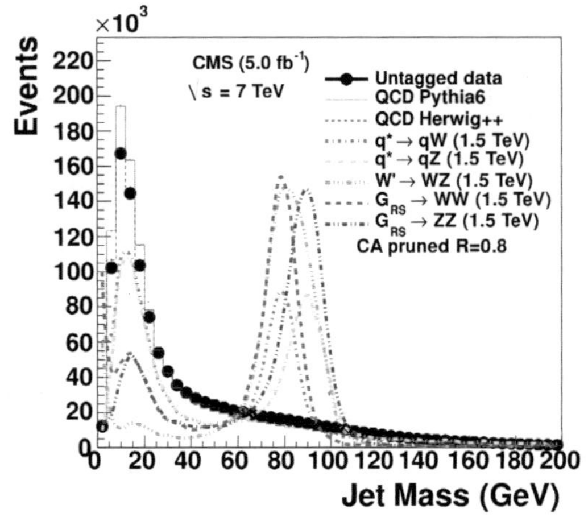

Figure 13. Pruned jet mass in data, signal and background simulations. All simulated distributions have been scaled to match the number of data events.

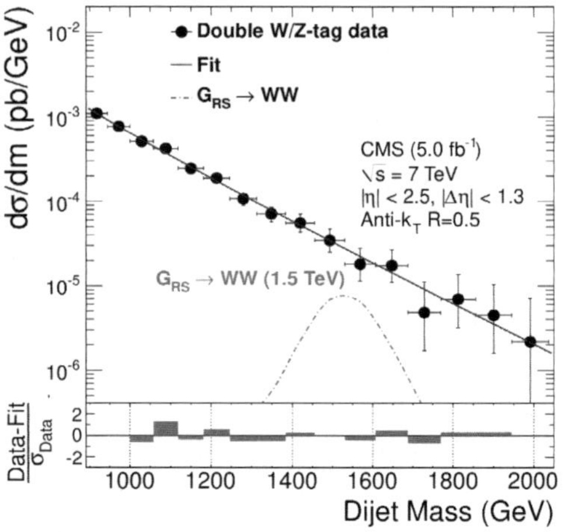

Figure 14. The double W/Z-tagged M_{jj} distributions in data fitted with the QCD background parametrization. Bottom pane: the corresponding pull distributions.

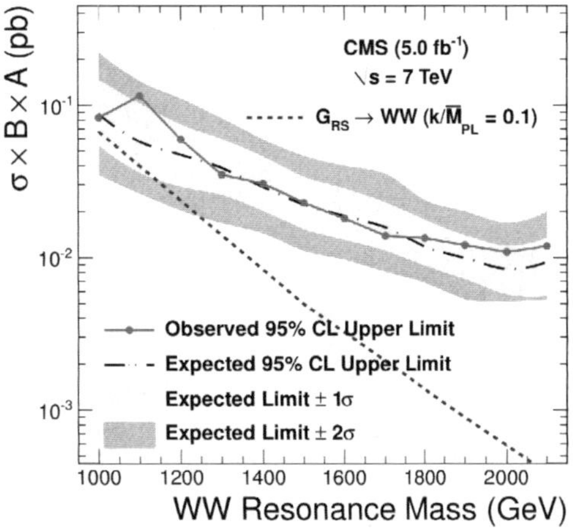

Figure 15. Expected and observed limits for WW resonances. The predicted cross sections as a function of resonance mass for the considered benchmark model is overlaid.

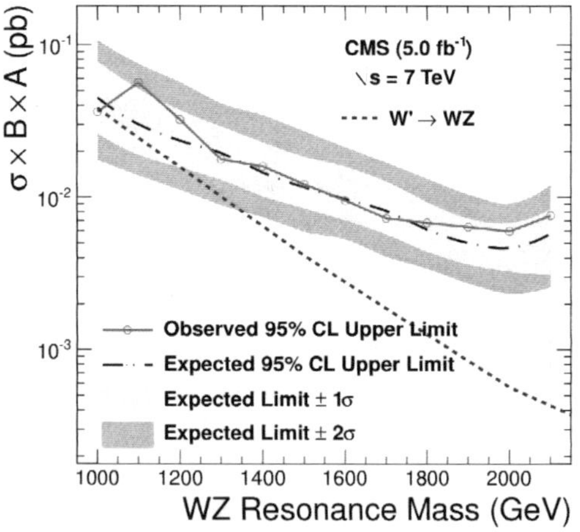

Figure 16. Expected and observed limits for WZ resonances. The predicted cross sections as a function of resonance mass for the considered benchmark model is overlaid.

6. Summary

The results of searches for new resonances decaying to a pair of massive vector bosons (WW, WZ, ZZ) were presented. All analyses used datasets of proton-proton collisions at $\sqrt{s} = 7$ TeV of center of mass energy, corresponding to 5.0 fb^{-1} of integrated luminosity collected by the Compact Muon Solenoid experiment at the Large Hadron Collider in 2011. The analyzed final states were WZ $\rightarrow 3\ell$+MET, ZZ $\rightarrow 2\ell$jj, WZ/ZZ $\rightarrow 2\ell/2\nu$+j and WW/WZ/ZZ $\rightarrow$ jj. No significant excesses compared to the standard model background expectation were observed, and upper limits at 95% confidence level were set on the production cross section times the branching fraction of hypothetical particles decaying to a pair of vector bosons. The results were interpreted in the light of several benchmark models, such as the Randall-Sundrum gravitons, the Sequential Standard Model W′, and Technicolor. Graviton resonances in the Randall-Sundrum model were excluded for masses smaller than 938 GeV/c^2, for coupling parameter $k/\overline{M}_{Pl} = 0.05$. For ADPS bulk Randall-Sundrum gravitons, masses smaller than 610 GeV/c^2 were excluded, for $k/\overline{M}_{Pl} = 0.05$, and smaller than 720 GeV/c^2, for $k/\overline{M}_{Pl} = 0.1$. Sequential Standard Model W′ with masses smaller than 1143 GeV/c^2 were excluded, as well as ρ_{TC} in the 167 – 687 GeV/c^2 mass range, in Low Scale Technicolor models with M(π_{TC}) = 3/4 M(ρ_{TC}) – 25 GeV/c^2. The most stringent upper limits to date on the cross section of double W/Z-tagged dijet events were set.

References

[1] CMS Collaboration 2008 The CMS experiment at the CERN LHC *JINST* **3** S08004

[2] CMS Collaboration 2012 Search for a W′ or techni-ρ decaying into WZ in pp collisions ar $\sqrt{s}$=7 TeV *Phys. Rev. Lett.* **109** 141801

[3] CMS Collaboration 2013 Search for a narrow, spin-2 resonance decaying to a pair od Z bosons in the $q\bar{q}\ell^+\ell^-$ final state *Phys. Lett.* B **4-5** 1208-1228

[4] CMS Collaboration 2013 Search for exotic resonances decaying into WZ/ZZ in pp collisions at $\sqrt{s}$=7 TeV *JHEP* **036**

[5] CMS Collaboration 2012 Search for heavy resonances in the W/Z-tagged dijet mass spectrum in pp collisions at 7 TeV *Preprint* arXiv:1212.1910

[6] Altarelli G, Mele B and Ruiz-Altaba M 1989 Searching for new heave vector bosons in p anti-p colliders *Z. Phys.* C **45** 109

[7] Randall L and Sundrum R 1999 Large mass hierarchy from a small extra dimension *Phys. Rev. Lett.* **83** 3370-3373

[8] Agashe K, Davoudiasl H, Perez G and Soni A 2007 Warped gravitons at the LHC and beyond *Phys. Rev.* D **76** 036006

[9] Eichten E and Lane K 2008 Low-scale technicolor at the Tevatron and LHC *Phys. Lett.* B **669** 235-238

[10] Read A L 2002 Presentation of search results: the CL(s) technique *J. Phys.* G **28** 2693-2704

[11] Alwall J, Herquet M, Maltoni F, Mattelaer O and Stelzer T 2011 MadGraph 5: going beyond *JHEP* **06** 128

[12] Sjostrand T, Mrenna S and Skands PZ 2006 Pythia 6.4 physics and manual *JHEP* **05** 026

[13] Gleisberg T, Hoeche S, Krauss F, Schonherr M, Schumann S, Siegert F and Winter J 2009 Event generation with Sherpa 1.1 *JHEP* **02** 007

[14] Bahr M et al. 2008 Herwig++ physics and manual *Eur. Phys. J.* C **58** 639-707

[15] Cacciari M, Salam GP and Soyez G 2008 The anti-kt jet clustering algorithm *JHEP* **04** 063

[16] Dokshitzer Yu L, Leder GD, Moretti S and Webber BR 1997 Better Jet Clustering Algorithms *JHEP* **9708** 001

[17] Campbell JM, Ellis RK and Williams C 2011 Vector boson pair production at the LHC *JHEP* **07** 018

[18] Gavin R, Li Y, Petriello F and Quackenbush S 2011 FEWZ 2.0: a code for hadronic Z production at next-to-next-to-leading order *Comput. Phys. Commun.* **182** 2388

[19] GEANT4 Collaboration 2003 GEANT4 - a simulation toolkit *Nucl. Instrum. Meth.* A **506** 250

[20] Gao Y, Gritsan AV, Guo Z, Melnikov K, Schulze M, Tran NV 2010 Spin determination of single-produced resonances at hadron colliders *Phys. Rev.* D **81** 075022

Journal of Physics: Conference Series **455** (2013) 012030 doi:10.1088/1742-6596/455/1/012030

Search for Higgs decaying to WW at CMS

Leonardo Di Matteo[1] ***on behalf of the CMS Collaboration***

[1] University and INFN of Milano-Bicocca

E-mail: leonardo.di.matteo@cern.ch

Abstract. A search for a standard model Higgs boson decaying into a W^+W^- pair is presented, using CMS proton-proton collision data samples both at 7 TeV and 8 TeV center-of-mass energy, corresponding to an integrated luminosity of 4.9 ± 0.2 fb^{-1} and 12.1 ± 0.5 fb^{-1} respectively. Both fully-leptonic and semi-leptonic W decay final states are considered. Upper limits are set on the Higgs boson production relative to the standard model expectation. The standard model Higgs boson is excluded at 95% confidence level in the mass ranges $[128 - 600]$ GeV and $[215 - 490] + [525 - 600]$ GeV using the fully-leptonic and semi-leptonic channels respectively. In the fully-leptonic final state, an excess of events is observed above background which is consistent with the expectations from a standard model Higgs boson of mass 125 GeV and has a statistical significance of 3.1 standard deviations.

1. Introduction

In the context of the Standard Model (SM) of particle physics, the origin of the masses of fundamental particles remains an open issue. The electroweak symmetry breaking mechanism, which causes vector bosons to acquire masses, rely on the existence of the Higgs field and its quantum, the Higgs boson, which is yet to be proven experimentally. One of the main goals of the CERN Large Hadron Collider (LHC) physics program is to investigate on the existence of the SM Higgs boson.

In particular, searches of the Higgs boson are being carried on in the two multipurpose experiments located on the LHC ring, ATLAS and CMS. In these proceedings, I report about two particular searches performed at CMS, both exploiting the Higgs decay into two W bosons: the first exploits the fully-leptonic final state, where each W boson decays in a lepton and a neutrino; the second explores the semi-leptonic final state, where one W decays leptonically and the other hadronically.

The fully-leptonic analysis, $H \to W^+W^- \to \ell\nu_\ell \ell\nu_\ell$ [1], is performed in the $110 - 600$ GeV Higgs mass range, while the semi-leptonic one, $H \to W^+W^- \to \ell\nu_\ell q\bar{q}$ [2], is restricted to the $170 - 600$ GeV range, due to its dependency on single lepton high p_T triggers. Both the analyses exploit the full 2011-2012 LHC pp collision dataset, corresponding to an integrated luminosity of 12.1 ± 0.5 fb^{-1}, at $\sqrt{s} = 8$ TeV, and 4.9 ± 0.2 fb^{-1}, at $\sqrt{s} = 7$ TeV.

2. Event reconstruction and physics objects

The CMS detector [3] is used to reconstruct proton-proton collisions. A full event reconstruction aimed at characterizing and identifying all stable particles in the event is obtained via a particle-flow (PF) technique [4]. This approach combines the information from all CMS sub-detectors to identify and reconstruct individual particles, classifying them into mutually exclusive categories:

charged hadrons, neutral hadrons, photons, electrons, and muons.

The electron reconstruction algorithm links clusters of energy deposits in the ECAL to trajectories in the inner tracker. To reconstruct the trajectories in the tracker, a dedicated model of electron energy loss is developed, and series of hits are fitted with a Gaussian sum filter. The electron candidates are filtered with an identification algorithm, relying on a multivariate technique that combines observables sensitive to the amount of bremsstrahlung along the electron trajectory, the geometrical and momentum matching between the electron trajectory and the associated clusters, and shower-shape observables.

The muon reconstruction algorithm exploits both the silicon tracker and the muon system. Segments in the muon spectrometer are matched with energy deposits in the calorimeters and tracks in the silicon detector, and muon candidates are built and filtered with quality criteria.

PF candidates are clustered together into Jets, using the anti-k_T algorithm [5] with distance parameter $R = 0.5$. Then, to account for the non-linear response of the calorimeters to the particle energies, and other instrumental effects, jet energy corrections are applied. The corrections are based on in-situ calibration using dijet and $\gamma +$ jet data samples. The pileup energy, computed as the median energy density in each event, is subtracted from each jet. Jets are additionally required to originate from the primary vertex, which is identified as the vertex with the highest summed p_T^2 of its associated tracks. Out of this collection, b-jets are tagged by looking at the jet displacement with respect to the primary vertex in the transverse direction. In addition, a multivariate selection is applied to separate jets from the primary interaction from those reconstructed due to energy deposits associated with pile-up, based on the differences in the jet shapes, in the relative multiplicity of charged and neutral components and in the different fraction of transverse momentum which is carried by the hardest components.

At the end of the reconstruction sequence, the missing transverse energy $E_T^{\rm miss}$ is found as the modulus of the negative vector sum of the transverse momenta of all reconstructed particles in an event.

Since charged leptons from W and Z boson decays are typically expected to be isolated from other activity in the event, lepton isolation is ensured by computing the sum of the transverse energies of all reconstructed particles, charged or neutral, within a cone of$\Delta\ R = \sqrt{(\Delta\eta)^2 + (\ \Delta\phi)^2} < 0.4$ around the lepton direction. In the computation, pileup is taken into account by subtracting the median energy density estimated on an event-by-event basis.

At trigger level, depending on the decay mode, events have to contain a pair of electrons or muons, one lepton with $p_T > 17$ GeV and the other with $p_T > 8$ GeV, or a single electron (muon) with $p_T > 27(24)$ GeV.

Efficiencies for trigger selection, reconstruction, identification, and isolation of leptons are measured from data, using the tag-and-probe technique [6], based on a pure sample of Z events. These measurements are performed in bins of p_T^ℓ and $|\eta^\ell|$. The overall efficiency for selecting electrons in the ECAL barrel (endcaps) vary from around 82% (73%) at $p_T^e \simeq 10$ GeV to 90% (89%) for $p_T^e \simeq 20$ GeV. It drops to about 85% in the ECAL barrel-endcap transition region, at $1.44 < |\eta^e| < 1.57$. Muon offline efficiencies are greater than ~98% in the whole $|\eta^\mu| < 2.4$ range. The muon trigger efficiency for events selected for this analysis ranges from 96% to 99%.

3. The $\mathrm{H} \to \mathrm{W^+W^-} \to \ell\nu_\ell\ell\nu_\ell$ search

The search strategy for the $\mathrm{H} \to \mathrm{W^+W^-} \to \ell\nu_\ell\ell\nu_\ell$ analysis is based on a signature with two isolated, oppositely charged, high p_T leptons (electrons or muons) and large missing transverse momentum $E_T^{\rm miss}$, due to the undetected neutrinos.

To improve the signal sensitivity, events are classified by jet multiplicity into three mutually exclusive categories, characterized by different signal yields and signal-to-background ratios. In the following, these will be referred to as the 0-jet, 1-jet and 2-jet samples, where the selected jets are required to have $E_T > 30$ GeV and $|\eta| < 4.7$. Events with more than three jets are discarded, while events with exactly three jets are incorporated in the 2-jet category if the lowest-p_T jet is not between the two other ones in pseudo-rapidity. Additionally, the signal candidates are splits into three final states: e^+e^-, $\mu^+\mu^-$, $e^\pm\mu^\mp$.

Tag leptons are required to have opposite charge, p_T greater than 20 GeV for the leading lepton ($p_T^{\ell,\max}$) and p_T greater than 10 GeV for the trailing lepton ($p_T^{\ell,\min}$), and to originate from the primary vertex of the event (chosen as the vertex with the highest $\sum p_T^2$). Only electrons (muons) with $|\eta| < 2.5(2.4)$ are considered in the analysis.

In addition to high momentum isolated leptons and low jet activity, missing transverse momentum is present in signal events but generally not in background. In this analysis, a *projected* E_T^{miss} variable is devised. It is defined as the component of E_T^{miss} transverse to the nearest lepton if the difference in azimuth (ϕ) between this lepton and the E_T^{miss} vector is less than $\pi/2$. If no leptons closer than $\pi/2$ with respect to the direction of E_T^{miss} in ϕ are present, then the *projected* E_T^{miss} is equal to E_T^{miss}. Since the *projected* E_T^{miss} resolution is degraded by pile-up, the minimum of two E_T^{miss} observables is used: the first includes all reconstructed particles in the event, while the second uses only the charged particles associated with the primary vertex. Events with *projected* E_T^{miss} above 20 GeV are selected for the analysis.

Top-quark background is suppressed by means of a top tagging technique based on soft-muon and b-jet tagging. The first method is designed to veto events containing soft-muons produced by b-quarks resulting from top-quark decays. The second one exploits the b-jet tagging algorithm, which searches for jet tracks with large impact parameter.

W+jets background is reduced by requiring a minimum dilepton transverse momentum ($p_T^{\ell\ell}$) of 45 GeV.

To reduce the background from WZ production, any event with a third lepton passing the identification and isolation requirements is rejected.

The contamination from Wγ production, where the photon is misidentified as an electron, is reduced by about 90% in the e^+e^- state by γ conversion rejection requirements.

The background from low mass resonances is rejected by requiring a dilepton mass ($m_{\ell\ell}$) greater than 12 GeV.

The huge Drell-Yan process, which greatly affects same-flavour final states, is dealt with a more complex strategy. First of all, the resonant component of the Drell-Yan production is rejected by requiring a dilepton mass outside a 30 GeV window centered on the Z mass pole. Then, the remaining off-peak contribution is suppressed by exploiting various E_T^{miss}-based approaches depending on the number of jets. In the 0/1-jet categories, Drell-Yan events are more difficult to separate from the signal, therefore, in this case, a dedicated multivariate selection, combining missing transverse momentum with kinematic and topological variables, is used to reject background events and maximize the surviving signal yield. In the 2-jet category, the dominant source of fake E_T^{miss} is the mis-measurement of the hadronic recoil, therefore the optimal performance is obtained by requiring E_T^{miss} greater than 45 GeV. Additionally, for the 2-jet category, the momenta of the dilepton system and of the most energetic jet must not be back-to-back in the transverse plane. The combination of these selections reduce the Drell-Yan background by three orders of magnitude, while rejecting less than 50% of signal events.

Furthermore, to enhance the sensitivity to a Higgs boson signal, a cut-based approach is devised for the final Higgs selection in all categories. Due to different signal kinematics for different Higgs mass hypotheses, separate optimizations are performed for different values of m_H in a cut-based analysis. In addition, a two-dimensional shape analysis technique is also pursued for the $e^{\pm}\mu^{\mp}$ final state in the 0/1-jet categories, since it allows for a direct physical interpretation of the observed data with a sensitivity comparable to other more complex techniques.

The optimized cut-based approach consists in extra requirements placed on $p_T^{\ell,\max}$, $p_T^{\ell,\min}$, $m_{\ell\ell}$, $\Delta\phi_{\ell\ell}$ and the transverse mass m_T, defined as $\sqrt{2p_T^{\ell\ell}E_T^{\rm miss}(1-\cos\Delta\,\phi_{E_T^{\rm miss}\ell\ell})}$, where$\Delta\,\phi_{E_T^{\rm miss}\ell\ell}$ is the azimuthal difference between $E_T^{\rm miss}$ and the dilepton system.

Also for the 2-d analysis, which exploits the independent variables m_T and $m_{\ell\ell}$, an additional loose set of requirements is applied: m_T must be greater than 80 GeV and smaller than 280 (600) GeV for m_H hypotheses smaller or equal than 250 GeV (greater than 250 GeV), while $m_{\ell\ell}$ must be smaller than 200 (600) GeV for m_H hypotheses smaller or equal than 250 GeV (greater than 250 GeV). Finally, $p_T^{\ell,\max}$ is required to be larger than 50 GeV for m_H hypotheses greater than 250 GeV. The two-dimensional distributions for the $m_H = 125$ GeV and $m_H = 200$ GeV Higgs signal hypotheses, the background processes and the data, are shown in Figure 1 for the 0-jet sample. This specific binning size is chosen to avoid empty bins for the overall background contribution for the present Monte Carlo statistics, while still keeping enough cells to differentiate the signal shape from the background shape. All the 80 bins enter in a binned likelihood fit of the data to the signal and background hypotheses in this two-dimensional shape analysis.

The 2-jet category is mainly sensitive to the vector boson fusion (VBF) production mode, whose cross section is roughly ten times smaller than that of the gluon-gluon fusion mode, but it is characterized by a relatively low background environment and it probes a different production mechanism to test the compatibility of a signal with the SM Higgs boson hypothesis. The $H \to W^+W^-$ events from VBF production are characterized by a pair of energetic forward-backward jets and very little hadronic activity in the rest of the event. Therefore, 2-jets events are further required to have zero hard jets and both the tag leptons present in the pseudorapidity region between the two leading jets. Additionally, to reject the main background from top-quark decays, two additional requirements are applied to the two jets, j_1 and j_2: $\Delta\eta(j_1, j_2) > 3.5$ and $m_{j_1j_2} > 500$ GeV. Finally, the cut-based approach in the 2-jet category makes use of the same m_H dependent requirements, with the exception that m_T is required to be larger than 30 GeV.

3.1. Background determination

To evaluate the background contamination in the signal region after the Higgs selection, a combination of techniques is used. When possible, data are used to estimate directly background contributions, thus avoiding large uncertainties related to the limited Monte Carlo simulation. The remaining contributions taken from simulation are small.

The W + jets and QCD multi-jet backgrounds are due to leptonic decays of heavy quarks, hadrons misidentified as leptons, and electrons from photon conversion. Hence, their estimation is derived directly from data using a control sample of events in which one lepton passes the standard analysis criteria and the other does not, but instead satisfies a relaxed set of requirements, resulting in a tight-fail sample. Firstly, the efficiency for a jet selected by the loose selection to pass the tight selection is measured using data from an independent multi-jet event sample dominated by non-prompt leptons. Secondly, this efficiency is used to extrapolate the tight-fail counts into the signal region (tight-tight sample). Systematic uncertainties related to the efficiency measurement dominate the overall uncertainty of this method, which is estimated to be about 36%.

Top-quark background is also estimated from data, by counting the number of top-tagged events and by extrapolating into the signal region using the corresponding top-tagging efficiency, which is measured on data with one b-tagged jet. This method is limited by the statistical uncertainty of the control sample and from the systematic uncertainties related to

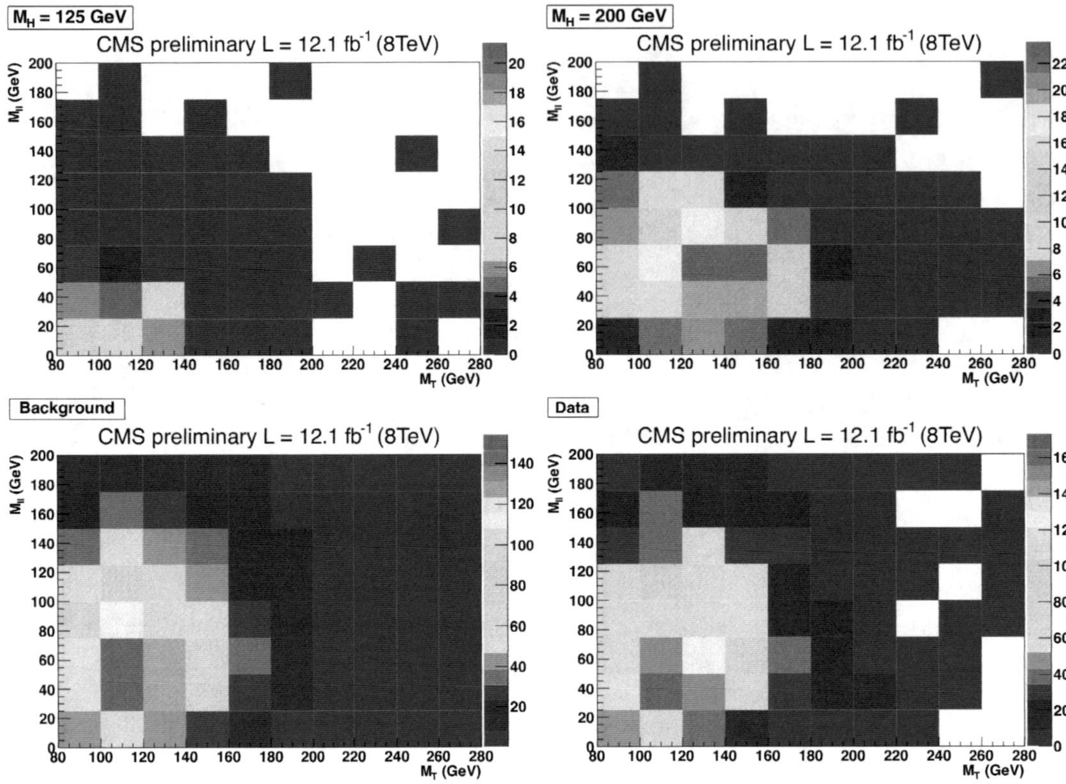

Figure 1. H $\rightarrow$ W$^+$W$^-$ $\rightarrow$ $\ell\nu_\ell\,\ell\nu_\ell$ search: ($m_{\rm T}$,$m_{\ell\ell}$) distributions in the 0-jet bin for the $m_{\rm H}$ = 125 GeV Higgs signal hypothesis (top left), the $m_{\rm H}$ = 200 GeV Higgs signal hypothesis (top right), the background processes (bottom left), and the data (bottom right).

the measurement of the tagging efficiency (about 20% in the 0-jet category and about 4% in the 1-jet category).

For the low-mass signal region, $m_{\rm H} \leq 200$ GeV, the non-resonant W$^+$W$^-$ contribution is estimated from data. Events with a dilepton mass larger than 100 GeV, where the Higgs boson signal contamination is negligible, are used to measure this contribution and the MC simulation is used to extrapolate into the signal region. The total uncertainty is about 10%. For larger Higgs boson masses the overlap between the non-resonant W$^+$W$^-$ and the Higgs boson signal becomes significant, and the sole simulation is used for the background estimation.

The Drell-Yan contribution to the same flavour final states is based on an extrapolation from the observed number of events with a dilepton mass within ± 7.5 GeV of the Z mass. The residual background in the control region is subtracted using e$^{\pm}\mu^{\mp}$ events. This method is limited by the statistical uncertainty of the control sample, which is about 20% to 50%. The Z/γ^* $\rightarrow$ $\tau\tau$ contamination is estimated using Z/γ^* $\rightarrow$ ee and Z/γ^* $\rightarrow$ $\mu\mu$ events selected in data, where the leptons are replaced with simulated τ decays, thus providing a better description of the experimental conditions with respect to the simulation.

The Wγ^* contamination, from asymmetric virtual photon decays where one lepton escapes detection, is determined using the MADGRAPH generator with dedicated cuts. The normalization scale of the simulated events is estimated through a control sample of high purity Wγ^* events with three reconstructed leptons. A measured factor of 1.6 ± 0.5 with respect to the leading order cross-section is found.

The remaining minor backgrounds, from WZ, ZZ (when the two selected leptons come from different bosons) and Wγ, are estimated from simulation.

For the two-dimensional analysis, the simulation is employed to build the templates, which are then cross-checked in data control samples. For the W + jets background the nominal shape is derived from the same control sample used to determine the normalization.

4. The $\mathrm{H} \to \mathrm{W}^+\mathrm{W}^- \to \ell\nu_\ell \mathrm{q}\overline{\mathrm{q}}$ search

The $\mathrm{W}^+\mathrm{W}^-$ semi-leptonic channel has two main advantages with respect to the fully-leptonic final state: it has a largest branching fraction and a reconstructible Higgs boson mass peak. On the other hand, it is contaminated by a large W + jets background. The precision on the estimation of this background largely determines the sensitivity of the analysis.

The reconstructed electrons (muons) are required to have $p_\mathrm{T} > 35(25)$ GeV, and are restricted to $|\eta| < 2.5(2.1)$. The jets are required to have $p_\mathrm{T} > 30\,\mathrm{GeV}$ and $|\eta| < 2.4$. Events with the electrons and muons, and with exactly two or three jets are analyzed separately, giving four categories in total. In addition, the electron(muon) categories are required to have $E_\mathrm{T}^\mathrm{miss} > 30(25)$ GeV.

The two highest-p_T jets are assumed to be the hadronic W decay products. According to the simulation, in the case of 2(3)-jet event categories, this criteria correctly identifies the jet pair at a rate which varies from 68(26)% for $m_\mathrm{H} = 200$ GeV to 88(84)% for $m_\mathrm{H} = 600$ GeV. Events with an incorrect dijet combination result in a broad non-peaking background in the m_WW spectrum.

The $(\ell, E_\mathrm{T}^\mathrm{miss})$ system is used to reconstruct the leptonic W candidate. To reduce the background from processes that do not contain $\mathrm{W} \to \ell\nu$ decays, requirements on $m_\mathrm{T}(\ell, E_\mathrm{T}^\mathrm{miss}) >$ 30 GeV and $|\Delta\phi_{\mathrm{leadingjet},\mathrm{E}_\mathrm{T}^\mathrm{miss}}| > 0.8$ (0.4) are imposed for electrons (muons). These selection criteria reduce the QCD multi-jet background, for which in many cases the $E_\mathrm{T}^\mathrm{miss}$ is generated by the mis-measurement of the hadronic recoil.

In addition to the listed cuts, a likelihood discriminant is built to enhance the signal sensitivity. The discriminant input variables comprise five angles between the Higgs decay products, fully describing the Higgs production kinematics [7], the p_T and rapidity of the $\mathrm{W}^+\mathrm{W}^-$ system, and the lepton charge. The likelihood discriminating power is optimized independently with dedicated simulation samples for several Higgs mass hypotheses, for each lepton flavor and for each jet multiplicity (2-jet, 3-jet): four different optimizations are therefore obtained per mass hypothesis. The figure of merit used for the optimization is the expected limit for the Higgs cross-section. For each of the event categories, events are retained if they survive a simple selection on the likelihood discriminant.

Finally, the m_WW resolution is improved by means of a kinematic fit, where both the W candidates are constrained to the W-boson mass to within its known width, with the longitudinal component of the neutrino momentum, $|p_\mathrm{z}|$, treated as a free parameter.

To estimate the normalizations of all background components in the signal region, an unbinned maximum likelihood fit is performed on the invariant mass distribution of the dijet system, m_{jj}. The fit is performed independently for each Higgs boson mass hypothesis. The signal region corresponding to the W mass window, 65 GeV $< m_{jj} <$ 95 GeV, is excluded from the fit.

The measured invariant mass of the $\mathrm{W}^+\mathrm{W}^-$ system, $\ell\nu_\ell jj$, is used as final discriminating variable. The binned distributions of $\ell\nu_\ell jj$ for the total background, signal and data for each mass hypothesis and event category are constructed and used as input for the limit-setting procedure. The $\ell\nu_\ell jj$ shape for the major background, W + jets, is determined from data as an interpolation of the shapes measured in two signal-free sideband regions of m_{jj} (55 GeV $< m_{jj} <$ 65 GeV, 95 GeV $< m_{jj} <$ 115 GeV). The relative fraction of the two sidebands into the signal region is found through simulation, separately for each Higgs boson mass hypothesis, by minimizing the χ^2 between the interpolated shape in the signal region and the expected one. In addition, to avoid statistical fluctuations due to the low event count, the interpolated distribution is fit with an exponential function. This function is then used for the limit-setting.

The m_{jj} and $m_{\ell\nu jj}$ distributions with final background estimates are shown in Figure 2, with selections optimized for a 500 GeV Higgs boson mass hypothesis, for the muon 2-jets category. All uncertainties arising from interpolation and fit procedures are propagated to the limit calculation as systematic uncertainties.

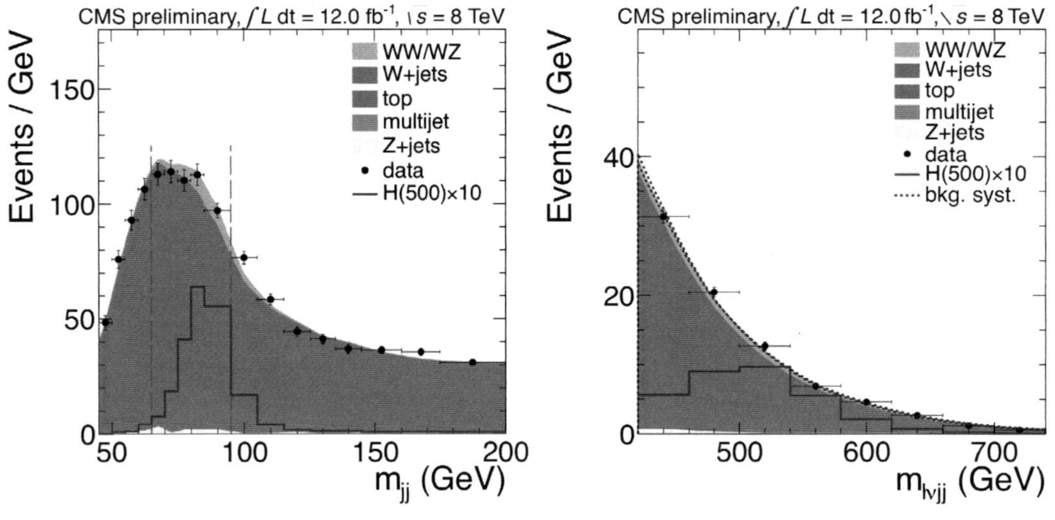

Figure 2. $\mathrm{H} \to \mathrm{W}^+\mathrm{W}^- \to \ell\nu_\ell \mathrm{q}\bar{\mathrm{q}}$ search: Invariant mass distributions for the $m_\mathrm{H} = 500$ GeV mass hypothesis, muon 2-jets category. On the left, the dijet invariant mass distribution with the major background contributions. The vertical lines corresponds to the signal region of this analysis 65 GeV $< m_{jj} <$ 95 GeV. On the right, the WW invariant mass distribution with the major background contributions in the signal region.

5. Results

An excess of data with respect to the total background prediction is observed in the $\mathrm{H} \to \mathrm{W}^+\mathrm{W}^- \to \ell\nu_\ell\ell\nu_\ell$ analysis for low Higgs mass selections. The significance of the excess with respect to the background only hypothesis is 3.1 standard deviations, while 4.1 standard deviations are expected assuming the SM prediction.

No other significant deviations are observed, and therefore upper limits on the Higgs boson production relative to the standard model Higgs expectation are derived. To compute the upper limits the modified frequentist construction CL_S [8] is used. The standard model Higgs boson is excluded in the mass range 128 – 600 GeV at 95% confidence level (CL), while the expected exclusion limit in the hypothesis of background only is 118 – 565 GeV.

With the $\mathrm{H} \to \mathrm{W}^+\mathrm{W}^- \to \ell\nu_\ell \mathrm{q}\bar{\mathrm{q}}$ analysis, the presence of a SM Higgs boson in the mass range 215 – 490 and 525 – 600 GeV is excluded at 95%, while the expected exclusion limit in the hypothesis of background only is 170 − 585 GeV.

Both the results are graphically summarized in Figure 3.

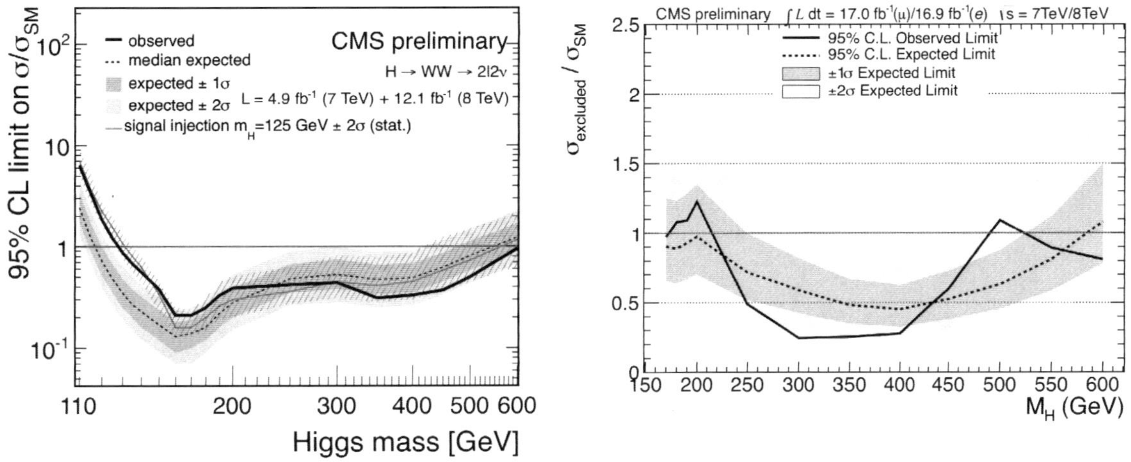

Figure 3. Expected and observed 95% CL upper limits on the cross section times branching fraction, relative to the SM Higgs expectation, for the $\mathrm{H} \to \mathrm{W}^+\mathrm{W}^- \to \ell\nu_\ell\ell\nu_\ell$ (left) and the $\mathrm{H} \to \mathrm{W}^+\mathrm{W}^- \to \ell\nu_\ell \mathrm{q}\overline{\mathrm{q}}$ (right) analyses. For the $\mathrm{H} \to \mathrm{W}^+\mathrm{W}^- \to \ell\nu_\ell\ell\nu_\ell$ search, the expected limits in the presence of the Higgs with $m_\mathrm{H} = 125$ GeV and its associated uncertainty are also shown.

[1] The CMS Collaboration 2012 CMS-PAS-HIG-12-042
[2] The CMS Collaboration 2012 CMS-PAS-HIG-12-046
[3] Chatrchyan S *et al.* (CMS) 2008 *JINST* **3** S08004
[4] The CMS Collaboration 2012 CMS-PAS-PFT-09-001
[5] Cacciari M, Salam G P and Soyez G 2008 *JHEP* **04** 063 (*Preprint* 0802.1189)
[6] Chatrchyan S *et al.* (CMS) 2011 *JHEP* **10** 132
[7] Gao Y *et al.* 2010 *Phys. Rev. D* **81** 075022 (*Preprint* 1001.3396)
[8] Read A L 2002 *Journal of Physics G: Nuclear and Particle Physics* **28** 2693 URL http://stacks.iop.org/0954-3899/28/i=10/a=313

Journal of Physics: Conference Series **455** (2013) 012031 doi:10.1088/1742-6596/455/1/012031

Studies of exotic quarkonium states at CMS

Alessandra Fanfani on behalf of CMS Collaboration

Dipartimento di Fisica ed Astonomia, Viale Berti/Pichat 6/2 , 40127 Bologna, Italy

E-mail: fanfani@bo.infn.it

Abstract.
Using large data samples of dimuon events in pp collisions at $\sqrt{s}$ = 7TeV, CMS has performed studies in the field of "exotic" quarkonium spectroscopy. The production of the X(3872) is studied measuring the ratio of the X(3872) and ψ(2S) cross sections times their branching fractions into $J/\psi\pi^+\pi^-$ as a function of transverse momentum (p_T). In addition, the fraction of X(3872) originating from B decays is determined. From these measurements the prompt X(3872) differential cross section times branching fraction as a function of p_T is extracted. The $\pi^+\pi^-$ mass spectrum of the $J/\psi\pi^+\pi^-$ system in the X(3872) decays is also investigated. Preliminary studies on structures in the $J/\psi\phi$ mass spectrum of the exclusive $B^+ \to J/\psi\phi K^+$ decays are reported.

1. Introduction

The observation of many new states, with masses above the open-charm threshold, that do not fit into the conventional quark model has renewed the interest in "exotic" quarkonium spectroscopy. There are several interpretations, such as molecular state , tetraquark state and hybrid mesons, however the theoretical picture is not clear and more experimental inputs are needed [1].

Using the pp collision data recorded by the Compact Muon Solenoid (CMS) experiment at the LHC in 2011, at a centre-of-mass energy of 7 TeV, the first targets of the "exotic" spectroscopy program in CMS have been the measurement of the X(3872) production and the search for structures in the $J/\psi\phi$ mass spectrum of the exclusive $B^+ \to J/\psi\phi K^+$ decay.

The main results from CMS are summarised as follows: after a brief description of the CMS detector in Section 2, measurements of the X(3872) production and its decay properties are presented in Section 3. In Section 4 the preliminary observation of structures in the $J/\psi\phi$ mass spectrum of the exclusive $B^+ \to J/\psi\phi K^+$ decay, is reported.

2. CMS detector

The central feature of the Compact Muon Solenoid (CMS) apparatus is a superconducting solenoid of 6 m internal diameter, providing a magnetic field of 3.8 T. Within the superconducting solenoid volume are a silicon pixel and strip tracker, a lead tungstate crystal electromagnetic calorimeter, and a brass/scintillator hadron calorimeter. Muons are measured in gas-ionization detectors embedded in the steel return yoke outside the solenoid. Extensive forward calorimetry complements the coverage provided by the barrel and endcap detectors. A two-level trigger system selects relevant pp collision events for offline reconstruction. The first level (L1) of the CMS trigger system is composed of custom hardware processors. The L1 trigger conditions are adjusted such as to limit the trigger rate to less than 100 kHz. The high-level trigger (HLT) runs on a processor farm to further reduce the rate to a few 100 Hz before data storage. A more detailed description can be found in Ref. [2].

3. X(3872) production cross-section measurements

The X(3872) was discovered by the Belle experiment in 2003 [3]. Despite a series of detailed studies at B-factories and Tevatron performed, the nature of the X(3872) still remains unknown. Quantitative predictions have been calculated for the differential production cross section of the X(3872) in $\mathrm{p\bar{p}}$ collisions at the Tevatron and pp collisions at the Large Hadron Collider (LHC) [4]. Measurement of the prompt production rate at the LHC as a function of transverse momentum provides a test of the NRQCD factorization approach to X(3872) production.

In this section the production of the X(3872) is studied in pp collisions at $\sqrt{s} = 7$ TeV, using the data recorded by the CMS experiment corresponding to an integrated luminosity of 4.8 fb^{-1}. The event selection is discussed in Section 3.1. The cross section measurement proceeds by determining the ratio of the X(3872) and ψ(2S) cross sections, where both states decay to $\mathrm{J}/\psi\pi^+\pi^-$, as described in Section 3.2. The fraction of nonpromptly produced X(3872) states is also reported in Section 3.3. In Section 3.4 the cross section for prompt X(3872) production is presented. Finally, Section 3.5 gives the $\pi^+\pi^-$ mass spectrum in X(3872) to $\mathrm{J}/\psi\pi^+\pi^-$ decays.

3.1. Event selection

The X(3872) is observed using the decays into $\mathrm{J}/\psi\pi^+\pi^-$, with the subsequent decay of the J/ψ into a pair of muons. The data are collected with requirements on the dimuon system imposed at the trigger level (HLT) such as rapidity within 1.25 and transverse momentum threshold initially 6.9 GeV, which was increased to 9.9 GeV near the end of data taking. The $\mathrm{J}/\psi\pi^+\pi^-$ system is reconstructed by combining the candidate muon tracks from each candidate J/ψ with pairs of oppositely charged tracks, which are assumed to be pions. Each $\mu^+\mu^-\pi^+\pi^-$ combination is refitted, constraining the four tracks to come from a common vertex and the muon-pair invariant mass to the J/ψ mass [5]. The event selection is described in detail in Ref. [6]. The analysis is performed in the kinematic range of p_T of the $\mathrm{J}/\psi\pi^+\pi^-$ system between 10 and 50 GeV and the rapidity within $|y| < 1.2$. Detailed event simulations are used to determine acceptances and efficiencies. Events containing X(3872) or ψ(2S) states are generated using PYTHIA [7], and decayed using EvtGen [8], with the signal resonances forced to decay into the $\mathrm{J}/\psi\pi^+\pi^-$ final state, as described in Ref. [6].

3.2. Measurement of the cross section ratio

The ratio of the cross section times the $\mathrm{J}/\psi\pi^+\pi^-$ branching fraction is obtained from the measured numbers of signal events for X(3872) and ψ(2S), $N_{\mathrm{X(3872)}}$ and $N_{\psi(\mathrm{2S})}$, correcting for the efficiency (ϵ) and acceptance (A) estimated from simulations, according to

$$R = \frac{\sigma(\mathrm{pp} \to \mathrm{X(3872)} + \mathrm{anything}) \cdot \mathcal{B}(\mathrm{X(3872)} \to \mathrm{J}/\psi\pi^+\pi^-)}{\sigma(\mathrm{pp} \to \psi(\mathrm{2S}) + \mathrm{anything}) \cdot \mathcal{B}(\psi(\mathrm{2S}) \to \mathrm{J}/\psi\pi^+\pi^-)} = \frac{N_{\mathrm{X(3872)}} \cdot A_{\psi(\mathrm{2S})} \cdot \epsilon_{\psi(\mathrm{2S})}}{N_{\psi(\mathrm{2S})} \cdot A_{\mathrm{X(3872)}} \cdot \epsilon_{\mathrm{X(3872)}}}. \quad (1)$$

The signal yields, $N_{\mathrm{X(3872)}}$ and $N_{\psi(\mathrm{2S})}$, are determined from unbinned maximum-likelihood fits to the invariant-mass spectra of the $\mathrm{J}/\psi\pi^+\pi^-$ system, separately for the X(3872) and ψ(2S), in the mass windows 3.75–4GeV and 3.6–3.8GeV, respectively. In the fits, the ψ(2S) resonance shape is parametrized using two Gaussian functions with a common mean, while a single Gaussian is used for the X(3872) signal. The nonresonant background is fitted with a second-order Chebyshev polynomial. The acceptances and efficiencies of the X(3872) and ψ(2S) final states are determined from the simulation. Studies are performed to verify the description of the data by the simulations and to determine the systematic uncertainties. A total systematic uncertainty of 5-13% on R is obtained, as described in Ref. [6]. The main systematic uncertainty is the dependence of the measurement on the p_T spectrum of the X(3872). The rapidly changing acceptance as a function of transverse momentum makes the R measurement very sensitive to the p_T spectrum, in particular for low transverse momentum and for the p_T-integrated result. The acceptance determination depends also on X(3872) quantum numbers and polarization. The results are obtained under the assumption that the X(3872) quantum numbers are $J^{PC} = 1^{++}$, as favoured by existing data [1], and that both X(3872) and ψ(2S) states are unpolarized. No systematic uncertainty is assigned to cover other cases.

The cross section ratio is determined in the rapidity region $|y| < 1.2$ as a function of the transverse momentum of the $J/\psi\pi^+\pi^-$ system as shown in Fig. 1(left). The results are also listed in Table 1. No significant dependence on transverse momentum is observed.

3.3. Measurement of the nonprompt fraction

The relative contribution to the total X(3872) yield resulting from decays of B hadrons, often referred to as the nonprompt fraction, is determined from the decay lifetime distribution. The "pseudo-proper" decay length ℓ_{xy} is defined in the plane transverse to the beam direction as the distance between the vertex formed by the four tracks of the $J/\psi\pi^+\pi^-$ system and the closest reconstructed primary vertex along the beam direction, corrected by the transverse Lorentz boost of the $J/\psi\pi^+\pi^-$ candidate. An event sample enriched in X(3872) candidates from B decays is selected by requiring that ℓ_{xy} be larger than 100 μm. The nonprompt fraction is then obtained from the ratio between the signal yields in this B-hadron-enriched sample and the signal yields in the inclusive sample, after correction for the efficiencies of the decay-length-selection criteria, as determined from simulations of prompt and nonprompt X(3872) states. The signal yields are extracted from fits to the $J/\psi\pi^+\pi^-$ invariant-mass spectrum, as described in Section 3.2. In the fits to the B-hadron-enriched sample, the fit parameters for the mass and width are fixed to those determined from the full sample.

Detailed studies are performed to verify the stability of the results and to determine the systematic uncertainties, as described in Ref. [6]. The measurement is dominated by its statistical uncertainty. A total systematic uncertainty of 6-10% is obtained.

The results are listed in Table 1 and shown in Fig. 1(right) as a function of p_T. The X(3872) nonprompt fraction reveals no significant dependence on transverse momentum and the integrated value is significantly smaller than that for the ψ(2S) [9]. The results are obtained ignoring effects related to the X(3872) polarization and no systematic uncertainty is assigned for polarization effects.

Table 1. The ratio R between the X(3872) and $\psi(2S)$ cross sections times branching fraction and the X(3872) nonprompt fractions, as a function of p_T, for the rapidity range $|y| < 1.2$. The uncertainties shown are statistical and systematic, respectively.

p_T(GeV)	R	X(3872) nonprompt fraction
10–13.5	$0.0727 \pm 0.0079 \pm 0.0097$	$0.272 \pm 0.057 \pm 0.016$
13.5–15	$0.0671 \pm 0.0072 \pm 0.0044$	$0.182 \pm 0.052 \pm 0.013$
15–18	$0.0687 \pm 0.0055 \pm 0.0051$	$0.246 \pm 0.043 \pm 0.015$
18–30	$0.0601 \pm 0.0042 \pm 0.0042$	$0.297 \pm 0.042 \pm 0.021$
30–50	$0.078 \pm 0.013 \pm 0.004$	$0.301 \pm 0.097 \pm 0.030$
10–50	$0.0656 \pm 0.0029 \pm 0.0065$	$0.263 \pm 0.023 \pm 0.016$

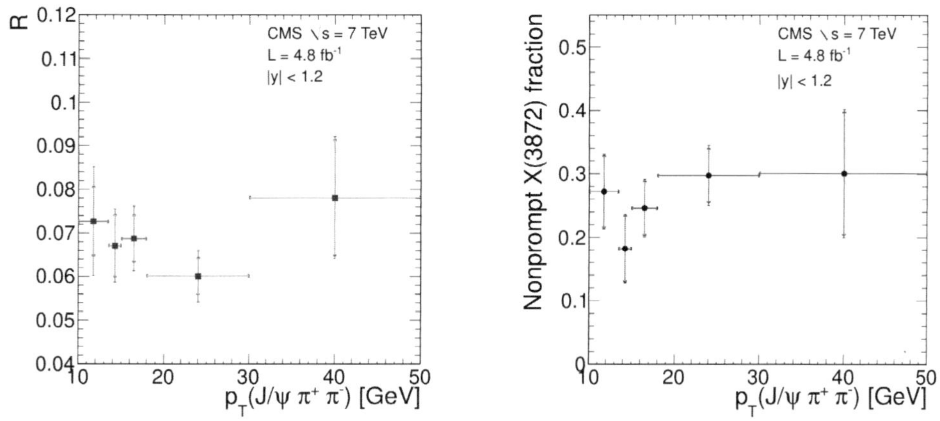

Figure 1. Ratio between the X(3872) and $\psi(2S)$ cross sections times branching fraction (left) and measured X(3872) nonprompt fraction (right), as a function of p_T. The inner error bars indicate the statistical uncertainty and the outer error bars represent the total uncertainty. The data points are placed at the centre of each p_T bin.

3.4. Determination of the prompt X(3872) production cross section

The cross section times branching fraction for prompt X(3872) production is determined from the measurement of the cross section ratio R and the nonprompt fraction, described above in Section 3.2 and 3.3, respectively, combined with a previous result of the prompt ψ(2S) cross section [9]. The latter measurement was performed using the $\psi(2S) \rightarrow \mu^+\mu^-$ decay mode and provides results as a function of transverse momentum up to 30 GeV and for the rapidity range $|y| < 1.2$. In the calculation, the branching fraction $\mathcal{B}(\psi(2S) \rightarrow J/\psi\pi^+\pi^-)$ is taken from Ref. [5], and $\mathcal{B}(\psi(2S) \rightarrow \mu^+\mu^-)$ is taken to be equal to the more precisely known $\mathcal{B}(\psi(2S) \rightarrow e^+e^-)$ [5]. No cancellation of systematic uncertainties is assumed in the combination. The main sources of systematic uncertainty are related to the measurement of the ratio R and the background lifetime fit in the measurement of the prompt ψ(2S) cross section [9]. The differential cross section for prompt X(3872) production times the branching fraction as a function of p_T, in the rapidity region $|y| < 1.2$, is listed in Table 2.

A calculation of the predicted differential cross section for prompt X(3872) production in pp collisions at $\sqrt{s} = 7$ TeV has been made using the NRQCD factorization formalism, assuming the X(3872) is formed from a $c\bar{c}$ pair with negligible relative momentum [4]. This calculation is normalized using Tevatron measurements [10] with the statistical uncertainty obtained from the experimental input data. The predictions from Ref. [4] were modified by the authors to match the phase-space of the measurement presented in this note. Comparisons of this prediction with the data, in Fig. 2, demonstrates that, while the shape is reasonably well described, the predicted cross section is much larger than observed in data.

The integrated prompt X(3872) cross section times branching fraction for the kinematic region $10 < p_T < 30$ GeV and $|y| < 1.2$ is also determined to be

$$\sigma^{\text{prompt}}(\text{pp} \rightarrow \text{X(3872)+anything}) \cdot \mathcal{B}(\text{X(3872)} \rightarrow J/\psi\pi^+\pi^-) = 1.06 \pm 0.11(\text{stat.}) \pm 0.15(\text{syst.}) \text{ nb.} \quad (2)$$

This result assumes that the X(3872) and ψ(2S) states are unpolarized. The NRQCD prediction for the prompt X(3872) cross section times branching fraction in the kinematic region of this analysis is 4.01 ± 0.88 nb [4], significantly above the measured value.

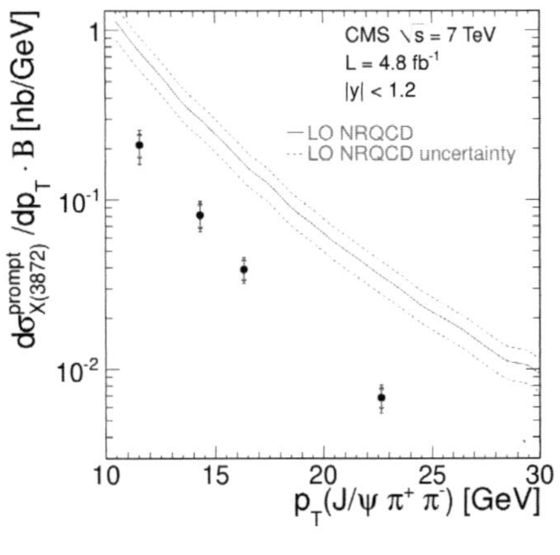

Figure 2. Measured differential cross section for prompt X(3872) production times branching fraction $X(3872) \to J/\psi\pi^+\pi^-$ (B) as a function of p_T. The inner error bars indicate the statistical uncertainty and the outer error bars represent the total uncertainty. The data points are placed where the value of the theoretical prediction is equal to its mean value over each bin, according to the prescription in Ref. [11]

Table 2. Prompt X(3872) differential cross section times branching fraction $\mathcal{B}(\mathrm{X}(3872) \to \mathrm{J}/\psi\pi^+\pi^-)$ as a function of p_T. The uncertainties shown are statistical and systematic, respectively.

p_T(GeV)	$d\sigma^{\mathrm{prompt}}_{\mathrm{X}(3872)}/dp_T \cdot \mathcal{B}(\mathrm{X}(3872) \to \mathrm{J}/\psi\pi^+\pi^-)$ (nb/GeV)
10–13.5	$0.211 \pm 0.034 \pm 0.035$
13.5–15	$0.081 \pm 0.013 \pm 0.010$
15–18	$0.0390 \pm 0.0054 \pm 0.0042$
18–30	$0.0068 \pm 0.0009 \pm 0.0009$

3.5. Measurement of the $\pi^+\pi^-$ invariant-mass distribution

The decay properties of the X(3872) are further investigated with a measurement of the $\pi^+\pi^-$ invariant-mass distribution from X(3872) decays to $\mathrm{J}/\psi\pi^+\pi^-$, within the kinematic range $10 < p_T < 50$ GeV and $|y| < 1.25$. To extract the dipion invariant-mass spectrum from X(3872) decays, the event sample is divided into twelve intervals of dipion invariant mass in the range $0.5 < m(\pi^+\pi^-) < 0.78$ GeV. In each interval, a maximum-likelihood fit to the $\mathrm{J}/\psi\pi^+\pi^-$ invariant-mass distribution is performed, where the signal is modelled with a single Gaussian. The position and width of the X(3872) signal are fixed to the values obtained in the fit to the full sample, except for the last interval, 0.765–0.78 GeV, where the mean and width of the Gaussian are left free to accommodate possible distortions of the signal shape near the upper kinematic limit. The X(3872) dipion invariant-mass distribution is extracted from the signal yields obtained from the fits to the data in each interval, after correction for detector acceptance and efficiencies, as estimated from the simulation. The systematic uncertainties on the signal extraction and on correction due to acceptance and efficiency are estimated, yielding to variations of 10-20% and 4-6%, respectively.

The resulting dipion invariant-mass spectrum, normalized to the total cross section in the interval $0.5 < m(\pi^+\pi^-) < 0.78$GeV, is presented in Fig. 3. The data are compared to X(3872) signal simulations with and without an intermediate ρ^0 in the $\mathrm{J}/\psi\pi^+\pi^-$ decay. The assumption of an intermediate ρ^0 decay gives better agreement with the data, confirming previous measurements [12, 13].

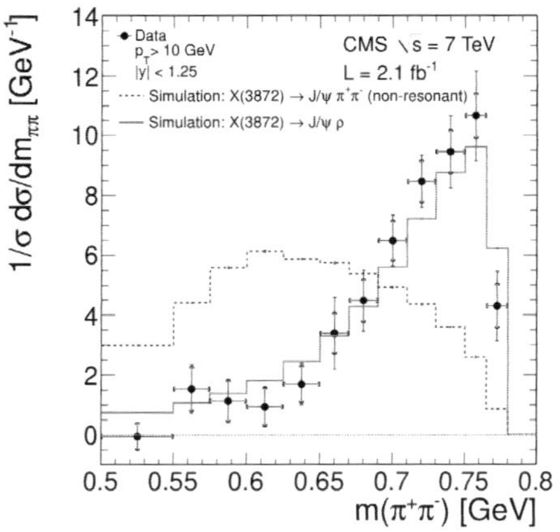

Figure 3. Dipion invariant-mass spectrum for $X(3872) \to J/\psi\pi^+\pi^-$ decays corrected for acceptance and efficiency. The distribution is normalized to unity by dividing by the total cross section for $0.5 < m(\pi^+\pi^-) < 0.78$ GeV. The inner error bars indicate the statistical uncertainty and the outer error bars represent the total uncertainty. The results are compared to results from simulation with (solid line) and without (dotted line) an intermediate ρ^0 decay .

4. Structures in the $J/\psi\phi$ mass spectrum of the exclusive $B^+ \to J/\psi\phi K^+$ decay

The CDF collaboration has reported an evidence for a narrow structure, called Y(4140), near the $J/\psi\phi$ threshold in the exclusive $B^+ \to J/\psi\phi K^+$ decay [14]. Recently, the LHCb collaboration reported a result in conflict with that of CDF using the same exclusive $B^+ \to J/\psi\phi K^+$ decays [15].

In CMS the study of the $J/\psi\phi$ mass spectrum from the $B^\pm \to J/\psi\phi K^\pm$ decays (when a decay mode is referenced, its charge conjugate is implied) is performed using an integrated luminosity of 5.2 fb^{-1}. The $\mu^+\mu^-$ channel is used to reconstruct J/ψ candidates that are selected by the HLT dimuon trigger. $B^\pm \to J/\psi\phi K^\pm$ candidates are reconstructed by combining three additional charged tracks pointing to the displaced J/ψ vertex with a total charge of ± 1 and with kaon mass assigned (kaon track). The p_T of the kaon tracks is required to be greater than 1 GeV. The five tracks, with $\mu^+\mu^-$ mass constrained to the J/ψ nominal value, are required to form a good 3D vertex with χ^2 probability greater than 1%. The reconstructed mass of the K^+K^- pair with the lower mass must satisfy $1.008 < m(K^+K^-) < 1.035$ GeV to be considered as a ϕ candidate. These selection requirements, which were designed to maximize the signal yield, were determined before the $J/\psi\phi$ mass spectrum was examined.

In order to search for possible structures in the $J/\psi\phi$ mass spectrum, the $J/\psi\phi K^+$ candidates are divided into bins of the mass difference, $\Delta M \equiv m(\mu^+\mu^-K^+K^-) - m(\mu^+\mu^-)$, and the $J/\psi\phi K^+$ mass distributions are fit to extract the B signal yield for each ΔM bin. A polynomial function is used for the combinatorial background and a double Gaussian for the B signal. The mean values of the two Gaussians are fixed to the nominal B mass, and the width values of the Gaussians, as well as their relative ratio are fixed to the values obtained from simulation.

The ΔM distribution is corrected for relative efficiency, estimated from simulation. The relative-efficiency-corrected ΔM distribution from the exclusive $B^+ \to J/\psi\phi K^+$ signal is shown in Fig. 4. The nonresonant decay of a real or quasi-particle state into three objects ($B^+ \to J/\psi\phi K^+$) can be described by a three-body phase space model. The ΔM continuum shape is also investigated with an event-mixing technique by applying the same kinematic constraint to J/ψ, ϕ, and K^+ each chosen from a different event. The three-body phase space background lies above the event-mixed background in all ΔM bins in the range of interest, as shown in Fig. 4(left). There are enhancements, relative to the phase-space continuum, at the $J/\psi\phi$ mass threshold and around $\Delta M = 1.2$ GeV. The observed structures, as shown in Fig. 4 (right), are modelled with S-wave relativistic Breit-Wigner functions convoluted with a Gaussian resolution function whose width is fixed to values obtained from simulation.

Various checks have been done to validate the robustness of the two structures. Each selection

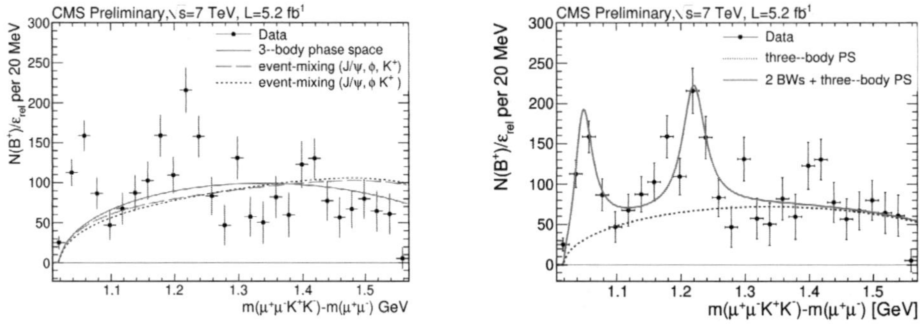

Figure 4. The relative-efficiency-corrected ΔM distribution from the exclusive $B^+ \to J/\psi\phi K^+$ signal. The background-only hypothesis fit are shown (left) with three different background shapes (solid, dashed, dotted). The binned χ^2 fit of the data with two S-wave relativistic Breit-Wigner functions convoluted with a Gaussian resolution function is shown (right) overlayed to the background-only fit contribution (dashed).

requirement has been varied. The analysis was repeated with tighter selection criteria that lowered the combinatorial background level by a factor of ten and retained 40% of the B signal events. The relative efficiencies for the first five bins are varied to confirm the robustness of the significance of the first structure. The *sPlot* formalism [16] projected to the ΔM distribution from the fit to the $J/\psi\phi K^+$ mass distribution is used to compare to the ΔM distribution obtained from previous fits to each ΔM bin. This is a background-subtraction technique by weighting each event based on the observed signal to background ratio. From these checks no indication of a possible bias is found.

Interpreting the two structures as $J/\psi\phi$ resonances with S-wave relativistic Breit-Wigner lineshapes over a three-body phase-space nonresonant component, a significance exceeding 5-sigma for the first structure is obtained. The fitted mass of the first structure is $\mathrm{m} = 4148.2 \pm 2.0(\mathrm{stat.}) \pm 4.6(\mathrm{sys.})$ MeV. This observation is consistent with a previous evidence for a narrow structure near $J/\psi\phi$ threshold by the CDF Collaboration [14]. There is evidence for a second structure with fitted mass $\mathrm{m} = 4316.7 \pm 3.0(\mathrm{stat.}) \pm 7.3(\mathrm{sys.})$ MeV.

5. Summary

The studies in the "exotic" quarkonium sector performed using large data samples of dimuon events in pp collisions at $\sqrt{s} = 7$TeV by the CMS experiment are summarised in this note.

The production of the X(3872) is studied measuring the ratio of the X(3872) and ψ(2S) cross sections times their branching fractions into $J/\psi\pi^+\pi^-$ as a function of transverse momentum (p_T). The fraction of X(3872) originating from B-hadron decays is $0.263 \pm 0.023(\mathrm{stat.}) \pm 0.016(\mathrm{sys.})$. No significant dependence on p_T is found. From these measurements, the cross section for prompt X(3872) production times branching fraction into $J/\psi\pi^+\pi^-$ has been extracted for the first time as a function of p_T. A value of $\sigma^{\mathrm{prompt}}(\mathrm{pp} \to \mathrm{X}(3872) + \mathrm{anything}) \cdot \mathcal{B}(\mathrm{X}(3872) \to J/\psi\pi^+\pi^-) = 1.06 \pm 0.11(\mathrm{stat.}) \pm 0.15(\mathrm{sys.})$ nb is found for the kinematic range $10 < p_T < 30$ GeV and $|y| < 1.2$. This result is made under the assumption that the X(3872) and ψ(2S) states are unpolarized. The NRQCD predictions for prompt X(3872) production at the LHC significantly exceed the measured value, while the p_T dependence is reasonably well described. The measured dipion mass spectrum for $\mathrm{X}(3872) \to J/\psi\pi^+\pi^-$ clearly favours the presence of an intermediate ρ^0 state.

Using the large $B^+ \to J/\psi\phi K^+$ sample collected by CMS preliminary results on observation of structures in the J/ψ mas spectrum are reported. A structure near $J/\psi\phi$ threshold is observed, consistent with a previous evidence by the CDF Collaboration, and there is evidence for a second structure around 4317 MeV.

Acknowledgments
We would like to thank Pierre Artoisenet and Eric Braaten for modifying their theoretical predictions [4] to match the phase-space of our measurement.

References

[1] Brambilla N *et al* 2011 Heavy quarkonium: progress, puzzles, and opportunities *Eur. Phys. J.* C **71** 1534
[2] CMS Collaboration 2008 The CMS experiment at the CERN LHC *JINST* **3** S08004
[3] Belle Collaboration 2003 Observation of a new narrow charmonium state in exclusive $B^{\pm} \rightarrow K^{\pm}\pi^{+}\pi^{-}\mathrm{J}/\psi$ decays *Phys. Rev. Lett.* **91** 262001
[4] Artoisenet P and Braaten E 2010 Production of the X(3872) at the Tevatron and the LHC *Phys. Rev.* D **81** 114018
[5] Particle Data Group 2012 Review of Particle Physics *Phys. Rev.* B **86** 010001
[6] CMS Collaboration 2013 Measurement of the X(3872) production cross section via decays to $\mathrm{J}/\psi\pi^{+}\pi^{-}$ in pp collisions at $\sqrt{s} = 7$ TeV *Preprint* arXiv:1302.3968
[7] Sjöstrand T, Mrenna S and Skands P Z 2006 PYTHIA 6.4 physics and manual *JHEP* **05** 026
[8] Lange D J 2001 The EvtGen particle decay simulation package *Nucl. Instrum. Meth.* A **462** 152
[9] CMS Collaboration 2012 J/ψ and $\psi(2S)$ production in pp collisions at $\sqrt{s} = 7$ TeV *JHEP* **02** 011
[10] CDF Collaboration 2005 The X(3872) at CDF II *Int. J. Mod. Phys.* A **20** 3765
[11] Lafferty G D and Wyatt T R 1995 *Nucl. Instrum. Meth.* A **355** 541
[12] CDF Collaboration 2006 measurement of the dipion mass spectrum in $\mathrm{X}(3872) \rightarrow \mathrm{J}/\psi\pi^{+}\pi^{-}$ decays *Phys. Rev. Lett.* **96**102002
[13] Belle Collaboration 2011 Bounds on the width, mass difference and other properties of X(3872) decays *Phys. Rev.* D **84** 052004
[14] CDF Collaboration 2009 Evidence for a narrow near-threshold structure in the $\mathrm{J}/\psi\phi$ mass spectrum in $B^{+} \rightarrow J/\psi\phi K^{+}$ decays *Phys. Rev. Lett.* **102** 242002
[15] LHCb Collaboration 2012 Search for the $X(4140)$ state in $B^{+} \rightarrow J/\psi\phi K^{+}$ decays *Phys.Rev.* D **85** 091103
[16] Pivk M and LeDiberder F R 2005 SPlot: A statistical tool to unfold data distributions *Nucl. Instrum. Meth.* A **555** 356

International Workshop on Discovery Physics at the LHC (Kruger2012) IOP Publishing
Journal of Physics: Conference Series **455** (2013) 012032 doi:10.1088/1742-6596/455/1/012032

Latest results on the search for $D^0 \rightarrow \mu\mu$ and $B^0_{s.d} \rightarrow \mu\mu$ decays from CMS

Fabrizio Palla[1]

INFN Sezione di Pisa, Largo Bruno Pontecorvo 3, 56126 Pisa (Italy)

E-mail: Fabrizio.Palla@cern.ch

Abstract.
The rare decays of $D^0 \rightarrow \mu\mu$, $B^0 \rightarrow \mu\mu$ and $B^0_s \rightarrow \mu\mu$ are sensitive to physics beyond the Standard Model as their branching fraction can be highly enhanced or suppressed in new physics models. The CMS experiment at the LHC has reported the studies of these rare decays in pp collisions at $\sqrt(s) = 7$ TeV using datasets collected in 2010 and 2011. In those searches, the number of observed events is consistent with the expectation from background plus Standard Model signal prediction. Results of the combination of the exclusion limits are presented.

Decays mediated by flavour-changing neutral currents (FCNC) are highly suppressed in the Standard Model (SM), and are sensitive to New Physics (NP) through higher order loop diagrams contributions. Charm and bottom mesons decays into muon pairs are in addition helicity suppressed, and the measurement of their decays are traditionally investigated to search for NP effects, which could enhance the small branching ratios (BR) of these decays.

We will present the status of the searches for the rare decays of $D^0 \rightarrow \mu\mu$, $B^0 \rightarrow \mu\mu$ and $B^0_s \rightarrow \mu\mu$ in CMS. In Section 1 we review the status of the searches for D^0 decays, and in Section 2 the one of the B^0 and B^0_s rare decays.

1. Searches for the rare decays $D^0 \rightarrow \mu\mu$

The rare decay $D^0 \rightarrow \mu^+\mu^-$ is highly suppressed by the GIM mechanism and by helicity. In the SM it proceeds via a W box diagram, also contributing to the $D^0 - \bar{D}^0$ mixing. Theoretical estimates of this branching ratio are approximately 10^{-18} from short range processes, increasing up to 10^{-13} when long-distance processes are included [1]. NP can enhance these estimates by several orders of magnitude [2].

CMS has searched for this decay in the data of pp collisions at the LHC, collected in 2011 at a centre-of-mass energy of 7 TeV, corresponding to an integrated luminosity of about 90 pb^{-1}[3]. The strategy of the analysis relies upon the determination of the ratio of the BR of the $D^{*+} \rightarrow D^0(\mu\mu)\pi^+$ over the BR of $D^{*+} \rightarrow D^0(K^-\mu^+\nu)\pi^+$ (otherwise explicited, charge conjugation is implied throughout the paper). In this way most of the systematic uncertainties cancel out. CMS has an efficient trigger for dimuon events, while for single-muons the thresholds on the muon transverse momentum (p_T) have been steadily increased with the luminosity, to cope with the constraints given by the allowed bandwidth. For this reason, the main challenge of the analysis is the reconstruction of the normalization decay, which has been affordable only

[1] On behalf of the CMS Collaboration

for the 2010 and 2011 data taking period, when the instantaneous luminosity did not exceeded about 4×10^{33} cm^{-2} s^{-1}. The trigger selection is the same for both decay modes of the D^0. It requires a high quality muon to be reconstructed in the muon chambers at the first level trigger (L1), and confirmed at the High Level Trigger (HLT) where the silicon tracker information is used. As the LHC instantaneous luminosity increased the thresholds increased as well, ranging from $p_T > 3$ GeV/c up to 15 GeV/c. Seven different taking periods were used: six in 2010 and one in 2011, the latter corresponding to an integrated luminosity of 54 pb^{-1}.

The semileptonic and the leptonic decays show the same final topology, with a displaced decay vertex with two charged tracks and a slow pion coming from the D^* decay occurring at the primary vertex. The primary vertex of the event is identified by the one with the largest sum of the tracks' p_T. Events without a primary vertex or events in which the selected primary vertex has a χ^2 confidence level (CL) less than 1% are discarded.

The $D^0 \to \mu^+\mu^-$ is reconstructed by pairing opposite sign muon candidates, reconstructed in the muon stations and in the silicon tracker, within the pseudorapidity range $|\eta| < 2.1$. One of the two muons is requested to "match" with the one that triggered the event (hence with a minimum p_T which depends on the data taking period), while the p_T of the other muon is required to be larger than 3 GeV/c. The two muon tracks are fit to a common vertex, and the pair retained if the vertex CL is larger than 1% and the primary vertex is recomputed excluding those muon tracks. The angle α between the momentum of the muon pair and the direction joining the primary and secondary vertices should satisfy $\cos\alpha > 0.99$. Finally the D^0 candidate is combined with a track (with $p_T > 0.6$ GeV/c) which is given the pion mass, and which must be compatible with originating from the primary vertex. Combinations with $\Delta M = M(\mu^+\mu^-\pi^+) - M(\mu^+\mu^-)$ smaller than 180 MeV/c^2 are retained as good D^{*+} candidates; if more than one candidate is found, the one with the ΔM closest to the nominal PDG mass difference is chosen.

The normalization channel, $D^{*+} \to D^0(K^-\mu^+\nu)\pi^+$, is reconstructed using a technique developed by E691 [4]. In that technique the neutrino momentum is reconstructed by imposing that the D^0 momentum is pointing to the primary vertex. The $K^-\mu^+\nu$ candidate is then combined with a π^+ to find the D^{*+} as done before. The muon candidate must satisfy the same cuts as the trigger muon of the $D^0 \to \mu^+\mu^-$ analysis, and the kaon candidate should have a $p_T > 0.8$ GeV/c and $|\eta| < 2.1$. The kaon and the muon are requested to come from the same vertex with a CL$>$ 1% and the neutrino momentum computed using the E691 technique. Finally the D^{*+} candidate is formed in a similar fashion as before. In order to evaluate the level of the background, also wrong sign (WS) combinations between the $K^-\mu^+\nu$ and the soft pion from the primary vertex (in which a pion with a charge opposite to the muon track) has been chosen, to be distinguished from the Right Sign (RS) combinations where the soft pion and muon tracks have identical charge.

In order to further reduce the backgrounds, both for the signal and normalization modes, the secondary vertex decay length is required to be larger than 3 times its computed uncertainty. The ΔM distribution of the WS and RS combinations is shown in Fig. 1 (Left). No evidence of peak is found for the WS and the shape of its distribution is taken to model the shape of the combinatorial backgrounds in the RS distribution. The number of $D^0 \to K^-\mu^+\nu$ candidates is extracted by a fit to the ΔM distribution of the RS combinations, in which the signal is described by two Gaussian functions. The fit returns 16458 ± 204 $D^0 \to K^-\mu^+\nu$ candidates. In order to select the $D^0 \to \mu^+\mu^-$ decay mode an additional cut is imposed: $|\Delta M - \Delta M_{PDG}| < 3$ MeV/c^2, where ΔM_{PDG} is the nominal ΔM mass difference between the D^{*+} and the D^0. The cut at 3 MeV/c^2 corresponds to about 3.6 times the expected Monte Carlo resolution. Figure **??**(Right) shows the dimuon invariant mass distribution after all cuts. The events in the signal region (shaded in red) are compatible with the background expectations, as evaluated by interpolating the sidebands (shaded in cyan).

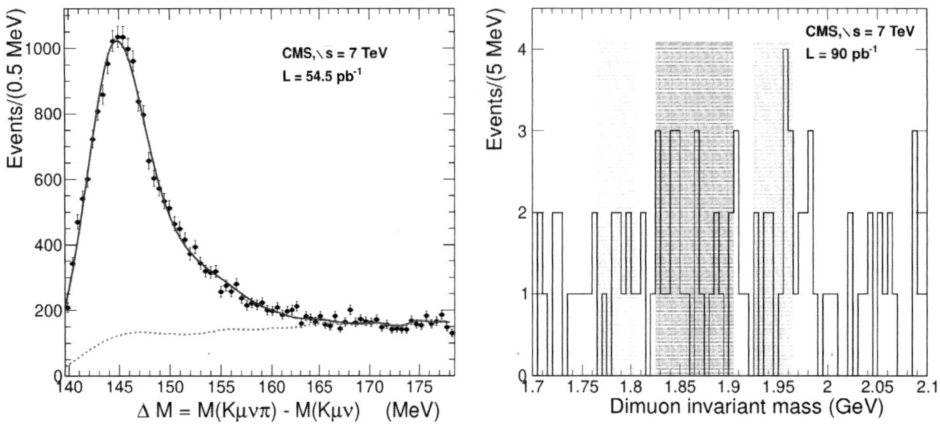

Figure 1. Left: Distribution of $\Delta M = M(K^-\mu^+\nu\pi^+) - M(K^-\mu^+\nu)$ for the 2011 RS sample with the result of the fit superimposed. The points are the data, the blue solid line shows the fit result. The background function modeled from the WS distribution is shown as the dashed red line. Right: The $\mu^+\mu^-$ invariant-mass distribution. The light-red band shows the signal region, while the two cyan bands show the sideband regions

An upper limit on the branching ratio $D^0 \to \mu^+\mu^-$ is determined using the following formula:

$$\mathcal{B}(D^0 \to \mu^+\mu^-) = \mathcal{B}(D^0 \to K^-\mu^+\nu) \times \frac{N_{\mu\mu}}{N_{K\mu\nu}} \times \frac{A_{\mu\mu}}{A_{K\mu\nu}} \times \frac{\epsilon_{\mu\mu}}{\epsilon_{K\mu\nu}} \quad (1)$$

where $\mathcal{B}(D^0 \to \mu^+\mu^-) = (3.30 \pm 0.13) \times 10^{-2}$ is the normalization branching ratio, taken from the PDG[5], $N_{\mu\mu}$ is the 90% CL upper limit on the $D^0 \to \mu^+\mu^-$ yield, $N_{K\mu\nu}$ the number of $D^0 \to K^-\mu^+\nu$ candidates, A and ϵ the acceptances and the efficiencies of the two decay modes, that are determined from the Monte Carlo simulation. The ratio of the acceptances (defined as the number of events in which both charged tracks are within the angular acceptance $|\eta| < 2.1$, divided by the total number of generated events) is 0.995 ± 0.006(stat). The efficiency has two components: the trigger and the selection efficiency. The former is defined as the number of events that pass the trigger requirements divided by the number of events in the acceptance. Given the three body decay of the normalization sample, the ratio of the trigger efficiencies between the signal and the normalization modes decreases with the increase of the p_T threshold of the muon, ranging from 0.149 ± 0.02 for the 3 GeV/c threshold down to 0.085 ± 0.003 for the 15 GeV/c threshold. The reconstruction efficiency is measured relative to the triggered events, and the ratio of the efficiencies between the signal and normalization channels varies from 2.2 ± 0.2 to 0.81 ± 0.05 from 3 to 15 GeV/c thresholds.

The systematics uncertainties on the acceptances and efficiencies are of the order of 15-20%, and are dominated by the one on the trigger efficiency, which depends on the correct modeling of the p_T spectrum of the muons from charm decays. This is measured by monitoring the luminosity weighted (and corrected for the efficiency) yield for the triggers in the 7 different periods of data taking.

The 90% CL upper limit is determined in each period of data taking, corresponding to different thresholds of the muon trigger, and then combined, giving

$$\mathcal{B}(D^0 \to \mu^+\mu^-) < 5.4 \times 10^{-7} \text{at } 90\ \%\ \text{CL}. \quad (2)$$

2. Searches for the rare decays $B_s^0 \to \mu\mu$ and $B^0 \to \mu\mu$

The rare decays $B_s^0 \to \mu^+\mu^-$ and $B^0 \to \mu^+\mu^-$ are highly suppressed in the SM due to their FCNC nature. These decays are forbidden at tree level and can proceed only through higher order diagrams, such as electroweak penguin and box diagrams. Furthermore, the decays are helicity suppressed and require an internal quark annihilation within the B meson. The SM predictions for the B_s^0 and B^0 branching ratios to dimuons are $(3.23 \pm 0.27) \times 10^{-9}$ and $(1.07 \pm 0.10) \times 10^{-10}$, respectively, with the main uncertainty resulting from the value of the B meson decay constant f_B [6]. The comparison with experimental results requires the inclusion of soft-photon radiation and $B_s^0 - \bar{B}_s^0$ oscillations effects each yielding O(10%) corrections to the predicted decay rate. Several extensions of the SM, such as supersymmetric models and models with a non-standard Higgs sector, predict enhancements or suppressions to the branching fractions for these rare decays. The rare nature of the processes and the rather precise SM predictions make these decays an excellent probe for physics beyond the SM.

The CMS experiment performed a simultaneous search for the rare decays $B_s^0 \to \mu^+\mu^-$ and $B^0 \to \mu^+\mu^-$ using an integrated luminosity of 5 fb^{-1} collected at the centre-of-mass energy of $\sqrt{s} = 7$ TeV. An event-counting experiment was performed in dimuon mass regions around the B_s^0 and B^0 masses and all selection criteria were established before observing the signal region. The analysis has been published in [7].

The main backgrounds are those from semileptonic decays of the B hadrons, in which the two muons are coming from the primary and secondary decays of the b and c hadrons, or those where the reconstructed muons are misidentified hadrons, or rare B decays. In particular, resonant decays of B hadrons in two hadrons where both are misidentified as muons are called "peaking background".

A normalization sample of events with $B^+ \to J/\psi K^+$ decays (where the J/ψ decays into a muon pair) was used to remove uncertainties related to the b-quark production cross section and the integrated luminosity. Combinatorial backgrounds were evaluated from the data in dimuon invariant mass sidebands while backgrounds from B decays were assessed with Monte Carlo (MC) simulation. The analysis was performed separately in two channels, barrel and endcap, and then combined for the final result. The barrel channel contained the candidates where both muons have $|\eta| < 1.4$ and the endcap channel included those where at least one muon had $|\eta| > 1.4$. $B \to \mu^+\mu^-$ candidates with traverse momentum above 6.5(8.5) GeV/c in the barrel(endcap) were formed by two oppositely-charged muons originating from a common vertex and with an invariant mass in the range $4.9 < m(\mu\mu) < 5.9$ GeV/c^2. The (sub)leading muon transverse momentum was required to be larger than (4.0) 4.5 GeV/c in the barrel and (4.2) 4.5 GeV/c in the endcap.

The primary vertex (PV) associated to the B candidate is chosen among all the reconstructed PVs in the event as the minimally separated PV from the B candidate along the z axis (where the z coordinate is along the beam direction). The PV is refit removing the tracks that already form the B candidate secondary vertex. Once the PV has been determined, more variables can be constructed, which distinguish the signal from the background :

- the angle between the B meson momentum direction and the one joining the primary and secondary vertex should be smaller than 50 mrad in the barrel and 30 mrad in the endcap.
- the three-dimensional impact parameter of the B meson with respect to the PV (δ_{3D}) and its uncertainty ($\sigma(\delta_{3D})$) are required to satisfy the following criteria: $\delta_{3D} < 80\mu$m and $\delta_{3D}/\sigma(\delta_{3D}) < 2$
- the three-dimensional decay length of the B meson should be larger than 13 (15) times its uncertainty in the barrel (endcap).

In order to suppress of the abundant semileptonic decays and be insensitive to the simultaneous pp collisions (pileup) (on average 8 events per bunch crossing), a set of isolation variables have been built:

- an isolation variable I
$$I = \frac{p_T(B)}{p_T(B) + \Sigma_i p_T(i)} \tag{3}$$
where the sum runs over the track momenta with a minimum p_T of 0.9 GeV/c, for tracks contained in a cone of $\Delta R < 0.7$ with respect to the B momentum. To reduce the sensitivity to the pileup, the sum is performed only on tracks coming either from the same PV or not associated to any PV (but having a distance of closest approach smaller than 0.05 cm). A good isolation is guaranteed by requiring $I > 0.80$.
- less than 2 tracks with $p_T > 0.5$ GeV/c and a distance of closest approach smaller than 0.03 cm with respect to the B candidate decay vertex
- finally, all tracks in the event that are either associated with the same PV as the B candidate or not associated to any other PV should have a distance of closest approach to the B vertex larger than 0.015 cm.

The signal selection has been validated using the normalization $B^+ \to J/\psi K^+$ and control $B_s^0 \to J/\psi\phi$ samples. Their reconstruction strategy is very similar to the signal one. The dimuon invariant mass should satisfy $3.0 < m(\mu\mu) < 3.2$ GeV/c^2, $p_T(\mu\mu) > 7$ GeV/c, $p_T(K) > 0.5$ GeV/c. For the $B_s^0 \to J/\psi\phi$ the ϕ selection requires $0.995 < m(KK) < 1.045$ GeV/c^2 and $\Delta R(KK) < 0.25$. The number of reconstructed B^+ mesons is $(82.7 \pm 4.2) \times 10^3$ in the barrel and $(23.8 \pm 1.2) \times 10^3$ in the endcap. The invariant mass distributions are fit with a double-Gaussian function for the signal and an exponential plus an error function for the background, as shown in Fig. 2.

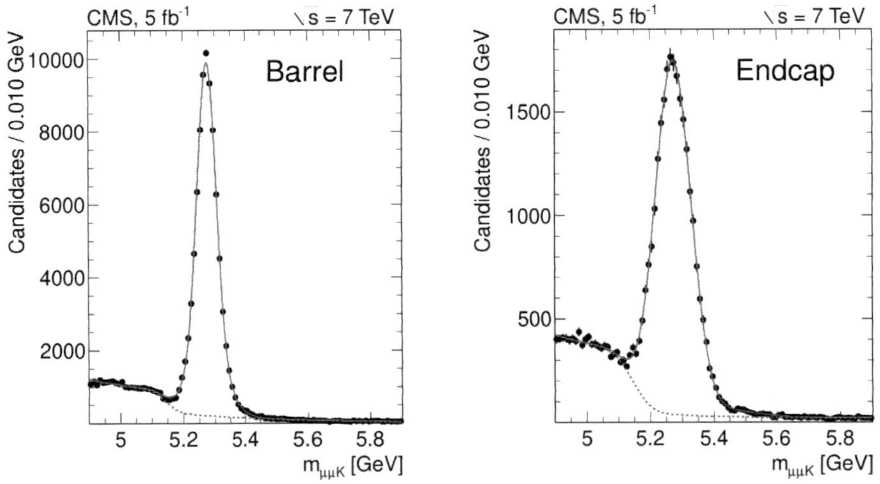

Figure 2. $B^+ \to J/\psi K^+$ invariant-mass distributions in the barrel (left) and endcap (right) channels. The solid (dashed) lines show the fits to the data (background).

The sideband subtracted data in the normalization and control samples have been compared to the MC simulation for the signal, for all the selection variables. The total difference in data and MC has been quantified by summing in quadrature the percentage difference in data and MC in each variable: this leads to 4% uncertainty for the normalization sample and to 3% for the control sample.

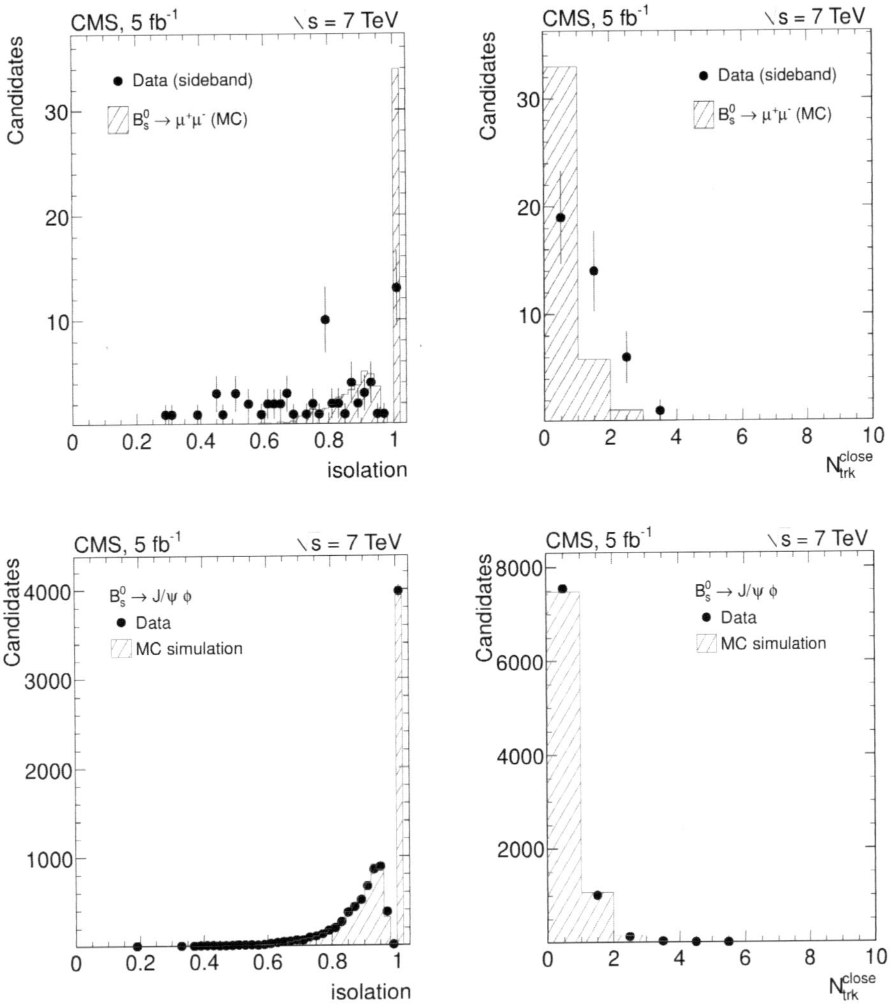

Figure 3. Comparison of simulated $B_s^0 \to \mu^+\mu^-$ decays and background (upper row) dimuon distributions as measured in the mass sidebands (upper row) or $B_s^0 \to J/\psi\phi$ (lower row), isolation variables: I and number of close tracks. The MC histograms are normalized to the number of events in the data.

The expected number of rare events is evaluated from the MC simulation and normalized to the measured B^+ yield, taking into account the muon misidentification probability measured from data in $D^{*+} \to D^0(K^-\pi^+)\pi^+$ and $\Lambda \to p\pi^-$ decays.

The branching fraction was estimated by using the following equation:

$$\mathcal{B}(B \to \mu^+\mu^-) = \frac{\mathrm{N_{sig}}}{\mathrm{N_{norm}}} \times \frac{f_u}{f_s} \times \frac{\epsilon_{\mathrm{norm}}}{\epsilon_{\mathrm{sig}}} \times \mathcal{B}(B^+ \to J/\psi K^+) \tag{4}$$

where $\mathrm{N_{norm}}$ is the number of reconstructed $B^+ \to J/\psi K^+$ decays, $\mathcal{B}(B^+)$ its branching ratio, ϵ_{norm} the total efficiency, $\mathrm{N_{sig}}$ is the number of observed candidates (after background subtraction) in the signal mass window, ϵ_{sig} the total signal efficiency, and $f_u/f_s = 0.267 \pm 0.021$ the ratio of the B^+ and B_s^0 production cross sections, is taken from [8] and its uncertainty taken as systematics.

The simulated dimuon mass resolution for signal events depends on the pseudorapidity of the B candidate and ranges from 37 MeV/c^2 for $\eta \sim 0$ to 77 MeV/c^2 for $|\eta| > 1.8$. The dimuon mass scale and resolution in the MC simulation are compared with the measured detector performance by studying $J/\psi \to \mu^+\mu^-$ and $\Upsilon(1S) \to \mu^+\mu^-$ decays. The residual differences between simulation and data are small and the uncertainty on the efficiency coming from these effects is estimated to be 3%.

Table 1 reports the results of the analysis, whereas Fig. 4 shows the measured dimuon invariant mass distributions after unblinding. In the sidebands the observed number of events is equal to six (seven) for the barrel (endcap) channel. Two (four) events are observed in the signal region in the barrel (endcap) for the $B_s^0 \to \mu^+\mu^-$ decay channel, while two (zero) events are observed in the barrel (endcap) channel for the $B^0 \to \mu^+\mu^-$.

In addition to the systematics already mentioned, the ones from the trigger and muon identification efficiencies vary between 3% and 8%, depending on the muon kinematics. Different production mechanisms would result in different acceptance, giving an uncertainty between 3.5% and 5%. The mass scale and resolution uncertainties give a 3% systematic uncertainty. Finally the combinatorial and rare decays background uncertainties are evaluated to be 4% and 20%, respectively.

Upper limits on the $B_s^0 \to \mu^+\mu^-$ and $B^0 \to \mu^+\mu^-$ branching ratios are determined using the CL method [9, 10] by combining the endcap and the barrel channels, and taking into account the statistical and systematic uncertainties. The combined upper limits for the barrel and endcap channels are $\mathcal{B}(B_s^0 \to \mu^+\mu^-) < 7.7 \times 10^{-9}$ (6.4×10^{-9}) and $\mathcal{B}(B^0 \to \mu^+\mu^-) < 1.8 \times 10^{-9}$ (1.4×10^{-9}) at 95% (90%) CL. The median expected upper limits at 95% CL are 8.4×10^{-9} (1.6×10^{-9}) for $B_s^0 \to \mu^+\mu^-$ ($B^0 \to \mu^+\mu^-$), where the number of expected signal events is based on the SM value. Including cross-feed between the $B^0$and B_s^0 decays, the background-only p value is 0.11 (0.24) for $B_s^0 \to \mu^+\mu^-$ ($B^0 \to \mu^+\mu^-$), corresponding to 1.2 (0.7) standard deviations. The p value for the background plus SM signal hypotheses is 0.71 (0.86) for$B_s^0 \to \mu^+\mu^-$ ($B^0 \to \mu^+\mu^-$).

Table 1. The event selection efficiency for signal events $\varepsilon_{\rm tot}$, the SM-predicted number of signal events $N_{\rm signal}^{\rm exp}$, the expected number of peaking background events $N_{\rm peak}^{\rm exp}$ and combinatorial background events $N_{\rm comb}^{\rm exp}$, and the number of observed events $N_{\rm obs}$ in the barrel and endcap channels for $B_s^0 \to \mu^+\mu^-$ and $B^0 \to \mu^+\mu^-$ The quoted errors include both, the statistical and the systematic uncertainties.

	Barrel		Endcap	
Variable	$B^0 \to \mu^+\mu^-$	$B_s^0 \to \mu^+\mu^-$	$B^0 \to \mu^+\mu^-$	$B_s^0 \to \mu^+\mu^-$
$\varepsilon_{\rm tot}$(%)	0.29 ± 0.02	0.29 ± 0.02	0.16 ± 0.02	0.16 ± 0.02
$N_{\rm signal}^{\rm exp}$	0.24 ± 0.02	2.70 ± 0.41	0.10 ± 0.01	1.23 ± 0.18
$N_{\rm peak}^{\rm exp}$	0.33 ± 0.07	0.18 ± 0.06	0.15 ± 0.03	0.08 ± 0.02
$N_{\rm comb}^{\rm exp}$	0.40 ± 0.34	0.59 ± 0.50	0.76 ± 0.35	1.14 ± 0.53
$N_{\rm total}^{\rm exp}$	0.97 ± 0.35	3.47 ± 0.65	1.01 ± 0.35	2.45 ± 0.56
$N_{\rm obs}$	2	2	0	4

3. Conclusions

The CMS collaboration has looked for rare decays of $B_{s,d}^0$ and D^0 mesons decaying into $\mu^+\mu^-$ in pp collisions at $\sqrt{s} = 7$ TeV. The data sample corresponds to an integrated luminosity of 5 fb^{-1} in 2011 for beauty mesons, and to 90 pb^{-1} for charm. The $B_{s,d}^0 \to \mu^+\mu^-$ result supersedes our

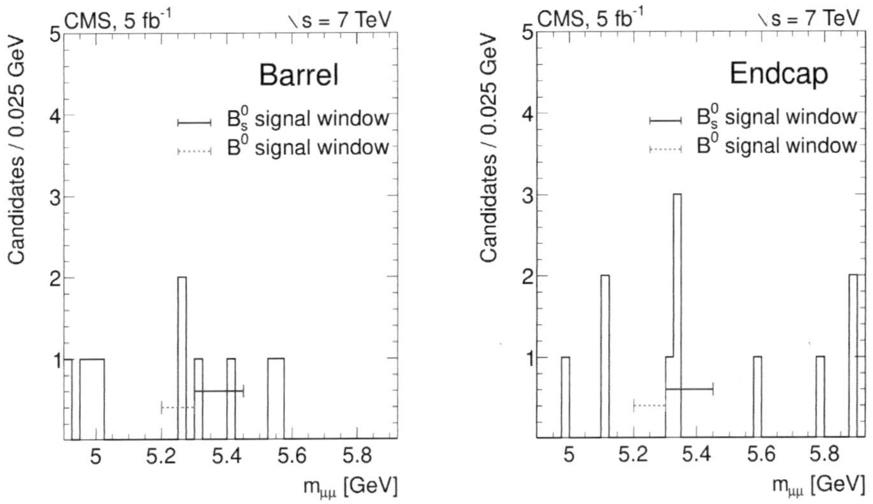

Figure 4. Dimuon invariant mass distributions in the barrel (left) and endcap (right) channels. The signal windows for B_s^0 and B^0 are indicated by horizontal lines.

previous measurements and gives a 95% CL upper limit on the branching fraction of 7.7×10^{-9} and $\mathcal{B}(B^0 \to \mu^+\mu^-) < 1.8 \times 10^{-9}$.

The $D^0 \to \mu^+\mu^-$ is normalized to $D^0 \to K^-\mu^+\nu_e$ decay mode, giving 5.4×10^{-7} upper limit at 90% CL. The main limitation in the analysis relies on the efficiency of the normalization sample

With the luminosity to be collected by the end of the 2012 LHC run (a factor 4 larger than the one used here), we will be having good chances to reach the $B_{s,d}^0$ standard model predicted BR.

References

[1] G. Burdnam et al., Phys. Rev. D 66, 014009 (2002).
[2] E. Golowich et al., Phys. Rev. D 79, 114030 (2009).
[3] CMS Collaboration, *"Search for the Flavour-Changing Neutral Current decay $D^0 \to \mu^+\mu^-$ with the experiment CMS at $\sqrt{s} = 7$ TeV"*, CMS-PAS-BPH-11-017 (2011).
[4] J. Anjos et al., *"A Study of the Semileptonic Decay Mode $D^0 \to K^-e^+\nu_e$"*, Phys. Rev. Lett. 62 (1989) 1587.
[5] K. Nakamura et al., *Review of Particle Physics*, J. Phys. G 37 (2010) 075021.
[6] A.J.Buras, J.Girrbach, D.Guadagnoli and G.Isidori, Eur. Phys. J. C 72 (2012) 2172.
[7] S. Chatrchyan et al., *"Search for $B_s^0 \to \mu^+\mu^-$ and $B^0 \to \mu^+\mu^-$ decays"*, JHEP 1204 (2012) 033.
[8] R. Aaij, et al., *"Measurement of b hadron production fractions in 7 TeV pp collisions"*, Phys.Rev. D85 (2012) 032008.
[9] A. L. Read, J. Phys. 28 (2002) 2693.
[10] T. Junk, Nucl. Instrum. Meth. A 434 (1999) 435.
[11] S.Chatrchyan et al. [CMS Collaboration], JHEP1204, (2012) 033.

Journal of Physics: Conference Series **455** (2013) 012033 doi:10.1088/1742-6596/455/1/012033

Studies on B hadron production, spectroscopy and decays at CMS

Xin Shi

National Taiwan University, Taiwan
On behalf of the CMS collaboration

E-mail: Xin.Shi@cern.ch

Abstract. We review several recent results of the CMS experiment in the field of B hadron productions, spectroscopy and decays. We performed a flavour-untagged measurement of the width difference $\Delta\Gamma_s$ in the $B_s \to J/\psi\phi$ decays. With an assumption of zero mixing phase, reconstructed B_s signal candidates are used to measure the polarization amplitudes in an angular and proper decay time analysis.

1. Introduction

The study of heavy-quark production in high-energy hadronic interactions plays a critical role in the testing next-to-leading order (NLO) quantum chromodynamics (QCD) calculations. Historically, the measured inclusive b-hadron production cross sections were significantly higher than the theory predictions [1], more data is needed to distinguish conclusively between the various theoretical approaches, particularly at large transverse momentum p_T [2]. Among the b-hadrons, the flavour mixing property of the B_s meson is particularly interesting, where the measurement of the lifetime difference ($\Delta\Gamma_s$) of the two mass eigenstates can be performed using the proper decay time distribution of B_s meson [3, 4].

2. The CMS detector

The CMS detector [5] is a general purpose detector at the LHC. It's inner tracker consists of silicon pixel and silicon strip layers. Muons are measured by drift tubes (DT), cathode strip chambers (CSC) and resistive plate chambers (RPC). The dimuon mass resolution is less than 1%, which makes it a powerful tool for B-physics study.

3. B-Hadron Production

The B-hadron production rates as a function of the hadron transverse momentum has been measured using the 7 TeV data on CMS shown in Figure 1, where the large normalization uncertainties for Λ_b and B_s are dominated by the poorly measured branching fractions for the decay channels used in the analysis. The B-hadron cross section are also measured and compared with theory predictions as shown Figure 2. Moreover, the first direct observation of a new strange heavy b baryon Ξ_b strongly decaying to a Ξ_b^- baryon and a charged pion is reported in [6].

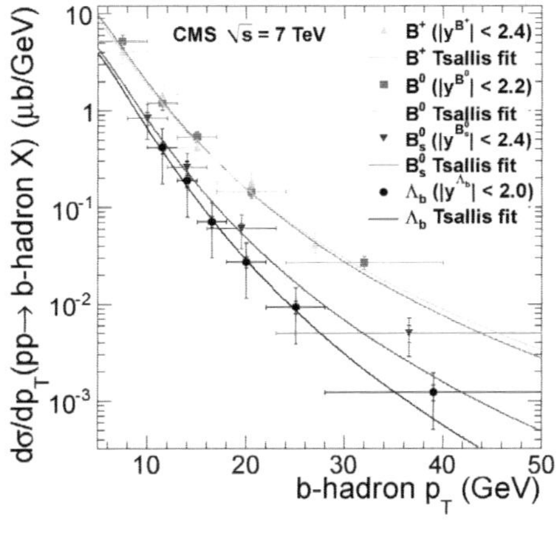

Figure 1. Comparison of b-hadron production rates versus hadron transverse momentum, where the inner error bars correspond to the bin-to-bin uncertainties, while the outer error bars represent the bin-to-bin plus normalization uncertainties added in quadrature.

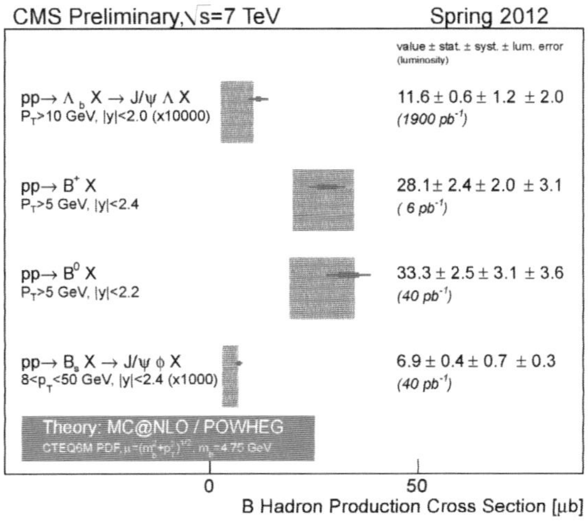

Figure 2. Summary of B hadron cross section measurements performed by CMS with 7 TeV p-p collisions at LHC. The inner error bars of the data points correspond to the statistical uncertainty, while the outer (thinner) error bars correspond to the quadratic sum of statistical and systematic uncertainties. The outermost brackets correspond to the total error, including a luminosity uncertainty which is also added in quadrature. Theory predictions at NLO are obtained using "MC@NLO".

4. Measurement of the B_s lifetime difference

The decay of a B_s meson is characterized by the possibility of the mixing between its two flavour eigenstates ($B_s - \bar{B}_s$) [7]. In the decay $B_s \to J/\psi\phi$ with $J/\psi \to \mu^+\mu^-$ and $\phi \to K^+K^-$, the final state is an admixture of the CP-even and CP-odd eigenstates. Since B_s is a pseudo-scalar meson, while J/ψ and ϕ are vector mesons, the orbital angular momentum can have the values $L = 0, 1, 2$. To measure the lifetime difference ($\Delta\Gamma_s$) for the decay rates of the two B_s mass eigenstates [3, 4], an analysis is needed to disentangle the two CP eigenstates.

The decay topology is described by three angles $\Theta = (\theta_T, \psi_T, \varphi_T)$ in the transversity basis [8], as shown in Figure 3. The decay of vector mesons is further described by the time evolution of three different amplitudes with different angular dependencies. The amplitudes at time $t = 0$ are defined using the longitudinal component $A_0(0)$ for $L = 0$, which is CP-even, and the transverse components $A_\perp(0)$ for $L = 1$ and $A_{||}(0)$ for $L = 2$ which are CP-odd and CP-even, respectively. In addition, the two strong phases are denoted by $\delta_{||}$ and $\delta_\perp$.

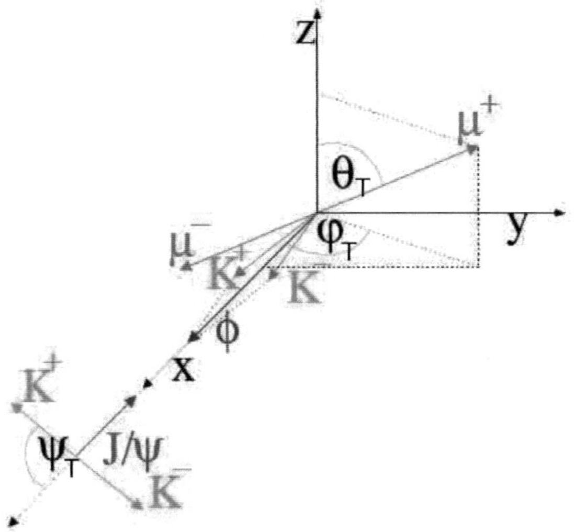

Figure 3. Definition of the three angles, (θ_T, ψ_T and φ_T), used for the description of the decay topology.

The differential decay rate can be represented [8] as:

$$\frac{d^4\Gamma(B_s(t))}{d\Theta\, dt} = f(\Theta, t; \alpha) = \sum_{i=1}^{6} O_i(\alpha, t).g_i(\Theta) \,, \tag{1}$$

where O_i are kinematics-independent observables, g_i are the angular distributions, and α denotes a set of physics parameters of interest (Γ_s, $\Delta\Gamma_s$, $|A_0|^2$, $|A_\perp|^2$, $\delta_{||}$). The kinematic observables are described using the following equations:

$$\begin{aligned}
O_1 &= |A_0(t)|^2 = |A_0(0)|^2 e^{-\Gamma_s t}[\cosh(\Delta\Gamma_s t/2) - \cos\phi_s \sinh(\Delta\Gamma_s t/2)] \\
O_2 &= |A_{||}(t)|^2 = |A_{||}(0)|^2 e^{-\Gamma_s t}[\cosh(\Delta\Gamma_s t/2) - \cos\phi_s \sinh(\Delta\Gamma_s t/2)] \\
O_3 &= |A_\perp(t)|^2 = |A_\perp(0)|^2 e^{-\Gamma_s t}[\cosh(\Delta\Gamma_s t/2) + \cos\phi_s \sinh(\Delta\Gamma_s t/2)] \\
O_4 &= Im(A_{||}^*(t)A_\perp(t)) = |A_{||}(0)||A_\perp(0)|e^{-\Gamma_s t}[-\cos(\delta_\perp - \delta_{||})\sin\phi_s \sinh(\Delta\Gamma_s t/2)] \\
O_5 &= Re(A_0^*(t)A_{||}(t)) = |A_0(0)||A_{||}(0)|\cos\delta_{||} e^{-\Gamma_s t}[\cosh(\Delta\Gamma_s t/2) - \cos\phi_s \sinh(\Delta\Gamma_s t/2)] \\
O_6 &= Im(A_0^*(t)A_\perp(t)) = |A_0(0)||A_\perp(0)|e^{-\Gamma_s t}[-\cos\delta_\perp \sin\phi_s \sinh(\Delta\Gamma_s t/2)] \;,
\end{aligned} \tag{2}$$

where ϕ_s is the mixing phase. Constraining ϕ_s to zero allows the terms containing $O_4(\alpha, t)\cdot g_4(\Theta)$ and $O_6(\alpha, t)\cdot g_6(\Theta)$ to be omitted. The individual angular distributions are given by the following equations:

$$\begin{aligned}
g_1 &= 2\cos^2(\psi_T)(1 - \sin^2(\theta_T)\cos^2(\varphi_T)), \\
g_2 &= \sin^2(\psi_T)(1 - \sin^2(\theta_T)\sin^2(\varphi_T)), \\
g_3 &= \sin^2(\psi_T)\sin^2(\theta_T), \\
g_4 &= -\sin^2(\psi_T)\sin^2(2\theta_T)\sin(\varphi_T), \\
g_5 &= \frac{1}{\sqrt{2}}\sin(2\psi_T)\sin^2(\theta_T)\sin(2\varphi_T), \\
g_6 &= \frac{1}{\sqrt{2}}\sin(2\psi_T)\sin(2\theta_T)\sin(\varphi_T) \;.
\end{aligned} \tag{3}$$

The theoretical expectation for the ratio $\Delta\Gamma_s/\Gamma_s$ is 0.12 ± 0.06 [7].

The events are selected by requiring two oppositely charged muons and two oppositely charged tracks. Then the B_s are build from 4-track vertex fit, where the dimuon invariant mass is constrained to the known J/ψ mass [9] and the invariant mass of a track pair is required to be within 10 MeV of the world average $\phi(1020)$-meson mass [9].

The overall efficiency includes: detector acceptance, the trigger conditions, and the selection cuts. No sizable correlation between the proper time and the angular variables is found. The correlation amongst the angular observables is also negligible. The proper decay length efficiency is almost flat in the range [0.02-0.3]cm.

An unbinned maximum likelihood fit to the data is performed by including the information on the invariant mass (m), proper decay time (t) and the three decay angles (Θ) of the reconstructed B_s candidates. From this five-dimensional fit, the physics parameters of interest $\Delta\Gamma_s$, Γ_s, $|A_\perp|^2$, $|A_0|^2$ and $\delta_{||}$ are determined, assuming that the mixing phase ϕ_s is zero.

The event likelihood function $\mathcal{L}$ can be represented as

$$\begin{aligned}
\mathcal{L} &= L_{\text{signal}} + L_{\text{background}} \;, \\
L_{\text{signal}} &= (f(\Theta, t; \alpha) \times G(t, \kappa, \sigma(t))]) \cdot M(m) \cdot \epsilon(t)\epsilon(\Theta) \;, \\
L_{\text{background}} &= b(\Theta, t, m) \,,
\end{aligned} \tag{4}$$

where L_{signal} is the PDF that describes the $B_s \to J/\psi\phi$ signal model, $L_{\text{background}}$ describes the background contributions, and $f(\Theta, t; \alpha)$ is the differential decay rate function. Here G is a Gaussian resolution function which makes use of the per event proper decay time uncertainty $\sigma(t)$ scaled by a factor κ, $\epsilon(t)$ is the proper decay time efficiency function, $\epsilon(\Theta)$ is the angular efficiency function, and $b(\Theta, t, m)$ describes the background model. The signal mass PDF $M(m)$ is given by the sum of two Gaussian functions.

Journal of Physics: Conference Series **455** (2013) 012033 doi:10.1088/1742-6596/455/1/012033

To extract the physics parameters of interest, a five-dimensional fit procedure is taken with several steps. First a one-dimensional fit of B_s mass to get the mean and the smaller of the two Gaussian function widths. Then fit the sideband region for the angular background shapes. Finally fit the full mass range. The mass distribution of the reconstructed B_s candidates is shown in Figure 5, together with the overlaid five-dimensional fit projection. The measured signal yield is 14456 ± 140, with a fitted mass mean of $5366.8 \pm 0.1\,\mathrm{MeV}$. The full fit was finally performed in the full mass range, and the one-dimensional projections on the distributions of the proper decay length distributions and the angular variables are shown in Figure 6 and Figure 7 respectively.

The total estimated systematic uncertainties are presented in Table 1, where the main contribution sources are background angular model, and proper time resolution.

Table 1. Systematic uncertainties associated to the quantities measured in the analysis.

Uncertainty source	$\Delta\Gamma_s$ [ps^{-1}]	$c\tau$ [cm]	$\lvert A_0\rvert^2$	$\lvert A_\perp\rvert^2$	$\delta_{\parallel}$ [rad]
Signal PDF modeling					
Signal mass model	0.00072	0.00012	0.0022	0.0006	0.039
Proper time resolution	0.00170	0.00006	0.0007	0.0000	0.007
ϕ_s approximation	0.00000	0.00001	0.0000	0.0000	0.002
S-wave assumption	0.00109	0.00001	0.0130	0.0066	0.056
Background PDF modeling					
Background mass model	0.00019	0.00000	0.0000	0.0001	0.003
Background lifetime model	0.00040	0.00000	0.0001	0.0002	0.003
Peaking B^0 background	0.00025	0.00006	0.0002	0.0022	0.050
Background angular model	0.00175	0.00003	0.0001	0.0064	0.161
Limited simulation statistics					
Angular efficiency parameters	0.00019	0.00002	0.0057	0.0055	0.037
Temporal efficiency parameters	0.00000	0.00005	0.0000	0.0000	0.000
Temporal efficiency parametrization	0.00181	0.00014	0.0005	0.0007	0.001
Angular efficiency parametrization	0.00063	0.00003	0.0021	0.0086	0.007
Likelihood function bias	0.00000	0.00004	0.0004	0.0000	0.014
Total uncertainty	0.00341	0.00022	0.0146	0.0140	0.187

The B_s decay width difference, B_s mean lifetime, transversity amplitudes ($|A_\perp|^2$ and $|A_0|^2$) and the strong phase ($\delta_{||}$) are measured to be

$$\begin{aligned}
\Delta\Gamma_s &= 0.048 \pm 0.024\,(\mathrm{stat.}) \pm 0.003\,(\mathrm{syst.})\,\mathrm{ps}^{-1}\,,\\
\tau_{B_s} &= 0.04580 \pm 0.00059\,(\mathrm{stat.}) \pm 0.00022\,(\mathrm{syst.})\,\mathrm{cm}\,,\\
|A_0|^2 &= 0.528 \pm 0.010\,(\mathrm{stat.}) \pm 0.015\,(\mathrm{syst.})\,,\\
|A_\perp|^2 &= 0.251 \pm 0.013\,(\mathrm{stat.}) \pm 0.014\,(\mathrm{syst.})\,,\\
\delta_{||} &= 2.79 \pm 0.14\,(\mathrm{stat.}) \pm 0.19\,(\mathrm{syst.})\,\mathrm{rad}\ .
\end{aligned}$$

the comparison of the decay width difference with other experiments is illustrated in Figure 4.

5. Summary

CMS is a powerful detector for studying B physics because of its excellent tracking and lepton identification. Several recent results of the CMS experiment in the field of B hadron productions, spectroscopy and decays are reviewed. A study of the $B_s \to J/\psi\phi$ decay with $J/\psi \to \mu^+\mu^-$ and $\phi \to K^+K^-$, yield $\Delta\Gamma_s = 0.048 \pm 0.024\,(\mathrm{stat.}) \pm 0.003\,(\mathrm{syst.})\,\mathrm{ps}^{-1}$. We look forward to the new results from CMS.

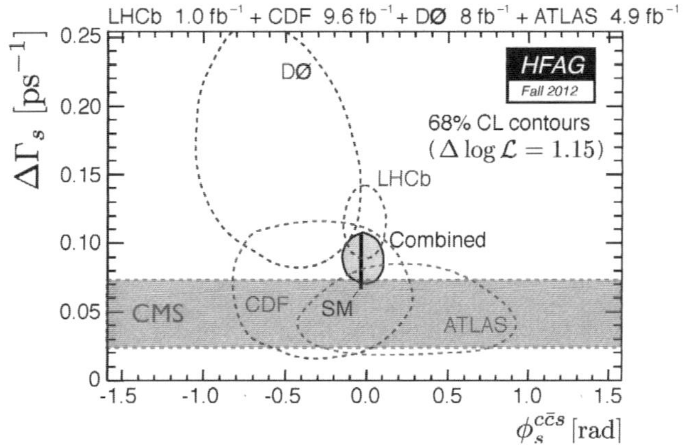

Figure 4. Comparison with other experiments. The blue region (CMS) was not in the original version by HFAG [10].

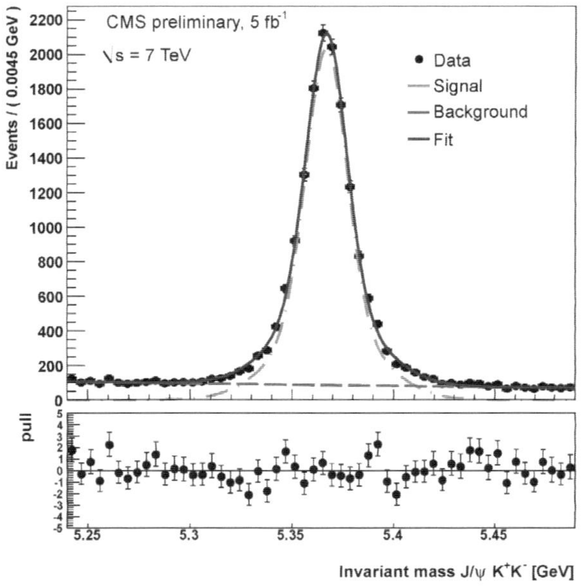

Figure 5. Mass projection of the five-dimensional maximum likelihood fit to the data. The points are the data distribution, the solid blue line shows the full fit, the green dash-dotted line is the signal model component, the red long-dashed line is the background component. The pull between the mass distribution and the fit is shown in the histogram below.

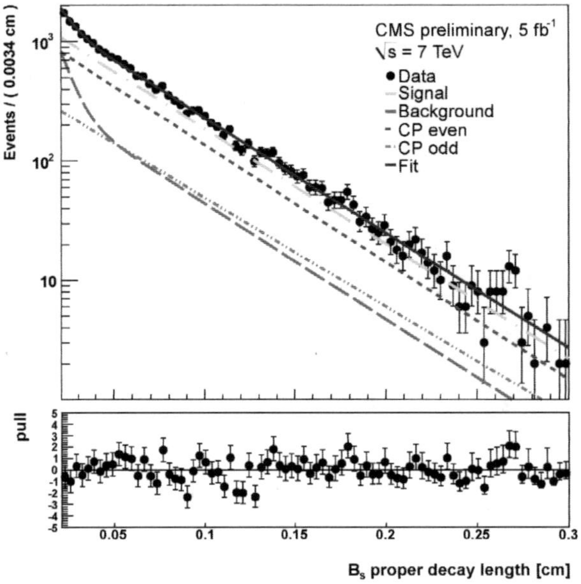

Figure 6. Proper decay length projection of the five-dimensional maximum likelihood fit to the data. The points are the data distribution, the solid blue line shows the full fit, the green dash-dotted line is the signal model component, with the CP-even (odd) component shown in blue short dashed line (magenta dash-dot-dotted line), and the red long dashed line is the background component. The pull between the proper decay length distribution and the fit is shown in the histogram below.

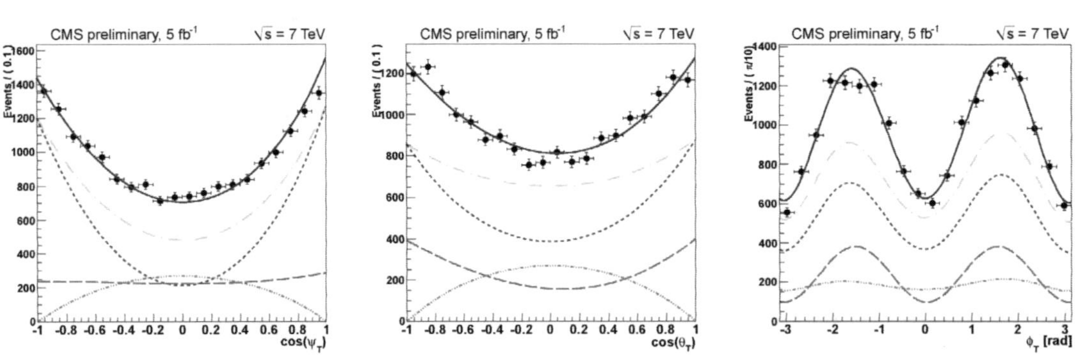

Figure 7. The angular projection of the five-dimensional maximum likelihood fit to the data. The plots show the projections on the angular variables $\cos(\psi_T)$, $\cos(\theta_T)$ and the angle ϕ. The various fit components are represented as in Figure 6.

Journal of Physics: Conference Series **455** (2013) 012033 doi:10.1088/1742-6596/455/1/012033

References

[1] Nason P, Dawson S and Ellis R K 1988 *Nucl. Phys.* **B303** 607
[2] Kniehl B A, Kramer G, Schienbein I and Spiesberger H 2008 *Phys. Rev.* **D77** 014011 (*Preprint* 0705.4392)
[3] Giurgiu G (For the CDF) 2010 *PoS* **ICHEP2010** 236 (*Preprint* 1012.0962)
[4] Uwer U (LHCb) 2011 *PoS* **BEAUTY2011** 051
[5] Chatrchyan S *et al.* (CMS Collaboration) 2008 *JINST* **3** S08004
[6] Chatrchyan S *et al.* (CMS Collaboration) 2012 *Phys.Rev.Lett.* **108** 252002 (*Preprint* 1204.5955)
[7] Dunietz I, Fleischer R and Nierste U 2001 *Phys.Rev.* **D63** 114015 (*Preprint* hep-ph/0012219)
[8] IDighe I and RFleischer 1999 *Eur.Phys.J.C* **6**
[9] Nakamura K *et al.* (Particle Data Group) 2010 *J.Phys.G* **G37** 075021
[10] Amhis *et al.* (Heavy Flavor Averaging Group) 2012 (*Preprint* 1207.1158)

International Workshop on Discovery Physics at the LHC (Kruger2012) IOP Publishing
Journal of Physics: Conference Series **455** (2013) 012034 doi:10.1088/1742-6596/455/1/012034

Search for Single and Pair-Production of Dijet Resonances with the CMS Detector

(Proceedings submitted to the Kruger 2012 Conference)

Kai Yi for the CMS Collaboration
Physics and Astronomy Department, University of Iowa, Iowa, USA
E-mail: yik@fnal.gov

Abstract.
Searches for new physics in the single and paired dijet mass spectrum are performed using data collected by the CMS experiment at the LHC at a collision energy of $\sqrt{s}$=7 or $\sqrt{s}$=8 TeV. No evidence for new physics is found and upper limits are set for various models. At 95% confidence level, a string resonance in the single dijet spectrum is excluded for masses between 1 and 4.7 TeV and, for the first time, a coloron in the paired dijet spectrum is excluded for masses between 250 and 740 GeV.

1. Introduction

Many beyond-the-standard-model models predict the existence of new massive strongly produced objects that decay into quarks (q) and gluons (g). These states may show up as narrow resonances in the leading (highest p_T) jet spectrum. We consider the following models: string resonance (qg, $q\bar{q}$ and gg) [1, 2]; E_6 diquark ($\bar{q}\bar{q}$ and qq) predicted from grand unified theory based on the E_6 gauge group [3]; excited quarks q^* (qg) expected from composite quark model [4, 5]; axial vector particles called axigluons A ($q\bar{q}$) [6, 7]; color-octet vector C ($q\bar{q}$) [8]; color-octet scalar S_8 (gg) [9]; new gauge bosons W' and Z' from new gauge symmetries ($q\bar{q}$) [10]; massive gravitons G ($q\bar{q}$ and gg) from the Randall-Sundrum (RS) model of extra dimensions [11]. The color-octet vector is predicted by the flavor-universal coloron model, which embeds the SU(3) symmetry of QCD in a larger gauge group; the color-octet scalar is included in many dynamical electroweak symmetry breaking models, such as technicolor.

In addition to singly produced particles, many models predict the pair production of massive colored particles [12]. The searches of new narrow resonance in the leading dijet mass spectrum is not optimized for pair-produced particles for the following reason: the probability that one of the particles decays into the two leading dijet is only about 10%. A dedicated search is needed for pair-produced particle. We use coloron production as our benchmark model ($gg \to CC \to q\bar{q}q\bar{q}$). We also consider the possibility that colorons decay into color-octet scalars (S_8). Furthermore, we consider an R-Parity violating SUSY model [13] in which pair-produced top squark (stops) each decay to $q\bar{q}$ as a third possibility.

Journal of Physics: Conference Series **455** (2013) 012034 doi:10.1088/1742-6596/455/1/012034

2. Jets Reconstruction at CMS

Jets are reconstructed using a particle flow technique [14] at CMS [15]. The CMS particle-flow algorithm combines calorimeter information with reconstructed tracks to identify photons, leptons, and both neutral and charged hadrons within jets. The particle-flow objects serve as input for jet reconstruction, performed using the anti-k_T algorithm [16] with a cone radius $R = \sqrt{(\Delta\eta)^2 + (\Delta\phi)^2}$ of 0.5 or 0.7. Jet energy scale corrections are derived from Monte Carlo (MC) simulation [17] as well as data.

3. Dijet Resonances Searches

We search for narrow resonances in the dijet mass spectrum using the two leading jets in the data. The data sample is collected with multijet high-level triggers based on H_T, the scalar sum of the transverse momenta of all jets in the event with $p_T > 40$ GeV and $|\eta| < 2.5$. The H_T threshold are 650 GeV and 750 GeV, depending on the running period. The trigger efficiency is measured from the data to be larger than 99.9% for dijet masses reconstructed offline above 890 GeV. Jets that pass the standard jet quality requirements are combined into "wide jets" [18] if their separation $R = \sqrt{(\Delta\eta)^2 + (\Delta\phi)^2} \leq 1.1$. The combining is done before determining the leading dijet mass spectrum.

The data used in this study correspond to an integrated luminosity of 4.0 fb^{-1} at a collision energy of $\sqrt{s}$=8 TeV [19]. The dijet mass background shape is modeled by a smooth parameterization. The dijet resonance shapes for qq, qg and gg partons are dominated by experimental resolution and are obtained from full simulation. No significant excess is observed. The upper limits at 95% CL set on the cross section times branching ratio times acceptance of centrally ($|\Delta\eta| < 1.3$ and $|\eta| < 2.5$) produced dijet mass resonances for (qq, qg, gg) are shown in Fig. 1. We set lower limits on the mass of string resonances, excited quarks, axigluons, colorons, S_8 resonances, E_6 diquarks, W$'$ and Z$'$ bosons, and RS gravitons in the 1–4.7 TeV range as listed in Table 1. Many of the values in the table represent significant improvements on exclusion limits from previous dijet mass searches.

Model	Final State	Obs. Mass Excl. [TeV]	Exp. Mass Excl. [TeV]
String Resonance (S)	qg	[1.0, 4.69]	[1.0,4.64]
Excited Quark (Q*)	qg	[1.0, 3.19]	[1.0,3.43]
E_6 Diquark (D)	qq	[1.0, 4.28]	[1.0,4.12]
Axigluon (A)/Coloron (C)	q$\bar{\text{q}}$	[1.0, 3.28]	[1.0,3.55]
S_8 Resonance (S_8)	gg	[1.0, 2.66]	[1.0,2.53]
W' Boson (W')	q$\bar{\text{q}}$	[1.0, 1.74] [1.97, 2.12]	[1.0,1.92]
Z' Boson (Z')	q$\bar{\text{q}}$	[1.0, 1.60]	[1.0,1.50]
RS Graviton (RSG)	q$\bar{\text{q}}$+gg	[1.0, 1.36]	[1.0,1.20]

Table 1. Observed and expected 95% CL exclusions on the mass of various resonances. Systematic uncertainties are included.

4. Searches for Pair-produced Dijet Resonances in a Four Jet Final State

We search for new narrow resonances in the paired dijet mass spectrum. The data sample is collected with triggers that require the presence of at least four jets, based on information from the calorimeters. Each jet must have $|\eta| < 3.0$ and $p_T > 70$ or 80 GeV, depending on the run period. The trigger is 99.5% efficient for events with our selection–four leading jets, each with a p_T exceeding 110 GeV with $|\eta| < 2.5$. The two jets in every possible pairs is required to have

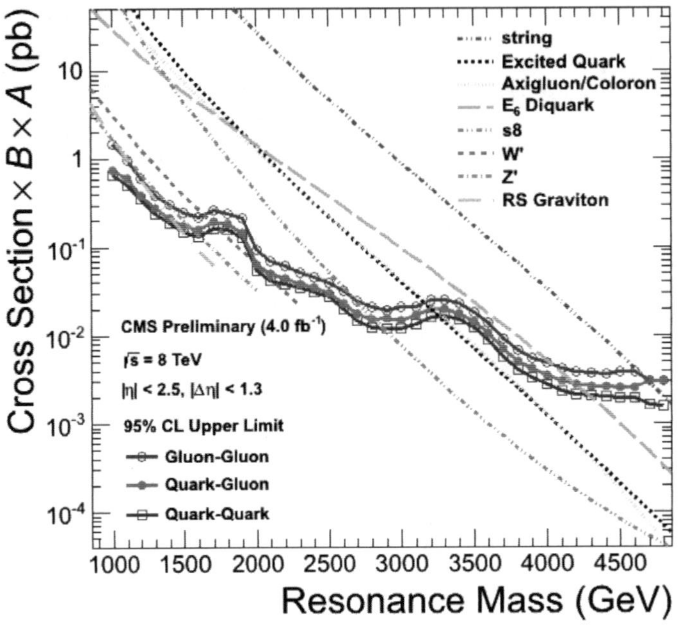

Figure 1. The observed 95% CL upper limits from the high-mass analysis on $\sigma \times B \times A$ for dijet resonances decaying through gluon-gluon (open circles), quark-gluon (solid circles), and quark-quark (open boxes) channels compared with theoretical predictions for string resonances [1, 2], E_6 diquarks [3], excited quarks [4, 5], axigluons [6, 7], colorons [8], S_8 resonances [9], new gauge bosons W′ and Z′ [10], and Randall-Sundrum gravitons [11].

a separation on $\Delta R_{jj} = \sqrt{(\Delta\eta)^2 + (\Delta\phi)^2} \geq 0.7$ to ensure a negligible overlap among the jets. There are three possible dijet pairs from the leading four jets, and each dijet pair is required to have $\Delta m < 0.15 m_{avg}$, where Δm is the mass difference between the two dijets and m_{avg} is their average mass. Thus the difference is required to be less than three times the dijet mass resolution (4.5%).

We require $\Delta > 25$ GeV for each dijet, where $\Delta = \Sigma_{i=1,2}(p_T)_i - m_{avg}$ is the scalar sum of the transverse momenta of the two jets in the dijet and the average pair mass in the event. The Δ requirement removes a broad structure around 300 GeV from QCD events and ensures a smoothly falling dijet mass background shape. It also reduces the QCD background by more than an order of magnitude while retaining approximately 25% of the signal.

The data used in this study correspond to an integrated luminosity of 5.0 fb^{-1} at a collision energy of $\sqrt{s}$=7 TeV [20]. No evidence of new physics is found in the data. The expected and observed 95% CL upper limits as a function of mass are presented in Fig. 2. For the first time, at 95% CL, the pair production of colorons is excluded for coloron masses between 250 and 740 GeV assuming that a coloron decays 100% into $q\bar{q}$, or between 250 and 580 GeV assuming that coloron decays into $q\bar{q}$ compete with decays into $S_8 S_8$. With the current integrated luminosity there is some sensitivity to a SUSY model for pair-produced stops.

5. Conclusion

We performed searches for new physics in the singly-produced and pair-produced dijet mass spectrum. No sign for new physics is found. Upper limits are set for various models and

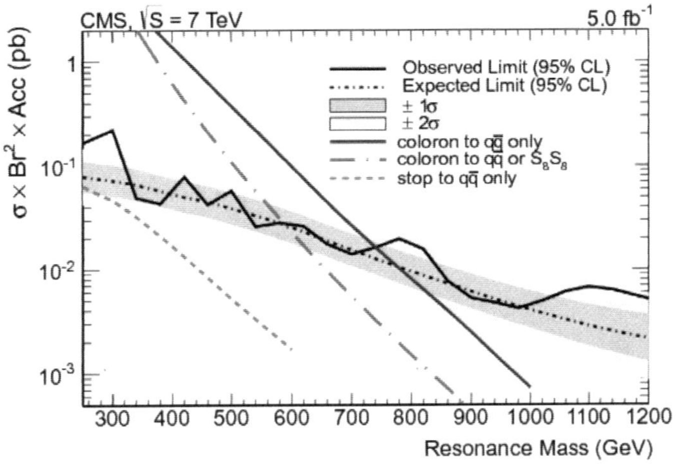

Figure 2. The observed and expected 95% CL limits on the product of the resonance pair production cross section, the square of the branching fraction to dijets, and the detector acceptance, given by the solid and dot-dashed black curves, respectively. The shaded regions indicate the 1σ and 2σ bands around the expected limits. Predictions of a coloron model and a SUSY model are also shown.

significant parameters space are excluded. We exclude the string resonance between 1 TeV and 4.7 TeV in the searches from the single dijet mass spectrum, and exclude the coloron with a mass between 250 GeV and 740 GeV for the first time in the searches from the paired dijet mass spectrum.

References

[1] L. A. Anchordoqui et al., "Jet signals for low mass strings at the Large Hadron Collider", Phys. Rev. Lett. 101 (2008) 241803, doi:10.1103/Physrevlett.100.171603.
[2] S. Cullen, M. Perelstein, and M. E. Peskin, "TeV strings and collider probes of large extra dimensions", Phys. Rev. D 62 (2000) 055012, doi:10.1103/PhysRevD.62.055012.
[3] J. L. Hewett and T. G. Rizzo, "Low-energy phenomenology of superstring-inspired E6 models", Phys. Rept. 183 (1989) 193, doi:10.1016/0370-1573(89)90071-9.
[4] U. Baur, I. Hinchliffe, and D. Zeppenfeld, "Excited Quark Production at Hadron Colliders", Int. J. Mod. Phys. A 2 (1987) 1285, doi:10.1142/S0217751X87000661.
[5] U. Baur, M. Spira, and P. M. Zerwas, "Excited Quark and Lepton Production at Hadron Colliders", Phys. Rev. D 42 (1990) 815, doi:10.1103/PhysRevD.42.815.
[6] P. H. Frampton and S. L. Glashow, "Chiral color: An alternative to the standard model", Phys. Lett. B 190 (1987) 157, doi:10.1016/0370-2693(87)90859-8.
[7] R. Chivukula, A. Farzinnia, E. H. Simmons et al., "Production of Massive Color-Octet Vector Bosons at Next-to-Leading Order", Phys.Rev. D85 (2012) 054005, doi:10.1103/PhysRevD.85.054005.
[8] E. H. Simmons, "Coloron phenomenology", Phys. Rev. D 55 (1997) 1678, doi:10.1103/PhysRevD.55.1678. References 11
[9] T. Han, I. Lewis, and Z. Liu, "Colored Resonant Signals at the LHC: Largest Rate and Simplest Topology", JHEP 1012 (2010) 085, doi:10.1007/JHEP12(2010)085, arXiv:1010.4309.
[10] E. Eichten et al., "Supercollider physics", Rev. Mod. Phys. 56 (1984) 579, doi:10.1103/RevModPhys.56.579.
[11] L. Randall and R. Sundrum, "An alternative to compactification", Phys. Rev. Lett. 83 (1999) 4690, doi:10.1103/PhysRevLett.83.4690.
[12] B. A. Dobrescu, K. Kong, and R. Mahbubani, Phys. Lett. B 670 (2008) 119; C. Kilic, S. Schumann, and M. son, JHEP 04 (2009) 128; Y. Bai and B. A. Dobrescu, JHEP 07 (2011) 100.
[13] C. Brust, A. Katz, and R. Sundrum, JHEP 08 (2012) 069; J. A. Evans and Y. Kats, arXiv:1209.0764.
[14] The CMS Collaboration, CMS Physics Analysis Summary CMS-PAS-PFT-10-001, (2010).
[15] The CMS Collaboration, JINST 3, S08004 (2008).

[16] Cacciari, Matteo and Salam, Gavin P. and Soyez, Gregory, Journal of High Energy Physics 2008 (2008), no. 04, 063.
[17] The CMS Collaboration, CMS PAS JME-10-010.
[18] M. Cacciari, J. Rojo, G. P. Salam et al., JHEP 0812 (2008) 032, arXiv:0810.1304; D. Krohn, J. Thaler, and L.-T. Wang, JHEP 02 (2010) 084; A. Abdesselam et al., Eur. Phys. J. C 71 (2011) 1661.
[19] The CMS collaboration, Phy. Rev. Lett. 105, 211801 (2010); The CMS collaboration, Phy. Lett. B 704, 123-142 (2011). The CMS Collaboration, CMS Physics Analysis Summary CMS-PAS-EXO-12-016, (2012). arXiv:1302.4794 [hep-ex].
[20] S. Chatrchyan *et al.* [CMS Collaboration], arXiv:1302.0531 [hep-ex].

Journal of Physics: Conference Series **455** (2013) 012035 doi:10.1088/1742-6596/455/1/012035

Testing the Standard Model with rare decays at the LHCb.

Eugeni Graugés, *on behalf of the LHCb Collaboration*

Departament d'Estructura i Constituents de la Matèria, Institut de Ciències del Cosmos, Universitat de Barcelona, Martí i Franquès 1, E-08028 Barcelona, Catalonia, Spain.

E-mail: eugeni.grauges@cern.ch

Abstract. The study of rare decays of B, D and K mesons (as well as τ leptons) can provide either very stringent tests for the Standard Model or excellent opportunities to search for hints of new physics beyond the SM. A review of the recent results of studies of such rare decays from analysis of *pp* collision data recorded at the LHCb experiment at CERN, is presented. All decays channels studied so far are in agreement with SM predictions.

1. Introduction

The LHCb experiment [1] at CERN's Large Hadron Collider (LHC) is focused on the study of b-quark hadrons to perform precision measurements of CP violating observables and rare decays that could unravel possible contributions from new physics (NP) beyond the Standard Model (SM). At the LHC proton-proton (*pp*) collision energies, charmed mesons and tau leptons are also copiously produced (at the level of b-hadrons or higher), and therefore, the corresponding equivalent observables can also be studied. These proceedings concentrate on key measurements of rare decays, while the measurements on CP violation observables are described on separate proceedings contribution [2]. The LHCb detector, described elsewhere [1], has been performing extremely well in the first LHC runs. With around 99% of operational readout channels and approximately a 95% of data taking efficiency, the LHCb experiment has been able to fully exploit the provided luminosity delivered by the LHC accelerator. The measurements presented here, unless mentioned otherwise, have been performed using the data set, corresponding to an integrated luminosity of 1.0 fb^{-1} of *pp* collisions collected at a centre-of-mass energy of 7 TeV during 2011. Roughly an additional 2 fb^{-1} of data at a centre-of-mass energy of 8 TeV have been collected during 2012 and, therefore, some of the analyses here presented might reach an increased sensitivity, once the entire data set is analysed.

The study of very suppressed, within the SM, beauty and charmed meson decays searching for any possible unpredicted enhancement of their respective branching fractions (BF) , is an excellent place to look for hints for NP processes, beyond the SM, that could provide extra contributions to the otherwise very suppressed decay amplitudes. Those extra contributions could also affect the Lorentz structure of the decay amplitudes and therefore the angular distributions of the decay products become also a very sensitive observable. Processes like $B_d \to K^{*0}\gamma$, $B_s \to \phi\gamma$, $B \to K^*\mu^+\mu^-$, $B^+ \to K^+\mu^+\mu^-$, $B_s \to \phi\mu^+\mu^-$, $B^+ \to \pi^+\mu^+\mu^-$, $B^0/B_s \to \mu^+\mu^-$, $K_s \to \mu^+\mu^-$ and $D \to \mu^+\mu^-$ are examples of flavour-changing neutral current processes (FCNC) which are expected to be very rare, since they cannot occur at tree level in the SM.

Additionally, searching for processes that are directly forbidden by the SM, such as those violating baryon (BNV) or lepton number conservation (or both) like $\tau \to p\mu^+\mu^-$ or $B^+ \to h\mu^+\mu^+$, would, if found, be unequivocally a signature of physics beyond the SM. Another example would be the τ lepton decay $\tau^- \to \mu^-\mu^+\mu^-$, which would correspond to lepton-flavor violating (LFV) decays, which are essentially forbidden in the SM because they can occur only through lepton mixing.

2. Rare B, D and K meson decays.

The SM decays of $B_d \to K^{*0}\gamma$ and $B_s \to \phi\gamma$ occur through a FCNC transition involving loop diagrams. In LHCb, the relative ratio of BFs of the former decay channels, that only differ in the B meson spectator quark, is measured to be [3]

$$\frac{BF(B_d \to K^{*0}\gamma)}{BF(B_s \to \phi\gamma)} = 1.23 \pm 0.06 \text{ (stat.)} \pm 0.04 \text{ (syst.)} \pm 0.10 \; (f_s/f_d),$$

where the first uncertainty is statistical, the second systematic, and the third is associated with the ratio of fragmentation fractions f_s/f_d. This result is in good agreement with the theoretical prediction of 1.0 ± 0.2 [4].

Since the initial B meson flavour in the $B_d \to K^{*0}\gamma$ decay channel is self tagged through the ulterior decay process $K^* \to K^+\pi^-$, the direct CP asymmetry of this decay can be easierly measured. Even though, in the SM, this CP asymmetry is predicted to be below 1% level [5], some NP models [6] predict it to be enhanced up to values around 15%. In LHCb, the most precise measurement to date of this CP asymmetry results in [7]

$$A_{CP}(B^0 \to K^{*0}\gamma) = (\ 0.8 \pm 1.7 \text{ (stat.)} \pm 0.9 \text{ (syst.)}\ )\ \%,$$

in agreement with the SM expectation [5]. Similarly the CP asymmetry of the $B^0 \to K^{*0}\mu^+\mu^-$ decay channel is measured to be [8]

$$A_{CP}(B^0 \to K^{*0}\mu^+\mu^-) = -\ 0.072 \pm 0.040 \text{ (stat.)} \pm 0.005 \text{ (syst.)},$$

which is also consistent with the SM prediction [9].

Studying the same former B meson decay channel, it is possible to access other observables sensitive to possible NP effects. For instance the forward-backward asymmetry defined in bins of invariant mass squared of the dimuon system $A_{FB}(q^2 = m^2_{\mu^+\mu^-})$ flips sign, according to the SM, at a precisely predicted value of q^2 between 4 - 4.3 GeV^2/c^4 [10]. The zero-crossing point of the forward-backward asymmetry is measured in LHCb and it is determined to be $q_0^2 = 4.9^{+1.1}_{-1.3}$ GeV^2/c^4. Also, the measured differential branching fractions (in q^2 bins) and other angular observables sensitive to NP are measured to be consistent with the SM predictions [11]. The same observables have been studied and measured in the similar, but charged B meson decay, $B^+ \to K^+\mu^+\mu^-$ channel [12], and no significant inconsistency with the SM predictions [13] has been found. Comparing the (neutral and charged) B meson decays $B^0 \to K^{(*)0}\mu^+\mu^-$ and $B^+ \to K^{(*)+}\mu^+\mu^-$, it is possible to measure the so-called isospin asymmetry, defined as

$$\mathrm{A}_I = \frac{\mathbf{BF}(B^0 \to K^{(*)0}\mu^+\mu^-) - (\tau_0/\tau_+)\mathbf{BF}(B^+ \to K^{(0)+}\mu^+\mu^-)}{\mathbf{BF}(B^0 \to K^{(*)0}\mu^+\mu^-) + (\tau_0/\tau_+)\mathbf{BF}(B^+ \to K^{(0)+}\mu^+\mu^-)},$$

in bins of q^2. The LHCb measurements for the $B \to K^*\mu^+\mu^-$ channels are in agreement with the SM expectations, while the isospin asymmetry measurement for the $B \to K\mu^+\mu^-$ channels, if integrated over the whole q^2 range, show a disgreement at the level of 4.4 standard deviations from the SM prediction of a 0 value [14]. These results are nevertheless consistent with previous measurement by other experiments [15].

Another rare B meson decay studied is the $B_s \to \phi\mu^+\mu^-$, whose BF is measured [16], normalized to the decay of $B_s \to \phi\ J/\psi$ (where $J/\psi \to \mu^+\mu^-$) , to be

$$\mathrm{BF}(B_s \to \phi\mu^+\mu^-) = (\ 0.78 \pm 0.10\ (\mathrm{stat.}) \pm 0.06(\mathrm{syst.}) \pm 0.28\ (\mathrm{BF})\ ) \times 10^{-6},$$

where the first error is statistical, the second error is systematic and the third error is due to the uncertainty on branching fractions, in this case $\mathrm{BF}(J/\psi \to \mu^+\mu^-)$.

The rarest B meson decay ever observed, that is, the $B^+ \to \pi^+\mu^+\mu^-$ decay, was first observed in LHCb. This decay corresponds to a FCNC transition from a b to a d quark type. Its BF was measured in LHCb to be $\mathrm{BF}(B^+ \to \pi^+\mu^+\mu^-) = (\ 2.4 \pm 0.6\ (\mathrm{stat.}) \pm 0.2(\mathrm{syst.})\ ) \times 10^{-8}$, in agreement with the SM prediction [17].

One of the key LHCb measurements and an excellent B meson decay channel to search for hints of NP beyond the SM is the $B^0/B_s \to \mu^+\mu^-$. These decays are doubly suppressed since they correspond to a FCNC trasition and the amplitudes of these processes are helicity supressed. The SM prediction [18] for these decay channels are of the order of 10^{-9} - 10^{-10}. Therefore, if any significant signal is found, implying an enhancement of the predicted BFs, it could be a very strong hint for NP. The search for these decay channels has been carried out using a data set of pp collisions of approximately 2fb^{-1}, collected by the LHCb experiment during 2011 and 2012 with a centre of mass energy of 7 TeV and 8 TeV respectively. Evidence for signal, with a significance of 3.5 standard deviations, is found [19] in the $B_s \to \mu^+\mu^-$ decay and the corresponding BF is measured to be

$$\mathrm{BF}(B_s \to \mu^+\mu^-) = (\ 3.2\ {}^{+1.5}_{-1.2}\ ) \times 10^{-9},$$

in impressive agreement with the precise SM predictions. No significant signal yield is found in the $B^0 \to \mu^+\mu^-$ decay channel search [19], and the corresponding upper limit (UL) at 95 % confidence level (CL) is set at

$$\mathrm{BF}(B^0 \to \mu^+\mu^-) < 9.4 \times 10^{-10}.$$

Similar analysis techniques have been used to look for other meson decays into a muon pair, that are as supressed as the B meson decays. Particularly the search for the decays of the K^0_s and D^0 mesons has been carried out in LHCb. The $\mathrm{K}^0_s \to \mu^+\mu^-$ decay has a BF predicted by the SM [20] of the order of 10^{-12} while the $\mathrm{D}^0 \to \mu^+\mu^-$ BF is expected to be of the order of 10^{-11} [21]. No significant signal above the background level has been found in both decays [22] [23] and the corresponding UL on the BF's at 95% CL have been set at

$$\mathrm{BF}\ (\mathrm{K}^0_s \to \mu^+\mu^-) < 11 \times 10^{-9}$$
$$\mathrm{BF}\ (\mathrm{D}^0 \to \mu^+\mu^-) < 1.3 \times 10^{-8}$$

The result for the K^0_s decay is a improvement of a factor 30 with respect to previous measurements, whilst in the result on the D^0 decay, an order of magnitude improvement is achieved with respect from the previous best limit [24].

3. Search for lepton number and flavor violation decays.

Lepton flavor violation is allowed in the SM when accomodating the neutrino oscillations. Nevertheless the BF for the LFV decay $\tau^- \to \mu^-\mu^+\mu^-$ predicted by the SM is beyond the actual experimental scope (some estimates consider it of the order of 10^{-54}). That might not be the case if NP models are considered, since some of them predict enhancements in its BF up to the 10^{-10} - 10^{-8} range. With a production cross section of τ leptons of approximately 22μb within the LHCb acceptance, nearly 10^{11} τ leptons are produced per fb^{-1} of collected pp collision data at a centre of mass energy of 7 TeV (mostly from D^+_s decays). No signal has been observed as a result of analysing this data set, and the corresponding UL on the BF at 95% CL has been set at [25]

$$\mathrm{BF}(\tau^- \to \mu^-\mu^+\mu^-) < 7.8 \times 10^{-8}.$$

Additionally a search is performed for the lepton and baryon number violating decays $\tau \to p\mu^+\mu^-$ and $\tau \to \bar{p}\mu^+\mu^-$. If found, this kind of process could provide clues for extra (beyond SM) sources of BNV, that could shed light into the explanation for the matter/antimatter univers asymmetry. Unfortunately, no signal is found in the data [26] and the corresponding UL at 95% CL are set to

$$\mathrm{BF}(\tau \to p\mu^+\mu^-) < 4.5 \times 10^{-7}$$
$$\mathrm{BF}(\tau \to \bar{p}\mu^+\mu^-) < 6.0 \times 10^{-7}.$$

Finally, a search for Majorana type neutrinos in B^- decays in final states containing hadrons plus a $\mu^-\mu^-$ pair have been performed using 0.41 fb^{-1} of data collected with LHCb detector in *pp* collisions at a centre of mass energy of 7 TeV. No signals are found and the UL are set on Majorana neutrino production as a function of mass, and also on B^- decay branching fractions [27].

4. Summary and Outlook.

The LHCb experiment is revealing itself as a fantastic experiment to look ar rare B, D and K meson decays that are very sensistive NP. So far, the analysis of the data have resulted in measurements, all of them consistent with the SM predictions and, therefore, in establishing very strong constraints for NP models beyond the SM. No evidence for LFV neither for LNV(BNV) has been found in analyzing specific τ lepton and B meson decays. Nevertheless, the precision of the current measurements is bound to be improved with the addition in the physics analysis of data set of *pp* collisions at a centre of mass energy of 8 TeV gathered in LHCb during 2012. The energy of the LHC is expected to increase up to 13 TeV, after a shutdown period of about two years (2013-2014), and an extra 5fb^{-1} are expected to be collected, allowing for a further improvement of the measurements precision and physics reach. An upgrade of the LHCb detector is foreseen during the period 2018-2019, just after a second long shut down of the LHC accelerator. The main goal of that upgrade is to operate the experiment at higher instantaneous luminosity (increasing the data collected) and to improve further the trigger efficiency for heavy quark decays to purely hadronic final states.

References

[1] A. A. Alves, Jr. *et al.* [LHCb Collaboration], JINST **3** (2008) S08005.
[2] Olaf Steinkamp *on behalf of the LHCb collaboration*, LHCb-PROC-2013-021
[3] R. Aaij *et al.* [LHCb Collaboration], Phys. Rev. D **85**, 112013 (2012)
[4] A. Ali, B. D. Pecjak, and C. Greub, Eur. Phys. J. C **55**, 577 (2008)
[5] M. Matsumori, A.I. Sanda, Y.Y. Keum, Phys. Rev. D **72** (2005) 014013, arXiv:hep-ph/0406055.
[6] C. Dariescu, M.-A. Dariescu, arXiv:0710.3819;
M. Aoki, G.-C. Cho, N. Oshimo, Phys. Rev. D **60** (1999) 035004, arXiv:hep-ph/9811251;
M. Aoki, G.-C. Cho, N. Oshimo, Nucl. Phys. B **554** (1999) 50, arXiv:hep-ph/9903385;
A.L. Kagan, M. Neubert, Phys. Rev. D **58** (1998) 094012, arXiv:hep-ph/9803368.
[7] R. Aaij *et al.* [LHCb Collaboration], Nuc. Phys. B **867** (2013) 118
[8] R. Aaij *et al.* [LHCb Collaboration], Phys. Rev. Lett. **110**, 031801 (2013)
[9] C. Bobeth, G. Hiller, and G. Piranishvili, JHEP **07** (2008) 106, arXiv:0805.2525.
W. Altmannshofer et al., JHEP **01** (2009) 019, arXiv:0811.1214
[10]] C. Bobeth, G. Hiller, D. van Dyk, and C. Wacker, JHEP **01** (2012) 107, arXiv:1111.2558.
M. Beneke, T. Feldmann, and D. Seidel, Eur. Phys. J. C **41** (2005) 173, arXiv:hep-ph/0412400
[11] R. Aaij *et al.* [LHCb Collaboration], LHCb-CONF-2012-008
[12] R. Aaij *et al.* [LHCb Collaboration], JHEP **02** (2013) 105 arxiv:1209.4284
[13] Christoph Bobeth, Gudrun Hiller, Danny van Dyk, JHEP **07** (2011) 067
C. Bobeth, G. Hiller, D. van Dyk and C. Wacker, JHEP **01** (2012) 107
[14] R. Aaij *et al.* [LHCb Collaboration], JHEP **07** (2012) 133
[15] BaBar collaboration, Phys.Rev. D **86** (2012) 032012, arXiv:1204.3933.
CDF collaboration, T. Aaltonen et al., Phys. Rev. Lett. **107** (2011) 201802, arXiv:1107.3753.
[16] R. Aaij *et al.* [LHCb Collaboration], LHCb-CONF-2012-003

[17] R. Aaij *et al.* [LHCb Collaboration], JHEP **12** (2012) 125, arxiv:1210.2645 and references therein.
[18] Buras, Isidori: arXiv:1208.0934
De Bruyn et al., Phys. Rev. Lett. **109**, 041801 (2012) using LHCbCONF2012002
[19] R. Aaij et al. [LHCb Collaboration] Phys. Rev. Lett. **110**, 021801 (2013), arxiv:1211.2674
[20] G. Ecker and A. Pich, Nucl. Phys. **B366** (1991) 189.
G. Isidori and R. Unterdorfer, JHEP **01** (2004) 009, arXiv:hep-ph/0311084.
[21] G. Burdman, E. Golowich, J. L. Hewett, and S. Pakvasa, Phys. Rev. **D66** (2002) 014009, arXiv:hep-ph/0112235.
[22] R. Aaij *et al.* [LHCb Collaboration], JHEP **01** (2013) 090
[23] R. Aaij *et al.* [LHCb Collaboration], LHCb-CONF-2012-005
[24] M. Petric *et al.* [The Belle Collaboration], Phys. Rev. D **81** (2010) 091102
[25] R. Aaij *et al.* [LHCb Collaboration], LHCb-CONF-2012-015
[26] R. Aaij *et al.* [LHCb Collaboration], LHCb-CONF-2012-027
[27] R. Aaij *et al.* [LHCb Collaboration], Phys. Rev. **D 85** (2012) 112004

Journal of Physics: Conference Series **455** (2013) 012036 doi:10.1088/1742-6596/455/1/012036

CP violation at LHCb

Olaf Steinkamp ***on behalf of the LHCb Collaboration***
Physik-Institut der Universität Zürich, Winterthurerstrasse 190, 8057 Zürich, Switzerland
E-mail: olafs@physik.uzh.ch

Abstract. Precision measurements of *CP* violating observables in the mixing and decay of *B* mesons provide excellent opportunities to search for possible contributions from New Physics beyond the Standard Model. In these proceedings, key measurements pursued by the LHCb collaboration at CERN's Large Hadron Collider (LHC) are described. Results obtained from the analysis of data collected in 2011 already match or even surpass the measurements of previous experiments in their precision. All results obtained so far are compatible with Standard Model predictions, hints for possible deviations reported by previous experiments are not confirmed. All results are dominated by statistical uncertainties and significant improvements are expected from analysing larger data sets. The total integrated luminosity accumulated by LHCb has already been tripled in 2012 and is expected to more than double again in the data taking period following the first long shutdown of the LHC. A comprehensive upgrade of the LHCb apparatus is then foreseen for the second long shutdown of the LHC accelerator in 2018/2019. The main goals of the LHCb upgrade are to operate the experiment at higher instantaneous luminosities and to improve further the trigger efficiency for heavy quark decays to purely hadronic final states.

1. Introduction

The main goal of the LHCb experiment [1] at CERN's Large Hadron Collider (LHC) is the search for possible contributions from physics beyond the Standard Model ("New Physics") by performing precision measurements of *CP* violating observables and rare decays of hadrons containing a b quark or a c quark. The LHCb detector and its performance in the first LHC runs, as well as key measurements of rare b and c decays, are described in another contribution to these proceedings [2]. Here, we will concentrate on measurements of *CP* violating observables. Unless mentioned otherwise, the measurements have been performed using the full 2011 data set, corresponding to an integrated luminosity of 1.0 fb^{-1} of pp collisions collected at a centre of mass energy of 7 TeV.

The only source of *CP* violation in the Standard Model of particle physics is a single complex phase in the Cabbibo-Kobayashi-Maskawa (CKM) matrix [3, 4], which parametrises the mixing between the quark families in charged current interactions. For the case of three quark families, the CKM matrix is fully described by four independent real parameters, namely three rotation angles and one complex phase. A non-zero value of this complex phase can induce *CP* violating asymmetries in processes in which two or more amplitudes with different weak phases interfere. Often, one of the interfering amplitudes involves loop diagrams with internal quark lines, such as box or penguin diagrams. Heavy new particles, which are predicted by most New Physics models, can then appear in the internal lines of these loop diagrams and modify the observed asymmetries with respect to Standard Model predictions. Precision measurements of these asymmetries therefore test the Standard Model and can reveal the presence of New Physics. Since these measurements probe the effect of virtual particles in internal loop diagrams, their sensitivity extends to much higher mass scales than direct searches for new particles, which are limited to the centre of mass energy of the accelerator at which the measurement is performed. Moreover, the pattern of observed deviations from Standard Model predictions can hint at the underlying dynamics of the New Physics at play. The B meson systems ($B^{\pm}$, $B^0/\overline{B}^0$ and $B_s^0/\overline{B}_s^0$) give rise to a rich phenomenology with a wide range of observables for which precise Standard Model predictions are available. Particularly high sensitivity to possible contributions from New Physics can be achieved in processes where the predicted asymmetry in the Standard Model is small.

The CKM picture gives rise to three mechanisms that can lead to CP violating asymmetries:

***CP* violation in mixing** in the $B^0\overline{B}^0$ and $B_s^0\overline{B}_s^0$ systems is caused by the interference of box diagrams with different internal quark lines. The mixing phase in the $B_s^0\overline{B}_s^0$ system is predicted with good precision to be close to zero in the Standard Model and is therefore particularly sensitive to possible contributions from New Physics.

***CP* violation in decay** is caused by the interference of decay diagrams with different weak phase and different strong phase leading to the same final state. This is the only possible source of CP violation for charged B mesons. In general, the poorly known strong phases limit the precision with which the weak phase can be extracted from the measured asymmetry. There are, however, cases in which the weak phase can be cleanly extracted using a combination of several related decay modes and exploiting symmetries of the strong interaction between these modes.

***CP* violation in the interference of mixing and decay** of B^0 or B_s^0 mesons to a final state that is accessible to both the $B_{(s)}^0$ meson and its antiparticle is caused by the interference of the direct decay amplitude of a $B_{(s)}^0$ meson to the given final state and the process where the $B_{(s)}^0$ meson first mixes to a $\overline{B}_{(s)}^0$, which then decays to the same final state. A prominent example is the "golden" decay mode $B^0 \to J/\psi K_S^0$, which allowed the precise measurement of the mixing phase $\sin 2\beta$ in the $B^0\overline{B}^0$ system by the B factories BaBar [5] and Belle [6].

In the following, we will encounter examples of all three types of CP violation.

2. Semileptonic Asymmetry

Second-order weak transitions (box diagrams) induce particle-antiparticle mixing in the $K^0\overline{K}^0$, $D^0\overline{D}^0$, $B^0\overline{B}^0$ and $B_s^0\overline{B}_s^0$ systems. The time evolution of the coupled particle–antiparticle system can be described by an effective 2×2 Schrödinger equation of the form

$$-i\frac{\partial}{\partial t}\begin{pmatrix} B_s^0 \\ \overline{B}_s^0 \end{pmatrix} = H \cdot \begin{pmatrix} B_s^0 \\ \overline{B}_s^0 \end{pmatrix} = \begin{pmatrix} M_{11} - \frac{i}{2}\Gamma_{11} & M_{12} - \frac{i}{2}\Gamma_{12} \\ M_{12}^{\star} - \frac{i}{2}\Gamma_{12}^{\star} & M_{22} - \frac{i}{2}\Gamma_{22} \end{pmatrix} \cdot \begin{pmatrix} B_s^0 \\ \overline{B}_s^0 \end{pmatrix}$$

(and similar for the other neutral meson–antimeson systems). The eigenstates of the effective Hamiltonian H are the heavy and light mass eigenstates, B_H and B_L, which have well defined masses, m_H and m_L, and lifetimes, Γ_H and Γ_L. Relevant observables are the mass difference $\Delta m_s \equiv m_H - m_L$ and the lifetime difference $\Delta\Gamma_s \equiv \Gamma_H - \Gamma_L$. Note that Δm_s is positive by definition, whereas $\Delta\Gamma_s$ can be positive or negative. Particle-antiparticle mixing is described by the off-diagonal elements of H. A relative phase ϕ_M between M_{12} and Γ_{12} induces CP violation in mixing, implying that the probability for the transition $B_s^0 \to \overline{B}_s^0$ is different from the probability for the transition $\overline{B}_s^0 \to B_s^0$. This CP asymmetry can be measured in the time-integrated semileptonic asymmetry

$$a_{\rm sl}^s \equiv \frac{\Gamma(\overline{B}_s^0 \to \ell^+ X) - \Gamma(B_s^0 \to \ell^- X)}{\Gamma(\overline{B}_s^0 \to \ell^+ X) + \Gamma(B_s^0 \to \ell^- X)} = \frac{\Delta\Gamma_s}{\Delta m_s}\tan\phi_M.$$

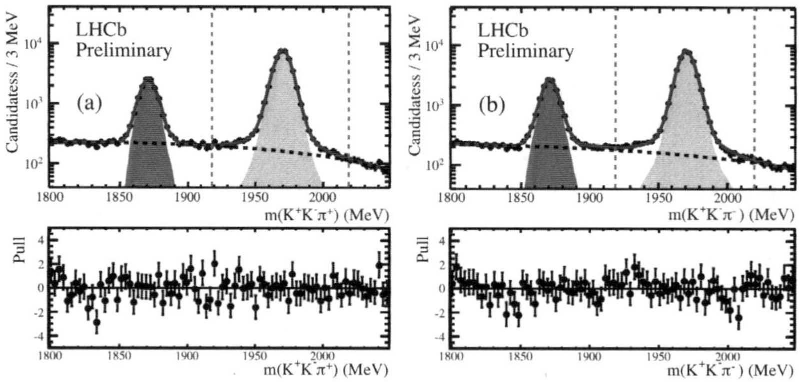

Figure 1. Invariant mass distributions for (a) $K^+K^-\pi^+$ candidates and (b) $K^+K^-\pi^-$ candidates for magnet up polarity, with $m(K^+K^-)$ within ± 20 MeV$/c^2$ of the ϕ meson mass.

In the Standard Model, the phase ϕ_M in the $B_s^0\overline{B}_s^0$ system is small and the semileptonic asymmetry is predicted to be $a_{\rm sl}^s = (1.9 \pm 0.3) \times 10^{-5}$ [7]. The corresponding asymmetry in the $B^0\overline{B}^0$ system is predicted to be $a_{\rm sl}^d = (4.1 \pm 0.6) \times 10^{-4}$ [7]. Measurements of $a_{\rm sl}^d$ reported by the B factories [8, 9, 10] and by D0 [11] agree with the Standard Model prediction. The D0 collaboration, however, also reported a measurement of the asymmetry in the rate of like-sign muon pairs, $A_{\mu\mu} \approx 0.6 \cdot a_{\rm sl}^d + 0.4 \cdot a_{\rm sl}^s$, which yielded an anomalously large value [12].

LHCb has reported a preliminary measurement of $a_{\rm sl}^s$ [13] using the decay $B_s^0 \to D_s^-\mu^+X$ with $D_s^- \to \phi\pi^-$ and $\phi \to K^+K^-$. No flavour tagging was employed to determine the initial flavour of the B_s^0 meson at its production. The measured raw asymmetry is then

$$A_{\rm raw} \equiv \frac{N(D_s^-\mu^+X) - N(D_s^+\mu^-X)}{N(D_s^-\mu^+X) + N(D_s^+\mu^-X)} = \frac{a_{\rm sl}^s}{2} + \kappa \cdot \left[a_{\rm p} - \frac{a_{\rm sl}^s}{2}\right],$$

where $a_{\rm p}$ is the $B_s^0 - \overline{B}_s^0$ production asymmetry and

$$\kappa \equiv \frac{\int e^{-\Gamma_s t} \cos(\Delta m_s t)\, \epsilon(t)\, dt}{\int e^{-\Gamma_s t} \cosh(\Delta\Gamma_s t/2)\, \epsilon(t)\, dt}$$

is a dilution factor due to $B_s^0 - \overline{B}_s^0$ oscillation, with $\epsilon(t)$ describing the reconstruction efficiency as a function of the decay time. The large oscillation frequency leads to a very small value of κ in the $B_s^0\overline{B}_s^0$ system, with $\kappa \approx 2 \times 10^{-3}$ for the LHCb decay time acceptance. Therefore, even a production asymmetry $a_{\rm p}$ of a few percent would be washed out to a negigible level compared to the current statistical precision of the measurement.

Candidates were selected using a trigger that first required a muon with significant transverse momentum, followed by two inclusive selections that made use either of the topology of a displaced decay vertex or the kinematic reconstruction of $\phi \to K^+K^-$ decays. To minimize and control potential detector asymmetries, the polarity of the LHCb dipole magnet was reversed several times throughout the data taking period. The complete analysis was performed separately for these two subsets of the data and the two results were then averaged to obtain the final result. Detection asymmetries largely cancel due to this averaging. Residual detection asymmetries were estimated from data, using dedicated control samples. Combinatorial background was subtracted statistically by fitting the $K^+K^-\pi^+$ invariant mass distributions as shown in Fig. 1 for candidates collected with one of the two magnet polarities. Other backgrounds are small and accounted for as part of the systematic uncertainties.

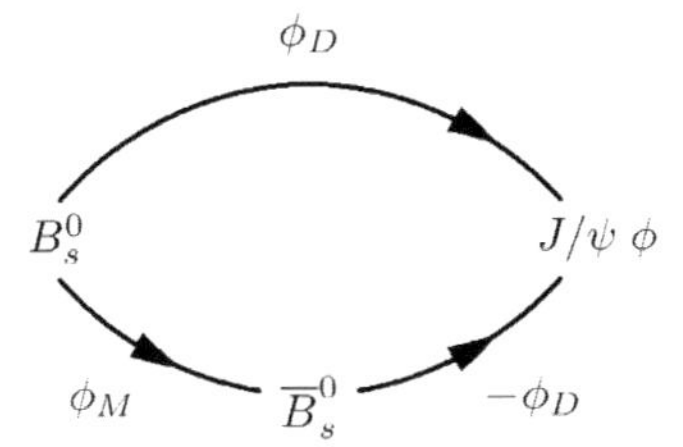

Figure 2. Illustration of weak phases in the interference of mixing and decay for $B_s^0 \to J/\psi\phi$. See the text for further explanations.

The measured asymmetry is $a_{\rm sl}^s = (0.24 \pm 0.54({\rm stat}) \pm 0.33({\rm syst})) \times 10^{-2}$ and agrees with the Standard Model prediction. The precision of the measurement is limited by the statistical uncertainty. The systematic uncertainty is dominated by the statistical uncertainty on the determination of efficiency ratios and can be expected to decrease with larger control samples.

3. *CP* violation in $B_s^0 \to J/\psi\phi$ and $B_s^0 \to J/\psi\pi^+\pi^-$

The final state $J/\psi\phi$ is accessible to both B_s^0 and $\overline{B}_s^0$. As illustrated in Fig. 2, the interference between the decay amplitudes and the $B_s^0 - \overline{B}_s^0$ mixing amplitude can give rise to a *CP* violating phase

$$\phi_s = \phi_M - 2\phi_D,$$

where ϕ_M is the $B_s^0 - \overline{B}_s^0$ mixing phase discussed in the previous section and ϕ_D is the weak decay phase. In the Standard Model, ϕ_D is very small, resulting in a precise prediction of $\phi_s = 0.036 \pm 0.002$ rad [15]. A measurement of ϕ_s is interesting since it is very sensitive to possible contributions from New Physics in $B_s^0 - \overline{B}_s^0$ mixing.

The experimental determination of ϕ_s proceeds through the measurement of the time-dependent *CP* asymmetry

$$A_{CP}(t) \equiv \frac{\frac{\partial\Gamma}{\partial t}(\overline{B}_s^0 \to J/\psi\phi) - \frac{\partial\Gamma}{\partial t}(B_s^0 \to J/\psi\phi)}{\frac{\partial\Gamma}{\partial t}(\overline{B}_s^0 \to J/\psi\phi) + \frac{\partial\Gamma}{\partial t}(B_s^0 \to J/\psi\phi)} = \frac{-\eta_f\,\sin\phi_s\,\sin(\Delta m_s t)}{\cosh(\Delta\Gamma_s t/2) - \eta_f\,\cos\phi_s\,\sinh(\Delta\Gamma_s t/2)},$$

where B_s^0 and $\overline{B}_s^0$ depict the flavour of the B meson at production, η_f is the *CP* eigenvalue of the final state, and Δm_s and $\Delta\Gamma_s$ are the mass and lifetime differences between the heavy and light mass eigenstates in the $B_s^0\overline{B}_s^0$ system as introduced in the previous section.

Reconstructing the two intermediate resonances in their decays $J/\psi \to \mu^+\mu^-$ and $\phi \to K^+K^-$ provides for a clear trigger signature and clean event samples. Several factors, however, make this a challenging measurement:

- The flavour of the B meson at production has to be derived using information from the remainder of the event. This flavour tagging is not perfect and the associated mis-tag fraction dilutes the oscillation amplitude. Moreover, the dilution factor has to be known precisely in order to extract ϕ_s.
- An excellent decay time resolution is required to resolve the rapid $B_s^0 - \overline{B}_s^0$ oscillation. The finite decay time resolution leads to another dilution of the oscillation amplitude and has therefore to be known precisely in order to extract ϕ_s.
- $B_s^0 \to J/\psi\phi$ being a decay of a pseudoscalar to two vector mesons, the two final state particles can be produced with a relative angular momentum of $L = 0$, 1 or 2. The final state is not a *CP* eigenstate, $L = 1$ being *CP* odd while $L = 0$ and 2 are *CP* even. A time-dependent angular analysis of the four final state particles can statistically separate the *CP* odd and *CP* even contributions but requires large statistics.

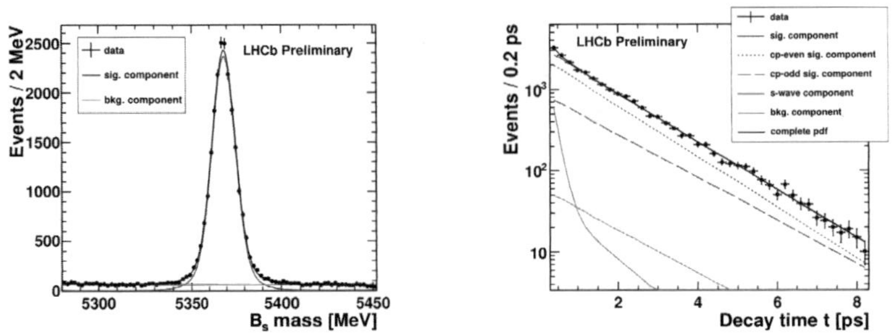

Figure 3. Left: $J/\psi K^+K^-$ invariant mass distribution of the event sample used for the ϕ_s measurement. Right: decay time distribution with fit projections overlayed.

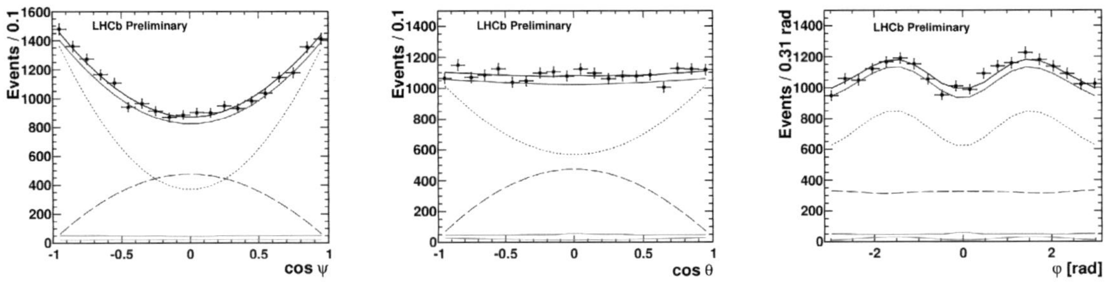

Figure 4. Distribution in the three helicity angles Ψ, θ and φ (see Ref. [23]), overlayed with projections of the time-dependent angular fit for ϕ_s. See the right plot in Fig. 3 for a legend.

- The value of $\Delta\Gamma_s$ has not been precisely measured previously and has to be determined simultaneously with ϕ_s.

Early measurements of ϕ_s performed by CDF and D0 [16, 17] indicated a possible tension with the Standard Model prediction. This tension, however, decreased and then disappeared when larger data samples were analysed [18, 19]. LHCb has performed a preliminary measurement of ϕ_s and $\Delta\Gamma_s$ using an event sample of about 21 000 $B_s^0 \to J/\psi\phi$ candidates extracted from the full 2011 data set [20][1]. The $J/\psi K^+K^-$ invariant mass spectrum is shown in Fig. 3.

A combination of opposite side tagging (OST) algorithms, described in Ref. [22], was employed to derive the initial flavour of the B meson. The OST exploits the fact that b quarks are mostly produced in quark-antiquark pairs. Flavour specific signatures from the decay of the other b hadron in the event are used to tag its flavour and it is then assumed that the flavour of the signal B meson is opposite to that of the tagging b hadron. The tagging algorithm returns a tagging decision as well as a mis-tag probability for each event, which is used to assign weights to signal events in the fit for ϕ_s. The event-by-event mis-tag probability was calibrated using a large sample of $B^\pm \to J/\psi K^\pm$ decays for which the true flavour of the signal B meson is known from the charge of the final state kaon. A global, linear calibration of the mis-tag probability is included in the fit for ϕ_s in order to absorb possible small differences between the tagging performances in $B_s^0 \to J/\psi\phi$ and in the calibration channel.

The decay time resolution was estimated event by event, using error propagation from the reconstructed momentum and decay length uncertainties. An overall calibration factor was then applied, which was determined from data using a large sample of prompt combinatorial

[1] While completing these proceedings, the measurement has been submitted for publication, see Ref. [21].

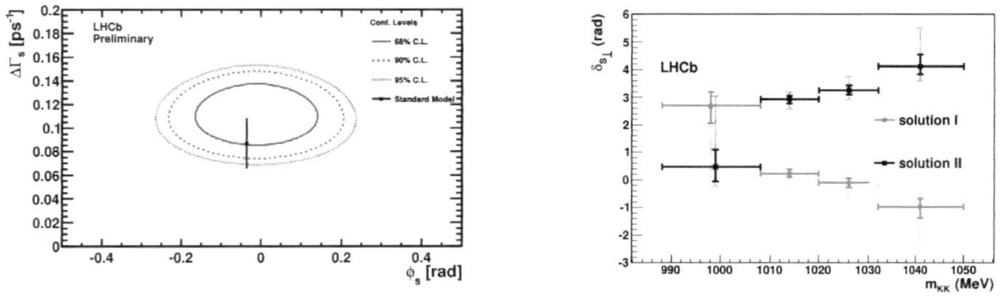

Figure 5. Left: Confidence regions in the (ϕ_s,$\Delta\Gamma_s$) plane, where only statistical uncertainties are included. Right: Measured phase difference $\delta_S - \delta_\perp$ for selected $B_s^0 \to J/\psi K^+K^-$ candidates in four bins of the K^+K^- invariant mass around the ϕ mass. Full blue circles show the result of the fit corresponding to $\Delta\Gamma_s > 0$ and ϕ_s close to zero, full black squares show the solution corresponding to $\Delta\Gamma_s < 0$ and ϕ_s close to π.

background events formed by a true $J/\psi \to \mu^+\mu^-$ and two random kaon candidates. The effective average decay time resolution was found to be 45 fs and is therefore of the same order of magnitude as the $B_s^0 - \overline{B}_s^0$ oscillation period of ≈ 56 fs. In the fit for ϕ_s, only events with decay time larger than 0.3 fs were considered. This cut significantly reduces combinatorial background and hardly reduces the statistical sensitivity to ϕ_s.

The time-dependent angular fit was performed in the transversity basis of the $J/\psi\phi$ system as defined in Ref. [23]. The full fit function is defined in Ref. [24]. Fig. 4 shows the angular distributions in the three transversity angles, together with the fit projections separating the *CP* even and *CP* odd components. The fit projection onto the decay time distribution is shown in Fig. 3. The different lifetimes of the *CP* even and *CP* odd components are clearly visible, indicating a non-zero value of $\Delta\Gamma_s$.

The result of the fit in the (ϕ_s,$\Delta\Gamma_s$) plane is shown in Fig. 5. It presents the most precise measurement of ϕ_s and $\Delta\Gamma_s$ to date and the first observation with more than five standard deviations of a non-zero value of $\Delta\Gamma_s$. The result is in good agreement with the Standard Model prediction. The fit function is invariant under a simultaneous variable transformation $(\phi_s, \Delta\Gamma_s, \delta_\parallel, \delta_\perp) \leftrightarrow (\pi - \phi_s, -\Delta\Gamma_s, 2\pi - \delta_\parallel, -\delta_\perp)$, where $\delta_\parallel$ and $\delta_\perp$ are the strong phases of two of the transversity amplitudes of the $J/\psi\phi$ system. This ambiguity was resolved in an LHCb analysis using about 1/3 of the 2011 data set, in which the evolution of the strong phases was studied as a function of the K^+K^- invariant mass around the ϕ resonance [25]. As shown in Fig. 4, the angular fit to the $B_s^0 \to J/\psi\phi$ sample reveals a small but not negligible S wave component, for which the K^+K^- system is not produced via the ϕ resonance. The strong phase δ_S for this S wave component is expected to vary only slowly across the ϕ resonance, whereas the strong phases of the resonant $\phi \to K^+K^-$ components are expected to go through a rapid positive phase shift at the location of the resonance. Fig. 5 shows for each of the two ambiguous solutions of the fit function the resulting strong phase difference $\delta_{S\perp} = \delta_S - \delta_\perp$ in four bins of the K^+K^- invariant mass around the ϕ resonance. Solution I shows the expected decreasing trend and is selected as the physical solution with a statistical significance of 4.7 standard deviations. This is the solution corresponding to a positive value of $\Delta\Gamma_s$ and ϕ_s close to zero, i.e. the solution which is in agreement with the Standard Model prediction.

LHCb has performed an independent measurement of ϕ_s using the decay $B_s^0 \to J/\psi\pi^+\pi^-$, with 755 MeV/$c^2$ < $m(\pi^+\pi^-)$ <1550 MeV/c^2 [26]. A modified Dalitz plot analysis of this decay has shown that the $\pi^+\pi^-$ system is produced predominantly in an S wave state and that the *CP* odd fraction of the $J/\psi\pi^+\pi^-$ final state is larger than 97.7 % at 95 % confidence level [27]. This simplifies the fit for ϕ_s since no angular analysis is required to separate *CP* odd

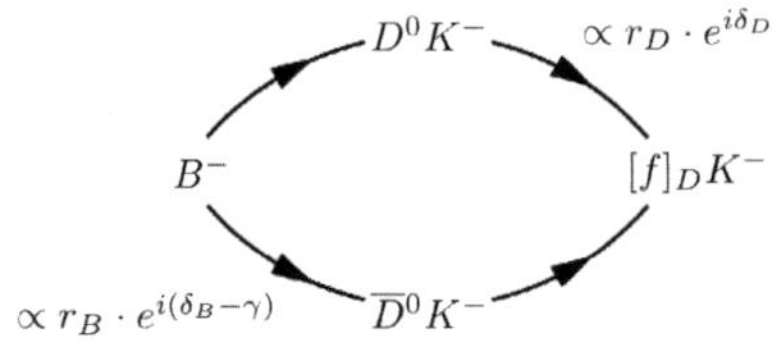

Figure 6. Illustration of the phases and ratios of magnitudes between the interfering amplitudes in tree decays $B^- \to DK^- \to [f]_D K^-$. Note that for the GLW method $r_D \equiv 1$ and $\delta_D \equiv 0$. See the text for further explanations.

and *CP* even components. On the other hand, the branching fraction for this decay is smaller than that for $B_s^0 \to J/\psi\phi$. The analysed event sample consisted of about 7400 signal candidates. A combination of the two ϕ_s analyses yields the preliminary result

$$\phi_s = -0.002 \pm 0.083(\text{stat}) \pm 0.027(\text{syst}) \text{ rad}.$$

This is the most precise determination of ϕ_s to date[2]. The precision of the measurements is limited by the statistical uncertainty. The dominant contribution to the systematic uncertainty is the assumption of no direct *CP* violation in the decay. The largest experimental uncertainty is due to angular acceptance corrections and can be further reduced with larger control channel samples.

4. Measurement of the CKM phase γ from $B^\pm \to DK^\pm$ and $B^\pm \to D\pi^\pm$ decays

The unitarity of the CKM matrix gives rise to six orthogonality relations, which can be visualised as triangles in the complex plane. The triangle derived from multiplying the first and third columns of the CKM matrix is of particular interest since its three sides are of comparable length. Its three angles are closely related to *CP* violating observables and the lengths of its sides can be measured from *CP* conserving observables. The overconstraint determination of this Unitarity Triangle provides a powerful consistency test of the CKM picture of *CP* violation.

The least well determined parameter of the Unitarity Triangle is the angle $\gamma \equiv \arg\left(-(V_{ud}V_{ub}^*)/(V_{cd}V_{cb}^*)\right)$. The CKMfitter group [28] quotes $\gamma = (66 \pm 12)^\circ$ from a global fit of CKM parameters, while the UTfit group quotes $\gamma = (72.2 \pm 9.2)^\circ$ [29].

A theoretically clean determination of γ can be derived from a combined measurement of time integrated decay rates for different $B^\pm \to DK^\pm$ tree decays. If the decay is observed in a final state $[f]_D$ that is accessible to both D^0 and $\overline{D}^0$ decays, the interference of

$$B^\pm \to D^0K^\pm \to [f]_D K^\pm \text{ and } B^\pm \to \overline{D}^0K^\pm \to [f]_D K^\pm,$$

as illustrated in Fig. 6, leads to a *CP* asymmetry that is sensitive to γ. Since the decays involve no loop diagrams, the extracted value for γ should be unaffected by possible contributions from New Physics. However, the measurements are experimentally challenging since they involve purely hadronic final states, which are difficult to trigger, and require excellent K/π separation. Moreover, branching fractions for the interesting decays are small.

Several methods using different combinations of final states $[f]_D$ have been proposed. They are usually named after the theoreticians who first proposed them.

The GLW method [30, 31] uses *CP* eigenstates such as $D^0 \to h^+h^-$, with h denoting a kaon or pion. The experimental observables are the ratios and asymmetries of time integrated decay

[2] The final result of the combined analysis, published in Ref. [21], is $\phi_s = 0.01 \pm 0.07(\text{stat}) \pm 0.01(\text{syst})$ rad.

rates

$$R_{CP+} \equiv \frac{\Gamma(B^- \to [h^+h^-]_D K^-) + \Gamma(B^+ \to [h^+h^-]_D K^+)}{\frac{1}{2}\{\Gamma(B^- \to [K^+\pi^-]_D K^-) + \Gamma(B^+ \to [K^-\pi^+]_D K^+)\}} = 1 + r_B^2 + 2\, r_B \cos\delta_B \cos\gamma$$

$$A_{CP+} \equiv \frac{\Gamma(B^- \to [h^+h^-]_D K^-) - \Gamma(B^+ \to [h^+h^-]_D K^+)}{\Gamma(B^- \to [h^+h^-]_D K^-) + \Gamma(B^+ \to [h^+h^-]_D K^+)} = \frac{2\, r_B \cos\delta_B \cos\gamma}{R_{CP+}},$$

where r_B and δ_B are the ratio of magnitudes and the relative strong phase between the two interfering decay amplitudes, respectively. The disadvantage of this method is that the decay $B^- \to \overline{D}^0 K^-$ is both CKM and colour suppressed with respect to the decay $B^- \to D^0 K^-$, leading to a small value of $r_B \approx 0.1$ and limiting the sensitivity to γ. For decays $B^\pm \to D\pi^\pm$, the CKM suppression factor is another factor 20 larger than for the case of $B^\pm \to DK^\pm$ and no sensitivity to γ is expected in these decays.

The ADS method [32, 33] tries to overcome the limitations of the GLW method by looking at quasi flavour-specific final states such as $K^-\pi^+$. The decay to this final state is Cabibbo favoured for $\overline{D}^0$ but doubly Cabibbo suppressed for D^0. The ratio of magnitudes between the D^0 and $\overline{D}^0$ decay amplitudes is $r_D \approx 0.05$, compensating for the small value of r_B and leading to similar total magnitudes for the two $B^- \to [f]_D K^-$ decay paths. Large interference is therefore possible, giving better sensitivity to γ. The price to pay is a very small total branching fraction for the doubly Cabibbo suppressed final states and the introduction of two additional parameters that have to be determined experimentally, namely r_D and the strong phase δ_D between the D^0 and $\overline{D}^0$ decay amplitudes. The experimental observables are again ratios and asymmetries of time integrated decay rates,

$$R_{ADS} \equiv \frac{\Gamma(B^- \to [K^+\pi^-]_D K^-) + \Gamma(B^+ \to [K^-\pi^+]_D K^+)}{\Gamma(B^- \to [K^-\pi^+]_D K^-) + \Gamma(B^+ \to [K^+\pi^-]_D K^+)} = r_B^2 + r_D^2 + 2\, r_B r_D \cos(\delta_B + \delta_D) \cos\gamma$$

$$A_{ADS} \equiv \frac{\Gamma(B^- \to [K^+\pi^-]_D K^-) - \Gamma(B^+ \to [K^-\pi^+]_D K^+)}{\Gamma(B^- \to [K^+\pi^-]_D K^-) + \Gamma(B^+ \to [K^-\pi^+]_D K^+)} = \frac{2\, r_B r_D \sin(\delta_B + \delta_D) \sin\gamma}{R_{ADS}}$$

The GGSZ or Dalitz-plot method [34, 35] exploits interference patterns across the Dalitz plot of self-conjugate three-body decays $D^0 \to K^0_S\pi^+\pi^-$. A rich resonance structure is observed, giving rise to large interference effects and good sensitivity to γ. However, variations of the strong phase δ_D across the Dalitz plot can be large and have to be taken into account in the extraction of γ. The associated systematic uncertainties ultimately limit the achievable precision of the measurement.

All three methods are being pursued in LHCb. Invariant mass distributions for the K^+K^- GLW mode and the suppressed ADS modes are shown in Fig. 7. Similar plots for the $\pi^+\pi^-$ GLW modes can be found in Ref. [36], where this analysis is described. The *CP* asymmetry in the $B^\pm \to DK^\pm$ GLW modes is clearly visible. Combining the two modes, the statistical significance of the asymmetry is 4.5 standard deviations. As expected, no asymmetry is seen in the $B^\pm \to D\pi^\pm$ GLW modes. Clear signals are also seen in the suppressed ADS modes and this measurement constitutes the first observation of the mode $B^\pm \to [\pi^\pm K^\mp]_D K^\pm$, with a statistical significance of about ten standard deviations. This mode displays evidence for *CP* asymmetry with a statistical significance of 4.0 standard deviations. A hint of an asymmetry is also seen in the mode $B^\pm \to [\pi^\pm K^\mp]_D \pi^\pm$, the statistical significance here being 2.4 standard deviations. The extracted results for the ratios R_{CP+} and R_{ADS} and the asymmetries A_{CP+} and A_{ADS} are listed in Ref. [36]. They are the world's most precise determinations of these parameters to date. All results are limited by statistical uncertainties. The leading systematic uncertainty on the ratios is due to the understanding of the π/K identification for the bachelor hadron, while systematic uncertainties on the asymmetries are dominated by uncertainties on production and

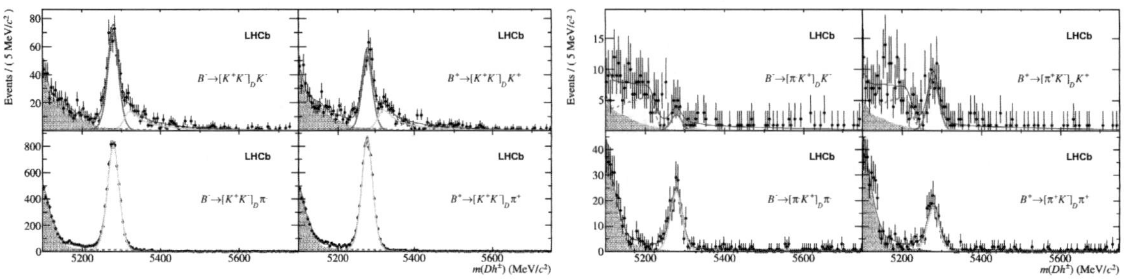

Figure 7. Invariant mass distributions for (left) the GLW modes $B^{\pm} \to [K^+K^-]_D h^{\pm}$ and (right) the suppressed ADS modes $B^{\pm} \to [\pi^{\pm}K^{\mp}]_D h^{\pm}$, where $h = K, \pi$.

detection efficiencies. These sources of uncertainties can be further reduced with larger control samples.

LHCb also performed a first GGSZ analysis [37] using D decays to $K^0_S\pi^+\pi^-$ and $K^0_S K^+K^-$. A model independent approach was chosen to avoid systematic uncertainties associated with the modelling of intermediate resonances in the D decay. Using measurements of the strong phase δ_D published [38] by the CLEO-c collaboration, the Dalitz plots were subdivided into regions (bins) in which the measured value of δ_D is approximately constant. The chosen binning scheme for the final state $K^0_S\pi^+\pi^-$ is shown in Fig. 8. The numbers $N^-_{\pm i}$ of B^- decays and $N^+_{\pm i}$ of B^+ decays in bin $\pm i$ are given by:

$$
\begin{aligned}
N^-_{\pm i} &\propto K_{\pm i} + (x_-^2 + y_-^2)\, K_{\mp i} + 2\sqrt{K_i K_{-i}}\,(x_- \cos(\delta_D)_{\pm i} \pm y_- \sin(\delta_D)_{\pm i}) \\
N^+_{\pm i} &\propto K_{\mp i} + (x_+^2 + y_+^2)\, K_{\pm i} + 2\sqrt{K_i K_{-i}}\,(x_+ \cos(\delta_D)_{\pm i} \mp y_+ \sin(\delta_D)_{\pm i}),
\end{aligned}
$$

where $K_{\pm i}$ is the number of flavour tagged D decays in bin $\pm i$ and the sensitivity to γ enters through the Cartesian parameters

$$x_{\pm} \equiv r_B \cos(\delta_B \pm \gamma) \quad \text{and} \quad y_{\pm} \equiv r_B \sin(\delta_B \pm \gamma).$$

The difference in B^- and B^+ signal yields for each bin is shown in Fig. 8 as well as the result of the fit in the ($x_{\pm}$,$y_{\pm}$) plane. The CKM angle γ is half the angle between the two vectors joining the origin and the best fit values for (x_+,y_+) and (x_-,y_-), respectively.

A combined analysis of all considered $B^{\pm} \to DK^{\pm}$ decay modes [39] yields preliminary results of

$$\gamma = (71.1^{+16.6}_{-15.7})^{\circ} \quad \text{and} \quad r_B = 0.092 \pm 0.008.$$

These results are in good agreement, and competitive, with the averages recently published by the BaBar [40] and Belle [41] collaborations.

5. *CP* violation in charmless *B* decays

The interference of $b \to u$ tree diagrams and $b \to s(d)$ penguin diagrams induces direct *CP* violation in two-body charmless $B^0_{(s)}$ decays. In the Standard Model, the weak phase in these decays is equal to the CKM phase γ. Possible contributions from New Physics can enter through the loop diagrams and could be revealed by comparing the weak phase measured in these decays with the value of γ measured in $B^{\pm} \to DK^{\pm}$ decays as described in the previous section. Hadronic uncertainties can be controlled by invoking U-spin symmetry (symmetry of the strong interaction under the exchange of the d and s spectator quarks) between related B^0 and B^0_s decays [42].

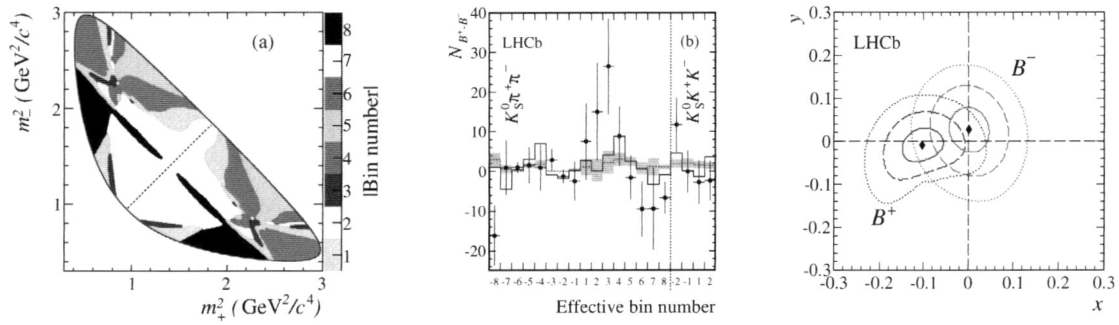

Figure 8. Left: Binning scheme chosen for the Dalitz plots in $D \to K_S^0\pi^+\pi^-$ decays. Bins above the dotted diagonal line are assigned positive bin numbers, bins below this line are assigned negative bin numbers. Middle: Difference of B^+ and B^- event yields in each bin of the Dalitz plot. The dashed line and grey shading indicate the expectation and uncertainty for the hypothesis of no *CP* violation. Right: One (solid), two (dashed) and three (dotted) standard deviation confidence levels for (x_+,y_+) (blue) and (x_-,y_-) (red). The points represent the best fit central values.

These measurements require large statistics of B_s^0 decays, efficient triggers for purely hadronic final states, and excellent K/π identification to separate the different final states. LHCb is the first experiment that has the potential to apply this technique. Two approaches are pursued: measurements of the time-dependent *CP* asymmetries in decays $B_{(s)}^0 \to h^+h^-$ $(h = K, \pi)$ and measurements of the time-integrated *CP* asymmetry in flavour-specific decays $B_{(s)}^0 \to K^\pm\pi^\mp$. While statistical uncertainties are currently still too large to achieve sensitivity to γ, important first steps have been taken in both approaches from analyses of parts of the 2011 data set.

In the first approach, the experimental observables are the time dependent *CP* asymmetries

$$A_{CP}(t) \equiv \frac{\frac{\partial\Gamma}{\partial t}(B_{(s)}^0 \to h^+h^-) - \frac{\partial\Gamma}{\partial t}(\overline{B}_{(s)}^0 \to h^+h^-)}{\frac{\partial\Gamma}{\partial t}(B_{(s)}^0 \to h^+h^-) + \frac{\partial\Gamma}{\partial t}(\overline{B}_{(s)}^0 \to h^+h^-)} = \frac{A_{h^+h^-}^{\rm dir}\cos(\Delta m_{(s)}t) + A_{h^+h^-}^{\rm mix}\sin(\Delta m_{(s)}t)}{\cosh(\frac{\Delta\Gamma_{(s)}}{2}t) - A_{h^+h^-}^{\Delta\Gamma}\sinh(\frac{\Delta\Gamma_{(s)}}{2}t)},$$

where the sensitivity to γ enters through the direct *CP* violation amplitude $A_{h^+h^-}^{\rm dir}$. The analysis is based on about 2/3 of the 2011 data sample [43]. Similar flavour tagging algorithms as in the measurement of ϕ_s were used to determine the flavour of the $B_{(s)}^0$ at production. Samples of flavour specific decays $B^0 \to K^+\pi^-$ were used to calibrate the estimated mis-tag probabilities. The values of Δm_d and Δm_s and the sign of $\Delta\Gamma_s$ were fixed to previous LHCb measurements. The observed raw asymmetries as a function of the decay time are shown in Fig. 9. They constitute the first measurement of the $B^0 \to \pi^+\pi^-$ asymmetry at a hadron collider and the first measurement of the $B_s^0 \to K^+K^-$ asymmetry ever.

In the second approach, the observables are the time-integrated *CP* asymmetries to flavour-specific final states,

$$A_{CP} \equiv \frac{\Gamma(B_{(s)}^0 \to K^\pm\pi^\mp) - \Gamma(\overline{B}_{(s)}^0 \to K^\mp\pi^\pm)}{\Gamma(B_{(s)}^0 \to K^\pm\pi^\mp) + \Gamma(\overline{B}_{(s)}^0 \to K^\mp\pi^\pm)}.$$

The invariant mass distributions obtained from an analysis of about 1/3 of the 2011 data set [44] are shown in Fig. 10. Non-zero raw asymmetries are clearly visible in both cases. Production and detection asymmetries are small and are corrected for using control channels. The measurements exclude the hypotheses of no *CP* violation with statistical significances of more than six standard deviations in the decay $B^0 \to K^+\pi^-$ and 3.2 standard deviations in the decay $B_s^0 \to K^-\pi^+$.

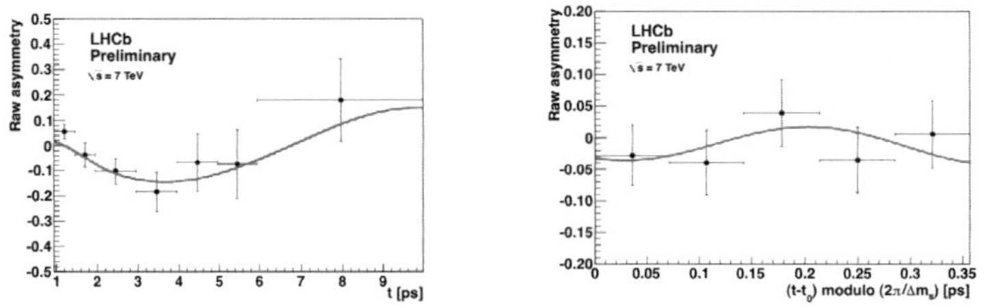

Figure 9. Time dependent *CP* asymmetry in (left) $B^0 \to \pi^+\pi^-$ and (right) $B_s^0 \to K^+K^-$.

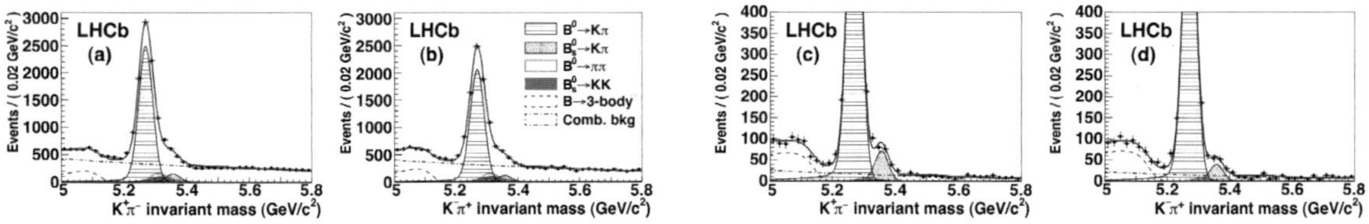

Figure 10. Invariant mass spectra for (a),(c) $K^+\pi^-$ and (b),(d) $K^-\pi^+$. Plots (a) and (b) show event selections optimized for sensitivity on $A_{CP}(B^0 \to K\pi)$, plots (c) and (d) show event selections optimized for sensitivity on $A_{CP}(B_s^0 \to K\pi)$.

These results are the first observation of *CP* violation with more than five standard deviations at a hadron collider and the first evidence with more than three standard deviations for *CP* violation in the B_s^0 system, respectively.

6. Summary and Outlook: The LHCb upgrade

The LHCb experiment has performed oustandingly well during the first three years of LHC operation. Many competitive and world-best measurements have been obtained from the analysis of the 2011 data set, corresponding to an integrated luminosity of 1.0 fb^{-1} of *pp* collisions collected at a centre of mass energy of 7 TeV. All results are dominated by statistical uncertainties and in most cases the leading systematic uncertainties can be further reduced by using larger samples of control channels. An additional 2 fb^{-1} of good quality data have been collected in 2012 and are being analysed. After the first long shutdown (LS1) of the LHC, the machine will resume operation at a centre of mass energy of 13 TeV. This will roughly double the heavy quark production cross sections. Assuming that the LHC will deliver about 5 fb^{-1} to LHCb during the run following LS1, the accumulated statistics will have increased by more than a factor ten by the time of the second long LHC shutdown (LS2), which is scheduled for 2018-2019.

The LHCb collaboration is preparing a comprehensive upgrade of the detector and its readout, to be installed and commissioned during LS2. The physics case and details of the upgrade plans are described in Ref. [47] and in the Letter of Intent [48] and the Framework Technical Design Report [49] for the upgrade. The upgrade plan contains two main components: to prepare the detector for operation at a five times increased instantaneous luminosity and to read out the full detector at the LHC bunch crossing rate of 40 MHz. By overcoming the limitations of the current hardware based trigger level, the 40 MHz readout will gain a factor of two in trigger efficiencies for *b* hadron decays to fully hadronic final states. Together with the increased heavy quark production cross section, the expected gains in event yields per year, compared to 2011, are a factor 10 in channels involving muons in the final state and a factor 20 in channels to fully hadronic final states. In total, the upgraded experiment is expected to collect an integrating luminosity of 50 fb^{-1} over 10 years of operation.

References

[1] A. A. Alves, Jr. *et al.* [LHCb Collaboration], JINST **3** (2008) S08005.
[2] E. Grauges Pous *on behalf of the LHCb Collaboration*, "Testing the Standard Model with rare decays at the LHCb," LHCb-PROC-2013-020.
[3] N. Cabibbo, Phys. Rev. Lett. **10** (1963) 531.
[4] M. Kobayashi and T. Maskawa, Prog. Theor. Phys. **49** (1973) 652.
[5] B. Aubert *et al.* [BaBar Collaboration], Phys. Rev. D **79** (2009) 072009 [arXiv:0902.1708 [hep-ex]].
[6] I. Adachi, H. Aihara, D. M. Asner, V. Aulchenko, T. Aushev, T. Aziz, A. M. Bakich and A. Bay *et al.*, Phys. Rev. Lett. **108** (2012) 171802 [arXiv:1201.4643 [hep-ex]].
[7] A. Lenz, arXiv:1205.1444 [hep-ph].
[8] B. Aubert *et al.* [BaBar Collaboration], Phys. Rev. Lett. **96** (2006) 251802 [hep-ex/0603053].
[9] B. Aubert *et al.* [BaBar Collaboration], hep-ex/0607091.
[10] E. Nakano *et al.* [Belle Collaboration], Phys. Rev. D **73** (2006) 112002 [hep-ex/0505017].
[11] V. M. Abazov *et al.* [D0 Collaboration], Phys. Rev. D **86** (2012) 072009 [arXiv:1208.5813 [hep-ex]].
[12] V. M. Abazov *et al.* [D0 Collaboration], Phys. Rev. D **84** (2011) 052007 [arXiv:1106.6308 [hep-ex]].
[13] LHCb Collaboration, LHCb-CONF-2012-022.
[14] R. Aaij *et al.* [LHCb Collaboration], Phys. Lett. B **713** (2012) 186 [arXiv:1205.0897 [hep-ex]].
[15] J. Charles, O. Deschamps, S. Descotes-Genon, R. Itoh, H. Lacker, A. Menzel, S. Monteil and V. Niess *et al.*, Phys. Rev. D **84** (2011) 033005 [arXiv:1106.4041 [hep-ph]].
[16] The CDF Collaboration, CDF public note CDF/ANAL/BOTTOM/PUBLIC/9458.
[17] V. M. Abazov *et al.* [D0 Collaboration], Phys. Rev. Lett. **101** (2008) 241801 [arXiv:0802.2255 [hep-ex]].
[18] T. Aaltonen *et al.* [CDF Collaboration], Phys. Rev. Lett. **109** (2012) 171802 [arXiv:1208.2967 [hep-ex]].
[19] V. M. Abazov *et al.* [D0 Collaboration], Phys. Rev. D **85** (2012) 032006 [arXiv:1109.3166 [hep-ex]].
[20] LHCb Collaboration, LHCb-CONF-2012-002.
[21] R. Aaij *et al.* [LHCb Collaboration], arXiv:1304.2600 [hep-ex].
[22] R. Aaij *et al.* [LHCb Collaboration], Eur. Phys. J. C **72** (2012) 2022 [arXiv:1202.4979 [hep-ex]].
[23] A. S. Dighe, I. Dunietz, H. J. Lipkin and J. L. Rosner, Phys. Lett. B **369** (1996) 144 [hep-ph/9511363].
[24] R. Aaij *et al.* [LHCb Collaboration], Phys. Rev. Lett. **108** (2012) 101803 [arXiv:1112.3183 [hep-ex]].
[25] R. Aaij *et al.* [LHCb Collaboration], Phys. Rev. Lett. **108** (2012) 241801 [arXiv:1202.4717 [hep-ex]].
[26] R. Aaij *et al.* [LHCb Collaboration], Phys. Lett. B **713** (2012) 378 [arXiv:1204.5675 [hep-ex]].
[27] R. Aaij *et al.* [LHCb Collaboration], Phys. Rev. D **86** (2012) 052006 [arXiv:1204.5643 [hep-ex]].
[28] J. Charles *et al.* [CKMfitter Group Collaboration], Eur. Phys. J. C **41** (2005) 1 [hep-ph/0406184], updated results and plots available at: http://ckmfitter.in2p3.fr.
[29] D. Derkach [UTfit Collaboration], arXiv:1301.3300 [hep-ph].
[30] M. Gronau and D. London, Phys. Lett. B **253** (1991) 483.
[31] M. Gronau and D. Wyler, Phys. Lett. B **265** (1991) 172.
[32] D. Atwood, I. Dunietz and A. Soni, Phys. Rev. Lett. **78** (1997) 3257 [hep-ph/9612433].
[33] D. Atwood, I. Dunietz and A. Soni, Phys. Rev. D **63** (2001) 036005 [hep-ph/0008090].
[34] A. Giri, Y. Grossman, A. Soffer and J. Zupan, Phys. Rev. D **68** (2003) 054018 [hep-ph/0303187].
[35] A. Poluektov *et al.* [Belle Collaboration], Phys. Rev. D **70** (2004) 072003 [hep-ex/0406067].
[36] R. Aaij *et al.* [LHCb Collaboration], Phys. Lett. B **712** (2012) 203 [Erratum-ibid. B **713** (2012) 351] [arXiv:1203.3662 [hep-ex]].
[37] R. Aaij *et al.* [LHCb Collaboration], Phys. Lett. B **718** (2012) 43 [arXiv:1209.5869 [hep-ex]].
[38] J. Libby *et al.* [CLEO Collaboration], Phys. Rev. D **82** (2010) 112006 [arXiv:1010.2817 [hep-ex]].
[39] LHCb Collaboration, LHCb-CONF-2012-032.
[40] J. P. Lees *et al.* [BaBar Collaboration], Phys. Rev. D **87** (2013) 052015 [arXiv:1301.1029 [hep-ex]].
[41] K. Trabelsi [Belle Collaboration], arXiv:1301.2033 [hep-ex].
[42] R. Fleischer, Eur. Phys. J. C **52** (2007) 267 [arXiv:0705.1121 [hep-ph]].
[43] LHCb Collaboration, LHCb-CONF-2012-007.
[44] R. Aaij *et al.* [LHCb Collaboration], Phys. Rev. Lett. **108** (2012) 201601 [arXiv:1202.6251 [hep-ex]].
[45] LHCb Collaboration, LHCb-CONF-2012-018.
[46] LHCb Collaboration, LHCb-CONF-2012-028.
[47] R. Aaij *et al.* [LHCb Collaboration], arXiv:1208.3355 [hep-ex].
[48] LHCb Collaboration CERN/LHCC-2011-001.
[49] LHCb Collaboration, CERN/LHCC-2012-007.

Journal of Physics: Conference Series **455** (2013) 012037 doi:10.1088/1742-6596/455/1/012037

The RHIC Beam Energy Scan program in STAR and what's next ...

Grazyna Odyniec

Lawrence Berkeley National Laboratory, Nuclear Science Division, Berkeley, CA 94720, USA

E-mail: G_Odyniec@lbl.gov

Abstract. In 2010, the Relativistic Heavy Ion Collider (RHIC) launched a multi-step experimental program to investigate the phase diagram of strongly interacting nuclear matter. The exploratory phase I of the Beam Energy Scan (BES) program was completed in 2011 with data sets ranging from 39 GeV down to 7.7 GeV. This, together with larger data sets at 62, 130 and 200 GeV, allowed for an initial look into the uncharted territory of QCD phase diagram. This diagram is considered to be the single most important graph of our field. All measurements taken by the STAR (Solenoidal Tracker At RHIC) detector at $\sqrt{s_{NN}}$ below the RHIC injection energy are affected by large statistical errors, steeply increasing with decreased energy. Nevertheless, they allowed for the first time a direct study of the anticipated critical point and phase transition signatures. The plans and the preparation for phase II of the BES program, with one order of magnitude larger statistics, are discussed.

1. Introduction

The first decade of RHIC (Relativistic Heavy Ion Collider at Brookhaven National Laboratory, US) operation showed that nuclear matter with partonic degrees of freedom (Quark Gluon Plazma) was formed in the early stages of Au+Au collision at $\sqrt{s_{NN}} = 200$ GeV. The preliminary results revealed that this new kind of matter is very complicated and has unique and unexpected properties. In spite of the huge recent progress in understanding the QGP phase, there are still questions to be addressed in the coming runs. In particular, the nature of the transformation from hadronic degrees of freedom to the partonic ones, and back to the state of hadron gas, is totally unknown.

So far, the theory gives very few insights into this problem. The generally accepted schematic representation of the QCD phase diagram is shown in Figure 1. Only the "edges" of this diagram are believed to be somehow understood: the lattice QCD calculations [1, 2] predict a smooth cross-over transition at the top RHIC energies, i.e. a critical temperature of 150-170 MeV and chemical potential $\mu_B \sim 0$ (top left in Fig.1), while several QCD based calculations [3, 4] show that at lower T and higher baryon chemical potential (right in Fig.1) a first-order phase transition may take place.

The end of a first-order phase transition line marks the QCD Critical Point (CP) (the red circle in Fig.1). Although a position of the CP as well as a location of a first-order phase transition are not known, there is enough information coming from the lattice calculation to suggest that the most probable location of CP would be in the mid-values of μ_B range, somewhere between 150 and 500 MeV.

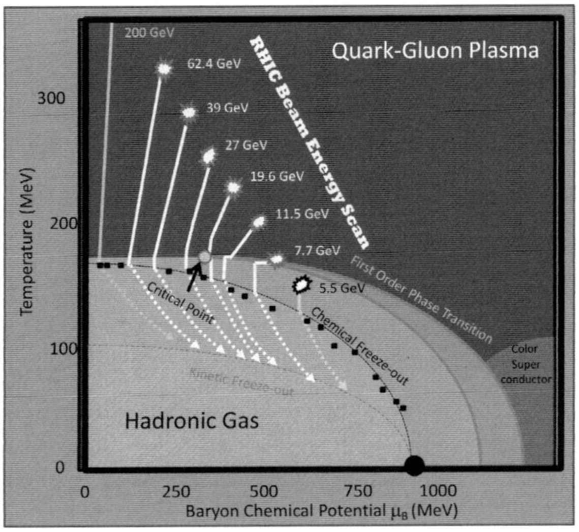

Figure 1. The picture of the QCD Phase Diagram with marked coverage of the RHIC BES program. Yellow trajectories represent schematics of the collision evolution at different energies of the BES program. The red circle symbolizes the critical point. Note, that yellow lines and the red circle are for illustrative purpose only.

The Au+Au collisions at lower, i.e. below the injection point, energies, available presently at RHIC, should be able to reach a chemical potential region of interest, and therefore allow for the exploration of the first-order phase transition, and possibly of the CP.

In 2010 and 2011 RHIC completed phase I of the BES program with data sets at 7.7, 11.5, 19, 27 and 39 GeV. This is complemented by the data collected earlier at higher energies (62, 130 and 200 GeV). Together they cover the μ_B interval from 20 to 450 MeV, which is believed to contain the range associated with the first order phase transition and CP.

In the following sections a short discussion of BES program specific goals, together with a few preliminary results of analyzed phase I data, will be presented.

2. Beam Energy Scan Program at RHIC

Presently the BES program is focused around three goals.

The first one, and the easiest, is to scan the phase diagram with variable collision energy (different μ_B and T) to find whether (and where in the $\sqrt{s_{NN}}$) the key QGP signatures observed at the top RHIC energy have been turned off. This may suggest that system is back to the hadron gas phase. The disappearance of a single signature would not be enough to claim an onset of deconfinement, because there are other phenomena not related to deconfinement which may cause a similar effect. However, the modification or disappearance of several signatures simultaneously would definitely provide a compelling case. The particular observables identified as the essential drivers of this part of the run are: constituent quark number scaling, hadron suppression in central collisions characterized by R_{cp}, untriggered pair correlations in the space of pair separation in azimuth and pseudorapidity and local parity violation in strong interactions.

The second goal is to find critical fluctuations, associated with a strong increase in the susceptibilities, which are expected in the vicinity of CP. However, because the finite size effects could wash out the critical behavior, the search for evidence of the softening of EOS as a system enters a mixed phase region implicitly associated with crossing first-order phase transition was proposed as a third goal of BES program. Essential observables in this search are: elliptic and directed flow of charged particles and of identified protons and pions, azimuthally-sensitive femtoscopy, and fluctuations indicated by large jumps in baryon, charge, and strangeness susceptibilities, as a function of system temperature.

Fig.1 shows the reach of RHIC's BES program in the (T, μ_B) plane. The yellow lines represent possible reaction trajectories at energies $\sqrt{s_{NN}}$ = 5, 7.7, 11.5, 19.6, 27 and 39 GeV. All energies, with the exception of 5 GeV, were run in 2010 and 2011. This choice of energies provides almost uniform coverage of the unknown (T, μ_B) territory and may allow us to narrow down an area of interest for further study.

3. RHIC and STAR

RHIC has a special and unique role in the exploration of the QCD phase diagram. Even with LHC operational, RHIC remains the only facility in the world capable of providing the necessary collision energies to execute the BES program.

Out of two large experiments presently on the RHIC floor, the STAR detector with its very large and uniform acceptance in rapidity, transverse momentum and azimuth, and excellent particle identification capabilities due to the addition of TOF in 2010, is particularly well suited for the challenges discussed in the previous sections. Moreover, as a collider experiment, the STAR detector acceptance at different energies remains the same allowing for a direct comparison between energies without the loss of statistics. It is particularly important for the measurements near the lower end of RHIC's range of operating energies, where luminosity drops dramatically and many analyses are statistics-limited. This situation will be greatly improved in the future phase II of the BES program.

4. Results

In the following sections BES selected results are presented. They are grouped to represent examples of each of three goals of the program, discussed earlier. The ones discussed here have the best resolving power given the constraints of the experiment (in phase I) and the total beam time. The details of these analyses and discussion of the other results can be found in [5, 17] and also in the recent Proceedings of Quark Matter Conferences [11, 12], Strange Quark Matter Conference [23], and CPOD 2011 Conference [13].

4.1. The "environment": freeze-out parameters

First, one needs to confirm that the environment created in heavy ion collisions at BES energies is suitable for this study.

Particle multiplicities observed in heavy ion and elementary collisions have been well described by statistical-thermal models [17, 7], which use experimentally measured particle ratios to extract two parameters: the temperature at the chemical freeze-out T_{ch}, and chemical potential μ_B.

Fig.2 shows the centrality dependence of the chemical freeze-out temperature as a function of chemical potential based on THERMUS model [8] with Grand Canonical ensemble for Au+Au collisions from 200 GeV to the lowest measured energy point at 7.7 GeV. Note, that while T_{ch} decreases only slightly (from 165 MeV at 200 GeV to 150 MeV at 7.7 GeV), there is a significant change in the value of the μ_B. With the lowest BES energy so far of 7.7 GeV, the RHIC μ_B range was extended to 400 GeV (with about 5% error). The critical region in μ_B has been predicted to span on the order of 100 MeV [10], which suggests that the overall program of BES measurements to date will offer reasonable coverage also below $\mu_B \sim 400$ MeV when the two recent energy points (19.6 and 27 GeV) are included in the analysis. The centrality dependence is still under investigation.

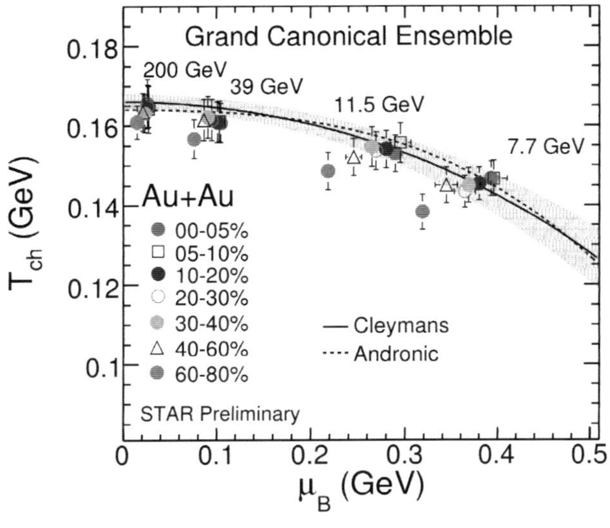

Figure 2. The variation of T_{ch} with μ_B, fitted with the Grand Canonical approach in THERMUS Model [8] using all available particle ratios, including strangeness.The lines represent systematic trends from Refs.[8, 9].

4.2. Onset of the QGP

RHIC results at top energies indicate that the passage through the phase transition to the partonic phase took place [11, 12]. Following the evolution of the established partonic signatures with $\sqrt{s_{NN}}$ from 200 to 7.7 GeV may bring us to the value of $\sqrt{s_{NN}}$ where these signatures change or even disappear completety. One of the key results that has been accepted as evidence of partonic degrees of freedom at RHIC has been the observation that the elliptic flow (expressed by anisotropy parameter v_2) scales with the number of constituent quarks in a given hadron. This indicates that the flow is established early on, where quarks are the relevant degrees of freedom (if the flow were to have been established during a hadronic phase, then the magnitude of v_2 for a given hadron would scale with its mass). Figure 3 shows the v_2 of protons and anti-protons in minimum bias collisions as a function of p_t for the five energies: 7.7, 11.5, 19.6, 27 and 39 GeV. The v_2 is calculated as $\langle \cos 2(\phi - \psi_2) \rangle$, where ϕ denotes the azimuthal angle of the produced particles and ψ_2 denotes the orientation of the second-order event plane. The v_2 gradually increases with energy. At 39 GeV the proton v_2 is only slightly higher than the anti-proton v_2, but this difference increases significantly with decreasing energy. The difference in particle and anti-particle v_2 suggests that the number of constituent quark (NCQ) scaling observed at $\sqrt{s_{NN}} = 200$ GeV [13] may no longer be valid at lower energies. The lower the energy the stronger the violation of NCQ scaling is observed.

More insights are provided by analysis of the nuclear modification factor (R_{cp}), another key signature measurement for the discovery of the QGP at RHIC. In high energy nuclear collisions the created strongly interacting medium is opaque to the fast moving partons, resulting in their measurable energy loss [14]. The suppression of high pt hadrons observed in A+A is quantified by the variable R_{cp}, i.e. the properly scaled (by the number of binary collisions) ratio of the number of particles in the central A+A collisions to the number of particles in peripheral A+A (where a QGP is not expected to be formed). The R_{cp} would equal to 1 if A-A system is a simple superposition of N-N. Any deviation from unity will imply a contribution from nuclear

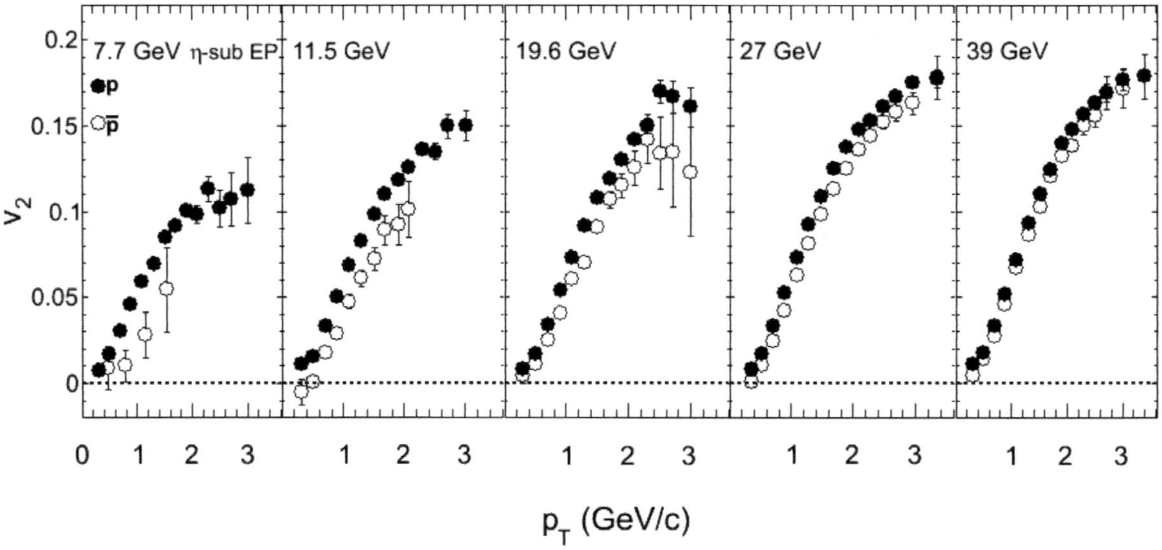

Figure 3. The v_2 of protons and anti-protons in minimum bias Au+Au collisions as a function of p_t for the five energies: 7.7, 11.5, 19.6, 27 and 39 GeV.

or medium effects. Early results on R_{cp} from 200 GeV Au+Au collisions at RHIC found a clear suppression of high p_t particles (see opened circles in Fig.4) which is attributed to the parton energy loss in the dense medium of quark gluon plasma [14]. The disappearance of this suppression would imply the absence of the partonic energy loss caused by the presence of quark gluon plasma, and would suggest that the hadronic interactions become dominant (replacing partonic ones). Thus, the study of the R_{cp} evolution with energy becomes very important. The results on R_{cp} of charged particles from Au+Au collisions at $\sqrt{s_{NN}}$ = 7.7, 11.5, 19.6, 27 and 39 GeV, together with the points for 62.4 and 200 GeV, are presented in Fig.4. The p_t reach of the charged hadron spectra extends beyond 3 GeV/c for all energies. The figure shows dramatic change in the suppression pattern. There is clear evidence of suppression at $\sqrt{s_{NN}}$ = 39, 62.4 and 200 GeV, which then appears to turn off between 19.6 and 39 GeV, becoming an enhancement at 7.7 and 11.5 GeV.

Those observations together with other ones [11, 12], suggest clearly that the change of dominant degrees of freedom, from partonic to hadronic, takes place somewhere in the energy range between 11.5 and 39 GeV. This interval of $\sqrt{s_{NN}}$ becomes the main focus phase II of the BES program, which is scheduled to start in 2017.

4.3. Critical Point (CP)

The location of the critical point is the most anticipated discovery of the BES program. Lattice QCD [16] shows the divergence of susceptibilities of conserved quantities such as baryon number, charge, and strangeness (B, Q, S) at the critical point (a similar critical behavior is known from classic thermodynamics), which translate into critical fluctuations in the multiplicity distributions that can be studied experimentally. The key observation is a change of the observable as a function of chemical potential μ_B. Therefore the experimental strategy amounts

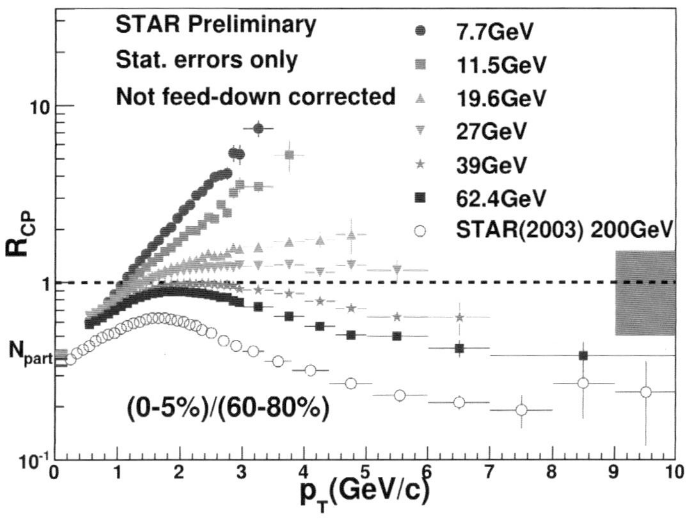

Figure 4. The STAR preliminary results of the energy dependence of charged particles R_{cp} [15]. The error bars on points reflect statistical errors. The systematic error (gray box) is independent of energy and p_t, and combines error on N_{coll} scaling and on the yields due to the background.

to small changes in energy with careful measurement of all aspects of the fluctuations. The STAR experiment studies of event-by-event fluctuation of proton, pion and kaon multiplicity ratios ($K/\pi, p/\pi$ and K/p) in Au+Au collisions at $\sqrt{s_{NN}}$ = 7.7, 11.5, 19.6, 39 and 200 GeV do not show "prescribed" anomalous behavior. All measured ratios change smoothly with energy or there is no observed change at all.

It has been noted that the multiplicity distribution moment analysis would be more suitable to these studies. Typically, experimental investigations are limited to the second moments, which are proportional to the square of the correlation length. Those, however, are estimated to be quite small in the vicinity of the critical point ($\sim$ 2-3 fm) [17]. Therefore, the higher moments of event-by-event multiplicity distribution are used; the higher the order of the moment the more sensitive it is to the correlation length of the system, e.g. the third moment (skewness) $s \sim \xi^{4.5}$ and the fourth moment (kurtosis) $\kappa^2 \sim \xi^7$ [17].

The measurements of higher moments of event-by-event identified particle multiplicity distributions and their variation with centrality and beam energy provide the very first direct link between experimental observables and Lattice QCD calculations. The measurements of higher moments of the net-proton distribution (proxy for net-baryon) carried out by the STAR collaboration are presented in Fig.5, with the $S\sigma$ and $\kappa\sigma^2$ of net-proton distributions in four centrality bins: 0-5 %, 5-10%, 30-40% and 70-80% in Au+Au collisions [18, 19, 24]. In the bottom panel, the ratios of $S\sigma$ over the corresponding Poisson function are shown. As one can see in the middle and bottom panel, the $\kappa\sigma^2$ and normalized $S\sigma$ values are close to the Poisson expectation for Au+Au collisions above $\sqrt{s_{NN}}$ = 39 GeV. At lower energies the results from most central collisions, 0-5 %, seem to deviate from Poisson, especially at $\sqrt{s_{NN}}$ = 19.6 GeV central collisions (middle and bottom panel of Fig.5). At lower beam energy, larger statistical errors preclude any decisive conclusions at the present time.

The energy dependent results from transport model UrQMD [20], dot-dashed line, which represent "non-CP effects" (like baryon number conservation and hardon scattering etc.) do not the reproduce data. They show a smooth decreasing trend from high to low energy.

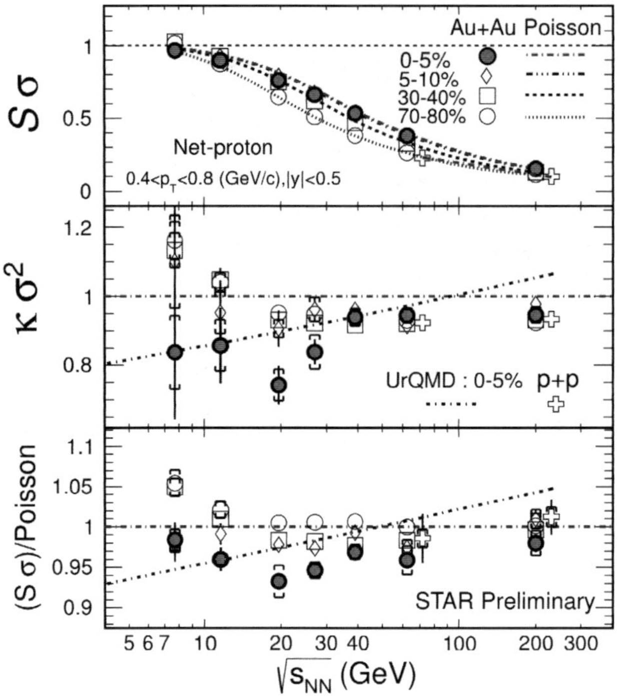

Figure 5. STAR preliminary results of the energy and centrality dependence of $S\sigma$ (top panel), $\kappa\sigma^2$ (middle panel) and $S\sigma/Poisson$ (bottom panel) of net-proton distributions from Au+Au collisions at RHIC [18]. Results from p+p collisions at $\sqrt{s_{NN}} = 62.4$ GeV and 200 GeV are shown as crosses. The results are compared to UrQMD model calculations for the top 5% central collisions shown as the dotted line. Dot-dashed lines are expectations from Poisson distributions.

4.4. First-order phase transition

Direct flow (v_1) probes the early stage of the collisions and is sensitive to the compression. Therefore, it is expected to be also sensitive to the first-order phase transition. The change of sign in the slope of dv_1/dy for protons has been proposed, based on hydrodynamic model calculations [21] to be a probe to the softening of the equation of state (EOS) and/or the first order phase transition. According to these calculations a first-order phase transition leads to a softening of the EOS, and this in turn leads to a local minimum in directed flow slope as a function of energy. This phenomenon is referred in the literature as a "collapse of proton flow" and the minimum is expected to correspond to the "softest point" in the EOS, where a relatively long-lived mixed phase occurs.

The direct flow excitation function measured by STAR for protons, antiprotons and pions near mid-rapidity is presented in Fig.6. The plotted quantity is dv_1/dy' , where the primed quantity y' refers to normalized rapidity y/y_{beam}. The lower panel in Fig.6 reveals that the inferred v_1 slope for net protons (the net protons signal is obtained by subtracting properly scaled antiprotons from protons) becomes negative with good statistical significance at 11.5 and 19.6 GeV, while it is consistent with zero at 27 GeV and positive at 7.7 GeV and above 27 GeV, including at 200 GeV. Indeed, the prominent dip and its associated double sign change bears a striking resemblance to the predicted signatures of a softening of the EOS associated with the first order phase transition [21]. In contrast, there is no hint of this remarkable behavior in the

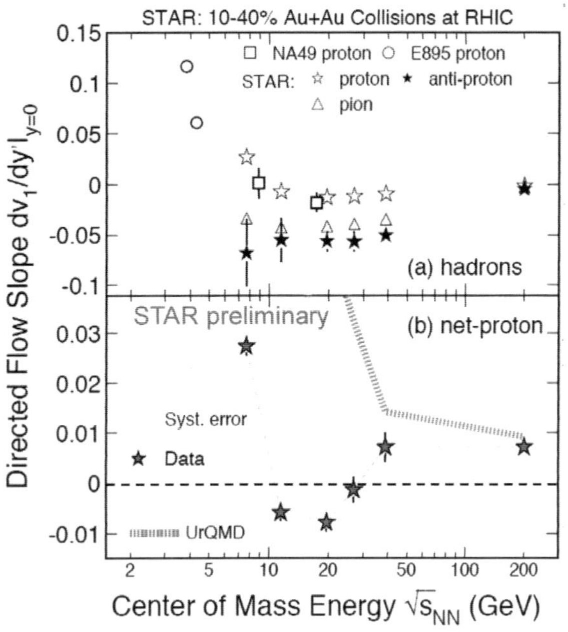

Figure 6. (a) - slope of directed flow vs rapidity for protons, antiprotons, and pions as a function of beam energy. (b) - proton signal with antiproton signal subtracted after appropriate scaling. The prediction of the UrQMd transport model is also plotted.

UrQMD transport model (purple line in Fig.6) which shows a positive slope at all energies for this observable. To better understand this observation and the possible role and relevance of stopping, one needs to make measurements of net proton v_1 as a function of centrality [22]. This will be possible in BES phase II part of the program which will deliver an order of magnitude more statistics.

5. Summary

Phase I of the BES program allowed to extend the μ_B reach of RHIC facility to values of approximately 400 MeV. Both the collider and experiments performance were excellent. As expected, a number of very interesting results was reported. It was observed that hadronic interactions in the produced nuclear matter become dominant over partonic interactions, when the collision energy decreases, especially below $\sqrt{s_{NN}} = 11.5$ GeV. In the most central Au+Au collisions, a deviation of the products of moments of net-protons distribution from the Poissonian expectations around μ_B, corresponding to beam energies of 19.6 - 27 GeV was found. The first hints of the softening of the EOS were noticed: the behavior of the directed flow, v_1, of charged hadrons at $\sqrt{s_{NN}} = 7.7$ - 200 GeV resembles striking similarity to the hydro-model predictions with the first-order phase transition built in. The implications of these measurements to understand the QCD phase structure are not yet resolved. The statistics collected during phase I are not sufficient to complete those studies. Therefore, STAR proposes the BES phase II program of precision measurements to map out a QCD phase diagram with an order of magnitude increase in data samples. BES phase II is planed for 2017.

5.1. BES phase II
The installation of electron cooling in RHIC will provide a significant improvement in luminosity for collisions below $\sqrt{s_{NN}}$ of 20 GeV. Simulation results indicate that at 7 and 20 GeV, the expected increasing factors are about 3-5 and 10, respectively [25]. Recently [26], it was shown that an additional improvement in luminosity may be possible by operating with longer bunches at the space-charge limit in a collider, which further extends benefits from cooling at low energies. Overall improvement from cooling could be about 10-fold gain in total luminosity.

5.2. Fixed target mode
While the electron cooling in RHIC is planned only for 2017, the extension of the BES program to the lower energies via the fixed target mode will be tested very soon.

Running in fixed target mode would allow for extension of the range of accessible baryon chemical potential from the current maximum of about $\mu_B \sim 400$ MeV up to ~ 800 MeV, corresponding to the collision energy $\sqrt{s_{NN}} \sim 2.5$ GeV, in the QCD phase diagram. The physics motivations of the fixed target mode focuses on the physics goal that is expected at the lowest energies - i.e. evidence of the first-order phase transition through identification of the softest point.

The fixed target will be mounted inside RHIC beam pipe. The fixed target data-taking can proceed with minimal or no impact on RHIC operations in collider mode. Collisions between injection-energy gold ions and a fixed gold target will allow the study of $\sqrt{s_{NN}} = 4.5$ GeV. The already developed sub-injection energy beams at RHIC, 11.5 and 7.7 GeV, will provide fixed target collisions at energies of $\sqrt{s_{NN}} = 3.5$ and 3.0 GeV, respectively. And the proposed for 2013 run at 14.6 GeV will provide $\sqrt{s_{NN}} = 4.0$ GeV fixed target collisions. The installation of 1% Au target in the STAR experiment is planned for the summer 2013 shutdown. It will be located about 2 meters from the center of the detector. During gold running, fixed-target events will be taken at injection energy at the beginning of each fill. This will happen concurrently with Au+Au data-taking at 200 GeV (i.e. during every store prior to accelerating Au ions to 200 GeV). The BES phase II program, which is discussed above, requests Au+Au collisions at 19.6, 14.6, 11.5 and 7.7 GeV after electron cooling upgrade. The gold target will be in place for these runs, allowing fixed target collisions at 4.5, 4.0, 3.5, and 3.0 GeV, respectively. These events can be triggered upon and recorded without reducing the number of collider events recorded as the expected luminosities will not fill the DAQ bandwidth.

6. Concluding remarks
RHIC sits at a "sweet spot" in energy, in which rapid changes occur in a number of signatures for energies up to approximately 30 GeV, while remaining surprisingly stable beyond that over the two orders of magnitude to the LHC.

With BES phase II the most exciting physics is still to come ...

References
[1] F.Karsch et al., Nucl. Phys. Proc. Suppl. 129, 614 (2004)
[2] M.Cheng et al., Phys. Rev.D 79, 074505 (2009) and references therein
[3] Y.Aoki et al. Nature 443, 675 (2006)
[4] S.Ejiri, Phys. Rev. D78, 074507 (2008), E.S.Bowwman and J.I.Kapusta, Phys. Rev. C79, 015202 (2009)
[5] http://drupal.star.bnl.gov/STAR/starnotes/public/sn0494
[6] B.I.Abelev et al., (STAR Coll.), Phys. Rev. C79, 034909 (2009); Phys. Rev. C81, 024911 (2010)
[7] J.Adams et al., (STAR Coll.), Nucl. Phys. A757, 102 (2005); P.Braun-Munzinger et al., Phys. Lett. B344, 43 (1995)
[8] S.Wheaton, J.Cleymans, and M.Hauer, Comp. Phys. Communications 180, 84 (2009)
[9] A.Andronic et al., Nucl. Phys. A 834, 237 (2010)

[10] R.V.Gavai and S.Gupta, Phys. Rev. D78, 114503 (2008); Phys. Rev. D71, 114014 (2005); arXiv:1001.3796; S.Gupta, arXiv:0909.4630; M.Asakawa et al., Phys. Rev. Lett. 101, 122302 (2008); P.Costa et al., Europhys. Lett. 86, 31001 (2009); Phys. Lett. B647, 431 (2007)
[11] Proceedings of Quark Matter 2011 Conference, Annecy, France, May 2011, J.Phys.G: Nucl.Part.Phys. 38 (2011)
[12] Proceedings of Quark Matter 2012 Conference, August 2012, Washington DC, US.
[13] L.Adamczyk et al., (STAR Coll.), arXiv: 1301.2348, submitted to PRC (in press), arXiv:1301.2347, submitted to PRL (in press); A.Schmah et al., (STAR Coll.), J.Phys.G: Nucl. Part. Phys. 38, 124045 (2011), Central Eur.J.Phys. 10, 1238 (2012), arXiv:1202.2389; S.Shi et al., (STAR Coll.),Central Eur. J. Phys. 10,(2012)
[14] X.N.Wang and M.Gyulassy, Phys. Rev. D44, 3501 (1991)
[15] E.Sangaline et al., (STAR Coll.), Quark Matter 2012 Conference Proceedings, Washington DC, August, 2012.
[16] M.A.Stephanov, Phys. Rev. Lett. 102, 032301 (2009); Phys. Rev. Lett. 107, 052301 (2001); M.Cheng et al., Phys. Rev. D79, 074505 (2009); R.V.Gavai and S.Gupta, Phys.Lett. B696, 459 (2011).
[17] M.A. Stephanov, arXiv:0809.3450.
[18] X.F.Luo et al., (STAR Coll.) Quark Matter 2012 Conference Proceedings, Washington DC, August, 2012.
[19] X.Dong et al.,(STAR Coll.) Quark Matter 2012 Conference Proceedings, Washington DC, August, 2012; D.Cebra at al.,(STAR Coll.) Quark Matter 2012 Conference Proceedings, Washington DC, August, 2012.
[20] UrQMD: http://urqmd.org/
[21] H.Stocker, Nucl. Phys. A750, 121 (2005); L.P.Csernai and D.Rohrich, Phys. Lett. B 458, 454 (1999); J.Brachmann et al., Phys. Rev. C61, 24909 (2000); D.R.Rischke et al., Heavy Ion Phys. I, 309 (1995)
[22] D. Keane et al., (STAR Coll.) Proc.Xth Quark Confinement and the Hadron Spectrum, Munich, October 2012, PoS (in press).
[23] S.Kabana et al., (STAR Coll.), Acta Phys. Pol. 2011, arXiv:1203.1814.
[24] M.M.Aggarwal et al., (STAR Coll.), Phys. Rev. Lett. 105 (2010) 022302.
[25] A.Fedotov and W.Fischer, RHIC CA-D, private communications, 2012.
[26] A.Fedotov and /m.Blaskiewicz, BNL C-AD Tech Note: C-A/AP/449 (February, 2012)

Journal of Physics: Conference Series **455** (2013) 012038 doi:10.1088/1742-6596/455/1/012038

Quarkonia production at the STAR detector

Christopher Beresford Powell[1,2] (STAR Collaboration)

[1] Lawrence Berkeley National Laboratory, Berkeley, California, USA

[2] University of Cape Town, Cape Town, Western Cape, South Africa

E-mail: cpowell001@gmail.com

Abstract. In this article we report on the results of quarkonia production in p+p, d+Au, and Au+Au collisions at midrapidity via the dielectron decay channel at $\sqrt{s_{NN}} = 200$ GeV from STAR. Results from J/ψ production in p+p collisions for $p_T < 14$ GeV/c are presented to provide a baseline for studying suppression in heavy ion collisions and are also used to understand the quarkonium production mechanism. The nuclear modification factor for J/ψ in d+Au collisions for $p_T < 5$ GeV/c and Au+Au collisions for $p_T < 10$ GeV/c is reported, along with J/ψ elliptic flow v_2 in Au+Au collisions. The results from Υ(1S+2S+3S) production in p+p and Au+Au collisions are also provided, and the Upsilon nuclear modification factor in Au+Au collisions is presented.

1. Introduction

The production of heavy quarkonia has been extensively used to probe the hot and dense medium created in heavy ion collisions, as they are expected to be suppressed in a deconfined medium due to the Debye screening of the heavy quark potential [1]. Because of their large mass, heavy quarks are primarily created in the initial hard scattering of the collision and provide information about the early stages and the evolution of the system. There are however various modifications other than color-screening to quarkonium production in heavy ion collisions, such as the recombination of charm quarks [2] into bound-state charmonium, formation time effects where high transverse momentum (p_T) particles may escape from the suppression region [3], and modifications to feeddown and sequential melting of excited states [4] There are further modifications to quarkonium production in heavy ion collisions from Cold Nuclear Matter effects (CNM) [5], such as modifications to parton distribution functions (PDFs) inside the nucleus (shadowing) [6], and nuclear absorption [7]. To disentangle all of these effects, a quantitative understanding of J/ψ and Υ production in p + A, d + A, and A+A is required.

In this article, the results for J/ψ production in p+p, d+Au, and Au+Au collisions at $\sqrt{s_{NN}} = 200$ GeV from the STAR detector are reported. The J/ψ p_T spectrum and nuclear modification factor at midrapidity ($|y| < 1$) for $p_T < 10$ GeV/c in $0 - 60\%$ centrality collisions are presented, along with the J/ψ elliptic flow v_2 in $20 - 60\%$ central Au+Au collisions. The Υ production in p+p and Au+Au collisions at $\sqrt{s_{NN}} = 200$ GeV are presented, and the nuclear modification factor in Au+Au collisions is discussed. The following sections in this article will describe the experimental setup and data used in this analysis, as well as a description of the analysis methods. This is followed by a discussion of the results for J/ψ and Υ production from STAR. Finally, a summary is provided.

2. Experiment and Analysis

The STAR experiment is a large acceptance multi-purpose detector which covers full azimuth and pseudorapidity of $|\eta| < 1$ [8]. The p+p, d+Au and Au+Au data were recorded in 2009, 2008, and 2010, respectively. The data were obtained using a minimum bias trigger by selecting on coincidences in the Vertex Position Detector (VPD) [9] located 5.7 m from the interaction point. The d+Au trigger also required a neutron to strike the East (gold facing) Zero Degree Calorimeter (ZDC) [10], which is located outside of the STAR magnet at 18 m from the interaction point. An additional trigger was also used in Au+Au collisions to identify central events by requiring a high occupancy in the detector. To improve statistics, the Upsilon analysis in p+p, d+Au, and Au+Au collisions, and the J/ψ analysis at high-p_T in p+p and Au+Au collisions used a high tower (HT) trigger, which required a minimum energy deposit in a single tower ($\Delta\eta \times \Delta\phi = 0.05 \times 0.05$) of the Barrel Electromagnetic Calorimeter (BEMC) [11]. A threshold energy of $E_T = 4.3$ GeV was used for the HT trigger in Au+Au collisions. Two HT triggers were used in p+p collisions with energy thresholds of $E_T = 2.6$ GeV and $E_T = 6.0$ GeV. Furthermore, the p+p dataset used for the Upsilon analysis was also required to satisfy a specific Upsilon trigger. This used the energy deposited in the calorimeter to reconstruct the candidate Υ particles assuming that any large energy deposits came from electrons, and events containing reconstructed particles with a mass near that of the Υ ($5 <$ GeV/$c^2 < 20$) were recorded.

The reconstruction of J/ψ and Υ(1S+2S+3S) has been performed via the dielectron decay channel, J/ψ (Υ) $\rightarrow e^+ + e^-$ with a branching ratio (B) of 5.9% (2.4%). The primary detectors used in this analysis are the Time Projection Chamber (TPC) [12], the Time of Flight (TOF) [13], and the BEMC. These detectors provide tracking and particle identification in $|\eta| < 1$ and $0 < \phi < 2\pi$. The TPC provides tracking and particle identification via the ionization energy loss per unit length (dE/dx) of charged particles. The TOF, which measures the flight time and velocity of particles, was used in the analysis of J/ψ in p+p and Au+Au data for low-p_T electron identification ($p_T \lesssim 1.5$ GeV/c) to improve electron-hadron discrimination provided by the TPC, especially where the electron and hadron dE/dx overlap. The TOF was not available for the d+Au data recording, and was not used for the Υ and high-p_T J/ψ analyses as it is less effective at separating electrons from hadrons at higher p_T. The BEMC is a lead-scintillator calorimeter and has been used to improve the particle identification at high p_T, in addition to providing a trigger mechanism for the HT and Upsilon trigger. The energy obtained from the BEMC has been combined with the TPC momentum to obtain the energy-to-momentum ratio (E/p) which can be used to distinguish electrons from hadrons.

3. J/ψ production in p+p collisions

The invariant cross section for inclusive J/ψ production in p+p collisions is shown as a function of p_T in the left panel of Fig. 1. The low-p_T results ($p_T < 3$ GeV/c) are combined with the high-p_T data ($2 < p_T < 14$ GeV/c) [14] to extend the J/ψ coverage at STAR to $0 < p_T < 14$ GeV/c. Also shown are the PHENIX results for $|y| < 0.35$ [15]. The data are compared to the Color Evaporation Model (CEM) [16] for prompt J/ψ production in p+p collisions, which includes initial (direct) production and contributions from excited states. The CEM is able to describe the data well for the entire range of transverse momentum, however it does not include contributions from B decay. The data are also compared to the Non-Relativistic QCD (NRQCD), next-to-leading order (NLO) Color Singlet and Color Octet (CS+CO) [17] model for prompt and direct J/ψ production, which agrees with the data well at high p_T, but does not provide a prediction at low p_T. Also shown are the next-to-next-to-leading order (NNLO*) Color Singlet (CS) [18] model for direct J/ψ production in p+p collisions. These under-estimate the yield for $p_T > 3$ GeV/c, however they do not include feeddown contributions from the decay of B meson or excited charmonium states. The fraction of inclusive J/ψ in p+p collisions from $B \rightarrow$ J/ψ decays is obtained by fitting the azimuthal angle correlation between high-p_T J/ψ and charged

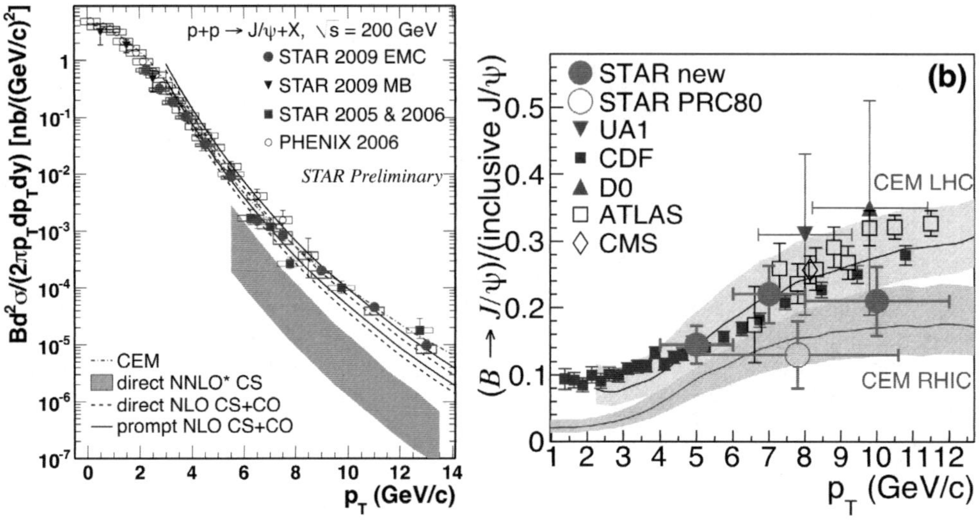

Figure 1. (Color online) (left) The invariant cross section versus transverse momentum for J/ψ in p+p collisions from STAR. The results are compared to theoretical predictions and other measurements. (right) The fraction of $B \rightarrow$ J/ψ for inclusive production in p+p collisions from STAR, compared to the results from other experiments and FONLL+CEM model calculations [14].

hadrons at STAR with the simulated correlation functions for J/ψ from prompt production and B feeddown. The results are shown in the right panel of Fig. 1, and the fraction of inclusive J/ψ from B feeddown was found to be $10-25\%$, increasing with increasing p_T [14]. The results are consistent to similar measurements at various experiments and with FONLL+CEM model predictions [19, 20].

4. J/ψ production in d+Au collisions

The nuclear modification factor in d+Au collisions (R_{dA}) for J/ψ with $p_T < 5$ GeV/c and $|y| < 1$ is shown in the left panel of Fig. 2 versus the number of nucleon-nucleon collisions (N_{coll}). The normalization uncertainty from the statistical and systematic uncertainty on the J/ψ p+p cross section, the uncertainty on the inelastic cross section, and the uncertainty on N_{coll} are indicated on the vertical axis. The J/ψ nuclear modification factor is compared to model predictions for the cold nuclear matter effects on J/ψ production in d+Au collisions. The CNM effects include the modification of nuclear parton distribution functions obtained from the EPS09 [21] parameterization. The shadowing from the nPDFs is with a J/ψ nuclear absorption cross section (σ_{abs}) [22], which is determined from a fit to the STAR data, and a value of $\sigma_{\text{abs}} = 2.8^{+3.5}_{-2.6}$ (stat.) $^{+4.0}_{-2.8}$ (syst.) $^{+1.8}_{-1.1}$ (EPS09) mb was obtained. The R_{dA} calculated using the EPS09 nPDFs combined with an absorption cross section of $\sigma_{\text{abs}} = 3$ mb is shown in Fig. 2, and the bands indicate the uncertainty of the EPS09 nPDFs. The model predictions are able to describe the data well.

The p_T dependence of the J/ψ nuclear modification factor in $|y| < 1$ for minimum bias d+Au collisions at $\sqrt{s_{NN}} = 200$ GeV collisions is shown in the right panel of Fig. 2. The grey band represents the statistical uncertainty on the J/ψ p+p cross section. The normalization uncertainties from the systematic uncertainty on the J/ψ p+p cross section and the uncertainty on N_{coll} are indicated on the vertical axis. The results are compared to PHENIX data in $|y| < 0.35$, and are in agreement within the uncertainties [23, 24].

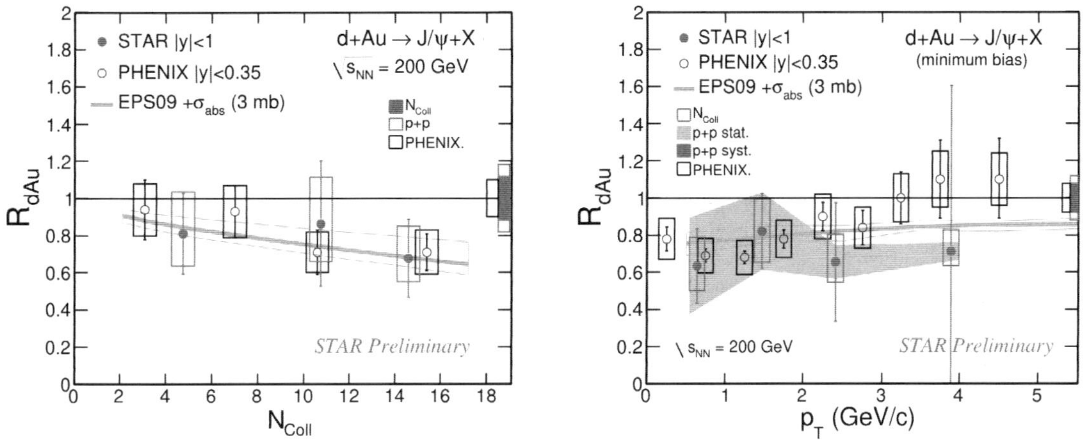

Figure 2. (Color online) (left) The J/ψ nuclear modification factor versus $N_{\rm coll}$ for $p_T < 5$ GeV/c in d+Au collisions. (right) The J/ψ nuclear modification factor versus p_T in d+Au collisions. The results are compared to shadowing calculations based on the EPS09 nPDFs combined with a J/ψ absorption cross section of $\sigma_{\rm abs} = 3$ mb, and the band indicates the uncertainty on the EPS09 predictions [21, 22].

5. J/ψ production in Au+Au collisions

The J/ψ invariant yield in Au+Au collisions at $\sqrt{s_{NN}} = 200$ GeV are shown in the left panel of Fig. 3 as a function of p_T and collision centrality [14]. The results are consistent with PHENIX measurements for $p_T < 5$ GeV/c and $|y| < 0.35$ [25]. The J/ψ invariant yield in Au+Au collisions is compared to a Tsallis Blast Wave (TBW) model assuming that the J/ψ flows like lighter hadrons (solid line) [26, 27]. The agreement with the data is good for $p_T > 2$ GeV/c, but underestimates the yield below this. The STAR data are also fitted with a TBW model in $0 < p_T < 10$ GeV/c that assumes a zero radial flow, $\beta = 0$ (dashed line) [26]. This improves the agreement with the data at low-p_T, suggesting that the J/ψ has a small radial flow, or that there may be contributions from recombination at low-p_T.

The J/ψ yield in Au+Au collisions is shown on a linear scale in the right panel of Fig. 3 for various collision centralities. The results are compared to predictions from viscous hydrodynamics using a J/ψ decoupling temperature of $T = 120$ MeV (dotted line) and $T = 165$ MeV (solid line) [34]. The predictions assume a zero chemical potential for J/ψ at kinetic freeze-out, and the scale of the predictions is determined from a fit to the data in $p_T < 5$ GeV. The data favor the higher decoupling temperature and are well described for $2 < p_T < 5$ GeV/c. However, the hydrodynamic calculations fail to describe the low p_T J/ψ yield ($p_T < 2$ GeV/c). The data are also compared to theoretical predictions based on the suppression of J/ψ due to color-screening and the statistical regeneration of charm quarks in Au+Au collisions, including B meson feeddown and formation time effects [28] (dashed line). The contribution from initial production dominates in peripheral events, with regeneration becoming more significant in central events and at low p_T. The predictions are able to describe the data across the measured transverse momentum range.

The J/ψ elliptic flow (v_2) in 20−60% centrality Au+Au collisions is shown in Fig. 4. The data are consistent with zero within the uncertainties, disfavoring the case where J/ψ is produced dominantly by coalescence from thermalized charm quarks for $p_T > 2$ GeV/c in mid-central collisions. The results are compared to various theoretical predictions [29] - [34]. The data are well described by models with J/ψ production from initial production only, or from a mixture between initial production and coalescence. This is consistent with the results shown in Fig. 3,

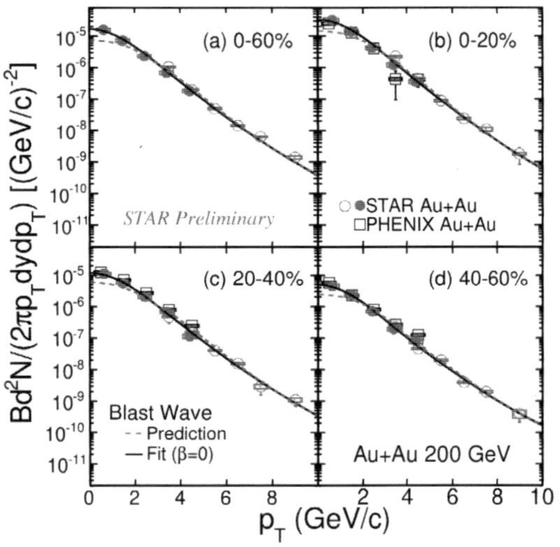

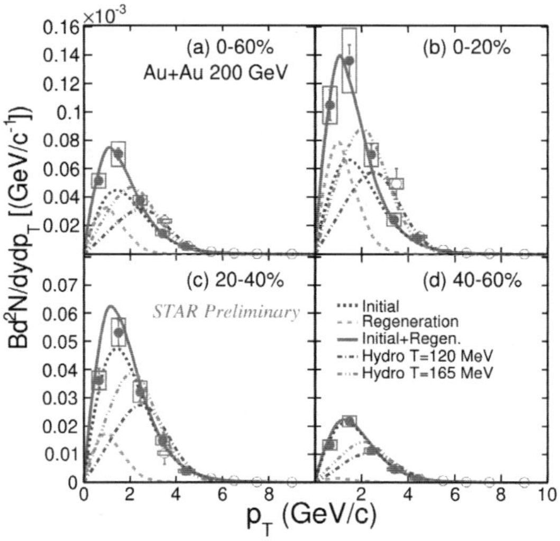

Figure 3. (Color online) (left) The J/ψ invariant yield versus transverse momentum and collision centrality in Au+Au collisions. The results are compared to previous measurements and TBW model predictions [14, 25, 26, 27]. (right) The J/ψ yield versus transverse momentum and collision centrality in Au+Au collisions. The results are compared to theoretical predictions [28, 34].

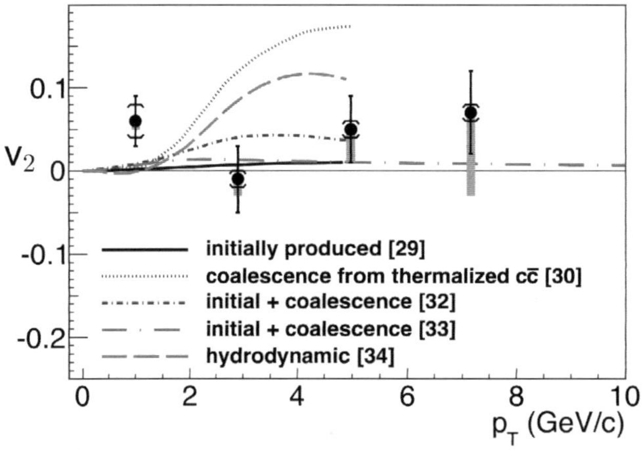

Figure 4. (Color online) The J/ψ elliptic flow v_2 in $20-60\%$ central Au+Au collisions. The data are compared to various model predictions.

which indicate that the contribution from coalescence is only significant at low p_T in central collisions.

The transverse momentum dependence of the J/ψ nuclear modification factor in Au+Au collisions (R_{AA}) is shown in the left panel of Fig. 5 for various collision centralities [14]. The uncertainty from $N_{\rm coll}$ and the inelastic cross section in p+p collisions are indicated by the boxes on the vertical axis, and the shaded bands indicate the statistical uncertainty on the J/ψ cross section in p+p collisions. The data are consistent with PHENIX results [25] in $|y| < 0.35$. The J/ψ R_{AA} is compared to theoretical calculations including contributions from prompt production and statistical charm quark regeneration in Au+Au collisions (dashed line [28] and solid line [35]). The latter also includes B feeddown and formation time effects to J/ψ production. Both models are able to reproduce the trend of the data, and these cannot be distinguished given the current data. A significant suppression is observed for $p_T < 3$ GeV/c in Au+Au collisions. The data exhibit an increase in R_{AA} for $p_T > 1$ GeV/c for all centralities, with R_{AA} approaching unity for $p_T > 5$ GeV/c in peripheral collisions.

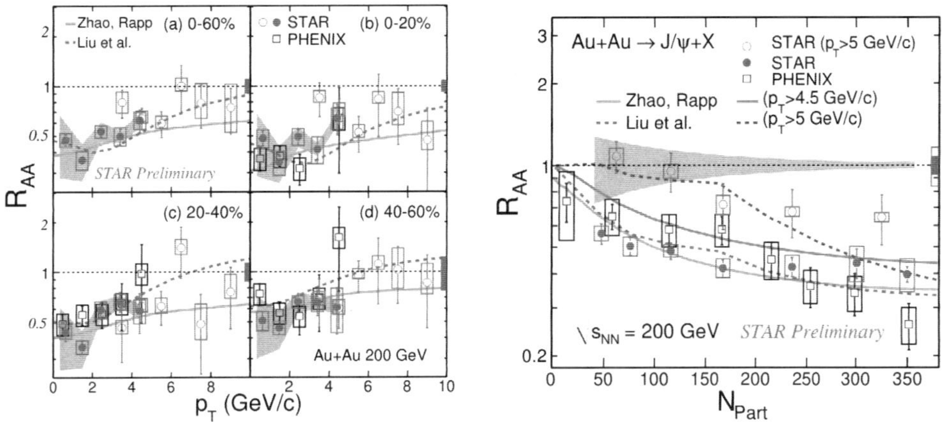

Figure 5. (Color online) (left) The J/ψ nuclear modification factor versus transverse momentum and collision centrality in Au+Au collisions. (right) The J/ψ nuclear modification factor versus N_{part} for J/ψ in Au+Au collisions [14, 36]. The data are compared to previous measurements and theoretical predictions (solid line [35] and dashed line [28]).

The centrality dependence of the J/ψ nuclear modification factor in Au+Au collisions is shown in the right panel of Fig. 5 as a function of N_{part} for $p_T < 5$ GeV/c (closed circles) and $p_T > 5$ GeV/c (open circles). The relative uncertainty on N_{coll} in Au+Au collisions is indicated by the shaded band. The uncertainty on the J/ψ cross section in p+p collisions and the uncertainty on the inelastic cross section in p+p collisions is indicated by the box on the vertical axis. The data are compared to the PHENIX results for $p_T < 5$ GeV/c in $|y| < 0.35$ [25] and are consistent within errors. The data exhibit a significant suppression in central collisions, decreasing in more peripheral collisions. There is less suppression observed at high p_T across the collision centrality range, with R_{AA} approaching unity in peripheral collisions. The results are compared to theoretical predictions for J/ψ production in Au+Au collisions including the suppression of J/ψ due to color-screening and the statistical regeneration of charm quarks [28, 35]. The predictions are able to describe the data reasonably well across the collision centrality range, and cannot be distinguished by the current low-p_T data.

6. Υ production in p+p collisions

The Υ(1S+2S+3S) yield and cross section in p+p collisions for $|y| < 0.5$ are shown in the left panel of Fig. 6. The cross section is compared to NLO pQCD calculations for Upsilon production using the CEM (boxes) [16] and CS (dotted line) [37] models. The data favor the CEM calculations. The Υ cross section is compared to the world data trend as a function of $\sqrt{s}$ in the right panel of Fig. 6. The results are also compared to CEM calculations [16], which are able to describe the data well across a wide range of collision energy.

7. Υ production in Au+Au collisions

The Υ invariant mass spectrum after like-sign background subtraction in 0−60% central Au+Au collisions is shown in the left panel of Fig. 7. This is fitted with a function which includes

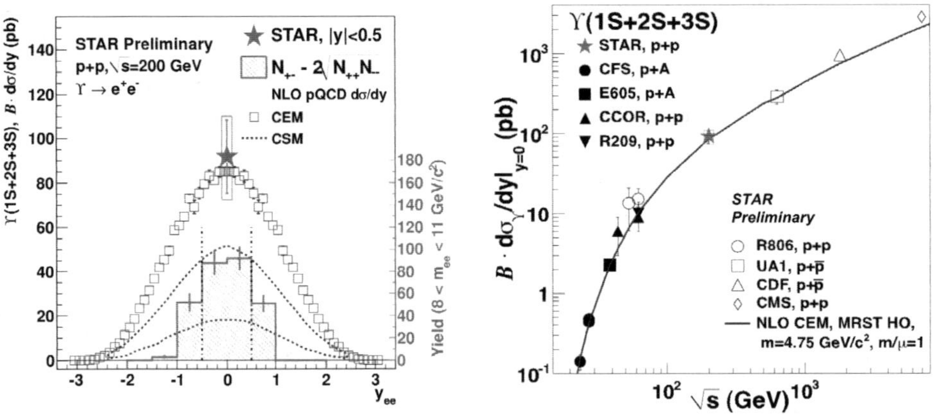

Figure 6. (Color online) (left) The Υ(1S+2S+3S) counts and cross section in p+p collisions. The results are compared to theoretical calculations [16, 37]. (right) The Υ cross section as a function of collision energy compared to the world data trend and CEM predictions [16].

the Υ(1S+2S+3S) signal shape and contributions from Drell-Yan and $b\bar{b}$. The shape of the Drell-Yan and $b\bar{b}$ contributions are determined using PYTHIA and CEM calculations, and the normalization is determined in the combined fit to the signal and background. The Upsilon yield is determined from the counts after subtracting the background. The Υ nuclear modification factor in Au+Au collisions is shown as a function of N_{part} in the right panel of Fig. 7. A significant suppression is observed in central events. The results are compared to lattice-based QCD calculations for the Υ melting, and are combined with a hydrodynamic model which is used to describe the expansion of the surrounding partonic medium [38]. The results favor the internal energy potential, which predicts an almost complete suppression of Υ(3S) in central collisions, with less suppression observed for Υ(2S) and Υ(1S).

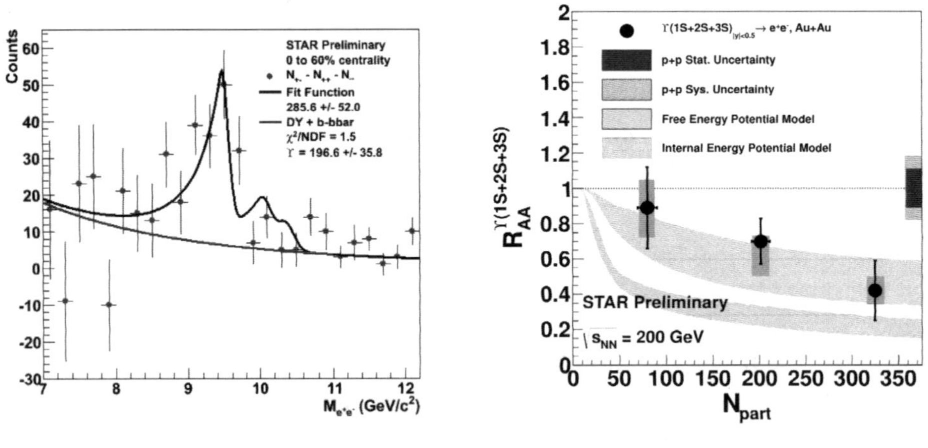

Figure 7. (Color online) (left) The Υ invariant mass spectrum in $0-60\%$ central Au+Au collisions after like-sign background subtraction. (right) The Υ(1S+2S+3S) nuclear modification factor versus N_{part} in Au+Au collisions. The results are compared to theoretical predictions [38].

8. Summary

The production of quarkonia at the STAR detector is described in this article. The J/ψ invariant cross section is presented and compared to various theoretical calculations, and the fraction of inclusive J/ψ production from B meson feeddown is calculated. The J/ψ nuclear modification factor in d+Au collisions is compared to the EPS09 calculations for shadowing and used to determine the J/ψ nuclear absorption cross section. The J/ψ invariant yield, elliptic flow, and nuclear modification factor in Au+Au collisions are compared to various theoretical calculations. Predictions based on the suppression and regeneration of J/ψ are able to describe the data reasonably well, and predict that initial production dominates mid-central and peripheral collisions, while the contribution from coalescence may become significant at low p_T in central collisions. The Υ(1S+2S+3S) invariant cross section in p+p collisions is compared to the world data trend and theoretical calculations, and can be described reasonably well by the color evaporation model. The Υ nuclear modification factor is presented, and exhibits a significant suppression in central collisions. The results are consistent with a model based on lattice QCD calculations of Upsilon melting in a hot medium and indicate an almost complete melting of the Υ(3S) state.

[1] Matsui T and Satz H 1986 *Phys Lett. B* **178** 416
[2] Braun-Munzinger P and Stachel J 2000 *Phys. Lett. B* **490** 196
[3] Karsch F and Petronzio P 1988 *Z. Phys. C* **37** 627
[4] Rafelski J, Thews R L and Schroedter M 2009 *Phys. Lett. B* **678** 72
[5] Vogt R 2005 *Phys. Rev. C* **71** 054902
[6] Noble J V 1981 *Phys. Rev. Lett.* **46** 412
[7] Tram V and Arleo F 2009 *Eur. Phys. J. C* **61** 847
[8] Ackermann K H *et al.* 2003 *Nucl. Instr. Meth. A* **499** 624
[9] Llope W J *et al.* 2004 *Nucl. Instr. Meth. A* **522** 252
[10] Adler C *et al.* 2003 *Nucl. Instr. Meth. A* **499** 433
[11] Beddo M *et al.* 2003 *Nucl. Instr. Meth. A* **499** 725
[12] Anderson M *et al.* 2003 *Nucl. Instr. Meth. A* **499** 659
[13] Bonner K *et al.* 2003 *Nucl. Instr. Meth. A* **499** 681
[14] Adamczyk L *et al.* 2011 (*Preprint* `arXiv:1208.2736v1`)
[15] Adare A *et al.* 2010 *Phys. Rev. D.* **82** 012001
[16] Frawley A, Ullrich T and Vogt R 2008 *Phys. Rept.* **462** 125
[17] Ma Y Q, Wang K and Chao K T 2011 *Phys. Rev. D* **84** 114001
[18] Artoisenet P *et al.* 2008 *Phys. Rev. Lett.* **101** 152001
[19] Bedjidian M *et al.* 2003 (*Preprint* `arXiv:hep-ph/0311048v1`)
[20] Cacciari M, Nason P and Vogt R 2005 *Phys. Rev. Lett.* **95** 122001
[21] Eskola K, Paukkunena H and Salgado C 2009 *Nucl. Phys. A* **830**(1-4) 599
[22] Vogt R 2010 *Phys. Rev. C* **81** 044903
[23] Adare A *et al.* 2008 *Phys. Rev. C* **77** 024912 erratum-ibid.C79:059901,2009
[24] Adare A *et al.* 2012 (*Preprint* `arXiv:1204.0777`)
[25] Adare A *et al.* 2007 *Phys. Rev. Lett.* **98** 232301
[26] Tang Z *et al.* 2009 *Phys. Rev. C* **79** 051901
[27] Tang Z *et al.* 2011 (*Preprint* `arXiv:1101.1912`)
[28] Liu Y, Qu Z, Xu N and Zhuang P 2009 *Phys. Lett B.* **678** 72
[29] Yan L, Zhuang P and Xu N 2006 *Phys. Rev. Lett.* **97** 232301 - v_2 model (1), (4)
[30] Greco V, Ko C M and Rapp R 2004 *Phys. Rev. Lett.* **B595** 202 - v_2 model (2)
[31] Ravagli L and Rapp R 2007 *Phys. Lett.* **B655** 126 - v_2 model (3)
[32] Zhao X and Rapp R 2008 - v_2 model (5) (*Preprint* `arXiv:0806.1239`)
[33] Liu Y, Xu N and Zhuang P 2010 *Nucl. Phys.* **A834** 317 - v_2 model (6)
[34] Heinz U W and Shen C 2011 private communication - v_2 model (7)
[35] Zhao X and Rapp R 2010 *Phys. Rev. C* **82** 064905
[36] Powell C B 2011 *PoS (EPS-HEP2011)* 485 (*Preprint* `arXiv:1111.6944v2`)
[37] Brodsky S and Lansberg J P 2010 *Phys. Rev. D* **81** 051502
[38] Strickland M and Bazow D 2012 (*Preprint* `arXiv:1112.2761v4`)

Journal of Physics: Conference Series **455** (2013) 012039 doi:10.1088/1742-6596/455/1/012039

Evidence for an excess of $B \to D^{(*)}\tau\nu$ decays

Bob Kowalewski, on behalf of the BaBar Collaboration

Department of Physics and Astronomy, University of Victoria, Victoria BC V8N2X3, Canada

E-mail: kowalews@uvic.ca

Abstract. The full data sample from the BaBar experiment is used to measure the ratios of branching fractions $\mathcal{R}(D) = \mathcal{B}(\bar{B} \to D\tau\bar{\nu}_\tau)/\mathcal{B}(\bar{B} \to D\ell\bar{\nu}_\ell)$ and $\mathcal{R}(D^*) = \mathcal{B}(\bar{B} \to D^*\tau\bar{\nu}_\tau)/\mathcal{B}(\bar{B} \to D^*\ell\bar{\nu}_\ell)$, where ℓ refers to an electron or muon. These ratios are sensitive to the presence of new physics in charged current decays and are sensitive probes for charged Higgs bosons. These ratios are predicted in the standard model with small theoretical uncertainties, and many experimental uncertainties cancel in the ratio. The measured values, $\mathcal{R}(D) = 0.440 \pm 0.058 \pm 0.042$ and $\mathcal{R}(D^*) = 0.332 \pm 0.024 \pm 0.018$, exceed standard model expectations by 2.0σ and 2.7σ, respectively. Accounting for correlations, the combined departure from the standard model is 3.4σ. The measured values are not compatible with a charged Higgs boson in type-II two-Higgs-doublet models, but can be accommodated in more general models involving charged mediators.

1. Introduction

This talk reports on a recent BaBar measurement published in Ref. [1]. This study is a search for lepton flavour violation in charged current decays of B mesons. Semileptonic decays form a large part of the total B decay width in the standard model (SM). The equality of coupling strengths for semileptonic B decays proceeding through electrons and muons has been tested at the few percent level in $b \to c\ell\nu$ transitions[1], and these decays are expected to be essentially free from new physics effects. However, semileptonic decays proceeding through τ leptons are less well constrained. There also exist well-motivated scenarios in which the coupling to third-generation leptons is modified; in particular, models involving charged Higgs bosons [2, 3, 4, 5, 6] provide a tree-level contribution to these decays that can lead to an $\mathcal{O}(1)$ change in the decay rate.

The purely leptonic decay $\bar{B} \to \tau\nu$ is also sensitive to new charged mediators. While that decay is interesting in its own right, the $\bar{B} \to D^{(*)}\tau\nu$ decay provides several advantages: it's far less rare ($b \to c$ versus $b \to u$ transition), has a smaller theoretical uncertainty on the SM expectation, and, as a three-body decay, allows for the measurement of decay distributions that are sensitive to the structure of the current.

The $\bar{B} \to D^*\tau\bar{\nu}_\tau$ decay rate in the SM can be written in terms of helicity amplitudes as follows [7]

$$\frac{d\Gamma}{dq^2} = \frac{G_F^2 \left|V_{cb}\right|^2 \left|\mathbf{p}^*_{D^{(*)}}\right| q^2}{96\pi^3 m_B^2} \left(1 - \frac{m_\tau^2}{q^2}\right)^2 \left[\left(|H_+|^2 + |H_-|^2 + |H_0|^2\right)\left(1 + \frac{m_\tau^2}{2q^2}\right) + \frac{3m_\tau^2}{2q^2}|H_s|^2\right] \quad (1)$$

[1] Throughout this paper, ℓ refers to an electron or a muon, and charge-conjugate decays are implied.

where $|\mathbf{p}^*_{D^{(*)}}|$ is the magnitude of the 3-momentum of the $D^{(*)}$ in the B rest frame. The helicity amplitudes $H_\pm$ do not contribute to $\bar{B} \to D(\ell/\tau)\nu$ decays, and the H_s amplitude does not contribute to $\bar{B} \to D^{(*)}\ell\nu$ decays since the coefficient of this term is proportional to the square of the charged lepton mass. The helicity amplitudes are constrained experimentally except for H_s, which is taken from an HQET-based prediction [6]. Many of the factors in Eq. 1 are common for e, μ and τ and associated uncertainties (e.g. on helicity amplitudes) partially cancel in the ratios $\mathcal{R}(D^{(*)})$. Experimental uncertainties on efficiencies and acceptances are also reduced in these ratios. The SM predictions for the ratios $\mathcal{R}(D^{(*)}) \equiv \mathcal{B}(\bar{B} \to D^{(*)}\tau\bar{\nu}_\tau)/\mathcal{B}(\bar{B} \to D^{(*)}\ell\bar{\nu}_\ell)$, using the latest input on measured form factors, are

$$\mathcal{R}_{SM}(D) = 0.297 \pm 0.017 \quad (2)$$
$$\mathcal{R}_{SM}(D^*) = 0.252 \pm 0.003 \quad (3)$$

The uncertainty due to the unmeasured helicity amplitude H_s has a smaller impact on $R_{SM}(D^*)$ than on $R_{SM}(D)$ due to the presence of additional helicity states in the $\bar{B} \to D^*$ transition. The uncertainties on the measured value are large compared to the uncertainty in the SM predictions.

2. Measurement strategy

The data were collected from e^+e^- collisions at a centre-of-mass (CM) energy of 10.58 GeV, where the $\Upsilon(4S)$ resonance decays to $B\bar{B}$ pairs with a cross-section of about 1 nb and the underlying $q\bar{q}$ continuum has a cross-section of about 3.3 nb. The beam energies are unequal but known, allowing measured quantities to be boosted into the CM frame of the colliding beams. BaBar collected a sample of 471 M $B\bar{B}$ events; in these events the two B mesons are produced near threshold, and there is insufficient energy to produce any additional particles. The energies of the B mesons therefore sum to the precisely-known beam energy, and the decay products of the B mesons overlap in the detector. The BaBar detector has silicon and gas chamber tracking for charged particles, a ring-imaging Cherenkov detector, and a CsI(Tl) calorimeter inside a 1.5T solenoid, and muon chambers interspersed in the return yoke; it is described in detail in Ref. [8].

The measurement involves the reconstruction of two related decay modes: the signal decay, $\bar{B} \to D^{(*)}\tau\bar{\nu}_\tau$ with $\tau \to \ell\nu_\tau\bar{\nu}_\ell$, and the normalization decay, $\bar{B} \to D^{(*)}\ell\bar{\nu}_\ell$. The detected particles in each case are the same. The signal decay involves three neutrinos; as a result, the proper assignment of particles to the B and $\bar{B}$ mesons requires one of them to be fully reconstructed in a hadronic decay mode. This reconstruction sums over approximately 1700 distinct decay chains and selects hadronic B_{tag} candidates based on the quantities $\Delta E \equiv E_{\text{tag}} - E_{\text{beam}}$ and $m_{ES} \equiv \sqrt{E^2_{\text{beam}} - |\mathbf{p}_{\text{tag}}|^2}$, where E_{tag} and $|\mathbf{p}_{\text{tag}}|$ are the energy and the magnitude of the 3-momentum of the B_{tag} candidate; these quantities are calculated in the CM frame. Given a B_{tag} candidate, a reconstructed $D^{(*)}$ meson and an identified electron or muon, two quantities are used to discriminate between signal and background. The missing mass squared is given by $M^2_{\text{miss}} = (p_{e^+e^-} - p_{\text{tag}} - p_{D^{(*)}} - p_\ell)^2$, where each p represents the corresponding 4-vector. This quantity peaks sharply at zero for the normalization decays, where there is only one missing neutrino, and is spread over the range 0 to $\sim 9\,\text{GeV}^2$ for signal decays. The second quantity used is the momentum of the lepton candidate $|\mathbf{p}^*_\ell|$ in the B rest frame; the leptons from the signal decay have a softer distribution than those from the normalization mode. Electron (muon) candidates with laboratory momenta above 300 (200) MeV can be identified as leptons; low-momentum muons are distinguished from pions based on dE/dx and energy deposition in the calorimeter.

Further requirements are imposed to suppress backgrounds. The signal B and the B_{tag} are required to have the expected charge correlation for coming from an $\Upsilon(4S)$ decay. Events with unassigned charged tracks are vetoed. A boosted decision tree (BDT) classifier is used to further separate correctly-reconstructed candidates from combinatorial background; the inputs to the BDT are the unassigned neutral electromagnetic cluster energy (E_{extra}), ΔE, the reconstructed masses of all intermediate D candidates, the mass differences for all $D^* \to D\pi$ candidates, the charged multiplicity of the B_{tag} and the angle between the thrust axes of the B_{tag} and the signal B candidates. The reconstruction of B_{tag} candidates and the identification of low-momentum leptons have been improved significantly since the previous BaBar analysis of these decays, resulting in an increase in the signal efficiency by more than a factor of three.

3. Backgrounds and control samples

The normalization decays, $\bar{B} \to D^{(*)}\ell\nu$, are much more numerous than the signal decays; normalization decays that are mis-reconstructed are a source of background and need to be well understood. On the other hand, normalization decays in the region $q^2 < 4\,\text{GeV}^2$ are essentially free of any signal contribution and serve as an excellent control sample for understanding the modelling of detector response. Other control samples are also available. The region $5.20 < m_{ES} < 5.26\,\text{GeV}$ allows the simulation of the combinatorial B_{tag} background to be corrected. A sample collected at e^+e^- collision energies just below $B\bar{B}$ threshold allows the simulation of $q\bar{q}$ continuum events to be corrected. Sidebands in E_{extra} allow further tests of the simulation.

An intrinsic source of background that overlaps significantly with the signal decays in both M^2_{miss} and $|\mathbf{p}^*_\ell|$ are decays of the type $\bar{B} \to D^{**}\ell\nu$ where D^{**} is a charmed state that decays to $D^{(*)}\pi$ and the π is undetected. The properties of this class of semileptonic decay are not well measured; furthermore, known decays of the type $\bar{B} \to X_c\ell\nu$ do not saturate the inclusive $b \to c\ell\nu$ decay rate, leaving room for other potential sources of background. A novel feature of this analysis relative to previous measurements of $\bar{B} \to D^{(*)}\tau\nu$ is the use of a dedicated control sample sensitive to these contributions, namely a sample where an additional neutral pion is reconstructed: $\bar{B} \to D^{(*)}\pi^0\ell$. Decays that proceed through any intermediate charm state (resonant or not) that ends up as $D^{(*)}\pi^0$ will produce a peak at $M^{2\prime}_{\text{miss}} = 0$ where $M^{2\prime}_{\text{miss}} = (p_{e^+e^-} - p_{\text{tag}} - p_{D^{(*)}} - p_{\pi^0} - p_\ell)^2$, providing experimental sensitivity to these decays. A second BDT, trained to select $\bar{B} \to D^{**}\ell\nu$ decays and reject $\bar{B} \to D^{(*)}(\ell/\tau)\nu$ decays and $B\bar{B}$ background, is used to improve the signal-to-noise in these control samples. The yield measured in these $D^{(*)}\pi^0\ell$ control samples constrains the amount of D^{**} background in the $D^{(*)}\ell$ samples.

4. Fit model and results

The fit takes as input events reconstructed in the four samples sensitive to the signal, $D^0\ell$, $D^{*0}\ell$, $D^+\ell$ and $D^{*+}\ell$, and in the four control samples, $D^0\pi^0\ell$, $D^{*0}\pi^0\ell$, $D^+\pi^0\ell$ and $D^{*+}\pi^0\ell$. The value of M^2_{miss} (or $M^{2\prime}_{\text{miss}}$) and $|\mathbf{p}^*_\ell|$ in each sample are input to an unbinned maximum-likelihood fit. In the fit, the continuum and $B\bar{B}$ combinatorial backgrounds are fixed, as is the background from charge cross-feed (where neutral B mesons are reconstructed as charge B mesons or vice-versa). The fit has 22 free parameters; 4 yields for the signal modes (one per sample), 4 yields for the normalization modes, 2 parameters for the feed-down from D^* to D when a transition particle is missed, and 12 parameters for the $D^{(*)}\pi^0\ell$ control samples (yield, feed-up from $D^{(*)}$, and $B\bar{B}$ background). The fit enforces constraints between related parameters, e.g. the D^{**} background in the signal samples and the corresponding yields in the $D^{(*)}\pi^0\ell$ control samples, and feed-up/feed-down probabilities in the signal and normalization modes.

There are 56 individual probability distribution functions (PDFs) used to describe the components of the fit. They are determined using large Monte Carlo samples with equivalent luminosities of nine times the data for $B\bar{B}$ events and twice the data for $q\bar{q}$ continuum events. The PDFs are constructed using non-parametric kernel estimation, and the uncertainties on the PDF shapes are obtained by creating 1000 distributions for each based on a bootstrapping technique (sampling with replacement).

Table 1. Fit results. The last two lines show the results of the isospin-constrained fit. The columns give the decay mode, the yields of signal and normalization events, the efficiency ratio for signal decays to normalization decays, the ratios $\mathcal{R}(D^{(*)})$, and the significance of the measured signal yield. Where two uncertainties are listed the first is statistical and the second, systematic.

Decay	$N_{\rm sig}$	$N_{\rm norm}$	$\epsilon_{\rm sig}/\epsilon_{\rm norm}$	$\mathcal{R}(D^{(*)})$	$\Sigma_{\rm tot}$
$B^- \to D^0\tau^-\bar{\nu}_\tau$	314 ± 60	1995 ± 55	0.368 ± 0.011	$0.429 \pm 0.082 \pm 0.052$	4.7
$B^- \to D^{*0}\tau^-\bar{\nu}_\tau$	639 ± 62	8766 ± 104	0.227 ± 0.004	$0.322 \pm 0.032 \pm 0.022$	9.4
$\bar{B}^0 \to D^+\tau^-\bar{\nu}_\tau$	177 ± 31	986 ± 35	0.384 ± 0.014	$0.469 \pm 0.084 \pm 0.053$	5.2
$\bar{B}^0 \to D^{*+}\tau^-\bar{\nu}_\tau$	245 ± 27	3186 ± 61	0.217 ± 0.005	$0.355 \pm 0.039 \pm 0.021$	10.4
$\bar{B} \to D\tau^-\bar{\nu}_\tau$	489 ± 63	2981 ± 65	0.372 ± 0.010	$0.440 \pm 0.058 \pm 0.042$	6.8
$\bar{B} \to D^*\tau^-\bar{\nu}_\tau$	888 ± 63	11953 ± 122	0.224 ± 0.004	$0.332 \pm 0.024 \pm 0.018$	13.2

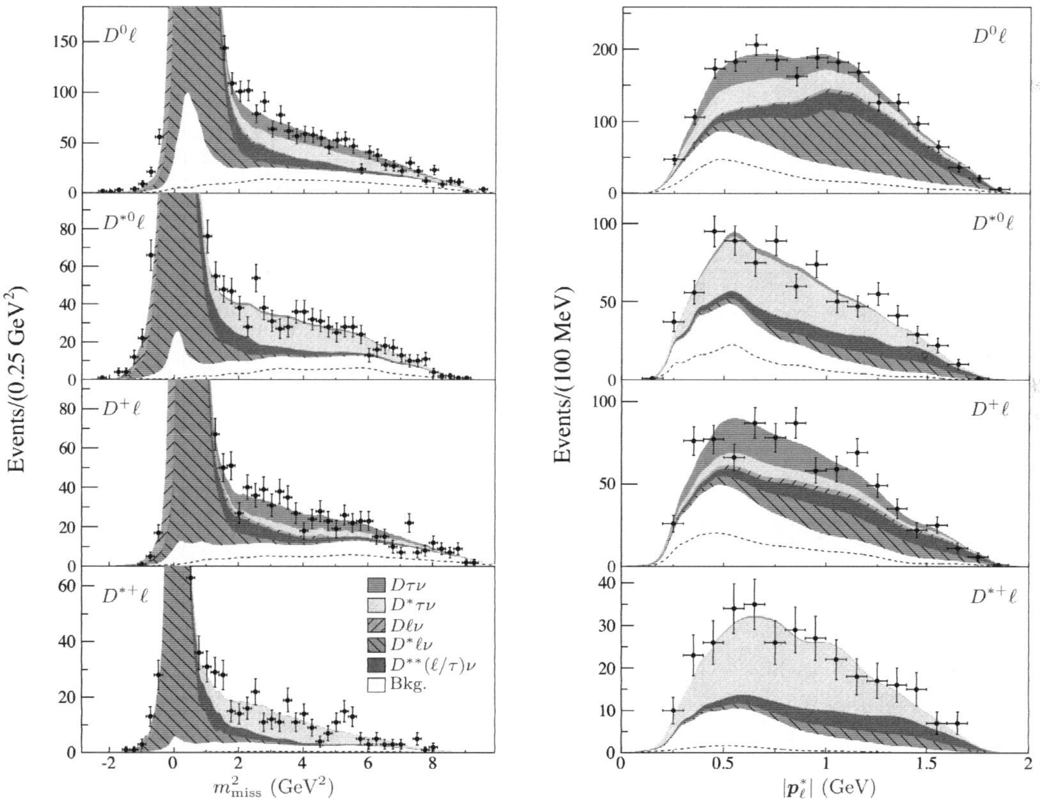

Figure 1. Projections of the fit for the four $D^{(*)}\ell$ samples in the two fit variables. The large peak from the normalization decays is cut off to emphasize the region populated by the signal decays. The dashed line in the yellow background component separates $q\bar{q}$ continuum from $B\bar{B}$ combinatorial background.

The fit results are shown in table 1 and the corresponding projections onto the fit variables are shown in figure 1 for the signal samples and in figure 2 for the $D^{(*)}\pi^0\ell$ control samples.

The results of an isospin-constrained fit are also given. The significance of the results, including systematic uncertainties, is well above 5σ for both $\bar{B} \to D\tau\nu$ and $\bar{B} \to D^*\tau\nu$. The measured values for $\mathcal{R}(D)$ and $\mathcal{R}(D^*)$ are consistent with previous measurements [9, 10, 11] but are significantly more precise.

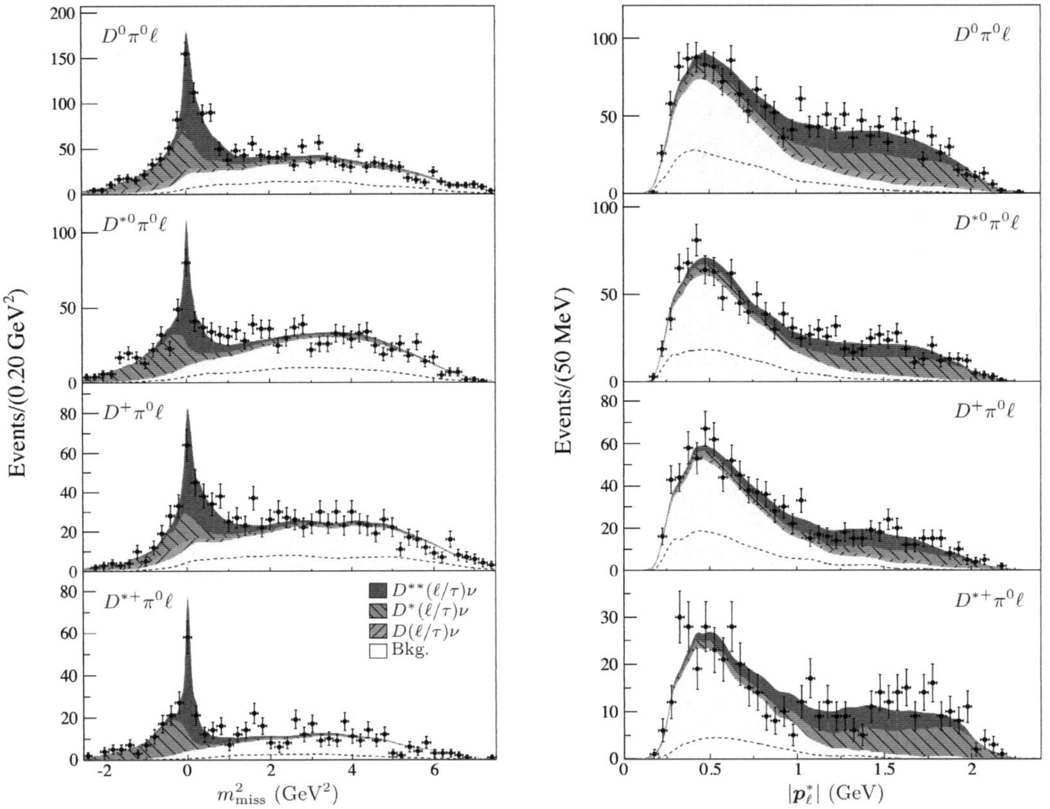

Figure 2. Projections of the fit for the four $D^{(*)}\pi^0\ell$ samples in the two fit variables. The components labelled (ℓ/τ) include contributions from all charged lepton flavours, but are dominated by contributions from e and μ.

The results for the e and μ channels are consistent with each other, and the results from the 4-channel fit are consistent with isospin symmetry. The results are stable across a wide range of BDT cuts, over which the number of events in the signal region range from 30% to 300% of the nominal sample, and the signal to background varies by a factor of 5.

An extensive set of systematic uncertainties were evaluated. They arise from imprecise knowledge of branching fractions, form factors, detector response and the statistics of the simulation samples and data control samples. These uncertainties were recorded for each channel as were their correlations between channels. The most significant uncertainties come from the modelling of the D^{**} background, which affects the constraints between the $D^{(*)}\ell$ and $D^{(*)}\pi^0\ell$ samples, the limited statistics of the MC samples, and the uncertainties on the fixed $q\bar{q}$ and $B\bar{B}$ samples; these are listed in table 2 along with the total uncertainties. The large negative correlation in the statistical uncertainty comes from the feed-down of the D^* decays into the $D\ell$ samples.

Journal of Physics: Conference Series **455** (2013) 012039 doi:10.1088/1742-6596/455/1/012039

Table 2. Leading sources of systematic uncertainty, total uncertainties and their linear correlation coefficient (ρ).

Source	Uncertainty (%) $\mathcal{R}(D)$	$\mathcal{R}(D^*)$	ρ
$D^{**}\ell\nu$ background	5.8	3.7	0.62
MC statistics	5.0	2.5	-0.48
$q\bar{q}$ and $B\bar{B}$ bkg.	4.9	2.7	-0.30
$\epsilon_{\rm sig}/\epsilon_{\rm norm}$	2.6	1.6	0.22
Systematic uncertainty	9.5	5.3	0.05
Statistical uncertainty	13.1	7.1	-0.45
Total uncertainty	16.2	9.0	-0.27

5. Discussion

The measured values for $\mathcal{R}(D)$ and $\mathcal{R}(D^*)$ exceed standard model predictions by 2.0σ and 2.7σ, respectively. Taking these measurements together and accounting for the correlation between them, their joint departure from the SM expectations[2] is 3.4σ. The comparison of these results with the Type-II two-Higgs-doublet model (2HDM) requires a re-evaluation of the signal PDFs and efficiency, since the presence of a spin-0 mediator changes the signal shape in $M^2_{\rm miss}$ and $|\mathbf{p}^*_\ell|$ as a function of $\tan\beta/m_H$. Figure 3 shows that, while each measurement can be accommodated individually, no value of $\tan\beta/m_H$ provides a good description of both $\mathcal{R}(D)$ and $\mathcal{R}(D^*)$, and the Type-II 2HDM is excluded over the full parameter space[3] at 99.8% confidence level. More general models with charged mediators are compatible with these measurements.

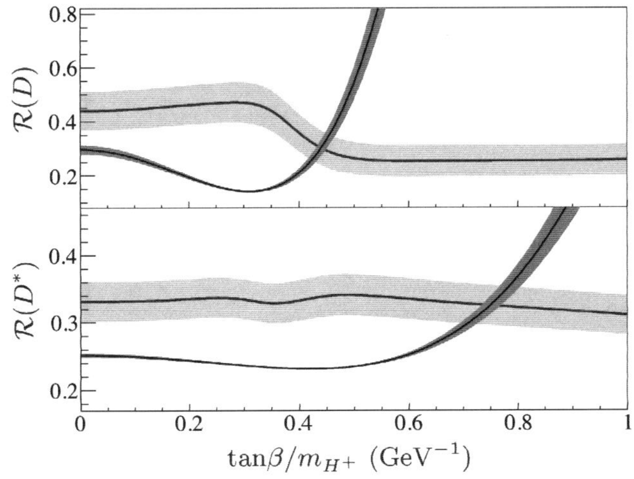

Figure 3. Comparisons of $\mathcal{R}(D)$ and $\mathcal{R}(D^*)$ as a function of $\tan\beta/m_H$ in the type-II 2HDM. The blue band shows the measured value, which varies as the signal kinematics change, and the red band shows the expected value in the 2HDM; the standard model corresponds to $\tan\beta/m_H = 0$.

[2] Using alternative calculations of the SM expectation the smallest departure is 3.2σ.

[3] The exclusion from this measurement is valid for $m_H > 15\,\text{GeV}$, but lighter m_H values have already been excluded by $b \to s\gamma$.

References

[1] Lees J *et al.* (BaBar Collaboration) 2012 *Phys. Rev. Lett.* **109** 101802
[2] Tanaka M 1995 *Z. Phys.* C **67** 321
[3] Itoh H, Komine S and Okada Y 2005 *Prog. Theor. Phys.* **114** 179
[4] Nierste U, Trine S and Westhoff S 2008 *Phys. Rev.* D **78** 015006
[5] Tanaka M and Watanabe R 2010 *Phys. Rev.* D **82** 034027
[6] Fajfer S, Kamenik JF and Nišandžić I 2012 *Phys. Rev.* D **85** 094025
[7] Korner JG and Schuler GA 1990 *Z. Phys.* C **46** 93
[8] Aubert B *et al.* (BaBar Collaboration) 2002 *Nucl. Instrum. Meth.* **A479** 1
[9] Matyja A *et al.* (Belle Collaboration) 2007 *Phys. Rev. Lett.* **99** 191807
[10] Aubert B *et al.* (BaBar Collaboration) 2008 *Phys. Rev. Lett.* **100** 021801
[11] Bozek A *et al.* (Belle Collaboration) 2010 *Phys. Rev.* D **82** 072005

International Workshop on Discovery Physics at the LHC (Kruger2012) IOP Publishing
Journal of Physics: Conference Series **455** (2013) 012040 doi:10.1088/1742-6596/455/1/012040

Searches for a light Higgs and dark photons at *BABAR*

Rocky So
Department of Physics and Astronomy
University of British Columbia
Vancouver, BC, CANADA
Representing the *BABAR* Collaboration

E-mail: rockyso@physics.ubc.ca

Abstract. A light CP-odd Higgs Boson (A^0) is predicted in extensions to the Standard Model such as Next-to-Minimal Supersymmetry Standard Model (NMSSM). We search for $\Upsilon(1S) \to \gamma A^0$; $A^0 \to \mu^+\mu^-$ or $\tau^+\tau^-$ and $\Upsilon(2S,3S) \to \gamma A^0$; $A^0 \to hadrons$ at *BABAR* using a sample of 122M Υ(3S), 99M Υ(2S), and 23M Υ(1S). We see no evidence of an NMSSM A^0. We also search for dark photons A' in $e^+e^- \to A'h'; h' \to A'A'$. We see no evidence of an A' with 516 pb^{-1} of data.

1. Introduction

A light CP-odd Higgs Boson (A^0) is predicted in extensions to the Standard Model such as Next-to-Minimal Supersymmetry Standard Model (NMSSM) [1]. An A^0 with mass less than two bottom quarks is not excluded by LEP constraints. We search for $\Upsilon(1S) \to \gamma A^0$; $A^0 \to \mu^+\mu^-$ or $\tau^+\tau^-$, and $\Upsilon(2S,3S) \to \gamma A^0$; $A^0 \to hadrons$ at *BABAR* using a sample of 122M Υ(3S), 99M Υ(2S), and 23M Υ(1S). We also search for dark photons A' in $e^+e^- \to A'h'; h' \to A'A'$.

2. Results

We see no evidence of an NMSSM A^0. Combining our previous results in $\Upsilon(2S,3S) \to \gamma A^0$; $A^0 \to \mu^+\mu^-$ [2] and $\tau^+\tau^-$ [3], we exclude some parameters space of NMSSM as shown in Figure 1 for $\Upsilon(1S) \to \gamma A^0$; $A^0 \to \mu^+\mu^-$ [4] and $\tau^+\tau^-$ [5]. We set 90% upper limits on product branching fraction $\mathcal{B}(\Upsilon(2S,3S) \to \gamma A^0) \cdot \mathcal{B}\,(A^0 \to hadrons)$ as shown in Figure 2.

We see no evidence of an A' with 516 pb^{-1} of data. We set 90% upper limits on $\alpha_D\epsilon^2$ as a function of A' and h' mass, where $\alpha_D = g_D^2/4\pi$ and g_D is dark sector gauge coupling [8]. ϵ is the mixing strength between the Standard Model and dark sector. This is shown in Figure 3.

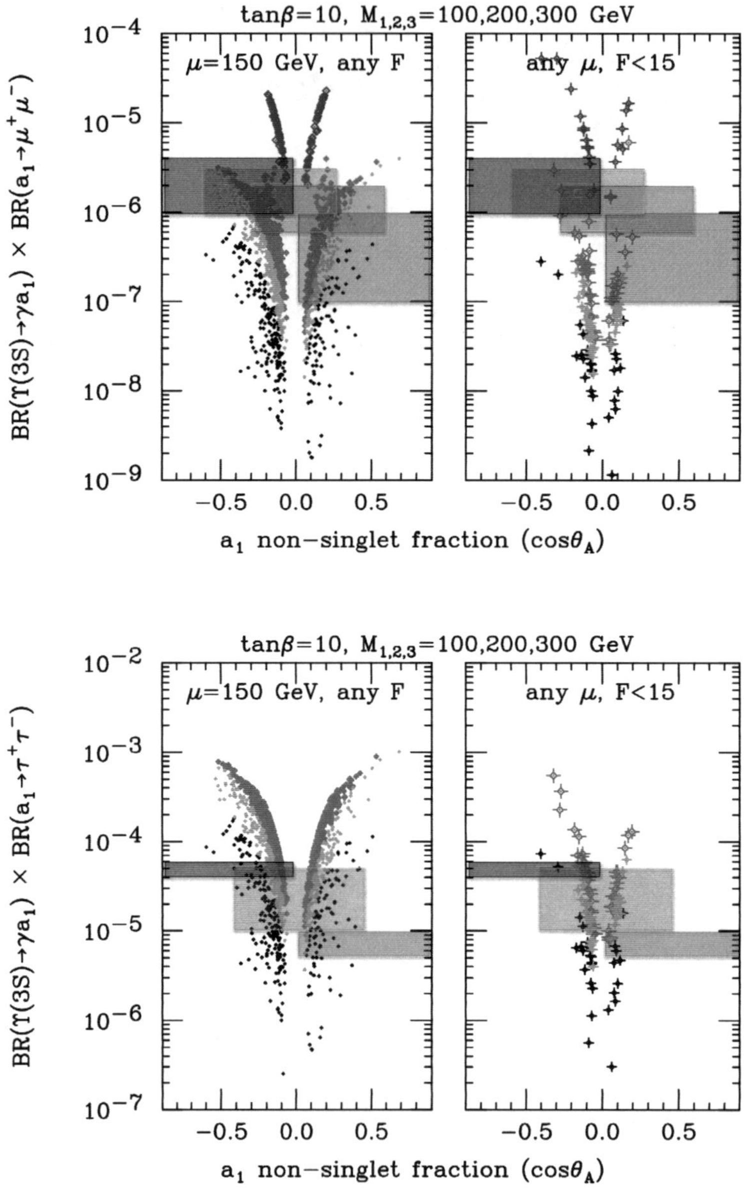

Figure 1. NMSSM parameters space excluded by *BABAR* data for A^0 goes to $\mu^+\mu^-$ (upper) and $\tau^+\tau^-$ (lower). Dots are branching fraction predictions. Boxes are our ranges of exclusion at different masses. The horizonal positions of different coloured boxes are separated for clarity. Blue is for A^0 mass less than $\tau^+\tau^-$, red is for A^0 mass less than $7.5\,\mathrm{GeV}/c^2$, green is for A^0 mass less than $8.8\,\mathrm{GeV}/c^2$, and black is for A^0 mass less than $9.2\,\mathrm{GeV}/c^2$. We exclude parameters space above the boxes.

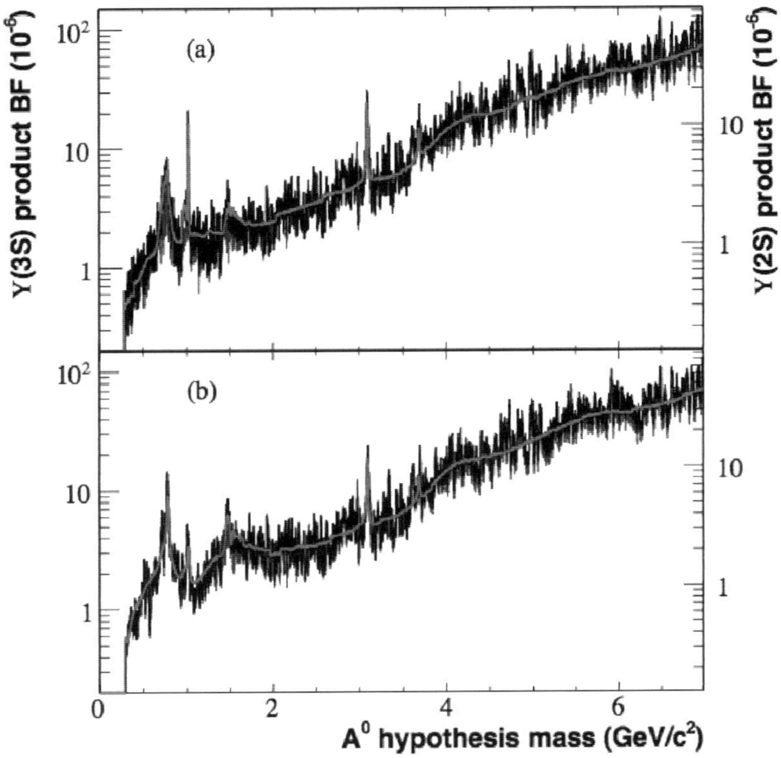

Figure 2. 90% confidence interval upper limits for $\mathcal{B}(\Upsilon(2S, 3S) \to \gamma A^0) \cdot \mathcal{B}\ (A^0 \to hadrons)$ for CP unspecified A^0 (a) CP-odd A^0 (b).

Journal of Physics: Conference Series **455** (2013) 012040 doi:10.1088/1742-6596/455/1/012040

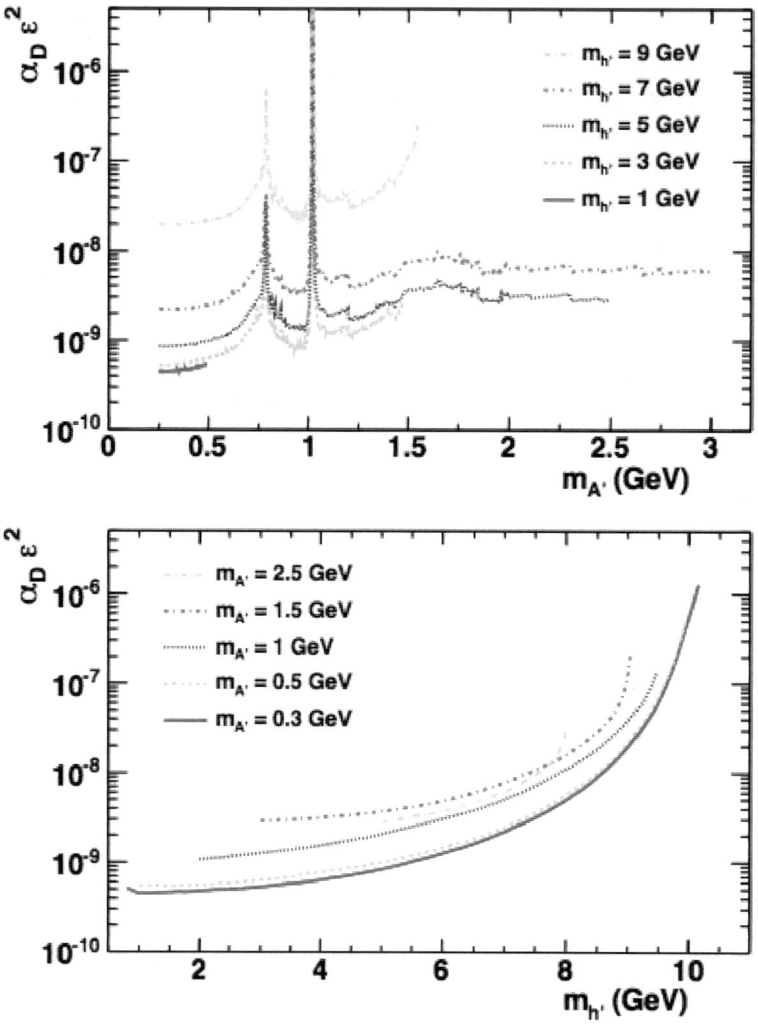

Figure 3. 90% confidence interval upper limits on $\alpha_D\epsilon^2$ as a function of A' mass (top) and h' mass (bottom).

References

[1] R. Dermisek and J. F. Gunion, Phys. Rev. D **81**, 075003 (2010).
[2] B. Aubert *et al.* (*BABAR* Collaboration), Phys. Rev. Lett. **103**, 081803 (2009).
[3] B. Aubert *et al.* (*BABAR* Collaboration), Phys. Rev. Lett.**103** , 181801 (2009).
[4] J.P. Lees *et al.* (*BABAR* Collaboration), preprint SLAC-PUB-15252, arXiv:1210.0287 [hep-ex] (2012).
[5] J.P. Lees *et al.* (*BABAR* Collaboration), preprint SLAC-PUB-15264, arXiv:1210.5669 [hep-ex] (2012).
[6] J.P. Lees *et al.* (*BABAR* Collaboration), Phys. Rev. Lett. **107**, 221803 (2011).
[7] J.P. Lees *et al.* (*BABAR* Collaboration), Phys. Rev. Lett. **108**, 211801 (2012).
[8] B. Batell, M. Pospelov, and A. Ritz, Phys. Rev. D. **79**, 115008 (2009).

Journal of Physics: Conference Series **455** (2013) 012041 doi:10.1088/1742-6596/455/1/012041

Solving nuclear shape conundrum at HIE-ISOLDE

J.N. Orce

Department of Physics and Astronomy, University of the Western Cape, P/B X17, Bellville, ZA-7535 South Africa

E-mail: jnorce@uwc.ac.za

Abstract.
It is the purpose of this paper to illustrate some of the low-energy nuclear physics that we want to pursue at the new HIE-ISOLDE radioactive-ion-beam facility at CERN. The University of the Western Cape leads an experimental proposal and co-authors two Letters of Intent, in collaboration with European institutions, at HIE-ISOLDE. Timely topics such as the "*Exploration of K-isomerism using unique high-K isomeric beams - CERN-INTC-I-101*" and "*Shape changes and proton-neutron pairing around the $N = Z$ line - CERN-INTC-I-102*" are addressed in these Letters of Intent. Our experimental proposal aims at performing a multi-step Coulomb-excitation of radioactive ^{70}Se ion beams using the ^{208}Pb(^{70}Se,^{70}Se*)^{208}Pb* reaction at a bombarding energy of 5.5 MeV/u. The physics goal is a precise measurement of the $\langle 2_1^+ \parallel \hat{E}2 \parallel 2_1^+ \rangle$ diagonal matrix element, related to the spectroscopic quadrupole moment, in ^{70}Se. Full simulations presented in this work show distinct angular distributions for plausible values of the spectroscopic quadrupole moment; with a predicted uncertainty of approximately ± 0.1 eb. Additional diagonal and transitional matrix elements will also be obtained. These results will shed light onto the origin of rarely-found oblate shapes and shape coexistence in this region of rapidly-changing shell structure.

1. Motivation

A remarkable feature of atomic nuclei is their ability to adopt different shapes for a small cost in energy compared to their total binding energy. Deformed nuclei may adopt axially-symmetric quadrupole shapes, which can either take prolate (cigar shape) or oblate (disk shape) forms. The vast majority of nuclei present prolate shapes; with only a handful of oblate-shaped exceptions. Even doubly closed-shell nuclei, in principle spherical in their ground state, can adopt a strongly-deformed shape in an excited state when Cooper pairs are broken. These strongly-deformed shapes were originally observed in doubly-magic ^{16}O and explained by Morinaga as 4 particles (p), with the remaining 4 holes (h), excited across the $N = Z = 8$ (N for neutrons and Z for protons) shell closure [1]. Mixing calculations between the spherical shell-model ground-state and deformed $2p-2h$ and $4p-4h$ configurations supported Morinaga's idea [2, 3]. These findings led to the phenomenon of shape coexistence [4], now generally observed throughout the nuclear chart except for light nuclei. These strongly-deformed configurations are generally accepted to arise from the residual proton-neutron interaction [5]; the physical origin or mathematical form of which remains unknown.

Particularly, the neutron single-particle level energies calculated by the deformed nuclear shell-model for the $A \approx 70$ region (where $A = N + Z$) are shown in Fig. 1. A similar plot is expected for proton single-particle energies. The $A \approx 70$ region is one of the best testing grounds for rapidly-changing nuclear shapes because of the subshell gaps at proton and neutron numbers 34, 36 and 38. In addition, triaxial shapes, i.e., non axially-symmetric quadrupole shapes, have also been invoked in this region to explain extremely deformed prolate shapes, triaxial (or γ) vibrations and the enhancement of nuclear collectivity. Among the different shapes a nucleus can take, the rare occurrence of oblate shapes remains to be elucidated and requires special attention.

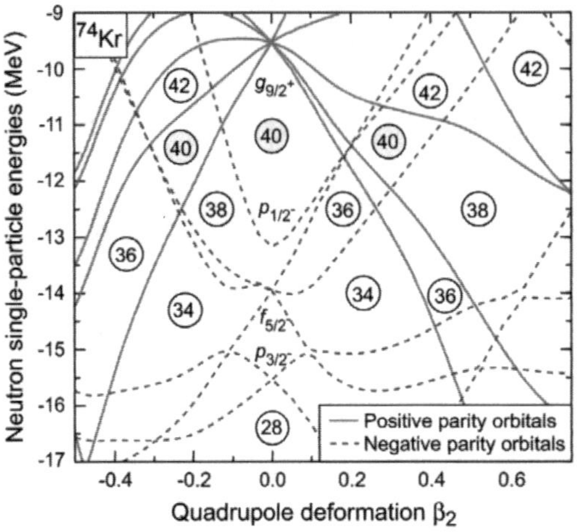

Figure 1. Deformed single-particle energies for the orbitals of interest. Figure taken from [6].

A key tool for measuring nuclear deformation is the reorientation effect in Coulomb excitation measurements. This effect is a second-order perturbation that generates a time-dependent hyperfine splitting of the nuclear levels and changes the population of the different magnetic substates; hence, modifying the Coulomb-excitation cross section as a function of the magnitude and sign of the spectroscopic quadrupole moment, Q_S, i.e., the quadrupole shape of the nucleus in the laboratory frame [7]. The advantage of Coulomb-excitation studies is that, unlike the nuclear interaction, the electromagnetic interaction is well understood. Nuclear contributions can simply be eliminated by choosing a 'safe' bombarding energy well below the Coulomb barrier. Moreover, Coulomb-excitation experiments provide measurements of matrix elements. Assuming an ideal rotor, for a $J^\pi = 2^+$ state, $Q_S(2_1^+)$ is related to the diagonal matrix element, $\langle 2_1^+ \| E2 \| 2_1^+ \rangle$ by,

$$Q_S(2_1^+) = \sqrt{\frac{16\pi}{5}} \frac{1}{\sqrt{2J+1}} \langle JJ20 \mid JJ \rangle \langle 2_1^+ \| E2 \| 2_1^+ \rangle = 0.75793 \langle 2_1^+ \| E2 \| 2_1^+ \rangle. \quad (1)$$

A reorientation-effect measurement of ^{70}Se beams ($Z = 34$) has recently been carried out at REX-ISOLDE [8]. Such a study, when combined with recent accurate lifetime measurements using the recoil-distance Doppler shift method [9], suggests an oblate shape for the ground state of ^{70}Se [9]. This is illustrated in Figure 2, where a positive $Q_S(2_1^+)$ value is deduced from the crossing of the Coulomb-excitation (black) sloping curve with the accurate $B(E2; 2_1^+ \rightarrow 0_1^+)$ value (red solid line) obtained from the lifetime measurement [9]. An earlier, and much shorter, lifetime measurement by Heese and collaborators suggests instead a prolate shape [10], as indicated by the overlap of the Coulomb-excitation curve and the horizontal dashed lines in

Fig. 2. Experimental and theoretical results in the light Se and Kr nuclei seems to suggest the emergence of oblate shapes as one approaches $N = Z$ [9]. However, the sudden decrease of the $B(E2; 2_1^+ \to 0_1^+)$ value from the NNDC accepted value of 44(9) W.u. [11] to the recent determination of 20.0(1.2) W.u. [9] cannot be accommodated theoretically (green asterisk in Fig. 2). A large uncertainty in the $Q_S(2_1^+)$ value prevents a more detailed comparison with theory. Aided by the higher yields and beam energies available at the new HIE-ISOLDE facility, the main goal of our experimental proposal is the accurate determination of the $\langle 2_1^+ \parallel \hat{E2} \parallel 2_1^+ \rangle$ diagonal matrix element, which, as shown in Fig. 2, currently spans in the range $-0.20 \lessapprox \langle 2_1^+ \parallel \hat{E2} \parallel 2_1^+ \rangle \lessapprox +1.0$ eb.

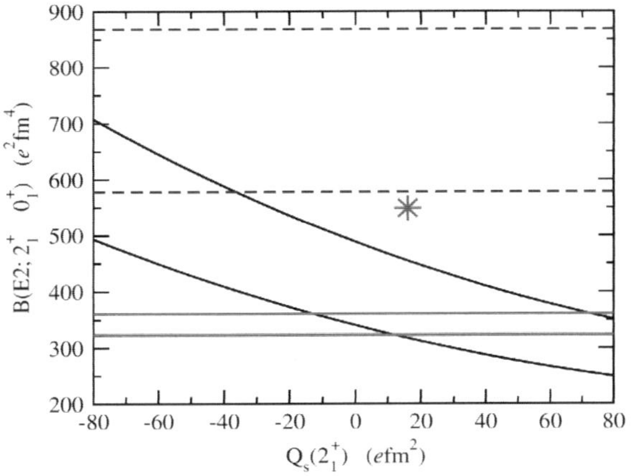

Figure 2. The $B(E2; 2_1^+ \to 0_1^+)$ value as a function of $Q_S(2_1^+)$ for the 2_1^+ state in ^{70}Se. The asterisk indicates the theoretical value. The large uncertainty in the Coulomb-excitation curve prevents a conclusive $Q_S(2_1^+)$ value. Figure taken from [9].

2. Experimental Details and Full Simulations

The HIE-ISOLDE project is a major upgrade to the existing REX-ISOLDE radioactive-ion-beam (RIB) facility at CERN, with the imminent objective to provide 5.5 MeV/u post-accelerated RIBs. This bombarding energy is ideal for multi-step Coulomb-excitation studies. We have made full use of this potential and proposed the study of the ^{208}Pb(^{70}Se,^{70}Se*)^{208}Pb* reaction at 5.5 MeV/u. Pure ^{70}Se beams, required for Coulomb-excitation measurements, have previously been delivered at REX-ISOLDE [8] using SeCO$^+$ molecules and breaking this molecule in the electron beam ion source (EBIS) [8]. The same procedure should be available at HIE-ISOLDE. The bombarding energy of 5.5 MeV/u (or 385 MeV) is well below the Coulomb barrier, at around 427 MeV, and assuming Cline's prescribed 5.0 fm separation between nuclear surfaces for heavy-ion reactions [12], safe for laboratory scattering angles $\theta_{lab} \leq 83°$. A large Sommerfeld parameter of $\eta = 196 \gg 1$ validates the semiclassical approximation [13] and a small adiabaticity parameter of $\xi = 0.35$ enhances the population of the 2_1^+ state in ^{70}Se through Coulomb excitation.

The powerful particle-γ coincidence technique will be utilised in this experiment. The de-excited γ-rays will be detected with the high-efficient MINIBALL array, composed of eight large cluster detectors and shown in Fig. 3, and the scattered ^{70}Se ions with a forward double-sided CD-type silicon detector covering $[17°, 52°]$ scattering angles in the laboratory frame. Background γ rays from the experimental hall can be suppressed by requiring a particle-γ coincidence condition. The particle-γ coincidence condition is satisfied by requiring a Ge hit and a hit in both the θ ring and ϕ sector of the Si detector within a time window. Events with Ge hits outside of this window are considered to be random coincidences and are background substracted.

Figure 3. The MINIBALL HPGe γ-ray array (seeing from above), comprising eight large HPGe cluster detectors, at the ISOLDE facility.

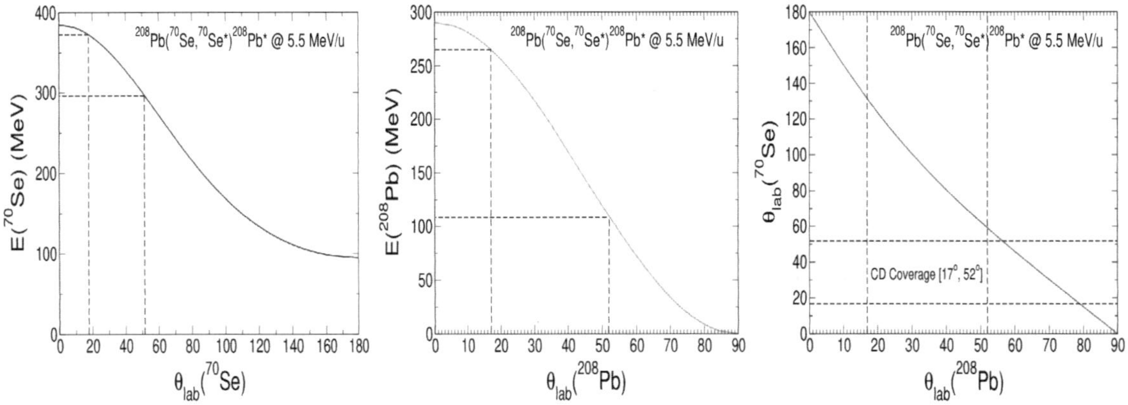

Figure . Reaction kinematics for the ^{70}Se ions and ^{208}Pb recoils.

The state-of-art silicon detector provides kinematic characterization of scattered ions for Doppler corrections to the γ-rays emitted from the Coulomb-excited beam nuclei. Reaction-kinematics plots shown in the first two panels of Fig. 4 illustrate the energies of ^{70}Se ions and ^{208}Pb recoils as a function of scattering laboratory angles. Their distinct energies for the $[17°, 52°]$ angular range permits a separation in the particle energy spectra. The right panel of Fig. 4 guarantees the selection of either ^{70}Se ions and ^{208}Pb recoils for the $[17°, 52°]$ angular range.

Using matrix elements extracted from Refs. [9, 11], coupled-channel GOSIA calculations [14], based on the semiclassical approximation of Coulomb-excitation theory [13], have been performed in this work. The results are presented in Fig. 5 for the population of the 2_1^+ state in ^{70}Se as a function of scattering angle in the laboratory frame and at plausible $Q_S(2_1^+)$ values given in Fig. 2. The differential cross section, shown in the left panel of Fig. 5, strongly depends on $Q_S(2_1^+)$, being stronger as the shape becomes more oblate, and peaks at $\theta_{lab} \approx 40°$. The right panel of Fig. 5 shows counts per five days of beam time as a function of scattering angle, assuming a yield of 10^4 ions/s and a 2-mg/cm^2 ^{208}Pb foil. The integrated yields for six sets of angular ranges have been summed up and illustrates the sensitivity to measure $Q_S(2_1^+)$. Five days of beam time (or fifteen shifts) will provide a measurement of $Q_S(2_1^+)$ with a ± 0.1 eb statistical uncertainty given by the error bars. This uncertainty could be smaller with the increasing yields expected with the upgrade to 2 GeV proton energies planned at HIE-ISOLDE. Comparatively, a yield of 10^4 ions/s for the previous ^{70}Se reorientation-effect measurement at 2.94 MeV/u yielded an area of 139(13) counts for the 2_1^+ peak [8]. Assuming the same yield and $Q_S(2_1^+) = +0.8$ eb, we approximately achieve 650 counts/day at 5.5 MeV/u; 500 counts/day for $Q_S(2_1^+) = 0$ eb. Experimental integrated yields will be normalised to the well-known $B(E2, 2_1^+ \rightarrow 0_1^+) = 342(19)$ e^{2}fm^4 value in ^{70}Se to account for experimental unknowns in the setup, such as the systematic uncertainties in the absolute beam energy, target thickness, particle detection efficiency and dead time of the data acquisition. Uncertainties in the γ-ray efficiency, the ϕ asymmetry of the cluster detector and the 5.5% error in the $B(E2, 2_1^+ \rightarrow 0_1^+)$ value will also be considered; although the statistical uncertainty will likely dominate the quoted error on this measurement.

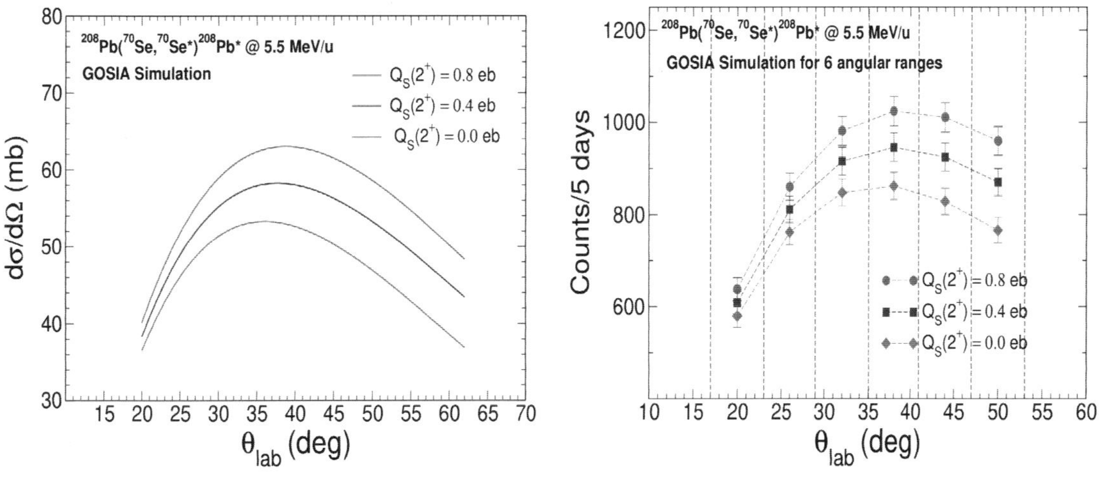

Figure . Simulated differential cross sections (left) and count rates per five days of beam time (right) as a function of scattering laboratory angle for the population of the 2_1^+ state in ^{70}Se. The dash lines in the right panel indicate the different angular ranges covered for the count-rate data points. The distinct angular distributions for plausible $Q_S(2_1^+)$ values indicate that a precise measurement of $Q_S(2_1^+)$ is feasible. The error bar accounts for an uncertainty of ≈ 0.1 eb.

Another question lies in understanding the excitation mechanism of higher-lying levels shown in Fig. 6. With a total cross section of tens of mb, we expect to obtain information on transitional matrix elements relating the 2_2^+, 4_1^+, 0_2^+, 4_2^+ and 6_1^+ states. About 250 counts calculated for the population of the 2_2^+ state may provide information on the sign of $\langle 2_2^+ \mid\mid \hat{E}2 \mid\mid 2_2^+ \rangle$ [8, 15].

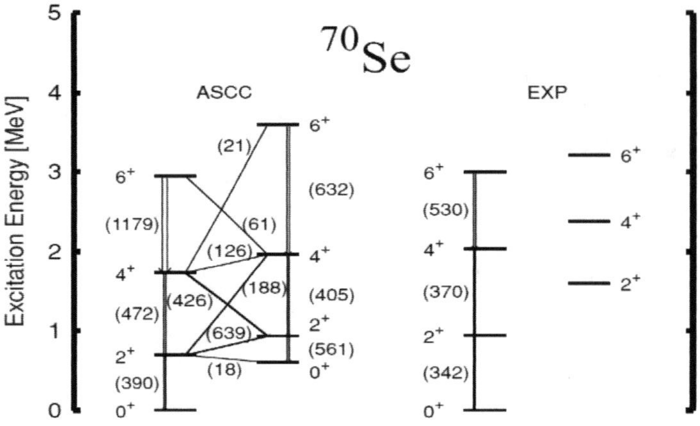

Figure . Theoretical (left) and experimental (right) excitation spectra and $B(E2)$ values (in units of e^{2}fm^4) for low-lying states in ^{70}Se [9, 16]. Noticeable predictions are the large $B(E2; 2_2^+ \rightarrow 2_1^+) = 639$ e^{2}fm^4 value and the low-lying 0_2^+ state.

3. Concluding emarks

It is plausible that shape coexistence may be enhanced in $N \approx Z$ nuclei by the occupation of the same orbitals for protons and neutrons and the resulting neutron-proton interaction [4, 5]. In fact, low-lying excited 0^+ states have been observed in other neutron-deficient Se [17, 18, 19] and Kr [20, 21, 22] isotopes and associated with shape coexistence. Figure 7 indicates the presence of an anomalous rotational behaviour for the 2_1^+ state in the neutron-deficient Se, Kr and Sr isotopes. The anomalous rise at $J = 2$ is particularly pronounced for the Se isotopes; being the largest for ^{70}Se. The general interpretation is that coexisting 0^+ configurations mix and result in a lowering of the excitation energy for the ground state; hence, the anomalous high excitation energy of the 2_1^+ state. Experimentally, the only candidate for a low-lying 0_2^+ state has been suggested at a higher ≈ 2 MeV excitation [17]; but a lower-lying 0^+ excitation below the 2_2^+ state could be identified in future β^+ decay measurements of ^{70}Br. The indication of shape coexistence in ^{70}Se may point at the existence of a low-lying 0^+ excitation.

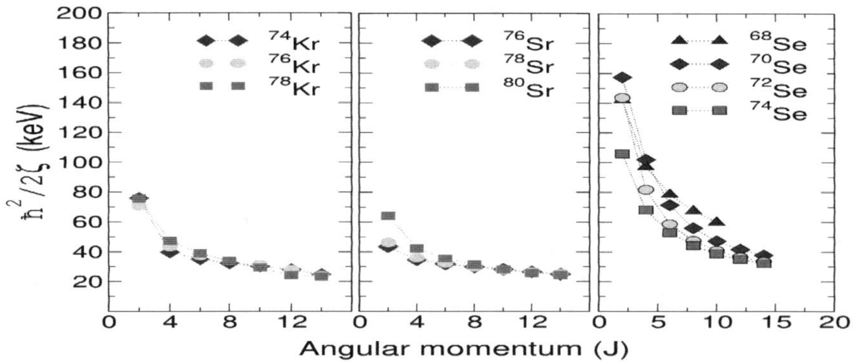

Figure . The experimental rotational parameter $\hbar^2/2\Im$ for the neutron-deficient Kr (left panel), Sr (central panel) and Se (right panel) isotopes. The anomalous rise at $J = 2$ is particularly pronounced for ^{70}Se.

The mixing of prolate-oblate shapes within rotational bands has been studied using various mean-field calculations [23, 24]. However, more realistic calculations can be expected from recent state-of-art beyond mean-field calculations, which include the interactions of multi-particle-multiple-hole excitations, and have, for instance, successfully described the structure of low-lying 0^+ states in ^{16}O in terms of self-consistent $0p-0h$, $2p-2h$, and $4p-4h$ Hartree-Fock states [25]. Similar beyond mean-field calculations will be desirable for the $A \approx 70$ region.

Furthermore, neutron-deficient nuclei in the $A \approx 70$ region are of astrophysical interest as these are produced in type I X-ray bursts. A better knowledge of the low-lying nuclear structure in this region is crucial for rp process network calculations and the understanding of luminosity curves through β decay and creation of elemental abundances after the burst [26]. Interestingly, comparison of the experimental Gamow-Teller strength distribution $B(GT)$ with mean-field calculations [27, 28] suggests that nuclei β decay depending on their shape!

This experimental proposal has been defended at the INTC meeting at ISOLDE last November 2012 and the referee's report is attached below:

CE N-INTC-2 12- , INTC-P-3 , Solving the shape conundrum in ^{70}Se.

"It is proposed to determine the sign and the magnitude of the spectroscopic quadrupole moment of the first 2^+ state in ^{70}Se, with the aim to provide a definite answer on the shape of this nucleus. The proposed method is the multi-step safe Coulomb excitation with the MINIBALL setup. Besides that, other transition matrix elements of higher-lying states can be obtained, to yield more spectroscopic information of this nucleus. This information is important to understand shape coexistence phenomena in this region, especially that discrepancies between experiment and theory are present. The physics case was found interesting and relevant. It was, however, not clear, whether the proposed experiment can provide definitive values for the $E2$ matrix elements from which the shape ^{70}Se shape can be determined. Therefore, full simulations of the experiment should be performed and a Letter of Clarification should be provided clarifying this point, before the shifts can be awarded. Thus, at this point, the Committee has not recommended any shifts for the approval of the Research Board."

The requested full simulations are shown in Figs. 4 and 5 and described in detail in section 2, and will be presented at the next INTC meeting this June 2013. The distinct angular distributions simulated in this work and shown in Fig. 5 permit a measurement of $Q_S(2_1^+)$ with an uncertainty of approximately ± 0.1 eb. Additional experimental proposals will be submitted to the INTC committee at HIE-ISOLDE in the near future. Complementary reorientation-effect measurements using multi-step Coulomb-excitation reactions with stable beams are scheduled at iThemba LABS end of 2013 and beginning of 2014. Such experiments will prepare our students with hands-on and data-analysis skills for similar RIB measurements at HIE-ISOLDE. Moreover, the prospect of a RIB facility at iThemba LABS is a strong motivation for capital development at HIE-ISOLDE. J.N.O. acknowledges D. Jenkins, N. Warr and the HIE-ISOLDE collaboration for fruitful physics discussions. J.N.O. and the UWC nuclear physics group acknowledge travel funding from the South Africa-CERN (SA-CERN) collaboration.

eferences

[1] Morinaga H 1956 Phys. Rev. **101** 254.
[2] Brown G E 1964 *Comptes Rendus du Congres International de Physique Nucleaire, (Vol. 1, Paris)* 129.
[3] Brown G E and Green A M 1966 Nucl. Phys. **75** 401.
[4] Heyde K and Wood J L 2011 Rev. Mod. Phys **83** 1467.
[5] de Shalit A and Goldhaber M 1953 Phys. Rev. **92** 1211.
[6] Bender M *et al.* 2006 Phys. Rev. C **73** 034322.
[7] Häusser O 1974 *Nuclear Spectroscopy and Reactions C, edited by J. Cerny (Academic, New York).*
[8] Hurst A M *et al.* 2007 Phys. Rev. Lett. **98** 072501.

[9] Ljungvall J *et al.* 2008 Phys. Rev. Lett. **100** 102502.
[10] Heese J *et al.* 1986 Z. Phys. A **325** 45.
[11] http://www.nndc.bnl.gov. (*NNDC database*)
[12] Cline D 1986 Ann. Rev. Nucl. Part. Sci. **36** 683.
[13] K. Adler, A. Bohr, T. Huus, B. Mottelson, and A. Winther, Rev. Mod. Phys. **28**, 432 (1956).
[14] Czosnyka T *et al.* 1983 Bull. Am. Phys. Soc. **28** 745.
[15] Orce J N *et al.* 2012 Phys. Rev. C **86** 041303(R).
[16] Rainovski G *et al.* 2002 J. Phys. G **28** 2617.
[17] Ahmed A *et al.* 1981 Phys. Rev. C **24** 1486.
[18] Hamilton J H *et al.* 1974 Phys. Rev. Lett. **32** 239.
[19] Mylaeus T *et al.* 1989 J. Phys. G**15** L135.
[20] Clément E *et al.* 2007 Phys. Rev. C **75** 054313.
[21] Bouchez E *et al.* 2003 Phys. Rev. Lett. **90** 082502.
[22] Becker F *et al.* 2006 Nucl. Phys. A**770** 107.
[23] Petrovici A, Schmid K W, Faessler A 2003 Nucl. Phys. A **728** 396.
[24] Hinohara N, Nakatsukasa T, Matsuo M and Matsuyanagi K 2009 Phys. Rev. C **80** 014305.
[25] Bender M and Heenen P -H 2003 Nucl. Phys. A**713** 390.
[26] Savory J *et al* 2009 Phys. Rev. Lett. **102** 132501.
[27] Nácher E *et al.* 2004 Phys. Rev. Lett. **92** 232501.
[28] Sarriguren P 2009 Phys. Rev. C **79** 044315.

Journal of Physics: Conference Series **455** (2013) 012042 doi:10.1088/1742-6596/455/1/012042

Omega Electroproduction

D A Unwuchola[1], S H Connell[1], M Aurousseau[1] , M M Dalton on behalf of the Jefferson Lab, Hall C, Baryon Collaboration.

[1] University of Johannesburg, Johannesburg, South Africa

E-mail: doomnull@gmail.com

Abstract. The differential cross section for $p(e, e'\omega)p$ has been studied at $Q^2 \sim 5.5$ (GeV/c)^2. Here Q^2 represents the four momentum squared of the virtual photon in the excitation of baryonic resonances by an electron projectile. In order to extract the ω-meson differential cross section from the JLAB data, the data was compared to a full Monte Carlo simulation of the detector based on events generated for omega production in a way that the production cross section was varied to achieve a match to the data. The bin selected for this procedure takes into account the measure of robustness of the stripping of the ω peak from the multi-pion background as well as the statistics in the measured data and the Monte Carlo simulation of the signal and background physics. An error estimation technique for the cross section was based on determining the dependence of the extracted cross section parameters on the experimental set-up (including parameters for the spectrometer, target beam geometeries and performance). We compare our results with a Regge-based model for hadronic content in the t-channel exchange of a photon in Q^2 region of overlap. There is an extension of this data into a completely new region, which is the highest yet measured.

1. Introduction

The extraction of the measured exclusive ω differential cross section at the highest achievable Q^2 values in the valence quark region has been performed. The importance of our study is that this average Q^2 of 5.5 (GeV/c)^2 falls in the region where the transition from low Q^2, low t physics, where soft, non-pertubative QCD processes characterised by constituent quarks dominate, to the high Q^2, high t regime where hard processes characterised by current quark correlations are expected to play an increasingly important role [1]. This paper describes the experiment, the extraction of the cross section from the data, the bin selection and the error estimation. Finally the data are discussed.

2. Experimental setup

The E01-002 experiment in Hall C at the Thomas Jefferson National Accelerator Facility (JLab) consists of an electron beam energy of 5.75 GeV incident on a hydrogen target, two spectrometers, associated electronics and software for reconstruction of events. The scattered electrons are detected using the Short Orbit Spectrometer (SOS) [2, 3],which is a resistive $QD\overline{D}$ (quadrupole, dispersive dipole, anti-dispersive dipole) spectrometer. The recoil protons are detected using the High Momentum Spectrometer (HMS), which is a $QQQD$ spectrometer. The ω particles were identified by the missing mass method which involves using event reconstruction and conservation laws. The instrumentation layout of the experiment can be found in Ref. [1].

3. Selection of data

The electron spectrometer was fixed in both angle and momentum therefore defining a central three-momentum transfer **q**. The direction of a boosted decay cone of protons is determined in turn by the vector **q**. In order to capture as much of this decay cone possible, the proton spectrometer was stepped in angle and momentum, with kinematics chosen such that adjacent settings overlapped in angle and momentum [4]. With such an approach, the systematic uncertainties associated with the imperfect knowledge of the spectrometer acceptance is reduced. The kinematic settings for the experiment is shown in Table 1.

Electron Arm P_{SOS} [GeV/c]	Electron Arm θ_{SOS} [degrees]	Proton Arm P_{HMS} [GeV/c]	Proton Arm θ_{HMS} [degrees]
1.74	47.5	2.13	13.5, 16.5, 19.5, 22.5
1.74	47.5	2.23	18.0, 12.0, 15.0, 21.0
1.74	47.5	2.57	11.2, 13.5, 16.5, 19.5, 22.5
1.74	47.5	2.69	12.0, 15.0, 18.0, 21.0, 24.0
1.74	47.5	3.10	11.2, 13.5, 16.5, 19.5, 22.5
1.74	47.5	3.24	12.0, 15.0, 18.0, 21.0, 24.0
1.74	47.5	3.73	11.2, 13.5, 16.5, 19.5, 22.5
1.74	47.5	3.90	12.0, 15.0, 18.0, 21.0
1.74	47.5	4.50	11.2, 13.5, 16.5, 19.5
1.74	47.5	4.70	15.0, 18.0

Table 1. Spectrometer settings for the $Q^2 \simeq 5.36$ GeV$^2/c^2$, ($E = 5.5$ GeV) data

The SOS was used to separate the electrons from the negatively charged pions. This was done by using both a threshold gas Čerenkov detector and a lead-glass calorimeter. In the HMS, protons were separated from positively charged pions using a combination of coincidence time (the difference between the trigger times of two the spectrometers) and particle velocity, or equivalently the time of flight β_{tof}.

The Hall C event builder code called ENGINE [1] was used for the offline replay of the collected raw signals. ENGINE reads each of the events, reconstructs trajectories, and generates kinematical and particle identification information for each event. In essence, the replay ENGINE converts raw data into calibrated physical quantities on an event-by-event basis. The analysis codes operates on the ENGINE ouput files (HBOOK) and converts them to ROOT files. From this point on the data is analysed within the ROOT framework using `C++` code.

In the process $p(e, e'p)\omega$, the kinematics of the undetected particle is reconstructed by calculating the missing mass squared m_x^2 from the conservation of the four-momentum. The extraction of the ω was accomplished by applying a cut on the m_x^2 spectrum around the omega peak and subtracting the background. To ensure the quality of the measurement, events of interest are selected by applying the set of 'standard' cuts to the data as listed in Table 2. Cuts such as the particle momentum deviation at both the SOS and HMS (δ_h and δ_s) are applied to ensure that only particles within the well understood region of the spectrometer momentum acceptance are used. Also used, but not listed in Table 2, are cuts on the collimators at both spectrometers [5]. These are to ensure that the path of the reconstructed track of a detected particle can be traced back through acceptable regions of the collimator slits.

For the purpose of extraction of the cross section as indicated in Table 3, the data are binned in W, $\cos\theta_\omega$, ϕ_{cm} and m_ω^2, where W is the invariant mass of the hadronic system, θ_ω is the polar angle between the direction of the ω and the three-momentum transfer vector $\vec{q}$ in the centre-of-mass of the resonance, ϕ_{cm} is the azimuthal angle of the ω with respect to the electron

Table 2. Set of 'standard' cuts applied to the data and to the simulation where applicable. † The Particle Identification cuts were not applied to the simulation.

Quantity	Cut	Purpose
†Coincidence time	$\mid t_{\text{coin}} - t_{\text{cent}} \mid \leqslant 1.5$	Selecting proton
HMS particle momentum deviation, $\delta_h = \frac{p - p_{\text{HMS}}}{p_{\text{HMS}}}$	$\mid \delta_h \mid \leqslant 9\%$	HMS acceptance
SOS particle momentum deviation, $\delta_s = \frac{p - p_{\text{SOS}}}{p_{\text{SOS}}}$	$-15\% \leqslant \delta_s \leqslant +20\%$	SOS acceptance
SOS x position focal plane, $X_{\text{SOS,f.p}}$	-20 cm $\leqslant X_{\text{SOS,f.p}} \leqslant +22$ cm	SOS acceptance
†SOS shower counter sum, E_{norm}	$E_{\text{norm}} \geqslant 0.7$	selecting electron
†SOS Čerenkov number of photons, $N_{p.e.}$	$N_{p.e.} \geqslant 0.5$	selecting electron
Missing mass squared m_x^2	$0.56 \text{ GeV}^2 \leqslant m_x^2 \leqslant 0.66 \text{ GeV}^2$	selecting ω particle

scattering plane, and m_ω^2 is the square of the missing mass for $p(e, e^{'}p)\omega$.

Variable	Range	Bins
W (GeV)	$1.72 \leq W \leq 1.92$	10
$\cos\theta_\omega$	$-1.0 \leq \cos\theta_\omega \leq 1.0$	5
ϕ_ω (rad)	$0 \leq \phi_\omega \leq 2\pi$	3

Table 3. ω analysis binning showing the ranges of W, $\cos\theta_\omega$ and ϕ_ω

For the ω analysis, the Monte Carlo ratio method is used to extract the cross section. The Hall C Monte Carlo simulation package (SIMC) included simulations of variety of effect as well as models for electroproduction cross section. In practice, the experimentally measured differential cross section is distorted by several systematic and random effects such as radiative processes, multiple scattering and other nuclear reactions, spectrometer acceptance and efficiency issues, to mention a few. These effects can not be reliably deconvoluted from the data. Instead, events generated from a model of the theoretical differential cross section are subjected to a Monte Carlo procedure which simulates the same systematic and random effects of the actual experiment (spectrometers, target, beam). The method used is outlined in four basic steps. Firstly, we obtain the data yield. Secondly, we generate the synthetic the Monte Carlo data. Thirdly, the ratio of the data yield and the Monte Carlo is calculated. Fourthly, the measured experimental cross section is scaled from the ratio applied to the theoretical cross section which had input to the Monte Carlo.

The results generated from the Monte Carlo where adjusted to be reasonably close to the data such that the same computational treatment could be used to analyze both the results from the Monte carlo and the data. For each run the data and the Monte Carlo results were binned in (W, cosθ_ω and ϕ_ω). A missing mass squared (m_x^2) cut which contains the ω particle peak is applied to the data and the Monte Carlo results, and the results integrated over the m_x^2. For the individual bins, the experimental differential cross-section is determined by the calculation of the ratio of the number of data events to the number of simulated Monte Carlo events, then this ratio is used to to scale the model cross-section. $\sigma^{data} = \sigma^{mod} \frac{N_{data}}{N_{MC}}$, where N_{data} is the number of ω particles in the data obtained by multiplying the data yield by the measured experimental charge; while N_{MC} is the number of ω particle generated from SIMC using a model input cross section σ^{mod}.

The background and signal contributions are estimated with a two-parameter fit in each (W, cosθ_ω, ϕ_ω) bin. Assuming the shape given by the model is correct, the parameters are the normalization of the signal and background. The fit minimized the χ^2 difference between the total Monte Carlo (signal + background) and the data. However,in some cases (out-of-plane ϕ bins), the multipion background and the m_x^2 of the ω spectra are too similar to have a reliable result from the fit. The is due to the decrease in acceptance, and in these cases it is assumed that the background normalization parameter is constant with respect to ϕ. This constraint was applied to the fit.

The main source of background is from events with more than one undetected particle, mainly multiple pions. The treatment of the background was done by simulating the m_x^2 spectra of the background using SIMC with a model of the largest contributing reactions (mainly pions). These yielded our approximation to the shape of the multipion background to a normalisation factor.

For local analysis, we retain bins in (W, cosθ_ω, $\phi_{\rm cm}$) if they passed the following three criteria. For the first criterion, considering the number of entries in the bins in the region of the ω missing mass squared peak, at least two of these must have at least 10 events. Secondly, a threshold for non-overlap of signal (S) and background (B) contributions of at least 10 %. This ensures that the fit applied to the data is reliable from the physics point of view. This second criterion also includes a rejection of the case when there is no background or no signal. Equation 1 indicates the metric used for the seperation of the S and the B.

$$\frac{1}{2}\int \frac{(S-B)^2}{(S+B)}\mathrm{dx} \tag{1}$$

The third criterion is to ensure that there is reliable amount of statistics for the fit. This is neccesary as there is lack of statistics from background simulation to give good results at the highest W bins of our measurement. The efficiency of the three criteria is 52 %. There is an overlap of 26 % between the first criterion and second criterion which together contribute 94 % of the efficiency. A typical missing mass squared plot showing a bin selected for passing through the three criteria and subsequently used for the extraction of the cross-section is displayed in Fig. 1.

4. Data Corrections

In order to extract reliable results from our data, corrections were made on a run-by-run basis, on the track reconstruction inefficiencies, dead times and also on offsets. All the corrections applied to the data to acquire physical quantities are shown in Table 4. The entire set of kinematic offsets that were use during the replay of our present data is listed in Table 5.

A 4 % correction due to proton absorption is done for the trigger efficiency. The condition for a proton to cause a trigger in the HMS is that it had to deposit enough energy to create above-threshold signals in at least three out of four scintillator planes in the detector stack. A trigger inefficiency for proton detection in the HMS is produced by protons which are not

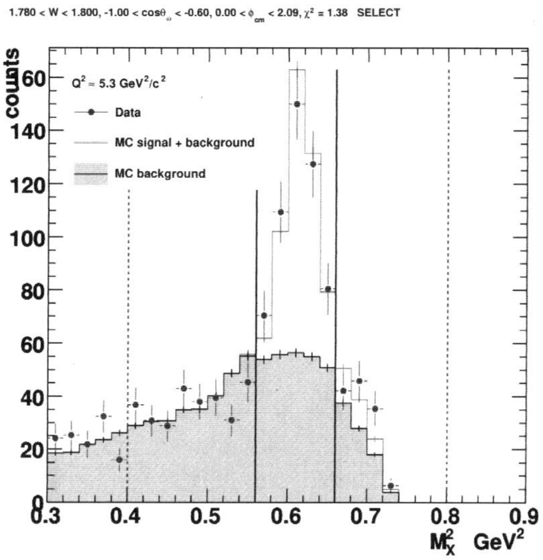

Figure 1. (Colour online) Plot representing one of the bins from which the cross-section is extracted as it is selected after passing through the three criteria. The (blue) points are the data while the solid line (red) is the sum of the simulations. The multipion background component is the green filled area and the ω production simulation component is within the area of the red line. The dashed blue lines show the region within which the background fit is done while the solid dark lines show the region within which the ω model cross-section is extracted.

Table 4. Corrections applied to the data. Indicated by the parentheses are the range of correction sizes applied on a run-by-run (†) or a bin-by-bin basis (‡).

Effects	Correction in %
Proton absorption	$+4 \pm 1$
†Computer dead time	+(1.0 – 19.1)
†HMS tracking	+(2.3 – 14.3)
†SOS tracking	+(0.3 – 0.9)
†Electronic dead time	+(0.0 – 2.4)
‡Random coincidence	-(0.0 – 7.6)

Table 5. Kinematic offsets applied to the data during the replay phase.

Quantity	HMS	SOS
θ	0.0 ± 0.5 mrad	0.0 ± 0.5 mrad
ϕ	$+1.1 \pm 0.5$ mrad	$+3.2 \pm 0.5$ mrad
p	-0.13 ± 0.05 %	-1.36 ± 0.05 %
E_e	0.00 ± 0.05 %	0.00 ± 0.05 %

detected in their interaction with the scintillator and by protons that do not pass through all the scintillators due to absorption.

The radiative corrections for our experiment were calculated within SIMC which is a Monte Carlo framework for applying radiative corrections in $(e, e'p)$ coincidence reactions at

GeV energies [1]. The size of the radiative corrections implemented by SIMC is done by running the full simulation with and without including radiative effects. The uncertainty in the radiative corrections was estimated to be 2%.

5. Systematic Error Analysis and Uncertainties Estimates

For a meaningful result, the statistical uncertainties must be correctly calculated and the estimates for systematic uncertainties should be reasonable. The determination of the statistical uncertainty on the cross-section is done by using the uncertainties in N_{data} and N_{MC}. For the N_{data}, the uncertainty is given by the statistical uncertainty (Poisson) in the number of measured real events and the contribution from accidental coincidnece events and events from the cell walls. As the Monte Carlo is done on for high statistics, the relative statistical error $R = \dfrac{N_{data}}{N_{MC}}$ is dominated by the uncertainty in the number of real events.

The concept and procedure used in the systematic studies in this work corresponds to that applied in Ref. [1] to the $p(e, e'p)\eta$ data. Depending on their source, estimates of the systematic uncertainties are accounted for in one of two ways. On one hand, we applied the assumed global uncertainty to the overall data to error sources that are independent of the binned kinematic variable (W, $\cos\theta_\omega$, ϕ_ω) of the extracted data.

On the other hand, for the systematic uncertainties that are expected to be kinematic-variable dependent are treated on a bin-by-bin basis. The estimation of such uncertainties are done by comparing for each bin the value of the cross section extracted with a nominal set of variables used in the Monte Carlo simulation to that extracted from using adjusted set of variables. The nominal set of variables are the best set of variables used for the 'standard' analysis from which the final ω differential cross-section is calculated. One at a time, each set of variables (variable) are altered and the analysis redone up to the stage the differential cross-section is acquired.

Table 6 lists various sources of kinematicly-dependent systematic errors considered in the analysis. The values of these variables are either not known precisely or the determination of the cross-section can be sensitive to their changes. As a fact, the target position in the beam direction, z_{targ}, is only known within 3 mm. The z_{targ} defines the entrance angles to the spectrometers which, in other words, could affect the spectrometer quantities. For the standard analysis, an offset of 1.5 mm from the nominal centre is used. The variaton chosen is 1.5 mm from either side of the nominal value. The Short Orbit Spectrometer (SOS) is found to be out-of-plane, but the exact amount is unknown [1]. Through a recent survey, a nominal value $x'_{\rm sos}$ = 2.62 mr is used in the extraction and offsets of 1.5 mr and 3.5 mr are used for the sysyematic analysis. The dependence of the missing mass squared, m_x^2, cut is checked by widening and subsequently tightening the cuts over reasonable limits on the region of backgroud fits and the region of the extraction of the model cross-section.

The expression for the systematic uncertainty in the differential cross section which results in a change in a given variable in the ith bin is given as : $\delta_i^v = \dfrac{\mid x_i - y_i^v \mid}{2}$ where x_i is the differential cross-section obtained from the nominal value of the variable in the main analysis, and y_i^v is the differential cross section obtained from the analysis with the change v applied to that variable.

The mean systematic error for all bins, weighted by the statistical error of the measurement in each bin is : $\langle\delta^v\rangle = \dfrac{\Sigma_i \delta_i^v/\sigma_i^2}{\Sigma 1/\sigma_i^2}$ where σ_i is the statistical error of the differential cross-section in bin i. The calculation of the total systematic uncertainty for the ith bin, δ^{tot}, is done by the addition in quadrature of the systematic uncertainty for each variation, δ_i^v, and the global systematic uncertainties, δ_{glo} : $\delta^{tot} = \sqrt{\Sigma_v(\delta_i^v)^2 + \Sigma\delta_{glo}^2}$

Source		Nominal value	Systematic variation	$\langle\delta^v\rangle$
z_{targ} offset	(mm)	1.5	0.0	2.2%
		1.5	3.0	2.8%
$x'_{\rm sos}$ offset	(mr)	2.62	1.5	1.6%
		2.62	3.5	1.4%
$m^2_{x,\rm background\text{-}fit\ region}$	(GeV^2)	$0.4 \leq m_x^2 \leq 0.8$	$0.3 \leq m_x^2 \leq 0.9$	2.3%
			$0.5 \leq m_x^2 \leq 0.7$	1.4%
$m^2_{x,\rm cross\text{-}section\ region}$	(GeV^2)	$0.56 \leq m_x^2 \leq 0.66$	$0.46 \leq m_x^2 \leq 0.76$	0.3%
			$0.59 \leq m_x^2 \leq 0.63$	0.1%

Table 6. Sources of kinematicly-dependent errors, values used for the main analysis (nominal values), their systematic variations and the weighted mean systematic error for all bins $\langle\delta^v\rangle$.

6. Results and Conclusion

The centre-of-mass differential cross-sections for the ω are extracted at an average Q^2 of 5.36 GeV^2 at an invariant mass range of 1.72 GeV $< W <$ 1.92 GeV at angle range -1.0$<$ cosθ $<$ 1.0, with full $\phi_{\rm cm}$ coverage. We compute the cross-sections in five cos$\theta_{\rm cm}$ bins and three $\phi_{\rm cm}$ bins. Shown in Fig. 2 are the computed cross-sections against the invariant mass W. For lower W bins where we have sufficient statistics, there is a strong correlation between our measured ω cross-sections (black dots) and the input model cross-sections (red triangle) [1]. The agreement between our extracted cross-sections and the input model cross-sections decreases for higher W bins due to lack of statistics from our data.

We also compare our measured cross-section with Regge-based model [6, 7] for hadronic content in the t- channel exchange of a photon in a similar Q^2 region. As shown in Fig 3, in the region of invariant mass 1.72 GeV $< W <$ 1.92 GeV and $Q^2 \sim 5$ GeV^2 where there is an overlap with our data, there is a good agreement between the model and our measurement.

Our present data do not have sufficient angular coverage at higher W to comment comprehensively on the presence of resonance mass region. The use of our data which is characterized by its very backward angular coverage can further be used to test additional models if more statistics at high invariant mass and scattering angle are made available. This data extends into a completely new region, which is the highest yet measured.

Future work includes a multipion background simulation at higher invariant mass and scattering angles. This will concurrently include the ρ meson which also has a similar mass but broader peak than the ω particle.

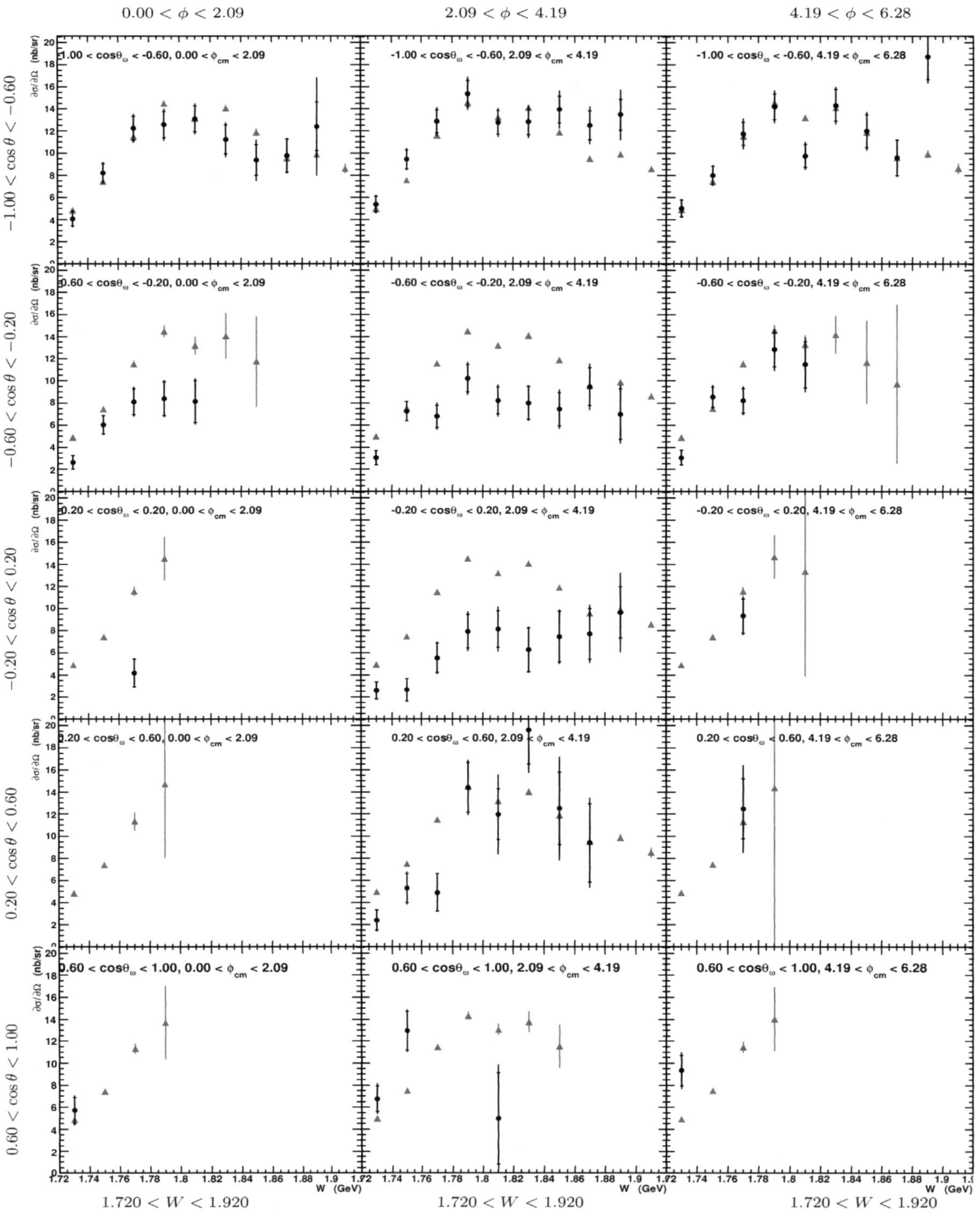

Figure 2. (Colour online) Extracted $ep \rightarrow ep\omega$ differential cross-sections against invariant mass W and flat in ϕ. The red triangles represent the input model cross-section and black dots represent the extracted differential cross-sections. The inner error bars are statistical and the outer error bars are the quadrature sum of the statistical and systematic errors.

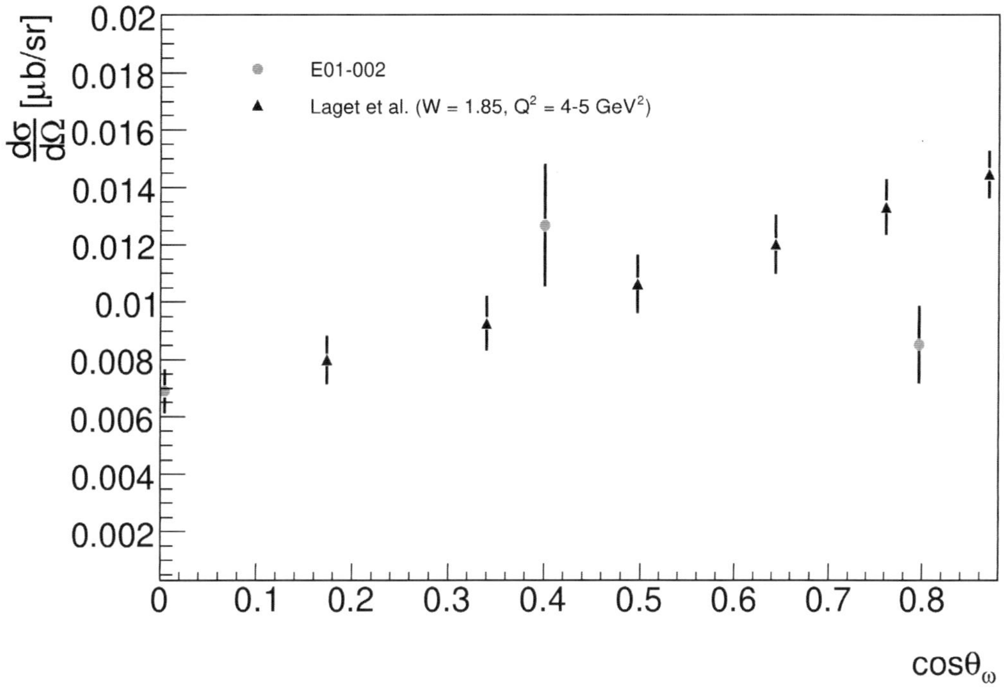

Figure 3. (Colour online) Measured cross-section (green dots) and model (black triangle) used in Refs. [6, 7] as a function of the scattering angle.

7. Acknowledgement

I would like to acknowledge the support of the staff and management at Jefferson Lab, and the staff of physics department in the University of Johannesburg.

References

[1] Dalton M M *et al* 2009 *Phys. Rev. C* **80** 015205
[2] Mohring R M 1999 Ph.D. thesis University of Maryland
[3] Arrington J R 1998 Ph.D. thesis California Institute of Technology
[4] Bosted P 2003 Kinematic settings and singles rates for E01-002 Hall C JLab
[5] Villano A N *et al* 2009 *Phys. Rev. C* **80** 035203
[6] Laget J M 2004 *Phys. Rev. D* **70** 054023
[7] Guidal M, Laget J M and Vanderhaeghen M 2003 *Phys. Rev. C* **68** 058201

Journal of Physics: Conference Series **455** (2013) 012043 doi:10.1088/1742-6596/455/1/012043

Lessons from LHC8

Gian Francesco Giudice
CERN, Theory Division, 1211 Geneva 23, Switzerland

I want to focus here on the lessons that we have learned from preliminary LHC results on Higgs searches. I will concentrate mostly on the Higgs boson, rather than on new physics, not only because the mechanism of electroweak (EW) symmetry breaking is one of the priorities in the LHC program, but also because we have new data on the Higgs and it is exciting to think about where this leads us to.

The phenomenon of EW symmetry breaking had already been established before the LHC. After LEP we had ample evidence for gauge structure in interactions (including triple gauge bosons couplings γWW and ZWW) and for the existence of longitudinal components of W and Z. Combining this information with knowledge of gauge boson masses, we conclude that propagating particles do not share the full symmetry of interactions, and thus that the EW symmetry is spontaneously broken. This means that every known phenomenon in particle physics (at least before December 13, 2011) can be described by the Lagrangian

$$L = -\frac{1}{4}\mathrm{Tr}F_{\mu\nu}F^{\mu\nu} + i\bar{f}\gamma^\mu D_\mu f + \frac{v^2}{4}\mathrm{Tr}D_\mu\Sigma^\dagger D^\mu\Sigma - \frac{v}{\sqrt{2}}\bar{f}_L\Sigma\lambda_f f_R + \mathrm{h.c.} \qquad (1)$$

$$\Sigma \equiv \exp\left(\frac{iT^a\pi^a}{v}\right) , \qquad (2)$$

where π^a are the longitudinal polarizations of W and Z, and $v = 246$ GeV. The first two terms in eq. (1) describe the kinetic terms and gauge interactions of the SM particles. The last two terms contain the effect of the longitudinal polarizations and the mass terms that arise when gauge symmetry is realized non-linearly.

Although the Lagrangian in eq. (1) was fully satisfying from the experimental point of view, even before Dec 13 every theorist knew that it could not be the full story. Scattering amplitudes of longitudinal gauge bosons grow like $(E/4\pi v)^2$, signaling loss of perturbative unitarity, and thus the onset of new phenomena, at $E \approx 4\pi v = 3$ TeV. One of the goals of the LHC was to discover what is the new phenomenon. It is well known that the simplest option is given by a single real scalar field h, which forms a complete $SU(2)$ doublet together with π^a. So 3/4 of the Higgs had already been found and the LHC discovered the missing 1/4. However, there is no strong motivation, other than simplicity, for choosing a single h and nature may have good reasons to make different choices. In this respect, hunting for the Higgs is not just looking for the last missing piece of the SM, but it means exploring an unknown territory and identifying the nature of the new force responsible for EW breaking, which I will call the *fifth force*.

When we examine the SM, we note that almost all of its open problems originate from Higgs interactions. The flavor problem comes from Yukawa couplings, the hierarchy problem from the Higgs bilinear, the stability problem from the Higgs quartic coupling (and one can add the cosmological constant problem from a constant term in the scalar potential). The crux of these puzzles is that the fifth force is not a gauge force. Therefore, it lacks the properties of uniqueness, robustness against deformations, and predictivity, which are characteristic of gauge theory.

In order to discuss what we have learned from Higgs data, I will identify two fundamental questions. The first question is: *What is the fifth force?* We want to know if it is weak or strong; if it is a gauge force or associated with a fundamental scalar. The answer to this question will come from precise measurements of Higgs couplings. These measurements will play the role that precision EW data played at the time of LEP. An important difference is that, in the case of Higgs couplings, deviations from the SM expectation could be large (not necessarily of one-loop size) and thus show up even at an early stage. Actually, the more natural the Higgs boson is, the more its properties must deviate from the SM. This is because a natural theory must give large corrections to the Higgs two-point function (to cure the hierarchy problem). These large corrections must also modify the Higgs production rate at the LHC and some of its decay channels, as can be easily seen by inserting two gluons (or two photons) in the Feynman diagram of the Higgs two-point function. This expectation is fully confirmed in all the examples of natural theories known to us. So measuring the Higgs couplings is the way to probe the fifth force and may be the first way for new physics to show up.

The second question is: *Is the Higgs natural?* This is not an idle question. Its importance goes beyond EW symmetry breaking and its answer will influence the strategy for future directions in particle physics. Naturalness is a concept fully linked to the use of effective field theories (EFT). EFT is the tool that we use to implement an intuitive notion: separation of scales. In simple words, separation of scales means that we don't need to know the motion of every atom inside the moon to compute its orbit. Or we don't need to know about quarks to describe physics at the atomic scale. We build a stack of EFT, one on top of the other, just like a matryoshka doll with one layer inside the other. Each EFT is appropriate to describe a certain energy regime, but it is connected to the next in the sense that free parameters in one layer can be computed in the next layer. One of the most remarkable results of modern physics has been the discovery that at each layer we find simpler physical laws, larger symmetry, unification of concepts that seemed unrelated in the previous layer. It is amazing that nature works this way, but it is just an empirical fact. We can use the criterion of naturalness in EFT to infer the energy at which the validity of one layer ends and a new layer must set in. Whenever a next layer exists, this procedure gives a reasonable answer, as shown by various examples (electron self-energy, pion mass difference, neutral kaon mass difference). When applied to the Higgs boson mass, this criterion gives a maximum scale for new physics at about 500 GeV.

Since we have not yet found any new physics at the LHC, one may wonder what is the fate of naturalness. The issue is not yet settled. It is quite possible that new physics is just around the corner. After all, the LHC has entered the territory of naturalness, but the exploration is far from complete. The alternative is that the idea of naturalness does not apply to the Higgs because there is a failure of the EFT approach. After all, dark energy could already be taken as evidence for failure of EFT, since the naturalness of the cosmological constant suggests new physics around 10^{-3} eV. Holography, gauge-gravity duality, and the AdS-CFT correspondence show that some theories are much richer than what a single Lagrangian can capture. The best we can do to describe the physical content is to resort to two different Lagrangians, two dual versions. Maybe this is an indication that our theoretical tools are failing, that an EFT Lagrangian is not able to catch all the underlying physics. There could be connections between small and large scales. A numerological curiosity is that, if we combine the largest possible scale (the Hubble length $H^{-1} = 10^{26}$ m) with the smallest (the Planck length $M_P^{-1} = 10^{-35}$ m), we can reproduce the scale of the cosmological constant ($\Lambda_{CC} = \sqrt{HM_P} = 5 \times 10^{-3}$ eV) and the weak scale ($\Lambda_{EW} = \sqrt{\Lambda_{CC} M_P} = 5$ TeV). If behind this numerical curiosity there is some theoretical IR/UV connection, we will never be able to catch it with an EFT.

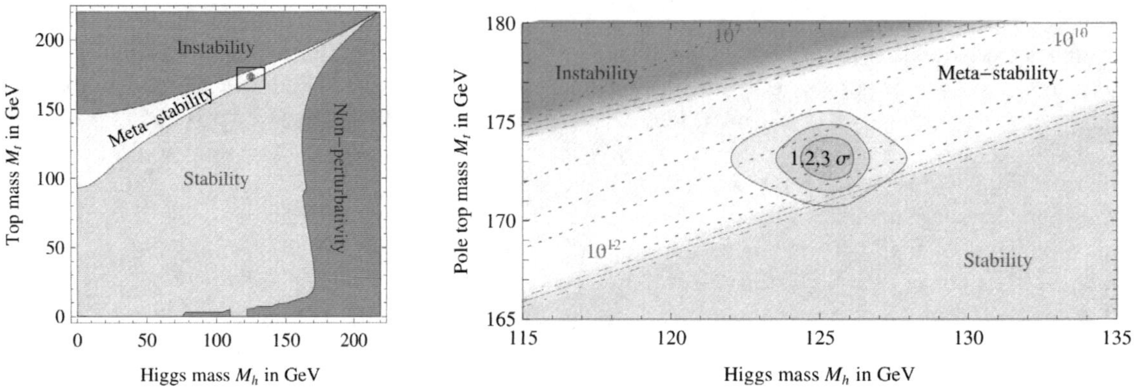

Figure 1. Regions of absolute stability, meta-stability and instability of the SM vacuum in the M_t–M_h plane. The frame on the right zooms in the region of the preferred experimental range of M_h and M_t (the gray areas denote the allowed region at 1, 2, and 3σ). The three boundaries lines correspond to $\alpha_s(M_Z) = 0.1184 \pm 0.0007$, and the grading of the colors indicates the size of the theoretical error. The dotted contour-lines show the instability scale Λ in GeV assuming $\alpha_s(M_Z) = 0.1184$.

Another approach which would invalidate naturalness is the idea of the multiverse. Out of the process of eternal inflation, a multitude of universes are created, each with its own values of the fundamental constants and its own physical laws. Anthropic arguments then select the kind of universe in which we live in or, in other words, the physical laws that govern nature. It may sound like a crazy idea to some (bordering on science fiction), but at present the multiverse yields the most convincing explanation of the cosmological constant. A lesson from the multiverse is that some of the questions that we thought to be fundamental may actually be just the result of environmental conditions, and carry no more significance than the shape of continents or the emergence of a particular animal species in Darwinian evolution.

So physics has reached a branching path and the LHC will tell us which way we have to follow. One path follows the road that guided us towards the extraordinary successes of particle physics in the last 100 years or so: a new layer, new symmetry, more unification that bring us closer to a single governing principle of nature. The other path is marked by the failure of naturalness, the collapse of the picture of a multi-layered matryoshka doll hiding a single final truth. We have only vague ideas of where this path leads to, and maybe the multiverse is the most concrete construction along this path. It is clear that establishing the fate of Higgs naturalness has far-reaching consequences for particle physics, well beyond the problem of EW breaking.

The LHC will teach us which path we have to follow. The first path is very familiar to particle physicists and promises new discoveries on the road towards final unification. The other path may mean finding the Higgs and nothing else at the LHC. It will force us to abandon naturalness and EFT, and look for new paradigms in a holistic vision, where nature should be seen as a whole and cannot always be reduced to its smaller components. But physics is a natural science and we want to find answers with the experimental method. While along the first path there are new phenomena to be studied and the goals are clear, how will we be able to make progress if the second path turns out to be true? This is a difficult question and I don't have an answer. However, I want to show how the measured value of the Higgs mass gives us some indirect hints.

For the sake of argument, let me assume that the SM with a single Higgs with mass around 125–126 GeV fully describes physics up to a very large energy scale. In this case, as shown in

Fig. 1, we learn that our universe lives in a very critical condition, at the verge of a cosmic catastrophe. It is a remarkable coincidence that we happen to live just at the boundary between two phases: the ordinary Higgs phase and a region where the Higgs field slides to very large values. The condition for absolute stability is

$$M_h \, [\text{GeV}] > 129.4 + 1.4 \left(\frac{M_t \, [\text{GeV}] - 173.1}{0.7} \right) - 0.5 \left(\frac{\alpha_s(M_Z) - 0.1184}{0.0007} \right) \pm 1.0_{\text{th}} \, .$$

Precise experimental information on the Higgs and top mass is needed to ascertain the ultimate fate of the universe and whether we live in a stable vacuum or not. The Higgs mass can be measured very precisely at the LHC. Then the top mass will become the largest source of uncertainty and every GeV in M_t counts as a shift of 2 GeV in the Higgs mass. A reduction in the error on the top mass may become the best telescope to peek into the future of our universe.

Also the hierarchy problem can be interpreted as a sign of near criticality between two phases. The coefficient m^2 of the Higgs bilinear in the scalar potential is the order parameter that describes the transition between the symmetric phase ($m^2 > 0$) and the broken phase ($m^2 < 0$). In principle, m^2 could take any value between $-M_P^2$ and $+M_P^2$, but quantum corrections push m^2 away from zero towards one of the two end points of the allowed range. The hierarchy problem is the observation that in our universe the value of m^2 is approximately zero or, in other words, sits near the boundary between the symmetric and broken phases. Therefore, if the LHC result is confirmed, we must conclude that both m^2 and λ, the two parameters of the Higgs potential, happen to be near critical lines that separate the EW phase from a different (and inhospitable) phase of the SM. Is criticality just a capricious numerical coincidence or is it telling us something deep?

The occurrence of criticality could be the consequence of symmetry. For instance, supersymmetry implies $m^2 = 0$. If supersymmetry is marginally broken, m^2 would remain near zero, solving the hierarchy problem. But if no new physics is discovered at the LHC, we should turn away from symmetry and look elsewhere for an explanation of the near-criticality. It is known that statistical systems often approach critical behaviors as a consequence of some internal dynamics or are attracted to the critical point by the phenomenon of self-organized criticality. As long as no new physics is discovered, the lack of evidence for a symmetry explanation of the hierarchy problem will stimulate the search for alternative solutions. The observation that both parameters in the Higgs potential are quasi-critical may be viewed as evidence for an underlying statistical system that approaches criticality. The multiverse is the most natural candidate to play the role of the underlying statistical system for SM parameters. If this vision is correct, it will lead to a new interpretation of our status in the multiverse: our universe is not a special element of the multiverse where the parameters have the peculiarity of allowing for life, but rather our universe is one of the most common products of the multiverse because it lies near an attractor critical point. In other words, the parameter distribution in the multiverse, instead of being flat or described by simple power laws (as usually assumed) could be highly peaked around critical lines because of some internal dynamics. Rather than being selected by anthropic reasons, our universe is simply a very generic specimen in the multitude of the multiverse. If you complained that string theory was making no predictions about our universe, do not rejoice in this. Now the situation may become even worse: the multiverse is making predictions, but about other universes.

What does a Higgs mass of 125-126 GeV tell us about natural theories? A Higgs mass smaller than 120 GeV would have been perfect for natural supersymmetry, while a mass larger than 130 GeV would have excluded the simplest scenarios. If the Higgs mass is really 125 GeV, right in the middle, then it looks like nature wants to tease theoretical physicists. In supersymmetry, a Higgs mass of 125 GeV can be reached, but only for extreme values of the parameters, especially those of the stop. Therefore certain natural setups, where parameters are correlated, are in bad

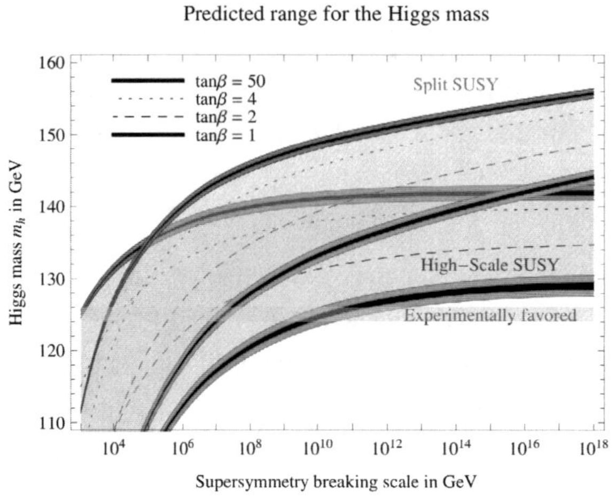

Figure 2. NNLO prediction for the Higgs mass M_h in High-Scale Supersymmetry (blue, lower) and Split Supersymmetry (red, upper) for $\tan\beta = \{1, 2, 4, 50\}$. The thickness of the lower boundary at $\tan\beta = 1$ and of the upper boundary at $\tan\beta = 50$ shows the uncertainty due to the present 1σ error on α_s (black band) and on the top mass (larger colored band).

shape (for instance gauge mediation), but the idea of low-energy supersymmetry is not killed.

As shown in Fig. 2, a Higgs mass of 125 GeV rules out the idea of Split Supersymmetry with a high scale, say larger than 10^8 GeV. However, it fits very well with Split Supersymmetry with a low scale. Actually the simplest model of Split Supersymmetry, based on anomaly mediation, predicts a hierarchy between scalars and fermions of a one-loop factor and thus looks like a very satisfactory solution.

The indication for a Higgs mass in the range 125–126 GeV is the most exciting result from the LHC so far. It has important consequences for supersymmetry and other theories beyond the SM, but the most puzzling (and surprising) message that we obtained from preliminary LHC data on the Higgs is the apparent near-criticality of the parameters entering the SM Higgs potential.

Journal of Physics: Conference Series **455** (2013) 012044 doi:10.1088/1742-6596/455/1/012044

Towards the Little Bang Standard Model

Ulrich Heinz
Physics Department, The Ohio State University, Columbus, Ohio 43210, USA
E-mail: heinz@mps.ohio-state.edu

Abstract. I review recent progress in developing a complete dynamical model for the evolution of the Little Bang fireballs created in relativistic heavy-ion collisions, and using the model to extract the transport properties and initial density fluctuations of the liquid quark-gluon plasma state of matter of which makes up these Little Bangs during the first half of their lives.

1. The Big Bang and the Little Bangs

Ultra-relativistic heavy-ion collisions at RHIC and the LHC produce fireballs made of extraordinarily hot matter, at initial energy densities (at the time when the matter reaches approximate local thermal equilibrium) that exceed the energy density of atomic nuclei in their ground states by two to three orders of magnitude. Due to enormous pressure gradients between the fireball center and the surrounding vacuum, these fireballs undergo explosive collective expansion, cooling down rapidly through several different states of matter, finally fragmenting into thousands of free-streaming hadrons whose energy and momentum distributions can be detected in the detectors set up around the collider rings. The evolution history of these "Little Bangs" has much similarity with the Big Bang that created our Universe (Fig. 1): Both undergo Hubble-like expansion,[1] feature a hierarchy of decoupling processes that are driven by the expansion dynamics (and not by the finite geometric size of the Little Bang), with chemical decoupling of the finally observed particle abundances[2] preceding kinetic decoupling,[3] and with initial-state quantum fluctuations[4] (shown in Figures 2 and 3) imprinting themselves[5] onto the experimentally observed final state.

[1] The relative velocity between two matter elements increases roughly linearly with their relative distance.

[2] In the Big Bang the chemical composition is frozen during the process of primordial nucleosynthesis at an age of about 3 minutes, in the Little Bang this happens during hadrosynthesis at the quark-hadron transition at an age of about 20-30 ioctoseconds [5].

[3] In the Big Bang, the formation of neutral atoms by charge recombination at an age of about 380,000 yr makes the universe transparent to light and freezes the thermal Bose-Einstein energy distribution of the Cosmic Microwave Background radiation. When the Little Bang reaches an age of about 40-50 ioctoseconds, the final-stage hadron gas becomes so dilute that strong interactions between the hadrons cease and their energies and momenta are "frozen out".

[4] The initial wave function of the universe, stretched by cosmic inflation, seeds density fluctuations in the Big Bang. In the Little Bang, the initial quark and gluon wave functions inside the nucleons within the colliding nuclei control its initial density distribution.

[5] In the Big Bang the initial density fluctuations evolve under the action of gravity, described by Einstein's general theory of relativity, into the observed Cosmic Microwave temperature fluctuation spectrum shown in Figs. 2 (left) and 3 (top), and ultimately into today's distributions of stars, galaxies, and galaxy clusters and superclusters (see top panel of Fig. 1). In the Little Bang, they evolve through viscous hydrodynamics into the measured anisotropic collective flow patterns and their event-by-event fluctuations that are the subject of this overview.

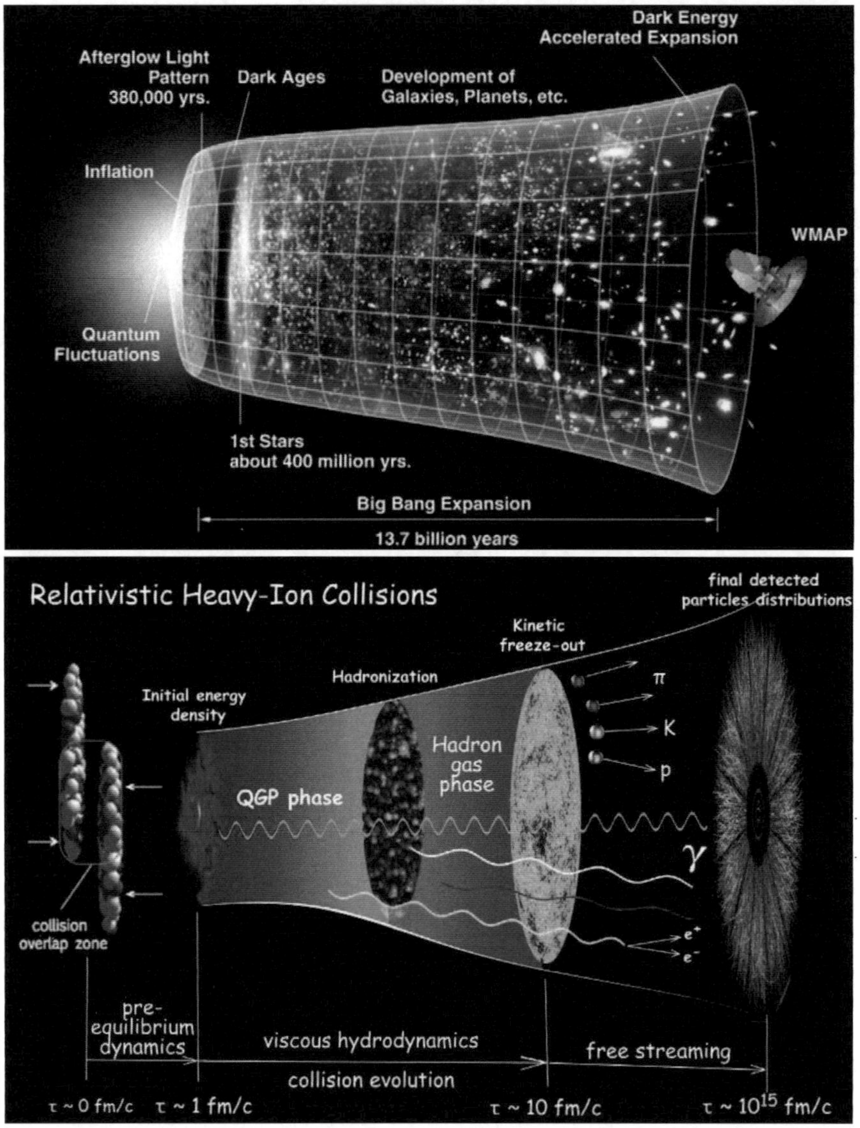

Figure 1. Artist's conception of the evolution of the Big Bang (top – credit: NASA) and the Little Bang (bottom – credit: Paul Sorensen and Chun Shen).

Of course, the Big and Little Bangs are quite different in other aspects: Their expansion rates differ by about 18 orders of magnitude; the Little Bang's expansion is 3-dimensional and driven by pressure gradients, not 4-dimensional and controlled by gravity; Little Bangs evolve on time scales of ioctoseconds, not billions of years; distances are measured in femtometers rather than light years. Most importantly, the Little Bang Standard Model is still under construction. This overview discusses recent progress of the edifice.

2. Eccentricity fluctuations, anisotropic flows, and flow fluctuations

We can observe only one Big Bang (the one that produced our universe), but at the Relativistic Heavy-Ion Collider (RHIC) and Large Hadron Collider (LHC) we have experimentally created and studied billions of Little Bangs. Each Little Bang is different: Highly successful phenomenology based on hydrodynamic evolution models [8, 4] has taught us that the initially very dense quark-gluon matter created in heavy-ion collisions reaches approximate local thermal

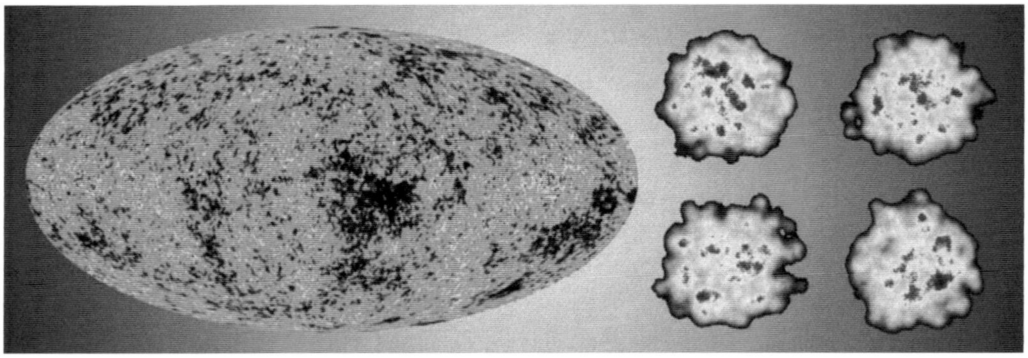

Figure 2. Temperature fluctuation spectrum of the Big Bang at age 380,000 yr (left), as measured through the Cosmic Microwave Background radiation by WMAP, and of four typical Little Bangs created in central Pb+Pb collisions at the LHC (right), at age 0.2 fm/c, as calculated from the IP-Glasma model [1]. Figure taken from [2].

equilibrium on a very short time scale [9] of order 1 fm/c (3 ioctoseconds), after which it evolves according to the macroscopic laws of relativistic viscous fluid dynamics. Pressure gradients in the fluctuating density profile (see Fig. 4 for a specific example) are the hydrodynamic forces that accelerate the fluid and cause it to expand and dilute. Spatial anisotropies and inhomogeneities in the initial density profile transverse to the beam direction lead to corresponding anisotropies in the final transverse expansion flow velocity profile that are imprinted on the momenta of the experimentally observed particles emitted from the collision. Each Little Bang features its own final flow velocity profile which (within hydrodynamics) is a deterministic classical response to the initial conditions that fluctuate from collision to collision due to quantum fluctuations in the initial nuclear wave function. We now know that the quark-gluon liquid that makes up the matter of the Little Bang during the first half of its life has very small viscosity, behaving like an almost ideal fluid [8]. This is a fantastic gift of Nature since it allows us to study experimentally the spectrum of initial-state quantum fluctuations through the final-state anisotropic flow fluctuations. Had the quark-gluon plasma (QGP) turned out to be highly viscous, all initial-state fluctuations would have been wiped out by dissipation before final decoupling of the emitted particles, thereby closing this observation window on the initial state of the Little Bang and on the quantum nature of the initial energy deposition process.

The Little Bang pressure p is related to its energy density e by its equation of state (EOS), and the anisotropy of the initial pressure gradients can be characterized by a series of harmonic eccentricity coefficients

$$\varepsilon_1 e^{i\Phi_1} \equiv -\frac{\int r\,dr\,d\varphi\,r^3 e^{i\varphi}\,e(r,\varphi)}{\int r\,dr\,d\varphi\,r^3 e(r,\varphi)}, \qquad \varepsilon_n e^{in\Phi_n} \equiv -\frac{\int r\,dr\,d\varphi\,r^n e^{in\varphi}\,e(r,\varphi)}{\int r\,dr\,d\varphi\,r^n e(r,\varphi)}\ (n>1), \quad (1)$$

where $e(r,\varphi)$ is the initial energy density distribution in the plane transverse to the beam direction. ε_n characterizes the magnitude of the nth harmonic deformation coefficient, and the angle Ψ_n gives the direction of the corresponding deformation component in the lab frame. Both the magnitudes ε_n and their orientations Ψ_n fluctuate from event to event. The mean values $\langle\varepsilon_n\rangle$ averaged over many events from three popular initial energy deposition models (see [4] for a description and original references) are shown as a function of harmonic index n in the left bottom panel of Fig. 3. They represent the primordial temperature fluctuation spectrum of the Little Bang. We see that each collision centrality generates a different class of Little Bangs, with its own $\langle\varepsilon_n\rangle$ power spectrum, and that the three models shown give quite different results for these power spectra. If we can somehow measure the centrality dependence of the initial $\langle\varepsilon_n\rangle$ power spectrum, we have a powerful constraint on the initial nuclear quark and gluon wave functions on our hands. This is reminiscent of the crucial role the CMB temperature power

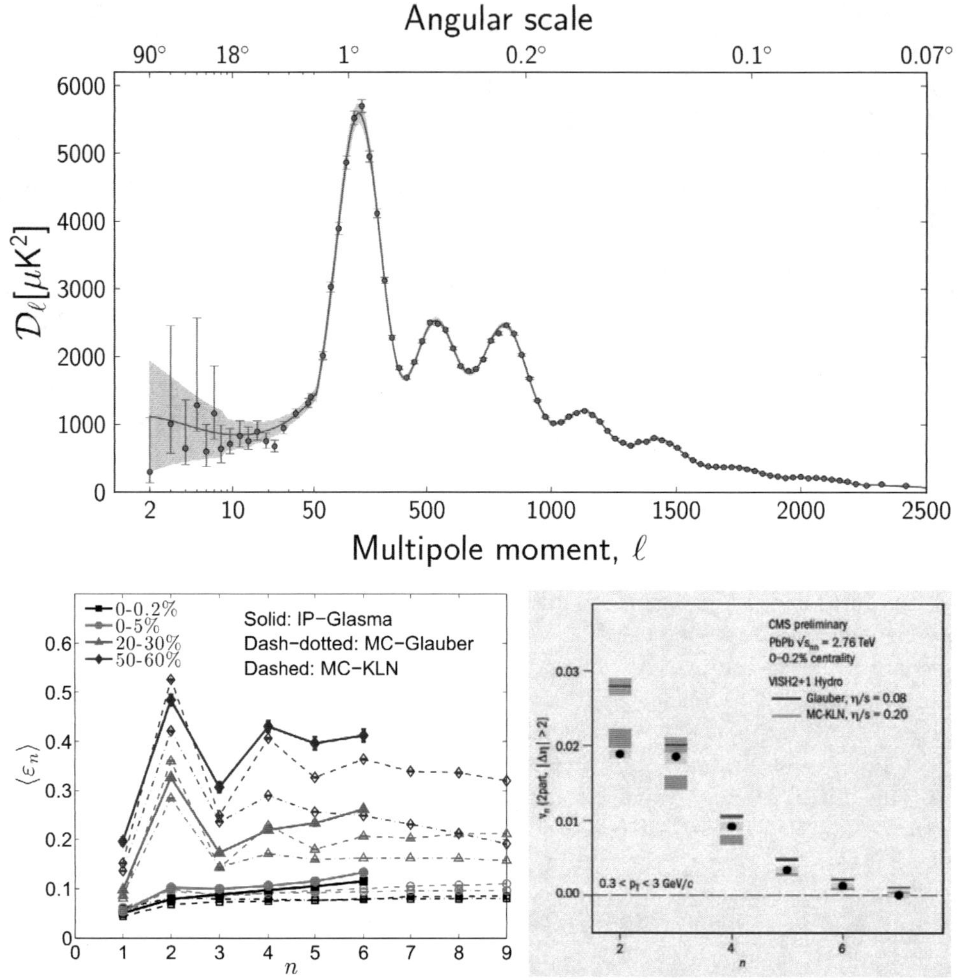

Figure 3. *Top:* Temperature power spectrum of the Big Bang at age 380,000 yr, as measured by the Planck satellite [3]. *Bottom left:* Primordial eccentricity power spectrum of the Little Bangs created in 2.76 A TeV Pb+Pb collisions of different centralities, at $\tau = 0$, from the energy density distributions of three different initial-state models (IP-Glasma, MC-Glauber, MC-KLN) [4]. *Bottom right:* The final Little Bang flow power spectrum for ultracentral (0-0.2% centrality) Pb+Pb collisions at $\sqrt{s} = 2.76\,A\,\mathrm{TeV}$ at the LHC, measured by the CMS Collaboration (Wei Li, *Quark Matter 2012*) and calculated with viscous hydrodynamics `VISH2+1` [6] using MC-Glauber and MC-KLN initial conditions with the indicated specific shear viscosities. Figure taken from [7].

spectrum (shown in the top panel of Fig. 3) has played in nailing down the parameters of the Standard Model of Big Bang Cosmology.

The hydrodynamic response to the initial ε_n-spectrum of the Little Bang is reflected in the observed transverse momentum distributions $dN_i/(dy p_T dp_T d\phi)$ of the various emitted hadronic species i, and in correlations among them. The transverse momentum spectra can again be characterized by a set of complex harmonic coefficients

$$V_n = v_n e^{in\Psi_n} := \frac{\int p_T dp_T d\phi \, e^{in\phi} \frac{dN}{dy p_T dp_T d\phi}}{\int p_T dp_T d\phi \frac{dN}{dy p_T dp_T d\phi}} \equiv \{e^{in\phi}\}, \tag{2}$$

$$V_n(p_T) = v_n(p_T) e^{in\Psi_n(p_T)} := \frac{\int d\phi \, e^{in\phi} \frac{dN}{dy p_T dp_T d\phi}}{\int d\phi \frac{dN}{dy p_T dp_T d\phi}} \equiv \{e^{in\phi}\}_{p_T}. \tag{3}$$

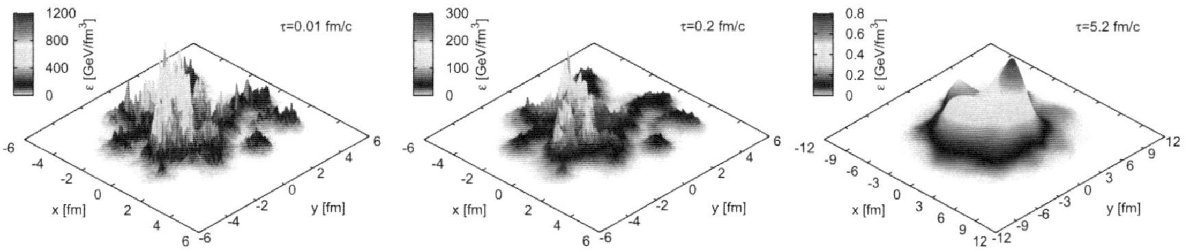

Figure 4. Typical transverse energy density profiles $e(x, y)$ from the IP-Glasma model [1] for a semiperipheral ($b = 8$ fm) Au+Au collision at $\sqrt{s} = 200\,A$ GeV, at times $\tau = 0.01$, 0.2, and 5.2 fm/c. From $\tau = 0.01$ fm/c to 0.2 fm/c the fireball evolves out of equilibrium according to the Glasma model [10]; at $\tau = 0.2$ fm/c the energy momentum tensor from the IP-Glasma evolution is Landau-matched to ideal fluid form (for technical reasons [11] the viscous pressure components are set to zero at the matching time) and henceforth evolved with viscous Israel-Stewart fluid dynamics, assuming $\eta/s = 0.12$ for the specific shear viscosity. The pre-equilibrium Glasma evolution is seen to somewhat wash out the large initial energy density fluctuations. The subsequent dissipative hydrodynamic evolution further smoothes these fluctuations. The asymmetric pressure gradients due to the prominent dipole asymmetry in the initial state of this particular event (visible as a left-right asymmetry of the density profile in the left panel) is seen to generate a dipole ("directed flow") component in the hydrodynamic flow pattern that pushes matter towards the right during the later evolution stages.

Here ϕ is the azimuthal angle around the beam direction of the particle's transverse momentum $\boldsymbol{p}_\mathrm{T}$, and the curly brackets denote the average over particles from a single collision. Each particle species i has its own set of anisotropic flow coefficients V_n. Eq. (2) defines the flow coefficients and associated flow angles for the entire event, whereas Eq. (3) is the analogous definition for the subset of particles in the event with a given magnitude of the transverse momentum p_T. I suppress the dependence of both types of flow coefficients on the rapidity y. v_n are known as the "integrated" anisotropic flows, $v_n(p_\mathrm{T})$ are called "differential" flows. By definition, both v_n and $v_n(p_\mathrm{T})$ are positive definite. Hydrodynamic simulations show that in general the flow angles Ψ_n depend on p_T, and that, as a function of p_T, $\Psi_n(p_\mathrm{T})$ wanders around the "average angle" Ψ_n that characterizes the integrated flow v_n of the entire event [12].

For each collision system, centrality class, and collision energy, the fluctuating initial state of the Little Bang is characterized by a distinct probability distribution $P(\varepsilon_n, \Phi_n)$, one particular moment of which is the $\langle\varepsilon_n\rangle$ power spectrum shown in Fig. 3. For different n, the angles Φ_n are more or less correlated with the direction of the impact parameter $\boldsymbol{b}$, but this direction cannot be measured experimentally. The viscous hydrodynamic evolution relates the probability distribution for the initial complex eccentricity coefficients deterministically to probability distributions $P(V_n)$ and $P(V_n(p_\mathrm{T}))$ for the integrated and differential final complex harmonic flow coefficients; these distributions are particle species specific. As will be discussed in the next section, the relation between the final complex flow coefficients V_n and the initial eccentricity coefficients $\varepsilon_n e^{in\Phi_n}$ (and thus between the corresponding flow and eccentricity probability distributions characterizing each class of collisions) depends on the viscosity of the Little Bang matter. One goal of the relativistic heavy-ion program is to both constrain the QGP viscosity and identify the correct theory for the initial-state quantum fluctuations by performing a complete experimental reconstruction of the final multi-dimensional distributions $P(V_n)$ and $P(V_n(p_\mathrm{T}))$.

Due to limited statistics arising from the finite number of particles emitted by each Little Bang, neither the magnitudes v_n nor the flow angles Ψ_n can be accurately determined for a single event. Experimental flow measures therefore involve angle correlations between two or more particles; for example, $\langle\{e^{in\phi_1}\}_{p_{T1}}\{e^{-in\phi_2}\}_{p_{T2}}\rangle = \langle v_n(p_{T1})v_n(p_{T2})\cos[n(\Psi_n(p_{T1})-\Psi_n(p_{T2}))]\rangle$ where the first particle from the event has transverse momentum $\boldsymbol{p}_{T1} = (p_{T1}, \phi_1)$ and the second particle has $\boldsymbol{p}_{T2} = (p_{T2}, \phi_2)$, the average $\{\ldots\}$ is over all such particles in the event and $\langle\ldots\rangle$ indicates the average of the result over many Little Bangs of the selected class. One sees that the event-by-event fluctuations and transverse momentum dependences of both the anisotropic flow magnitudes v_n and their associated flow angles Ψ_n affect these experimental observables.

Different such observables correspond to different correlation functions between the v_n's and Ψ_n's all of which can, for a given initial energy deposition model with probability distribution $P(\varepsilon_n, \Phi_n)$, be computed from the hydrodynamically predicted probability distributions $P(V_n)$ and $P(V_n(p_T))$. One can define experimental observables that separate the fluctuations of the anisotropic flow magnitudes $v_n(p_T)$ from those in the flow angles $\Psi_n(p_T)$ [12]. These should be powerful discriminators between different $P(V_n)$ and $P(V_n(p_T))$ distributions, and thus between different initial-state fluctuation models. Experimental studies of these v_n and Ψ_n fluctuations and their p_T-dependences have just gotten under way; the results will be interesting and should significantly advance the construction of the Little Bang Standard Model.

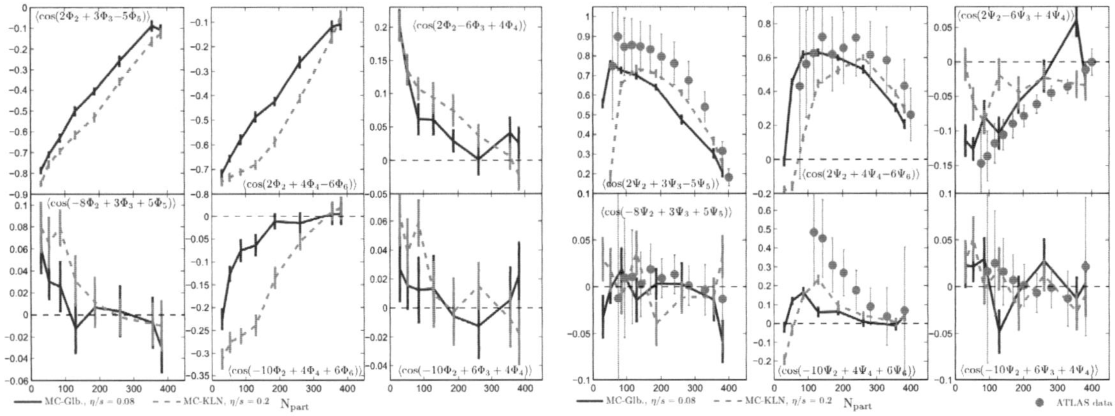

Figure 5. Three-plane correlations between the initial-state participant planes (left 6-panel) and the final-state flow planes (right 6-panel), for 2.76 A TeV Pb+Pb collisions as functions of collision centrality, indicated by the number of participant nucleons $N_{\rm part}$ on the horizontal axes. The solid (dashed) lines are for MC-Glauber (MC-KLN) initial conditions evolved event-by-event with viscous hydrodynamics using $\eta/s = 0.08$ (0.2) for the specific shear viscosity [14]. Filled circles show the experimental values measured by the ATLAS Collaboration [16].

The angles Φ_n associated with the initial deformation parameters ε_n are not only correlated with the (unmeasurable) direction of the impact parameter $\boldsymbol{b}$, but also with each other, in ways that are predicted by (and thus depend on) the initial energy deposition model [13, 14]. These so-called participant-plane correlations in the initial state are translated by hydrodynamic response into final-state flow angle correlations. These flow angle correlations were measured in Pb+Pb collisions at the LHC by the ALICE [15] and ATLAS [16] Collaborations. All measured two- and three-plane flow angle correlations [16] are qualitatively reproduced by viscous hydrodynamic calculations, both in their magnitudes and centrality dependences [14] (right panels in Fig. 5). They differ, however, qualitatively from the initial participant-plane correlations [14], often even in sign (left panels in Fig. 5). Hydrodynamic evolution is nonlinear and leads to mode-coupling between different harmonics [17]. For example, elliptic and triangular deformations ε_2 and ε_3 provide a nonlinear contribution to pentangular flow V_5 which, for large impact parameters where the elliptic deformation ε_2 is big, overwhelms the linear response to ε_5 and completely decorrelates the pentangular flow plane Ψ_5 from the angle Φ_5 of the initial pentangular density deformation [17]. Only by accounting for mode-coupling, either directly by following the non-linear hydrodynamic evolution [14] or through a non-linear response analysis that keeps at least second-order terms [18], can the experimental data be reproduced. These flow-plane correlations thus represent an *experimentum crucis* in support of the hydrodynamic paradigm for the Little Bang; dynamical models without a large degree of local thermalization and hydrodynamic collective flow will not be able to describe the dynamical change of character between the initial-state participant-plane and final-state flow-plane correlations [14].[6]

[6] The strength of the flow-plane correlations depends on the QGP shear viscosity [14]. A careful quantitative

3. QGP shear viscosity from anisotropic flow measurements

The efficiency of the fluid to convert spatial anisotropies in the pressure gradients into anisotropic flows is degraded by shear viscosity. The "conversion efficiency" v_n/ε_n is therefore a measure for the specific shear viscosity η/s of the expanding fluid. This is shown in Fig. 6. Higher harmonics, reflecting variations on smaller spatial scales, are suppressed more strongly by shear viscosity than lower flow harmonics.

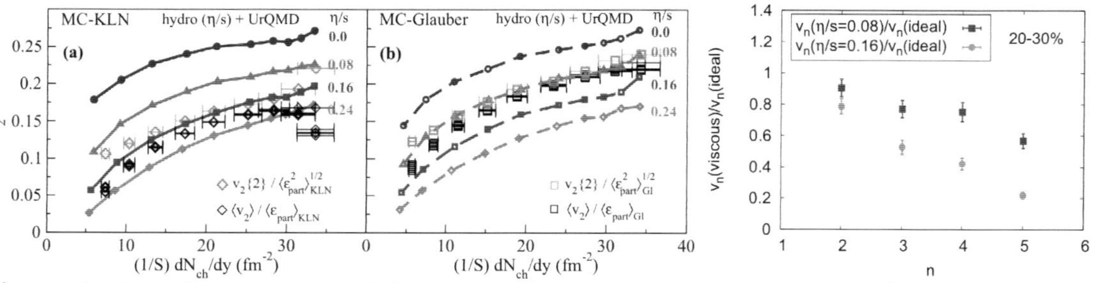

Figure 6. *Left:* The eccentricity-scaled integrated elliptic flow of all charged hadrons, $v_2^{\rm ch}(\eta/s)/\varepsilon$, as a function of total charged hadron multiplicity density per unit overlap area, $(1/S)(dN_{\rm ch}/dy)$. The experimental data points show two measures for the elliptic flow ($\langle v_2 \rangle$ [19] and $v_2\{2\}$ [20]) from 200 *A* GeV Au-Au collisions at RHIC, measured by the STAR Collaboration. Both panels use the same sets of data, but use different average initial eccentricities $\langle \varepsilon \rangle$ and overlap areas $\langle S \rangle$ (obtained from the MC-Glauber and MC-KLN models) to normalize the vertical and horizontal axes. The theoretical curves were computed with the `VISHNU` model [21], for different (temperature-independent) choices of the specific QGP shear viscosity $(\eta/s)_{\rm QGP}$. *Right:* Viscous suppression of v_n for $\eta/s = 0.08$ (squares) and 0.16 (circles), as function of harmonic index n, for Au+Au collisions at 20-30% centrality [22].

The azimuthally symmetric part of the collective transverse flow, called radial flow, boosts particles from low to high transverse momentum in proportion to their masses; it is thus responsible for the distribution of the hydrodynamically generated momentum anisotropies over the various particle species and in $p_{\rm T}$ [25]. To correctly describe the differential anisotropic flows $v_n(p_{\rm T})$, for all charged hadrons or for specific identified hadron species, thus requires that the total hydrodynamic momentum anisotropy (which "measures" η/s), the radial flow (which is also sensitive to bulk viscosity that suppresses radial flow [26]), and the chemical composition of the system at final decoupling (which reflects the kinetics of chemical freeze-out) are *all* correctly described by the model. By analyzing instead the total flow anisotropies $v_n^{\rm ch}$ of *all* charged hadrons, integrated over $p_{\rm T}$, one can strongly reduce the model sensitivity to bulk viscosity and final chemical composition and obtain a much more robust estimate of η/s [21, 25]. The left panel in Fig. 6 shows such an extraction of η/s from elliptic flow data. The `VISHNU` model used for this extraction couples the viscous hydrodynamic evolution of the QGP phase to a microscopic kinetic evolution of the hadronic phase after hadronization, thereby eliminating many uncertainties in earlier simplified calculations that described the hadronic phase macroscopically, too. The different theoretical lines show that, for a given initial energy deposition model, $v_2^{\rm ch}/\varepsilon_2$ is *only* sensitive to the specific QGP shear viscosity. Once $(\eta/s)_{\rm QGP}$ has been adjusted to the measured total charged hadron elliptic flow as in Fig. 6 (resulting in $(\eta/s)_{\rm QGP} \simeq 0.2$ for MC-KLN and $(\eta/s)_{\rm QGP} \simeq 0.08$ for MC-Glauber initial conditions), one finds that the model also correctly describes the $p_{\rm T}$-spectra and differential elliptic flows $v_2(p_{\rm T})$ of all charged hadrons together, as well as for specific identified hadron species (pions, kaons, protons), for all collision centralities [25]. Furthermore, it correctly *predicted* the analogous observables for Pb+Pb collisions at the LHC [6, 27, 28], in particular an increased mass splitting between the differential elliptic flows for light and heavy hadrons [6, 29].

study of these correlations by more precise experiments and more systematic theoretical analyses should be able to separate the effects arising from the initial-state fluctuation spectrum and from dissipative transport.

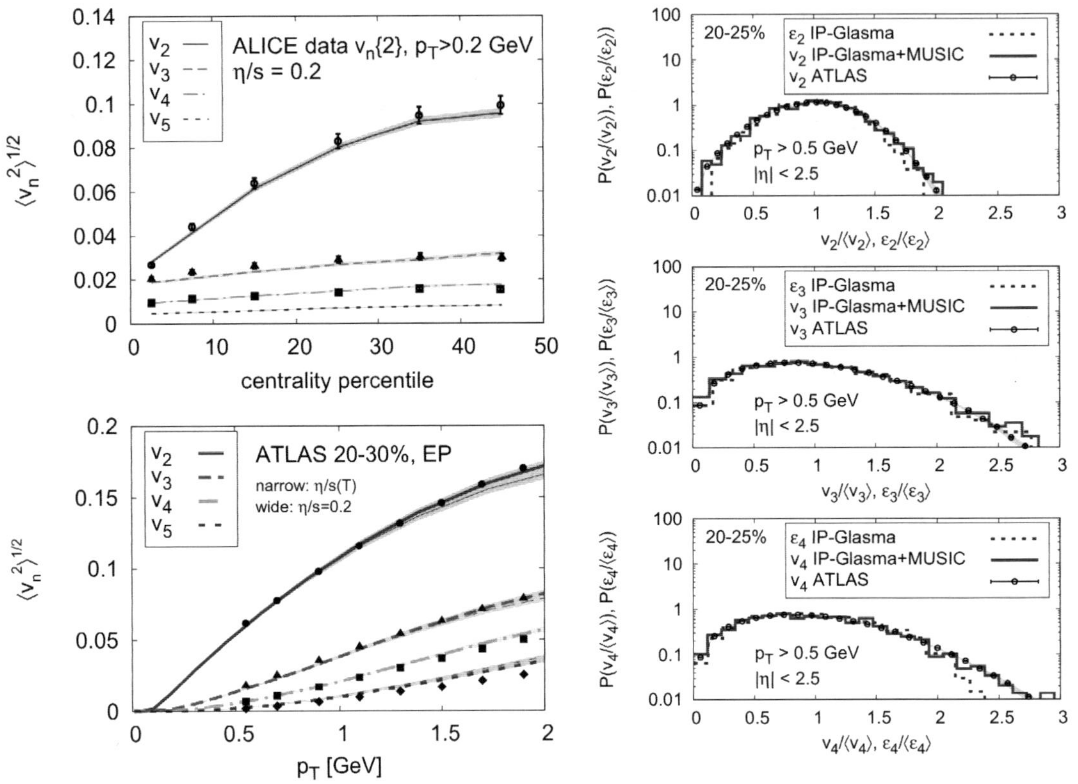

Figure 7. *Top left:* The centrality dependence of $v_n\{2\}$ from 2.76 ATeV Pb+Pb collisions measured by ALICE [15] compared to viscous hydrodynamic model calculations [11]. *Bottom left:* Comparison of $v_n(p_T)$ for the same collision system at 20−30% centrality from ATLAS [23] with hydrodynamical calculations, using both a constant average and a temperature dependent η/s [11]. *Right:* Scaled distributions of $v_{2,3,4}$ (from top to bottom) from viscous hydrodynamics with IP-Glasma initial conditions [11] compared with experimental data from ATLAS [24] and with the scaled distributions of the corresponding initial eccentricities $\varepsilon_{2,3,4}$. Nonlinear hydrodynamic evolution causes slightly larger variances for the v_n distributions compared to those of ε_n. The data in both panels are from Pb+Pb collisions at the LHC.

Unfortunately, the two left panels in Fig. 6 also show that different initial energy deposition models lead to very different estimates for $(\eta/s)_{\rm QGP}$, due to their different initial eccentricities ε_2 (see Fig. 3). In [30] it was shown that this model ambiguity can be resolved by analyzing simultaneously elliptic and triangular flow (v_2 and v_3) data. This analysis eliminated the MC-KLN model with $(\eta/s)_{\rm QGP} = 0.2$ as a viable candidate, and thereby demonstrated the power of a comprehensive set of anisotropic flow data to (over)constrain the dynamical evolution model. Unfortunately, the MC-Glauber model did not enjoy a much longer life itself: The v_n data ($n = 2, \dots, 7$) from ultra-central Pb+Pb collisions shown by CMS at *Quark Matter 2012* (see the right bottom plot in Fig. 3) and the event-by-event v_n probability distributions ($n = 2, \dots, 4$) measured by ATLAS and shown at the same conference [24] (see the right panel in Fig. 7) can not simultaneously be described by viscous hydrodynamics for *any* choice of $(\eta/s)_{\rm QGP}$ if initial fluctuation spectra from either of these two models (MC-KLN and MC-Glauber) are used.

Fortunately, a new and much more successful initial-state model saved the day: the IP-Glasma model [31, 1], based on the Color Glass Condensate idea [10], implements gluonic field fluctuations inside nucleons [33, 1] as well as gluon saturation effects [10, 32]. As shown in Fig. 7, these IP-Glasma initial conditions (represented by the solid lines in Fig. 3, lower left panel), when (after a short initial pre-equilibrium stage modeled by classical Yang-Mills evolution) evolved with viscous fluid dynamics, reproduce the entire measured spectrum of charged hadron anisotropic flow coefficients v_n, both integrated over and differential in p_T, for all collision centralities, as well as the measured [24] event-by-event distribution of v_2, v_3, and v_4, again for a range of collision centralities [11]. This is shown in Fig. 7 for Pb+Pb collisions at the LHC but also holds for Au+Au collisions at RHIC [11]. The only difference is that for LHC energies the effective specific shear viscosity of the QGP fluid must be chosen somewhat larger ($(\eta/s)_{\rm LHC} \simeq 0.2$) than at top RHIC energy ($(\eta/s)_{\rm RHIC} \simeq 0.12$). Both data sets are consistent with a temperature dependent specific shear viscosity $(\eta/s)(T)$ proposed in [34] (the thin lines in the lower left panel in Fig. 7) that has a minimum value around $1/(4\pi) = 0.08$ at the pseudocritical temperature $T_{\rm cr}$ where the QGP hadronizes and then rises to about 5 times that minimal value at $2T_{\rm cr}$. While this temperature-dependence of η/s is not yet tightly constrained by the available analyses of RHIC and LHC data, the need for a somewhat larger effective shear viscosity at the higher temperatures probed at the LHC, already noted in earlier work [35, 34], appears to solidify, and the early evidence from RHIC data for a very low specific shear viscosity near $T_{\rm cr}$ [36], perhaps as low as the KSS bound [37] of $1/(4\pi)$, is strongly supported by the work in [11].

4. Summary: status of and prospects for the Little Bang Standard Model

The Little Bang Standard Model is still under construction, but its key features are showing through the scaffolding: **1.** Every heavy-ion collision system, centrality class, and collision energy generates a different class of Little Bangs, each with its own characteristic initial density fluctuation and final anisotropic flow fluctuation spectrum. **2.** The initial fluctuation spectrum can be computed from the Color Glass Condensate (CGC) theory using e.g. the IP-Glasma model; gluon field fluctuations inside the nucleons within the colliding nuclei play an essential role in this spectrum. **3.** After a very short pre-equilibrium evolution stage, not lasting much longer than $1\,{\rm fm}/c$, which is best described using classical Yang-Mills field dynamics corrected for quantum fluctuations, the Little Bangs are in the QGP phase which undergoes viscous hydrodynamic evolution until it has cooled down to the pseudo-critical temperature for hadronization, $T_{\rm cr} \simeq 160\,{\rm MeV}$. The specific QGP shear viscosity is small, around $1/(4\pi)$ near $T_{\rm cr}$, rising modestly at higher temperatures; its effective value for $200\,A\,{\rm GeV}$ Au+Au collisions at RHIC is around $0.12 = 1.5/(4\pi)$, for $2.76\,A\,{\rm TeV}$ Pb+Pb collisions it is around $0.2 = 2.5/(4\pi)$. This small shear viscosity allows flow anisotropies to build in response to the initial fluctuations and anisotropies in the pressure gradients, to values that are large enough to be experimentally observed in the final state but still visibly attenuated by shear viscous effects, following an η/s-dependent pattern where higher harmonic flow coefficients are suppressed more strongly than lower harmonics. The spectrum of flow anisotropies, their magnitudes, directions, particle species and p_T-dependences, and event-by-event fluctuations provide a rich menu of experimental observables from which the QGP shear viscosity and initial fluctuation spectra can be reconstructed. **4.** After hadronization, the Little Bang continues to evolve as a dilute, highly dissipative hadron resonance gas that is highly inefficient in generating any additional momentum anisotropies but also doesn't erase those established earlier during the QGP phase. This makes it possible to extract from hadronic final state observables quantitative information about the earlier QGP and CGC phases.

In this relatively simple form, the Little Bang Standard Model applies to Little Bangs created at top RHIC and LHC energies. At lower collision energies, our understanding of the initial conditions becomes less reliable and the lifetime of the hydrodynamic QGP stage shrinks, making a quantitative theory of the Little Bang more challenging, but not less interesting.

Acknowledgments

This work was supported by the U.S. Department of Energy under Grants No. DE-SC0004286 and (within the framework of the JET Collaboration) DE-SC0004104. I thank my (former) students Huichao Song, Chun Shen, and Zhi Qiu for their important contributions to the construction of the LBSM.

References

[1] Schenke B, Tribedy P, and Venugopalan R 2012 *Phys. Rev. Lett.* **108** 252301; and *Phys. Rev. C* **86** 034908

[2] Tribble R (chair), Burrows A *et al.* 2013 *Implementing the 2007 Long Range Plan*, Report to the Nuclear Science Advisory Committee, January 31, 2013. Available at `http://science.energy.gov/np/nsac/reports/`.

[3] Ade PAR *et al.* (Planck Collab.) 2013 *Preprint* 1303.5075 [astro-ph.CO]

[4] Heinz U and Snellings R 2013 *Annu. Rev. Nucl. Part. Sci.* **63** in press (*Preprint* 1301.2826 [nucl-th]).

[5] Heinz U 1999 in *Strong and Electroweak Matter '98*, J. Ambjorn et al. (eds.), World Scientific, Singapore, p. 81-100 (*Preprint* arXiv:hep-ph/9902424); and *Nucl. Phys.* **A661** 140

[6] Shen C, Heinz U, Huovinen P, and Song H 2011 *Phys. Rev. C* **84** 044903

[7] Harris JW, Kharzeev D, and Ullrich T 2012 *CERN Courier* **52** (no. 9) 17

[8] Kolb PF and Heinz U 2004 in *Quark-Gluon Plasma 3*, Hwa RC *et al.* (eds.), World Scientific, Singapore, p. 634-714 (*Preprint* arXiv:nucl-th/0305084).

[9] Heinz U and Kolb PF 2002 *Nucl. Phys.* **A702** 269

[10] Kovner A, McLerran LD, and Weigert H 1995 *Phys. Rev. D* **52** 6231; Kovchegov YV and Rischke DH 1997 *Phys. Rev. C* **56** 1084; Krasnitz A and Venugopalan R 1999 *Nucl. Phys.* **B557** 237; Krasnitz A and Venugopalan R 2000 *Phys. Rev. Lett.* **84** 4309; Krasnitz A and Venugopalan R 2001 *Phys. Rev. Lett.* **86** 1717; Lappi T 2003 *Phys. Rev. C* **67** 054903; Lappi T and McLerran LD 2006 *Nucl. Phys.* **A772** 200

[11] Gale C, Jeon S, Schenke B, Tribedy P, and Venugopalan R 2013 *Phys. Rev. Lett.* **110** 012302; and *Preprint* 1210.5144 [hep-ph]

[12] Heinz U, Qiu Z, and Shen C 2013 *Phys. Rev. C* **87** 034913

[13] Jia J and Mohapatra S 2012 *Preprint* 1203.5095 [nucl-th]; Jia J and Teaney D 2012 *Preprint* 1205.3585 [nucl-ex]

[14] Qiu Z and Heinz U 2012 *Phys. Lett. B* **717** 261

[15] Aamodt K *et al.* (ALICE Collab.) (2011) *Phys. Rev. Lett.* **107** 032301; Bilandzic A *et al.* (ALICE Collab.) 2012 *Preprint* 1210.6222 [nucl-ex]

[16] Jia J *et al.* (ATLAS Collab.) 2012 *Preprint* 1208.1427 [nucl-ex]

[17] Qiu Z and Heinz U 2011 *Phys. Rev. C* **84** 024911

[18] Teaney D and Yan L 2012 *Phys. Rev. C* **86**, 044908; and *Preprint* 1210.5026 [nucl-th]

[19] Ollitrault J-Y, Poskanzer AM, and Voloshin SA 2009 *Phys. Rev. C* **80** 014904

[20] Adams J *et al.* (STAR Collab.) 2005 *Phys. Rev. C* **72** 014904

[21] Song H, Bass SA, and Heinz U 2011 *Phys. Rev. C* **83** 024912; Song H, Bass SA, Heinz U, Hirano T, and Shen C 2011 *Phys. Rev. Lett.* **106**, 192301

[22] Schenke B, Jeon S, and Gale C 2012 *Phys. Rev. C* **85** 024901

[23] Aad G *et al.* (ATLAS Collab.) 2012 *Phys. Rev. C* **86** 014907

[24] Jia J *et al.* (ATLAS Collab.) 2012 *Preprint* 1209.4232 [nucl-ex]; ATLAS Collab 2012 *ATLAS note ATLAS-CONF-2012-114*. `http://cds.cern.ch/record/1472935/files/ATLAS-CONF-2012-114.pdf`

[25] Song H, Bass SA, Heinz U, Hirano T, Shen C 2011 *Phys. Rev. C* **83** 054910

[26] Song H and Heinz U 2010 *Phys. Rev. C* **81** 024905

[27] Heinz U, Shen C, and Song H 2012 *AIP Conf. Proc.* **1441** 766

[28] Abelev B *et al.* (ALICE Collab.) 2012 *Phys. Rev. Lett.* **109** 252301; and 2013 *Preprint* 1303.0737 [hep-ex]

[29] Muller B, Schukraft J, and Wyslouch B *Annu. Rev. Nucl. Part. Sci.* **62** 361

[30] Qiu Z, Shen C, and Heinz U (2012) *Phys. Lett. B* **707** 151

[31] Schenke B, Tribedy P, and Venugopalan R 2012 *Phys. Rev. C* **86** 034908

[32] Bartels J, Golec-Biernat KJ, and Kowalski H (2002) *Phys. Rev. D* **66** 014001; Kowalski H and Teaney D 2003 *Phys. Rev. D* **68** 114005

[33] Tribedy P and Venugopalan R 2011 *Nucl. Phys.* **A850** 136

[34] Niemi H, Denicol GS, Huovinen P, Molnar E, and Rischke DH 2011 *Phys. Rev. Lett.* **106** 212302; and 2012 *Phys. Rev. C* **86** 014909

[35] Luzum M 2011 *Phys. Rev. C* **83** 044911; Song H, Bass SA, and Heinz U 2011 *Phys. Rev. C* **83** 054912

[36] Lacey RA and Taranenko A 2006 *Proceedings of Science* **CFRNC2006** 021

[37] Policastro G, Son DT, and Starinets AO 2001 *Phys. Rev. Lett.* **87** 081601; Kovtun P, Son DT, and Starinets AO 2005 *Phys. Rev. Lett.* **94** 111601

International Workshop on Discovery Physics at the LHC (Kruger2012) IOP Publishing
Journal of Physics: Conference Series **455** (2013) 012045 doi:10.1088/1742-6596/455/1/012045

The Deconfinement of Charm and Beauty

Helmut Satz
Fakultät für Physik, Universität Bielefeld
Postfach 100 131, D-33501 Bielefeld, Germany

Abstract:

We discuss the spectral analysis of quarkonium states in a hot medium of deconfined quarks and gluons, and we show that such an analysis provides a way to determine the thermal properties of the quark-gluon plasma.

1. Introduction

Since more than thirty years it is known that besides the almost massless *up* and *down* quarks of the everyday world, and the still relatively light *strange* quarks required to account for the strange mesons and hyperons observed in hadron collisions, there are heavy quarks at the other end of the scale, whose bare masses alone are larger than those of most of the normal hadrons. These heavy quarks first showed up in the discovery of the J/ψ meson [1], of mass of 3.1 GeV; it is a bound state of a *charm* quark (c) and its antiquark ($\bar{c}$), each having a mass of some 1.2–1.5 GeV. On the next level there is the Υ meson [2], with a mass of about 9.5 GeV, made up of a *bottom* or *beauty* quark-antiquark pair ($b\bar{b}$), with each quark here having a mass around 4.5 - 4.8 GeV. Both charm and bottom quarks can of course also bind with normal light quarks, giving rise to open charm (D) and open beauty (B) mesons. The lightest of these 'light-heavy' mesons have masses of about 1.9 GeV and 5.3 GeV, respectively.

The bound states of a heavy quark Q and its antiquark $\bar{Q}$ are generally referred to as quarkonia. Besides the initially discovered vector ground states J/ψ and Υ, both the $c\bar{c}$ and the $b\bar{b}$ systems give rise to a number of other *stable* bound states of different quantum numbers. They are stable in the sense that their mass is less than that of two light-heavy mesons, so that strong decays are forbidden. The measured stable charmonium spectrum contains the $1S$ scalar η_c and vector J/ψ, three $1P$ states χ_c (scalar, vector and tensor), and the $2S$ vector state ψ', whose mass is just below the open charm threshold. There are further charmonium states above the ψ'; these can decay into $D\bar{D}$ pairs, and we shall here restrict our considerations only to quarkonia stable under strong interactions.

The study of quarkonia has played and continues to play a major role in many aspects of QCD: in spectroscopy, in production and decay, and last but not least as probe of hot QCD media [3]. Quarkonia are rather unusual hadrons. The masses of the light hadrons, in particular those of the non-strange mesons and baryons, arise almost entirely from the interaction energy of their nearly massless quark constituents. In contrast, the quarkonium masses are largely determined by the bare charm and bottom quark masses. These large quark masses allow a very straightforward calculation of many basic quarkonium properties, using non-relativistic potential theory. It is found that, in particular, the ground states and the lower excitation levels of quarkonia are very much smaller than the normal hadrons, and that they are very tightly bound. Now deconfinement is a matter of scales: when the separation between normal hadrons becomes much less than the size of these hadrons, they melt to form the quark-gluon plasma. What happens at this point to the much smaller quarkonia? When do they become dissociated? That is the main question we want to address here. We shall show that the disappearance of specific quarkonia signals the presence of a deconfined medium of a specific temperature [4]. Thus the study of the quarkonium spectrum in a given medium is akin to the spectral analysis of stellar media, where the presence or absence of specific excitation lines allows a determination of the temperature of the stellar interior.

We had defined quarkonia as bound states of heavy quarks which are stable under strong decay, i.e., $M_{c\bar{c}} \leq 2M_D$ for charmonia and $M_{b\bar{b}} \leq 2M_B$ for bottomonia. Since the quarks are heavy, with $m_c \simeq 1.2-1.5$ GeV for the charm and $m_b \simeq 4.5-4.8$ GeV for the bottom quark, quarkonium spectroscopy can be studied quite well in non-relativistic potential theory. The simplest ("Cornell") confining potential [5] for a $Q\bar{Q}$ at separation distance r has the form

$$V(r) = \sigma\, r - \frac{\alpha}{r} \tag{1}$$

with a string tension $\sigma \simeq 0.2$ GeV^2 and a Coulomb-like term with a gauge coupling $\alpha \simeq \pi/12$. The corresponding Schrödinger equation

$$\left\{2m_c - \frac{1}{m_c}\nabla^2 + V(r)\right\}\Phi_i(r) = M_i\Phi_i(r) \tag{2}$$

then determines the bound state masses M_i and the wave functions $\Phi_i(r)$, and with

$$\langle r_i^2\rangle = \int d^3r\; r^2|\Phi_i(r)|^2 \Big/ \int d^3r\; |\Phi_i(r)|^2. \tag{3}$$

the latter in turn provide the (squared) average bound state "radii", here defined as the $Q\bar{Q}$ separation for the state in question. The solution of eq. (2) gives in fact a very good account of the full (spin-averaged) quarkonium spectroscopy, as shown in more detail in [6]. Let us seen what happens to this in a strongly interacting medium.

2. Interaction Range and Colour Screening

Consider a colour-singlet bound state of a heavy quark Q and its antiquark $\bar{Q}$, put into the medium in such a way that we can measure the energy of the system as function of the $Q\bar{Q}$ separation r (see Fig. 1). The quarks are assumed to be heavy so that they are static and any energy changes indicate changes in the binding energy. We consider first the case of vanishing baryon density; at $T=0$, the box is therefore empty.

In vacuum, i.e., at $T=0$, the free energy of the $Q\bar{Q}$ pair is assumed to have the string form [5]

$$F(r) \sim \sigma r \tag{4}$$

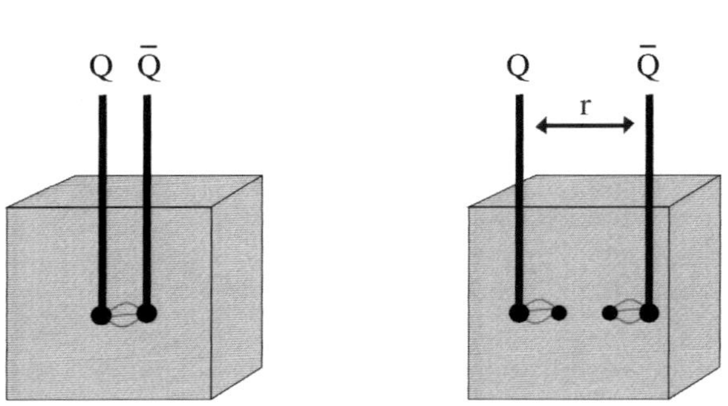

Figure 1: String breaking for a $Q\bar{Q}$ system

where $\sigma \simeq 0.16$ GeV^2 is the string tension as determined in the spectroscopy of heavy quark resonances (charmonium and bottomonium states). Thus $F(r)$ increases with separation distance; but when it reaches the value of a pair of dressed light quarks (about the mass of a ρ meson), it becomes energetically favorable to produce a $q\bar{q}$ pair from the vacuum, break the string and form two light-heavy mesons $(Q\bar{q})$ and $(\bar{Q}q)$. These can now be separated arbitrarily far without changing the energy of the system (Fig. 1).

The string breaking energy for charm quarks is found to be

$$F_0 = 2(M_D - m_c) \simeq 1.2 \text{ GeV}; \tag{5}$$

for bottom quarks, one obtains the same value,

$$F_0 = 2(M_B - m_b) \simeq 1.2 \text{ GeV}, \tag{6}$$

using 1.3 GeV and 4.7 GeV for charm and bottom quark mass, respectively. Hence the onset of string breaking is evidently a property of the vacuum as a medium. It occurs when the two heavy quarks are separated by a distance

$$r_0 \simeq 1.2 \text{ GeV}/\sigma \simeq 1.5 \text{ fm}, \tag{7}$$

independent of the mass of the (heavy) quarks connected by the string.

If we heat the system to get $T > 0$, the medium begins to contain light mesons, and the large distance $Q\bar{Q}$ potential $F(\infty, T)$ decreases, since we can use these light hadrons to achieve an earlier string breaking through a kind of flip-flop recoupling of quark constituents [7], resulting in an effective screening of the interquark force (see Fig. 2). Near the deconfinement point, the hadron density increases rapidly, and hence the recoupling dissociation becomes much more effective, causing a considerable decrease of $F(\infty, T)$.

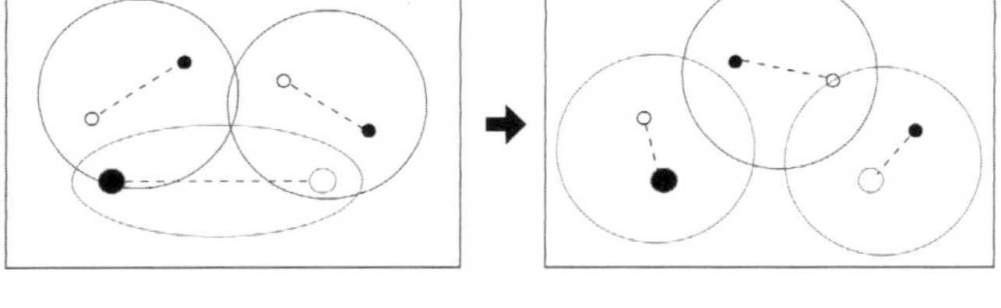

Figure 2: In-medium string breaking through recoupling

A further increase of T will eventually bring the medium to the deconfinement point T_c, at which chiral symmetry restoration causes a rather abrupt drop of the light quark dressing (equivalently, of the constituent quark mass), increasing strongly the density of constituents. As a consequence, $F(\infty, T)$ now continues to drop sharply. Above T_c, light quarks and gluons become deconfined colour charges, and this quark-gluon plasma leads to a colour screening, which limits the range of the strong interaction. The colour screening radius r_D, which determines this range, is inversely proportional to the density of charges, so that it decreases with increasing temperature. As a result, the $Q\bar{Q}$ interaction becomes more and more short-ranged.

In summary, starting from $T = 0$, the $Q\bar{Q}$ probe first tests vacuum string breaking, then a screening-like dissociation through recoupling of constituent quarks, and finally genuine colour screening. In Fig. 3, we show the behaviour obtained in full two-flavour QCD for the colour-singlet $Q\bar{Q}$ free energy as a function of r for different T [8].

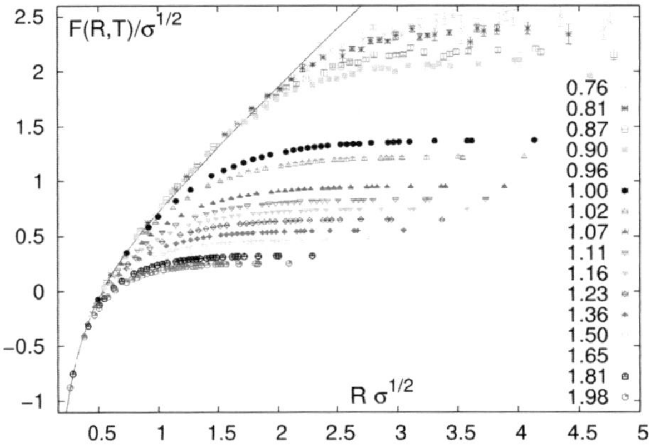

Figure 3: The colour singlet $Q\bar{Q}$ free energy $F(r, T)$ vs. r at different T [8]

It is evident in Fig. 3 that the asymptotic value $F(\infty, T)$, i.e., the energy needed to separate the $Q\bar{Q}$ pair, decreases with increasing temperature, as does the separation distance at which "the string breaks". For the moment we consider the latter to be defined by the point beyond which the free energy remains constant within errors, returning in section 4.3 to a more precise definition. The behaviour of both quantities is shown in Fig. 4. Deconfinement is thus reflected very clearly in the temperature behaviour of the heavy quark potential: both the string breaking energy and the interaction range drop sharply around T_c. The latter decreases from hadronic size in the confinement region to much smaller values in the deconfined medium, where colour screening is operative.

The in-medium behaviour of heavy quark bound states thus does serve quite well as probe of the state of matter in QCD thermodynamics. We had so far just considered $Q\bar{Q}$ bound states in general. Let us now turn to a specific state such as the J/ψ. What happens when the range of the binding force becomes smaller than the radius of the state? Since the c and the $\bar{c}$ can now no longer see each other, the J/ψ must dissociate for temperatures above this point. Hence the dissociation points of the different quarkonium states provide a way to measure the temperature of the medium. The effect is illustrated schematically in Fig. 5, showing how with increasing temperature the different charmonium states

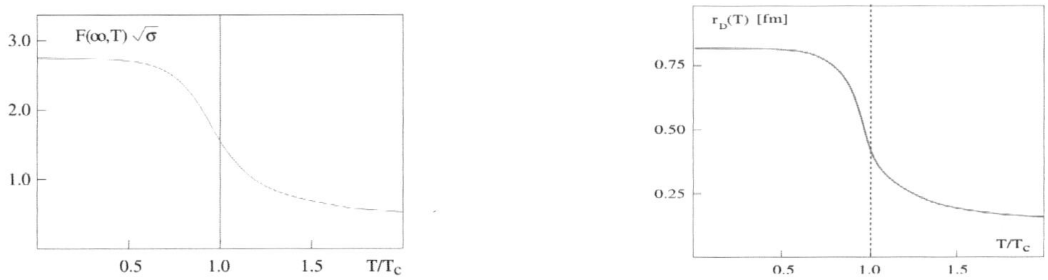

Figure 4: String breaking potential and interaction range at different temperatures

"melt" sequentially as function of their binding strength; the most loosely bound state disappears first, the ground state last.

Moreover, since finite temperature lattice QCD also provides the temperature dependence of the energy density [9], the melting of the different charmonia or bottomonia can be specified as well in terms of ϵ. In Fig 6, we illustrate this, combining $\epsilon(T)$ with the force radii shown in Fig. 4. It is evident that although ψ' and χ_c are both expected to melt around T_c, the corresponding dissociation energy densities could still differ.

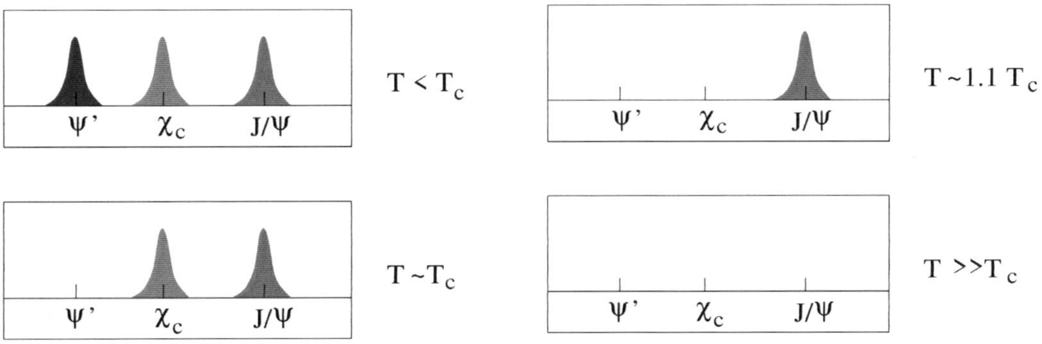

Figure 5: Charmonium spectra at different temperatures

To make these considerations quantitative, we thus have to find a way to determine the in-medium melting points of the different quarkonium states. This problem has been addressed in three different approaches:

- Model the heavy quark potential as function of the temperature, $V(r,T)$, and solve the resulting Schrödinger equation (2).
- Determine the internal energy $U(r,T)$ of a $Q\bar{Q}$ pair at separation distance r from lattice results for the corresponding free energy $F(r,T)$, using the thermodynamic relation

$$U(r,T) = -T^2 \left(\frac{\partial[F(r,T)/T]}{\partial T}\right) = F(r,T) - T\left(\frac{\partial F(r,T)}{\partial T}\right), \tag{8}$$

  and solve the Schrödinger equation with $V(r,T) = U(r,T)$ as the binding potential.
- Calculate the quarkonium spectrum directly in finite temperature lattice QCD.

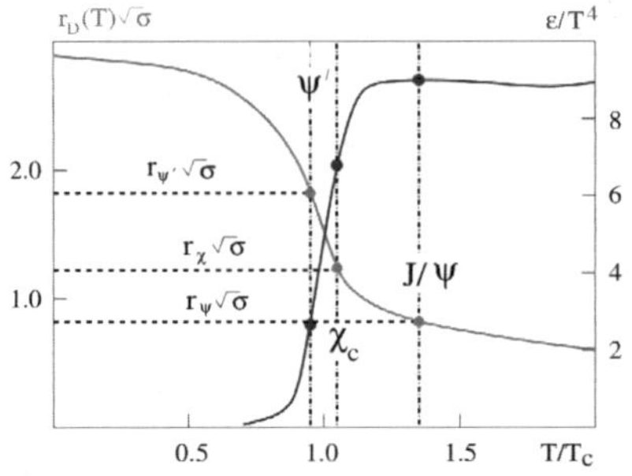

Figure 6: Charmonium dissociation vs. temperature and energy density

Clearly the last is the only model-independent way, and it will in the long run provide the decisive determination. However, the direct lattice study of charmonium spectra has become possible only quite recently, and so far most results are obtained in quenched QCD (no dynamical light quarks); corresponding studies for bottomonia are still more difficult. Hence much of what is known so far is based on Schrödinger equation studies with different model inputs. To illustrate the model-dependence of the dissociation parameters, we first cite some early work using different models for the temperature dependence of $V(r,T)$, then some recent studies based on lattice results for $F(r,T)$, and finally summarize the present state of direct lattice calculations of charmonia in finite temperature media.

3. Potential Model Studies

The first quantitative studies of finite temperature charmonium dissociation [10] were based on screening in the form obtained in one-dimensional QED, the so-called Schwinger model. The confining part of the Cornell potential (1), $V(r) \sim \sigma r$, is the solution of the Laplace equation in one space dimension. In this case, Debye-screening leads to [11]

$$V(r,T) \sim \sigma r \left\{ \frac{1 - e^{-\mu r}}{\mu r} \right\}, \tag{9}$$

where $\mu(T)$ denotes the screening mass (inverse Debye radius) for the medium at temperature T. This form reproduces at least qualitatively the convergence to a finite large distance value $V(\infty,T) = \sigma/\mu(T)$, and since $\mu(T)$ increases with T, it also gives the expected decrease of the potential with increasing temperature. Combining this with the usual Debye screening for the $1/r$ part of eq. (1) then leads to

$$V(r,T) \sim \sigma r \left\{ \frac{1 - e^{-\mu r}}{\mu r} \right\} - \frac{\alpha}{r} e^{-\mu r} = \frac{\sigma}{\mu} \left\{ 1 - e^{-\mu r} \right\} - \frac{\alpha}{r} e^{-\mu r} \tag{10}$$

for the screened Cornell potential. In [10], the screening mass was assumed to have the form $\mu(T) \simeq 4\ T$, as obtained in first lattice estimates of screening in high temperature $SU(N)$ gauge theory. Solving the Schrödinger equation with these inputs, one found that both the ψ' and the χ_c become dissociated essentially at $T \simeq T_c$, while the J/ψ persisted up to about $1.2\ T_c$. Note that as function of the energy density $\epsilon \sim T^4$, this meant that the J/ψ really survives up to much higher ϵ.

This approach has two basic shortcomings:

- The Schwinger form (9) corresponds to the screening of σr in one space dimension; the correct result in three space dimensions is different [11].
- The screening mass $\mu(T)$ is assumed in its high energy form; lattice studies show today that its behaviour near T_c is quite different [12].

While the overall behaviour of this approach provides some first insight into the problem, quantitative aspects require a more careful treatment.

When lattice results for the heavy quark free energy as function of the temperature first became available, an alternative description appeared [13]. It assumed that in the thermodynamic relation (8) the entropy term $-T(\partial F/\partial T)$ could be neglected, thus equating binding potential and free energy,

$$V(r,T) = F(r,T) - T(\partial F/\partial T) \simeq F(r,T). \tag{11}$$

Using this potential in the Schrödinger equation (2) specifies the temperature dependence of the different charmonium masses. On the other hand, the large distance limit of $V(r,T)$ determines the temperature variation of the open charm meson D,

$$2M_D(T) \simeq 2m_c + V(\infty, T) \tag{12}$$

In fig. 7, we compare the resulting open and hidden charm masses as function of temperature. It is seen that the ψ' mass falls below 2 M_D around 0.2 T_c, that of the χ_c at about 0.8 T_c; hence these states disappear by strong decay at the quoted temperatures. Only the ground state J/ψ survives up to T_c and perhaps slightly above; the lattice data available at the time did not extend above T_c, so further predictions were not possible.

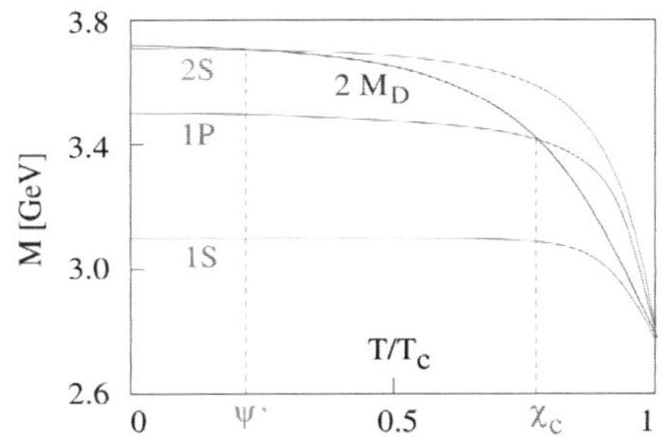

Figure 7: Temperature dependence of open and hidden charm masses [13]

The main shortcoming of this approach is quite evident. The neglect of the entropy term in the potential reduces $V(r,T)$ and hence the binding. As a result, the D mass drops faster with temperature than that of the charmonium states, and it is this effect which leads to the early charmonium dissociation. Moreover, in the lattice studies used here, only the colour averaged free energy was calculated, which leads to a further reduction of the binding force.

We conclude from these attempts that for a quantitative potential theory study, the free energy has to be formulated in the correct three-dimensional screened Cornell form, and it then has to be checked against the space- and temperature-dependence of the corresponding colour singlet quantity obtained in lattice QCD [12, 14]

4. Screening Theory

The modification of the interaction between two charges immersed in a dilute medium of charged constituents is provided by Debye-Hückel theory, which for the Coulomb potential in three space dimensions leads to the well-known Debye screening,

$$\frac{1}{r} \rightarrow \frac{1}{r}\epsilon^{-\mu r}, \tag{13}$$

where $r_D = 1/\mu$ defines the screening radius [15]. Screening can be evaluated more generally [11] for a given free energy $F(r) \sim r^q$ in d space dimensions, with an arbitrary number q. We shall here apply this to the two terms of the Cornell form, with $q = 1$ for the string term, $q = -1$ for the gauge term, in $d = 3$ space dimensions [12].

We thus assume that the screening effect can be calculated separately for each term, so that the screened free energy becomes

$$F(r,T) = F_s(r,T) + F_c(r,T) = \sigma r f_s(r,T) - \frac{\alpha}{r} f_c(r,T). \tag{14}$$

The screening functions $f_s(r,T)$ and $f_c(r,T)$ must satisfy

$$\begin{aligned} f_s(r,T) &= f_c(r,T) = 1 \ \text{ for } T \rightarrow 0, \\ f_s(r,T) &= f_c(r,T) = 1 \ \text{ for } r \rightarrow 0, \end{aligned} \tag{15}$$

since at $T = 0$ there is no medium, while in the short-distance limit $T^{-1} \gg r \rightarrow 0$, the medium has no effect. The resulting forms are [11]

$$F_c(r,T) = -\frac{\alpha}{r}\left[e^{-\mu r} + \mu r\right] \tag{16}$$

for the gauge term, and

$$F_s(r,T) = \frac{\sigma}{\mu}\left[\frac{\Gamma(1/4)}{2^{3/2}\Gamma(3/4)} - \frac{\sqrt{\mu r}}{2^{3/4}\Gamma(3/4)} K_{1/4}[(\mu r)^2]\right] \tag{17}$$

for the string term. The first term in eq. (17) gives the constant large distance limit due to colour screening; the second provides a Gaussian cut-off in $x = \mu r$, since $K_{1/4}(x^2) \sim \exp\{-x^2\})$; this is in contrast to the exponential cut-off given by the Schwinger form (10), but in accord with string breaking arguments.

At temperatures $T > T_c$, when the medium really consists of unbound colour charges, we thus expect the free energy $F(r,T)$ to have the form (14), with the two screened terms given by eqs. (16) and (17). In Fig. 8a, it is seen that the results for the colour singlet free energy calculated in two-flavour QCD [16] are indeed described very well by this form, with $m_c = 1.25$ GeV and $\sqrt{\sigma} = 0.445$ GeV, as above. The only parameter to be determined, the screening mass μ, is shown in Fig. 9, and as expected, it first increases rapidly in the transition region and then turns into the perturbative form $\mu \sim T$.

The behaviour of $Q\bar{Q}$ binding in a plasma of unbound quarks and gluons is thus well described by colour screening. Such a description is in fact found to work well also for $T < T_c$, when quark recombination leads to an effective screening-like reduction of the interaction range, provided one allows higher order contributions in $x = \mu r$ in the Bessel

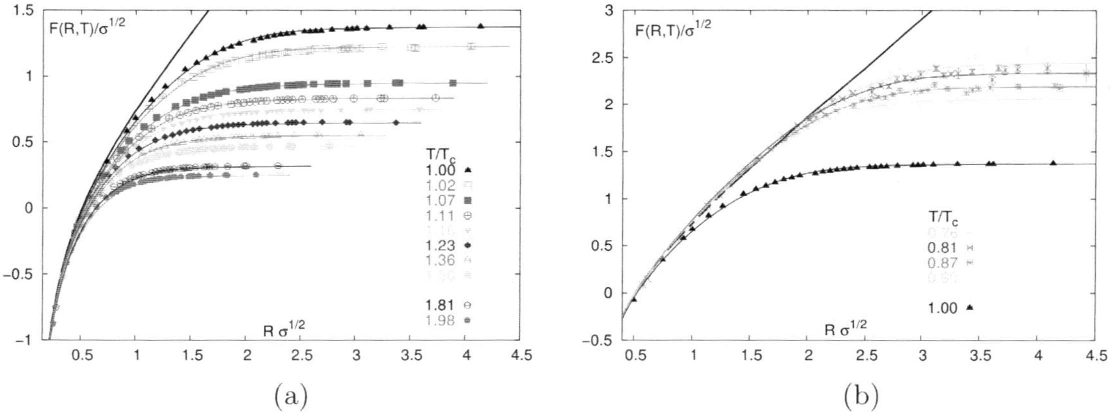

Figure 8: Screening fits to the $Q\bar{Q}$ free energy $F(r,T)$ for $T \geq T_c$ (left) and $T \leq T_c$ (right) [12]

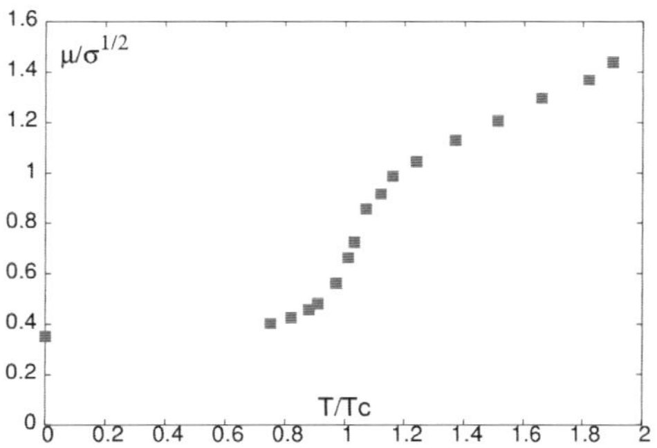

Figure 9: The screening mass $\mu(T)$ vs. T [12]

function $K_{1/4}(x^2)$ governing string screening [12]. The resulting fit to the two-flavour colour singlet free energy below T_c is shown in Fig. 8b, using $K_{1/4}(x^2 + (1/2)x^4)$ in eq. (17); the corresponding values for the screening mass are included in Fig. 9.

With the free energy $F(r,T)$ of a heavy quark-antiquark pair given in terms of the screening form obtained from Debye-Hückel theory, the internal energy $U(r,T)$ can now be obtained through the thermodynamic relation (8), and this then provides the binding potential $V(r,T)$ for the temperature dependent version of the Schrödinger equation (2). The resulting solution then specifies the temperature-dependence of charmonium binding as based on the correct heavy quark potential [12].

Let us briefly comment on the thermodynamic basis of this approach. The pressure P of a thermodynamic system is given by the free energy, $P = -F = -U + TS$; it is determined by the kinetic energy TS at temperature T and entropy $S(T)$, reduced by the potential energy $U(T)$ between the constituents. In our case, all quantities give the difference between a thermodynamic system containing a $Q\bar{Q}$ pair and the corresponding system without such a pair. The question now is which thermodynamic quantity constitutes the correct potential term in the two-body Schrödinger equation. First studies had used the color-averaged free energy for $V(r,T)$, but subsequently lattice results for the color-singlet free energy became available, leading to somewhat stronger binding. Eventually it was argued that the correct potential is given by the internal energy $U(r,T)$. The proposals of the last years now cover the whole range, with the general form $xU(r,T) + (1-x)F(r,T)$, where $0 < x < 1$. As is evident from what was said here, the effective binding increases with x. The internal energy $U(r,T)$, as the expectation value of the difference of the Hamiltonians with and without the pair, includes the indirect "cloud-cloud" binding in addition to the direct $Q\bar{Q}$ interaction and hence provides a stronger binding. We believe that the indirect binding cannot be neglected, so that the correct form of the potential should indeed be $U(r,T)$. To determine the dissociation points for the different quarkonium states, we then have to solve the Schrödinger equation (2) with $V(r,T) = U(r,T)$.

At this point, we add that recently the $Q\bar{Q}$ binding problem has been studied quite extensively in the weak-coupling limit [17–21], corresponding largely to the QED case. Here one can in fact show that the potential energy is given by the free energy $V(r,T)$. In this case, however, the behavior of the free energy as well as that of the screening mass is very different from that found in the strong coupling regime of interest to us here, for $T_c \leq T \leq 2\,T_c$. Nevertheless, these results do throw some doubt on potential studies using the full internal energy as potential. In addition, the imaginary part of the potential, here so far neglected, has to be taken into account; for a review of recent developments, see [22]

From eqs. (16) and (17) we get for the $Q\bar{Q}$ potential

$$V(r,T) = V(\infty,T) + \tilde{V}(r,T), \tag{18}$$

with

$$V(\infty,T) = c_1\frac{\sigma}{\mu} - \alpha\mu + T\frac{d\mu}{dT}[c_1\frac{\sigma}{\mu^2} + \alpha], \tag{19}$$

and $c_1 = \Gamma(1/4)/2^{3/2}\Gamma(3/4)$, and where $\tilde{V}(r,T)$ contains the part of the potential which vanishes for $r \to \infty$. The behaviour of $V(\infty,T)$ as function of the temperature is shown in Fig. 10. It measures (twice) the energy of the cloud of light quarks and gluons around an isolated heavy quark, of an extension determined by the screening radius, relative to the energy contained in such a cloud of the same size without a heavy quark. This energy difference arises from the interaction of the heavy quark with light quarks and gluons of the medium, and from the modification of the interaction between the light constituents themselves, caused by the presence of the heavy charge. – The behaviour of $\tilde{V}(r,T)$ is shown in Fig. 11 for three different values of the temperature. It is seen that with increasing T, screening reduces the range of the potential.

The relevant Schrödinger equation now becomes

$$\left\{-\frac{1}{m_c}\nabla^2 + \tilde{V}(r,T)\right\}\Phi_i(r) = \Delta E_i(T)\Phi_i(r) \tag{20}$$

where

$$\Delta E_i(T) = M_i - 2m_c - V(\infty,T) \tag{21}$$

is the binding energy of charmonium state i at temperature T. When it vanishes, the bound state i no longer exists, so that $\Delta E_i(T) = 0$ determines the dissociation temperature T_i for that state. The temperature enters only through the T-dependence of the screening mass $\mu(T)$, as obtained from the analysis of the lattice results for $F(r,T)$. –

Journal of Physics: Conference Series **455** (2013) 012045 doi:10.1088/1742-6596/455/1/012045

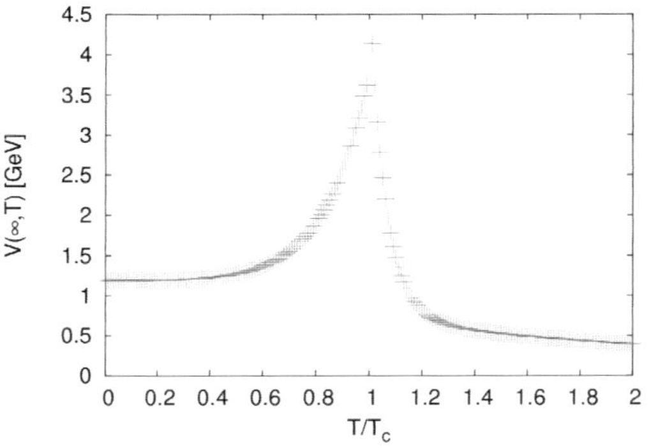

Figure 10: The large distance limit of the quarkonium potential $V(r,T)$ [12]

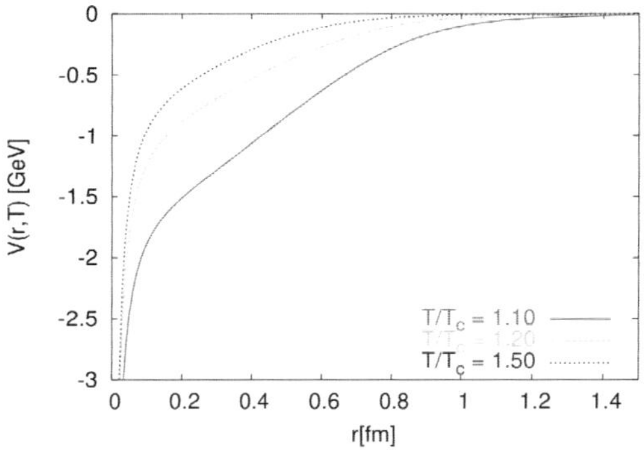

Figure 11: Variation of $\tilde{V}(r,T)$ with r for different T [12]

Equivalently, the divergence of the binding radii, $r_i(T) \to \infty$, can be used to define the dissociation points T_i.

The same formalism, with $m_b = 4.65$ GeV replacing m_c, leads to the bottomonium dissociation points. The combined quarkonium results are listed in Table 1. They agree quite well with those obtained in a very similar study based on corresponding free energies obtained in quenched lattice QCD [23], indicating that above T_c gluonic effects dominate. Using a parametrically generalized screened Coulomb potential obtained from lattice QCD results also leads to very compatible results for the $N_f = 2$ and quenched cases [24]. We recall here that the main underlying change, which is responsible for the much higher dissociation temperatures for the quarkonium ground states, is the use of the full internal energy (8), including the entropy term, as potential in the Schrödinger equation: this makes the binding much stronger.

We should note, however, that in all such potential studies it is not so clear what binding energies of less than a few MeV or bound state radii of several fermi can mean in a medium whose temperature is above 200 MeV and which leads to screening radii of less than 0.5 fm. In such a situation, thermal activation [25] can easily dissciate the bound state.

state	$J/\psi(1S)$	$\chi_c(1P)$	$\psi'(2S)$	$\Upsilon(1S)$	$\chi_b(1P)$	$\Upsilon(2S)$	$\chi_b(2P)$	$\Upsilon(3S)$
T_d/T_c	2.10	1.16	1.12	> 4.0	1.76	1.60	1.19	1.17

Table 1: Quarkonium Dissociation Temperatures [6]

5. Charmonium Correlators

The direct spectral analysis of charmonia in finite temperature lattice has come within reach only in very recent years [26]. It is possible now to evaluate the correlation functions $G_H(\tau, T)$ for hadronic quantum number channels H in terms of the Euclidean time τ and the temperature T. These correlation functions are directly related to the corresponding spectral function $\sigma_H(M,T)$, which describe the distribution in mass M at temperature

T for the channel in question. In Fig. 12, schematic results at different temperatures are shown for the J/ψ and the χ_c channels. It is seen that the spectrum for the ground state J/ψ remains essentially unchanged below 1.5 T_c. At 3 T_c, however, it has disappeared; the remaining spectrum is that of the $c\bar{c}$ continuum of J/ψ quantum numbers at that temperature. In contrast, the χ_c is already absent at 1.1 T_c, with only the corresponding continuum present.

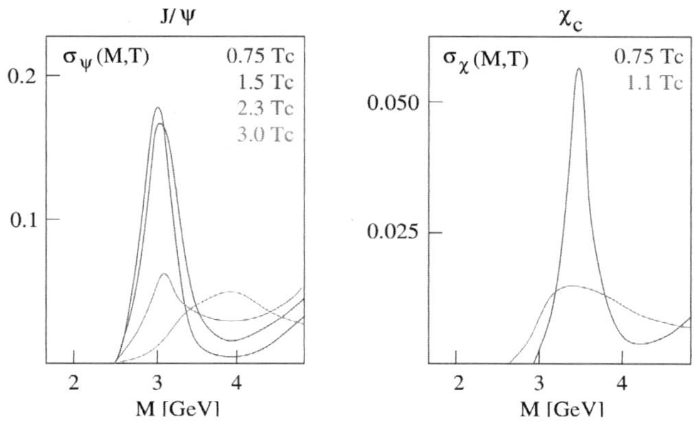

Figure 12: J/ψ and χ_c spectral functions at different temperatures

These results are clearly very promising: they show that in a foreseeable future, the dissociation parameters of quarkonia can be determined *ab initio* in lattice QCD. For the moment, however, they remain indicative only, since the underlying calculations were generally performed in quenched QCD, i.e., without dynamical quark loops; for a report on recent extensive studies on this level, see [27]. Since quark loops are crucial in the break-up of quarkonia into light-heavy mesons, final results require calculations in full QCD. Some first calculations in two-flavour QCD have appeared [28] and support the late dissociation of the J/ψ. The widths of the observed spectral signals are at present determined by the precision of the lattice calculations; to study the actual physical widths, much higher precision is needed. At present, further large-scale lattice studies are underway, both for charmonia and for bottomonia; for a survey, see [29]. So far, however, they have not led to results much different from the earlier work discussed here.

Finally, one has so far only first signals at a few selected points; a temperature scan also requires higher performance computational facilities. Since the next generation of computers, in the multi-Teraflops range, is presently going into operation, the next years should bring the desired results. So far, in view of the mentioned uncertainties in both approaches, the results from direct lattice studies and those from the potential model calculations of the previous section appear quite compatible.

6. Conclusions

The theoretical analysis of the in-medium behaviour of quarkonia has greatly advanced in the past decade. Potential model studies based on lattice results for the colour-singlet free energy appear to be converging to results from direct lattice calculations, and within a few more years, the dissociation temperatures for the different quarkonium states should be known precisely. Through corresponding calculations of the QCD equation of state, these temperatures provide the energy density values at which the dissociation occurs.

We conclude with some conceptual comments on high energy heavy ion collisions in general and on the role of quarkonium analysis for this field in particular. These remarks are addressed especially to the younger players in the game, in order to explain to them a little our motivation in doing what we are doing.

When they first became available as tools of research, high energy accelerators were employed simply to study what happens when two particles collide at relativistic energies; the experimental aims were largely exploratory. In the following stages, specific objectives came into play. The $p - \bar{p}$ collider at CERN was built to find the heavy weak interaction bosons Z_0 and $W^{\pm}$, and it did, thereby providing crucial support for the unified electroweak theory. The aim of LEP was to test the predictions of perturbative QCD to high accuracy, and it did, confirming that QCD was indeed the basic theory of strong interactions. The LHC finally was the Higgs-search machine, and it (very likely) has now also found the elusive boson, required to account for the inherent constituent masses in the standard model. Similarly, the early heavy ion accelerators at Berkeley and at GSI were constructed to check what happened in the collisions of two heavy nuclei, how they fragmented and gave rise to collective features of nuclear matter. But in this field as well, the next generation of high energy heavy ion experiments was initiated, at CERN and at Brookhaven, with a specific well-defined charge: to produce and study in the laboratory the deconfined state of matter, the quark-gluon plasma, predicted by quantum chromodynamics.

The investigation of nuclear collisions is a very multi-faceted enterprise. It involves initial state aspects, parton structure and its limits as color glass, multiple parton interactions, non-equilibrium evolution, thermalization questions, hydrodynamic expansion, viscosity and flow, and much more. The production of deconfined matter through such collisions is a rather speculative issue, and the view of sceptics was perhaps best summarized by Richard Feynman when he said: "if I throw my watch against the wall, I get a broken watch, not a new state of matter". The mentioned non-equilibrium issues are difficult, if not impossible, to account for fully in terms of first principle QCD calculations. Thus there seems to be only one way to assure that high energy heavy ion collisions indeed produce the quark-gluon plasma of QCD: to calculate in equilibrium statistical QCD some observable features and then show that these features indeed occur in high energy nuclear collisions. The fundamental theory of strongly interacting matter is quantum chromodynamics, and so our final aim must be to connect first principle QCD calculations with high energy heavy ion data. Then, and only then have we really carried out the assigned charge mentioned above.

In heavy ion studies, there is a wealth of additional, preliminary and alternative issues to be addressed, much interesting physics to be carried out. The challenge to model these features has in recent years been strongly emphasized, and sometimes has even led to claims that a direct quantitative comparison between *ab initio* QCD predictions and data, without proper account for the complications arising from the small fireball size and lifetime and its rapid dynamical evolution in heavy-ion collisions, is not meaningful and can yield very misleading results. It seems to me that this is just another way of stating Feynman's objection: heavy ion collisions will basically remain a study of the fragmentation of his watch when it hits the wall.

Many in the field, however, both in theory and in experiment, continue to believe that it is a possible to compare *ab initio* results obtained in statistical QCD to data from high energy heavy ion collisions. The calculation and the measurement of quarkonium dissociation thresholds can provide such a chance and thus help to establish that nuclear collisions indeed produce the new state of matter we are looking for.

References

[1] J. J. Aubert et al., Phys. Rev. Lett. 33 (1974) 1404;
J. E. Augustin et al., Phys. Rev. Lett. 33 (1974) 1406.

[2] S. W. Herb et al., Phys. Rev. Lett. 39 (1977) 252.

[3] For a recent survey, see
N. Brambilla et al., *Heavy Quarkonium Physics*, Yellow Report CERN-2005-005.

[4] T. Matsui and H. Satz, Phys. Lett. B 178 (1986) 416.

[5] E. Eichten et al., Phys. Rev. D17 (1978) 3090; Phys. Rev. D 21 (1980) 203.

[6] See e.g. H. Satz, J. Phys. G32 (2006) R25.

[7] H. Miyazawa, Phys. Rev. D 20 (1979) 2953;
V. Goloviznin and H. Satz, Yad. Fiz. 60N3 (1997) 523

[8] O. Kaczmarek et al., Prog. Theor. Phys. Suppl. 153 (2004) 287;
O. Kaczmarek and F. Zantow, hep-lat/0503017.

[9] F. Karsch, Lect. Notes Phys. 583 (2002) 209.

[10] F. Karsch, M.-T. Mehr and H. Satz, Z. Phys. C 37 (1988) 617

[11] V. V. Dixit, Mod. Phys. Lett. A5 (1990) 227.

[12] S. Digal et al., Eur. Phys. J. C 43 (2005) 71.

[13] S. Digal, P. Petreczky and H. Satz, Phys. Lett. B 514 (2001) 57.

[14] A. Mocsy and P. Petreczky, Eur. Phys. J. C 43 (2005) 77;
A. Mocsy and P. Petreczky, hep-ph/0512156

[15] L. D. Landau and E. M. Lifshitz, *Statistical Physics*, Pergamon Press, London, 1958.

[16] O. Kaczmarek and F. Zantow, Eur. Phys. J. C 43 (2005) 63.

[17] N. Brambilla et al., Phys. Rev. D 78 (2008) 014017.

[18] M. Laine et al., JHEP 0703 (2007) 054.

[19] Y. Burnier, M. Laine and M. Vepsalainen, JHEP 0801 (2008) 043.

[20] O. Philipsen, arXiv:0810.4685.

[21] A. Beraudo, J.-P. Blaizot and C. Ratti, Nucl. Phys. A 806 (2008) 312; arXiv:0812.1130.

[22] M. Laine, arXiv:1108.5965.

[23] C.-Y. Wong, hep-ph/0408020

[24] W. Alberico et al., hep-ph/0507084;
C.-Y. Wong, hep-ph/0509088.

[25] D. Kharzeev, L. D. McLerran and H. Satz, Phys. Lett. B 356 (1995) 349.

[26] T. Umeda et al., Int. J. Mod. Phys. A16 (2001) 2215;
M. Asakawa and T. Hatsuda, Phys. Rev. Lett. 92 (2004);
S. Datta et al., Phys. Rev. D 69 (2004) 094507;
H. Iida et al., hep-lat/0509129

[27] H.-T. Ding et al., arXiv:1204.4945

[28] R. Morrin et al., hep-lat/0509115.

[29] O. Kaczmarek, arXiv:1208.4075

International Workshop on Discovery Physics at the LHC (Kruger2012) IOP Publishing
Journal of Physics: Conference Series **455** (2013) 012046 doi:10.1088/1742-6596/455/1/012046

Collisional and Radiative Energy Loss of Heavy Quarks

J. Aichelin [1]**, P.B. Gossiaux, T. Gousset**

SUBATECH, Laboratoire de Physique Subatomique et des Technologies Associées, Université de Nantes - IN2P3/CNRS - Ecole des Mines de Nantes
4 rue Alfred Kastler, F-44072 Nantes, Cedex 03, France

E-mail: aichelin@subatech.in2p3.fr

Abstract. Heavy quarks are one of the key tools to study the properties of the expanding plasma created in ultrarelativistic heavy ion collisions. In this context we develop a framework to study the $Qq \rightarrow Qqg$ collsions in scalar QCD and provide the results for different key quantities. We investigate the radiative energy loss and compare the results to the the collisional energy loss. We conclude that radiative and collisional energy loss together allow for a description of the recently published ALICE data.

1. Introduction

The study of the properties of a plasma of quarks and gluons (QGP) which existed some milliseconds after the Big Bang and which can very probably be produced in the collisions of ultrarelativistic heavy ions is the main objective of the presently ongoing experiments at RHIC (Brookhaven/United States) and CERN (Switzerland & France) in which a couple of thousand physicists are involved. The challenge is to find this state of matter and to identify its properties although its lifetime is extremely short (10-23 s) and the spatial extension is of the order of 10-14 m. Then a transition from partons to the hadrons occurs and only the hadrons can be finally observed in the detectors. At the phase transition the measured multiplicity of light hadrons is well described by statistical models, but presently it is not obvious if this statistical equilibrium is achieved already when the QGP is created or only shortly before the hadonization takes place or even by means of the hadronization process itself. The answer to this question is essential for understanding the plasma properties.

Quantum Chromo Dynamics (QCD), the underlying theory of the strong interaction, predicts that two and three particle correlations, the precursors of hadrons, are already formed above the critical temperature Tc in the QGP phase and the few calculations of the viscosity in a QGP indicate that only close to the phase transition the viscosity is sufficiently small to bring or maintain the expanding QGP to local equilibrium. On the other hand ideal hydrodynamical models which are based on the assumption that the local equilibrium is established already at the beginning of the QGP expansion and that the phase transition is sudden at a critical energy density (Cooper Frye assumption), can predict many observables. Presently it is not clear whether this is a true verification of the assumption of a local equilibrium during the expansion

[1] invited speaker

or whether the rather unknown initial conditions of the plasma expansion is at the origin of this agreement. The fact that the centrality dependence of several observables cannot be described and, even in central collisions, a couple of observables cannot be correctly predicted, may also be considered as a hint for non-equilibrium effects.

In order to asses directly the properties of the plasma during the expansion of the plasma it is necessary to concentrate on those probes which do not come to an equilibrium with the plasma. Heavy quarks and jets are such probes. Especially heavy quarks (charm and beauty) are well suited for that - due to their large mass they are produced dominantly by hard processes during the early stage of the reaction when the QGP is formed. Their initial creation in a hard process is predicted to be very similar as in proton-proton collisions and is therefore directly accessible by experiment. While traversing the plasma the heavy quarks do not come to equilibrium with the environment since their interaction with the partons of the plasma is not strong enough. However, due to the exchange of the momentum in the partonic interactions, the final momentum distribution of heavy quarks (and, as a consequence of the observed heavy mesons) contains information on the number of scattering partners as well as on the momentum transfer during the QGP phase. Both depend on whether the dynamical evolution of the expanding system occurs in equilibrium or not. The heavy mesons are therefore a decisive tool to study the nature of the expansion and the properties of the plasma.

Heavy quarks are produced in hard binary initial collisions between the incoming protons. Their production cross sections are known from pp collisions and can as well be calculated in pQCD calculations. Therefore the initial transverse momentum distribution of the heavy quarks is known. Comparing this distribution with that measured in heavy ion collisions allow to define $R_{AA} = (d\sigma_{AA}/dp_t^2) \;/\; (N_c\, d\sigma_{pp}/dp_t^2)$, where N_c is the number of the initial binary collisions between projectile and target. The deviation of R_{AA} from one measures the interaction of the heavy quark with the plasma because the hadron cross sections of heavy mesons are small. The heavy quark does not come to thermal equilibrium with the QGP therefore R_{AA} contains the information on the interaction of the heavy quark while it traverses the plasma.

Unfortunately the experimental results depend not only on the elementary interaction but also on the description of the expansion of the QGP [1]. Therefore the ultimate aim is to control the expansion by results on the light meson sector. This has not been achieved yet for the LHC and therefore it is difficult to asses the influence of the expansion on the observables. We use here the approach from Kolb and Heinz which has reasonably well described the midrapidity light mesons at RHIC [2]. We adjust only the charged particle multiplicity to the value measured at LHC.

The R_{AA} of 0.2 values observed for large p_t heavy mesons are much smaller than originally expected. Early theoretical approaches based on perturbative QCD (pQCD) calculation gave much larger values and it has been doubted, whether pQCD is the right tool to describe this interaction. This early calculation, however, used ad hoc assumptions on the coupling constant α_s and the infrared regulator μ. With a standard choice μ and α_s an artificial K factor, an overall multiplication factor of the elastic cross section of around 10 [3, 4] had to be introduced to match the experimental data.

A while ago we advanced an approach for the collisional energy loss of heavy quarks in the QGP [5, 6, 7]. Our matrix element has the same form as PQCD calculations in first order Born approximation with a infrared regulator μ. The infrared regulator is, however, not taken as constant but proportional to the Debye mass with a constant of proportionality which is chosen in that way that we obtain the same energy loss as calculations which employ a hard thermal loop for scattering with a low momentum transfer. In addition we employ a coupling constant determined by the sum rule advanced by Dokshitzer and later used by Peshier. Both these improvements increase the cross section especially for small momentum transfers and reduced therefore the necessary K factor to 2. For details we refer to [5, 6, 7]

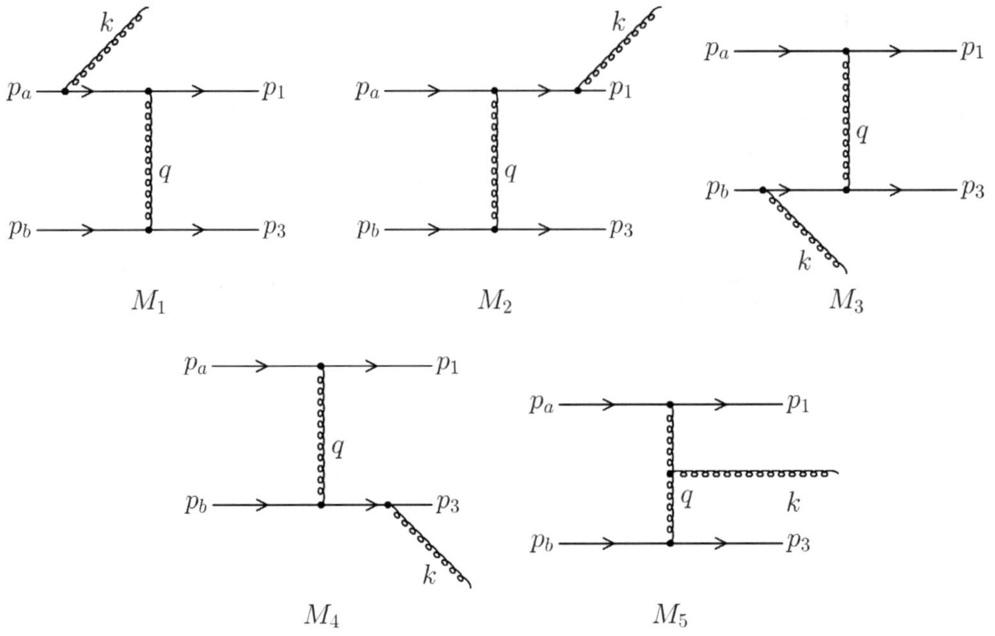

Figure 1. (Color online) The five matrix elements which contribute to the gluon bremsstrahlung.

In addition to the energy loss by elastic collisions the particle loose energy as well by radiation. The is the dominant form of energy loss for light mass quarks but plays as well an important role for heavy quarks. It is the purpose of these proceedings to study these inelastic collisions in detail and to investigate their influence for the energy loss in heavy ion collisions [8, 9].

2. adiative Energy Loss

Extending our collisional energy loss approach to radiative energy loss we have to consider the following 5 processes whose matrix elements are displayed in fig.1. The commutation relation

$$T^bT^a = T^aT^b - if_{abc}T^c \tag{1}$$

allows us to regroup the 5 matrix elements into 3 combinations, each of them being independently gauge invariant:

$$\begin{aligned} iM^{QED}_{h.q.} &= C_a i(M_1 + M_2) \\ iM^{QED}_{l.q.} &= C'_a i(M_3 + M_4) \\ iM^{QCD} &= C_c i(M_1 + M_3 + M_5). \end{aligned} \tag{2}$$

h.q. (l.q.) mark the emission of the gluon from the heavy (light quark) line. C_a, C'_a and C_c are the color algebra matrix elements. The matrix elements labeled as QED are the bremsstrahlung diagrams already observed in Quantum Electrodynamics (QED), whereas that labeled QCD is the genuine diagram of Quantum Chromodynamics (QCD). The QCD diagram is the main objet of interest here because it dominates the energy loss of heavy quarks.

We evaluate the matrix elements in scalar QCD (see ref.[13]). They are given by

$$iM_1^{SQCD} = C_A(ig)^3\frac{(p_b+p_3)^\mu}{(p_3-p_b)^2}D_{\mu\nu}[p_3-p_b](\frac{(p_a+p_1-k)^\nu(2p_a-k)\epsilon}{(p_a-k)^2-m^2} - \epsilon^\nu)$$

$$\begin{aligned} iM_5^{SQCD} &= C_c(ig)^3 D^{\mu\mu'}[p_3-p_b]D^{\nu\nu'}[p_1-p_a]\left[g_{\mu'\nu'}(p_a-p_1+p_3-p_b)_\sigma + \right. \\ & \left. g_{\nu'\sigma}(p_1-p_a-k)_{\mu'} + g_{\sigma\mu'}(p_b-p_3+k)_{\nu'}\right]\epsilon^\sigma \\ & \cdot \frac{(p_3+p_b)^\mu(p_a+p_1)^\nu}{(p_3-p_b)^2(p_1-p_a)^2} \end{aligned} \tag{3}$$

M_3 is obtained by replacing $p_a \to p_b$ and $p_1 \to p_3$ in M_1. To discuss the physics we adopt the following light cone vectors

$$\begin{aligned} p_a &= \{\sqrt{s-m^2}, \frac{m^2}{\sqrt{s-m^2}}, 0, 0\} \\ p_b &= \{0, \sqrt{s-m^2}, 0, 0\} \\ k &= \{x\sqrt{s-m^2}, 0, \vec{k_t}\} \\ p_1 &= p_a + q - k = \{p_a^+(1-x) - \frac{q_t^2}{p_b^-}, \frac{(\vec{k_t}-\vec{q_t})^2+m^2}{(1-x)p_a^+}, \vec{q_t}-\vec{k_t}\} \\ p_3 &= p_b - q = \{\frac{q_t^2}{p_b^-}, p_b^- - \frac{(1-x)k_t^2 - x(\vec{k_t}-\vec{q_t})^2 + m^2x^2}{p_a^+(1-x)x}, -\vec{q_t}\} \end{aligned} \tag{4}$$

Here we study only the dominant QCD terms.

Using light cone gauge and integrating over seven phase space variables the cross section can be written as [9]

$$\frac{d\sigma^{Qq\to Qgq}}{dxd^2k_td^2q_t} = \frac{1}{2(s-m^2)}|M|^2 \frac{1}{4(2\pi)^5\sqrt{\Delta}}\Theta(\Delta). \tag{5}$$

The calculation of this cross section is straight forward but not transparent. $|M|^2$ diverges for small momentum transfer q_t^2. Therefore we have to introduce a multiplicative infrared regulator

$$G(q_t^2, \mu) = \frac{q_t^4}{(q_t^2+\mu^2)^2}. \tag{6}$$

The regulated cross section

$$\frac{d\tilde{\sigma}^{Qq\to Qgq}}{dxd^2k_td^2q_t} = G(q_t^2,\mu)\cdot \frac{d\sigma^{Qq\to Qgq}}{dxd^2k_td^2q_t} \tag{7}$$

can only be solved numerically and we employ a Monte Carlo procedure. This solution we call model I. In the limit

$$s, xs >> |q_t^2|, \vec{k}_t^2 \tag{8}$$

$|M|^2$ simplifies and we obtain [9]

$$|M|^2 = 16\pi^3 x(1-x)|M_{el}|^2 \cdot P_g(m, \vec{q_t}, \vec{k_t}, x) \tag{9}$$

with the regularized elastic matrix element

$$|M_{el}|^2 = \frac{2}{9}\frac{g^2 4s^2}{(q_t^2+\mu^2)^2} \tag{10}$$

and

$$P_g(m, \vec{q_t}, \vec{k_t}, x) = \frac{C_A\alpha_S(1-x)}{x\pi^2}\left(\frac{\vec{k_t}}{\vec{k}_t^2 + x^2m^2} - \frac{\vec{k_t}-\vec{q_t}}{(\vec{k_t}-\vec{q_t})^2+x^2m^2}\right)^2 \tag{11}$$

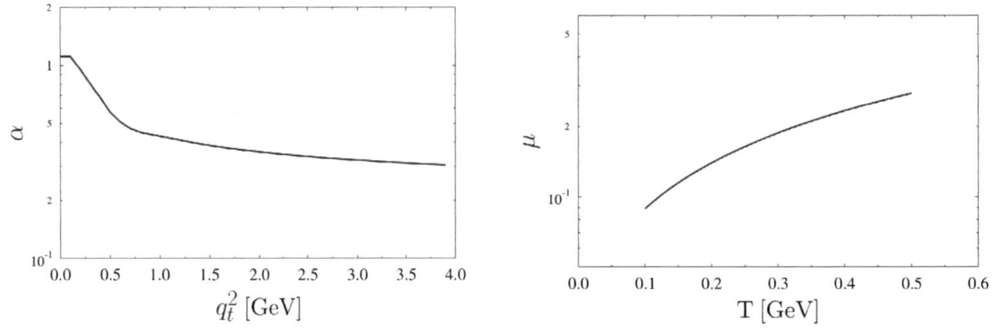

Figure 2. (Color online)Left: the running coupling constant as a function of the momentum transfer q_t^2. Right: The infrared regulator $\mu(T)$ as a function of the temperature.

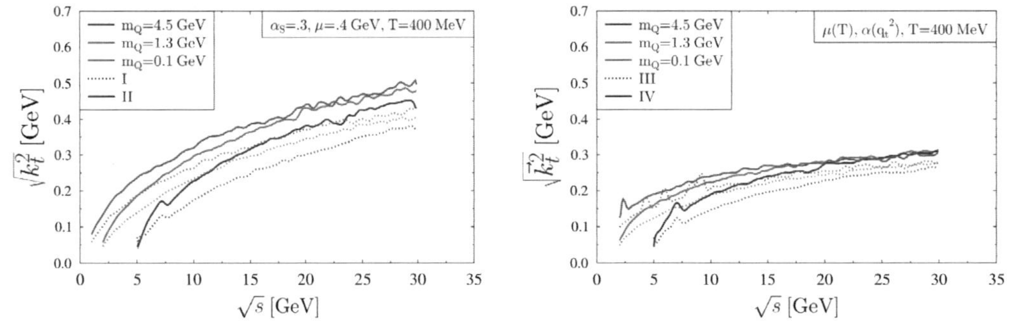

Figure 3. (Color online) Average transverse momentum of the emitted gluon, $\sqrt{< k_t^2 >}$, as a function of $\sqrt{s}$ for different masses of the heavy quark. On the left hand side we display the results of the models I and II, on the right hand that of the models III and IV.

is the distribution function of the produced gluons. In the limit of vanishing heavy quark masses this model corresponds to the Gunion Bertsch approach ref. [10]. This model we call model II.

For the expanding plasma we go one step further and adopt the results of refs. [5, 6, 7] by introducing in eq. 6 a running coupling constant at the vertices of the exchanged gluon and a temperature depending infrared regulator. Both are displayed in fig. 2. For the gluon emission we keep a constant coupling constant $\alpha_S = 0.3$. These features are introduced by multiplying the square of the matrix elements of the models I and II by

$$\frac{\alpha_S^2(q_t^2)(q_t^2 + \mu^2)^2}{\alpha_S^2 \cdot (q_t^2 + \mu(T)^2)^2}. \tag{12}$$

The modified full solution we call model III and the high energy approximation model IV. These 4 models we use now to study some key quantities which characterize the reaction. Fig.3 displays the average transverse momentum $\sqrt{< k_t^2 >}$ of the emitted gluon as a function of $\sqrt{s}$. The left hand side shows this observable for the models I and II, the right hand side that for the models III and IV. We display the results for three different heavy quark masses, m= .1, 1.3 and 4.5 GeV. We see an increase of $\sqrt{< k_t^2 >}$ with increasing $\sqrt{s}$ because more energetic collisions

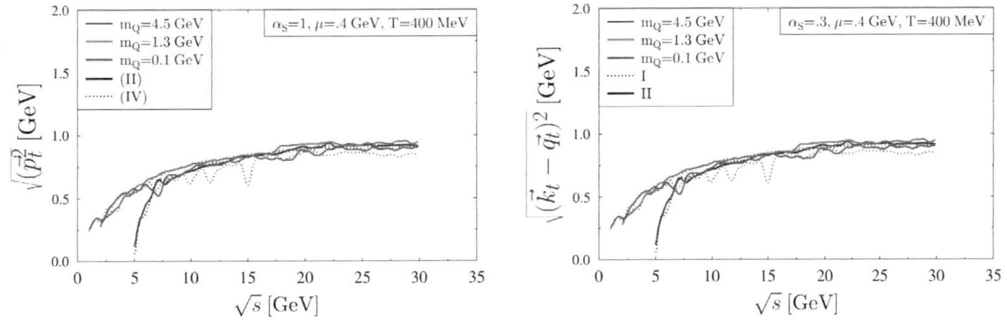

Figure . (Color online) Average transverse momentum transfer to the heavy quark $\sqrt{< p_t^2 >} = \sqrt{< (\vec{k}_t - \vec{q}_t)^2 >}$ as a function of $\sqrt{s}$ for different masses of the heavy quark. On the left hand side we display the results of the models I and II, on the right hand that of the models III and IV.

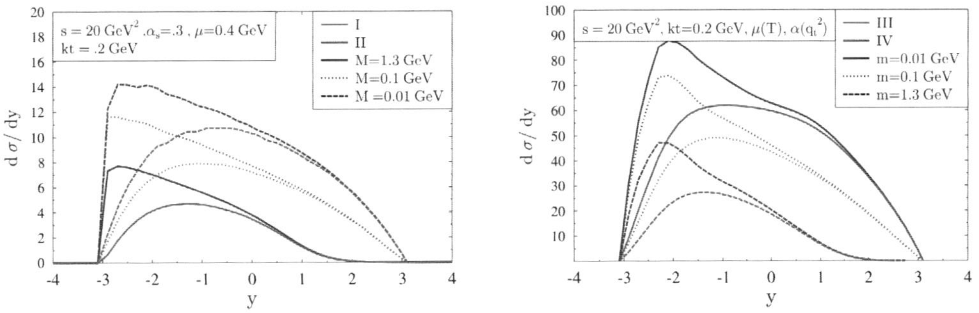

Figure . (Color online) Rapidity dependence of the cross section for different masses of the heavy quark. On the left hand side we display the results for the models I and II, on the right hand that for the models III and IV.

yield a larger momentum transfer between projectile and target quark. Clearly the running coupling constant favors low momentum transfers, q_t^2, which yields consequently lower values of $\sqrt{< k_t^2 >}$. Also the increase of $\sqrt{< k_t^2 >}$ with $\sqrt{s}$ is more moderate. The full solution yields smaller $\sqrt{< k_t^2 >}$ values than the high energy approximation. It is remarkable that despite of the dead cone effect [11] the $\sqrt{< k_t^2 >}$ of heavy quarks is smaller than that of light quarks.

Another quantity of interest is the final transverse momentum given to the heavy quark. This quantity is displayed in fig. 4 for different masses of the heavy quark. The left hand side shows this observable for the models I and II, the right hand side for the models III and IV. $\sqrt{< (\vec{k}_t - \vec{q}_t)^2 >}$ is almost independent of the heavy quark mass and the high energy approximation reproduces the full solution very well. As already observed for $\sqrt{< k_t^2 >}$ we find as well for $\sqrt{< (\vec{k}_t - \vec{q}_t)^2 >}$ that a running coupling constant suppresses a large momentum transfer and hence the final transverse momenta of the outgoing particles is smaller. This is

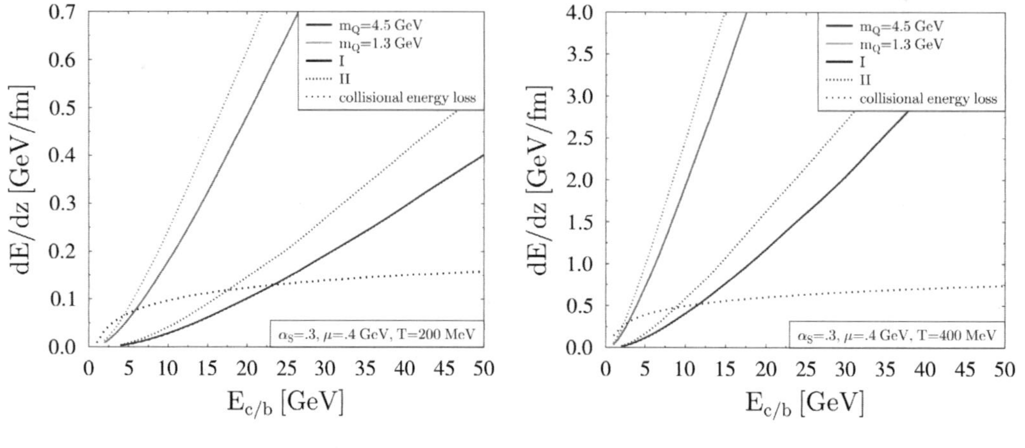

Figure . (Color online) Radiative and collisional energy loss as a function of the energy of the heavy quark for the models I and II, left for a temperature of T=200 MeV, right for T=400 MeV.

clearly seen by comparing the upper and the lower red lines of the right figure which compare directly $\sqrt{<(\vec{k}_t-\vec{q}_t)^2>}$ of the 4 models. In contradistinction to $\sqrt{<k_t^2>}$, $\sqrt{<(\vec{k}_t-\vec{q}_t)^2>}$ is -besides threshold effects - for a running coupling constant (almost) independent of $\sqrt{s}$ and also independent of the mass of the heavy quark.

The rapidity dependence of the cross section

$$\frac{d\sigma^{Qq\to Qgq}}{dyd^2k_td^2q_t} \propto x\frac{d\sigma^{Qq\to Qgq}}{dxd^2k_td^2q_t} \tag{13}$$

is shown in fig. 5, on the left hand side for the models I and II, on the right hand for the models III and IV. We see how the finite heavy masses shift the center of the distribution from $y=0$ zu negative y values. We observe as well that the high energy approximation is valid for positive y values only whereas for negative y values, corresponding to $x<\frac{\sqrt{k_t^2}}{\sqrt{s}}$, the conditions for the approximations are not anymore satisfied and strong deviations appear. The running coupling constant and the introduction of $\mu(T)$ increase the cross section by a factor up to seven, especially for small x values.

The collisional and radiative energy loss due to the scattering of the heavy quark with light quarks are compared in fig.6 for the models I and II. The collisional energy loss in the scattering of a heavy quark with a light quark of the plasma is calculated as [12]

$$\frac{dE}{dz} = \frac{4\pi}{3}a_S^2T^2\frac{n_f}{6}log(\frac{ET}{\mu^2}) \tag{14}$$

with a coupling constant of $\alpha_s=.3$, an infrared cut off of μ=.4 GeV and a temperature of T=200 MeV (left) and 400 MeV (right). E is the incident energy of the heavy quark and the number of flavours, n_f, is taken as three. The radiative energy loss is obtained by the approximate formula

$$\frac{dE}{dz} = \int\int\frac{d^3k}{(2\pi)^3}\rho(k)Ex\frac{d\sigma_{rad}}{dx}(\sqrt{s})dx \tag{15}$$

assuming that $\sqrt{s}=\sqrt{m^2+6ET}$. We see that for low energies of the heavy quark the collisional energy loss is larger than the radiative energy loss but already for moderate energies the radiative

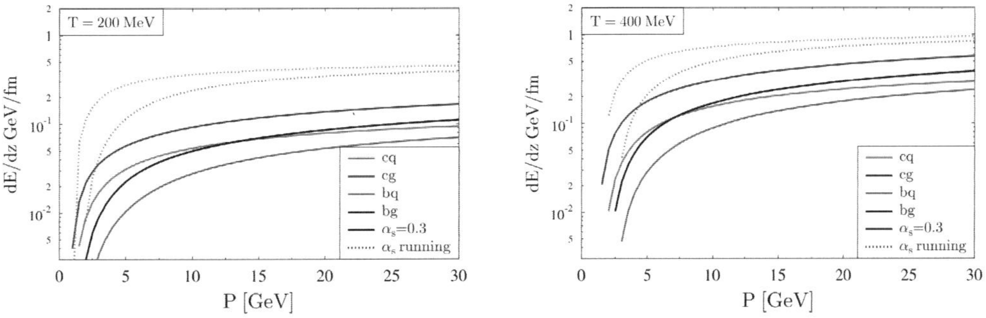

Figure . (Color online) The collisional energy loss for fixed α_S and μ as compared to running α_S and $\mu(T)$ for the temperatures T=200MeV (left) and T=400 MeV (right).

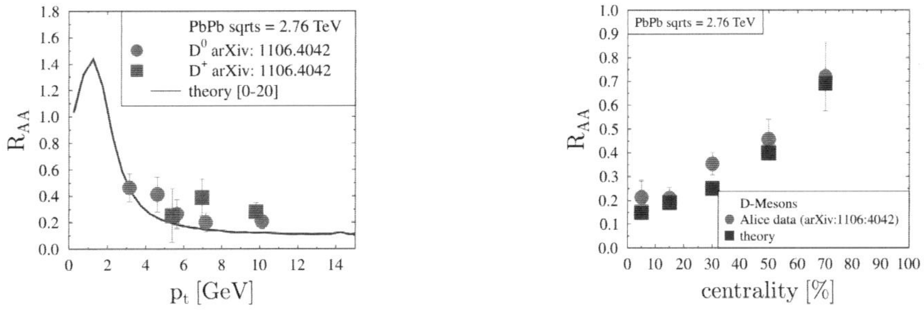

Figure . (Color online) Left: The calculated $R_{AA}(p_t)$ spectrum in comparison with the experimental data. Right: The p_t integrated $R_{AA}(p_t)$ spectrum for different centralities.

energy loss dominates. Finally we compare in fig.7 the collisional energy loss for fixed α_S and μ with running α_S and $\mu(T)$ for the temperatures T=200MeV (left) and T=400 MeV (right). cq, cg, bq and bg refer to charm-light quark, charm-gluon, bottom-light quark and bottom-gluon collisions, respectively. As for the radiative cross section we see also for the elastic cross section a strong enhancement and consequently a larger energy loss when α_S is running and μ becomes a function of the temperature $\mu(T)$.

If we include now both, the radiative energy loss and the collisional energy loss, calculated with running coupling constant and temperature dependent $\mu(T)$, in our calculation of the heavy quark energy loss in an expanding plasma in which the plasma is modeled by the hydrodynamical calculations of Kolb and Heinz [2] we describe [14] the recently measured $R_{AA}(p_t)$ values (the pt spectrum measured in heavy ion collisions divided by that in pp collisions properly normalized by the total number of elementary collisions in heavy ion reactions) of the Alice collaboration almost quantitatively. This is shown in fig. 8, where on the left hand side $R_{AA}(p_t)$ is compared to the Alice data, whereas on the right hand side the centrality dependence of the p_t integrated spectrum is displayed.

In conclusion, we have studied how radiative collisions between a heavy and a light quark change the momentum of the heavy quark. We have shown the rapidity spectra of the emitted gluons and their average transverse momentum as well as the average transverse momentum transfer to the heavy quark. In these studies the full SQMD matrix element has been employed and compared to an approximate high energy formula which reduces to the Gunion Bertsch approach for vanishing heavy quark quark masses. We have demonstrated how an effective coupling constant and a temperature dependent infrared regulator, which we have developed in the framework of elastic collisions between heavy and light quarks, can be introduced in the radiation cross section and have studied how this changes the cross section. Finally we have shown that with these ingredients the measured R_{AA} values can be reproduced when the model is employed in an expanding plasma scenario.

[1] P. B. Gossiaux, S. Vogel, H. van Hees, J. Aichelin, R. Rapp, M. He and M. Bluhm, arXiv:1102.1114 [hep-ph].
[2] P. F. Kolb and U. W. Heinz, arXiv:nucl-th/0305084. published in Qurak Gluon Plasma, World Scientific
[3] G. D. Moore and D. Teaney, Phys. Rev. C **71** 064904 (2005).
[4] B. Svetitsky, Phys. Rev. D **37** (1988) 2484.
[5] P. B. Gossiaux, J. Aichelin Phys. Rev. **C78**, 014904 (2008), arXiv:0802.2525 [hep-ph]
[6] P. B. Gossiaux, R. Bierkandt and J. Aichelin arXiv:0901.0946 [hep-ph]
[7] P. B. Gossiaux and J. Aichelin arXiv:0901.2462 [nucl-th]
[8] P. B. Gossiaux, J. Aichelin, T. Gousset and V. Guiho, J. Phys. G **37**, 094019 (2010) [arXiv:1001.4166 [hep-ph]].
[9] J. Aichelin, P.B. Gossiaux and T. Gousset, to be published
[10] J. F. Gunion and G. Bertsch, Phys. Rev. D **25** (1982) 746.
[11] . L. Dokshitzer and D. E. Kharzeev, Phys. Lett. B **519** (2001) 199 [arXiv:hep-ph/0106202].
[12] S. Peigne and A. V. Smilga, Phys. Usp. **52** (2009) 659 [Usp. Fiz. Nauk **179** (2009) 697] [arXiv:0810.5702 [hep-ph]].
[13] E. Meggiolaro, Phys. Rev. D **53** (1996) 3835 [arXiv:hep-th/9506043].
[14] J. Aichelin, P. B. Gossiaux and T. Gousset, Acta Phys. Polon. B **43** (2012) 655 [arXiv:1201.4192 [nucl-th]].

Journal of Physics: Conference Series **455** (2013) 012047 doi:10.1088/1742-6596/455/1/012047

Higgs to four leptons searches at the LHC : constraints on Abelian Hidden sector models

Mathieu Aurousseau
University of Johannesburg, Johannesburg, South Africa
E-mail: mathieu.aurousseau@cern.ch

Abstract. Recent results on Higgs searches at the LHC by the ATLAS and CMS collaborations claim the observation of a new boson with properties compatible with the Standard Model (SM) Higgs boson, at a mass of about 126 GeV. The production rate of the new particle seems compatible with the SM prediction. In some models beyond the SM, an Abelian Hidden sector is coupled to the SM, and the resulting exotic Higgs and new gauge (Z') fields are allowed to mix with the corresponding SM fields. This admits new processes such as the decay of the lightest Higgs boson (after mixing) to a pair of new gauge bosons, and the decay of the new gauge bosons to pairs of leptons : $H \to Z'Z' \to 4\ell$. This modifies the rate of 4-lepton events sought for in the usual SM $H \to ZZ^{(*)} \to 4\ell$ search. Assuming the observed particle is a Higgs boson of a mass of 126 GeV, we investigate the consequences of the addition of this new process to the search for a SM Higgs boson decaying to 4 leptons, and the resulting constraints on the Abelian Hidden sector model parameters. The constraints are drawn in terms of Higgs mixing angle and new gauge boson mass.

1. Introduction

The observation of a new particle consistent with the Standard Model (SM) Higgs (H) boson [1, 2] opens a new era for the LHC experiments, which are switching from the discovery mode to measuring the properties of the new particle. Several quantities are of particular interest: spin, CP state, mass. A precise measurement of these quantities will require refined analyses. Another quantity, which is in fact the first to be measured experimentally because it is the one leading to the discovery (or observation) claim, is the production rate. This number is predicted in the SM, as a function of the H mass M_H, and experiments usually quote the signal strength $\mu = \sigma_{\rm obs}/\sigma_{\rm SM}$, the ratio between the observed rate and the SM predicted rate. Current experimental results show that μ is consistent with 1, although still with large uncertainties, which shall shrink as experiments analyze more data and refine their analyses.

A particularly interesting scenario is if μ becomes significantly different from 1, because it then requires extensions of the SM. An example of such extensions is the existence of a Hidden sector connected to the SM, and in particular a case where the Higgs domains of the two sectors are mixing. This is a general concept, that can be applied in several models [3, 4, 5, 6], and it has beed studied in general in [7, 8]. Applications to a specific model will require measurements of precise numbers such as the Higgs couplings, observation of new gauge bosons, etc. In this article, we focus on a particular model, the Hidden Abelian Higgs Model (HAHM) proposed by Wells *et al.* [3, 4]. It is a particularly interesting model since it allows processes such as $H \to Z'Z' \to 4\ell$ and $H \to ZZ' \to 4\ell$. These processes can be sought for directly, using minimal changes in the standard $H \to ZZ^{(*)} \to 4\ell$ searches carried on by ATLAS or CMS. Another interesting aspect is that for a light Z' boson (around 50 or 60 GeV) and a light Higgs boson (around 125 GeV), a $H \to Z'Z' \to 4\ell$ event would pass the usual cuts applied for the $H \to ZZ^{(*)} \to 4\ell$ analyses, so that we may have direct access to the "hidden" part of the Higgs width. This provides two interesting observables for the new model :

(*i*) a modified rate of 4ℓ-events compared to the SM prediction; and

(*ii*) a narrow resonance in the "secondary" dilepton invariant mass distribution (the one containing the off-shell Z).

It has to be noted that for lighter Z' (typically below 50 GeV), $H \to Z'Z'$ events would be mostly cut out by analysis cuts on the invariant mass, but $H \to ZZ'$ events may survive. This of course depends strongly on the analysis cuts that are applied and may be different between the various experiments.

The present study focuses on the first item (*i*) and investigates how the rate of observed 4ℓ-events is changed assuming Wells *et al.* model (HAHM) instead of the SM.

2. Hidden Abelian Higgs Model (HAHM)

2.1. Model parameters

The model is presented in [3, 4], where its parameters are defined. Three of the parameters can be considered as free: the Higgs boson mass M_H, the new gauge boson mass $M_{Z'}$, and the mixing angle between the SM and Hidden Higgs sectors θ_h. For convenience, we will use the sine-square of the mixing angle, $\sin^2 \theta_h$, which we will denote s_h^2, and similarly, the cosine-square of this angle will be denoted c_h^2.

2.2. Couplings

2.2.1. Higgs couplings In the model, the Higgs boson is coupled to SM particles in a similar way as in the SM, with the following features :

(*i*) decays involving new gauge bosons ($H \to ZZ'$, $H \to Z'Z'$) are allowed;

(*ii*) $H \to ZZ^{(*)}$ couplings are non-trivially modified with respect to SM; and

(*iii*) all other couplings are suppressed by a factor c_h^2, in particular the effective coupling to gluons for the Higgs boson production by gluon fusion.

To the first order, the $H \to ZZ'$ process is suppressed compared to the other decays. In the following, we will consider only $H \to ZZ^{(*)}$ and $H \to Z'Z'$ decays.

2.2.2. Coupling to SM fermions The gauge bosons Z and Z' are coupled to SM fermions. Couplings depend on many model parameters. To the first order, the $Z \to \ell^+\ell^-$ coupling in the HAHM is the same as in the SM, and we will consider that the new gauge boson Z' decays only to leptons with a 100% branching fraction.

3. Experimental results from ATLAS and CMS in the $H \to ZZ^{(*)} \to 4\ell$ search

3.1. Higgs boson mass and signal strength

In this study, we assume that the new boson observed by the ATLAS and CMS experiments [1, 2] is a Higgs boson of mass 126 GeV. The experimental results on the signal strength μ (ratio of the observed cross-section to the SM expectation) are shown in Table 1 with the corresponding uncertainty σ_μ. These numbers are in the 4ℓ channel only.

The approach of this study is to assess how the HAHM parameters are constrained by these experimental results, if we assume that $H \to Z'Z'$ events are allowed and may be observed.

	μ	σ_μ
ATLAS	1.4	± 0.6
CMS	0.7	$^{+0.4}_{-0.3}$

Table 1. Experimental results from ATLAS [1] and CMS [2] in the 4ℓ channel.

3.2. Experimental analysis cuts

The ATLAS and CMS experiments have different analysis strategies, but both analyses are cut-based. The kinematics of the leptons from $H \to Z'Z' \to 4\ell$ events is similar to that of the $H \to ZZ^{(*)} \to 4\ell$, therefore we will assume that all efficiencies and acceptance are unchanged. Both analyses are based on selecting events with 4 leptons and forming lepton pairs. One pair has a mass m_{12} close to the SM Z boson mass, and the other has a mass m_{34} on which the cut is relaxed to account for off-shell Z boson. The cuts on these quantities are:

- ATLAS: $m_{12} > 50$ GeV, $m_{34} > 17.5$ GeV for a 4-lepton mass of about 126 GeV;
- CMS: $m_{12} > 40$ GeV, $m_{34} > 12$ GeV.

In order to have an approach that is consistent with both experiments and avoids boundary effects, we will consider the following cases:

(i) if $20 \le M_{Z'} < 35$ GeV, the $H \to Z'Z'$ events cannot be detected by the experiments; and
(ii) if $55 \le M_{Z'} < 80$ GeV, the $H \to Z'Z'$ events can be detected by the experiments.

The cases of $M_{Z'} < 20$ GeV and $35 \le M_{Z'} < 55$ GeV are discarded because of the m_{34} and m_{12} cuts respectively. We also do not consider the case where the Z' boson would be heavier than the Z boson.

4. Constraints on HAHM parameters

We draw limits on hidden sector parameters by calculating the event rate predicted by the new model, compared to the SM, and by checking the compatibility with the experimental results.

In the HAHM, the hidden width is known as a function of the three free parameters: M_H, $M_{Z'}$ and s_h^2. The Higgs boson mass being fixed to 126 GeV, results will be presented in a $(M_{Z'}, s_h^2)$ plane.

4.1. Method

Our approach is to calculate, using only the theoretical (phenomenological) considerations, the expected production rate of $H \to 4\ell$ events, compared to the SM prediction. In other words, we can calculate, if we consider only gluon fusion (dominant production mode at the LHC) :

$$\frac{\sigma_{\rm HAHM}}{\sigma_{\rm SM}} = \frac{\sigma_{\rm HAHM}(gg \to H) \cdot {\rm BR}_{\rm HAHM}(H \to 4\ell)}{\sigma_{\rm SM}(gg \to H) \cdot {\rm BR}_{\rm SM}(H \to 4\ell)} \quad . \tag{1}$$

We compare the above quantity in a $(M_{Z'}, s_h^2)$ plane, with the measured value of the production rate divided by the SM prediction (μ). This approach is similar to the one used in [7], although applied to a specific model (HAHM). Of course, in the HAHM model, the Hidden decay width is linked to the Z' mass. The reason we chose to look specifically at the Z' mass is that in the usual approach [7, 8], it is considered that the decays to Hidden particles is not accessible (invisible), and therefore not sought for. We want to place ourselves in the case of Hidden particles (Z') decaying to SM fermions, therefore potentially adding signal to the SM search $H \to ZZ^{(*)} \to 4\ell$.

In Eq. 1, the denominator is given by the SM values, and depends only on the (SM) H mass. The term ${\rm BR}_{\rm SM}(H \to 4\ell)$ involves only $H \to ZZ^{(*)} \to 4\ell$ events, and therefore is a constant of the SM (since the Z mass is known). In the numerator, however, values come from the HAHM model and depend on the free parameters M_H, $M_{Z'}$ and s_h^2. The cross-section term depends on the Higgs mass and the mixing angle, and we assume the universal suppression :

$$\frac{\sigma_{\mathrm{HAHM}}(gg \to H)}{\sigma_{\mathrm{SM}}(gg \to H)} = c_h^2 = (1 - s_h^2) \quad . \tag{2}$$

The term $\mathrm{BR}_{\mathrm{HAHM}}(H \to 4\ell)$ can be decomposed into three components:

$$\mathrm{BR}_{\mathrm{HAHM}}(H \to 4\ell) = \mathrm{BR}_{\mathrm{HAHM}}(H \to ZZ^{(*)} \to 4\ell) + \mathrm{BR}_{\mathrm{HAHM}}(H \to Z'Z' \to 4\ell) + \mathrm{BR}_{\mathrm{HAHM}}(H \to ZZ' \to 4\ell) \tag{3}$$

As stated before, $H \to Z'Z' \to 4\ell$ events are considered as detected if $M_{Z'} > 55$ GeV, and as not detected if $M_{Z'} < 35$ GeV. Besides $H \to ZZ' \to 4\ell$ are not taken into account. Therefore, the formulae that are used to evaluate the rate of events in the HAHM model are :

$$\frac{\sigma_{\mathrm{HAHM}}}{\sigma_{\mathrm{SM}}} = \begin{cases} (1 - s_h^2) \cdot \frac{\mathrm{BR}_{\mathrm{HAHM}}(H \to ZZ^{(*)})}{\mathrm{BR}_{\mathrm{SM}} H \to ZZ^{(*)}} & \text{if} \quad M_{Z'} < 35 \text{ GeV} \\ (1 - s_h^2) \cdot \frac{\mathrm{BR}_{\mathrm{HAHM}}(H \to ZZ^{(*)}) \cdot [\mathrm{BR}_{\mathrm{SM}}(Z \to 2\ell)]^2 + \mathrm{BR}_{\mathrm{HAHM}}(H \to Z'Z')}{\mathrm{BR}_{\mathrm{SM}} H \to ZZ^{(*)} \cdot [\mathrm{BR}_{\mathrm{SM}}(Z \to 2\ell)]^2} & \text{if} \quad M_{Z'} > 55 \text{ GeV} \end{cases} , \tag{4}$$

where we have used that the branching fraction $Z' \to \ell^+\ell^-$ is 100%. An immediate consequence, is that in the first case ($M_{Z'} < 35$ GeV), the ratio is always less than 1, since the branching fractions in the HAHM model are lower than in the SM. This means that for a low-mass Z', the HAHM model predicts less events than the SM, given the experimental analysis cuts that are used. On the other hand, in the second case ($M_{Z'} > 55$ GeV), the ratio can be either larger or smaller than 1.

4.2. Results

Results are presented first for the low-mass Z' (20 GeV $< M_{Z'} <$ 35 GeV), and then for the larger-mass Z' (55 GeV $< M_{Z'} <$ 80 GeV). We show the distribution of $\sigma_{\mathrm{HAHM}}/\sigma_{\mathrm{SM}}$ in the $(M_{Z'}, s_h^2)$ plane, and we superimpose the contour corresponding to experimental results (ATLAS, CMS) μ, and 68% and 95% C.L.. The values used have already been introduced previously, in Table 1.

4.2.1. Low-mass Z' (20 GeV$< M_{Z'} <$ 35 GeV)

In the low-mass region, we assume that the exotic decays do not pass the experimental analysis cuts, and therefore Eq. 4 (top) is used. Figure 1 shows the distribution of $\sigma_{\mathrm{HAHM}}/\sigma_{\mathrm{SM}}$ in the $(M_{Z'}, s_h^2)$ plane, with the μ, and 68% and 95% C.L. contours.

Only the 95% C.L. contour (dot-dashed curve) is visible on the left plot (ATLAS). This is because the ATLAS result μ is larger than 1, while the HAHM predicts less events than the SM. The yellow region below the dot-dashed curve is compatible with the ATLAS result at 95% C.L.. On the other hand, CMS result μ is smaller than 1, therefore we see the μ contour (solid line), and we also see the 68% C.L. contour (dashed line).

4.2.2. High-mass Z' (55 GeV$< M_{Z'} <$ 80 GeV)

In the high-mass region, we assume that the exotic decays pass the experimental analysis cuts, and therefore Eq. 4 (bottom) is used. Making this assumption also implies that we assume that the kinematics of the leptons (which are the objects reconstructed experimentally) is the same as for standard $H \to ZZ^{(*)} \to 4\ell$ events. This assumption has been checked using events generated with MadGraph. Figure 2 shows the distribution of $\sigma_{\mathrm{HAHM}}/\sigma_{\mathrm{SM}}$ in the $(M_{Z'}, s_h^2)$ plane, with the μ, and 68% and 95% C.L. contours.

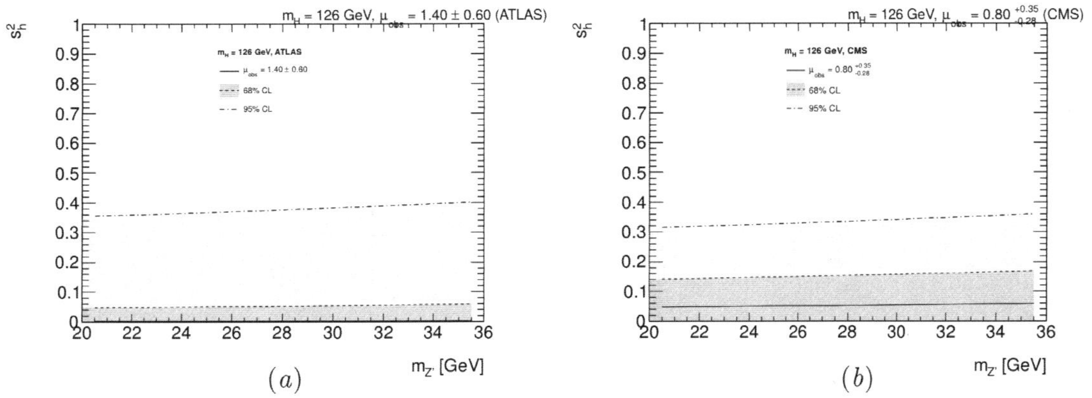

Figure 1. (a) $\mathrm{BR}_{\mathrm{SM}}(H \to ZZ)$ in the $(M_{Z'}, s_h^2)$ plane, (b) $\mathrm{BR}_{\mathrm{SM}}(H \to Z'Z')$ in the $(M_{Z'}, s_h^2)$ plane.

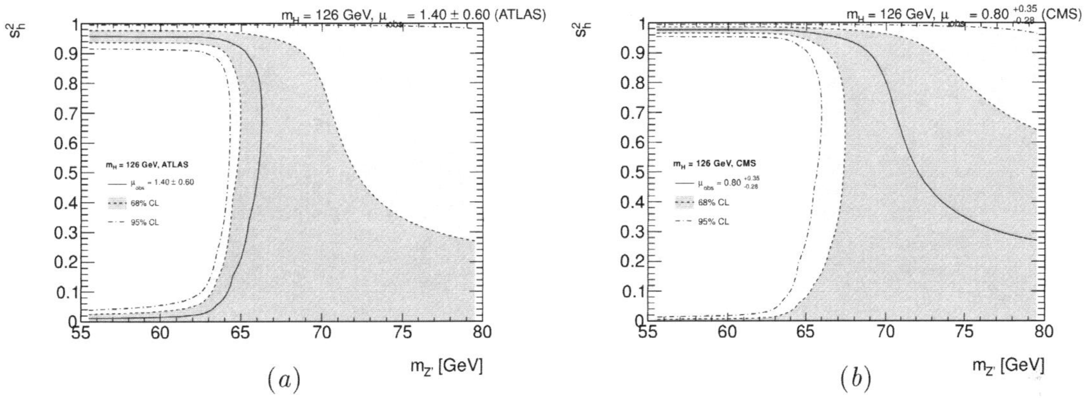

Figure 2. (a) $\mathrm{BR}_{\mathrm{SM}}(H \to ZZ)$ in the $(M_{Z'}, s_h^2)$ plane, (b) $\mathrm{BR}_{\mathrm{SM}}(H \to Z'Z')$ in the $(M_{Z'}, s_h^2)$ plane.

The same colors and styles are used for the contours: the solid line is the μ contour, the dashed line is the 68% C.L. contour, and the dot-dashed line the 95% C.L. contour. As already stated, $\sigma_{\mathrm{HAHM}}/\sigma_{\mathrm{SM}}$ in this case can be smaller or larger than 1, therefore more contours are visible. The green [yellow] region between the two dashed [dot-dashed] lines is the part of the parameter space that is compatible with the experimental results at 68% C.L. [95% C.L.]. We see that in both cases, a large region of the parameter space is excluded (not compatible at 95% C.L.), below $M_{Z'} \sim 63$ GeV, and between $s_h^2 \in [0.02, 0.94]$ approximately. Indeed, for $M_{Z'} < M_H/2$, the $H \to Z'Z'$ decay with two on-shell bosons is kinematically allowed, and will significantly increase the number of expected 4ℓ-events. It should be reminded here that the branching fraction of $Z' \to 2\ell$ has been taken as 100%, which overestimates the number of expected events with the HAHM model. Lower branching fraction would lead to a larger region compatible at 68 or 95% C.L. .

For $M_{Z'} > M_H/2$, since both $H \to ZZ^{(*)}$ and $H \to Z'Z'$ are suppressed ($H \to Z'Z'$ is suppressed, and $H \to ZZ^{(*)}$ gets the SM value, so we are almost in the SM case), most of the parameter space is compatible at least at 95% C.L., with the experimental results.

Journal of Physics: Conference Series **455** (2013) 012047 doi:10.1088/1742-6596/455/1/012047

5. Conclusion

Assuming that the newly-observed particle is a Higgs boson of mass 126 GeV, we show that the measured event rate is compatible with a large part of the HAHM parameter space. Refined analyses from the ATLAS and CMS experiments, with larger statistics and improved systematic uncertainties treatment, are expected soon and will improve the limits set on the Higgs mixing angle and new gauge boson mass. A direct search for $H \to Z'Z' \to 4\ell$ events by the experiments would complement this study.

References

[1] The ATLAS Collaboration 2012 *Physics Letters B*
[2] The CMS Collaboration 2012 *Physics Letters B*
[3] Wells J 2008 *arXiv preprint arXiv:0803.1243*
[4] Gopalakrishna S, Jung S and Wells J 2008 *Physical Review D* **78** 055002
[5] Schabinger R and Wells J 2005 *Physical Review D* **72** 093007
[6] Wells J 2002 *arXiv preprint hep-ph/0205328*
[7] Bock S, Lafaye R, Plehn T, Rauch M, Zerwas D and Zerwas P 2010 *Physics Letters B* **694** 44–53
[8] Englert C, Plehn T, Rauch M, Zerwas D and Zerwas P 2012 *Physics Letters B*

The QGP dynamics in relativistic heavy-ion collisions

E. L. Bratkovskaya[1,2], **W. Cassing**[3], **V. P. Konchakovski**[3,4], **O. Linnyk**[3], **V. Ozvenchuk**[2], **V.D. Toneev**[5,2] **and V. Voronyuk**[5,2,4]

[1] Institute for Theoretical Physics, University of Frankfurt, Frankfurt, Germany
[2] Frankfurt Institute for Advanced Study, Frankfurt, Germany
[3] Institute for Theoretical Physics, University of Giessen, Giessen, Germany
[4] Bogolyubov Institute for Theoretical Physics, Kiev, Ukraine
[5] Joint Institute for Nuclear Research, Dubna, Russia

Abstract. The dynamics of partons and hadrons in relativistic nucleus-nucleus collisions is analyzed within the novel Parton-Hadron-String Dynamics (PHSD) transport approach, which is based on a dynamical quasiparticle model for the partonic phase (DQPM) including a dynamical hadronization scheme. The PHSD model reproduces a large variety of observables from SPS to LHC energies, e.g. the quark-number scaling of elliptic flow, transverse mass and rapidity spectra of charged hadrons, dilepton spectra, open and hidden charm production, collective flow coefficients etc., which are associated with the observation of a sQGP. The 'highlights' of the latest results on collective flow are presented and open questions/perspectives are discussed.

1. Introduction

The dynamics of the early universe in terms of the 'Big Bang' may be studied experimentally by ultrarelativistic nucleus-nucleus collisions at Relativistic-Heavy-Ion-Collider (RHIC) or Large-Hadron-Collider (LHC) energies in terms of 'tiny bangs' in the laboratory. The Power Spectrum extracted from the Cosmic Microwave Background Radiation has some analogy to the Fourier components of particles in the azimuthal angular distribution [1]. The discovery of large azimuthal anisotropic flow at RHIC has provided conclusive evidence for the creation of dense partonic matter in ultra-relativistic nucleus-nucleus collisions. With sufficiently strong parton interactions, the medium in the collision zone can be expected to achieve local equilibrium and exhibit approximately hydrodynamic flow [2, 3, 4]. The momentum anisotropy is generated due to pressure gradients of the initial "almond-shaped" collision zone produced in noncentral collisions [2, 3]. The azimuthal pressure gradient extinguishes itself soon after the start of the hydrodynamic evolution, so the final flow is only weakly sensitive to later stages of the fireball evolution. The pressure gradients have to be large enough to translate an early asymmetry in density of the initial state to a final-state momentum-space anisotropy. In these collisions a new state of strongly interacting matter is created, being characterized by a very low shear viscosity η to entropy density s ratio, η/s, close to a nearly perfect fluid [5, 6, 7]. Lattice QCD (lQCD) calculations [8, 9] indicate that a crossover region between hadron and quark-gluon matter should have been reached in these experiments.

An experimental manifestation of this collective flow is the anisotropic emission of charged particles in the plane transverse to the beam direction. This anisotropy is described by the different flow parameters defined as the proper Fourier coefficients v_n of the particle distributions in azimuthal angle ψ with respect to the reaction plane angle Ψ_{RP}. At the highest RHIC collision energy of $\sqrt{s_{NN}}$ = 200 GeV, differential elliptic flow measurements $v_2(p_T)$ have been reported for a broad range of centralities or number of participants N_{part}. For N_{part} estimates, the geometric fluctuations associated with the positions of the nucleons in the collision zone serve as the underlying origin of the initial eccentricity fluctuations. These data are found to be in accord with model calculations that an essentially locally equilibrated quark gluon plasma (QGP) has little or no viscosity [10, 11, 12]. Collective flow continues to play a central role in characterizing the transport properties of the strongly interacting matter produced in heavy-ion collisions at RHIC and even LHC and shed some light on the scale of initial state fluctuations.

The Beam-Energy-Scan (BES) program proposed at RHIC [13] covers the energy interval from $\sqrt{s_{NN}}$ = 200 GeV, where partonic degrees of freedom play a decisive role, down to the AGS energy of $\sqrt{s_{NN}} \approx 5$ GeV, where most experimental data may be described successfully in terms of hadronic degrees-of-freedom, only. Lowering the RHIC collision energy and studying the energy dependence of anisotropic flow allows to search for the possible onset of the transition to a phase with partonic degrees-of-freedom at an early stage of the collision as well as possibly to identify the location of the critical end-point that terminates the cross-over transition at small quark-chemical potential to a first order phase transition at higher quark-chemical potential [14, 15].

This contribution aims to summarize excitation functions for different harmonics of the charged particle anisotropy in the azimuthal angle at midrapidity in a wide transient energy range, *i.e.* from the AGS to the top RHIC energy. The first attempts to explain the preliminary STAR data with respect to the observed increase of the elliptic flow v_2 with the collision energy have failed since the traditional available models did not allow to clarify the role of the partonic phase [16]. In this contribution we investigate the energy behavior of different flow coefficients, their scaling properties and differential distributions (cf. Ref. [17, 18]). Our analysis of the STAR/PHENIX RHIC data – based on recent results of the BES program – will be performed within the Parton-Hadron-String Dynamics (PHSD) transport model [19] that includes explicit partonic degrees-of-freedom as well as a dynamical hadronization scheme for the transition from partonic to hadronic degrees-of-freedom and vice versa. For more detailed descriptions of PHSD and its ingredients we refer the reader to Refs. [20, 21, 22, 23].

2. Results for collective flows

We directly continue with the results from PHSD in comparison with other approaches and the available experimental data.

2.1. Elliptic flow

The largest component, known as elliptic flow v_2, is one of the early observations at RHIC [24]. The elliptic flow coefficient is a widely used quantity characterizing the azimuthal anisotropy of emitted particles,

$$v_2 =< cos(2\psi - 2\Psi) >=< \frac{p_x^2 - p_y^2}{p_x^2 + p_y^2} > \ , \tag{1}$$

where Ψ_{RP} is the azimuth of the reaction plane, p_x and p_y are the x and y component of the particle momenta and the brackets denote averaging over particles and events. This coefficient can be considered as a function of centrality, pseudorapidity η and/or transverse momentum p_T. We note that the reaction plane in PHSD is given by the $(x - z)$ plane with the z-axis in beam direction. The reaction plane is defined as a plane containing the beam axes and the impact parameter vector.

We recall that at high bombarding energies the longitudinal size of the Lorentz contracted nuclei becomes negligible compared to its transverse size. The forward shadowing effect then becomes negligible and the elliptic flow fully develops in-plane, leading to a positive value of

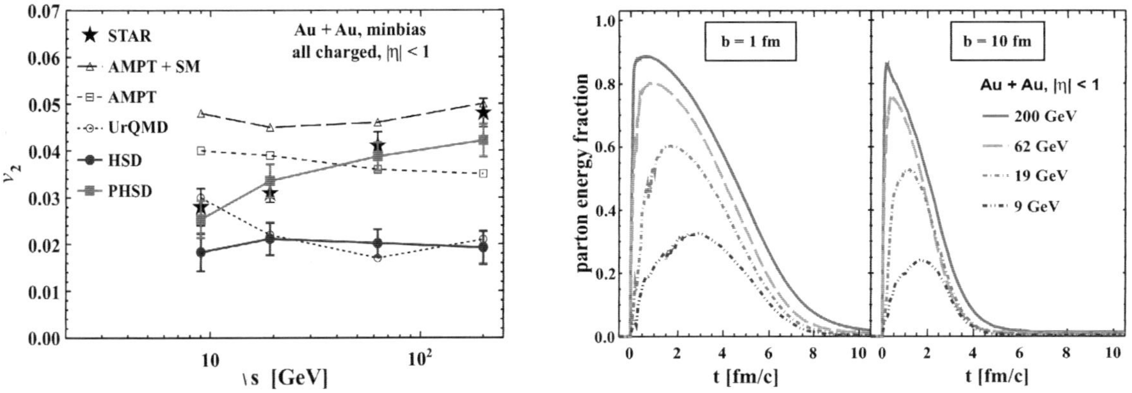

Figure 1. (l.h.s.) The average elliptic flow v_2 of charged particles at midrapidity for minimum bias collisions at $\sqrt{s_{NN}}$ = 9.2, 19.6, 62.4 and 200 GeV (stars) is taken from the data compilation of Ref. [16]). The corresponding results from different models are compared to the data and explained in more detail in the text. (r.h.s.) Evolution of the parton fraction of the total energy density at midrapidity (from PHSD) for different collision energies at impact parameters b =1 fm and 10 fm.

the average flow v_2 since no shadowing from spectators takes place. In Fig. 1 (l.h.s.) the experimental v_2 data compilation for the transient energy range is compared to the results from various models: PHSD, HSD as well as from UrQMD amd AMPT as included in Ref. [16]. The centrality selection is the same for the data and the various models.

The HSD [27, 28, 29] and UrQMD (Ultra relativistic Quantum Molecular Dynamics) [25, 26] are the hadron-string models and, thus, essentially provide information on the contribution from the hadronic phase [30]. As seen in Fig. 1, being in agreement with data at the lowest energy $\sqrt{s_{NN}}$ = 9.2 GeV, the HSD and UrQMD model results then either remain approximately constant or decrease slightly with increasing $\sqrt{s_{NN}}$ and do not reproduce the rise of v_2 with the collision energy as seen experimentally.

The AMPT (A Multi Phase Transport model) [31, 32] uses initial conditions of a perturbative QCD (pQCD) inspired model which produces multiple minijet partons according to the number of binary initial nucleon-nucleon collisions. The string melting (SM) version of the AMPT model (labeled in Fig. 1 as AMPT-SM) is based on the idea of 'melting' of hadrons or strings above the critical energy density of $\varepsilon \sim 1$ GeV/fm^3 to massless partons. The subsequent scattering of the quarks are based on a parton cascade with (adjustable) effective cross sections which are significantly larger than those from pQCD [31, 32]. Once the partonic interactions terminate, the partons hadronize through the mechanism of parton coalescence.

We find from Fig. 1 that the interactions between the minijet partons in the AMPT model indeed increase the elliptic flow significantly as compared to the hadronic models UrQMD and HSD. An additional inclusion of interactions between partons in the AMPT-SM model gives rise to another 20% of v_2 bringing it into agreement (for AMPT-SM) with the data at the maximal collision energy. So, both versions of the AMPT model indicate the importance of partonic contributions to the observed elliptic flow v_2 but do not reproduce its growth with $\sqrt{s_{NN}}$. The authors address this result to the partonic-equation-of state (EoS) employed which corresponds to a massless and noninteracting relativistic gas of particles. This EoS deviates severely from the results of lattice QCD calculations for temperatures below 2-3 T_c. Accordingly, the degrees-of-freedom are propagated without self-energies and a parton spectral function.

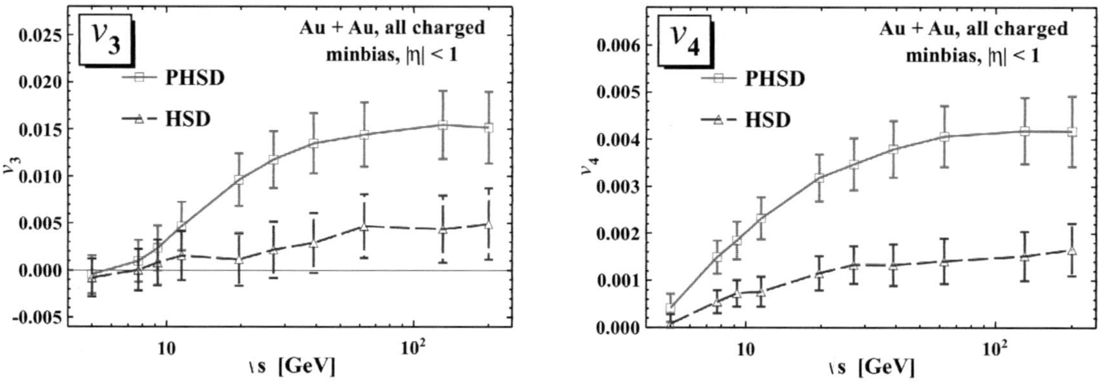

Figure 2. Average anisotropic flows v_3 (l.h.s.) and v_4 (r.h.s.) of charged particles at mid-pseudorapidity for minimum bias Au + Au collisions calculated within the PHSD (solid lines, red) and HSD (dashed lines, blue) models.

The PHSD approach incorporates the latter medium effects in line with a lQCD equation-of-state and also includes a dynamical hadronization scheme based on covariant transition rates. As has been demonstrated in Refs. [17, 18] and explicitly shown in Fig. 1 (l.h.s.), the elliptic flow v_2 from PHSD (red line) agrees with the data from the STAR collaboration and clearly shows an increase with bombarding energy.

An explanation for the increase in v_2 with collision energy is provided in Fig. 1 (r.h.s.) where the partonic fraction of the energy density is shown with respect to the total energy where the energy densities are calculated at mid-rapidity. As discussed above the main contribution to the elliptic flow is coming from an initial partonic stage at high $\sqrt{s}$. The fusion of partons to hadrons or, inversely, the melting of hadrons to partonic quasiparticles occurs when the local energy density is about $\varepsilon \approx 0.5$ GeV/fm^3. As follows from Fig. 1, the parton fraction of the total energy goes down substantially with decreasing bombarding energy while the duration of the partonic phase is roughly the same. The maximal fraction reached is the same in central and peripheral collisions but the parton evolution time is shorter in peripheral collisions. One should recall again the important role of the repulsive mean-field for partons in the PHSD model that leads to an increase of the flow v_2 with respect to HSD predictions (cf. also Ref. [33]). We point out in addition that the increase of v_2 in PHSD relative to HSD is also partly due to the higher interaction rates in the partonic medium because of a lower ratio of η/s for partonic degrees-of-freedom at energy densities above the critical energy density than for hadronic media below the critical energy density [34, 35, 36]. The relative increase in v_3 and v_4 in PHSD essentially is due to the higher partonic interaction rate and thus to a lower ratio η/s in the partonic medium which is mandatory to convert initial spacial anisotropies to final anisotropies in momentum space [37].

2.2. Higher-order flow harmonics

Depending on the location of the participant nucleons in the nucleus at the time of the collision, the actual shape of the overlap area may vary: the orientation and eccentricity of the ellipse defined by the participants fluctuates from event to event. Note, however, that by averaging over many events an almond shape is regained for the same impact parameter.

Recent studies suggest that fluctuations in the initial state geometry can generate higher-order flow components [1, 10, 38, 39]. The azimuthal momentum distribution of the emitted

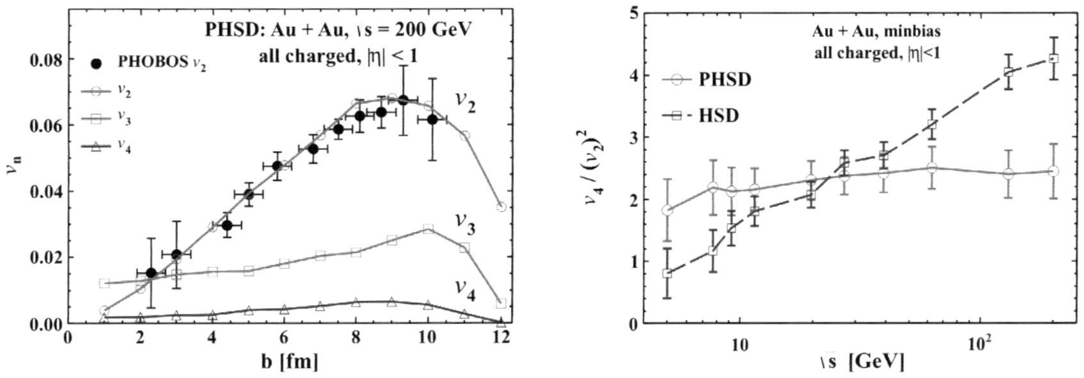

Figure 3. (l.h.s.) Impact parameter dependence of anisotropic flows of charged particles at mid-pseudorapidity for minimum bias collisions of Au+Au at $\sqrt{s_{NN}}$ = 200 GeV. Experimental points are from Ref. [43]. (r.h.s.) Beam energy dependence of the ratio $v_4/(v_2)^2$ for Au+Au collisions. The solid and dashed curves are calculated within the PHSD and HSD models, respectively.

particles is commonly expressed in the form of a Fourier series as

$$E\frac{d^3N}{d^3p} = \frac{d^2N}{2\pi p_T dp_T dy}\left(1 + \sum_{n=1}^{\infty} 2v_n(p_T)\cos(n(\psi - \Psi_n))\right), \tag{2}$$

where v_n is the magnitude of the n-th order harmonic term relative to the angle of the initial-state spatial plane of symmetry Ψ_n. The anisotropy in the azimuthal angle ψ is usually characterized by the even order Fourier coefficients with the reaction plane $\Psi_n = \Psi_{RP}$: $v_n = \langle \exp(\imath n(\psi - \Psi_{RP}))\rangle$ $(n = 2, 4, ...)$, since for a smooth angular profile the odd harmonics vanish. For the odd components, e.g. v_3, one should take into account event-by-event fluctuations with respect to the participant plane $\Psi_n = \Psi_{PP}$. We calculate the v_3 coefficients with respect to Ψ_3 as: $v_3\{\Psi_3\} = \langle\cos(3[\psi - \Psi_3])\rangle/Res(\Psi_3)$. The event plane angle Ψ_3 and its resolution $Res(\Psi_3)$ are calculated as described in Ref. [40] via the two-sub-events method [41, 42].

In Fig. 2 we display the PHSD and HSD results for the anisotropic flows v_3 and v_4 of charged particles at mid-pseudorapidity for Au+Au collisions as a function of $\sqrt{s_{NN}}$. The pure hadronic model HSD gives $v_3 \approx 0$ for all energies. Accordingly, the results from PHSD (dashed red line) are systematically larger than from HSD (dashed blue line). Unfortunately, our statistics are not good enough to allow for more precise conclusions. The hexadecupole flow v_4 stays almost constant in the energy range $\sqrt{s_{NN}} \geq 10$ GeV; at the same time the PHSD gives noticeably higher values than HSD which we attribute to the higher interaction rate in the partonic phase.

Alongside with the integrated flow coefficients v_n the PHSD model reasonably describes their distribution over centrality or impact parameter b. A specific comparison at $\sqrt{s_{NN}} = 200$ GeV is shown in Fig. 3 for v_2, v_3 and v_4. While v_2 increases strongly with b up to peripheral collisions, v_3 and v_4 are only weakly sensitive to the impact parameter. The triangular flow is always somewhat higher than the hexadecupole flow in the whole range of impact parameters b.

2.3. Ratios of different harmonics

Different harmonics can be related to each other. In particular, hydrodynamics predicts that $v_4 \propto (v_2)^2$ [44]. The simplest prediction that $v_4 = 0.5(v_2)^2$ is given for a boosted thermal freeze-out distribution of an ideal fluid, Ref. [45]. In this work it was noted also that v_4 is largely

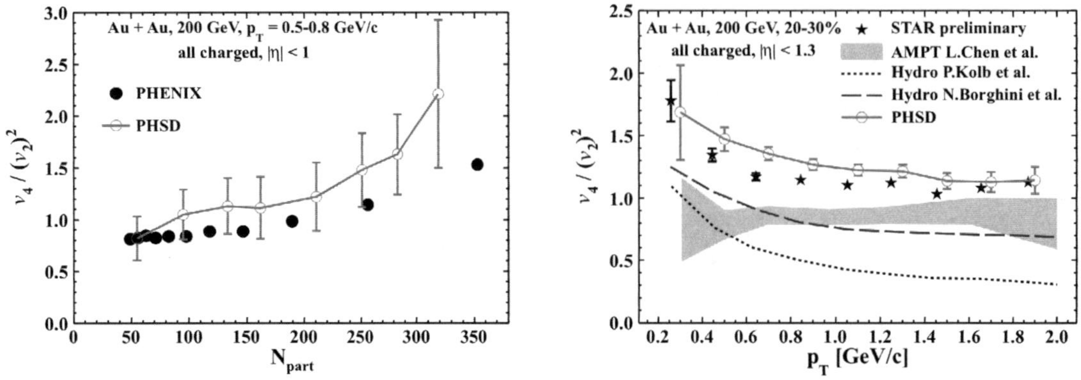

Figure 4. (l.h.s.) Participant number dependence of the $v_4/(v_2)^2$ ratio of charged particles for Au+Au ($\sqrt{s_{NN}}$ = 200 GeV) collisions. The experimental data points for 0.5< p_T <0.8 GeV/c are from Ref. [46]. (r.h.s.) Transverse momentum dependence of the ratio $v_4/(v_2)^2$ of charged particles for Au+Au (at $\sqrt{s_{NN}}$ = 200 GeV) collisions. The dashed and dot-dashed lines are calculated within the hydrodynamic approaches from Refs. [44] and [45], respectively. The shaded region corresponds to the results from the AMPT model [48]. The experimental data points are from the STAR Collaboration [47].

generated by an intrinsic elliptic flow (at least at high p_T) rather than the fourth order moment of the fluid flow. This is a motivation for studying the ratio $v_4/(v_2)^2$ rather than v_4 alone. As is seen in Fig. 4 (r.h.s.), indeed the ratio calculated within the PHSD model is practically constant in the whole range of $\sqrt{s_{NN}}$ considered but significantly deviates from the ideal fluid estimate of 0.5. In contrast, neglecting dynamical quark-gluon degrees-of-freedom in the HSD model, we obtain a monotonous growth of this ratio.

The dependence of the $v_4/(v_2)^2$ ratio versus the number of participants N_{part} is shown in Fig. 4 for charged particles produced in Au + Au collisions at $\sqrt{s_{NN}}$ = 200 GeV. The PHSD results are roughly in agreement with the experimental data points from Ref. [47] but overshoot them for $N_{part} \sim 250$.

As pointed out before, the ratio $v_4/(v_2)^2$ is sensitive to the microscopic dynamics. In this respect we show the transverse momentum dependence of the ratio $v_4/(v_2)^2$ in Fig. 4 for charged particles produced in Au+Au collisions at $\sqrt{s_{NN}}$ = 200 GeV (20-30% centrality). The PHSD results are quite close to the experimental data points from Ref. [47], however, overestimate the measurements by up to 20%. The hydrodynamic results – plotted in the same figure – significantly underestimate the experimental data and noticeably depend on viscosity. The partonic AMPT model [48] discussed above also predicts a slightly lower ratio than the measured one, however, being in agreement with both hydrodynamic models for p_T 0.8 GeV/c. Our interpretation of Fig. 4 (r.h.s.) is as follows: the data are not compatible with ideal hydrodynamics and a finite shear viscosity is mandatory (in viscous hydrodynamics) to come closer the experimental observations. The kinetic approaches AMPT and PHSD perform better but either overestimate (in AMPT) or slightly underestimate the scattering rate of soft particles (in PHSD). An explicit study of the centrality dependence of these ratios should provide further valuable information.

3. Conclusions

In summary, relativistic collisions of Au+Au from $\sqrt{s_{NN}}$ = 5 to 200 GeV have been studied within the PHSD approach which includes the dynamics of explicit partonic degrees-of-freedom as well as dynamical local transition rates from partons to hadrons and also the final hadronic scatterings. Whereas earlier studies have been carried out for longitudinal rapidity distributions of various hadrons, their transverse mass spectra and the elliptic flow v_2 as compared to available data at SPS and RHIC energies [19, 20], here we have focussed on the PHSD results for the collective flow coefficients v_2, v_3 and v_4 in comparison to experimental data in the large energy range from the RHIC Beam-Energy-Scan (BES) program as well as different theoretical approaches ranging from hadronic transport models to ideal and viscous hydrodynamics. We mention explicitly that the PHSD model from Ref. [20] has been used for all calculations performed in this study and no tuning (or change) of model parameters has been performed.

We have found that the anisotropic flows – elliptic v_2, triangular v_3, hexadecapole v_4 – are reasonably described within the PHSD model in the whole transient energy range naturally connecting the hadronic processes at lower energies with ultrarelativistic collisions where the quark-gluon degrees of freedom become dominant. The smooth growth of the elliptic flow v_2 with the collision energy demonstrates the increasing importance of partonic degrees of freedom. Other signatures of the transverse collective flow, the higher-order harmonics of the transverse anisotropy v_3 and v_4 change only weakly from $\sqrt{s_{NN}} \sim 7$ GeV to the top RHIC energy of $\sqrt{s_{NN}} = 200$ GeV, roughly in agreement with experiment. As shown in this study, this success is related to a consistent treatment of the interacting partonic phase in PHSD whose fraction increases with the collision energy.

The analysis of correlations between particles emitted in ultrarelativistic heavy-ion collisions at large relative rapidity has revealed an azimuthal structure that can be interpreted as solely due to collective flow [49, 50, 51, 52]. This interesting new phenomenon, denoted as triangular flow, results from initial state fluctuations and a subsequent hydrodynamic-like evolution. Unlike the usual directed flow, this phenomenon has no correlation with the reaction plane and should depend weakly on rapidity. Event-by-event hydrodynamics [53] has been a natural framework for studying this triangular collective flow but it has been of interest also to investigate these correlations in terms of the PHSD model. We have found the third harmonics to increase steadily in PHSD with bombarding energy. The coefficient v_3 is compatible with zero for $\sqrt{s_{NN}} > 20$ GeV in case of the hadronic transport model HSD which does not develop 'ridge-like' correlations. In this energy range PHSD gives a positive v_3 due to dominant partonic interactions.

Different harmonics can be related to each other and in particular, hydrodynamics predicts that $v_4 \propto (v_2)^2$ [44]. In this work it was noted also that v_4 is largely generated by an intrinsic elliptic flow (at least at high p_T) rather than the fourth order moment of the fluid flow. Indeed, the ratio $v_4/(v_2)^2$ calculated within the PHSD model is approximately constant in the whole considered range of $\sqrt{s_{NN}}$ but significantly deviates from the ideal fluid estimate of 0.5. In contrast, neglecting dynamical quark-gluon degrees-of-freedom in the HSD model, we obtain a monotonous growth of this ratio.

The transverse momentum dependence of the ratio $v_4/(v_2)^2$ at the top RHIC energy has given further interesting information (*cf.* Fig. 4) by comparing the various model results to the data from STAR which are interpreted as follows: the STAR data are not compatible with ideal hydrodynamics and a finite shear viscosity is mandatory (in viscous hydrodynamics) to come closer the experimental ratio observed. The kinetic approaches AMPT and PHSD perform better but either overestimate (in AMPT) or slightly underestimate the scattering rate of soft particles (in PHSD).

We recall that our present PHSD calculations employ 'naturally' Glauber type initial state fluctuactions which appear compatible with experimental observations up to top RHIC energies. On the other hand one might expect that at LHC energies an initial 'glasma' phase might play a sizeable role and that the gluon-field fluctuations - of lower scale - could show up in the Fourier decomposition of the azimuthal angular distribution. It will be interesting to compare the coefficients v_n for high multiplicity pp, $p+Pb$ and $Pb+Pb$ reactions as a function of transverse momentum p_T (and centrality).

The authors acknowledge financial support through the "HIC for FAIR" framework of the "LOEWE" program.

References

[1] Mishra A P *et al.* 2008 *Phys. Rev.* C **77** 064902
[2] Ollitrault J Y 1992 *Phys. Rev.* D **46** 229
[3] Heinz U and Kolb P 2002 *Nucl. Phys.* A **702** 269
[4] Shuryak E V 2009 *Prog. Part. Nucl. Phys.* **62** 48
[5] Shuryak E V 2005 *Nucl. Phys.* A **750** 64
[6] Gyulassy M and McLerran L 2005 *Nucl. Phys.* A **750** 30
[7] Peshier A and Cassing W 2005 *Phys. Rev. Lett.* **94** 172301
[8] Cheng M *et al.* 2008 *Phys. Rev.* D **77** 014511
[9] Aoki Y *et al.* 2009 *JHEP* **0906** 088
[10] Adare A *et al.* 2007 *Phys. Rev. Lett.* **98** 172301
[11] Romatschke P and Romatschke U 2007 *Phys. Rev. Lett.* **99** 172301
[12] Xu Z, Greiner C and Stöcker H 2008 *Phys. Rev. Lett.* **101** 082302
[13] Abelev B I *et al.* 2010 *Phys. Rev.* C **81** 024911
[14] Lacey R A *et al.* 2007 *Phys. Rev. Lett.* **98** 092301
[15] Aggarwal M M *et al.* 2010 *arXiv:1007.2613*
[16] Nasim M, Kumar L, Netrakanti P K and Mohanty B 2010 *Phys. Rev.* C **82** 054908
[17] Konchakovski V P *et al.* 2012 *Phys. Rev.* C **85** 011902
[18] Konchakovski V P *et al.* 2012 *Phys. Rev.* C **85** 044922
[19] Cassing W and Bratkovskaya E L 2009 *Nucl. Phys.* A **831** 215
[20] Bratkovskaya E L, Cassing W, Konchakovski V P and Linnyk O 2011 *Nucl. Phys.* A **856** 162
[21] Cassing W 2007 *Nucl. Phys.* A **791** 365
[22] Cassing W 2007 *Nucl. Phys.* A **795** 70
[23] Cassing W 2009 *Eur. J. Phys.* **168** 3
[24] Ackermann K H *et al.* 2001 *Phys. Rev. Lett.* **86** 402
[25] Cassing W *et al.* 1990 *Phys. Rep.* **188** 363
[26] Bass S A *et al.* 1998 *Prog. Part. Nucl. Phys.* **41** 255
[27] Bratkovskaya E L, Cassing W and Stöcker H 2003 *Phys. Rev.* C **67** 054905
[28] Ehehalt W and Cassing W 1996 *Nucl. Phys.* A **602** 449
[29] Cassing W and Bratkovskaya E L 1999 *Phys. Rep.* **308** 65
[30] Bratkovskaya E L *et al.* 2004 *Phys. Rev.* C **69** 054907
[31] Lin Z W and Ko C M 2002 *Phys. Rev.* C **65** 034904
[32] Lin Z W, Ko C M, Li, B A Zhang B and Pal S 2005 *Phys. Rev.* C **72** 064901
[33] Cassing W and Bratkovskaya E L 2008 *Phys. Rev.* C **78** 034919
[34] Mattiello S and Cassing W 2010 *Eur. Phys. J.* C **70** 243
[35] Demir N and Bass S A 2009 *Phys. Rev. Lett.* **102** 172302
[36] Ozvenchuk V *et al.*, *arXiv:1212.5393 [hep-ph]*
[37] Petersen H, Coleman-Smith C, Bass S A and Wolpert R 2011 *J. Phys.* G **38** 045102
[38] Petersen H and Bleicher M 2010 *Phys. Rev.* C **81** 044906
[39] Alver B and Roland G 2010 *Phys. Rev.* C **81** 054905
[40] Adare A *et al.* 2011 *Phys. Rev. Lett.* **107** 252301
[41] Poskanzer A M and Voloshin S A 1998 *Phys. Rev.* C **58** 1671
[42] Bilandzic A, Snellings R and Voloshin S A 2011 *Phys. Rev.* C **83** 044913
[43] Back B B *et al.* 2005 *Phys. Rev.* C **72** 051901
[44] Kolb P F 2003 *Phys. Rev.* C **68** 031902(R)
[45] Borghini N and Ollitrault J Y 2006 *Phys. Lett.* B **642** 227
[46] Gong X Y *et al.* 2011 *J. Phys.* G **38** 124146
[47] Bai Y *et al.* 2007 *J. Phys.* G **34** S903
[48] Chen L W, Ko C M and Lin Z W 2004 *Phys. Rev.* C **69** 031901
[49] Xu J and Ko C M 2011 *Phys. Rev.* C **84** 014903
[50] Schenke B, Jeon S and Gale C 2011 *Phys. Rev. Lett.* **106** 042301
[51] Teaney D and Yan L 2011 *Phys. Rev.* C **83** 064904
[52] Luzum M, Gombeaud C and Ollitrault J Y 2010 *Phys. Rev.* C **81** 054910
[53] Gardim F G, Grassi F, Hama Y, Luzum M and Ollitrault J Y 2011 *Phys. Rev.* C **83** 064901

Journal of Physics: Conference Series **455** (2013) 012049 doi:10.1088/1742-6596/455/1/012049

The Tsallis Distribution for $p-p$ collisions at the LHC.

Jean Cleymans
UCT-CERN Research Centre and Department of Physics
University of Cape Town, Rondebosch, South Africa

E-mail: jean.cleymans@uct.ac.za

Abstract. The Tsallis distribution has been used widely in high energy physics to describe the transverse momentum distributions of particles produced in $p-p$ collisions.. In this talk we review some recent developments.

There exists a rich and wide variety of distributions covering a large range of applications [1, 2]. Those having a power law behaviour have attracted considerable attention in physics in recent years but there is a a long history in other fields such as biology and economics [3].

In high energy physics power law distributions have been applied by a very large number of scientists [4, 5, 7, 6, 8] to the description of transverse momenta of secondary particles produced in $p-p$ collisions. Indeed the available range of transverse momenta has expanded considerably with the advent of the Large Hadron Collider (LHC).

Collider energies up to 7 TeV are now available in $p-p$ collisions and transverse momenta of hundreds of GeV are now a common occurrence. In this presentation the focus will be on various forms of distributions first proposed by C. Tsallis about twenty-five years ago [9]. Applications of the Tsallis distribution to high energy $e^+ - e^-$ annihilation has been considered by Bediaga, Curado and de Miranda [10]. A recent review of power laws in elementary and heavy-ion collisions can be found in [11].

In the analysis of the new data, a Tsallis-like distribution gives excellent fits to the transverse momentum distributions as shown by the STAR [4] and PHENIX [5] collaborations at RHIC and by the ALICE [6], ATLAS [7] and CMS [8] collaborations at the LHC. In this talk we review recent work on the parameterization used by these groups and discuss a slightly different one which might lead to a more consistent interpretation and has the bonus of being thermodynamically consistent [12, 13].

For high energy physics a consistent form of Tsallis statistics (see e.g. [13] and references therein) for the particle number, energy density and pressure is given by the expressions given below

$$N = gV \int \frac{d^3p}{(2\pi)^3} \left[1 + (q-1)\frac{E-\mu}{T}\right]^{-\frac{q}{q-1}}, \tag{1}$$

$$\epsilon = g \int \frac{d^3p}{(2\pi)^3} E \left[1 + (q-1)\frac{E-\mu}{T}\right]^{-\frac{q}{q-1}}, \tag{2}$$

$$P = g \int \frac{d^3p}{(2\pi)^3} \frac{p^2}{3E} \left[1 + (q-1)\frac{E-\mu}{T}\right]^{-\frac{q}{q-1}}. \tag{3}$$

where T and μ are the temperature and the chemical potential, V is the volume and g is the degeneracy factor. A similar version has been presented in [14, 15, 16, 17] but with different power-law behaviour.

As is well-known the Tsallis distribution [9, 18] introduces a new parameter q which for transverse momentum spectra is always close to 1, typical values for the parameter q obtained from fits to the transverse momentum distribution are in the range 1.1 to 1.2. In this talk we always assume $q > 1$.

The expressions (1), (2) and (3) are thermodynamically consistent, e.g. it can be easily shown [13] that relations of the type

$$N = V \left.\frac{\partial P}{\partial \mu}\right|_T, \tag{4}$$

are satisfied. It has been verified that the proposed expressions have thermodynamic consistency [12, 13].

It follows from (1) that the momentum distribution is given by,

$$\frac{d^3N}{d^3p} = \frac{gV}{(2\pi)^3} \left[1 + (q-1)\frac{E-\mu}{T}\right]^{-q/(q-1)}, \tag{5}$$

or, expressed in terms of transverse momentum, p_T, transverse mass, m_T, and rapidity y

$$\frac{d^2N}{dp_T\, dy} = gV \frac{p_T m_T \cosh y}{(2\pi)^2} \left[1 + (q-1)\frac{m_T \cosh y - \mu}{T}\right]^{-q/(q-1)}. \tag{6}$$

At mid-rapidity $y = 0$ and for zero chemical potential $\mu = 0$ this reduces to

$$\left.\frac{d^2N}{dp_T\, dy}\right|_{y=0} = gV \frac{p_T m_T}{(2\pi)^2} \left[1 + (q-1)\frac{m_T}{T}\right]^{-q/(q-1)}. \tag{7}$$

This is the expression used in [12, 13] to fit the LHC transverse momentum spectra.

It is well-known since 1988 [9] that in the limit where the parameter q goes to 1 this reduces to the standard Boltzmann distribution:

$$\lim_{q\to 1} \frac{d^2N}{dp_T\, dy} = gV \frac{p_T m_T \cosh y}{(2\pi)^2} \exp\left(-\frac{m_T \cosh y - \mu}{T}\right). \tag{8}$$

The parameterization given in Eq. (7) is close to the one used e.g. by the STAR [4], PHENIX [5], ALICE [6], ATLAS [7] and CMS [8] collaborations where the following form is used :

$$\left.\frac{d^2N}{dp_T\, dy}\right|_{y=0} = p_T \frac{dN}{dy} \frac{(n-1)(n-2)}{nC(nC + m_0(n-2))} \left[1 + \frac{m_T - m_0}{nC}\right]^{-n} \tag{9}$$

$p-p$ 900 GeV		
Particle	q	T
π^+	1.166 ±0.004	0.0649 ±0.0010
π^-	1.162 ±0.003	0.0648 ± 0.0008
K^+	1.145 ±0.008	0.0776 ±0.0044
K^-	1.142 ±0.003	0.0769 ± 0.0013
K_S^0	1.135 ±0.002	0.0919 ±0.0015
p	1.109 ±0.004	0.0949 ± 0.0025
$\bar{p}$	1.110 ±0.018	0.0973 ±0.0015
Λ	1.106 ±0.004	0.0796 ± 0.0048
Ξ^-	1.081 ±0.004	0.1262 ± 0.0037

Table 1. Fitted values of the T and q parameters measured in $p-p$ collisions by the ALICE and CMS collaborations using the Tsallis form (7) for the momentum distribution.

where n, C and m_0 are fit parameters. Indeed, after substituting

$$n \to \frac{q}{q-1} \tag{10}$$

and

$$nC \to \frac{T + m_0(q-1)}{q-1}, \tag{11}$$

Eq. (9) becomes

$$\left.\frac{d^2N}{dp_T\,dy}\right|_{y=0} = p_T\frac{\mathrm{d}N}{\mathrm{d}y}\frac{(n-1)(n-2)}{nC(nC+m_0(n-2))} \left[\frac{T}{T+m_0(q-1)}\right]^{-q/(q-1)} \left[1+(q-1)\frac{m_T}{T}\right]^{-q/(q-1)}. \tag{12}$$

Which has the same dependence on the transverse momentum as (7) apart from an additional factor m_T on the right-hand side. It has to be pointed out that the inclusion of the rest mass in the substitution Eq. (11) is not in agreement with the Tsallis distribution as it breaks m_T scaling which is present in the Tsallis form (6) but not in Eq. (9).
The inclusions of the factor m_T leads to a more consistent interpretation of the variables q and T [12, 13].
The distribution (7) has been used to fit the data for identified particles, π, K and p for the ALICE [6] collaboration and K_s^0, Λ and Ξ for the CMS [8] collaboration in $p-p$ collisions at 900 GeV [12, 13]. The resulting values for q and T at 900 GeV beam energy are listed in Table 1. The transverse momentum distributions for the ALICE [6] data are shown in Fig. (1). The results obtained are comparable to the ones obtained recently in [20, 25].

Interesting results were obtained in Refs. [20, 21, 22, 23] where spectra for identified particles were analyzed and the resulting values for the parameters q and T were considered.

The results obtained for the values of q are shown in Fig. (2) for two different LHC beam energies. Note that the values for q at 7 TeV are slightly higher than those obtained at 0.9 TeV.

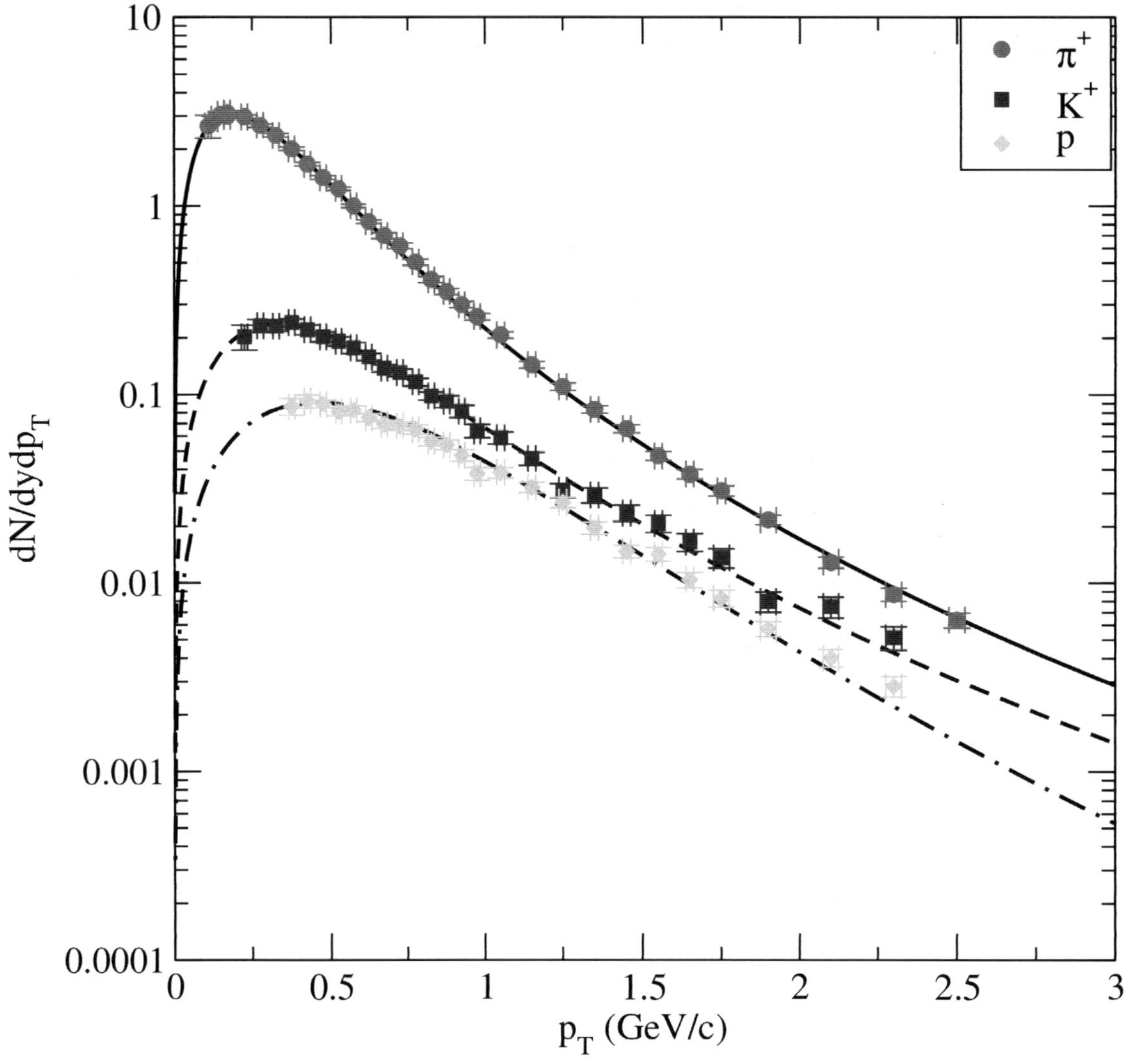

Figure 1. Fit to the π, K, p transverse momentum distributions in $p-p$ collisions as measured by the ALICE collaboration [6] using the Tsallis distribution function as given by (7).

This has been analyzed in a systematic way in [26] where the parameter q was extracted from results for charged hadrons at different beam energies. A similar analysis was recently made in [27, 28], the energy dependence is compatible but the values of q are systematically lower than those obtained in [26] due to a different power in the Tsallis distribution.

In conclusion we can say that the use of the Tsallis parameterization presented in (1) leads to a good description of identified particles in $p-p$ collisions at the LHC with a reasonably consistent set of parameters.

Acknowledgments

The results presented here were obtained in collaboration with G.I. Lykasov, A.S. Parvan, A.S. Sorin, O.V. Teryaev and D. Worku.

[1] M.E.J. Newman, Contemporary Physics **46**, 323 (2005).

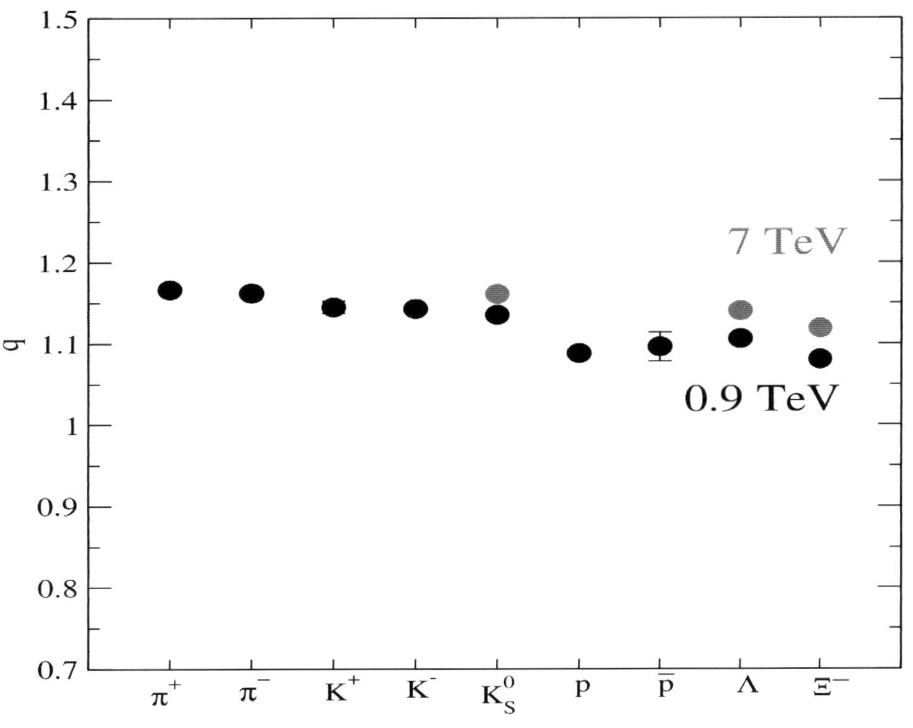

Figure 2. Values of the Tsallis parameter q for different species of hadrons. The values were extracted from results obtained by the ALICE [6] and CMS [8] collaborations.

[2] A. Clauset, C.R. Shalizi, M.E.J. Newman, SIAM Review **51**, 661, 2009.
[3] M. Mitzenmacher, Internet Mathematics, **1**, 226 (2003).
[4] B. I. Abelev et al. (STAR Collaboration), Phys. Rev. C **75**, 064901 (2007).
[5] A. Adare et al. (PHENIX Collaboration), Phys. Rev. C **83**, 052004, (2010); Phys. Rev. C **83**, 064903 (2011).
[6] K. Aamodt, et al. (ALICE Collaboration), Eur. Phys. J. C **71** 1655 (2011).
[7] G. Aad, et al. (ATLAS Collaboration), New J. Phys. **13** (2011) 053033.
[8] V. Khachatryan, et al. (CMS Collaboration), JHEP **05**, 064 (2011).
[9] C. Tsallis, J.Statist.Phys. **52**, 479 (1988); arxiv:hep-ph/9905255
[10] I. Bediaga, E.M.F. Curado, J.M. de Miranda, Physica A **286**, 156 (2000).
[11] G. Wilk and Z. Wlodarczyk, Eur. Phys. J. A**40**, 299 (2009); **48**, 161 (2012).
[12] J. Cleymans and D. Worku, J. Phys. **G**39 025006 (2012) .
[13] J. Cleymans and D. Worku, Eur. Phys. J. **A** (to be published), arXiv:1203.4343.
[14] A. Lavagno, Physics Letters **A** 301, 13, (2002).
[15] W.A. Alberico, A. Lavagno, Eur. Phys. J. **A** 40, 313 (2009).
[16] A. Drago, A. Lavagno, P. Quarati,, Physica **A**, 472 (2004).
[17] A. Lavagno, Phys. Rev. C **8**1, 044909 (2010).
[18] C. Tsallis, R. S. Mendes, A. R. Plastino, Physica A **261**, 534 (1998).
[19] Cheuk-Yin Wong, G. Wilk, Acta Physica Polonica, 43 (2012) 2047.
[20] L. Marques, E.Andrade-II, A. Deppman, arXiv:1210.1725[hep-ph]
[21] A. Deppman, arXiv:1212.0379[hep-ph].
[22] I. Sena, A. Deppman, arXiv:1208.2952[hep-ph]
[23] I. Sena, A. Deppman, arXiv:1209.2367[hep-ph]
[24] K. Ürmössy, arXiv:1212.0260[hep-ph].
[25] I. Sena, A. Deppman, arXiv:1208.2952.
[26] J. Cleymans, G.I. Lykasov, A.S. Parvan, A.S. Sorin, O.V. Teryaev, D. Worku, arXiv:1302.1970.
[27] T. Wibig, J. Phys. G: Nucl. Part. Phys. **37** 115009 (2010).
[28] T. Wibig, I. Kurp, JHEP **0**312 039 (2003).

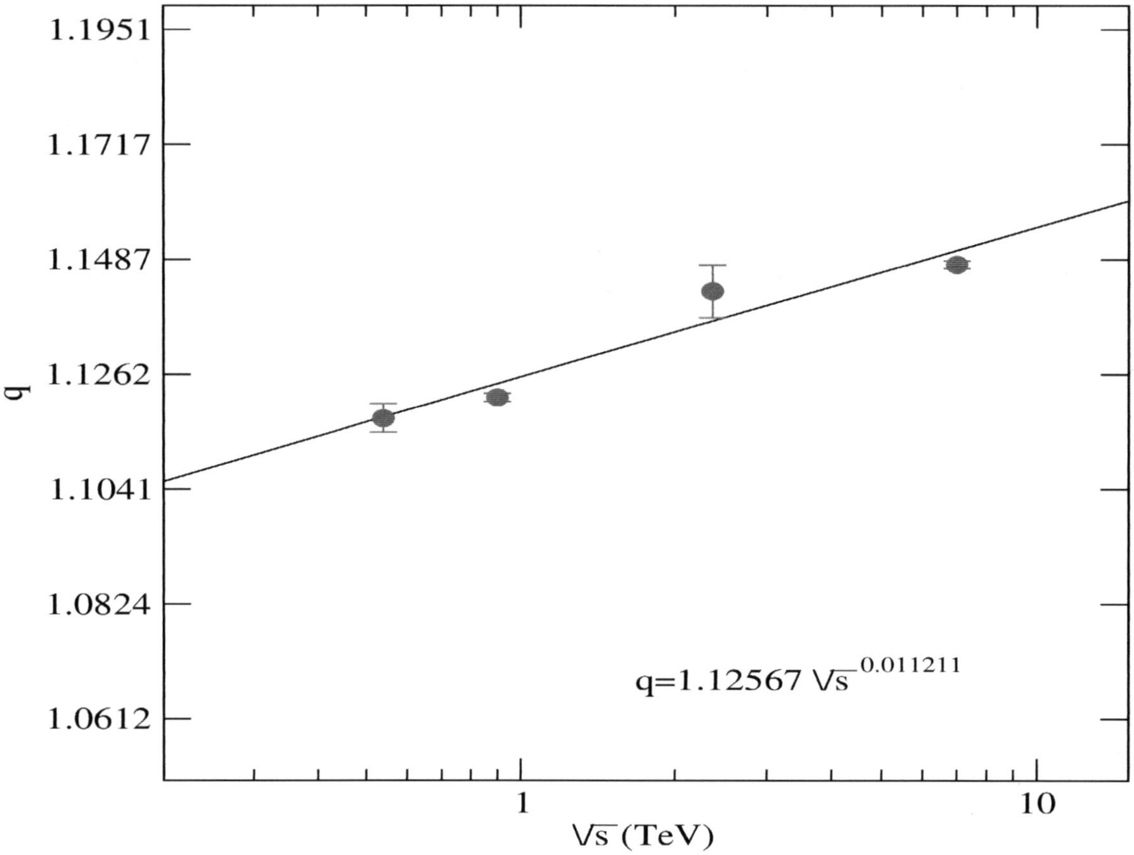

Figure 3. (Color online) Energy dependence of the Tsallis parameter q appearing in the Tsallis distribution [26].

Evolution of Yukawa Couplings and Quark Flavour Mixings in the 5D MSSM

Ammar Abdalgabar[1] **and A. S. Cornell**[2]

National Institute for Theoretical Physics; School of Physics, University of the Witwatersrand, Wits 2050, South Africa

E-mail: [1]ammar.abdalgabar@students.wits.ac.za, [2]alan.cornell@wits.ac.za

Abstract. The evolution equations of the Yukawa couplings and quark mixings are derived for the one-loop renormalization group equations in the 5D Minimal Supersymmetric Standard Model on an S^1/Z_2 orbifold. Different possibilities for the matter fields are discussed such as the cases of bulk propagating or brane localised fields. We discuss in both cases the evolution of the mass ratios and the implications for the mixing angles.

1. Introduction

A theory of fermion masses and the associated mixing angles provide an interesting puzzle and a likely window to physics beyond the Standard Model (SM), where one of the main issues in particle physics is to understand the fermion mass hierarchy and mixings. A clear feature of the fermion mass spectrum is [1]

$$m_u \ll m_c \ll m_t \, , \; m_d \ll m_s \ll m_b \, , \; m_e \ll m_\mu \ll m_\tau \, . \tag{1}$$

There have been many attempts to understand the fermion mass hierarchies and their mixings by making use of Renormalization Group Equations (RGEs) especially for Universal Extra Dimension (UED) models and their possible extensions (see Refs. [2, 3] and references therein). There are several versions of UED models, the simplest being the case of one flat extra dimension compactified on an S^1/Z_2 orbifold which has a size $1/R \approx 1$ TeV. This compactification will lead to a tower of new particle states in the effective 4-dimensional (4D) theory. As such, in the 4D effective theory there appears an infinite tower of massive Kaluza-Klein (KK) states, with a mass contribution inversely proportional to the radius of the extra-dimension. Due to the orbifolding mechanism the momentum is no longer conserved along the fifth dimension, and the symmetry is reduced to a KK parity conservation, $P = (-1)^n$, which is now an exact symmetry, with n referring to the KK level. That is, the UED models are effective theories in four dimensions with a cutoff Λ, with the consequence that the tree level spectrum is highly degenerate and where loop corrections to masses become important [4]. The phenomenology of these UED models will arise when their flat extra dimensions allow all the SM fields (or some subset) to propagate in the full space-time [5, 6].

There are many reasons to study such models, primarily as they provide a way to address the "hierarchy problem", that is, the question of why the Planck scale $M_{Pl} \sim 10^{19}$ GeV, is so much smaller than the weak scale 246 GeV, but also to provide a means of breaking the electroweak symmetry, the generation of fermion mass hierarchies, and in studying the CKM matrix and new sources of CP violation [7, 8]. Furthermore, TeV scale grand unification and sources of dark matter are also possible in these theories [9, 10]. To date much of the interest in UED models has been for its source of beyond the SM TeV-scale physics, largely arising from the tower of KK states approximately degenerate in mass at the scale set by the inverse of the compactification radius. KK parity and the 4D conservation of momentum imply that contributions to SM particle masses occur only for interactions at loop level, and that the lightest KK particle will be stable and a suitable dark matter candidate [11].

Recall that in extra-dimensional scenarios we are lead to a power law running of the gauge couplings due to the large number of KK states. These KK modes contribute the same way as the zeroth mode does, which we identify as the usual 4D Minimal Supersymmetric SM (MSSM) particles. In 5-dimensions, though, only Dirac fermions are allowed by the Lorentz algebra. As such there are eight supercharges which corresponds to the 4D view point of an $\mathcal{N} = 2$ supersymmetry, where the S^1/Z_2 orbifolding present in our UED models [3] will break the $\mathcal{N} = 1$ supersymmetry, for more details see Refs. [12, 9].

2. The 5D MSSM models

In the 5-dimensional (5D) MSSM, the Higgs superfields and gauge superfields always propagate into the fifth dimension. However, different possibilities of localisation for the matter superfields can be studied. We shall consider the two limiting cases of superfields with SM matter fields all in the bulk or all superfields containing SM matter fields restricted to the brane. When all fields propagate in the bulk, the action for the matter fields Φ_i is [12, 3]:

$$\begin{aligned} S_{matter} &= \int d^8 z dy \,[\bar{\Phi}_i \Phi_i + \Phi_i^c \bar{\Phi}_i^c + \Phi_i^c \partial_5 \Phi_i \delta(\bar{\theta}) - \bar{\Phi}_i \partial_5 \bar{\Phi}_i^c \delta(\theta) \\ &\quad + \tilde{g}(2\bar{\Phi}_i V \Phi_i - 2\Phi_i^c V \bar{\Phi}_i^c + \Phi_i^c \chi \Phi_i \delta(\bar{\theta}) + \bar{\Phi}_i \bar{\chi} \bar{\Phi}_i^c \delta(\theta))] \ . \end{aligned} \tag{2}$$

Similarly, when all superfields containing SM fermions are restricted to the brane, the part of the action involving only gauge and Higgs fields is not modified, whereas the action for the superfields containing the SM fermions becomes:

$$S_{matter} = \int d^8 z dy \delta(y) \,[\bar{\Phi}_i \Phi_i + 2\tilde{g}\bar{\Phi}_i V \Phi_i] \ . \tag{3}$$

For $\mathcal{N} = 1$, 4D supersymmetry the superfield formalism is well established: superfields describe quantum fields and their superpartners, as well as auxiliary fields, as a single object. This simplifies the notation and the calculation considerably. A similar formulation for a 5D vector superfield and the superfield formulation for matter supermultiplets has been developed in Refs.[12, 9]. Note that in our model the Yukawa couplings in the bulk are forbidden by the 5D $\mathcal{N} = 1$ supersymmetry. However, they can be introduced on the branes, which are 4D subspaces with reduced supersymmetry. We will write the following interaction terms, called brane interactions, containing Yukawa-type couplings:

$$S_{brane} = \int d^8 z dy \delta(y) \left(\frac{1}{6} \tilde{\lambda}_{ijk} \Phi_i \Phi_j \Phi_k \right) \delta(\bar{\theta}) + h.c. \ . \tag{4}$$

In 4D MSSM, the one-loop correction to the gauge couplings are given by

$$16\pi^2 \frac{dg_i}{dt} = b_i g_i^3 \ , \tag{5}$$

where $b_i = (\frac{33}{5}, 1, -3)$, $t = \ln(\mu/M_Z)$, and M_Z is the Z boson mass [13].

Table 1. The terms present in the various Yukawa evolution equations, see Eq.(7).

Scenarios	G_u	G_d	G_e
Bulk	$\frac{13}{15}g_1^2 + 3g_2^2 + \frac{16}{3}g_3^2$	$\frac{7}{15}g_1^2 + 3g_2^2 + \frac{16}{3}g_3^2$	$\frac{9}{5}g_1^2 + 3g_2^2$
Brane	$\frac{43}{30}g_1^2 + \frac{9}{2}g_2^2 + \frac{32}{3}g_3^2$	$\frac{19}{30}g_1^2 + \frac{9}{2}g_2^2 + \frac{32}{3}g_3^2$	$\frac{33}{10}g_1^2 + \frac{9}{2}g_2^2$
Scenarios	$T_d = T_e$	T_u	
Bulk	$(3Tr(Y_d^\dagger Y_d) + Tr(Y_e^\dagger Y_e))\pi S^2(t)$	$3Tr(Y_u^\dagger Y_u)\pi S^2(t)$	
Brane	$3Tr(Y_d^\dagger Y_d) + Tr(Y_e^\dagger Y_e)$	$3Tr(Y_u^\dagger Y_u)$	
Scenarios	F_u	F_d	F_e
Bulk	$(3Y_u^\dagger Y_u + Y_d^\dagger Y_d)\pi S^2(t)$	$(3Y_d^\dagger Y_d + Y_u^\dagger Y_u)\pi S^2(t)$	$(3Y_e^\dagger Y_e)\pi S^2(t)$
Brane	$(6Y_u^\dagger Y_u + 2Y_d^\dagger Y_d)S(t)$	$(6Y_d^\dagger Y_d + 2Y_u^\dagger Y_u)S(t)$	$6Y_e^\dagger Y_e S(t)$

In 5D MSSM, the one-loop corrections to gauge couplings are given by [3, 5]

$$16\pi^2 \frac{dg_i}{dt} = (b_i + (S(t) - 1)\tilde{b_i})g_i^3 \ , \qquad (6)$$

where $S(t) = e^t M_Z R$ is the number of KK states, $\tilde{b_i} = (\frac{66}{5}, 10, 6)$ for matter fields in the bulk and $\tilde{b_i} = (\frac{6}{5}, -2, -6)$ for matter fields localised to the brane. As can be seen in Fig. 1 (and Eq. (6)), the one-loop running of the gauge couplings changes with energy scale drastically and lowers the unification scale considerably. Specifically, for the compactification radii $R^{-1} = 1$, 4, 15 TeV, we find that the gauge couplings meet at around $\Lambda \approx 30$, 120, 430 TeV respectively. As such the Yukawa couplings also receive finite corrections at each KK level whose magnitudes depend on the cutoff energy scale. The evolution of the Yukawa couplings were derived using the standard techniques of Refs. [12, 3]. As such, the one-loop RGEs for Yukawa couplings in the 5D MSSM are:

$$16\pi^2 \frac{dY_i}{dt} = Y_i \left[T_i - G_i S(t) + F_i\right] \ , \qquad (7)$$

where $i = u, d, e$ and the values of G_i, F_i and T_i are given in table 1. That is, when the energy scale $\mu > \frac{1}{R}$, or when the energy scale parameter $t > \ln(\frac{1}{M_Z R})$ (where we have set M_Z as the renormalization point) we shall use Eq.(7), however, when the energy scale $M_Z < \mu < \frac{1}{R}$, the Yukawa evolution equations are dictated by the usual MSSM ones [13].

Yukawa coupling matrices can be diagonalized by using two unitary matrices U and V, where

$$UY_u^\dagger Y_u U^\dagger = diag(f_u^2, f_c^2, f_t^2) \ , \ VY_d^\dagger Y_d V^\dagger = diag(h_d^2, h_s^2, h_b^2) \ .$$

The CKM matrix appears as a result (upon this diagonalisation of quark mass matrices) of $V_{CKM} = UV^\dagger$. The variation of the CKM matrix and its evolution equation for all matter fields in the bulk is [14, 2, 6]:

$$16\pi^2 \frac{dV_{i\alpha}}{dt} = \pi S^2 \left[\sum_{\beta, j \neq i} \frac{f_i^2 + f_j^2}{f_i^2 - f_j^2} h_\beta^2 V_{i\beta} V_{j\beta}^* V_{j\alpha} + \sum_{j, \beta \neq \alpha} \frac{h_\alpha^2 + h_\beta^2}{h_\alpha^2 - h_\beta^2} f_j^2 V_{j\beta}^* V_{j\alpha} V_{i\beta} \right] \ . \qquad (8)$$

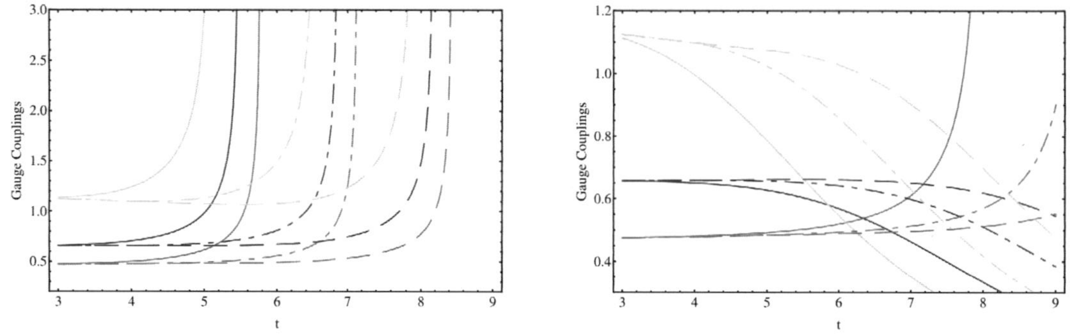

Figure 1. (Colour online) Gauge couplings (g_1 (red), g_2 (blue), g_3 (green) with: in the left panel, all matter fields in the bulk; and the right panel for all matter fields on the brane; for three different values of the compactification scales (1 TeV (solid line), 4 TeV (dot-dashed line), 15 TeV (dashed line))) as a function of the scale parameter t.

For all matter fields on the brane, the CKM evolution is the same as Eq.(8) but πS^2 is replaced by 2 S.

In making use of these equations for this current work, we recall that in the 4D MSSM the particle spectrum contains two Higgs doublets and the supersymmetric partners to the SM fields. After the spontaneous symmetry breaking of the electroweak symmetry, five physical Higgs particles are left in the spectrum. The two Higgs doublets H_u and H_d, with opposite hypercharges, are responsible for the generation of the up-type and down-type quarks masses respectively. The vacuum expectation values (vevs) of the neutral components of the two Higgs fields satisfy the relation $v_u^2 + v_d^2 = (\frac{246}{\sqrt{2}})^2 = (174GeV)^2$. The fermions mass matrices appear after the spontaneous symmetry breaking from the fermion-Higgs-Yukawa couplings. As a result, the initial Yukawa couplings are given by the ratios of the fermion masses to the appropriate Higgs vev as follows:

$$f_{u,c,t} = \frac{m_{u,c,t}}{v_u}, \; h_{d,s,b} = \frac{m_{d,s,b}}{v_d}, \; y_{e,\mu,\tau} = \frac{m_{e,\mu,\tau}}{v_d}, \tag{9}$$

where we define $tan\beta = \frac{v_u}{v_d}$, which is the ratio of vevs of the two Higgs fields H_u and H_d.

3. Numerical Results and Conclusions

For our numerical calculations we assume the fundamental scale is not far from the range of the LHC and set the compatification radii to be $R^{-1} = 1$ TeV, 4 TeV and 15 TeV. Only some selected plots will be shown and we will comment on the other similar cases not explicitly presented. We quantitatively analyse these quantities in the 5D MSSM with small, intermediate, and large $tan\beta$, though we observed similar behaviours for all values of $tan\beta$; the initial values we shall adopt at the M_Z scale are: for the gauge couplings $\alpha_1(M_Z) = 0.01696$, $\alpha_2(M_Z) = 0.03377$, and $\alpha_3(M_Z) = 0.1184$; for the fermion masses $m_u(M_Z) = 1.27$ MeV, $m_c(M_Z) = 0.619$ GeV, $m_t(M_Z)$ = 171.7 GeV, $m_d(M_Z) = 2.90$ MeV, $m_s(M_Z) = 55$ MeV, $m_b(M_Z) = 2.89$ GeV, $m_e(M_Z) = 0.48657$ MeV, $m_\mu(M_Z) = 102.718$ MeV, and $m_\tau(M_Z) = 1746.24$ MeV [3, 15].

As illustrated in Figs. 2 and 3 the mass ratios evolve in the usual logarithmic fashion when the energies are below 1 TeV, 4 TeV and 15 TeV respectively. However, once the first KK threshold is reached the contributions from the KK states become increasingly significant and the effective 4D MSSM couplings begin to deviate from their normal trajectories. They evolve faster and faster after that point, their evolution diverging due to the faster running of the gauge

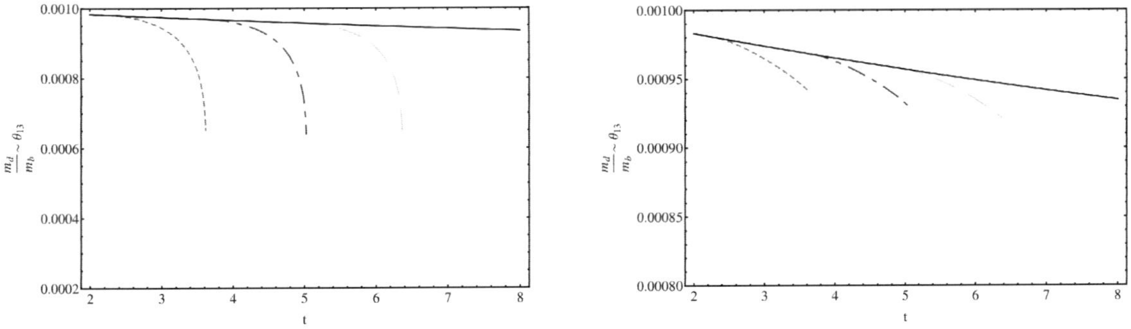

Figure 2. (Colour online) Evolution of the mass ratio $\frac{m_d}{m_b}$, with: in the left panel, all matter fields in the bulk; and the right panel for all matter fields on the brane. Three different values of the compactification raduis have been used 1 TeV (dotted red line), 4 TeV (dot-dashed blue line), 15 TeV (dashed green line), all as a function of the scale parameter t.

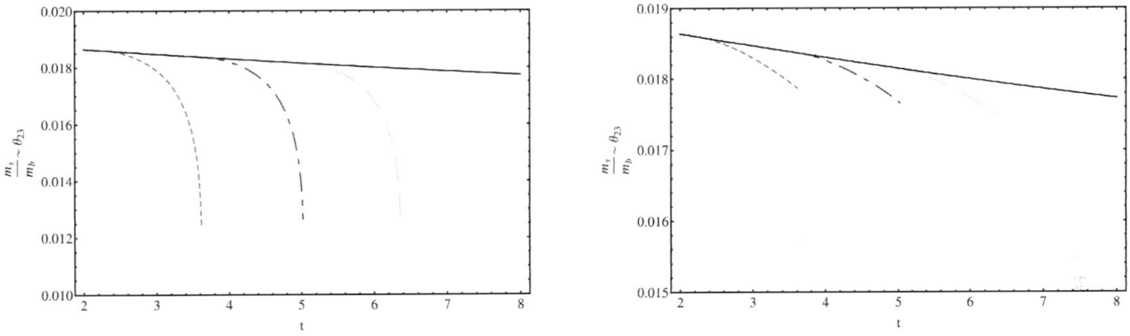

Figure 3. (Colour online) Evolution of mass ratio $\frac{m_s}{m_b}$, with the same notations as Fig.2.

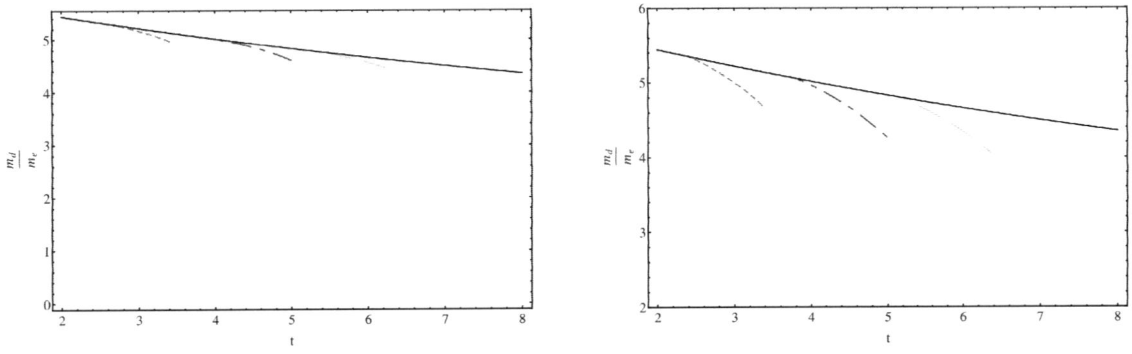

Figure 4. (Colour online) Evolution of mass ratio $\frac{m_d}{m_e}$, with the same notations as Fig.2.

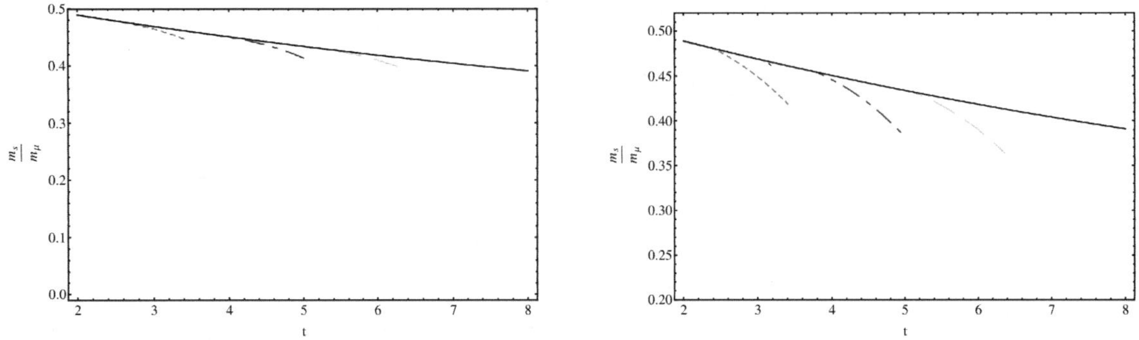

Figure 5. (Colour online) Evolution of mass ratio $\frac{m_s}{m_\mu}$, with the same notations as Fig.2.

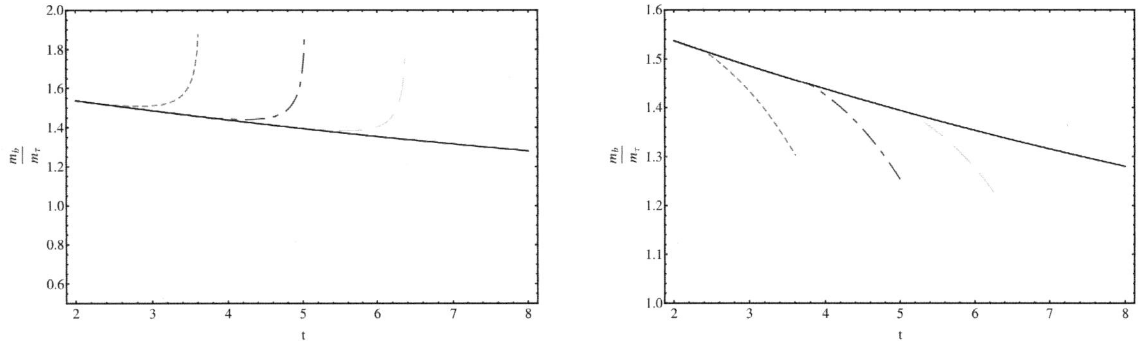

Figure 6. (Colour online) Evolution of mass ratio $\frac{m_b}{m_\tau}$, with the same notations as Fig.2.

couplings, where in approaching our cutoff for the effective theory, Λ, any new physics would then come into play, see Fig.1. As such, we have chosen cutoffs for our effective theory of; for the bulk case, where g_3 becomes large (and perturbation theory breaks down), and for the brane case, where $g_2 = g_3$ (and an expected mechanism for unification would take over). As such, the one-loop running of the gauge couplings changes with energy scale drastically and lowers the unification scale considerably. Specifically, for the compactification radii $R^{-1} = 1, 4, 15$ TeV, we find that for the brane localised matter fields case the cut-off $\Lambda \approx 30, 120, 430$ TeV respectively. For the bulk case the cut-off is around $\Lambda \approx 6, 30, 70$ TeV respectively.

On the other hand, in $SU(5)$ theory we have $m_d = m_e$, $m_s = m_\mu$ and $m_b = m_\tau$ at the unification scale, where in 5D MSSM, due to power law running of the Yukawa couplings, the renormalization effects on these relations can be large for $\frac{m_d}{m_e}$ and $\frac{m_s}{m_\mu}$, for both scenarios see Figs. 4 and 5. We have shown, by numerical analysis of the one loop calculation, that the mass ratios $\frac{m_d}{m_e}$, $\frac{m_s}{m_\mu}$ and $\frac{m_b}{m_\tau}$ decrease as energy increases. However, $\frac{m_b}{m_\tau}$ for matter fields in the bulk increases as energy increases see Fig. 6 left panel.

As depicted in Fig. 6, for the third generation the mass ratios increase rapidly as one crosses the KK threshold at $\mu = R^{-1}$ for the bulk case, resulting in a rapid approach to a singularity before the unification scale is reached. However, for the brane localised case the contribution from the gauge couplings may become significant, therefore the trajectory might change direction, and the mass ratios decrease instead of increasing.

In conclusion, in this paper the mass ratios in 5D MSSM on a S^1/Z_2 orbifold, for different possibilities of matter field localisation, were discussed. That is, where they are either bulk propagating or localised to the brane. We found that the 5D MSSM has substantial effects on the scaling of fermion masses for both cases, including both quark and lepton sectors. We quantitatively analysed these quantities in the 5D MSSM with small, intermediate, and large $tan\beta$ values, though we observed similar behaviours for all values of $tan\beta$. We have shown that the scale dependence is no longer logarithmic, having a power law behaviour. We also found that for both scenarios the theory is valid up to the unification scale receiving significant renormalization group corrections. Therefore the 5D MSSM promises exciting phenomenology for upcoming collider physics results.

References

[1] D. Falcone, Int. J. Mod. Phys. A **17**, 3981 (2002) [hep-ph/0105124]; L. -X. Liu, Int. J. Mod. Phys. A **25**, 4975 (2010) [arXiv:0910.1326 [hep-ph]].

[2] A. S. Cornell and L. -X. Liu, Phys. Rev. D **83**, 033005 (2011) [arXiv:1010.5522 [hep-ph]].

[3] A. S. Cornell, A. Deandrea, L. -X. Liu and A. Tarhini, Phys. Rev. D **85**, 056001 (2012) [arXiv:1110.1942 [hep-ph]].

[4] H. -C. Cheng, K. T. Matchev and M. Schmaltz, Phys. Rev. D **66**, 036005 (2002) [hep-ph/0204342].

[5] G. Bhattacharyya, A. Datta, S. K. Majee and A. Raychaudhuri, Nucl. Phys. B **821**, 48 (2009) [arXiv:0904.0937 [hep-ph]].

[6] L. -X. Liu and A. S. Cornell, PoS KRUGER **2010**, 045 (2010) [arXiv:1103.1527 [hep-ph]].

[7] G. -Y. Huang, K. Kong and S. C. Park, JHEP **1206**, 099 (2012) [arXiv:1204.0522 [hep-ph]].

[8] V. D. Barger, M. S. Berger and P. Ohmann, Phys. Rev. D **47**, 2038 (1993) [hep-ph/9210260].

[9] N. Arkani-Hamed, T. Gregoire and J. G. Wacker, JHEP **0203**, 055 (2002) [hep-th/0101233]; N. Marcus, A. Sagnotti and W. Siegel, Nucl. Phys. B **224**, 159 (1983); E. A. Mirabelli and M. E. Peskin, Phys. Rev. D **58**, 065002 (1998) [hep-th/9712214]; I. L. Buchbinder, S. J. Gates, Jr., H. -S. Goh, W. D. Linch, III, M. A. Luty, S. -P. Ng and J. Phillips, Phys. Rev. D **70**, 025008 (2004) [hep-th/0305169].

[10] K. R. Dienes, E. Dudas and T. Gherghetta, Nucl. Phys. B **537**, 47 (1999) [hep-ph/9806292].

[11] I. Antoniadis, Phys. Lett. B **246**, 377 (1990); C. Csaki, In *Shifman, M. (ed.) et al.: From fields to strings, vol. 2* 967-1060 [hep-ph/0404096]; Y. A. Kubyshin, hep-ph/0111027; V. A. Rubakov, Phys. Usp. **44**, 871 (2001) [Usp. Fiz. Nauk **171**, 913 (2001)] [hep-ph/0104152]; A. Perez-Lorenzana, J. Phys. Conf. Ser. **18**, 224 (2005) [hep-ph/0503177]; M. Quiros, hep-ph/0302189.

[12] A. Deandrea, J. Welzel, P. Hosteins and M. Oertel, Phys. Rev. D **75**, 113005 (2007) [hep-ph/0611172].

[13] A. S. Cornell and L. -X. Liu, Phys. Rev. D **84**, 036002 (2011) [arXiv:1105.1132 [hep-ph]]; A. S. Cornell, A. Deandrea, L. -X. Liu and A. Tarhini, arXiv:1209.6239 [hep-ph]; A. S. Cornell, A. Deandrea, L. -X. Liu and A. Tarhini, arXiv:1206.5988 [hep-ph].

[14] K. S. Babu, Z. Phys. C **35**, 69 (1987).

[15] Z. -z. Xing, H. Zhang and S. Zhou, Phys. Rev. D **77**, 113016 (2008) [arXiv:0712.1419 [hep-ph]].

International Workshop on Discovery Physics at the LHC (Kruger2012) IOP Publishing
Journal of Physics: Conference Series **455** (2013) 012051 doi:10.1088/1742-6596/455/1/012051

RHIC and LHC heavy ion collisions with pQCD Boltzmann transport

I. Bouras[1], O. Fochler[1], F. Reining[1], F. Senzel[1], J. Uphoff[1], C. Wesp[1], Z. Xu[2] and C. Greiner[1]

[1] Institut für Theoretische Physik, Johann Wolfgang Goethe-Universität, Max-von-Laue-Str. 1, D-60438 Frankfurt am Main, Germany

[2] Department of Physics, Tsinghua University, Beijing 100084, China

E-mail: carsten.greiner@th.physik.uni-frankfurt.de

Abstract. We discuss recent applications of the partonic perturbative QCD based cascade model BAMPS with focus on heavy-ion phenomeneology in hard and soft momentum range. First, the elliptic flow and suppression of charm and bottom quarks are studied at LHC energies. Thereafter, we compare in a detailed study the standard Gunion-Bertsch approximation of the matrix elements for inelastic processes to the exact results at leading order perturbative QCD. Since a disagreement is found, we propose an improved Gunion-Bertsch matrix element, which agrees with the exact result in all phase space regions. In addition to recent jet-quenching investigations we present results on the application of jet reconstruction algorithms in BAMPS. Finally, jet phenomena inducing collective effects such as Mach cones, are demonstrated in an expanding medium at relativistic heavy-ion collisions. We address the question, whether the signal of a Mach cones is the origin for a double peak structure in extracted two-particle correlations.

1. Introduction

In the last couple of years the heavy-ion community put a lot of effort in understanding the mechanism present in heavy-ion collisions at ultrarelativistic energies in the experiments at the Relativistic Heavy-Ion Collider (RHIC) and the Large Hadron Collider (LHC). Recent experimental and theoretical results indicate that a new state of matter,the quark-gluon plasma (QGP), with a very short life time is created. Although the QGP is not available for direct observation, its properties can be deduced from the measurement of the produced hadrons in the final state.

The large values of the measured hadronic elliptic flow v_2 [1], which is the second coefficient of the Fourier series of the azimuthal particle multiplicity, suggests that equilibration of quarks and gluons occurs on a very short time scale ≤ 1 fm/c. This also suggests that the shear viscosity over entropy density ratio η/s of the QGP is very small, which means that the QGP behaves like a nearly perfect fluid. All these conclusions can be drawn from comparison of experimental results with hydrodynamic calculations. However, an understanding of the mechanism of fast thermalization can not be achieved in the scope of hydrodynamic models. The early pre-equilibrium dynamics of the QGP must be studied in the scope of the microscopic theory.

In contrast to the hydrodynamic approach, kinetic transport theory is a microscopic theory and thus allows to study processes of soft and hard processes simultaneously. This is in particular important for detailed understanding of further properties of the quark-gluonic medium, such as the suppression of jets and heavy-quarks. Suppression of jets, also known as jet quenching, is quantified by comparing the hadron multiplicities measured in heavy-ion collisions with appropriately scaled multiplicities from $p+p$

collisions [2]. In addition, very exciting jet-associated particle correlations were observed [3], which might be the result of a conical emission off propagating shock waves in form of Mach Cones. These Mach Cones might be induced by high-energy partons traversing the expanding medium [4]. However, recent studies show that the most promising explanation for the appearing double-peak structure are fluctuating initial conditions[5]. Nevertheless, the issue about Mach cones is yet not clarified and demands a deeper investigation.

The kinetic transport model BAMPS (Boltzmann Approach to Multiparton Scatterings) [6] has been developed to provide a unified description of dynamics of the early QGP stage of heavy-ion collisions (HIC) including perturbative QCD based elastic and inelastic processes. BAMPS has been applied to provide explanations of fast thermalization on a very short time scale ≤ 1 fm/c [7] as well as a small value of $\eta/s \approx 0.08-0.2$ for $\alpha_s = 0.6-0.3$ [8, 9]. These results have been challenged in Ref. [10], which claims that a mis- or double-counting of symmetry factors when applying the Gunion-Bertsch (GB) matrix element to inelastic processes might lead to an overestimation of interaction rates in BAMPS by a factor of six. In Ref. [11] we addressed these issues in great detail by providing extensive numerical comparisons between the GB approximation and the exact leading order pQCD result for light partons. To this end, we proposed an improved version of the GB matrix element that cures some problems within the GB approximations and, thus, is valid in all regions of phase space. However, we want to emphasize that the deviations of rates computed in the original GB approximation compared to rates computed from the improved GB and the exact matrix element are caused deep within the approximations made by GB and are not given by simple symmetry factors as argued in Ref. [10]. The line of argumentation is outlined in Sec. 3. Recent BAMPS calculations with the original GB matrix element provide results on elliptic flow [12, 13] and jet quenching [14] at RHIC energies, which is for the first time done in a consistent and fully pQCD–based microscopic transport model. New calculations for various observables with the improved GB matrix element are underway. In addition, BAMPS has been used in certain works as a reference for hydrodynamic calculations. This opens the possibility to study hydrodynamic phenomena for arbitrary viscosity.

In these proceedings we demonstrate the application of BAMPS on several number of phenomena observed in the recent heavy-ion experiments. In Sec. 2 BAMPS results on elliptic flow and suppression of charm and bottom quarks for LHC energies are introduced. In the following Sec. 3 we show the comparison of the Gunion-Bertsch approximation to the exact result within leading order perturbative QCD. In Sec. 4 we introduce preliminary results on the application of jet reconstruction algorithms. Finally, in Sec. 5 we demonstrate the evolution of a jet inducing a shock wave in form of a Mach cone in relativistic heavy-ion collisions.

2. Elliptic flow and suppression of heavy quarks

Heavy quarks are a good probe to study the properties of the QGP. They are accurately calibrated due to their well known production process. Furthermore, heavy quarks are produced entirely in the early stage of the heavy ion collision due to their large mass [15] and are also tagged during hadronization due to flavor conservation. While heavy quarks at RHIC can only be measured indirectly via heavy flavor electrons, at LHC for the first time it is possible to reconstruct D mesons and, therefore, receive information only about charm quarks.

The elliptic flow v_2 and the nuclear modification factor R_{AA} are important observables for heavy quarks. Although those particles are rare probes, both observables are experimentally accessible for fragmentation and decay products of heavy quarks such as D mesons or heavy flavor electrons. The R_{AA} reflects how much energy heavy quarks loose in the QGP. The v_2 is large if heavy quarks interact often with the medium and pick up its collective flow.

All the calculations for heavy quarks in this section are done with a running coupling and an improved Debye screening. The latter means that the screening mass of the t channel of elastic scatterings is determined such that the energy loss matches the energy loss of a heavy quark calculated within the hard thermal loop approach.

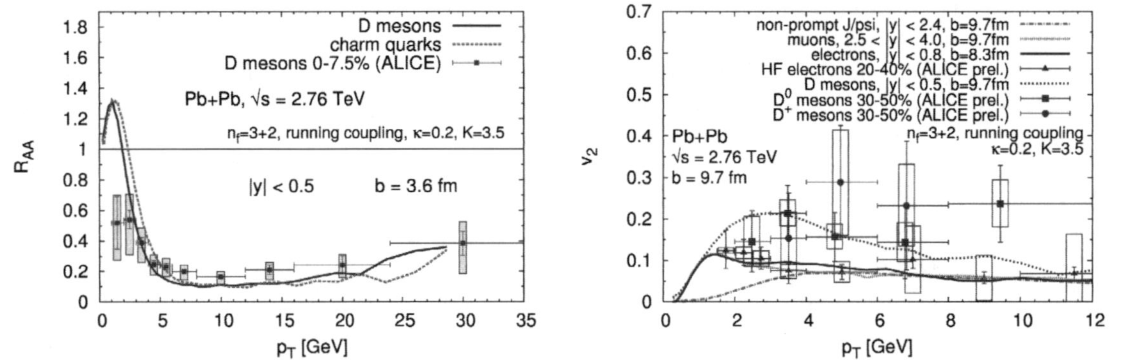

Figure 1. Left: Nuclear modification factor R_{AA} of D mesons for $b = 3.6\,\mathrm{fm}$ with data [22]. Right: Elliptic flow v_2 of D mesons, non-prompt J/ψ, muons, and electrons at LHC together with available data for D mesons and electrons [23, 24].

Detailed studies in BAMPS [16, 17, 18, 19, 20, 21] show that elastic energy loss of heavy quarks alone is not compatible with the experimental data at RHIC and LHC. However, elastic energy loss explains a significant portion of the overall suppression. If we employ a running coupling and improved Debye screening the experimental data for both v_2 and R_{AA} for both RHIC and LHC can be explained if the elastic cross section is multiplied with the artificial factor $K = 3.5$. This indicates that radiative energy loss should be about two times larger than the elastic energy loss. However, this must be checked in a forthcoming study, which will be carried out with the improved Gunion-Bertsch matrix element from Sec. 3 generalized to heavy quarks. On the left hand side of Fig. 1 the nuclear modification factor of D mesons for very central events at LHC is depicted. Our prediction with BAMPS is slightly smaller than the experimental data points. This is in agreement with the observation that we also underestimate the R_{AA} of non-prompt J/ψ, heavy flavor electrons, and muons [16].

The right hand side of Fig. 1 shows our predictions of the elliptic flow of D mesons, non-prompt J/ψ, heavy flavor electrons, and muons. For D mesons and electrons there is already data available which agrees well with our calculations.

3. Comparison of the Gunion-Bertsch cross section to the exact result

A commonly used approximation to the leading order perturbative QCD matrix element for partonic $2 \leftrightarrow 3$ processes is a result derived by GUNION and BERTSCH (GB) [25]. This approximation gives a comparatively simple expression for the gluon radiation amplitude in terms of the transverse momentum of the radiated gluon $k_\perp$ and the transverse exchanged momentum $q_\perp$.

When employing the matrix element to obtain rates or cross sections from phase space integration, the amplitude by GUNION and BERTSCH deviates from the exact result [26] in characteristic regions of the phase space [11], which leads to a sizeable deviation [10]. Therefore, we propose an improved version of the Gunion-Bertsch matrix element which agrees very well with the exact result in all phase space regions:

$$\left|\overline{\mathcal{M}}_{qq'\to qq'g}\right|^2 \simeq= 12g^2 \left|\overline{\mathcal{M}}_{qq'\to qq'}\right|^2_{\mathrm{sa}} (1-\bar{x})^2 \frac{q_\perp^2}{k_\perp^2 (\mathbf{q}_\perp - \mathbf{k}_\perp)^2} \tag{1}$$

where “sa” stands for small angle approximation, g is the strong coupling and $\bar{x}$ the symmetrized longitudinal momentum fraction of the emitted gluon,

$$\bar{x} = \frac{k_\perp}{\sqrt{s}} e^{|y|} \, . \tag{2}$$

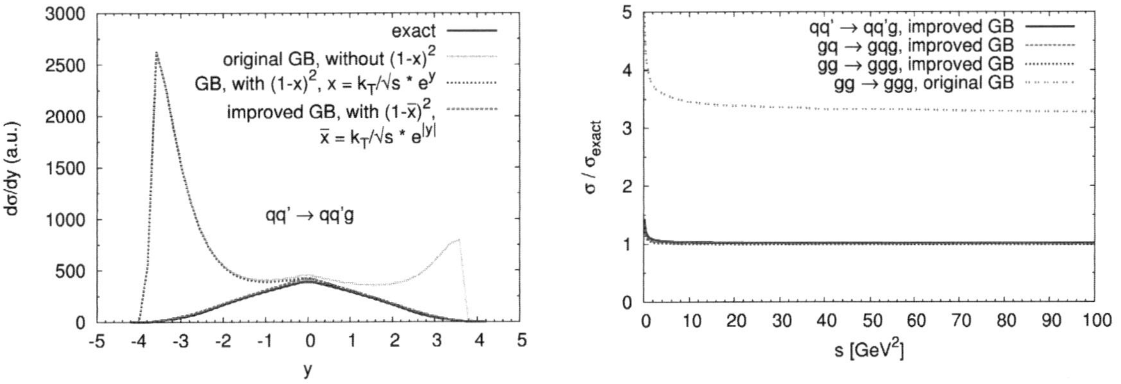

Figure 2. Left: Differential cross section $d\sigma/dy$ for the process $qq' \to qq'g$ calculated with the exact [26], original GB [25], GB with $(1-x)^2$, and improved GB with $(1-\bar{x})^2$ (Eq. (1)) matrix element. Right: Ratio of total cross section $\sigma_{2\to 3}$ of the improved and original GB matrix elements to the exact one for different processes. For details see Ref. [11].

The derivation is explicitly shown in Ref. [11]. The difference to the original GB matrix element consists of two parts, a) keeping a kinematic factor $(1-x)^2$, where x is the fraction of light cone momentum carried by the radiated gluon; and b) respecting the symmetry of the process by explicitly combining results from the two gauge choices $A^+ = 0$ and $A^- = 0$, restricting the emission of gluons to the respective forward direction. The latter is necessary since the GB approximations are not applicable in the respective backward regions for both gauges [11]. In contrast to the original GB matrix element, the improved result is not only valid at mid-rapidity, but also at forward and backward rapidity.

The left panel of Fig. 2 compares the rapidity spectrum of the emitted gluon as given by different approximations of the matrix element by depicting the differential cross section $d\sigma/dy$ for the process $qq' \to qq'g$. Although the original GB matrix element agrees very well with the exact result at mid-rapidity (the region in which GUNION and Bertsch were mainly interested in) large deviations can be seen at forward and backward rapidity. Based on a detailed analytic investigation of the underlying approximations, our proposed improved version of the GB matrix element agrees very well in all phase space regions with the exact result. This remarkable agreement holds for virtually all center of mass energies $\sqrt{s}$ as well as for the other radiative processes $qg \to qgg$ and $gg \to ggg$, which is depicted on the right hand side of Fig. 2.

The implementation of the new improved GB matrix element in our transport model BAMPS is currently under way. As part of this study we will also implement the running coupling for elastic and radiative processes as well as radiative processes for heavy quarks in Sec. 2.

4. Jet reconstruction within BAMPS

Another method for studying the parton energy loss inside a heavy-ion medium is the reconstruction of jets. The initial hard scattering processes of the approaching nucleons lead to back-to-back parton pairs, which gain a high amount of virtuality during these scattering processes. In the subsequent evolution of the partons, they try to decrease their virtuality by splitting processes like $q \to qg$ or $g \to gg$, which can be described by the DGLAP evolution equation [27, 28, 29]. These fragmentation processes lead to particle showers with a broad angle and momentum distribution. In order to provide a description of the energy loss mechanism inside the created medium, jet reconstruction methods [30, 31] are used. They combine single shower particles to a common "full jet" based on their distances $\Delta R = \sqrt{(\Delta y)^2 + (\Delta\phi)^2}$ to the jet axis. The idea behind reconstructing jets within heavy-ion collisions is that when subtracting appropriately any background medium contribution out of the jets, jet reconstruction allows the direct study of the in-medium energy loss of the initial high p_t partons.

In p+p collisions, where no medium creation is expected, these splitting processes already lead to an imbalance in the momenta of the reconstructed jets with the two highest transverse momenta. These jets are associated with the initial back-to-back parton pair and the momentum asymmetry is caused by stochastically distributed vacuum splitting processes out of the considered jet cone. Experimental results [32, 33, 34] in $\sqrt{s_{NN}} = 2.76\,\text{TeV}$ Pb+Pb collisions at the LHC show a significant enhancement of this momentum imbalance in central HIC in comparison to p+p-collisions. As a measure of this enhancement the momentum imbalance A_J,

$$A_J = \frac{p_{\text{t;Leading}} - p_{\text{t;Subleading}}}{p_{\text{t;Leading}} + p_{\text{t;Subleading}}}, \tag{3}$$

is defined, where $p_{\text{t;Leading}}$ ($p_{\text{t;Subleading}}$) is the reconstructed transverse momentum of the jet with the highest (second highest) transverse momentum. The suppression of balanced events in HIC is supposed to be the result of different in-medium energy loss of the two partons within the created bulk medium, which is a consequence of a non-central spatial production point of the initial di-jet pair.

In this section we present our preliminary results on the momentum imbalance A_J simulated within the transport model BAMPS with a constant coupling $\alpha_s = 0.3$. For the initial momentum spectra of the partons we use a distribution sampled according to a parametrized parton distribution [35], starting at $p_{t;0} = 100\,\text{GeV}$. Because BAMPS describes only scattering processes of particles on the mass-shell, it is necessary to model the initial splitting processes of the virtual partons properly for reproducing the findings in p+p collisions. Therefore the shower routines of the event generator PYTHIA [36] are used to model the virtual splitting processes. Because the medium modification of the created parton showers is to be evaluated within the BAMPS framework, it is necessary to switch off any hadronization processes and terminate the splitting processes within PYTHIA prematurely. For that reason, the standard PYTHIA global termination criterion in the virtuality $Q_0 = 1\,\text{GeV}$ is replaced by an energy-dependent minimum virtuality scale $Q_0 = \sqrt{\frac{E_{\text{parton}}}{\tau}}$ depending on the individual parton energy and a global shower time τ. Throughout this section the shower time is assumed as $\tau = 0.2\,\text{fm}$. Calculations within a static medium showed that the energy loss of the reconstructed jets is, for realistic values of τ, nearly independent of the used shower time. The initial spatial production points of the parton pairs are determined by a Glauber modeling of the initial nucleus-nucleus collisions based on a Woods-Saxon density profile.

The created parton showers are subsequently evolved within an offline recorded BAMPS background event. At every timestep the shower particles may interact with medium particles which afterwards become shower particles by their own. Only the so defined shower particles are used for reconstructing jets by the "anti-k_t"-algorithm [37] with resolution parameter $R = 0.3$ as implemented in the common "FastJet"-package [38]. As a first step for considering background effects, transverse momentum cuts $p_{t;cut}$ are used to separate shower particles from the underlying background medium. Therefore the in-medium scatterings processes of each shower particle are stopped individually when reaching $p_{t;cut}$.

Figure 3 (left) shows the calculated A_J distribution for central $\sqrt{s_{NN}} = 2.76\,\text{TeV}$ Pb+Pb collisions (0-10%, which corresponds to a mean impact parameter $b = 3.4\,\text{fm}$) in comparison with data by CMS for different $p_{t;cut}$ values. Experimental trigger conditions by CMS ($p_{\text{t;Leading}} > 120\,\text{GeV}$, $p_{\text{t;Subleading}} > 30\,\text{GeV}$, $\Delta\phi > \frac{2\pi}{3}$ and $|\eta_{\text{jet}}| < 2$) are employed and an effective modeling of the experimental jet energy resolution is used. For that an independent Gaussian smearing of the reconstructed jet momenta based on a comparison between PYTHIA simulations and p+p data by CMS [34] is applied.

The calculations show that the shape of the A_J distribution is highly sensitive to the momentum cut in the shower evolution. Simulations with an earlier termination of the scattering processes lead to a momentum imbalance which is peaked around $A_J \approx 0.6$ whose value is limited by the employed trigger conditions. For lower values of $p_{t;cut}$ and therefore a later cut in the shower evolution, the distribution gets shifted to more balanced events.

The used cut parameter $p_{t;cut}$ is obviously not sufficient to fulfill the assumed separation between the shower particles and the underlying medium particles. With a lower $p_{t;cut}$ parameter the further

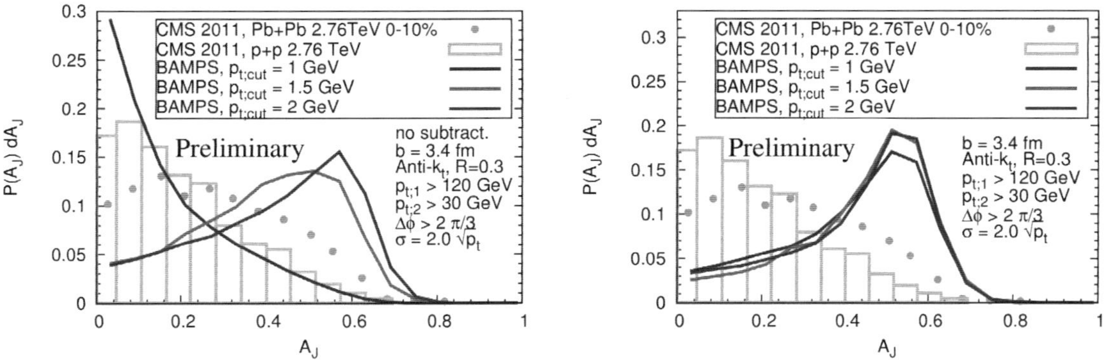

Figure 3. Momentum imbalance A_J in central Pb + Pb collisions at the LHC different $p_{t;cut}$ values [34]. Left: Jets are reconstructed based on only shower particles with no additional background subtraction employed. Right: Jets are reconstructed based on events with both shower and medium particles. Background contamination is subtracted by "CMS noise/pedestal"-background subtraction methods.

in-medium scattering processes lead to a higher contamination of medium momentum within the reconstructed jet momenta. This momentum contamination is similar to both the leading and subleading jet and at the same time greater than the in-medium momentum loss out of the reconstructed jets. Hence, the A_J distribution is shifted to events with balanced leading jet momenta.

To subtract the background contamination in the reconstructed jets more effectively, we have studied the sampling of appropriate background events in which the scattered medium particles are removed to prevent a double-counting of particles simultaneously in the shower and the medium list. These pruned events can afterwards be superposed with the simulated shower particles to get events with all shower and background particles. After reconstructing jets based on these events, common experimental subtraction algorithms [39, 34] can be used to subtract the background contamination within the reconstructed jets. However, the sampling of background events with removed scattered medium particles is not trivial: Because of the used test-particle method within BAMPS, the background medium events have different test-particle numbers than the overlayed shower events. Therefore we employ a local sampling method based on a grid in the $\eta - \phi$-plane. Detailed studies have shown that this sampling procedure leads to an uncertainty of $\approx 10\%$ in the transverse momentum of jets reconstructed in background events without additional shower particles.

Figure 3 (right) shows the calculated A_J distribution with the already described parameters and trigger conditions but with reconstructed jets based on shower as well as medium particles which are subtracted by a "CMS noise/pedestal"[34]-like background subtraction method. The distribution of A_J is not longer dependent on the used cut $p_{t;cut}$ in the shower evolution: It shows a strong peak structure at high A_J values independent from $p_{t;cut}$. This means, that the medium p_t contamination is successfully removed by the employed background subtraction. However, the enhancement and thereby the relative jet momentum loss within BAMPS is too strong in comparison with data. This is in qualitative agreement with former studies of the nuclear modification factor R_{AA} both at RHIC and LHC [20].

We have presented a procedure in which it is possible to simulate the evolution of the parton shower within a partonic medium while considering any recoil effects and further in-medium scattering processes. While considering these further in-medium scattering processes it is insufficient to use a trivial momentum cut in the in-medium shower evolution to separate between the shower and medium regime. Furthermore, we have shown that when studying reconstructed jets based on events with shower and medium particles together with a background subtraction, the momentum imbalance is independent from the employed cut in the shower evolution. Similar to the single inclusive hadron suppression, the momentum imbalance of reconstructed jet pairs within BAMPS is too strong in comparison with experimental data. However, recent studies [11] propose an improvement of the Gunion-Bertsch matrix element, which is employed for the radiative $2 \to 3$ processes within BAMPS. The implementation of the improved Gunion-Bertsch matrix element into BAMPS is currently under way and with it we expect much less partonic energy loss. Furthermore, it is planned to realize a stochastic modeling of the Landau-Pomeranchuk-Migdal effect for radiative processes of hard probes and to investigate its implications for the magnitude and shape of high-pt and jet observables.

5. Transition from ideal to viscous Mach cones in BAMPS

Highly energetic partons propagating through the hot and dense QGP rapidly lose their energy and momentum as the energy is deposited in the medium. Measurements of two- and three-particle correlations in heavy-ion collisions show a complete suppression of the away-side jet, whereas for lower p_T a double peak structure is observed in the two-particle correlation function [3]. One possible origin of these structures were assumed to be the interaction of fast partons with the soft matter which generates collective motion of the medium in form of Mach cones. [4, 40, 41, 42, 43, 44]. Although today the most promising explanation for this phenomenon is due to fluctuating initial conditions[5], we address here the question whether the jet-induced Mach cones can still be considered as possible candidates.

In a previous work [45] we demonstrated in BAMPS the transition of Mach cones from perfect fluid limit to the highly viscous regime by adjusting the shear viscosity over entropy density ratio η/s. For this purpose we used a simplified setup of a static box in order to investigate the pure evolution of a Mach cone. Using two different source terms with an infinite energy reservoir we have shown, that the double peak structure in extracted two-particle correlations appears only, if the source deposits only energy, but no momentum. However, this only holds if in addition the energy deposition is large enough. Furthermore, a finite shear viscosity over entropy density ratio η/s can destroy this double peak structure.

Although this source term with only energy but no momentum deposition was able to explain under some circumstances the double peak structure in two-particle correlations, the physical motivation for this source term is weak, since its properties does not fit in the usual picture of a jet assumed in heavy-ion collisions. Therefore, the origin of such a double peak structure observed in heavy-ion collision is most probably not the source alone, but maybe medium effect, such as the expansion of the medium. Furthermore, due to the infinite energy in our previous setups, the source never stopped, which also may have an effect of the final results.

In this proceeding we demonstrate first preliminary results using a more realistic setup in order to describe the evolution of jet-induced Mach cones in relativistic collisions of heavy-ions. For this purpose we use smoothed initial conditions for the initial setup and consider only central collisions, $\mathbf{b} = 0$, in order to avoid effects coming from elliptic flow v_2. For the initial single-particle distribution function we use the following parametrization:

$$f(\mathbf{x},\mathbf{p}) = K\frac{1}{E}\left(\frac{Q^n}{Q^n + p_T^n}\right)^m e^{\left(-y_{\rm rap}^2/\sigma_y^2\right)} e^{\left(-z^2/\sigma_z^2\right)} T_{\rm A}(\mathbf{x}_T - \mathbf{b}/2) T_{\rm B}(\mathbf{x}_T + \mathbf{b}/2) \tag{4}$$

where $T_{\rm A}$ and $T_{\rm B}$ are the nuclear thickness functions and $y_{\rm rap}$ is the longitudinal momentum rapidity. The parameters we use are $Q = 1.3\,\text{GeV}$, $n = 4$, $m = 1.5$, $\sigma_y = 1$ and $\sigma_z = 0.13\,\text{fm}$. For the normalization we use $K = 0.0135$. The parameters are chosen in such a way to fit the final spectra of recent heavy-ion collisions at RHIC. The shear viscosity over entropy density ratio is kept constant $\eta/s = 0.08$ in the whole simulation. We use only binary collisions with an isotropic angle distribution. The jet is initialized at midrapidity with $E = 50$ GeV and looses its energy only via binary scatterings. Due to the test particle method [6] the number of jets in BAMPS has to be scaled with the number of test particles, $N_{\rm test}$. This is required in order to preserve the correct relation between jet and medium.

On the left hand side of Fig. 4 we show a snapshot at $t = 8$ fm/c at midrapidity of the simulation in BAMPS. In order to demonstrate the induced shock wave of the jet, we set the initial position of the jet

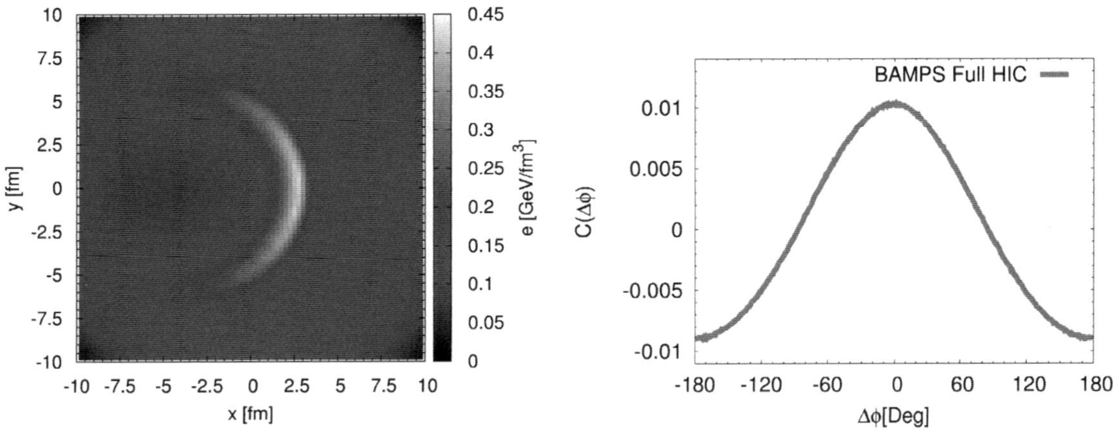

Figure 4. Left: Snapshot at $t = 8$ fm/c of a jet with initially $E = 50$ GeV propagating through an expanding medium at relativistic heavy-ion collisions inducing a Mach cone like structure. Right: Extracted two particle correlations (without any momentum cuts) of the corresponding simulation on the left panel at final time $t = 12$ fm/c.

to $x = -4$ fm and $y = 0$ fm. The jet initially propagates in positive x-direction. As expected, due to the sufficient small viscosity we observe the development of a strongly curved shock wave propagating through the medium. The strongly curved profile is a result of the loosing energy of the jet as well as the fact that the medium expands very fast. On the right hand side of Fig. 4 we show the corresponding two-particle correlations of the simulation at $t = 12$ fm/c. We clearly observe only a peak at $\Delta\phi = 0$, and not a double peak. This indicates, that we either have to use special momentum cuts in order to observe a double peak structure, or the expansion of the medium as well as the more realistic jet scenario are still not enough to obtain a double peak structure. Further investigations on this topic using the transport model BAMPS are currently under way.

Acknowledgements

The authors are grateful to the Center for the Scientific Computing (CSC) at Frankfurt for the computing resources. This work was supported by BMBF and the Helmholtz International Center for FAIR within the framework of the LOEWE program launched by the State of Hesse.

[1] Stephen Scott Adler et al. Elliptic flow of identified hadrons in Au + Au collisions at s(NN)**(1/2) = 200-GeV. *Phys. Rev. Lett.*, 91:182301, 2003.

[2] John Adams et al. Transverse momentum and collision energy dependence of high p(T) hadron suppression in Au + Au collisions at ultrarelativistic energies. *Phys. Rev. Lett.*, 91:172302, 2003.

[3] Fuqiang Wang. Measurement of jet modification at RHIC. *J. Phys.*, G30:S1299–S1304, 2004.

[4] Horst Stoecker. Collective Flow signals the Quark Gluon Plasma. *Nucl. Phys.*, A750:121–147, 2005.

[5] Bjorn Schenke, Sangyong Jeon, and Charles Gale. Elliptic and triangular flow in event-by-event (3+1)D viscous hydrodynamics. *Phys.Rev.Lett.*, 106:042301, 2011.

[6] Zhe Xu and Carsten Greiner. Thermalization of gluons in ultrarelativistic heavy ion collisions by including three-body interactions in a parton cascade. *Phys. Rev.*, C71:064901, 2005.

[7] Andrej El, Zhe Xu, and Carsten Greiner. Thermalization of a color glass condensate and review of the 'Bottom-Up' scenario. *Nucl. Phys.*, A806:287–304, 2008.

[8] Zhe Xu and Carsten Greiner. Shear viscosity in a gluon gas. *Phys. Rev. Lett.*, 100:172301, 2008.

[9] Andrej El, Azwinndini Muronga, Zhe Xu, and Carsten Greiner. Shear viscosity and out of equilibrium dissipative hydrodynamics. *Phys. Rev.*, C79:044914, 2009.

[10] Jiunn-Wei Chen, Jian Deng, Hui Dong, and Qun Wang. Shear and Bulk Viscosities of a Gluon Plasma in Perturbative QCD: Comparison of Different Treatments for the gg<->ggg Process. 2011.

[11] Oliver Fochler, Jan Uphoff, Zhe Xu, and Carsten Greiner. Radiative parton processes in perturbative QCD - an improved version of the Gunion and Bertsch cross section from comparisons to the exact result. 2013.

[12] Zhe Xu, Carsten Greiner, and Horst Stöcker. PQCD calculations of elliptic flow and shear viscosity at RHIC. *Phys. Rev. Lett.*, 101:082302, 2008.
[13] Zhe Xu and Carsten Greiner. Elliptic flow of gluon matter in ultrarelativistic heavy- ion collisions. *Phys. Rev.*, C79:014904, 2009.
[14] Oliver Fochler, Zhe Xu, and Carsten Greiner. Towards a unified understanding of jet-quenching and elliptic flow within perturbative QCD parton transport. *Phys. Rev. Lett.*, 102:202301, 2009.
[15] Jan Uphoff, Oliver Fochler, Zhe Xu, and Carsten Greiner. Heavy quark production at RHIC and LHC within a partonic transport model. *Phys. Rev.*, C82:044906, 2010.
[16] Jan Uphoff, Oliver Fochler, Zhe Xu, and Carsten Greiner. Open Heavy Flavor in Pb+Pb Collisions at $\sqrt{s} = 2.76$ TeV within a Transport Model. *Phys.Lett.*, B717:430–435, 2012.
[17] Jan Uphoff, Oliver Fochler, Zhe Xu, and Carsten Greiner. Open heavy flavor and J/ψ at RHIC and LHC. 2012.
[18] Alex Meistrenko, Andre Peshier, Jan Uphoff, and Carsten Greiner. Collisional energy loss of heavy quarks. 2012.
[19] Jan Uphoff, Oliver Fochler, Zhe Xu, and Carsten Greiner. Elliptic flow and energy loss of heavy quarks in ultra-relativistic heavy ion collisions. *Phys.Rev.*, C84:024908, 2011.
[20] Oliver Fochler, Jan Uphoff, Zhe Xu, and Carsten Greiner. Jet quenching and elliptic flow at RHIC and LHC within a pQCD-based partonic transport model. *J.Phys.G*, G38:124152, 2011.
[21] Jan Uphoff, Oliver Fochler, Zhe Xu, and Carsten Greiner. Heavy quarks at RHIC and LHC within a partonic transport model. *Nucl.Phys.*, A855:444–447, 2011.
[22] Alessandro Grelli. D meson nuclear modification factors in Pb-Pb collisions at $\sqrt{s_{NN}} = 2.76$ TeV with the ALICE detector. 2012.
[23] Giacomo Ortona. Open-Charm Meson Elliptic Flow Measurement in Pb-Pb Collisions at sqrt(sNN) = 2.76 TeV with ALICE at the LHC. 2012.
[24] Shingo Sakai. Measurement of the nuclear modification factor and v2 of electrons from heavy flavour decays in Pb-Pb collisions at 2.76 TeV with ALICE. 2012. Quark Matter 2012 proceedings.
[25] J.F. Gunion and G. Bertsch. Hadronization by color bremsstrahlung. *Phys.Rev.*, D25:746, 1982.
[26] Frits A. Berends, R. Kleiss, P. De Causmaecker, R. Gastmans, and Tai Tsun Wu. Single bremsstrahlung processes in gauge theories. *Phys. Lett.*, B103:124, 1981.
[27] V.N. Gribov and L.N. Lipatov. Deep inelastic e p scattering in perturbation theory. *Sov.J.Nucl.Phys.*, 15:438–450, 1972.
[28] Yuri L. Dokshitzer. Calculation of the Structure Functions for Deep Inelastic Scattering and e+ e- Annihilation by Perturbation Theory in Quantum Chromodynamics. *Sov.Phys.JETP*, 46:641–653, 1977.
[29] Guido Altarelli and G. Parisi. Asymptotic Freedom in Parton Language. *Nucl. Phys.*, B126:298, 1977.
[30] Gavin P. Salam. Towards Jetography. *Eur.Phys.J.*, C67:637–686, 2010.
[31] Matteo Cacciari, Juan Rojo, Gavin P. Salam, and Gregory Soyez. Jet Reconstruction in Heavy Ion Collisions. *Eur.Phys.J.*, C71:1539, 2011.
[32] Georges Aad et al. Observation of a Centrality-Dependent Dijet Asymmetry in Lead-Lead Collisions at $\sqrt{s_{NN}} = 2.77$ TeV with the ATLAS Detector at the LHC. *Phys.Rev.Lett.*, 105:252303, 2010.
[33] Serguei Chatrchyan et al. Observation and studies of jet quenching in PbPb collisions at nucleon-nucleon center-of-mass energy = 2.76 TeV. *Phys.Rev.*, C84:024906, 2011.
[34] Serguei Chatrchyan et al. Jet momentum dependence of jet quenching in PbPb collisions at $\sqrt{s_{NN}} = 2.76$ TeV. *Phys.Lett.*, B712:176–197, 2012.
[35] M. Gluck, E. Reya, and A. Vogt. Dynamical parton distributions of the proton and small x physics. *Z.Phys.*, C67:433–448, 1995.
[36] Torbjorn Sjostrand, Stephen Mrenna, and Peter Skands. PYTHIA 6.4 physics and manual. *JHEP*, 05:026, 2006.
[37] Matteo Cacciari, Gavin P. Salam, and Gregory Soyez. The Anti-k(t) jet clustering algorithm. *JHEP*, 0804:063, 2008.
[38] Matteo Cacciari, Gavin P. Salam, and Gregory Soyez. FastJet User Manual. *Eur.Phys.J.*, C72:1896, 2012.
[39] Olga Kodolova, I. Vardanian, A. Nikitenko, and A. Oulianov. The performance of the jet identification and reconstruction in heavy ions collisions with CMS detector. *Eur.Phys.J.*, C50:117–123, 2007.
[40] J. Casalderrey-Solana, E.V. Shuryak, and D. Teaney. Hydrodynamic flow from fast particles. 2006.
[41] Barbara Betz, Jorge Noronha, Giorgio Torrieri, Miklos Gyulassy, Igor Mishustin, et al. Universality of the Diffusion Wake from Stopped and Punch-Through Jets in Heavy-Ion Collisions. *Phys.Rev.*, C79:034902, 2009.
[42] Ioannis Bouras et al. Relativistic Shock Waves and Mach Cones in Viscous Gluon Matter. *J. Phys. Conf. Ser.*, 230:012045, 2010.
[43] I. Bouras et al. Relativistic shock waves in viscous gluon matter. *Phys. Rev. Lett.*, 103:032301, 2009.
[44] I. Bouras et al. Investigation of shock waves in the relativistic Riemann problem: A comparison of viscous fluid dynamics to kinetic theory. *Phys. Rev.*, C82:024910, 2010.
[45] I. Bouras, A. El, O. Fochler, H. Niemi, Z. Xu, et al. Transition From Ideal To Viscous Mach Cones In A Kinetic Transport Approach. *Phys.Lett.*, B710:641–646, 2012.

International Workshop on Discovery Physics at the LHC (Kruger2012) IOP Publishing
Journal of Physics: Conference Series **455** (2013) 012052 doi:10.1088/1742-6596/455/1/012052

Is It Hard Yet? The Qualitative Agreement of pQCD Energy Loss with RHIC and LHC Data

W. A. Horowitz

Department of Physics, University of Cape Town, Private Bag X3, Rondebosch 7701, South Africa

E-mail: wa.horowitz@uct.ac.za

Abstract. Heavy flavor research is a vigorous and active topic in high-energy QCD physics. Comparing theoretical predictions to data as a function of flavor provides a unique opportunity to tease out properties of quark-gluon plasma. We explicitly demonstrate this utility with energy loss predictions based on the assumption of 1) a weakly-coupled plasma weakly coupled to a high-p_T probe using pQCD and 2) a strongly-coupled plasma strongly coupled to a high-p_T probe using AdS/CFT; we find that while the former enjoys broad qualitative agreement with data, it is difficult to reconcile the latter with experimental measurements.

1. Introduction

Our goal as heavy ion physicists is to quantitatively extract experimentally and understand theoretically the properties of hot, dense nuclear matter; we wish to describe part of the phase diagram of the strong force. This is an extremely immodest goal. For example, detailed first-principles calculations for the phase diagram of hydrogen, the most simple QED system, is contemporary research [1]. We have a number of tools at our disposal for exploring the properties of quark-gluon plasma (QGP) from experimental measurements; we will focus on some of the observables related to high transverse momentum (high-p_T) processes.

High-p_T particles are especially interesting as they are the decay products of high-p_T partons, which are the most direct probe of the relevant degrees of freedom in a quark-gluon plasma [2, 3]. One is able in principle to learn about QGP by making an assumption regarding the physics of the QGP and comparing the necessary theoretical consequences of those assumptions to data. One hopes to use this approach to falsify certain assumed descriptions of the plasma and add evidence for others by requiring consistency in the description of measurements associated with energy loss. It turns out that this consistency is quite hard to achieve, and, as a result, energy loss provides us with a valuable window through which to actually measure the physics of quark-gluon plasma. To explicitly demonstrate the power of this method we will take two extreme, generic assumptions regarding the medium and how it couples to high-p_T probes and compare the results to data.

An energy loss calculation, in a broad sense, results in a probability of a parton losing some of its initial momentum, $P(\Delta p_T \,|\, p_T,\, L,\, T,\, M_Q,\, R)$, where L is the pathlength the parent parton travels, T is the temperature of the plasma, M_Q is the mass (or effective mass) of the parent parton, and R is the representation (i.e. is the parent parton a gluon or quark). Unfortunately one cannot directly alter these parameters experimentally to test energy loss theories; rather

Journal of Physics: Conference Series **455** (2013) 012052 doi:10.1088/1742-6596/455/1/012052

one changes, for instance, the collision species, the $\sqrt{s}$ of the collision, or—as is especially useful for investigating heavy flavor—the mass of the measured hadron, and compares to theoretical predictions.

2. Extracting Physics by Comparing to Data

2.1. Strongly-coupled Medium Strongly Coupled to a Probe

To begin our energy loss comparison to data using two extreme assumptions about the physics of the QGP, let's consider a strongly-coupled medium strongly coupled to the high-p_T probe. There are many reasons to believe that strong-coupling dynamics dominate the physics of the medium, and in particular, that AdS/CFT techniques provide valuable insight into these processes [3, 4]. For instance, running coupling calculations suggest that at $T \sim 250$ MeV—a not unreasonable placeholder for the QGP temperature—$g \sim 2$ and $\lambda = g^2 N_c \sim 12 \gg 1$; it's worth noting that in phenomenological applications T is never large compared to Λ_{QCD}. Also, for $T \gtrsim T_c$, lattice calculations nontrivially deviate from the Stefan-Boltzmann limit, and in such a way that is reasonably well described using AdS/CFT [4]. Finally, the viscosity to entropy ratio extracted from hydrodynamics calculations [5] suggest $\eta/s \sim 1/4\pi$, which is readily explained by AdS/CFT [6].

There have been calculations of the energy loss of both light and heavy quarks using the AdS/CFT correspondence. For heavy flavor, the energy loss is a drag, $dp_T/dt = -\mu p_T$, where $\mu = \pi\lambda^{1/2}T^2/2M_Q$ [7, 8]; this is similar to weakly-coupled energy loss in the Bethe-Heitler regime, but very different from the predictions of perturbative quantum chromodynamics (pQCD) in the deep Landau-Pomeranchuk-Migdal (LPM) region where $dp_T/dt \sim -LT^3 \ln(p_T/M_Q)$ [9]. Comparison between AdS/CFT calculations and data are difficult because there is no unique mapping from the parameters of QCD to those of $\mathcal{N} = 4$ SYM and $AdS_5 \times S^5$. Nevertheless, comparing over reasonable assumptions for parameter values in AdS/ CFT yields a quantitative agreement between theoretical predictions and data for non-photonic electron suppression at RHIC [10–12]. But we see again the power of comparing theoretical calculations to a wide range of data when we attempt to simultaneously describe the suppression of heavy flavor at LHC. Keeping all parameters fixed and only changing the temperature of the medium (which we do according to the measured $\sqrt{s}$ dependence of the multiplicity) we can calculate zero parameter predictions for LHC, shown in Fig. (1). While the B meson suppression is currently consistent with data within the large experimental and theoretical uncertainties, the AdS/CFT calculations significantly overpredict the suppression of D mesons. Before strong conclusions are drawn, however, it turns out that momentum fluctuations [13, 14] (especially longitudinal)—whose importance should only affect momenta parametrically large compared to the momenta at which the formalism breaks down, and were neglected in these calculations—likely play a significant role numerically.

One wants not only to simultaneously compare AdS/CFT predictions to data as a function of $\sqrt{s}$ but also as a function of parton species. Unfortunately the theory of light flavor energy loss [18–20] is less well understood in AdS/CFT than for heavy quarks. Added difficulties arise in the light sector due to the lack of 1) an analytic solution for falling string configurations and 2) a good working definition for the energy lost by the probe (in principle one can exactly compute $T^{\mu\nu}$ for the plasma and thus the energy lost by the probe, but this is an extremely difficult problem both in terms of the analytics and the numerics). Preliminary estimates, however, suggest that light flavor energy loss is also overpredicted by AdS/CFT: the thermalization time for light quarks in the medium is of the order of 3 fm; even when the 1D Hubble flow of the QGP is included the thermalization time is only increased to about 4 fm [21].

In addition to checking the flavor dependence, these AdS/CFT energy loss calculations may also be tested by changing collision species and looking at the suppression of heavy flavor at forward rapidities [22].

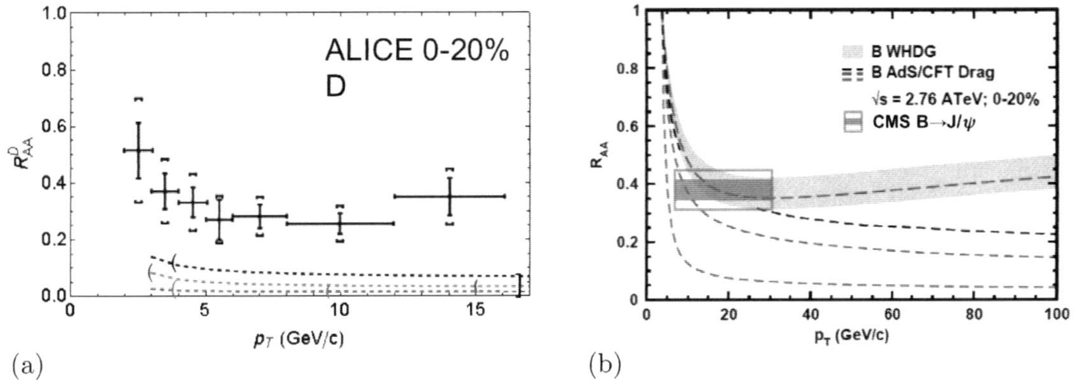

Figure 1. Comparison of (a) D [15] and (b) B [16] meson $R_{AA}(p_T)$ for 0-20% centrality collisions at LHC to AdS/CFT heavy quark drag predictions constrained by RHIC data [11, 12, 17].

2.2. *Weakly-coupled Medium Weakly Coupled to a Probe*

The assumption of the dominance of weakly-coupled dynamics in heavy ion collisions is also not unreasonable. For $T \sim 250$ MeV, $\alpha_s(2\pi T) = 0.3$. Also, multi-loop thermal field theory is in good agreement with lattice data for thermodynamic properties of QCD at a few times T_c, albeit with large uncertainties [23]. Finally, numerically daunting parton cascade calculations show that including $2 \to 3$ channels yields $\eta/s \sim \text{few}/4\pi$ [24].

Continuing with the assumption of a weakly-coupled plasma coupling weakly to a probe, the medium is described by two scales: the Debye screening length, given in terms of the Debye mass $\mu \sim gT$, and a mean free path for gluons, $\lambda^g_{mfp} \sim 1/g^2T$ (see [25] and references therein). When evaluated at temperature scales relevant for RHIC and LHC and with all the numerical coefficients, one finds[1] an ordering of scales $1/\mu \ll \lambda^g_{mfp} \ll L$, where again L is the pathlength travelled by the parent parton and is on the scale of the radius of the nucleus, $L \sim R_A$. Since $1/\mu \ll \lambda^g_{mfp}$, high-$p_T$ particles scatter off of well defined, separated medium quasi-particles. It is important to note however that for heavy quarks $L/\lambda_{mfp} \sim 4$ (and similarly even for gluons), and therefore energy loss models that assume a large number of collisions (and that thus the central limit theorem holds), such as those using Langevin or rates methods, likely require large corrections.

In pQCD with its quasi-particle picture one can distinguish between two types of energy loss: elastic and inelastic, otherwise known as collisional and radiative, respectively. There is a long history of pQCD-based elastic energy loss calculations (see [12] and references therein); see [2] for a review of pQCD-based radiative energy loss calculations. Leading order estimates of the size of elastic energy loss yield $dp_T^{el}/dt \sim -T^2 \ln(p_T/M_Q)$. Naively, at asymptotically large energies intuition based on classical electromagnetism leads to the conclusion that $\Delta E_{el} \ll \Delta E_{rad}$, but this is based on a Bethe-Heitler estimate of radiative energy loss in which subsequent collisions with medium particles yield incoherently summable emissions. However, there is one more important scale to qualitatively understand radiative energy loss, the formation time, τ_{form}, which characterizes the distance required for an emitted gluon to be resolved independently from the emitting parton. There is a large uncertainty in the size of τ_{form} but for emissions of large energy gluons in QGP, $\tau_{form} \gg \lambda^g_{mfp}$, and a single gluon emission is thus produced from coherent scatterings off of multiple in-medium quasi-particles. This reduction in the amount of emitted radiation is known as the Landau-Pomeranchuk-Migdal, or LPM, effect [2]. In the LPM limit $\Delta E_{rad} \propto -L\,T^3 \ln(p_T/M_Q)$. With this reduction in radiative energy loss it is possible for

[1] For $T_{RHIC} = 350$ MeV ($T_{LHC} = 450$ MeV), $1/\mu \simeq 0.3\,(0.2)$ fm and $\lambda^g_{mfp} \simeq 0.8\,(0.7)$ fm.

elastic energy loss to be important even at asymptotically high energies. We will be using the Wicks-Horowitz-Djordjevic-Gyulassy (WHDG) model of convolved radiative and elastic energy loss [25] for explicit comparison to experimental data; in this calculation which uses thermal field theory methods for computing the elastic energy loss and the Djordjevic-Gyulassy-Levai-Vitev (DGLV) derivation for the radiative, the elastic energy loss remains a significant contributor to total energy loss even for 250 GeV/c partons at LHC; see Fig. (2).

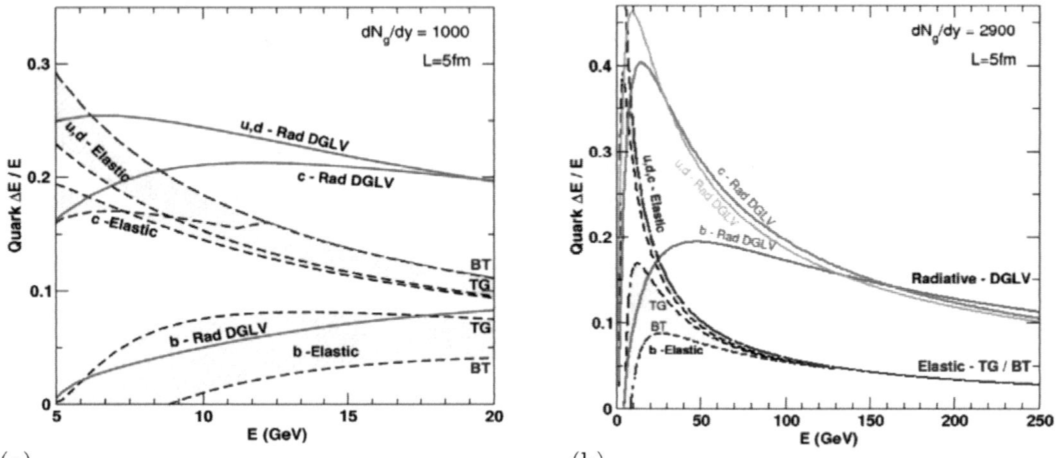

(a) (b)

Figure 2. Elastic and radiative energy loss for gluons and light, charm, and bottom quarks travelling a distance $L = 5$ fm through QGP at (a) RHIC and (b) LHC temperatures [12].

We would like to compare the WHDG model to as many observables as possible. Using the thermal field theoretic methods to relate μ and λ^g_{mfp} to temperature, assuming that the temperature profile is proportional to the Glauber participant density, and that all couplings are approximately fixed at $\alpha_s = 0.3$ there is only one free parameter in the theory, the proportionality constant relating the observed multiplicity to the entropy of the plasma. The PHENIX experiment rigorously extracted the best fit value of this parameter and its uncertainty, the rapidity density of gluons $dN_g/dy = 1400^{200}_{-375}$ [26], by comparing to their $R^{\pi^0}_{AA}(p_T)$ measurement in most central $\sqrt{s} = 200$ AGeV collisions. Before immediately comparing to the multitude of data from RHIC and LHC, it is worth noting that the lack of precision and accuracy inherent in pQCD due to both the complicated nature of the theory and the relatively large size of its coupling constant: even NLO calculations of production rates in hadronic collisions tend to be correct only within a factor of 2 of the data [27, 28]. With this in mind, the agreement between the LO WHDG energy loss theory and data shown in Fig. (3) over a range of centralities, collision energies, measurements, and flavors is surprisingly good.

There are a number of directions in which these perturbative calculations can be improved. For instance one might try to model the energy loss using a parton cascade, which trades a better treatment of multiple gluon emission for a less accurate treatment of the quantum mechanical formation time effects [33]. Or one might attempt a NLO ansatz for running coupling along with a more careful treatment of production spectra and time evolution [34].

2.3. Direct Comparison of the Pictures

Although the high-p_T physics evidence for a weakly-coupled plasma weakly coupled to a probe is strong and there are significant signs of disagreement between the predictions of a strongly-coupled plasma strongly coupled to a probe it is worth considering a measurement that shows

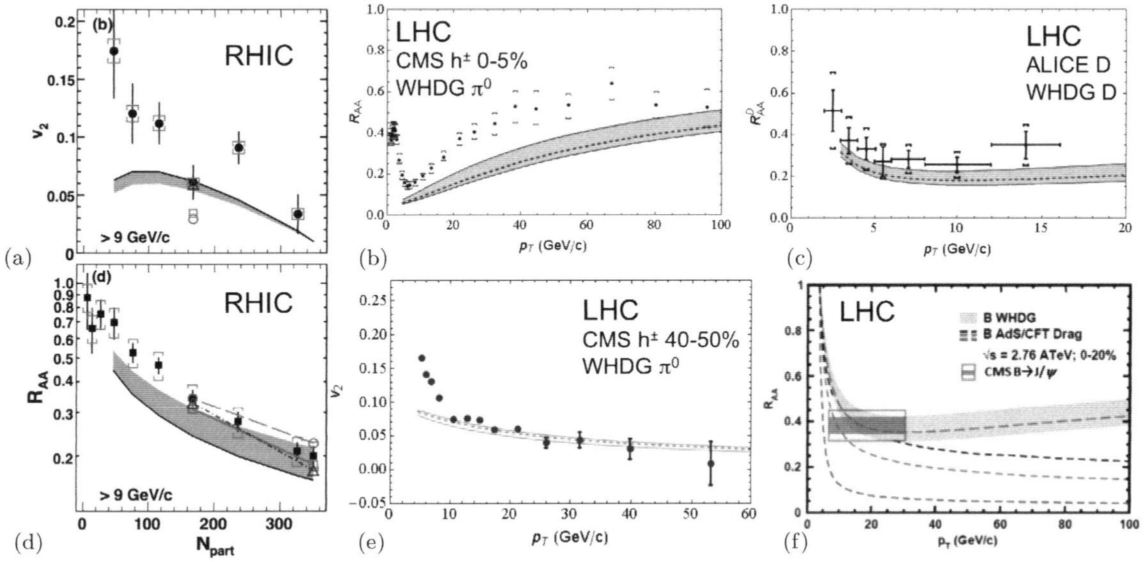

Figure 3. Constrained zero parameter WHDG predictions compared to data for (a) $v_2(N_{part})$ at RHIC [25, 29], (b) 0-5% centrality $R_{AA}(p_T)$ for light flavors at LHC [28, 30], (c) $R^D_{AA}(p_T)$ at 0-20% centrality at LHC [15, 31], (d) $R^{\pi^0}_{AA}(N_{part})$ at RHIC [25, 29], (e) $v_2(p_T)$ at LHC for light flavors at 40-50% centrality [31, 32], and (f) $R^B_{AA}(p_T)$ at 0-20% centrality at LHC [16, 17].

a qualitative difference between the two pictures. One may emphasize the different mass and momentum dependencies of the pQCD and AdS/CFT results by considering the double ratio of D to B meson R_{AA} as seen in Fig. (4). While the leading order AdS/CFT results are applicable (up to a speed limits indicated on the graph) the mass dependence of the energy loss remains; on the other hand the mass dependence drops out for the perturbative results at asymptotically large momenta.

3. Summary

We seek a coherent, consistent picture of the physics of QGP in our quest to understand its properties. The comparison of energy loss calculations to data provide a direct probe of the relevant degrees of freedom in a QGP and how this physics interacts with high-p_T particles. As we have known since the first anisotropy and heavy flavor results from RHIC, a simultaneous description of multiple observables related to energy loss physics is very hard to achieve. In particular, despite successes at RHIC, predictions for LHC based on the strong coupling physics of AdS/CFT do not appear to describe the data, although possibly important physics was neglected. However, LO pQCD results give a rather good qualitative description of a suite of observables including pion and heavy flavor suppression and anisotropy from RHIC to LHC. Should we find that these tentative conclusions hold, it becomes a very interesting question of how the strong-coupling low-p_T physics of the QGP medium as implied by hydrodynamics comparisons to data turn over to weak-coupling physics of the QGP medium as implied by the energy loss calculations.

4. Acknowledgments

Support from the National Research Foundation of South Africa and SA-CERN is gratefully acknowledged.

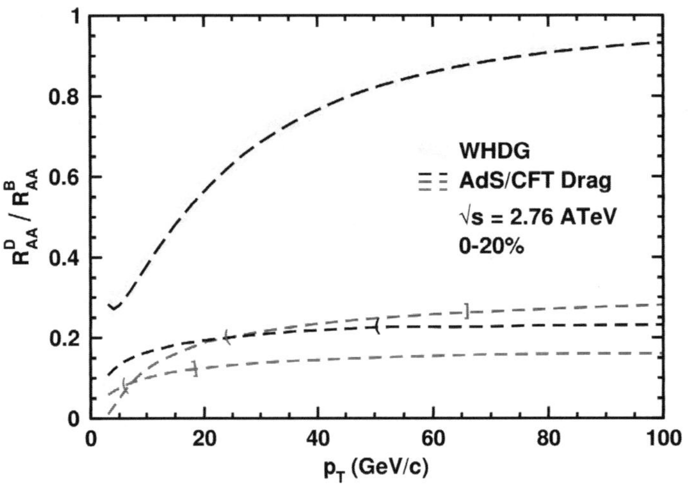

Figure 4. Comparison of the double ratio of D meson to B meson $R_{AA}(p_T)$ for 0-20% centrality collisions at LHC using the pQCD-based WHDG energy loss model [25, 30] and a model based on AdS/CFT drag energy loss [11, 17].

References

[1] Militzer B 2000 *Path Integral Monte Carlo Simulations of Hot Dense Hydrogen* Ph.D. thesis University of Illinois, Urbana-Champaign

[2] Majumder A and Van Leeuwen M 2011 *Prog.Part.Nucl.Phys.* **A66** 41–92 (*Preprint* 1002.2206)

[3] Casalderrey-Solana J, Liu H, Mateos D, Rajagopal K and Wiedemann U A 2011 (*Preprint* 1101.0618)

[4] Gubser S S 2009 *Nucl.Phys.* **A830** 657C–664C (*Preprint* 0907.4808)

[5] Heinz U, Shen C and Song H C 2012 *AIP Conf.Proc.* **1441** 766–770 (*Preprint* 1108.5323)

[6] Kovtun P, Son D and Starinets A 2005 *Phys.Rev.Lett.* **94** 111601 (*Preprint* hep-th/0405231)

[7] Herzog C, Karch A, Kovtun P, Kozcaz C and Yaffe L 2006 *JHEP* **0607** 013 (*Preprint* hep-th/0605158)

[8] Gubser S S 2006 *Phys.Rev.* **D74** 126005 (*Preprint* hep-th/0605182)

[9] Gyulassy M, Levai P and Vitev I 2001 *Nucl.Phys.* **B594** 371–419 (*Preprint* nucl-th/0006010)

[10] Akamatsu Y, Hatsuda T and Hirano T 2009 *Phys.Rev.* **C79** 054907 (*Preprint* 0809.1499)

[11] Horowitz W and Gyulassy M 2008 *Phys.Lett.* **B666** 320–323 (*Preprint* 0706.2336)

[12] Horowitz W A 2010 (*Preprint* 1011.4316)

[13] Gubser S S 2008 *Nucl.Phys.* **B790** 175–199 (*Preprint* hep-th/0612143)

[14] Casalderrey-Solana J and Teaney D 2007 *JHEP* **0704** 039 (*Preprint* hep-th/0701123)

[15] Abelev B *et al.* (ALICE Collaboration) 2012 *JHEP* **1209** 112 (*Preprint* 1203.2160)

[16] Chatrchyan S *et al.* (CMS Collaboration) 2012 *JHEP* **1205** 063 (*Preprint* 1201.5069)

[17] Horowitz W 2012 *AIP Conf.Proc.* **1441** 889–891 (*Preprint* 1108.5876)

[18] Gubser S S, Gulotta D R, Pufu S S and Rocha F D 2008 *JHEP* **0810** 052 (*Preprint* 0803.1470)

[19] Chesler P M, Jensen K, Karch A and Yaffe L G 2009 *Phys.Rev.* **D79** 125015 (*Preprint* 0810.1985)

[20] Ficnar A 2012 *Phys.Rev.* **D86** 046010 (*Preprint* 1201.1780)

[21] Morad R and Horowitz W A *in preparation*

[22] Horowitz W and Kovchegov Y V 2009 *Phys.Lett.* **B680** 56–61 (*Preprint* 0904.2536)

[23] Andersen J O, Leganger L E, Strickland M and Su N 2011 *JHEP* **1108** 053 (*Preprint* 1103.2528)

[24] El A, Muronga A, Xu Z and Greiner C 2009 *Phys.Rev.* **C79** 044914 (*Preprint* 0812.2762)

[25] Wicks S, Horowitz W, Djordjevic M and Gyulassy M 2007 *Nucl.Phys.* **A784** 426–442 (*Preprint* nucl-th/0512076)

[26] Adare A *et al.* (PHENIX Collaboration) 2008 *Phys.Rev.* **C77** 064907 (*Preprint* 0801.1665)

[27] Adare A *et al.* (PHENIX Collaboration) 2011 *Phys.Rev.* **C84** 044905 (*Preprint* 1005.1627)

[28] Chatrchyan S *et al.* (CMS Collaboration) 2012 *Eur.Phys.J.* **C72** 1945 (*Preprint* 1202.2554)

[29] Adare A *et al.* (PHENIX Collaboration) 2010 *Phys.Rev.Lett.* **105** 142301 (*Preprint* 1006.3740)

[30] Horowitz W and Gyulassy M 2011 *Nucl.Phys.* **A872** 265–285 (*Preprint* 1104.4958)

[31] Horowitz W and Gyulassy M 2011 *J.Phys.* **G38** 124114 (*Preprint* 1107.2136)

[32] Chatrchyan S *et al.* (CMS Collaboration) 2012 *Phys.Rev.Lett.* **109** 022301 (*Preprint* 1204.1850)

[33] Uphoff J, Fochler O, Xu Z and Greiner C 2012 (*Preprint* 1208.1970)

[34] Buzzatti A and Gyulassy M 2012 (*Preprint* 1207.6020)

Journal of Physics: Conference Series **455** (2013) 012053 doi:10.1088/1742-6596/455/1/012053

Implications of LHC Higgs results for supersymmetry

Sabine Kraml

Laboratoire de Physique Subatomique et de Cosmologie, UJF Grenoble 1, CNRS/IN2P3, INPG, 53 Avenue des Martyrs, F-38026 Grenoble, France

E-mail: sabine.kraml@lpsc.in2p3.fr

Abstract. The current LHC Higgs results may be used as a guide for where to look for SUSY. I discuss implications for the MSSM and NMSSM. Particularly interesting are NMSSM scenarios with large λ and small $\tan\beta$; they are characterized by light stops and light higgsinos, and offer the intriguing possibility of observing more than one light Higgs at the LHC.

1. Introduction

The recent discovery [1, 2] of a new particle with mass around 125 GeV and properties consistent with a Standard Model (SM) Higgs boson is a first triumph for the LHC physics program. However, while this discovery completes our picture of the SM, it still leaves many fundamental questions open. One of the most pressing issues is that the SM does not explain the value of the electroweak (EW) scale itself: Why is the Higgs boson so light when it is predicted to be driven to the scale of Grand Unified Theories ($M_{\rm GUT}$), or even the Planck scale, by radiative corrections? Either new physics appears at the EW scale, or the Higgs mass-squared is fine tuned at the 10^{-32} level.

New particles that couple to the Higgs can however also modify the Higgs couplings, and thus the production and decay rates in various channels. So on the one hand we expect physics beyond the SM (BSM) to explain the Higgs mass, on the other hand the measured mass and signal strengths provide significant constraints on concrete BSM realizations, see *e.g.* [3, 4]. Moreover, while the SM provides a reasonably good fit to to the current data, some new physics contributions to the Higgs couplings to gluons and photons are preferred, as shown in Fig. 1.[1]

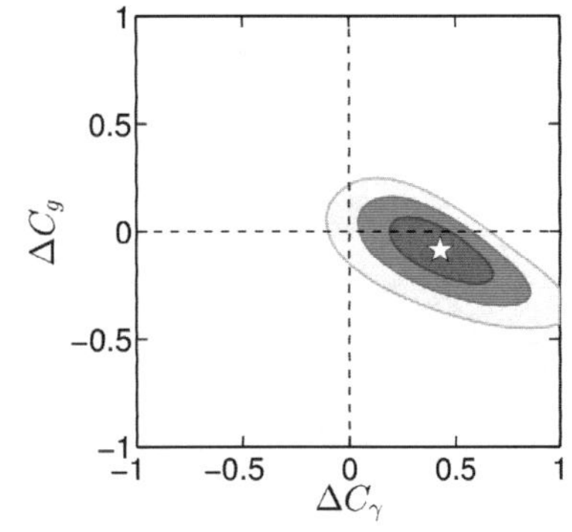

Figure 1. Global fit of additional loop contributions ΔC_g and ΔC_γ from new particles to the Higgs couplings to gluons and photons, assuming SM values for the couplings to W, Z and SM fermions. The fit includes all available Higgs signal strengths from ATLAS, CMS and the Tevatron experiments. The red, orange and yellow ellipses show the 68%, 95% and 99.7% CL regions, respectively. The white star marks the best-fit point $\Delta C_g = -0.086$, $\Delta C_\gamma = 0.426$. From [4].

[1] Based on the experimental results available at the end of 2012.

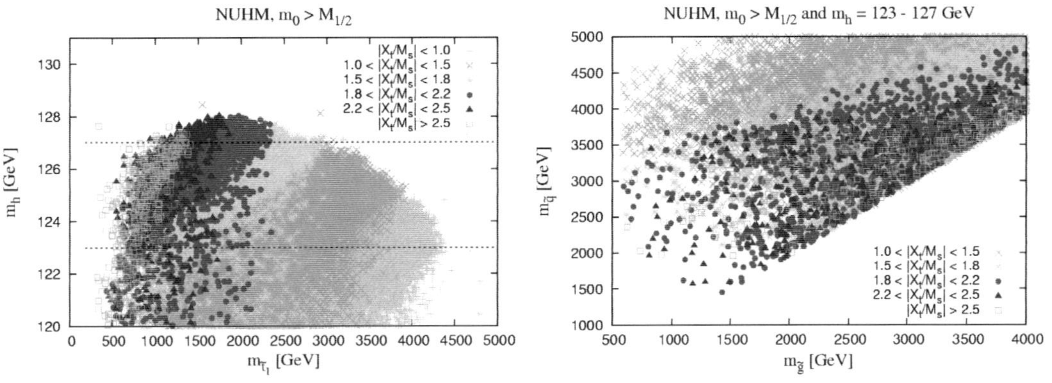

Figure 2. Left: dependence of m_h on $m_{\tilde{t}_1}$ in the NUHM model, with the amount of stop mixing indicated by a color code. Right: projection of the points with $m_h = 123-127$ GeV in the squark versus gluino mass plane. From [6].

The arguably best motivated extension of the SM is weak-scale supersymmetry (SUSY), introducing a new symmetry between fermions and bosons. SUSY solves the hierarchy problem provided SUSY particles exist at or around the TeV energy scale. The (Next-to-) Minimal Supersymmetric Standard Model, (N)MSSM, moreover predicts a light, often SM-like, Higgs boson with mass below $\approx$ 135 (140) GeV. This has always been regarded as an intriguing feature, and even more so with the actual observation of a Higgs-like state at 125 GeV. So far, however, SUSY searches at ATLAS and CMS show no signal whatsoever, and the mass limits in particular for squarks and gluinos have been pushed well into the TeV range [7, 8].

So the Higgs very likely has been found — but where is supersymmetry? In fact, the SUSY particles relevant for the Higgs sector and the hierarchy problem, stops and higgsinos, are much less constrained than light-flavor squarks and gluinos. "Natural" SUSY still lives on. In this talk I therefore discuss some implications of the LHC Higgs results for supersymmetric models.

2. Minimal Supersymmetric Standard Model

In the MSSM, $m_h^2 = m_Z^2 \cos^2 2\beta$ at tree level, where $\tan\beta = v_u/v_d$, $v = \sqrt{v_u^2 + v_d^2} = 174$ GeV. This quickly saturates to $m_h^2 \lesssim m_Z^2$ for $\tan\beta \gtrsim 5$. To further lift m_h^2 from $m_Z^2 = (91\text{ GeV})^2$ to around $(125\text{ GeV})^2$, radiative corrections nearly as large as the tree-level value are required. The leading one-loop correction comes from the top–stop sector and is given by [5]

$$\Delta m_h^2 = \frac{3}{4\pi^2}\frac{m_t^4}{v^2}\left(\log\frac{M_S^2}{m_t^2} + \frac{X_t^2}{M_S^2}\left(1 - \frac{X_t^2}{12\,M_S^2}\right)\right). \tag{1}$$

Here m_t is the running top-quark mass at the scale m_t, $M_S^2 = m_{\tilde{t}_1} m_{\tilde{t}_2}$ with $m_{\tilde{t}_{1,2}}$ the stop masses, and X_t is the stop mixing parameter, $X_t = A_t - \mu\cot\beta$, at the scale M_S. The contribution from the logarithmic term in Eq. (1) can be increased by simply raising M_S, but naturalness demands that the SUSY scale should be not too far above the EW scale. The X_t contribution is maximized at $|X_t/M_S| \simeq \sqrt{6} = 2.45$; this is called the maximal-mixing scenario. So $m_h \simeq 125$ GeV requires either (unnaturally) heavy stops, or maximal mixing. This is illustrated in Fig. 2 for a semi-constrained version of the MSSM with universal gaugino mass $M_{1/2}$, scalar mass m_0 and trilinear coupling A_0 all defined at $M_{\rm GUT}$, but non-universal Higgs mass parameters (NUHM model). As can be seen, a 125 GeV Higgs together with stops in the 0.5–1 TeV mass range indeed requires maximal mixing, *i.e.* very large $|A_t|$ (left plot). At the same time, gluinos and

1st/2nd generation squarks turn out to be heavy, with masses above 1–2 TeV (right plot), thus automatically avoiding the current LHC limits. The Higgs signal strengths in the $\gamma\gamma$ and ZZ channels are however $R \lesssim 1$ in this case, see [6].

3. Next-to-Minimal Supersymmetric Standard Model

The NMSSM differs from the MSSM by to the presence of the gauge singlet superfield $\hat{S}$ [9]. In the simplest Z_3 invariant realization of the NMSSM, the Higgs mass term $\mu \hat{H}_u \hat{H}_d$ in the superpotential $W_{\rm MSSM}$ of the MSSM is replaced by the coupling λ of $\hat{S}$ to $\hat{H}_u$ and $\hat{H}_d$ and a self-coupling κS^3. The superpotential $W_{\rm NMSSM}$ is given by:

$$W_{\rm NMSSM} = \lambda \hat{S} \hat{H}_u \cdot \hat{H}_d + \frac{\kappa}{3} \hat{S}^3 + \dots \,, \tag{2}$$

where hatted letters denote superfields, and the dots denote the MSSM-like Yukawa couplings of $\hat{H}_u$ and $\hat{H}_d$ to the quark and lepton superfields. Once the real scalar component of $\hat{S}$ develops a vev $\langle S \rangle$, the first term in $W_{\rm NMSSM}$ generates an effective μ-term, $\mu_{\rm eff} = \lambda \langle S \rangle$.

As compared to two independent parameters in the Higgs sector of the MSSM at tree level, often chosen as $\tan\beta$ and M_A, the Higgs sector is now described by

$$\lambda \,,\; \kappa \,,\; A_\lambda \,,\; A_\kappa ,\; \tan\beta \;= v_u/v_d \,,\; \mu_{\rm eff} \,. \tag{3}$$

The neutral Higgs sector of this model consists of three CP-even (h_1, h_2, h_3) and two CP-odd (a_1, a_2) states. The CP-even mass eigenstates are superpositions of the neutral CP-even components of H_u, H_d and S:

$$\begin{aligned} h_1 &= \mathcal{S}_{1,d}\, H_d + \mathcal{S}_{1,u}\, H_u + \mathcal{S}_{1,s}\, S \,, \\ h_2 &= \mathcal{S}_{2,d}\, H_d + \mathcal{S}_{2,u}\, H_u + \mathcal{S}_{2,s}\, S \,, \\ h_3 &= \mathcal{S}_{3,d}\, H_d + \mathcal{S}_{3,u}\, H_u + \mathcal{S}_{3,s}\, S \,, \end{aligned} \tag{4}$$

with the couplings to gauge bosons and fermions determined by the 3×3 mixing matrix $\mathcal{S}$, *e.g.* $g_{h_iVV}/g_{H_{\rm SM}VV} = \cos\beta\, \mathcal{S}_{i,d} + \sin\beta\, \mathcal{S}_{i,u}$.

An interesting feature is that the coupling $\lambda S \hat{H}_u \hat{H}_d$ in the superpotential leads to an extra tree-level contribution to the SM-like Higgs mass $m_{\tilde{h}}^2 = m_Z^2 \cos^2 2\beta + \lambda v^2 \sin^2 2\beta + \Delta m_{\tilde{h}}^2$. It is thus much easier to obtain $m_h \simeq 125$ GeV in constrained versions of the NMSSM then in their MSSM equivalents [10]. Moreover, as pointed out by Ellwanger [11, 12], for large λ (and small $\tan\beta$) doublet–singlet mixing can reduce the hbb coupling, thus enhancing the di-photon signal at the LHC. This works in fact for both, the lightest and the second-lightest scalar, h_1 and h_2, and either of them could be the observed state at 125 GeV [12, 10, 13].

For illustration, Fig. 3 shows the result of a scan of the "semi-constrained" NMSSM with universal m_0, $M_{1/2}$ and A_0 at the GUT scale, but the NMSSM-specific parameters of Eq. (3) treated as free parameters at the EW scale. The scan was performed with `NMSSMTools` [14]; all points have a neutralino as the lightest SUSY particle (LSP) and obey the current mass limits as well as the constraints on BR($B_s \to X_s\gamma$), ΔM_s, ΔM_d, BR($B_s \to \mu^+\mu^-$), BR($B^+ \to \tau^+\nu_\tau$) and BR($B \to X_s\mu^+\mu^-$) at 2σ. The light, medium and dark blue points have $\Omega h^2 \leq 0.136$ and obey the bounds on the spin-independent LSP–proton scattering cross section from XENON100. Light and medium red points have $0.094 \leq \Omega h^2 \leq 0.136$ and of course also pass the XENON100 bounds. (The shades of blue and red just help indicate the level of enhancement or suppression of the $\gamma\gamma$ signal.) The green points have $\Omega h^2 \leq 0.136$ and in addition explain Δa_μ within 2σ.

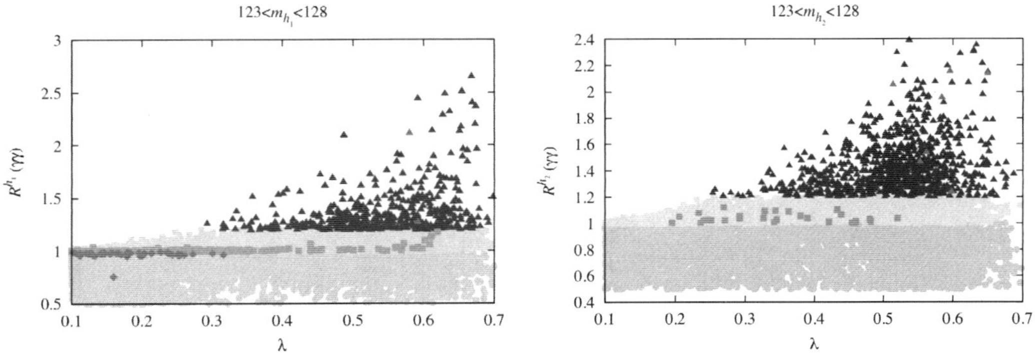

Figure 3. Signal strength (relative to SM) in the $h_i \to \gamma\gamma$ channel as function of λ from a scan over the semi-constrained NMSSM, on the left for the h_1 lying in the $123-128$ GeV mass range, on the right for the h_2 lying in the $123-128$ GeV range. See text for color code.

4. Two Higgs bosons at 98 and 125 GeV

If the h_2 of the NMSSM is responsible for the signal at 125 GeV, a particularly interesting question [15] is whether one could simultaneously explain the LHC signal and the small ($\sim 2\sigma$) LEP excess in $e^+e^- \to Zb\bar{b}$ in the vicinity of $M_{b\bar{b}} \sim 98$ GeV. We recall that the LEP excess is clearly inconsistent with a SM-like Higgs boson at this mass, being only about $10-20\%$ of the rate predicted for the $H_{\rm SM}$. Consistency with such a result for the h_1 is natural if the h_1 couples at a reduced level to ZZ, which, in turn, is automatic if the h_2 has substantial ZZ coupling, as required by the observed LHC signals.

As above, we perform a scan over the semi-constrained NMSSM. We compute the ratio of the gg or vector-boson fusion (VBF) induced Higgs cross section times the Higgs branching ratio to a given final state X, relative to the corresponding value for the SM Higgs boson, as[2]

$$R^{h_i}_{gg}(X) \equiv \frac{\Gamma(h_i \to gg)\ {\rm BR}(h_i \to X)}{\Gamma(H_{\rm SM} \to gg)\ {\rm BR}(H_{\rm SM} \to X)}, \quad R^{h_i}_{\rm VBF}(X) \equiv \frac{\Gamma(h_i \to WW)\ {\rm BR}(h_i \to X)}{\Gamma(H_{\rm SM} \to WW)\ {\rm BR}(H_{\rm SM} \to X)}, \tag{5}$$

where h_i is the ith NMSSM scalar Higgs, and $H_{\rm SM}$ is the SM Higgs boson, taking $m_{H_{\rm SM}} = m_{h_i}$.

To describe the LEP and LHC data the h_1 and h_2 must have $m_{h_1} \sim 98$ GeV and $m_{h_2} \sim 125$ GeV, respectively, with the h_1 being largely singlet and the h_2 being primarily doublet. Figure 4 shows $R^{h_1}_{\rm VBF}(b\bar{b})$ versus $R^{h_2}_{gg}(\gamma\gamma)$ for the scan points that pass LEP, B-physics and dark matter constraits as above and have in addition $m_{h_1} \in [96, 100]$ GeV and $m_{h_2} \in [123, 128]$ GeV. (These ranges take into account a 2–3 GeV theoretical error in the computation of the Higgs masses.) The points with $0.1 \leq R^{h_1}_{\rm VBF}(b\bar{b}) \leq 0.25$ would provide the best fit to the LEP excess. As can be seen, a large portion of these points have $R^{h_2}_{gg}(\gamma\gamma) > 1$ as preferred by LHC data.

In the following we thus require $m_{h_1} \in [96, 100]$ GeV with $0.1 \leq R^{h_1}_{\rm VBF}(b\bar{b}) \leq 0.25$, and $m_{h_2} \in [123, 128]$ GeV with $R^{h_2}_{gg}(\gamma\gamma) > 1$. We refer to this as the "$98 + 125$ GeV Higgs scenario". Points with $\Omega h^2 < 0.094$ are represented by blue circles and points with $\Omega h^2 \in [0.094, 0.136]$ (the "WMAP window") are represented by red/orange diamonds.

Two distinct WMAP-window regions appear. The red region has $R^{h_2}_{gg}(\gamma\gamma) \sim 1.6$ and corresponds $\mu_{\rm eff} \sim 120$ GeV and $\tan\beta \sim 2$; as can be seen in Fig. 5, it features a partly light spectrum with $m_{\tilde{\chi}^0_1} \sim 70-80$ GeV, $m_{\tilde{\chi}^\pm_1} \sim 105-110$ GeV and $m_{\tilde{t}_1} \sim 0.2-1$ TeV, while

[2] Note that $R^{h_1}_{\rm VBF}(b\bar{b})$ is equivalent to $R^{h_1}_{Vh_1}(bb)$ as relevant for LEP.

Journal of Physics: Conference Series **455** (2013) 012053 doi:10.1088/1742-6596/455/1/012053

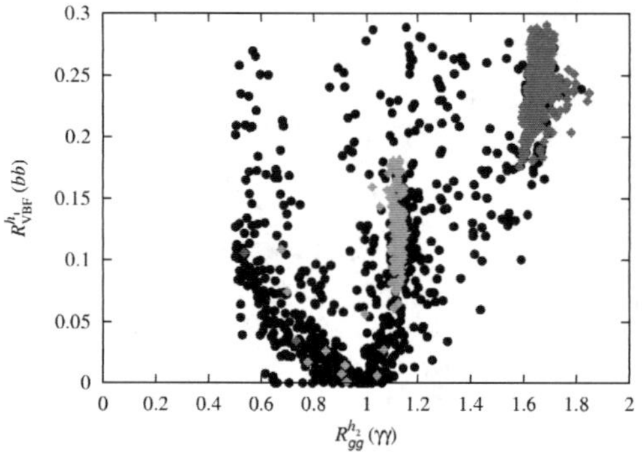

Figure 4. Signal strengths (relative to SM) $R^{h_1}_{\rm VBF}(b\bar{b})$ versus $R^{h_2}_{gg}(\gamma\gamma)$ for $m_{h_1} \in [96, 100]$ GeV and $m_{h_2} \in [123, 128]$ GeV in the semi-constrained NMSSM. Blue points have $\Omega h^2 < 0.094$ while red and orange points have $\Omega h^2 \in [0.094, 0.136]$. From [15].

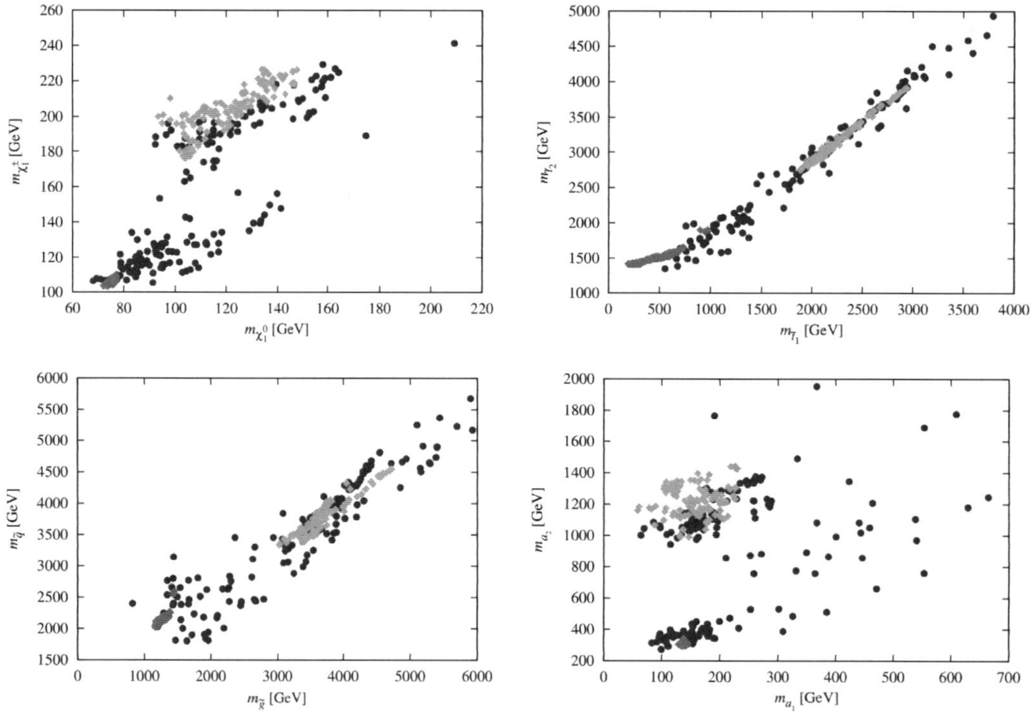

Figure 5. Expectations for sparticle and Higgs masses in the 98 + 125 GeV Higgs scenario. Blue points have $\Omega h^2 < 0.094$ while red and orange points have $\Omega h^2 \in [0.094, 0.136]$. From [15].

$m_{\tilde{g}} \gtrsim 1$ TeV and $m_{\tilde{q}} \gtrsim 2$ TeV. Again, LHC SUSY limits are automatically avoided by the Higgs-sector requirements! Moreover, the other Higgses are light, too, $m_{a_2} \sim 150$ GeV and $m_{h_3} \simeq m_{H^\pm} \simeq m_{a_2} \sim 300 - 400$ GeV. The orange region is quite different. It appears at $\mu_{\rm eff} \sim 200$ GeV and $\tan\beta \sim 5 - 8$ and has $R^{h_2}_{gg}(\gamma\gamma) \sim 1.1$. The overall mass scale is much heavier: $m_{\tilde{\chi}^0_1} \sim 90 - 150$ GeV and $m_{\tilde{t}_1} > 1.8$ TeV, see Fig. 5. Squarks and gluinos lie in the $3 - 5$ TeV mass range, above the reach of the 14 TeV LHC. The heavy Higgses also have masses above 1 TeV, only the a_1 is still light with $m_{a_1} \lesssim 250$ GeV.

The LSP decomposition and its expected spin-independent scattering cross section off protons are shown in Fig. 6. The prospects to test the 98 + 125 GeV Higgs scenario at the LHC and a future ILC are discussed in detail in [15].

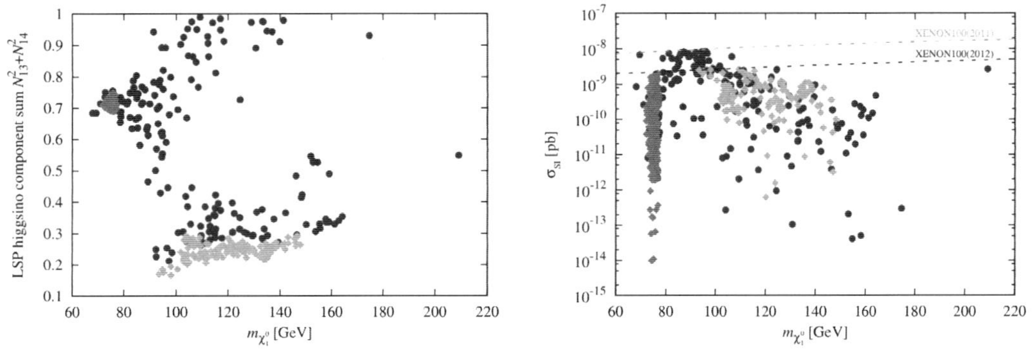

Figure 6. LSP higgsino component (left) and spin-independent scattering cross section (right) as function of the LSP mass for the 98 + 125 GeV Higgs scenarios. From [15].

5. Degenerate case: two Higgses hiding in the 125 GeV signal?

As mentioned, enhanced rates in the $\gamma\gamma$ channel arise in the NMSSM with large λ when the h_1 and h_2 are sufficiently close in mass that one Higgs, h_i, "steals" (through mixing) some of the $b\bar{b}$ width of the other Higgs, h_j. The state with the enhanced $\gamma\gamma$ signal and mass near 125 GeV can be either the h_1 or the h_2. It is however also possible that h_1 and h_2 both lie in the 123–128 GeV mass window [16]. In this case, a second mechanism for large $\gamma\gamma$ rates emerges — namely both h_1 and h_2 contribute significantly and their summed rate is enhanced even though their individual rates are more or less at, or even somewhat below, the SM level.

Figure 7 shows the correlation of $gg \to (h_1, h_2) \to \gamma\gamma$ signal strengths in the semi-constrained NMSSM when both h_1 and h_2 lie in the 123–128 GeV mass range. We see that often one Higgs dominates the signal, but it is also possible that both have $R_{gg}^{h_i}(\gamma\gamma) \gtrsim 0.5$ thus giving a combined signal larger than 1.

To go a step further, we take the net signal in given production and decay channels Y and X to simply be $R_Y^h(X) = R_Y^{h_1}(X) + R_Y^{h_2}(X)$, and we define the resulting "effective" Higgs mass as

$$m_h^Y(X) \equiv \frac{R_Y^{h_1}(X) m_{h_1} + R_Y^{h_2}(X) m_{h_2}}{R_Y^{h_1}(X) + R_Y^{h_2}(X)} \, . \tag{6}$$

Of course, the extent to which it is appropriate to combine the rates from the h_1 and h_2 depends upon the degree of degeneracy and the experimental resolution. It should be noted that the widths of the h_1 and h_2 are of the same order of magnitude as the width of a 125 GeV SM Higgs boson, *i.e.* they are very much smaller than this resolution.

In Fig. 8, we display in the left-hand plot the strong correlation between $R_{gg}^h(\gamma\gamma)$ and $R_{gg}^h(VV)$, $V = W, Z$. Note that if $R_{gg}^h(\gamma\gamma) \sim 1.5$, as suggested by current experimental results, then in this scenario $R_{gg}^h(VV) \geq 1.2$.

The scenario again prefers small μ_{eff}, which is very favorable in point of view of fine tuning, in particular if stops are also light. Indeed a good fraction of our points with degenerate h_1, h_2 and $R(\gamma\gamma) > 1$ features light stops with $m_{\tilde{t}_1} \in [300, 700]$ GeV and $M_{\text{SUSY}} = \sqrt{m_{\tilde{t}_1} m_{\tilde{t}_2}} \lesssim 1$ TeV. Because of the small μ_{eff}, the LSP is dominantly a light higgsino. A relic density of $\Omega h^2 \simeq 0.1$ can be achieved for LSP masses just below 80 GeV, see the right-hand plot in Fig. 8. The LSP is 70–80% higgsino in this case, with order 20% singlino admixture. More details on the consequences for collider and dark matter phenomenology can be found in [16].

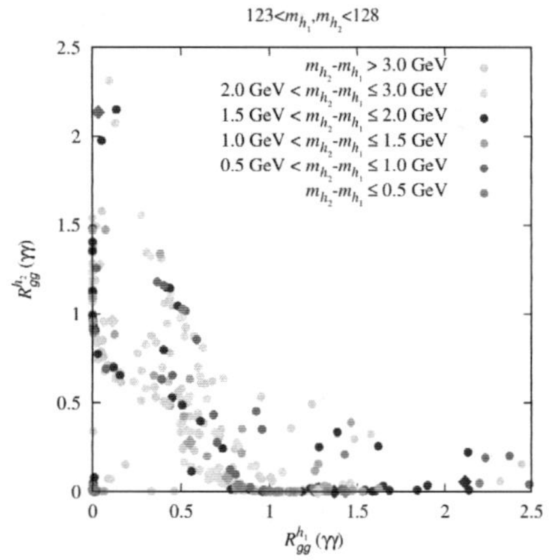

Figure 7. Correlation of $gg \to (h_1, h_2) \to \gamma\gamma$ signal strengths when both h_1 and h_2 lie in the 123–128 GeV mass range. Circular points have $\Omega h^2 < 0.094$, while diamond points have $0.094 \le \Omega h^2 \le 0.136$. Points are color coded according to $m_{h_2} - m_{h_1}$ as indicated on the figure. From [16].

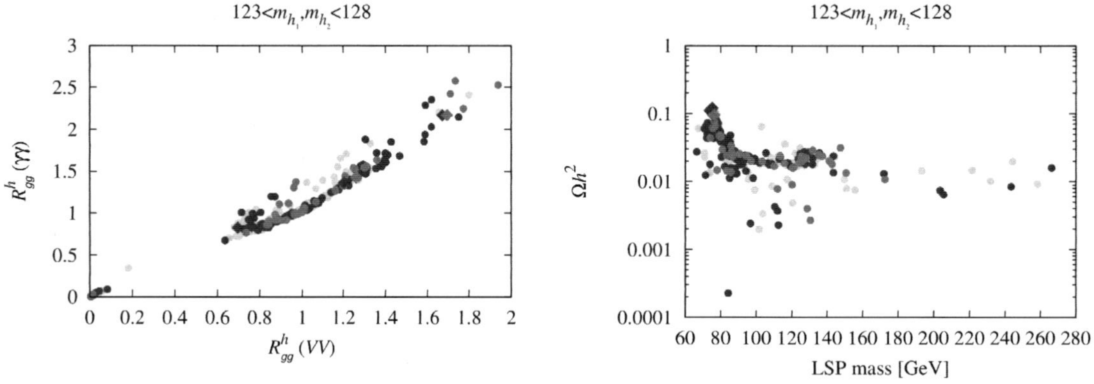

Figure 8. Correlation between the gg induced $\gamma\gamma$ and VV signal strengths (left) and relic density Ωh^2 versus LSP mass (right) for NMSSM points with quasi-degenerate h_1 and h_2 in the 123–128 GeV mass window. The green, blue and red points have $\Delta m = m_{h_2} - m_{h_1} =$ 2–3 GeV, $\Delta m =$ 1–2 GeV and $\Delta m \le 1$ GeV, respectively. From [16].

6. Diagnosing degenerate Higgs bosons

Two or more degenerate Higgs bosons will in general have different relative production rates in the VBF and gg fusion channels for one or more final states. In [17] we thus proposed double ratios of signal strengths as a useful diagnostic tool to reveal the existence of one ore more quasi-degenerate (but non-interfering in the small width approximation) Higgs states. For models with Higgs doublets+singlets, the relevant double ratios are:

$$\text{I)}: \frac{R^h_{\rm VBF}(\gamma\gamma)/R^h_{gg}(\gamma\gamma)}{R^h_{\rm VBF}(b\bar{b})/R^h_{gg}(b\bar{b})}, \quad \text{II)}: \frac{R^h_{\rm VBF}(\gamma\gamma)/R^h_{gg}(\gamma\gamma)}{R^h_{\rm VBF}(WW)/R^h_{gg}(WW)}, \quad \text{III)}: \frac{R^h_{\rm VBF}(WW)/R^h_{gg}(WW)}{R^h_{\rm VBF}(b\bar{b})/R^h_{gg}(b\bar{b})}, \tag{7}$$

each of which should be unity if only a single Higgs boson is present but are generally expected to deviate from 1 if two (or more) Higgs bosons are contributing to the net Higgs signals. Values obtained in the semi-constrained NMSSM are shown in Fig. 9.

Acknowledgements

I wish to thank Genevieve Belanger, Beranger Dumont, Ulrich Ellwanger, John F. Gunion and Yun Jiang for a most fruitful and rewarding collaboration, and the organizers of KRUGER2012

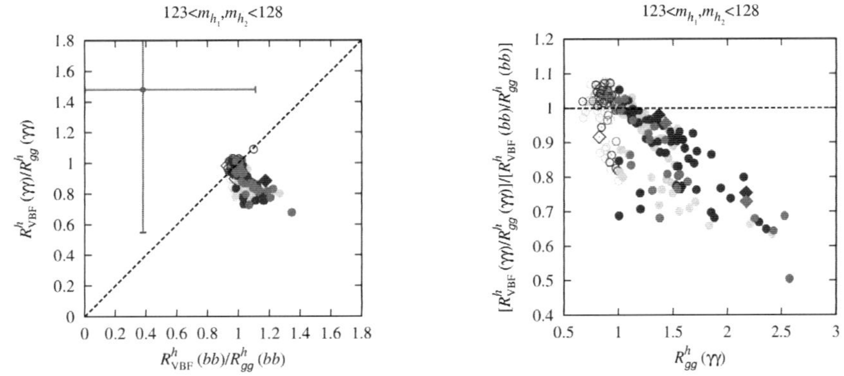

Figure 9. Illustration of the double ratio I) of Eq. (7) for degenerate h_1 and h_2 in the 123–128 GeV mass range in the semi-constrained NMSSM. The green, blue and red points have $\Delta m = m_{h_2} - m_{h_1} = 2$–$3$ GeV, $\Delta m = 1$–2 GeV and $\Delta m \leq 1$ GeV, respectively. From [17].

for the invitation to this extraordinary conference. The work presented here was supported in part by IN2P3 under contract PICS FR–USA No. 5872 and by a PEPS-PTI grant "Physique Théorique et ses Interactions".

References

[1] G. Aad *et al.* [ATLAS Collaboration], "Observation of a new particle in the search for the Standard Model Higgs boson with the ATLAS detector at the LHC," Phys. Lett. B **716** (2012) 1 [arXiv:1207.7214 [hep-ex]].
[2] S. Chatrchyan *et al.* [CMS Collaboration], "Observation of a new boson at a mass of 125 GeV with the CMS experiment at the LHC," Phys. Lett. B **716** (2012) 30 [arXiv:1207.7235 [hep-ex]].
[3] G. Cacciapaglia, A. Deandrea, G. D. La Rochelle and J. -B. Flament, "Higgs couplings beyond the Standard Model," JHEP **1303** (2013) 029 [arXiv:1210.8120 [hep-ph]].
[4] G. Belanger, B. Dumont, U. Ellwanger, J. F. Gunion and S. Kraml, "Higgs Couplings at the End of 2012," JHEP **1302** (2013) 053 [arXiv:1212.5244 [hep-ph]].
[5] A. Djouadi, "The Anatomy of electro-weak symmetry breaking. II. The Higgs bosons in the minimal supersymmetric model," Phys. Rept. **459** (2008) 1 [hep-ph/0503173].
[6] F. Brummer, S. Kraml and S. Kulkarni, "Anatomy of maximal stop mixing in the MSSM," JHEP **1208** (2012) 089 [arXiv:1204.5977 [hep-ph]].
[7] `https://twiki.cern.ch/twiki/bin/view/AtlasPublic/SupersymmetryPublicResults`.
[8] `https://twiki.cern.ch/twiki/bin/view/CMSPublic/PhysicsResultsSUS`.
[9] For a review, see *e.g.* U. Ellwanger, C. Hugonie and A. M. Teixeira, "The Next-to-Minimal Supersymmetric Standard Model," Phys. Rept. **496** (2010) 1 [arXiv:0910.1785 [hep-ph]].
[10] J. F. Gunion, Y. Jiang and S. Kraml, "The Constrained NMSSM and Higgs near 125 GeV," Phys. Lett. B **710** (2012) 454 [arXiv:1201.0982 [hep-ph]].
[11] U. Ellwanger, "Enhanced di-photon Higgs signal in the Next-to-Minimal Supersymmetric Standard Model," Phys. Lett. B **698** (2011) 293 [arXiv:1012.1201 [hep-ph]].
[12] U. Ellwanger, "A Higgs boson near 125 GeV with enhanced di-photon signal in the NMSSM," JHEP **1203** (2012) 044 [arXiv:1112.3548 [hep-ph]].
[13] U. Ellwanger and C. Hugonie, "Higgs bosons near 125 GeV in the NMSSM with constraints at the GUT scale," Adv. High Energy Phys. **2012** (2012) 625389 [arXiv:1203.5048 [hep-ph]].
[14] `http://www.th.u-psud.fr/NMHDECAY/nmssmtools.html`.
[15] G. Belanger, U. Ellwanger, J. F. Gunion, Y. Jiang, S. Kraml and J. H. Schwarz, "Higgs Bosons at 98 and 125 GeV at LEP and the LHC," JHEP **1301** (2013) 069 [arXiv:1210.1976 [hep-ph]].
[16] J. F. Gunion, Y. Jiang and S. Kraml, "Could two NMSSM Higgs bosons be present near 125 GeV?," Phys. Rev. D **86** (2012) 071702 [arXiv:1207.1545 [hep-ph]].
[17] J. F. Gunion, Y. Jiang and S. Kraml, "Diagnosing Degenerate Higgs Bosons at 125 GeV," Phys. Rev. Lett. **110** (2013) 051801 [arXiv:1208.1817 [hep-ph]].

Journal of Physics: Conference Series **455** (2013) 012054 doi:10.1088/1742-6596/455/1/012054

Heavy-quark binary scattering in the quark-gluon plasma

A. Peshier[⋆]

[⋆]University of Cape Town, Physics Department, Rondebosch 7700, South Africa

E-mail: `Andre.Peshier@uct.ac.za`

Abstract. I argue that the strong quenching and the sizable anisotropy of heavy quark flavors observed in heavy-ion collisions has a substantial component from binary scattering processes in the quark-gluon plasma. To arrive at this conclusion, I illustrate essential requirements for a more accurate theoretical description by considering first the average collisional energy loss of heavy quarks in a thermal background. On this basis, I describe a consistent transport approach, formulating the in-medium evolution of heavy quarks as a stochastic Markov process, which has been implemented for applications in heavy-ion phenomenology.

1. Introduction

Heavy quark flavors are distinguished probes for various properties of the strong interaction. In the context of heavy-ion phenomenology, in a quark-gluon plasma of temperature T, quarks of mass $M \gg T$ provide an additional scale, which may help to clarify an ongoing debate on the relevance of different in-medium parton energy loss mechanisms: The relative suppression, or 'quenching', of particle yields in nucleus-nucleus collisions has often been attributed *entirely* to radiative energy loss of *light* partons [1, 2] – a picture challenged by the observation that heavy quarks, which radiate less [4], are quenched almost as much as light quarks [5, 6]. This has revived a new interest (see [7] and references therein) for the collisional energy loss as a complementary suppression mechanism, in particular for moderately large momenta.[1] In any case, a reliable theoretical understanding of collisions of heavy quarks with thermal partons is necessary for a well-defined baseline for the analysis of the experimental data.

2. Mean energy loss

The average energy loss of a parton subject to binary collisions with gluons and n_f light quark flavors ($i = g, q$) reads

$$\frac{dE}{dx} = \frac{(dv)^{-1}}{2E} \sum_i \int_k \frac{n_i(k)}{2k} \int_{k'} \frac{\bar{n}_i(k')}{2k'} \int_{p'} \frac{1}{2E'} (2\pi)^4 \delta^{(4)}(P+K-P'-K') \, |\mathcal{M}_i|^2 \, \omega \,. \tag{1}$$

Here d is the degeneracy of the parton ($d = 6$ for quarks), $v = p/E$ is its velocity, $n(k) = 1/(e^{k/T} \mp 1)$ and $\bar{n} = 1 \pm n$ are the thermal distribution functions of the collision

[1] It seems interesting to recall that Bjorken [8] proposed *collisional* quenching of partons as a signature for the creation of a quark-gluon plasma, based on his adaption of the Bethe-Bloch formula to QCD.

and recoil partners; $\omega = E - E'$ is the energy transfer. The matrix elements $\mathcal{M}_i$ of the relevant processes (see Fig. 1) are squared and summed over color/spin states of all particles, and we use a shorthand notation $\int_k = \int d^3k/(2\pi)^3$. Bjorken [8] estimated dE/dx of energetic partons from

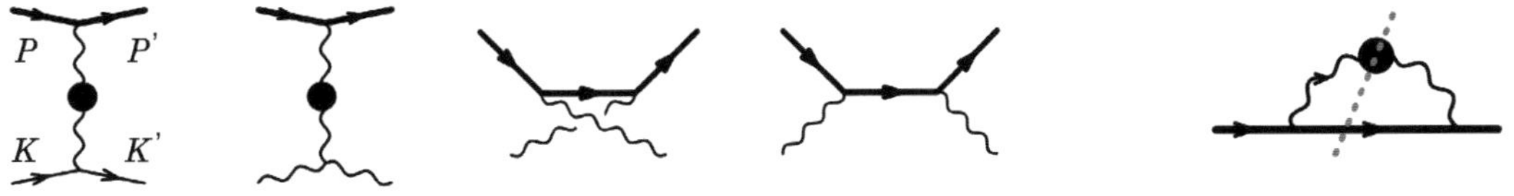

Figure 1. Heavy-quark scattering off light quarks (thin lines) and gluons (wavy lines). The t-channel contributions, which require screening (depicted by a blob), can be calculated from the cut *hard thermal loop* resummed self-energy (right diagram).

the two t-channel processes by

$$\frac{dE}{dx}^{\text{Bjorken}} = \frac{4\pi T^2}{3} \left(1 + \tfrac{n_f}{6}\right) \alpha^2 \left(\ln \frac{ET}{\mu^2} + c(n_f) + \dots \right), \tag{2}$$

imposing a cut-off $\mu \sim \sqrt{\alpha}T$ of the order of the Debye mass m_D. For phenomenological applications the 'leading' logarithm is not large, which motivated Braaten and Thoma [9] to calculate the constant $c(n_f)$, or in other words the cut-off μ in (2), for heavy quarks, and in particular in the limit $E \gg M^2/T$, using *hard thermal loop* (HTL) perturbation theory.

A principal shortcoming for phenomenological estimates is that neither (2) nor the HTL-improved result specify the value of the QCD coupling α. The common *prescription* in the literature is to *ad hoc* assign a value given by the running coupling at the order of the lowest Matsubara frequency, $\alpha \to \alpha(\sim (2\pi T)^2)$. This not only introduces sizable uncertainties, it is also inconsistent: Screening, which is required to render dE/dx finite, arises from resumming thermal contributions of loop corrections to the Born scattering amplitudes. The concurrent vacuum contributions of the loop corrections then make the coupling momentum dependent; the relevant scale(s) of the running coupling in the final result is thus determined unambiguously. I have shown in [10] that in the leading-log contribution from t-channel scattering in the large-E limit,

$$\left.\frac{dE}{dx}\right|^t \sim \int_k \frac{n(k)}{2k} \int_{-s}^{-m_D^2} dt \, \frac{d\sigma}{dt}(-t), \tag{3}$$

the differential cross section $d\sigma/dt \sim \alpha^2/t^2$ has to be evaluated with a running coupling $\alpha(Q^2)$ at the scale $Q^2 = t$. At leading order $\alpha(t) \sim 1/\ln(|t|/\Lambda^2)$, which then implies replacing $\alpha^2 \ln(ET/m_D^2) \to \alpha(m_D^2)\alpha(ET)\ln(ET/m_D^2)$ in (2). Evaluating one factor of α at the *soft scale* m_D^2 leads to larger quantitative estimates compared to previous 'fix-coupling guesses'. Interestingly, in the limit $E \to \infty$ the energy loss ceases to grow logarithmically and saturates, $dE/dx|^t \sim T^2\alpha(m_D^2) \sim m_D^2$.[2]

The running coupling calculation has been pushed beyond leading-log accuracy. Our derivation [7, 12] reveals also that the QCD Compton scattering processes (see Fig. 1) actually do contribute already at leading-log order due to a collinear enhancement, which results – besides a 'new' log term – in a revised value of the next-lo-leading log constant $c(n_f)$,

$$\frac{dE}{dx} = \frac{4\pi T^2}{3} \left[\left(1 + \tfrac{n_f}{6}\right) \alpha(m_D^2)\alpha(ET) \left(\ln \frac{ET}{m_D^2} + c'(n_f) \right) + \tfrac{2}{9}\, \alpha(M^2)\alpha(ET) \ln \frac{ET}{M^2} + \dots \right]. \tag{4}$$

[2] For a discussion of setting the scale in the running coupling for the Debye mass see [11].

Journal of Physics: Conference Series **455** (2013) 012054 doi:10.1088/1742-6596/455/1/012054

To estimate the running coupling for the time-like Compton processes, we continue the usual 1-loop perturbative expression according to [13],

$$\alpha(Q^2) = \frac{1}{\beta_0} \begin{cases} L_-^{-1} \\ \frac{1}{2} - \pi^{-1}\mathrm{atn}(L_+/\pi) \end{cases} \quad \text{for } Q^2 \lessgtr 0\,, \tag{5}$$

where $\beta_0 = (11 - \frac{2}{3}\,n_f)/(4\pi)$ with $n_f = 3$, and $L_\pm = \ln(\pm Q^2/\Lambda^2)$. We use $\Lambda \simeq 0.2\,\mathrm{GeV}$ [11], and also constrain the coupling strength in the far infra-red, $\alpha(Q^2) \lesssim 1$.[3]

In a recent study [14] we have furthermore investigated the collisional energy loss in a thermal background that flows with a given collective 4-velocity $u^\mu = \gamma(1, \boldsymbol{u})$, as described by Jüttner functions $n_u(k) = [\exp(K_\mu u^\mu/T) \pm 1]^{-1}$ which replace the thermal distributions in (1). It might be anticipated from the manifest covariant form of intermediate expressions like (3) that we should, in the final result for dE/dx, replace $ET \to P_\mu u^\mu T$, both in the kinematic logarithms and in the arguments of the running coupling. We have confirmed this expectation by an explicit calculation and find

$$\frac{dE}{dx} = \frac{4\pi T^2}{3}\alpha(P_\mu u^\mu T)\left[(1+\tfrac{n_f}{6})\,\alpha(m_D^2)\ln\frac{P_\mu u^\mu T}{m_D^2} + \tfrac{2}{9}\,\alpha(M^2)\ln\frac{P_\mu u^\mu T}{M^2} + \ldots\right] \tag{6}$$

at leading-log accuracy. For typical conditions expected in heavy-ion collisions, the flow of the background can have effects of the order of 50% on the mean energy loss of heavy quarks.

3. Markov evolution of heavy quarks

Our analytical considerations on the average energy loss illustrate the relevance of two essential requirements for a justified phenomenological understanding of in-medium heavy-quark collisions: taking into account the running of the QCD coupling, and a calculational accuracy beyond leading-log accuracy. On this basis we have developed a transport description of heavy quarks subject to binary collisions in a thermalized quark-gluon plasma. Rewriting their mean energy loss (1) as

$$\frac{dE(p)}{dx} = \int_p \frac{d^3\Gamma(\boldsymbol{p}', \boldsymbol{p})}{d(p')^3}\,\omega\,, \tag{7}$$

we can first easily obtain the heavy-quark interaction rate $d^3\Gamma(\boldsymbol{p}', \boldsymbol{p})/d(p')^3$, differential with respect to the outgoing momentum. We calculate this scattering rate beyond leading-log accuracy by using an *effective* mass cut-off [15] for t-channel scattering,

$$\mu^2(t) = \kappa \cdot 4\pi\,(1 + \tfrac{n_f}{6})\,\alpha(t)\,T^2\,, \tag{8}$$

which i) incorporates the running coupling and ii) parameterizes the complex HTL structure of soft electric gluon exchanges by a parameter κ. Its value

$$\kappa = (2e)^{-1} \approx 0.2 \tag{9}$$

is adjusted unambiguously by the requirement that the resulting mean energy loss (7) has to reproduce the analytic formula (4). The value (9) is considerably smaller than unity, which implies that our interaction rates are larger, in proportion, than previous estimates based on a cut-off using the Debye mass in the form $m_D^2 = 4\pi\,(1 + \frac{n_f}{6})\,\alpha\,T^2$.

With $d^3\Gamma/d(p')^3$ at hand, we can describe the stochastic propagation of a heavy quark by means of a statistical ensemble. Its momentum distribution evolves as a function of time τ according to

$$f(\boldsymbol{p}', \tau) \;\to\; f(\boldsymbol{p}', t + \delta\tau) = \int_p \mathcal{T}_{\delta\tau}(\boldsymbol{p}', \boldsymbol{p}) f(\boldsymbol{p}, \tau)\,, \tag{10}$$

[3] The numerical results presented below are largely insensitive to the precise value of the upper bound.

which is a first order Markov process with the kernel

$$\mathcal{T}_{\delta\tau}(\boldsymbol{p}', \boldsymbol{p}) = (1 - \delta\tau\, \Gamma(\boldsymbol{p}))\, \delta(\boldsymbol{p}' - \boldsymbol{p}) + \delta\tau\, \frac{d^3\Gamma(\boldsymbol{p}', \boldsymbol{p})}{d(p')^3}\,, \tag{11}$$

where $\Gamma(\boldsymbol{p})$ denotes the total rate for scattering out of momentum state $\boldsymbol{p}$. In discrete form, i. e. for binned momenta, (11) becomes a (positive definite) transition matrix satisfying $\sum_{p'} \mathcal{T}_{p'p} = 1$ for all $\boldsymbol{p}$. Such matrices are called *stochastic*; among their noteworthy properties (discussed in more detail in [15]) we emphasize here only one: They describe a Markov evolution (10) converging to a unique fix point, which in our case – due to detailed balance of the thermal scattering rate – is the thermal equilibrium distribution.

I underline that this basic criterion is *not* met by a number of heavy-quark transport approaches used in the literature to analyze heavy-ion experiments. E. g., the interaction rate $d^3\Gamma(\boldsymbol{p}', \boldsymbol{p})/d(p')^3$ is sometimes obtained by means of Fokker-Planck type equations whose validity, however, relies on the dominance of soft momentum exchanges. This is usually justified in non-relativistic situations, while in the present relativistic case momentum exchanges range from soft $\sim m_D^2$ to hard $\sim ET$, as evident e. g. from the argument of the 'Coulomb logarithm' in (2). Accordingly, in these Fokker-Planck approaches coefficients have to be *'tuned by hand'* in order to guarantee thermalization of the heavy quarks (after sufficiently long time) in a thermal bath.

4. Phenomenological implications and conclusions

In order to apply our formalism for the analysis of heavy-ion experiments, we need to model the dynamics of the quark-gluon plasma background. For the following results we use specific temperature and flow profiles, $T(\tau, \boldsymbol{r})$ and $\boldsymbol{u}(\tau, \boldsymbol{r})$, extracted either from the QCD Boltzmann cascade [16] or, alternatively, from the (3+1)-dimensional ideal hydrodynamics code [17], both initialized with typical RHIC conditions (see [14] for details). We follow standard procedures by considering charm and bottom quarks ($M_c \simeq 1.3\,\mathrm{GeV}$ and $M_b \simeq 4.6\,\mathrm{GeV}$) with initial spacial distributions according to the Glauber model [18], and with initial momenta sampled by a QCD event generator [19]. We neglect shadowing and the Cronin effect as irrelevant for heavy-quarks with momenta $p_t \gtrsim 2\,\mathrm{GeV}$. We also assume, when the local temperature of the background reaches $T_c \simeq 200\,\mathrm{GeV}$, a fragmentation [20] of the heavy quarks into heavy mesons, whose subsequent decay into electrons we compute with PYTHIA [21].

As a first observable we consider the transverse spectrum of electrons from heavy flavor decays, normalized to the corresponding spectrum in proton-proton collisions times the number of binary collisions N_{bc} according to the Glauber model,

$$R_{AA}(p_t, y) = \frac{d^2N_{AA}/(dp_t\, dy)}{N_{\mathrm{bc}}\, d^2N_{pp}/(dp_t\, dy)}\,. \tag{12}$$

Figure 2 shows R_{AA} calculated at mid-rapidity, $|y| < 0.35$, for a central AuAu-collision at RHIC with $\sqrt{s} = 200\,\mathrm{AGeV}$, compared to experimental findings in the centrality class of $0\% - 10\%$. Obviously, finer details of the background evolution models are not crucial. Our results clearly follow the trend of the data, with some room for effects due to radiative energy loss. Adapting a common attempt to estimate the relative importance of collisional vs. radiative effects we *ad hoc* scale the binary interaction rate by a constant factor which the experimental data constrain by $1 \lesssim K \lesssim 3$. A similar picture emerges in Figure 3 when comparing our numerical results for non-central collisions to PHENIX data, with a slightly smaller K-value which could be a consequence of the different path length dependence for collisional and radiative energy loss.

As a second observable we compute the elliptic anisotropy, defined by the ensemble average

$$v_2(p_t) = \left\langle \frac{p_x^2 - p_y^2}{p_t^2} \right\rangle\,, \tag{13}$$

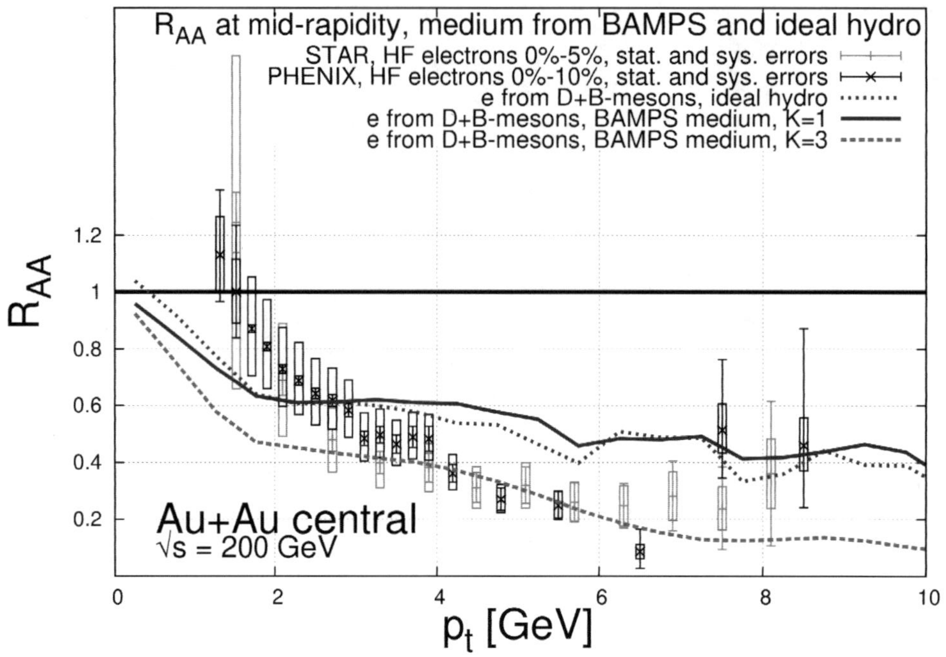

Figure 2. The nuclear modification factor (12) at mid-rapidity for central AuAu-collisions at RHIC; results from our Markov approach [14] vs. RHIC data [5, 22]. The lower dotted line shows the effect of scaling the binary interaction rate by a factor $K = 3$, see text.

for electrons from heavy-flavor decays, whose transverse momentum dependence is shown in Figure 4. Also in this case the binary interactions alone give a substantial contribution to the experimental results, and again the binary scattering rate had to be scaled by a factor $K \approx 3$ in order to saturate the data.[4] In this sense we can indeed conclude that collisional effects are of similar importance as radiative effects.

To summarize: We have reviewed analytic results for the mean energy loss of heavy flavors due to binary collisions in a thermalized quark-gluon plasma. We have outlined a consistent transport description within a framework that takes into account the momentum dependence of the QCD coupling and screens soft interactions by a next-to-leading log adjusted cut-off. We have demonstrated that resulting collisional effects are quantitatively as important as radiative effects and thus need be thoroughly considered in heavy-ion phenomenology.

Acknowledgments
I thank S. Peigné, A. Meistrenko, J. Uphoff and C. Greiner for their collaboration on the presented topics, and I acknowledge support of my research by the South African National Research Foundation as well as SA-CERN.

[4] We note that our approach, namely the value (9) for the screening parameter, is justified for larger heavy quark energies, $E \gtrsim M^2/T$, i.e possibly a few GeV for charm quarks. Results at smaller momenta should thus be understood as extrapolations (which may still give reasonable estimates).

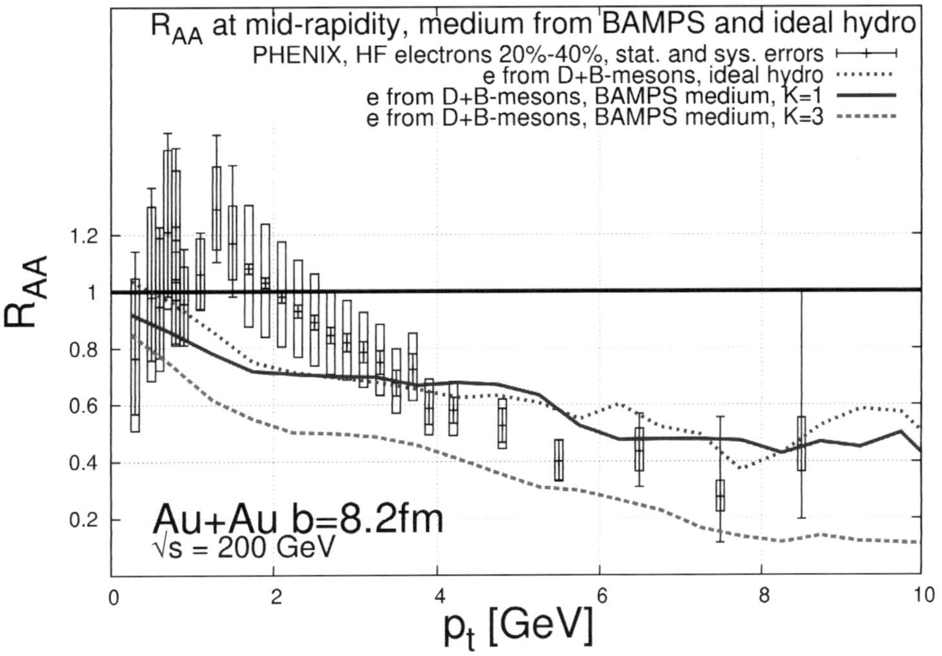

Figure 3. Same as Figure 2, but for non-central collisions with impact parameter $b = 8.2\,\mathrm{fm}$ vs. PHENIX data for the centrality class 20%–40%.

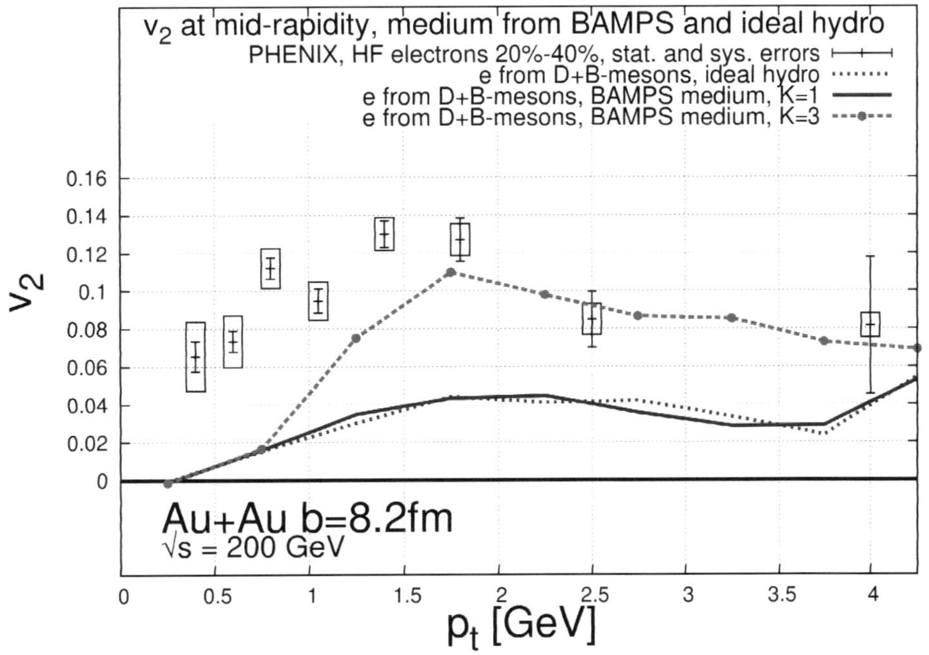

Figure 4. Same as Figure 3, but for the elliptic flow.

References

[1] R. Baier, Y. L. Dokshitzer, A. H. Mueller, S. Peigné and D. Schiff, Nucl. Phys. B **484**, 265 (1997) [arXiv:hep-ph/9608322].
[2] M. Gyulassy, P. Levai and I. Vitev, Phys. Rev. Lett. **85**, 5535 (2000) [arXiv:nucl-th/0005032].
[3] R. Baier and D. Schiff, JHEP **0609**, 059 (2006) [arXiv:hep-ph/0605183].
[4] Y. L. Dokshitzer and D. E. Kharzeev, Phys. Lett. B **519**, 199 (2001) [arXiv:hep-ph/0106202].
[5] B. I. Abelev *et al.* [STAR Collaboration], Phys. Rev. Lett. **98**, 192301 (2007) [arXiv:nucl-ex/0607012].
[6] A. Adare *et al.* [PHENIX Collaboration], Phys. Rev. Lett. **98**, 172301 (2007) [arXiv:nucl-ex/0611018].
[7] S. Peigné and A. Peshier, Phys. Rev. D **77**, 114017 (2008) [arXiv:0802.4364 [hep-ph]].
[8] J. D. Bjorken, FERMILAB-PUB-82-059-THY.
[9] E. Braaten and M. H. Thoma, Phys. Rev. D **44**, 2625 (1991).
[10] A. Peshier, Phys. Rev. Lett. **97**, 212301 (2006) [arXiv:hep-ph/0605294].
[11] A. Peshier, arXiv:hep-ph/0601119.
[12] S. Peigné and A. Peshier, Phys. Rev. D **77**, 014015 (2008) [arXiv:0710.1266 [hep-ph]].
[13] Y. L. Dokshitzer, G. Marchesini and B. R. Webber, Nucl. Phys. B **469**, 93 (1996). [arXiv:hep-ph/9512336].
[14] A. Meistrenko, J. Uphoff C. Greiner and A. Peshier, Nucl. Phys. A **901**, 51 (2013) [arXiv:1204.2397 [hep-ph]].
[15] A. Peshier, Nucl. Phys. A **888**, 7 (2012) [arXiv:0801.0595 [hep-ph]].
[16] Z. Xu and C. Greiner, Phys. Rev. C **71** (2005) 064901 [hep-ph/0406278].
[17] D. H. Rischke, S. Bernard and J. A. Maruhn, Nucl. Phys. A **595** (1995) 346 [nucl-th/9504018].
[18] M. L. Miller, K. Reygers, S. J. Sanders and P. Steinberg, Ann. Rev. Nucl. Part. Sci. **57** (2007) 205 [nucl-ex/0701025].
[19] S. Frixione, P. Nason and B. R. Webber, JHEP **0308** (2003) 007 [hep-ph/0305252].
[20] C. Peterson, D. Schlatter, I. Schmitt and P. M. Zerwas, Phys. Rev. D **27** (1983) 105.
[21] T. Sjostrand, S. Mrenna and P. Z. Skands, Comput. Phys. Commun. **178** (2008) 852 [arXiv: 0710.3820 [hep-ph]].
[22] A. Adare *et al.* [PHENIX Collaboration], Phys. Rev. C **84** (2011) 044905 [arXiv:1005.1627 [nucl-ex]].

International Workshop on Discovery Physics at the LHC (Kruger2012) IOP Publishing
Journal of Physics: Conference Series **455** (2013) 012055 doi:10.1088/1742-6596/455/1/012055

Forward physics with proton tagging at the LHC

Christophe Royon

CEA/IRFU/Service de physique des particules, CEA/Saclay, 91191 Gif-sur-Yvette cedex, France

Abstract. We present some inclusive and exclusive diffractive processes to be studied at the LHC, namely QCD, quartic anomalous couplings between γ and W/Z bosons and diffractive Higgs processes.

In this short report, we discuss some potential measurements to be performed using proton tagging detectors at the LHC. We can distinguish two kinds of measurements: the first motivation of these detectors is a better understanding of the structure of the colorless exchanged object, the Pomeron, in a domain of energy unexplored until today and the second motivation is to explore beyond standard model physics such as quartic anomalous couplings between photons and W/Z bosons and γ. We assume in the following intact protons to be tagged in dedicated detectors located at about 210 m for ATLAS (220 m for CMS) as described at the end of this report.

1. Inclusive diffraction measurement at the LHC

In this section, we discuss potential measurements at the LHC that can constrain the Pomeron structure, the colorless object which is exchanged in diffractive interactions. The Pomeron structure in terms of quarks and gluons has been derived from QCD fits at HERA and it is possible to probe this structure and the QCD evolution at the LHC.

1.1. Jet production in double Pomeron exchanges processes

The high energy and luminosity at the LHC allow the exploration of a completely new kinematical domain. Tagging both diffractive protons in ATLAS will allow the QCD evolution of the gluon and quark densities in the Pomeron to be probed and the gluon and quark densities to be measured using the dijet and $\gamma + jet$ cross-section measurements. The different diagrams of the processes that can be studied at the LHC are shown in Fig. 1, namely double pomeron exchange (DPE) production of dijets (left), of γ+jet sensitive respectively to the gluon and quark contents of the Pomeron, and the jet gap jet events (right). As an example, we show in Fig.2, the ratio of the number of events obtained for γ+jet and dijet production for a luminosity of 300 pb^{-1} when the protons are tagged in the ATLAS Forward Physics detectors (AFP) at 210 m. Fig. 2 shows the potential of this measurement to constrain further the quark density in the Pomeron. The QCD fits to the proton diffractive structure functions and to the dijet cross section performed at HERA do not allow to constrain the differences between the quark densities in the Pomeron, and the quark densities are assumed to be equal in the fit ($u = \bar{u} = d = \bar{d} = s = \bar{s}$). At the LHC, we will be able to test this hypothesis by constraining further the quark densities in the Pomeron as shown in Fig. 2. Let us notice that the best observable is the ratio of the

Journal of Physics: Conference Series **455** (2013) 012055 doi:10.1088/1742-6596/455/1/012055

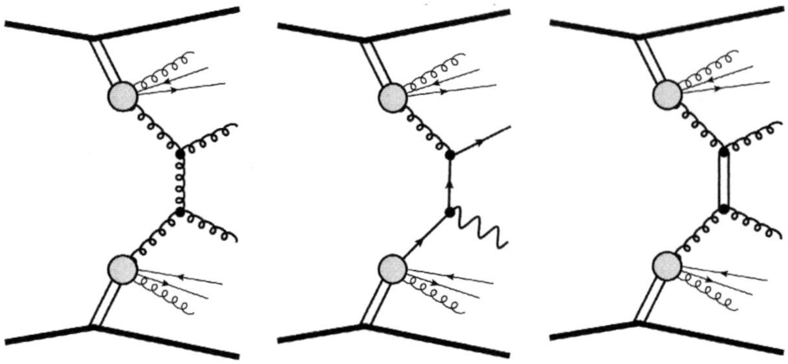

Figure 1. Inclusive diffractive diagrams. From left to right: jet production in inclusive double pomeron exchange, γ+jet production in DPE, jet gap jet events

total diffractive mass computed by measuring the two intact protons in the final state between γ+jet and dijet events since most systematics uncertainties cancel in the ratio. Fig. 2 displays the ratio obtained for a luminosity of 300 pb^{-1} which corresponds to three weeks of data at low luminosity at the LHC.

1.2. Jet gap jet production in double Pomeron exchanges processes
This process is illustrated in Fig 1, right [1, 2]. Both protons are intact after the interaction and detected in AFP, two jets are measured in the ATLAS central detector and a gap devoid of any energy is present between the two jets. This kind of event is important since it is sensitive to QCD resummation dynamics given by the Balitsky Fadin Kuraev Lipatov [3] (BFKL) evolution equation . This process has never been measured to date and will be one of the best methods to probe these resummation effects, benefitting from the fact that one can perform the measurement for jets separated by a large angle (there is no remnants which 'pollute' the event). As an example, the cross section ratio for events with gaps to events with or without gaps as a function of the leading jet p_T is shown in Fig 3 for 300 pb^{-1}.

2. Exclusive jet and Higgs boson production at the LHC
The Higgs and jet exclusive production in both Khoze Martin Ryskin [4] (KMR) and CHIDe [5] models have been implemented in the Forward Physics Monte Carlo (FPMC) [6], a generator that has been designed to study forward physics, especially at the LHC. It aims to provide a variety of diffractive processes in one common framework, *i.e.* single diffraction, double pomeron exchange, central exclusive production and two-photon exchange.

Different uncertainties are associated with the models of exclusive diffractive processes. There are three main sources of uncertainties: the gap survival probability which will be measured using the first LHC data (in this study we assume a value of 0.1 at the Tevatron and 0.03 at the LHC), the gluon density, which contains the hard and the soft part (contrary to the hard part, the soft one is not known precisely and originates from a phenomenological parametrisation), and finally the limits of the Sudakov integral which appear in the calculation of exclusive processes, which have not been all fixed by theoretical calculations (apart from the upper limit for the Higgs case) and thus are not known precisely [7].

In order to constrain the uncertainty on the Higgs boson cross section, we study the possible constraints using early LHC measurements of exclusive jets using an integrated luminosity of 100 pb^{-1}. In addition to the statistical uncertainties, we consider a conservative 3% jet energy

Journal of Physics: Conference Series **455** (2013) 012055 doi:10.1088/1742-6596/455/1/012055

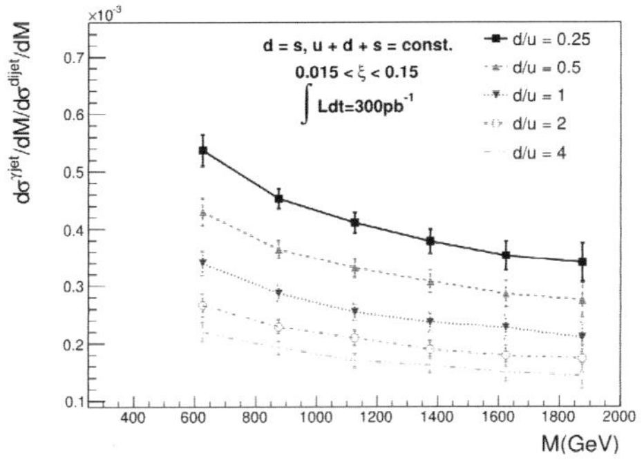

Figure 2. Ratio between the γ+jet and dijet productions and sensitivity on the quark structure in the Pomeron.

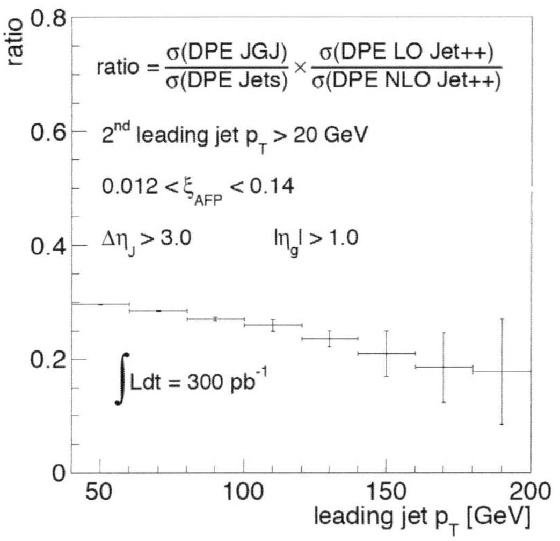

Figure 3. Ratio of DPE Jet gap jet events to standard DPE dijet events as a function of the leading jet p_T [1].

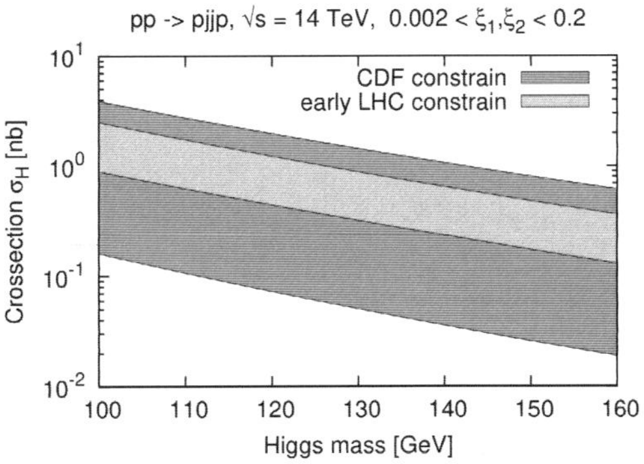

Figure 4. Total uncertainty for exclusive Higgs production at the LHC: c onstraint provided by the CDF measurements and the possible LHC measurement with a luminosity of 100 pb^{-1} [7].

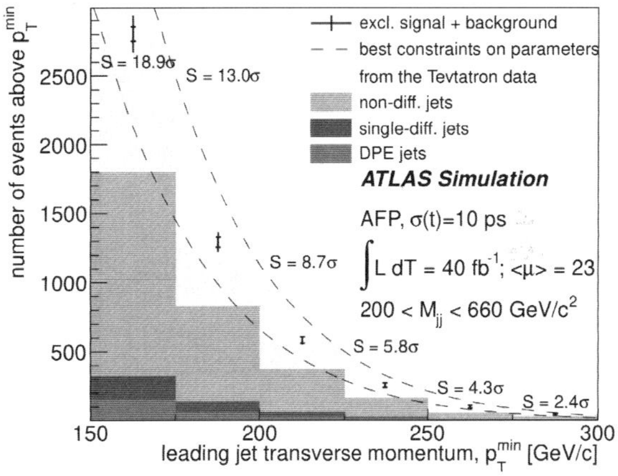

Figure 5. Measurement of the exclusive jet production at the LHC in the ATLAS experiment [8].

scale uncertainty as the dominant contribution to the systematic error. We obtain the prediction for Higgs boson prediction as shown in Fig. 4 (yellow band).

A possible measurement of exclusive jet cross section (see Fig. 6) at the LHC in the ATLAS experiment is shown in Fig. 5. The results of the measurement is shown as black points for a luminosity of 40 fb^{-1} and 23 pile up events, assuming the protons to be detected in the ATLAS Forward Physics detectors [8]. The expected contributions from background (non-diffractive, single diffractive with pile up and double pomeron exchange events) are shown as well as the exclusive jet one in yellow. The statistical significance of the measurement if up to 19σ.

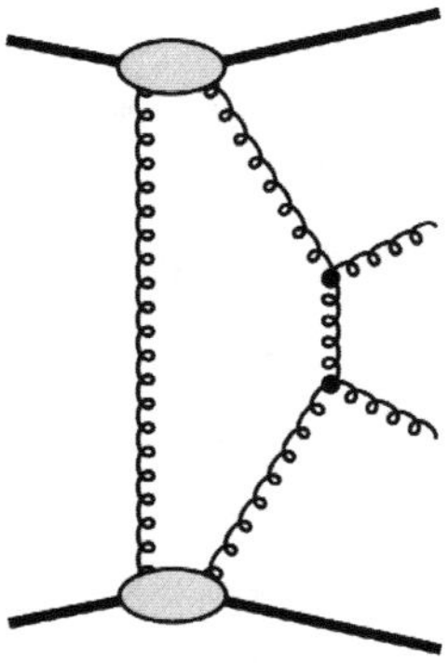

Figure 6. Exclusive jet production.

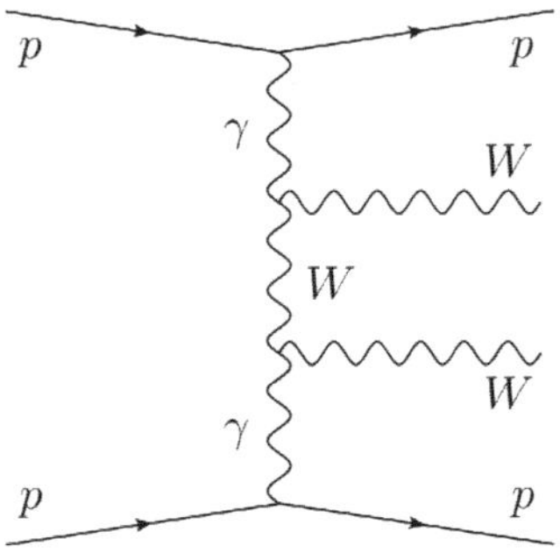

Figure 7. Diagram showing the two-photon production of W pairs.

3. Exclusive WW and ZZ production

In the Standard Model (SM) of particle physics, the couplings of fermions and gauge bosons are constrained by the gauge symmetries of the Lagrangian. The measurement of W and Z boson pair productions via the exchange of two photons allows to provide directly stringent tests of one of the most important and least understood mechanism in particle physics, namely the electroweak symmetry breaking.

3.1. Photon exchange processes in the SM

The process that we intend to study is the W pair production shown in Fig. 7 induced by the exchange of two photons [9]. It is a pure QED process in which the decay products of the W bosons are measured in the central detector and the scattered protons leave intact in the beam pipe at very small angles

After simple cuts to select exclusive W pairs decaying into leptons, such as a cut on the proton momentum loss of the proton ($0.0015 < \xi < 0.15$) — we assume the protons to be tagged in the ATLAS Forward Physics detectors [8] — on the transverse momentum of the leading and second leading leptons at 25 and 10 GeV respectively, on $\not{E}_T > 20$ GeV, $\Delta\phi > 2.7$ between leading leptons, and $160 < W < 500$ GeV, the diffractive mass reconstructed using the forward detectors, the background is found to be less than 1.7 event for 30 fb^{-1} for a SM signal of 51 events.

3.2. Quartic anomalous couplings

The parameterization of the quartic couplings based on [10] is adopted. The cuts to select quartic anomalous gauge coupling WW events are similar as the ones we mentioned in the previous section, namely $0.0015 < \xi < 0.15$ for the tagged protons, $\not{E}_T > 20$ GeV, $\Delta\phi < 3.13$ between the two leptons. In addition, a cut on the p_T of the leading lepton $p_T > 160$ GeV and on the diffractive mass $W > 800$ GeV are requested since anomalous coupling events appear at high mass. Fig 8 displays the p_T distribution of the leading lepton for signal and the different considered backgrounds. After these requirements, we expect about 0.7 background events for an expected signal of 17 events if the anomalous coupling is about four orders of magnitude lower than the present LEP limit ($|a_0^W/\Lambda^2| = 5.4\ 10^{-6}$) for a luminosity of 30 fb^{-1}. The strategy to select anomalous coupling ZZ events is analogous and the presence of three leptons or two like sign leptons are requested. Table 1 gives the reach on anomalous couplings at the LHC for luminosities of 30 and 200 fb^{-1} compared to the present OPAL limits [11]. We note that we can gain almost four orders of magnitude in the sensitivity to anomalous quartic gauge

Journal of Physics: Conference Series **455** (2013) 012055 doi:10.1088/1742-6596/455/1/012055

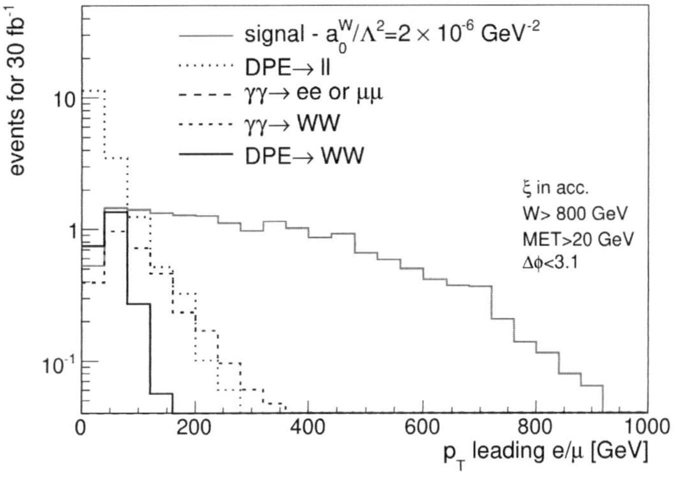

Figure 8. Distribution of the transverse momentum of the leading lepton for signal and background after the cut on W, $\not{E}_T$, and $\Delta\phi$ between the two leptons [9].

Couplings	OPAL limits [GeV^{-2}]	Sensitivity @ $\mathcal{L}$ = 30 (200) fb^{-1} 5σ	95% CL
a_0^W/Λ^2	[-0.020, 0.020]	5.4 10^{-6} (2.7 10^{-6})	2.6 10^{-6} (1.4 10^{-6})
a_C^W/Λ^2	[-0.052, 0.037]	2.0 10^{-5} (9.6 10^{-6})	9.4 10^{-6} (5.2 10^{-6})
a_0^Z/Λ^2	[-0.007, 0.023]	1.4 10^{-5} (5.5 10^{-6})	6.4 10^{-6} (2.5 10^{-6})
a_C^Z/Λ^2	[-0.029, 0.029]	5.2 10^{-5} (2.0 10^{-5})	2.4 10^{-5} (9.2 10^{-6})

Table 1. Reach on anomalous couplings obtained in γ induced processes after tagging the protons in AFP compared to the present OPAL limits. The 5σ discovery and 95% C.L. limits are given for a luminosity of 30 and 200 fb^{-1} [9]

couplings compared to LEP experiments, and it is possible to reach the values expected in extra-dimension models. The tagging of the protons using the ATLAS Forward Physics detectors is the only method at present to test so small values of quartic anomalous couplings.

As mentioned already, in order to achieve this experimental program, a proposal to add forward proton detectors located at 210 m (respectively 240) and 420 m from the interaction point was submitted tp the ATLAS (respectively CMS) collaborations. Both detectors allow a good acceptance in mass between 110 GeV and 1.4 TeV. By 2014, only the 210 m detectors are foreseen. They are sensitive to high diffractive masses and will allow to fulfill the program to look for anomalous couplings extra-dimension models with an unprecedented precision if these projects are approved by the ATLAS and CMS collaborations.

The search for quartic anomalous couplings between γ and W bosons was performed again after a full simulation of the ATLAS detector including pile up [8]. Integrated luminosities of 40 and 300 fb^{-1} with, respectively, 23 or 46 average pile-up events per beam crossing have been considered. The W pair is assumed to decay leptonically. The full list of background processes used for the ATLAS measurement of Standard Model WW cross-section was simulated, namely $t\bar{t}$, WW, WZ, ZZ, W+jets, Drell-Yan and single top events. In addition, the additional diffractive backgrounds mentioned in the previous paragraph were also simulated, The requirement of the presence of at least one proton on each side of AFP within a time window of 10 ps allows us to reduce the background by a factor of about 200 (50) for $\mu =$ 23 (46). The p_T of the leading lepton originating from the leptonic decay of the W bosons is required to be $p_T > 150$ GeV, and that of the next-to-leading lepton $p_T > 20$ GeV. Additional requirement of the dilepton mass to be above 300 GeV allows us to remove most of the Z and WW events. Since only leptonic decays of the W bosons are considered, we require in addition less than 3 tracks associated to the primary vertex, which allows us to reject a large fraction of the non-diffractive backgrounds (e.g. $t\bar{t}$, diboson productions, W+jet, etc.) since they show much higher track multiplicities. Remaining Drell-Yan and QED backgrounds are suppressed by requiring the difference in azimuthal angle between the two leptons $\Delta\phi < 3.1$. After these requirements, a similar sensitivity with respect to fast simulation without pile up was obtained.

Of special interest will be also the search for anomalous quartic $\gamma\gamma\gamma\gamma$ anomalous couplings which is now being implemented in the FPMC generator. Let us notice that there is no present existing limit on such coupling and the sensitivity using the forward proton detectors is expected to be similar as the one for $\gamma\gamma WW$ or $\gamma\gamma ZZ$ anomalous couplings. If discovered at the LHC, $\gamma\gamma\gamma\gamma$ quartic anomalous couplings might be related to the existence of extra-dimensions in the universe, which might lead to a reinterpretation of some experiments in atomic physics. As an example, the Aspect photon correlation experiments [12] might be interpreted via the existence of extra-dimensions. Photons could communicate through extra-dimensions and the deterministic interpretation of Einstein for these experiments might be true if such anomalous couplings exist. From the point of view of atomic physics, the results of the Aspect experiments would depend on the distance of the two photon sources.

4. Forward Proton Detectors in ATLAS and CMS

In this section, we describe the proposal to install the ATLAS Forward Proton (AFP) detector in order to detect intact protons at 206 and 214 meters on both side of the ATLAS experiment [8] (similar detectors will be installed around CMS by TOTEM/CMS). This one arm will consist of two sections (AFP1 and AFP2) contained in a special design of beampipe (or in more traditional roman pots). In the first section (AFP1), in one pocket of the beampipe, a tracking station composed by 6 layers of Silicon detectors will be deployed. The second station AFP2 will contain another tracking station similar to the one already described and a timing detector. The aim of this setup, mirrored by an identical arm placed on the opposite side of the ATLAS interaction point, will be to tag the protons emerging intact from the pp interactions so allowing ATLAS to exploit the program of diffractive and photoproduction processes described in the previous sections.

4.1. Movable beam pipes

The idea of movable Hamburg beam pipes is quite simple [13]: a larger section of the LHC beam pipe than the usual one can move close to the beam using bellows so that the detectors located at its edge (called pocket) can move close to the beam by about 2.5 cm when the beam is stable (during injection, the detectors are in parking position). In its design, the predominant aspect is the minimization of the thickness of the portions called floor and window (see Fig. 9). Minimizing the depth of the floor ensures that the detector can go as close to the beam as possible allowing us to detect protons scattered at very small angles, while minimizing the depth of the thin window is important to keep the protons intact and to reduce the impact of multiple interactions. Two configurations exist for the movable beam pipes: the first one at 206

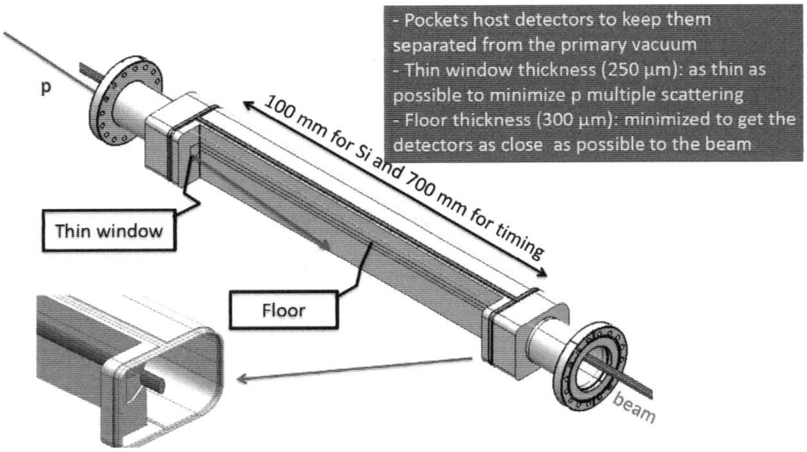

Figure 9. Scheme of the movable beam pipe.

m form the ATLAS interaction point hosts a Si detector (floor length of about 100 mm) and the second one (floor length of about 400 mm) the timing and the Si detectors.

4.2. 3D Silicon detectors

The purpose of the tracker system is to measure points along the trajectory of beam protons that are deflected at small angles as a result of collisions. The tracker when combined with the LHC dipole and quadrupole magnets, forms a powerful momentum spectrometer. Silicon tracker stations will be installed in Hamburg beam pipes (HBP) at $\pm$ 206 and $\pm$ 214 m from the ATLAS.

The key requirements for the silicon tracking system at 220 m are:

- Spatial resolution of $\sim$ 10 (30) μm per detector station in x (y)
- Angular resolution for a pair of detectors of about 1 μrad
- High efficiency over the area of 20 mm $\times$ 20 mm corresponding to the distribution of diffracted protons
- Minimal dead space at the edge of the sensors allowing us to measure the scattered protons at low angles
- Sufficient radiation hardness in order to sustain the radiation at high luminosity
- Capable of robust and reliable operation at high LHC luminosity

The basic building unit of the AFP detection system is a module consisting of an assembly of a sensor array, on-sensor read-out chip(s), electrical services, data acquisition and detector control system. The module will be mounted on the mechanical support with embedded cooling and other necessary services. The sensors are double sided 3D 50$\times$250 micron pixel detectors with slim-edge dicing built by FBK and CNM companies. The sensor efficiency has been measured to be close to 100% over the full size in beam tests. A possible upgrade of this device will be to use 3D edgeless Silicon detectors built in a collaboration between SLAC, Manchester, Oslo, Bergen... A new front-end chip FE-I4 has been developed for the Si detector by the Insertable B Layer (IBL) collaboration in ATLAS. The FE-I4 integrated circuit contains readout circuitry for 26 880 hybrid pixels arranged in 80 columns on 250 μm pitch by 336 rows on 50 μm pitch, and covers an area of about 19 mm $\times$ 20 mm. It is designed in a 130 nm feature size bulk CMOS process. Sensors must be DC coupled to FE-I4 with negative charge collection. The FE-I4 is very well suited to the AFP requirements: the granularity of cells provides a sufficient spatial resolution, the chip is radiation hard enough (up to a dose of 3 MGy), and the size of the chip is sufficiently large that one module can be served by just one chip.

The dimensions of the individual cells in the FE-I4 chip are 50 μm $\times$ 250 μm in the x and y directions, respectively. Therefore to achieve the required position resolution in the x-direction of $\sim$ 10 μm, six layers with sensors are required (this gives $50/\sqrt{12}/\sqrt{5} \sim 7$ μm in x and roughly 5 times worse in y). Offsetting planes alternately to the left and right by one half pixel will give a further reduction in resolution of at least 30%. The AFP sensors are expected to be exposed to a dose of 30 kGy per year at the full LHC luminosity of $10^{34}\text{cm}^{-2}\text{s}^{-1}$.

4.3. Timing detectors

A fast timing system that can precisely measure the time difference between outgoing scattered protons is a key component of the AFP detector. The time difference is equivalent to a constraint on the event vertex, thus the AFP timing detector can be used to reject overlap background by establishing that the two scattered protons did not originate from the same vertex as the the central system. The final timing system should have the following characteristics [15]:

- 10 ps or better resolution (which leads to a factor 40 rejection on pile up background)
- Efficiency close to 100% over the full detector coverage
- High rate capability (there is a bunch crossing every 25 ns at the nominal LHC)
- Enough segmentation for multi-proton timing
- Level trigger capability

Figure 10 shows a schematic overview of the first proposed timing system, consisting of a quartz-based Cerenkov detector coupled to a microchannel plate photomultiplier tube (MCP-PMT), followed by the electronic elements that amplify, measure, and record the time of the event along with a stabilized reference clock signal. The QUARTIC detector consists of an array of 8$\times$4 fused silica bars ranging in length from about 8 to 12 cm and oriented at the average Cerenkov angle. A proton that is sufficiently deflected from the beam axis will pass through a row of eight bars emitting Cerenkov photons providing an overall time resolution that is approximately $\sqrt{8}$ times smaller than the single bar resolution of about 30 ps, thus approaching the 10 ps resolution goal. Prototype tests have generally been performed on one row (8 channels) of 5 mm $\times$ 5 mm pixels, while the initial detector is foreseen to have four rows to obtain full acceptance out to 20 mm from the beam. The beam tests lead to a time resolution per bar of the order of 34 ps. The different components of the timing resolution are given in Fig. 11. The upgraded design of the timing detector has equal rate pixels, and we plan to reduce the the width of detector bins close to the beam, where the proton density is highest.

At higher luminosity of the LHC (phase I starting in 2019), higher pixelisation of the timing detector will be required. For this sake, a R&D phase concerning timing detector developments based on Silicon photomultipliers (SiPMs), avalanche photodiods (APDs), quartz fibers, diamonds has been started in Saclay. In parallel, a new timing readout chip has been developed. It uses waveform sampling methods which give the best possible timing resolution. The aim of this chip called SAMPIC [16] is to obtain sub 10 ps timing resolution, 1GHz input bandwidth, no dead time at the LHC, and data taking at 2 Gigasamples per second. The cost per channel is estimated to be of the order for \$10 which a considerable improvement to the present cost of a few \$1000 per channel, allowing us to use this chip in medical applications such as PET imaging detectors. The holy grail of imaging 10 picosecond PET detector seems now to be feasible: with a resolution better than 20 ps, image reconstruction is no longer necessary and real-time image formation becomes possible.

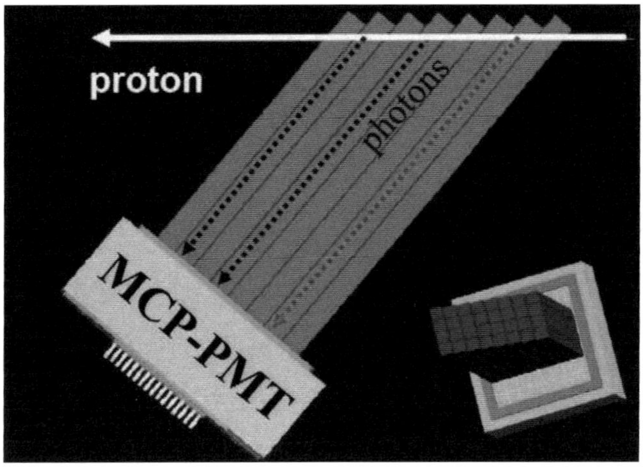

Figure 10. A schematic diagram of the QUARTIC fast timing detector.

Component	δt(ps) Current	δt(ps) Projected (8 ch +cable)	Improve ment	δt(ps) Phase 0 (8 channels)
Radiator (fused silica bar) ~10 pe's	22	22	Optimize radiator	17
MCP-PMT (64 channel 25 um Planacon)	20	20	10 um tube	15
CFD	5	5	-	5
HPTDC	16	16	-	15
Reference Clock	-	3	-	3
Total/bar	34	34		28
Cable		15%	retune CFD	5%
Total/ detector	14	14	-	10

Figure 11. Different components of the timing detector resolution [15].

5. References

[1] C. Marquet, C. Royon, M. Trzebinski, R. Zlebcik, arXiv:1212.2059, accepted by Phys. Rev. D; O. Kepka, C. Marquet, C. Royon, Phys. Rev. D**83** (2011) 034036.

[2] C. Marquet, C. Royon, Phys. Rev. D**79** (2009) 034028; O. Kepka, C. Royon, C. Marquet, R. Peschanski, Eur. Phys .J. C**55** (2008) 259-272; Phys. Lett. B**655** (2007) 236-240; H. Navelet, R. Peschanski, C. Royon, S. Wallon, Phys. Lett. B**385** (1996) 357;H. Navelet, R. Peschanski, C. Royon, Phys. Lett. B366 (1996) 329.

[3] V. S. Fadin, E. A. Kuraev, L. N. Lipatov, Phys. Lett. B**60** (1975) 50; I. I. Balitsky, L. N. Lipatov, Sov.J.Nucl.Phys. **28** (1978) 822;

[4] V. A. Khoze, A. D. Martin and M. G. Ryskin, Eur. Phys. J. C **23** (2002) 311.

[5] J. R. Cudell, A. Dechambre, O. F. Hernandez and I. P. Ivanov, Eur. Phys. J. C **61** (2009) 369.

[6] M. Boonekamp, A. Dechambre, V. Juranek, O. Kepka, M. Rangel, C. Royon, R. Staszewski, e-Print: arXiv:1102.2531; M. Boonekamp, V. Juranek, O. Kepka, C. Royon "Forward Physics Monte Carlo", "Proceedings of the workshop: HERA and the LHC workshop series on the implications of HERA for LHC physics," arXiv:0903.3861 [hep-ph].

[7] A. Dechambre, O. Kepka, C. Royon, R. Staszewski, Phys. Rev. D **83** (2011) 054013. 14; T. Coughlin, J. Forshaw, PoS DIS2010 (2010) 064.

[8] ATLAS Coll., CERN-LHCC-2011-012.

[9] E. Chapon, O. Kepka, C. Royon, Phys. Rev. **D81** (2010) 074003; O. Kepka and C. Royon, Phys. Rev. D **78** (2008) 073005; J. de. Favereau et al., preprint arXiv:0908.2020.

[10] G. Belanger and F. Boudjema, Phys. Lett. B **288** (1992) 201.

Journal of Physics: Conference Series **455** (2013) 012055 doi:10.1088/1742-6596/455/1/012055

[11] G. Abbiendi *et al.* [OPAL Collaboration], Phys. Rev. D **70** (2004) 032005 [arXiv:hep-ex/0402021].
[12] A. Aspect, P. Grangier, G. Roger, Phys. Rev. Lett., Vol. 49, no 2 (1982) 91-94; A. Aspect, J. Dalibard, G. Roger, Phys. Rev. Lett., Vol. 49, Iss. 25 (1982) 1804.
[13] K. Piotrzkowski, U. Schneekloth, Proc. of the ZEUS Collaboration meeting, March 1994, DESY, Hamburg.
[14] ATLAS IBL Coll., CERN-LHCC-2010-013, ATLAS-TDR-019 5/09/2010.
[15] A. Brandt, Microchannel Plate PMT Lifetime and Performance, DIRC2011: Workshop on Fast Cerenkov detectors, Giessen, Germany, April 4-6, 2011.
[16] E. Delagnes, D. Breton, F. Lugiez, R. Rahmanifard, Nucl. Science IEEE Transactions, Volume 54, Issue 5, Part 2, Oct 2007, 1735; E. Delagnes, patent WO 2008/050177, EP2076963.

*

An effective model of QCD thermodynamics[1]

L. Turko[1], D. Blaschke[1,2], D. Prorok[1] and J. Berdermann[3]
[1]Instytut Fizyki Teoretycznej, Uniwersytet Wrocławski, Poland,
[2]Bogoliubov Laboratory for Theoretical Physics, JINR Dubna, Russia,
[3]DES Zeuthen, Germany

E-mail: turko@ift.uni.wroc.pl, blaschke@ift.uni.wroc.pl, prorok@ift.uni.wroc.pl, j.berdermann@desy.de

PACS numbers: 12.38.Gc, 12.38.Mh, 12.40.Ee, 24.85.+p

Abstract. A combined effective model reproducing the equation of state of hadronic matter as obtained in recent lattice QCD simulations is presented. The model reproduces basic physical characteristics encountered in dense hadronic matter in the quark-gluon plasma (QGP) phase and the lower temperature hadron resonance gas phase. The hadronic phase is described by means of an extended Mott-Hagedorn resonance gas while the QGP phase is described by the extended PNJL model. The dissociation of hadrons is obtained by including the state dependent hadron resonance width.

1. Introduction

Simulations of lattice QCD (LQCD) are in practice the only reliable approach to QCD thermodynamics which covers the broad region of strongly interacting matter properties from the hadron gas at low temperatures to a deconfined quark gluon plasma phase at high temperatures. Recently, finite temperature LQCD simulations have overcome the difficulties of reaching the low physical light quark masses and approaching the continuum limit which makes this theoretical laboratory now a benchmark for modeling QCD under extreme conditions.

We are going to present a combined effective model reproducing the equation of state of hadronic matter as obtained in recent lattice QCD simulations [1, 2]. The model should reproduce basic physical characteristics of processes encountered in dense hadronic matter, from the hot QCD phase through the critical temperature region till the lower temperature hadron resonance gas phase. In medium properties of hadrons are different from those in the vacuum. The very notion of the mass shell should be modified, as was postulated quite long ago [3]. The interaction becomes effectively nonlocal due to the Mott effect and hadrons eventually gradually dissolve into quarks and gluons in the high temperature phase. Then, with the increasing temperature, quark masses are less and less important although the massless Stefan-Boltzmann limit would be eventually reached only at extremely high temperature.

It has been shown that the equation of state derived from that time QCD lattice calculation [4] can be reproduced by a simple hadron gas resonance model. The rapid rise of the number of degrees of freedom in lattice QCD data around the critical temperature $T_c \sim 150 - 170$ MeV, can be explained quantitatively by a resonance gas below the critical temperature T_c [5, 6].

[1] Talk presented at the Symposium: On Discovery Physics at the LHC – Kruger 2012, RPA, December 3 - 7, 2012

*

For higher temperatures the model is modified by introducing finite widths of heavy hadrons [7, 8] with a heuristic ansatz for the spectral function which reflects medium modifications of hadrons. This fits nicely the lattice data, also above T_c as is shown in Fig. 1.

This Mott-Hagedorn type model [9] has been constructed to fit nicely the lattice data, also above T_c where it does so because it leaves light hadrons below a mass threshold of $m_0 = 1$ GeV unaffected. The description of the lattice data at high temperatures is accidental because the effective number of those degrees of freedom approximately coincides with that of quarks and gluons. The QGP presence in this region is formally simulated here by the smart choice of the mass cut-off parameter such that cut-off defined stable light hadrons provide the same number of degrees of freedom as partonic components of the QGP.

The considered model is here gradually refined to take into account those physical processes present in the full QCD treatment. The uniform treatment of all hadronic resonances, without artificial stability island, is reached by a state-dependent hadron resonance width $_i(T)$ given by the inverse collision time scale in a resonance gas [10].

In order to remove this unphysical aspect of the otherwise appealing model one has to extend the spectral broadening also to the light hadrons and thus describe their disappearance due to the Mott effect while simultaneously the quark and gluon degrees of freedom appear at high temperatures due to chiral symmetry restoration and deconfinement.

In the present contribution we will report results obtained by introducing a unified treatment of all hadronic resonances with a state-dependent width $_i(T)$ in accordance with the inverse hadronic collision time scale from a recent model for chemical freeze-out in a resonance gas [10]. The appearance of quark and gluon degrees of freedom is introduced by the Polyakov-loop improved Nambu–Jona-Lasinio (PNJL) model [11, 12]. The model is further refined by adding perturbative corrections to $\mathcal{O}(\alpha_s)$ for the high-momentum region above the three-momentum cutoff inherent in the PNJL model. One obtains eventually a good agreement with lattice QCD data, comparable with all important physical characteristics taken into account.

2. Extended Mott-Hagedorn resonance gas

2.1. No quarks and gluons; hadronic spectral function with state-independent ansatz

We introduce the width of a resonance in the statistical model through the spectral function

$$A(M,m) = N_M \frac{\cdot m}{(M^2 - m^2)^2 + \; ^2 \cdot m^2}\,, \tag{1}$$

a Breit-Wigner distribution of virtual masses with a maximum at $M = m$ and the normalization factor

$$\begin{aligned} N_M &= \left[\int_{m_0^2}^{\infty} d(M^2) \frac{\cdot m}{(M^2 - m^2)^2 + \; ^2 \cdot m^2} \right]^{-1} \\ &= \frac{1}{\frac{\pi}{2} + \arctan\left(\frac{m^2 - m_0^2}{\Gamma \cdot m}\right)}\,. \end{aligned} \tag{2}$$

And the model ansatz for the resonance width is given by [7]

$$(T) = C_\Gamma \left(\frac{m}{T_H}\right)^{N_m} \left(\frac{T}{T_H}\right)^{N_T} \exp\left(\frac{m}{T_H}\right)\,, \tag{3}$$

where $C_\Gamma = 10^{-4}$ MeV, $N_m = 2.5$, $N_T = 6.5$ and the Hagedorn temperature $T_H = 165$ MeV.

The energy density of this model with zero resonance proper volume for given temperature T and chemical potentials: μ_B for baryon number and μ_S for strangeness, can be cast in the form

$$\varepsilon(T,\mu_B,\mu_S) = \sum_{i:\ m_i<m_0} g_i\ \varepsilon_i(T,\mu_i;m_i) + \sum_{i:\ m_i\geq m_0} g_i \int_{m_0^2}^{\infty} d(M^2)\ A(M,m_i)\ \varepsilon_i(T,\mu_i;M), \tag{4}$$

where $m_0 = 1$ GeV and the energy density per degree of freedom with a mass M is

$$\varepsilon_i(T,\mu_i;M) = \int \frac{d^3k}{(2\pi)^3} \frac{\sqrt{k^2+M^2}}{\exp\left(\frac{\sqrt{k^2+M^2}-\mu_i}{T}\right)+\delta_i}, \tag{5}$$

with the degeneracy g_i and the chemical potential $\mu_i = B_i\cdot\mu_B + S_i\cdot\mu_S$ of hadron i. For mesons, $\delta_i = -1$ and for baryons $\delta_i = 1$. According to Eq. (1) the energy density of hadrons consists of the contribution of light hadrons for $m_i < m_0$ and the contribution of heavier hadrons smeared with the spectral function for $m_i \geq m_0$.

For simplicity, we assume $n_S = 0$ for the strangeness number density and $n_B = 0$ for the baryon number density. Then $\mu_B = 0$ and $\mu_S = 0$ always, so the temperature is the only significant statistical parameter here. In such a case and for a fixed volume we have

$$\varepsilon + P = T\cdot s = T\cdot\frac{\partial P}{\partial T}, \tag{6}$$

where $P = P(T)$ and $s = s(T)$ are the pressure and entropy density, respectively. This is the first order ordinary differential equation for the pressure and the general solution reads

$$P(T) = \frac{T}{T_0}\cdot P_0 + T\int_{T_0}^{T} dT'\ \frac{\varepsilon(T')}{T'^2}, \tag{7}$$

where $P_0 = P(T_0)$. To have well-defined solution for the initial temperature $T_0 = 0$ one has to assume that $\lim_{T_0\to 0} P(T_0)/T_0 = s_0 < \infty$. Then

$$P(T) = s_0\cdot T + T\int_{0}^{T} dT'\ \frac{\varepsilon(T')}{T'^2}. \tag{8}$$

And the entropy density reads:

$$s(T) = \frac{\partial P}{\partial T} = s_0 + \int_{0}^{T} dT'\ \frac{\varepsilon(T')}{T'^2} + \frac{\varepsilon(T)}{T}, \tag{9}$$

where $s(0) = s_0$. We put $s_0 = 0$ as suggested by the Nernst postulate. The sound velocity squared is given by

$$c_s^2 = \frac{\partial P}{\partial \varepsilon}. \tag{10}$$

In Fig. 1 we show the results for the thermodynamic quantities (pressure, energy density and squared speed of sound) of the MHRG model at this stage. The nice correspondence with results from lattice QCD is not accidental for the temperature region $T \sim T_c \sim 200$ MeV, where has

Journal of Physics: Conference Series **455** (2013) 012056 doi:10.1088/1742-6596/455/1/012056

been shown in [1] that a hadron resonance gas perfectly describes the lattice QCD data. For $T > T_c$ the broadening of the spectral function (1) which affects at this stage of the model only the hadronic resonances with $m > m_0$ leads to the vanishing of their contribution at about $2T_c$ while the light hadrons with masses $m < m_0$ are not affected and gradually reach the Stefan-Boltzmann (SB) limit determined by their number of degrees of freedom. As has been noted in [13], this number ($\sum_{i=\pi,K,\eta,f_0,\rho,\omega,K^*,\eta',f_0,a_0} g_i = 3+4+1+1+9+3+6+1+1+3 = 32$) accidentally (or by duality arguments) coincides with that of the quarks and gluons ($\sum_{i=q,g} g_i = 7/8 * N_c * N_f * N_s * 2 + (N_c^2 - 1) * 2 = 31.5$) for $N_c = N_f = 3$. Therefore, imposing that all mesons lighter than $m_0 = 1$ GeV are stable provides us with a SB limit at high temperatures which fakes that of quarks and gluons in the case for three flavors.

Although providing us with an excellent fit of the lattice data, the high-temperature phase of this model is unphysical since it ignores the Mott effect for light hadrons. Due to the chiral phase transition at T_c, the quarks loose their mass and therefore the threshold of the continuum of quark-antiquark scattering states is lowered. At the same time the light meson masses, however, remain almost unaffected by the increase in the temperature of the system. Consequently, they merge the continuum and become unbound - their spectral function changes from a delta-function (on-shell bound states) to a Breit-Wigner-type (off-shell, resonant scattering states). This phenomenon is the hadronic analogue [14] of the Mott-Anderson transition for electrons in solid state physics (insulator-conductor transition).

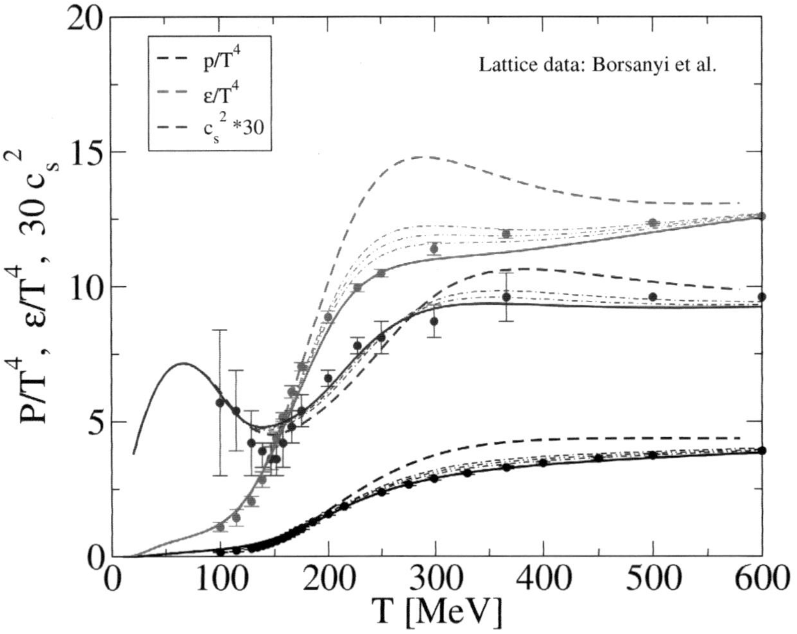

Figure 1. (Color online) Thermodynamic quantities for the old Mott-Hagedorn Resonance Gas model [7]. Different line styles correspond to different values for the parameter N_m in the range from $N_m = 2.5$ (dashed line) to $N_m = 3.0$ (solid line). Lattice QCD data are from Ref. [1].

It has been first introduced for the hadronic-to-quark-matter transition in [15]. Later, within the NJL model, a microscopic approach to the thermodynamics of the Mott dissociation of mesons in quark matter has been given in the form of a generalized Beth-Uhlenbeck equation of state [16], see also [17].

2.2. Hadronic spectral function with state-dependent ansatz

As a microscopic treatment of the Mott effect for all resonances is presently out of reach, we introduce an ansatz for a state-dependent hadron resonance width $\ _i(T)$ given by the inverse collision time scale recently suggested within an approach to the chemical freeze-out and chiral condensate in a resonance gas [10]

$$\ _i(T) = \tau^{-1}_{\mathrm{coll,i}}(T) = \sum_j \lambda \, \langle r_i^2 \rangle_T \langle r_j^2 \rangle_T \; n_j(T) \; , \tag{11}$$

which is based on a binary collision approximation and relaxation time ansatz using for the in-medium hadron-hadron cross sections the geometrical Povh-Hüfner law [18]. In Eq. (11) the coefficient λ is a free parameter, $n_j(T)$ is the partial density of the hadron j and the mean squared radii of hadrons $\langle r_i^2 \rangle_T$ obtain in the medium a temperature dependence which is governed by the (partial) restoration of chiral symmetry. For the pion this was quantitatively studied within the NJL model [19] and it was shown that close to the Mott transition the pion radius is well approximated by

$$r_\pi^2(T) = \frac{3}{4\pi^2} f_\pi^{-2}(T) = \frac{3M_\pi^2}{4\pi^2 m_q} |\langle \bar{q}q \rangle_T|^{-1} \; . \tag{12}$$

Here the Gell-Mann–Oakes–Renner relation has been used and the pion mass shall be assumed chirally protected and thus temperature independent.

For the nucleon, we shall assume the radius to consist of two components, a medium independent hard core radius r_0 and a pion cloud contribution $r_N^2(T) = r_0^2 + r_\pi^2(T)$, where from the vacuum values $r_\pi = 0.59$ fm and $r_N = 0.74$ fm one gets $r_0 = 0.45$ fm. A key point of our approach is that the temperature dependent hadronic radii shall diverge when hadron dissociation (Mott effect) sets in, driven basically by the restoration of chiral symmetry. As a consequence, in the vicinity of the chiral restoration temperature all meson radii shall behave like that of the pion and all baryon radii like that of the nucleon.

The resulting energy density behaviour is shown in Fig. 2. This part of the model we call Mott-Hagedorn-Resonance-Gas (MHRG). When all hadrons are gone at $T \sim 250$ MeV, we are clearly missing degrees of freedom!

3. Quarks, gluons and hadron resonances

We improve the PNJL model over its standard versions [11, 12] by adding perturbative corrections in $\mathcal{O}(\alpha_s)$ for the high-momentum region above the three-momentum cutoff Λ. In the second step, the MHRG part is replaced by its final form, using the state-dependent spectral function for the description of the Mott dissociation of all hadron resonances above the chiral transition. The total pressure obtains the form

$$P(T) = P_{\mathrm{MHRG}}(T) + P_{\mathrm{PNJL,MF}}(T) + P_2(T) \; . \tag{13}$$

where $P_{\mathrm{MHRG}}(T)$ stands for the pressure of the MHRG model, accounting for the dissociation of hadrons in hot dense, matter.

The $\mathcal{O}(\alpha_s)$ corrections can be split in quark and gluon contributions

$$P_2(T) = P_2^{\mathrm{quark}}(T) + P_2^{\mathrm{gluon}}(T) \; , \tag{14}$$

where P_2^{quark} stands for the quark contribution and P_2^{gluon} contains the ghost and gluon contributions. The total perturbative QCD correction to $\mathcal{O}(\alpha_s)$ is

$$P_2 = -\frac{8}{\pi}\alpha_s T^4 (I_\Lambda^+ + \frac{3}{\pi^2}((I_\Lambda^+)^2 + (I_\Lambda^-)^2)), \tag{15}$$

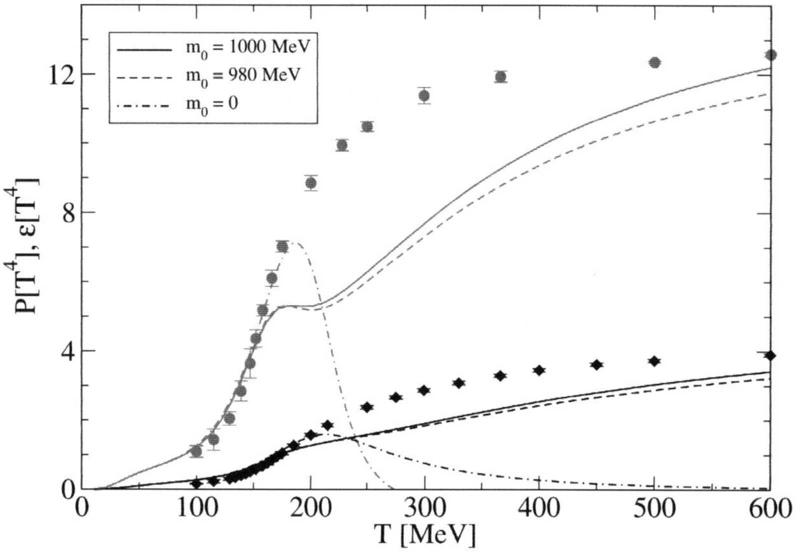

Figure 2. (Color online) Energy density (red lines and symbols) and pressure (black lines and symbols) for the state-dependent width model of Eq. (11) and three values of the mass threshold m_0: 1 GeV (solid lines), 980 MeV (dashed lines), 0 (dash-dotted lines). Lattice QCD data are from Ref. [1].

where $I_\Lambda^\pm = \int\limits_{\Lambda/T}^{\infty} \frac{dx\; x}{e^x \pm 1}$. The corresponding contribution to the energy density is given in standard way by Eq. (6).

We will now include an effective description of the dissociation of hadrons due to the Mott effect into the hadron resonance gas model by including the state dependent hadron resonance width (11) into the definition of the HRG pressure

$$P_{\rm MHRG}(T) = \sum_i \delta_i d_i \int \frac{d^3p}{(2\pi)^3} dM\, A_i(M) T \ln\left(1 + \delta_i e^{-\sqrt{p^2+M^2}/T}\right) . \quad (16)$$

From the pressure as a thermodynamic potential all relevant thermodynamical functions can be obtained. Combining the α_s corrected meanfield PNJL model for the quark-gluon subsystem with the MHRG description of the hadronic resonances we obtain the results shown in the right panel of Fig. 1 where the resulting partial contributions in comparison with lattice QCD data from Ref. [1] are shown.

We see that the lattice QCD thermodynamics is in full accordance with a hadron resonance gas up to a temperature of ~ 170 MeV which corresponds to the pseudocritical temperature of the chiral phase transition. The lattice data saturate below the Stefan-Boltzmann limit of an ideal quark-gluon gas at high temperatures. The PNJL model, however, attains this limit by construction. The deviation is to good accuracy described by perturbative corrections to $\mathcal{O}(\alpha_s)$ which vanish at low temperatures due to an infrared cutoff procedure. The transition region $170 \leq T[\text{MeV}] \leq 250$ is described by the MHRG model, resulting in a decreasing HRG pressure which vanishes at $T \sim 250$ MeV.

We have presented two stages of an effective model description of QCD thermodynamics at finite temperatures which properly accounts for the fact that in the QCD transition region it is dominated by a tower of hadronic resonances. To this end we have further developed a generalization of the Hagedorn resonance gas thermodynamics which includes the finite lifetime

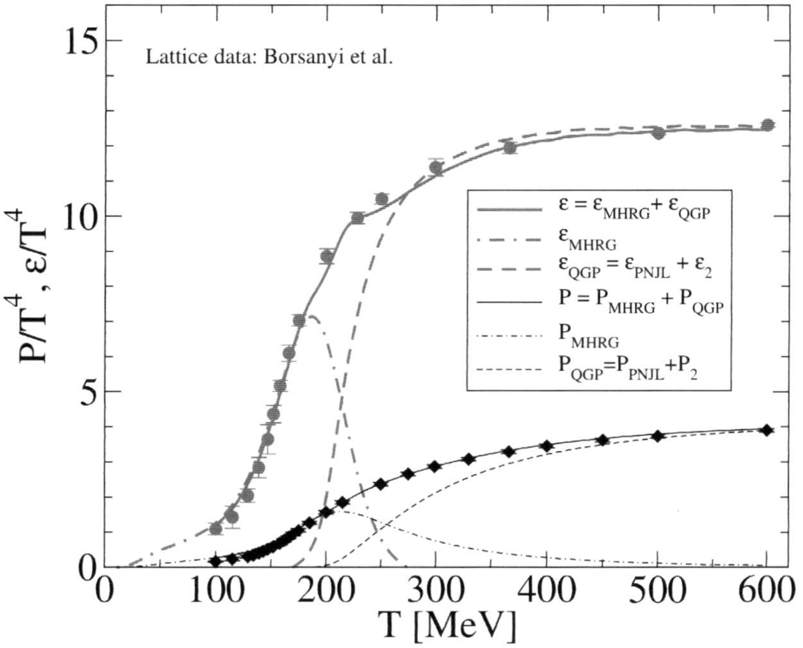

Figure 3. (Color online) Thermodynamic quantities for the new Mott-Hagedorn Resonance Gas where quark-gluon plasma contributions are described within the PNJL model including α_s corrections (dashed lines). Hadronic resonances are described within the resonance gas with finite width, as an implementation of the Mott effect (dash-dotted line). The sum of both contributions (solid lines) is shown for the energy density (thick lines) and pressure (thin lines) in comparison with the lattice data from [1].

of hadronic resonances in a hot and dense medium by a model ansatz for a temperature- and mass dependent spectral function.

. Conclusion and outlook

After having presented the MHRG-PNJL model with the state-dependent spectral function approach we show the summary of the thermodynamic quantities in Fig. 4. We have presented two stages of an effective model description of QCD thermodynamics at finite temperatures which properly accounts for the fact that in the QCD transition region it is dominated by a tower of hadronic resonances. In the first of the two stages of the developments we presented here, we have used the fact that the number of low-lying mesonic degrees of freedom with masses below ~ 1 GeV approximately equals that of the thermodynamic degrees of freedom of a gas of quark and gluons. In the second one we have further developed a generalization of the Hagedorn resonance gas thermodynamics which includes the finite lifetime of heavy resonances in a hot and dense medium by a model ansatz for a temperature- and mass dependent spectral function which is motivated by a model for the collision time successfully applied in the kinetic description of chemical freeze-out from a hadron resonance gas. A next step should take into account also the effects of continuum correlations in hadronic scattering channels in accordance with the Levinson theorem [20] as discussed recently for the example of pion dissociation within the PNJL model [21]. Our account for $\mathcal{O}(\alpha_s)$ corrections from quark and gluon scattering in the plasma may be seen from this perspective.

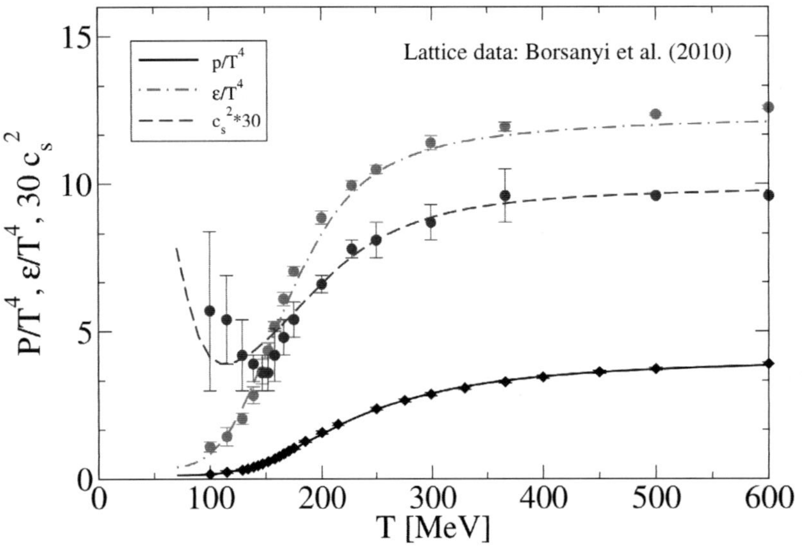

Figure . (Color online) Thermodynamic quantities as in Fig. 1 for the MHRG-PNJL model compared to lattice data from Ref. [1].

Acknowledgments
D.B. wants to thank K.A. Bugaev, K. Redlich and A. Wergieluk for discussions and collaboration. This work was supported in part by the Polish National Science Center (NCN) under contract No. N N202 0523 40 and the "Maestro" programme DEC-2011/02/A/ST2/00306 as well as by the Russian Foundation for Basic Research under grant No. 11-02-01538-a (D.B.).

[1] S. Borsanyi *et al.*, JHEP **1011**, 077 (2010).
[2] A. Bazavov *et al.*, Phys. Rev. D **80**, 014504 (2009).
[3] F. Karsch, K. Redlich and L. Turko, Z. Phys. C **60**, 519 (1993).
[4] F. Karsch, Lect. Notes Phys. **583**, 209 (2002); [hep-lat/0106019].
[5] F. Karsch, K. Redlich and A. Tawfik, Eur. Phys. J. C **29**, 549 (2003).
[6] C. Ratti *et al.* [Wuppertal-Budapest Collaboration], Nucl. Phys. A **855**, 253 (2011) [arXiv:1012.5215 [hep-lat]].
[7] D. B. Blaschke, K. A. Bugaev, Fizika B **13**, 491 (2004);
Phys. Part. Nucl. Lett. **2**, 305 (2005).
[8] D. B. Blaschke, K. A. Bugaev, Phys. Part. Nucl. Lett. **2**, 305 (2005). [Pisma Fiz. Elem. Chast. Atom. adra **2**, 69 (2005)].
[9] L. Turko, D. Blaschke, D. Prorok and J. Berdermann, Acta Phys. Polon. Supp. **5**, 485 (2012) [arXiv:1112.6408 [nucl-th]].
[10] D. B. Blaschke *et al.*, Phys. Part. Nucl. Lett. **8**, 811 (2011).
[11] K. Fukushima, Phys. Lett. B **591**, 277 (2004).
[12] C. Ratti, M. A. Thaler, W. Weise, Phys. Rev. D **73**, 014019 (2006).
[13] G. E. Brown, H. A. Bethe, P. M. Pizzochero, Phys. Lett. B **263**, 337 (1991).
[14] D. Zablocki, D. Blaschke and G. Röpke, *Metal-to-Nonmetal Transitions*, Springer Series in Materials Science, **132**, 161 (2010).
[15] D. Blaschke *et al.*, Phys. Lett. B **151**, 439 (1985).
[16] J. Hüfner *et al.*, Annals Phys. **234**, 225 (1994).
[17] A. E. Radzhabov *et al.*, Phys. Rev. D **83**, 116004 (2011).
[18] B. Povh, J. Hüfner, Phys. Lett. B **245**, 653 (1990).
[19] H. J. Hippe and S. P. Klevansky, Phys. Rev. C **52**, 2172 (1995).
[20] R. Dashen, S. -K. Ma and H. J. Bernstein, Phys. Rev. **187**, 345 (1969).
[21] A. Wergieluk, D. Blaschke, . .L. Kalinovsky and A. Friesen, arXiv:1212.5245 [nucl-th].

AUTHOR INDEX